PHAGOCYTOSIS: THE HOST

Volume 5 • 1999

ADVANCES IN
CELL AND MOLECULAR BIOLOGY
OF MEMBRANES AND ORGANELLES

DEDICATION

This volume is dedicated to the memory of
Zanvil A. Cohn, mentor and friend.

PHAGOCYTOSIS: THE HOST

Volume 5 • 1999

ADVANCES IN
CELL AND MOLECULAR BIOLOGY
OF MEMBRANES AND ORGANELLES

Series Editor: ALAN M. TARTAKOFF
Institute of Pathology
Case Western Reserve University

Volume Editor: SIAMON GORDON
Sir William Dunn School of Pathology
University of Oxford

JAI PRESS INC.
Stamford, Connecticut

100 Prospect Street
Stamford, Connecticut 06901

ISBN: 1-55938-999-0

Transferred to digital printing 2006

CONTENTS

SECTION III. SIGNALING

SECTION IV. THE PATHWAY

SECTION V. RESPONSES

LIST OF CONTRIBUTORS

Arie E. Abo	Department of Pathology and Microbiology Duke University Medical Center Durham, North Carolina
Alan Aderem	Department of Immunology University of Washington Seattle, Washington
Matthew L. Albert	Lab of Cellular Physiology and Immunology Rockefeller University New York, New York
Carmen Alvarez-Dominguez	Department of Cell Biology and Physiology Washington University School of Medicine St. Louis, Missiouri
Nina Bhardwaj	Lab of Cellular Physiology and Immunology Rockefeller University New York, New York
Scott D. Blystone	Department of Anatomy and Cell Biology State University of New York Syracuse, New York
Eric J. Brown	Infectious Diseases Division Washington University School of Medicine St. Louis, Missouri
Shamshad Cockcroft	Department of Physiology University College London London, England
Rosangela P. da Silva	Sir William Dunn School of Pathology University of Oxford Oxford, England

Lodewijk V. Dekker — Centre for Molecular Medicine
Rayne Institute
University College London Medical School

Peter Elsbach — Department of Medicine and Microbiology
New York University School of Medicine
New York, New York

R. Alan B. Ezekowitz — Department of Pediatrics
Massachusetts General Hospital
Boston, Massachusetts

Nathalie C. Franc — Cutaneous Biology Research Center
Massachusetts General Hospital
Charlestown, Massachusetts

Iain P. Fraser — Department of Pediatrics
Massachusetts General Hospital
Boston, Massachusetts

Wendy Garrett — Department of Cell Biology
Yale University School of Medicine
New Haven, Connecticut

Siamon Gordon — Sir William Dunn School of Pathology
University of Oxford
Oxford, England

Steven Greenberg — Departments of Medicine and Pharmacology
Columbia University
New York, New York

Sergio Grinstein — Division of Cell Biology
The Hospital for Sick Children
Toronto, Ontario, Canada

Philippe Gros — Department of Biochemistry
McGill University
Montreal, Quebec, Canada

Samantha Gruenheid — Department of Biochemistry
McGill University
Montreal, Quebec, Canada

Richard Haworth — Sir William Dunn School of Pathology
University of Oxford
Oxford, England

David J. Hackam	Division of Cell Biology The Hospital for Sick Children Toronto, Ontario, Canada
Kayo Inaba	Department of Zoology Graduate School of Science Kyoto University Kyoto, Japan
Luis Mayorga	Instituto de Histologia y Embriologia Facultad de Ciencias Medicas Universidad Nacional de Cuyo Mendoza, Argentina
Keng-Mean Lin	Department of Physiology Southwestern Medical Center at Dallas The University of Texas Dallas, Texas
Ira Mellman	Department of Cell Biology Yale University School of Medicine New Haven, Connecticut
Carl F. Nathan	Beatrice and Samuel Seaver Laboratory Cornell University Medical College New York, New York
Jean Pieters	Basel Institute for Immunology Basel, Switzerland
Nick Platt	Sir William Dunn School of Pathology University of Oxford Oxford, England
Ori D. Rotstein	Division of Cell Biology The Hospital for Sick Children Toronto, Ontario, Canada
Anthony W. Segal	Centre for Molecular Medicine Rayne Institute University College London Medical School London, England
Michael U. Shiloh	Beatrice and Samuel Seaver Laboratory Cornell University Medical College New York, New York

Emil Skamene Department of Biochemistry
McGill University
Montreal, Quebec, Canada

Philip D. Stahl Department of Cell Biology and Physiology
Washington University School of Medicine
St. Louis, Missiouri

Ralph M. Steinman Lab of Cellular Physiology and Immunology
Rockefeller University
New York, New York

Richard W. Stockley Centre for Molecular Medicine
Rayne Institute
University College London Medical School
London, England

Thomas P. Stossel Experimental Medical Division
Longwood Medical Center
Brigham and Women's Hospital
Boston, Massachusetts

Hui-Qiao Sun Department of Physiology
Southwestern Medical Center at Dallas
The University of Texas
Dallas, Texas

Joel A. Swanson Department of Anatomy and Cell Biology
University of Michigan Medical School
Ann Arbor, MI

Shannon Turley Department of Cell Biology
Yale University School of Medicine
New Haven, Connecticut

David M. Underhill Department of Immunology
University of Washington
Seattle, WA

Frans Wientjes Centre for Molecular Medicine
Rayne Institute
University College London Medical School
London, England

Debora Williams-Herman Department of Immunology
University of California
San Francisco, California

Zena Werb
Department of Anatomy
University of California
San Francisco, California

Masaya Yamamoto
Department of Physiology
Southwestern Medical Center at Dallas
The University of Texas
Dallas, Texas

Helen L. Yin
Department of Physiology
Southwestern Medical Center at Dallas
The University of Texas
Dallas, Texas

INTRODUCTION TO THE SERIES

The remarkable vigor and central importance of cell biology result from the realization that emphasis on structure/function relations at the cellular and subcellular levels is essential for a rigorous and satisfactorily complete understanding.

Unlike many subdivisions of biomedical science, cell biology is not linked to any one methodology. It often emphasizes topological or topographic questions, and it is concerned with the structure, biogenesis, and turnover of macromolecular structures; however, there is no limit to the techniques and conceptual approaches that it brings to bear on these issues. Indeed, this has even been true since the term *cell biology* was first used. Certain scientists trace its origins back to ultrastructure and histology; others consider E.B. Wilson's *The Cell in Development and Heredity* to epitomize the foundations of cell biology, while others consider cell biology as an outgrowth of somatic cell genetics or the extension of biophysics to objects of increasingly large size. This varied and often changing identity makes sense. It also suggests the rich intellectual mix that underlies today's successes in cell biology research.

This series aims to match the continuing evolution of cell biology—with particular emphasis on cell membranes—by treating coherent areas in multiauthored volumes. This approach allows a multifaceted coverage of topics that benefits from the unity of vision of the volume editor(s) but does not rely on one individual to synthesize an entire subject. Hopefully, the result will be more readable and intrinsically richer than lengthy review chapters, which attempt to be all encompassing.

Alan M. Tartakoff
Series Editor

PREFACE

The history of phagocytosis is intimately associated with the rise of immunology and growth in understanding of host resistance to infection. More recently, improved methods of cellular and molecular biology have shown the value of phagocytosis as a model system, to study membrane recognition and trafficking, signaling and altered gene expression. Above all, it has become evident that pathogens employ a marvelous range of strategies to penetrate host cells, where required for their own survival, and evade host anti-microbial mechanisms. The continuing medical importance of infectious agents and resistance to chemotherapeutic agents make it doubly important to study the basic mechanisms of host cell-pathogen interactions, and their role in pathogenesis of infectious disease, chronic inflammation and autoimmunity.

Phagocytosis is an ancient mechanism by which free-living individual cells interacted with their microbial and cellular neighbors. Its function in multicellular organisms included the clearance of naturally dying cells and of cell fragments, as part of tissue homeostasis and remodeling, as well as the response to injury, cell death and invasion by more or less -adapted symbiotic and parasitic invaders. Specialized host cells such as macrophages evolved as professional phagocytes, and their derivatives, dendritic cells, became uniquely proficient at regulating T lymphocyte responses to microbial antigens. Even in the absence of macrophages and polymorphonuclear leukocytes, other nonprofessional cell-types are able to ingest and destroy their dying neighbors, albeit at reduced efficiency; this clearly can suffice in organisms such as *Caenorhabitis elegans* and in mice that lack all myeloid cells after genetic ablation of the transcription factor PU-1. Such animals develop reasonably normally, but lack resistance to infection and die soon after birth unless they are rescued by a hemopoietic graft.

Several texts, some of which are listed below, review particular aspects of phagocytosis in relation to pathogen invasion. However, the treatment is usually from the perspective of either the host cell or the microbe—with little integration of the subject, adequate appreciation of the amazing diversity, or overlap of the cellular mechanisms employed by each participant. Evolutionary pressures are strong, to reward success or eliminate failure; novel, even unexpected strategies keep cropping up, often pointing to a neglected area of previous research. The present volume focuses mainly on the host aspect of phagocytosis. An accompanying volume (Volume 6) in this series presents strategies of cellular invasion from the viewpoint of the microbe.

This field of study is growing rapidly after a somewhat slow start over recent decades. This collection of invited chapters attempts to reflect current research, and brings together cell biologists, microbiologists and immunologists with disparate interests. However, there is a certain unity, even repetition of key themes, hopefully like a symphony rather than a boring catalogue. It will be evident that editorial bias favors intracellular parasitism and medically important organisms. The neutrophil is far more than a supporting player to the macrophage, and some attempt is made to remind the reader of some of its unique skills. To retain a manageable size, the emphasis is on relatively early events such as mutual recognition, cell entry, and response, rather than on longterm changes in gene expression by either host cell or pathogen. Viruses are excluded not because of lack of importance but because of somewhat different research approaches, although it is becoming increasingly clear that large viruses (e.g., vaccinia) and listeria monocytogenes, share common strategies in invasion and intercellular spread.

Other aspects of phagocytosis that deserve more attention than given here include turnover of cells in degenerative and neoplastic disease, targeting of cells via liposomes, potentially important in gene therapy, and avoidance of premature clearance of foreign DNA. There remains a great deal to discover regarding the immune response to non- or altered self ("danger") particulates, versus the silent clearance of apoptotic cells. Clinically, there is as yet little impact of phagocytosis research on diagnosis, genetic variation among individuals in their response to infectious agents, evaluation of phagocytic function *in vivo*, or management of disordered phagocytic capacity. Sudden, massive uptake of dying organisms following antibiotic treatment may have serious, if transient effects on the host. Finally, as we learn more about the subtleties of the mechanisms by which host and pathogen recognize each other, and the consequences of their encounter, we may acquire the skills to manipulate their interactions, boost the phagocytes where appropriate, or block their deleterious effects.

ACKNOWLEDGMENTS

Work in the author's laboratory is supported by the Medical Research Council, Wellcome Trust, and the Arthritis Research Campaign. I would like to thank Christine Holt for her invaluable help in all stages of preparing this book for publication.

SUGGESTIONS FOR FURTHER READING

de Duve, C. (1995). In: *Vital Dust. Life as a Cosmic Imperative.* New York: Basic Books. A rare discussion of primitive phagocytes, as molecules evolved into single and multicellular organisms.

Karnovsky, M.L. and Bolis, L. (Eds.) (1982). In: *Phagocytosis—Past and Future.* Academic Press. Based on the conference "A century since Metchnikoff" held in the province of Messina, Sicily, Italy. September 22–27, 1980.

Margolis, L. (1992). In: *Symbiosis in Cell Evolution.* San Francisco: W.H. Freeman. The endosymbiotic origin of mitochondria and chloroplasts.

Metchnikoff, E. (1905). In: *Immunity in Infective Diseases.* Cambridge: Cambridge University Press. A summing up of the labors of 25 years.

Shaw, B. *The Doctor's Dilemma.* A tragi-comedy first produced in 1906 and published in 1911. Current edition available from Penguin Books, London. Almost farcical portrayal of Sir Colenso Ridgeon, the enthusiast for opsonic phagocytosis, based on Almroth-Wright who coined the term.

Tauber, A.I. and Chernyak, L. (1991). In: *Metchnikoff and the Origins of Immunology. From Metaphor to Theory.* Oxford University Press. A synthesis by an American immunologist and a Russian philosopher, of embryologic roots to evolution and cellular biology.

Zwilling, B.S. & Eisenstein, T.K. (Eds.) (1994). Macrophage-pathogen interactions. New York: Marcel-Dekker Inc. Integrates macrophage immunobiology and interactions with a range of bacteria, parasites, fungi and viruses.

Individual chapters cite a wide range of review articles, as well as key papers of modern observations.

Siamon Gordon
Editor

SECTION I

CONTEXT

THE EARLY HISTORY OF PHAGOCYTOSIS

Thomas P. Stossel

I. INTRODUCTION

This book summarizes the state of knowledge of the science of phagocytosis as we enter the third millenium. Like the rest of biological science at this moment, phagocytosis is approaching a level of fundamental comprehension previously limited to the physical sciences, and the practical spinoffs of this knowledge have bright commercial and medical possibilities. Yet phagocytosis was a subject of interest to biologists long before the emergence of modern cell and molecular biol-

Advances in Cell and Molecular Biology of Membranes and Organelles
Volume 5, pages 3-18.

ISBN: 1-55938-999-0

ogy, indeed, two centuries before the molecular definition of a genetic code. So what?

Historians argue that a decent and rich life requires an understanding of ones precedents, and that a perception of how small any individual is in the vast tapestry of history affords the humility that engenders character. However, many dismiss history as an irrelevant graveyard at best and an instrument of political oppression at worst: the tyranny of "dead white males." Scientists tend to have such an attitude by default. Most academic institutions and certainly industry reward scientists for "what they've done for us (fill in the blank defining 'us'—the granting agency, the academic department, product development) lately." The profession seems to select for those who don't waste time asking "why am I doing what I do?" or "where did I get the idea?"

Ironically, the communication of science has a historicity of sorts, namely the hoarding of credit documented as attribution of previous work in papers and grant applications that reviewers police viciously, but (an assumption purely my own) usually to defend their own perceived credit territory. One might think that the incredible retrieval capacity of electronic databases would facilitate a broader generosity of attribution, but if anything, it enables references to disappear more readily, should one key article omit a citation, the absence of which is faithfully reproduced with the stroke of a key in subsequent scholarship based on that article.

Does it matter? I don't know, but, for my own amusement I have compiled a history of phagocytosis, including some obscure and, to my knowledge previously unattributed research. Hurried entrepreneurs can safely ignore it; others who are more relaxed may find it interesting.

Phagocytosis has a broad and a narrow definition. The former encompasses the spectrum of activities undertaken by a class of cells of the hematopoietic system in animal physiology related to tissue remodeling, immunity and inflammation. The latter is the attribute of particle engulfement that unifies this class of otherwise very diverse cells. Ilya Metchnikoff (1845–1916) unquestionably carried phagocytosis from its narrow to its broad conception. Moreover, the force of Metchnikoff's theory concerning the role of phagocytes in immunity and inflammation and the charm of the story by which he came to develop his theory have been so influential as to create the widespread belief that he *discovered* the narrower definition too, the fundamental cellular activity of phagocytosis. I cannot claim to have read Metchnikoff's publications exhaustively, but to the extent that I have surveyed his original work and historical accounts of others (Metchnikoff, E., 1882, 1887, 1891, 1905; Metchnikoff, O., 1920; Herrlinger, 1956; Heifets, 1982; Hirsch, 1982; Tauber, 1991), I found no references to earlier observations regarding phagocytosis per se. By contrast, Metchnikoff did make abundant citations of research supporting his own views or to those contradicting his ideas concerning interactions between leukocytes and blood vessel walls.

II. THE INITIAL DESCRIPTIONS OF PHAGOCYTOSIS

To my knowledge, the first article describing phagocytosis is entitled "Infusion animals that eat others," written by Johann August Ephraim Goeze (1777). Goeze (1731–1793), studied theology in Halle and became a Lutheran pastor in Aschersleben and Quedlinburg, towns in Saxony not far from the birthplace of Johann Sebastian Bach. The article on the discovery of phagocytosis appears on pages 373–384 in volume 3 of a journal called *Activities of the Berlin Society of Naturalist Companions* published in 1777. Becuase it is a historic milestone not generally known, I provide a translation of the article in its entirety. I do so also so the reader can appreciate Goeze's vivid and anthropomorphic depiction of the predatory behavior of his microorganisms. This attitude is particularly interesting, in light of Metchnikoff's later imaging of pitched battles between phagocytes and microbes, a metaphor about which much as been made regarding its influence on the way we think about health and disease.

> The following observation taught me that predation and compassion exist in the microscopic world as they do in ours. On November 8, 1776 I prepared two infusions in my room: one from hay and running water which had stood for a time in a copper pot and another from hay and well water and allowed both to become stale in the warm room. By November 18 the latter was teeming with life, the former lifeless. Is it possible that some green copper rust contaminated the water and hindered the generation of fusoria? This seemed likely to me, so I removed the old water from the first infusion, preserved the macerated hay and added stale well water which had been standing covered for a while in the warm room. In less than two days, this preparation was full of life forms. Here I must add a separate observation. If I make an infusion with cold well water, it yields no infusoria; therefore I might have concluded that this water was too hard and too full of nitrate. If, however, I leave the same well water standing in the room for a few days and then use it for the infusion, it supports infusoria. I cannot yet imagine how to explain these findings.
>
> The first infusion, the one with macerated hay and stale well water, which I will call A, contains two types of infusoria. The first variety was present in such abundance that the entire slimy surface at the top of water was a mass of these animals. They are the common elongated oval creatures with crooked heads (numbers 1–6) [The numbers all refer to figure 1]. Number 6 shows an attractive adult one, revealing its innards, which I surmise to be offspring. So does Mr. Müller [The Danish naturalist, O.F. Müller—I believe he was the first to refer to the innards of a protist as "gelatinous."]. You can also observe the radius of its circular movements, which proceed from left to right. If the large adults are mothers of my infusoria, as the Privy Councillor [Mueller] believes, then all the creatures are of one strain and their reproductive method is similar to *Volvox globator*. On other hand, I have seen medium-sized examples of these animals divide. But now I will abandon these very familiar animals.
>
> The second species was at least three-fold larger than the first but not so numerous. Their shape is broad, with bristles at the head and at the rear atop round protuberances. They move jerkily forward and backward in fits and starts. The bristles cause enormous glitter and even cause the water to swirl. I found single examples of these earlier in balm-mint, but I never previouly observed them in such numbers and size as in this infusion (b, see below).
>
> I very distinctly recorded the characteristics of this animal, as did the Privy Councillor: its mouth, its innards, the jerky movements which propelled it for enormous distances; if it flattens its bristles and uses them as feet, the body rolls up at both ends and, like a hollow tube, runs

quickly about a piece of slime. It is truly bizarre when it straightens up and extends its bristles like masts and touches objects with them like feelers or horns. The reader can find pictures of examples contorted into various shapes in numbers 7–12. The Privy Councillor did not cite an authority, but I believe it correct to see these animals in the figures of Joblot, volume 1, P. 11, tables 2.f.3 and 8.f.9.9, which he calls *la grosse Araignee aquatique* ["the fat water spider"] on p. 78. He also mentioned that they devoured *Cornemeuses* ["bagpipes"]. The figures, however, are a bit unnatural, as is typical for Joblot's pictures.

We can add a third feature to the usual characteristics of this animal, the first being *natans* (swimming), the second being *ambulans* (walking)—which was described by the Privy Councillor—and a new one, *devorans* (eating). This last is the spectacle I wish to describe in detail.

When I pipetted by first drop and examined it with the number 3 ocular and type A objective of my 'Composit' microscope, I saw with surprise that these 'hair bugs' (*Haarwanzen*) numbers 7–9) crowded among the small oval animals. Some they had already swallowed. Others were stuck in the predator's neck, while still others got free and took care not to approach the hair bugs again. Shortly thereafter took place a horrid hunt and massacre never previously seen in the microscopic world. At first I could not believe my eyes, because my mind recalled the works of Mr. Müller in "Histor. Verm. Vol. 1. p.2.p.58" declaring "*Nec ullus oculatior animalcula revera ab animalculis devorari vidit*" [...no observer ever saw infusoria devoured by others]. I applied the second, third and fourth drops and saw the same spectacle. I observed them with the number 1 objective in the same size as in numbers 12 and 13, and the predators devoured more and more prey. The following day I showed the scene to my friends without telling them anything in advance, and at the first glance they all exclaimed: "these are horrible predators that devour the other animals in half-dozens." To allay all doubts, I sent the flask with the predators to a local merchant who is a great nature lover and went for an eight-day winter journey. I told him nothing about the predators but only asked him to examine the water of this infusion with his Hoffman Composit (microscope), and to preserve the flask until my return. When I did so, he was extraordinarily excited and full of wonder about the strange predators which he found in the infusion and which devoured masses of other animals. I only mention these things to convince the reader that I was not deluded by illusions or optical artifacts.

Now I want to describe the predation scene in detail. As soon as the predator met one of the oval animals it suddenly dove at it and grasped it with the lips of its mouth which is located on its inferior side (number 8a). The captured animal defended itself as best it could. It struggled for a time in the jaws of the predator, especially if in an oblique position when seized. In that case the predator worked to turn the victim into a longitudinal position which was easier to accommodate to the tube which was its stomach. During such a battle the victim often got free. Generally, however, it was so exhausted that it swam to the nearest edge and expired.

If the predator (number 7) seized the victim properly so that one end stuck in its throat, one might expect it to slide easily into the stomach. This is not what happens, however, in that the predator begins to choke and jerks itself backwards until the prey is fully swallowed. During this choking process I noted several predators assuming different shapes (numbers 7–10). Often they bent sideways as in Figures 7 and 8, and in this manner the animals were swallowed. When one was engulfed, the chase and predation continued.

Numbers 12 and 13 show an astoundingly voracious predator. As I watched it, it swallowed three oval animals. One of them pushed the abdominal skin outward as if it had no room, as in the case of those hydra that overeat. The engulfed oval animals did not retain their oval shape for long. Within five minutes they rounded as were nine others already processed by digestion. In all, the predator had twelve victims internalized which became progressively smaller and finally were little dots, when their juices were digested. The same predator ingested the front part of an oval animal such that its cilium appeared obliquely cut off. The victim was stuffed deeper, and the cilium became visible again. Hence, the predator uses cilia for more than one purpose—how versatile the structure of such a single small animal!

> Often predators seize an oval animal in their jaws and move about the droplet before they devour the victim. Sometimes another predator appears and steals it from the first. Occasionally such an altercation results in the liberation of the victim which remains alive. Evidently the predators lack teeth or other organs of injury but rather simply use the edges of their lips to ensnare the prey (number 8a). When the predators were quite stuffed with prey they consumed little more but jerked about the droplet as usual. Observing them, you can easily mistake the ingested prey for offspring.
>
> I also noted that predation was unsuccessful when the oval animals were clustered as they are wont to do. Perhaps they instinctively do this to gain security from predators, as pigeons gather in a crowd to avoid attack from hawks. The moment I dispersed the cluster of oval animals with a sharp needle, the predation and digestion began anew.
>
> I must mention another experiment providing evidence that the hair bugs actually swallowed the oval animals. I applied a droplet to the lower plate of the recently devised Hoffman press with screws and small watchsprings—I will describe the device at another occasion—and superimposed the other plate, screwing it down until it touched the surface the drop. Then I gave the predators time to function and devour. One swallowed five small animals one by one, and they all were visible in its belly. Then I screwed the plate down further which caused the predator to become quiescent, whereas the swallowed animals still moved about with-in its body. By gentle further application of the screw, I gave them a final squeeze that caused the predator to burst, releasing the swallowed animals to freedom. The liberated animals immediately continued swimming in the liquid. I was delighted to be the deliverer of those swallowed victims, even in the microscopic world. This is convincing proof of the benefits of this excellent device for microscopical experiments which otherwise would be impossible, such as, for example, the delivery of wheel and globe animals, the convenient compression of vinegar eels, nematodes so forth.

Arguably Goeze's priority to the first description of phagocytosis is challengeable because his descriptions do not establish unequivocally that the method of "ingestion" truly resembles what happens when leukocytes engulf objects. Moreover, since I cannot identify the microorganisms participating in Goeze's drama, I cannot claim that subsequent research has established that they are phagocytic in the manner of leukocytes. Nevertheless, Goeze's allusion to cells using "the edges of their lips" certainly evokes the image of extended lamellar sheets we associate with the phagocytosis of erythrocytes or zymosan by macrophages. Furthermore, we are still learning to appreciate the complexity of engulfment mechanisms which are (at least to me) unexpectedly individualized to the type of cell doing the ingesting and the size, surface and nature of the prey being eaten. So I am inclined to give Goeze the benefit of the doubt. And the elegant experiment "releasing" the captive prey is up to standards that would justify publication in today's most prestigious biomedical journals.

Two years after Goeze's paper, Baron Friederich Wilhelm von Gleichen-Russworm (1717–1783) of Nuremberg, Germany, narrated how he became curious about objects he saw inside of certain protozoans (infusoria). He narrowed the possible explanations for these bodies to either infusoria eggs or ingested matter and designed a simple experiment to settle the question. He added carmine to the infusion and over a period of days documented that the red-colored carmine bits progressively collected inside of the little animals. Although he could not ascer-

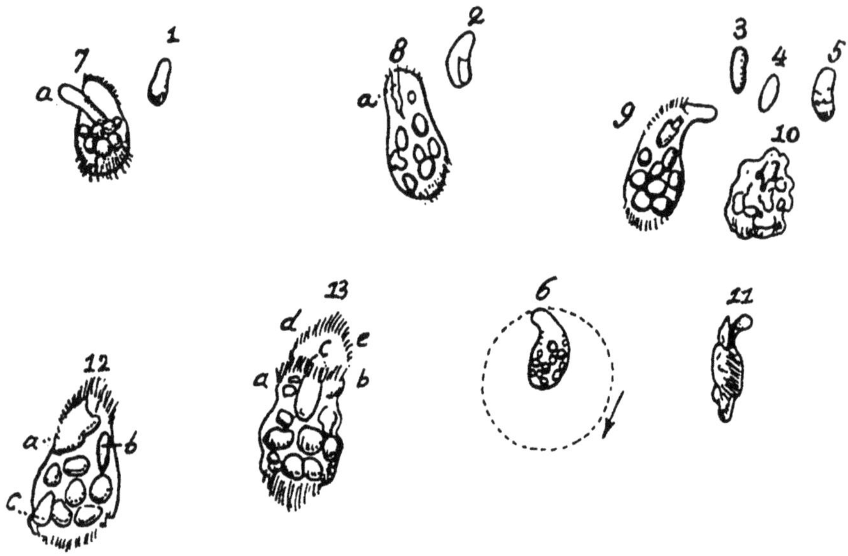

Figure 1. Goeze's drawings depicting infusoria being ingested by others.

tain how the creatures took up the dye, he thought they might have a mouth. This point, as I will discuss further, was to remain controversial for nearly half a century. Gleichen-Russworm, who cites Goeze's findings, pursued a military career in the service of the Markgrave of Bayreuth which he ended in 1748 to manage the estates inherited from his mother. He obviously managed to undertake scientific investigations on the side. Gleichen-Russworm's observations concerning phagocytosis appear in a chapter called "The Engulfing Infusion Animals," (pp 140–144) of a posthumously published monograph entitled *Studies of Germinating and Infusion Animals and How to Accomplish Microscopic Observations on the Germination of Animals and on Diverse Infusions*. Excerpts from that chapter, which I believe is an account of the first time an investigator actually fed things to cells, follow:

> I added carmine to a wheat infusion containing *Pandeloquen*, and 'small oval animals,' which had been present for months. By the next day my expectations were fulfilled in that I was convinced not only that the meal had been swallowed but that I could learn more about the viscera of these animals.
>
> This is all well and good, but the nature of the stained objects inside the animals is a more difficult question. At first glance as you watch the quiescent or slowly moving creatures, you cannot avoid the idea that the internal pellets are eggs, because they are surrounded by bright rings as seen around frog eggs, especially when separated in space as in c and d, not compressed as in a and b [The letters refer to Figure 2]. Even though they are stained, you might think they are animals within animals. Against this conclusion, however, is that I could not see the slightest motion of pellets in the water outside of the animals, even though I watched for hours. Even

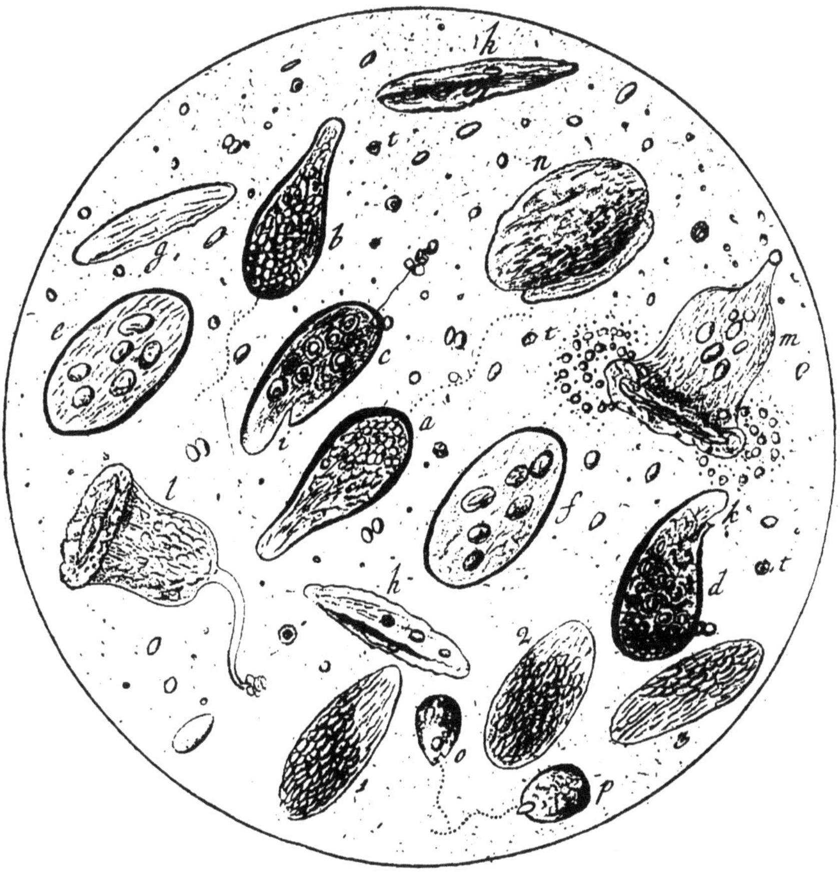

Figure 2. Drawings by Gleichen-Russworm (1786) of protists that have ingested carmine dye.

inside the animals, the pellets moved little and then only slowly. Once I saw them roll about rapidly in both parts of an animal that had recently undergone fission.

I persisted with these observations for four weeks and must say that of all those you find in this book none required so much time, patience and visual concentration as these to convince me that the pellets truly discharged from the animals. I could see them pop out of the rear and once from the sides (c and d) of an animal; I even saw two and three come out connected together.

Several times I also saw animals of this kind which dragged behind a bundle of the same pellets (c) only to lose it after a while. But I must notify anyone who wants to see such discharge of pellets to spare himself the time and effort waiting for such an event which is so rare that I only saw it 10 to 12 times in innumerable observations. Once and only once I even saw a completely quiescent *Pandeloquen* in an infusion which drew towards and then away from itself a very transparent bit of slime sprinkled with red pellets which could have been offspring that came to lie 5 or 6 lengths away from the animal. With each pull back towards the animal I noted the multiplication of small mobile black dots which seemed to appear from beneath the immotile animal and now rolled around what might been its spawn. Finally, the animal jerked and took off so fast that I immediately lost it from view. All this gave the impression of birth.

Henceforth I devoted all attention to the red pellets beneath my magnifier, ignoring all else. As soon as I noted drying of the water, I added fresh water with great caution not to sweep away pellets that I had observed. I repeated this process up to four times, attempting to prolong the observation period so as to find out whether the sessile pellets would move or grow. I detected neither, although I sometimes got the impression that tiny black spots appeared from their surroundings (t). This observation was unclear, nevertheless, as I used my highest power lens with which everything outside of an animal appears dark. Given the aforementioned doubts about the vitality of these pellets and the likelihood that they were eggs, I confess that I know not what they are if not excretions of the animals. But there are possible objections to this idea.

This infusion also contained those animals which I had seen in others, especially wheat pollen infusions. They are described in my *Microscopic Discoveries*, Table 12. They move forward with the broad part of their bodies followed by the narrower part, like a carp's tail, only much faster, shown in a and b. Although I never saw such movement of the *Pandeloquen* as in d and c, I believe they belong to the same species. Bright oval animals covered with bubbles (e and f) were new to me. I considered that they might have come from the teacup I always have beside the microscope to dilute the most concentrated infusion on the slide as I had done with this infusion. The water is no older than 36 hours, and I found a lot of these animals in the first drop. (This should serve as a warning to test any water used for this purpose). To find out whether they also would ingest colored water, I poured a bit from a teacup into the infusion. They remained white and transparent over a period of 14 days.

Finally I have to introduce two *Pandeloquen* (h) which already had released most of their pellets. A careful inspection of animals c and d showed in both animals an incision at their narrow side (i and k) which resembles so much a mouth in shape and position that I truly think it is one. This discovery I consider to be most fortuitous because this incision is only visible when it is open and this happens only in a few animals and then only for a few moments. I would have kept silent about it, had the animals not done me the favor of opening their mouths more than the first one I saw, and did so a bit more slowly, thereby allaying all my doubts.

There was an astonishing increase in my animals in the stained infusion, whereas they diminished in the unstained wheat or mother fusions at the same rate. And then at the end of March I discovered a bell-shaped animal (1). Its gait was not fast, but it turned round and round. On this occasion I could not find more than two animals of this species in the same drop. The next day, however, I noted them much multiplied, and there was one with a rather large red spot in its body and a small pellet at the sharp pointed head (m). The frontal part, which is the mouth, was dilated. The animal soon contracted, and the sides in both directions were pulled unevenly as if the perimeter of a wheel were pulled by the spokes on one side, giving the body an oval shape with only the mouth appearing beneath it, like a snail under its shell.

*Mr. Pastor Gotze had the good fortune to observe in a hay infusion masses of those infusoria which were described by Mr. Privy Councillor Müller and named *Trichoda cimex* (hairbugs), because they are trimmed with bristles both at the top and the rear parts of their bodies. As he described their eagerness to devour and their skill in catching other infusoria, these are real predators in the microscopic world which you might call 'infusion wolves.'

III. EARLY STUDIES OF THE MECHANISM OF PHAGOCYTOSIS

Evident in Gleichen-Russworm's description is the not unreasonable belief that protists are "complete animals," endowed with the organs of larger animals on a smaller scale. This view was especially championed by the German biologist Christian Gottfried Ehrenberg (1795–1896). In his observations reported on the behavior of amebas, he insisted that they ingested objects through a mouth, indeed with a trunk, and that the objects came to rest within a diminutive stomach and intestines (Ehrenberg, 1839).

Opposing this view were Félix Dujardin (1801–1867), a professor of biology at Rennes, France and also the French naturalist Jean Baptiste Lamarck, best known for advocating that acquired traits can be heritable. They asserted that amebas did not have external or internal organs, but rather imbibed nutrition through "pores." To account for the internalization of large solid objects, they allowed that openings in the surface might appear, but that these were transient (Dujardin, 1839, 1841; Rousseau and Laissus, 1972). This Franco-German dispute stalemated at the predictable level of mutual accusations about competence and adequacy of instrumentation.

Not until much later was the issue resolved, when Albert von Koelliker (1817–1905) of Würzburg, Germany caught the heliozoan, *Actinophrys sol*, in the act of ingesting another protozoan (von Koelliker, 1848). He described how the prey initially attached to one of the many spines which project from the entire circumference of the heliozoan's body, and how the spines then retracted so that the target became apposed to the cell body. Thereupon a "hole" opened in the cell periphery and closed over the object. Six years later, Edouard Claparède (1832–1871) of Geneva confirmed von Koelliker's fundamental finding and extended it by documenting that the peripheral cytoplasm of ameboid organisms appeared to flow around objects undergoing internalization. According to his description, something more active than the simple "opening of a hole" was taking place (Claparède, 1854).

> The amebas feed in a most remarkable way. They glide slowly along, attach themselves like snakes to the prey to be swallowed and, like a soft mist moving across a landscape, completely encircle it: one has the impression that the object still lies underneath, but it has already been enclosed within the body.

IV. DISCOVERY OF LEUKOCYTES AND THEIR TRAFFICKING

Although by mid-nineteenth century the phagocytic act had been seen only in unicellular microorganisms and had not yet been recognized as a property of metazoan cells, something was already known about the "colorless corpuscles," some of which were later to be defined as phagocytes. It is possible that leukocytes were first seen in 1678 by the father of microscopy, Antonie van Leeuwenhoek (1632–1743) of Delft, in human saliva (Dobell, 1960). They were more definitively described by Joseph Liétaud (1705–1780) and by Jean Baptiste de Sénac (1693–1770) of Versailles in 1750 as components of the blood and by William Hewson (1739–1774) of London in 1771 as occupants of the lymph (Hewson, 1846). Hewson proposed that leukocytes were the progenitors of erythrocytes in the blood. Now that we know that hematopoietic stem cells circulate, Hewson's proposal was correct!

The association between pus and inflammation was known to ancient physicians, but the composition of pus was not. Although corpuscles were observed in pus by de Sénac, a more definite relationship between blood leukocytes and pus was not established until the late 1830s. Friederich Gustav Jakob Henle (1809–1886) in Berlin, George Gulliver (1804–1882) in London, and Julius Vogel (1814–1880) in Giessen, Germany noted the similarity between the dimensions and granularity of "pus corpuscles" ("Eiterkörperchen") and certain blood corpuscles and the appearance of "pus cells" in the blood of animals and humans with suppurative wounds (Gulliver, 1838; Henle, 1838; Vogel, 1838; Hewson, 1846).

In 1843 William Addison (1802–1881), a physician of Malvern, England, reported that leukocytes moved from the circulation into the tissues, and that this movement increased during inflammation. In particular, he stated that leukocytes entered tubercules from the blood in cases of tuberculosis (Addison, 1844; Rather, 1972). Addison's claims were in keeping with the partially correct "blastema" theories favored by Jean Cruveilhier (1791–1874) of Paris and later by Rudolph Virchow (1821–1902), the founder of cellular pathology, which posited that tissue cells were replenished and nourished by liquid and the formed elements of the blood, and that this process was accelerated during inflammation. Where Addison failed to stake a clear claim on establishing the phenomenon of leukocyte emigration from the blood to the tissue, later to be namd diapedesis, was in his conviction that capillaries had no cellular walls. This detail was corrected three years later by Augustus Waller (1816–1870). Waller, also an Englishman but who did most of his work in France and Germany and who is best known for the discovery of neuronal Wallerian degeneration, set up an ingenious system for observing capillaries of the frog tongue with the microscope. He found that in response to local injury, leukocytes emigrated through capillaries which he clearly recognized had a cellular lining. Waller too fell short of a definitive description of diapedesis, because he concluded that the process was totally passive, and that the leukocytes did not actively migrate (Waller, 1847).

Waller's conclusion was not surprising, because it was not until the following year, 1847, that Thomas Wharton Jones (1808–1891), a London eye surgeon and physiologist, reported for the first time the ameboid movements of leukocytes (Wharton Jones, 1846). He also differentiated the leukocytes into non-granular, "finely granulated" and "coarsely granulated" types: later to be called lymphocytes, neutrophils, and eosinophils, respectively. The observation of leukocyte ameboid movement was independently made by Davaine (1812–1882) of Paris in 1850 (Davaine, 1850) and confirmed by Nathaniel Lieberkühn (1822–1867) in Germany in 1854 (Lieberkühn, 1854).

However, the ability of leukocytes, like amebas, to ingest prey was not recognized up to that time, and, indeed, Lieberkühn concluded that these cells were unable to engulf particles (Lieberkühn, 1854) . The error was corrected in 1862 when Ernst Haeckel (1834–1919), a noted German marine biologist and social Darwinist, described the ability of molluscan leukocytes to ingest India ink particles (Haeckel, 1862), and this attribute was confirmed by others, including Lieberkühn, for the granular mammalian blood leukocytes (Schultze, 1863; Von Recklinghausen, 1863).

Within a few years, details concerning the morphology of leukocyte movements and ingestion became clarified. Albert von Recklinghausen (1833–1910) a disciple of Virchow and a professor of pathology in Königsberg, Germany, noted for the first time the translocational locomotion of leukocytes (von Recklinghausen, 1863). "Ameboid" movements reported hitherto consisted of cellular shape changes and extensions of pseudopodia but not actual locomotion. von Recklinghausen also described the movements of cellular granules ("Molekularbewegung") at sites of pseudopod extensions and saw that the granules never quite moved to the hyaline tips of the extended pseudopodia. William Preyer (1841–1897), an English born naturalist working in Berlin and Max Schultze (1810-1902), a German biologist who coined the term "protoplasm," made descriptions of the morphology of leukocyte locomotion and phagocytosis which are nearly as complete as any made to this day (Preyer, 1864; Schultze, 1865).

In 1867, Julius Cohnheim (1839–1884), another pupil of Virchow's, concluded that leukocyte emigration through capillaries was an active process (Cohnheim, 1873), only to retract this claim six years later under the influence of the dominant theory that vascular pemeability changes alone were sufficient to account for the inflammatory process. According to this view, the hydrostatic pressure of the circulation forced fluid and cells out of capillaries, the permeability of which was increased by inflammation. Nevertheless, important phenomenological observations about leukocytes and inflammation were made around that time. In 1868, Hering found that leukocytes became adherent to vessels near sites of injury (cited by Metchnikoff, 1891), and in the following year Bilroth (1829–1894) noted that connective tissue destruction appeared to be associated with leukocyte collections in inflamed sites (Bilroth, 1869).

Except for Cohnheim's retraction, there were few notable events concerning phagocytes during the next decade. As early as 1847, Alexander Ecker (1816–1887) of Basel described erythrocytes inside of rabbit spleen cells. He speculated that the red cells may have been "jammed" into the larger cells by circulatory forces (Ecker, 1847). In 1870 Lieberkühn showed that leukocytes could ingest erythrocytes (Lieberkühn, 1870). William Osler, then studying at Oxford, found erythrocyte-containing leukocytes in the blood of humans with various diseases (Osler, 1882). Bizzozero (1846–1901), noted for the discovery of blood platelets, described in 1871 the clearance of carbon particles by cells of lymph nodes and also the internalization of small leukocytes by larger leukocytes in the spleen. He referred to these cells as "cellilliferous."

Thus, by 1881 the activities of phagocytes were for the most part known, including their movement from circulation to the tissues, their tendency to adhere to capillary surfaces, their translational movement, their ingestion ability, their appearance in pus and association with inflammation and tissue destruction. Yet, except for the blastema theory, there was no explanation to integrate all of this information, and the stage was set for Metchnikoff.

V. THE METCHNIKOFF SYNTHESIS

Metchnikoff was born in Kharkov, Russia where he obtained his university education. He studied marine biology in Italy and Germany, specifically focusing on the digestive processes of coelenterates, a function involving phagocytic uptake of particulate material (Schultze, 1865). Ironically, his first trip out of Russia was to Würzburg, Germany in an attempt to study with von Koelliker who had first observed the act of phagocytosis. Since the academic calendars in Germany and Russia differed, Metchnikoff arrived penniless in Würzburg during the vacation term, and his effort was in vain. He eventually took a teaching position at the University of Odessa in 1875. During an excursion to Messina, Sicily in 1879, he observed wandering phagocytes in the tissues of a variety of marine organisms which seemed to have no relation to their digestive tracts. He again departed for Messina in 1882 because of political repression resulting from the assassination of Czar Alexander II. This move was facilitated by the economic independence afforded by the death of his wealthy second wife's parents. Not needing other employment (or research grants!), he pursued his interests in marine biology. The story of how he completed his first experiment suggesting a role for phagocytes in inflammation has often been told. The idea that mobile cells might be important for defense occurred to him as he observed such cells within the transparent larva of a starfish in the microscope. He inserted a rose thorn into a larva and found the next day to his delight that the point of the thorn was covered with leukocytes.

In the ensuing years Metchnikoff promulgated the theory that there is a constant battle between microorganisms and phagocytes. If the battle is won, the organism survives, if not, it dies. In either case, the cost of the battle is inflammation which includes tissue destruction. The ability of the phagocytes to kill microorganisms and to degrade tissue derives from "ferments" within the leukocyte. In addition, Metchnikoff, stimulated by Carl Claus of Vienna, introduced the term "phagocyte" to replace the less dignified "eating cells" ("Fresszellen") and described the two principal types of phagocyte, the macrophage and the granulocyte (he called these microphages). He noted that the microphages were very motile, had lobulated nuclei and that they were eventually ingested by the less mobile macrophages.

Alexander von Humboldt wrote that the frequent sequence of reactions to an important discovery is first a denial of its veracity, then a denigration of its importance and finally usurpation of credit for it. The response to Metchnikoff was unexceptional in this regard. Although powerful scientists such as Virchow and Koch accepted his ideas rapidly, others resisted them. A major source of disbelief was the primacy of the vascular theory of inflammation which had led Cohnheim to retract his ideas concerning the active role of leukocytes in diapedesis. Metchnikoff's rebuttal to these objections is especially noteworthy because of his brilliant use of comparative pathology. He surveyed the response of organisms throughout the phylogenetic tree to injury pointing particularly to examples where phagocytes accumulated at injured sites in organisms lacking a vascular system, such as the starfish larva, Daphnia, sponges, jellyfish and worms. He also invoked the fact that diapedesis of leukocytes, but not of erythrocytes, occurred in recently killed animals with arrested blood circulations.

Metchnikoff cited the observation of an increased phagocytosis by leukocytes of microorganisms in immunized animals as evidence for his theory. But here he was on softer ground, since the role of immunization in phagocytosis was later shown to be indirect, and since his adversaries marshalled evidence that infectious diseases prevailed in instances where there was brisk phagocytosis. Interestingly, some found Metchnikoff's ideas about battles between microbes and purposeful phagocytes too anthropomorphic and attacked him on this ground (Baumgarten, 1889).

Metchnikoff's detractors also adumbrated others' work mentioned above as establishing prior to Metchnikoff a scavenger role for phagocytes in tissue remodeling at the same time that they doubted his contention that these cells were involved in antibacterial defense. The best claim for some priority in this regard could go to Müllendorf (1859–1879), who observed spirochetes of the species *Borellia recurrentes* enclosed within peripheral blood leukocytes of patients with recurrent fever and suggested that the leukocytes might be enaged in digesting the microorganisms ("Verzehrungsprozess") (Müllendorf, 1879) . Müllendorf's early death in the year that he reported this observation precluded his pursuing any priority claim.

Nevertheless, Metchnikoff's view of phagocytes as scavengers was rapidly accepted, and by 1890 the phagocyte theory of host defense and inflammation was well established and his priority as its champion assured. The place of phagocytosis in immunity had to be shared with humoral factors when they were discovered around that time. Strong support for Metchnikoff's ideas came from the work of Leber, published in 1888, which introduced the phenomenon of chemotaxis of leukocytes toward damaged tissues (Leber, 1981) . Metchnikoff joined the staff of the Institut Pasteur in Paris in 1888 where he remained until his death. He shared one of the first Nobel Prizes in Physiology and Medicine with Paul Ehrlich, the father of humoral immunity, in 1904.

In assessing what led Metchnikoff to synthesize the available facts into a coherent and powerful theory, we might speculate that the newly evolved and persuasive germ theory of disease became coupled in Metchnikoff's mind to his experimental observations of phagocytes voraciously ingesting such microorganisms, a phenomenon which, as has been described in this chapter, was well established.

Why did Metchnikoff not even cast a nod to those who made such observations? Possibly the relevant information was well known and required no attribution. Perhaps Metchnikoff was so focused on the big picture and fighting for his ideas that a mere detail such as who described the phagocytic process of cells was too trivial to countenance. If so, Metchnikoff is in the good company of many brilliant scientists who either are so self-absorbed in their creativity or else so dependent on external validation, that they minimize or totally ignore the often modest but essential accomplishments of others which may have been antecedents to their glory. Some scientists appreciate the subjectivity of their conclusions and decide that the generous attribution is a human value and that the appreciation of those credited is as "true" as the most venerated scientific fact. In the end, everybody's dead, so maybe it isn't important. But it isn't the sharing of credit, of which Metchnikoff got plenty, that kills you.

ACKNOWLEDGMENTS

I thank Ned Hiam for generously and consistently supporting my research program which led from scattered studies concerning phagocytosis to a long effort to understand at least some mechanisms by which cells crawl. These mechanisms may or may not finally explain how phagocytosis actually takes place, that relatively trivial detail in the broader concept of phagocytic function. I hope that other clinically important spinoffs will emerge from this research. My parents rummaged through libraries to find some of the references cited in this chapter. I hope they enjoyed the diversion, because since it took so long for me to get around to collating the information, they are no longer able to see it in an organized form.

REFERENCES

Addison, W. (1844). The actual process of nutrition in the living structure demonstrated by the microscope; and the renewal of the tissues and secretions, with the phenomena and products of inflammation. Trans. Prov. Med. Surg. Assn. 12, 235–306.

Baumgarten, P. (1889). Zur Kritik der Metschnikoff'sschen Phagocytentheorie. Z. f. Klin. Med. 15, 1–41.

Bilroth, C.A.T. (1869). Mancherlei über die morphologischen Vorgänge bei der Entzundung. Med. J. 18, 18.

Claparède, E. (1854). Ueber Actinophrys Eichhornii. Müllers Archiv 398–419.

Cohnheim, J. (1867). Ueber Enztundung und Eiterung. Virchows Arch. Path. Anat. Physiol. Klin. Med. 40,179–202.

Cohnheim, J. (1873). In: Neue Untersuchungen über die Entzundung. August Hirschwald, Berlin.

Davaine, C. (1850). Les globules blancs du sang. Societé de Biologie (Paris). C.R. hebdomdaires, Mem. II, 103–105.

Dobell, C. (1960). In: Anthony van Leeuwenhoek and his Little Animals. Dover Publications, New York.

Dujardin, M.F. (1838). Mémoire sur l'organisation des Infusoires. Ann. des Sciences Naturelles 2e Ser, Zoologie. 10, 230–315.

Dujardin, M.F. (1841). In: Histoire Naturelle des Zoophytes Infusoires, p. 228. Paris.

Ecker, A. (1847). Ueber die Veränderungen welche die Blutkörperchen in der Milz erleiden. Z. Rationelle Med. 6, 260–265.

Ehrenberg Chr. G. (1839). In: *Die Infusionstierchen als volkommene Organismen.* Vol. 1, p. 127. Voss, Leipzig.

Gleichen-Russworm, F.W. (1786). In: *Abhandlung über die Saamen und Infusionsthierchen.* pp. 140–144. A.W. Winterschmidt, Nürnberg.

Goeze, J.A.E. (1777). Infusionstierchen die Andre fressen. Beschäftigungen der berlinischen Gesellschaft naturforschender Freunde 3, 373–384.

Gulliver, G. (1838). Researches on suppuration. London & Edinburgh Phil. Mag. J. Sci. 13, 193–202.

Haeckel, E. (1862). In: *Die Radiolaren.* Vol. I, pp. 135–140. Karl Riemer, Leipzig.

Heifets, L. (1982). Centennial of Metchnikoff's discovery. J. Reticuloendothelial Soc. 31, 381–391.

Henle, F.G.J. (1838). Ueber Schleim und Eiterbildung und ihr Verhältnis zur Oberhaut. Hufeland Journ. Arzn. 5, 3–62.

Herrlinger, R. (1956). Die historische Entwicklung des Begriffes Phagocytose. Ergebn. d. Anatomie u. Entwicklungsgeschichte 35, 334–357j.

Hewson, W. (1846). An inquiry into the properties of the blood with remarks on some of its morbid appearances. In: *The Works of William Hewson,* pp. 78–82, 244. The Sydenham Society, London.

Hirsch, J.G., & Hirsch, B.T. (1982). Metchnikoff's life and scientific contributions in historical perspective. In: *Phagocytosis, Past and Future.* (Karnovosky, M.L. & Bolis, L., Eds.), Academic Press, New York.

Leber, T. (1981). *Die Enstehung der Entzundung und die Wirkung der Entzundungerregenden Schädlichkeiten nach Vorzugsweise am Auge angestellten Untersuchungen.* Wilhelm Engelmann, Leipzig.

Lieberkühn, N. (1854). Ueber die Psorospermien. Müllers Arch. f. Anat. Physiol. Wiss. Med. 1–24.

Lieberkühn, N. (1870). In: *Ueber Bewegungerscheinungen der Zellen.* pp 358–359. Marburg.

Metchnikoff, E. (1882). Ueber die intrazellulare Verdauung bei wirbellosen Tieren. Arb. aus dem Zool. Inst. zu Wien II, 141–158.

Metchnikoff, E. (1891). *Lectures on the Comparative Pathology of Inflammation.* Transl. P.H. Starling. E.H. Starling, Dover Publications, New York, 1968.

Metchnikoff, E. (1901). Immunity in Infective Diseases. Trans. F.G. Binnie, Cambridge.

Metchnikoff, O. (1920). Vie d'Elie Metchnikoff 1845–1916, pp. 106–111, Librairie Hachette, Paris.

Metschnikoff, E. (1887). Sur la lutte des cellules de l'organisme contre l'invasion des microbes. Ann. Inst. Pasteurl, 321–336.

Müllendorf, J. (1879). Febris recurrens nach Beobachtungen im Stadt. Krankenhaus Dresden. Diss. Leipzig.

Osler, W.M. (1882). Note on cells containing red blood corpuscles. Lancet 1, 238.

Preyer, W. (1864). Ueber amoeboide Blutkörperchen. Virchows Arch. Path. Anat. Physiol. Klin. Med. 30, 417–441.

Rather, L.J. (1972). Addison and the White Corpuscles: An Aspect of Nineteenth-Century Biology. University of California Press, Berkeley.

Rousseau, G., & Laissus, Y. (Eds.) (1972). In: *Inédits de Lamarck.* p. 23. Masson & Cie, Paris.

Schultze, M. (1863). *Das Protoplasma der Rhizopoden und der Pflanzenzellen.* Wilhelm Engelman Verlag, Leipzig.

Schultze, M. (1865). Ein heizbarer objekttisch und seine verwendung bei untersuchungen des blutes. Arch. f. Mikrosk. Anat. 1, 1–42.

Tauber, A.I., & Chernyak, L. (1991). *Metchnikoff and the Origins of Immunology. From Metaphor to Theory.* Oxford University Press, Oxford.

Vogel, J. (1838). In: *Physiologisch-pathologische Untersuchungen über Eiter, Eiterung und die damit verwandten Vorgänge*. J.J. Palm, S.E. Henke, Erlangen.
von Koelliker, A. (1848). Das Sonnentierchen, Actinophrys sol. Z. Wiss. Zool. 1, 198–217.
Von Recklinghausen, F. (1863). Ueber Eiter und Bindegewebskörperchen. Virchows Arch. Path. Anat. Physiol. Klin. Med. 28, 157–197.
Waller, A. (1847). Microscopic observations on the perforation of the capillaries by the corpuscles of the blood, and oil—the origin of mucus and pus-globules. London & Edinburgh Phil. Mag. J. Sci. 29(3rd ser), 397–405.
Wharton Jones, T. (1846). The blood-corpuscle considered in its different phases of development in the animal series. Phil. Trans. Roy. Soc. London. 63–87.

DROSOPHILA HEMOCYTES, PHAGOCYTOSIS, AND CROQUEMORT, A MACROPHAGE RECEPTOR

Nathalie C. Franc

Advances in Cell and Molecular Biology of Membranes and Organelles
Volume 5, pages 19-46.

ISBN: 1-55938-999-0

I. INTRODUCTION

Phagocytosis is a critical event in both immune defense and development, and is essential for survival of all multicellular organisms. Phagocytosis is the process of recognition and engulfment of particles such as pathogens or tissue debris that accumulate during infection, inflammation and wound repair (reviews in Unanue, 1984; Ezekowitz, 1989). It is also involved in the clearance of cells that undergo programmed cell death (PCD, also called apoptosis) (review in Savill, 1997b). Phagocytosis of apoptotic cells prevents the damage that these cells could induce if their cytoplasmic contents were released in the developing tissues (review in Wyllie et al., 1980; Raff, 1992). It also allows the elimination of the T lymphocytes during the resolution of a viral infection (Akbar et al., 1994; review in Savill, 1997a). In vertebrates, phagocytosis is mainly performed by migrating, bone marrow-derived cells called professional phagocytes. They are the neutrophils and monocytes that circulate in the blood, and the macrophages that reside in every tissue.

Phagocytosis requires sequential signal transduction events that lead to (i) the recruitment of the phagocytes at the site of infection or apoptosis, (ii) the recognition of the particles to be ingested, (iii) the movement of the cytoskeleton of the phagocyte that forms pseudopodia around the particle to be ingested, and (iv) the formation and maturation of a phagosome, a structure that results from the fusion of intracellular vesicles such as lysosomes with the plasma membrane around the particle. The phagosome is where degradation of the ingested particle occurs. Certainly the most important event in phagocytosis is the recognition. It is essential that phagocytes distinguish specifically between "self" and "non-self," or "abnormal self." Therefore, phagocytes express specific surface receptors that have evolved to recognize biochemical features of pathogens, or recognize host proteins deposited onto the pathogen in host defense (review in Brown, 1994). In addition, receptors have evolved to recognize biochemical changes at the surface of the cells undergoing apoptosis (review in Savill, 1997b). Because of the diversity of pathogens that infect mammals and the different origins of the dying cells, these mechanisms of recognition are multiple and intricate. This variety and complexity is well illustrated by the nature of the various macrophage phagocytic receptors that have been characterized so far in vertebrates, some of which are described in this volume.

Over the past decade, studies have demonstrated that the immune defense of insects resembles the innate immunity of vertebrates and relies on both humoral and cellular mechanisms that are tightly interconnected. The humoral response is characterized by a rapid and transient synthesis of antimicrobial peptides by the fat body, the functional analogue of the mammalian liver and by the hemocytes, or blood cells (reviews in Hultmark, 1993; Boman, 1995; Hoffmann et al., 1996; Hoffmann et al., 1999). In addition, there is induction of proteolytic cascades such as the coagulation cascade that leads to blood clotting at the site of injury, and the prophenoloxidase cascade that leads to melanization of large intruders, such as eggs of parasitic wasps (reviews in Rizki and Rizki, 1984; Nappi and Vass, 1993). The cellular reactions consist predominantly of phagocytosis and/or encapsulation of invading microorganisms (reviews in Gupta, 1979; Ratcliffe, 1993). *Drosophila* hemocytes are important effectors of these defense mechanisms, as they accumulate at the site of infection and participate in wound healing, encapsulation and phagocytosis of pathogens (reviews in Gupta, 1979; Ratcliffe, 1993). Moreover, *Drosophila* hemocytes also play an important role in removal of apoptotic cells during development as well as during metamorphosis (review in Wyllie et al., 1980; Tepass et al., 1994). Three *Drosophila* proteins have been recently identified which may be phagocytic receptors in immune defense and/or developmental events. Among them is croquemort, a CD36-related receptor (cluster of differentiation 36 antigen), which we have shown to function as a phagocytic receptor *in vivo* in both mammalian and insect systems (Franc et al., 1996).

This chapter provides an overview of the origins and multiple functions of *Drosophila* hemocytes, and a discussion of the role of croquemort as a macrophage (embryonic phagocytic hemocyte) receptor in phagocytosis of apoptotic cells during *Drosophila* embryogenesis.

II. ORIGINS AND FUNCTIONS OF *DROSOPHILA* HEMOCYTES

A. During Embryogenesis

Developmental Pattern of Embryonic Hemocytes and Conversion into Macrophages

In a very elegant study, Tepass and colleagues described the embryonic conversion of hemocytes into macrophages during *Drosophila* embryogenesis (Tepass et al., 1994). They clarified the origins of hemocytes in embryos, followed their developmental patterns and compared their distribution with that of cells fated to die. Hemocytes appear approximately two hours after gastrulation (stage 10) and derive from the procephalic mesoderm that overlies the neuroblasts in the head region. This mesoderm forms a separate mitotic domain where cells undergo four divisions during stages 8–11. In addition to the procephalic mesoderm, small groups of cells in the gnathal segments may also give rise to hemocytes. After the final division, the majority of hemocytes are scattered round or irregularly shaped cells that are free in the hemocoel (plasma). By stage 11, the complement of 700 hemocytes is reached and this number will remain constant throughout embryogenesis. Hemocytes spread throughout the embryo, moving anteriorly and ventrally to populate the clypeolabrum and gnathal segments of the head, and posteriorly to populate the tail end of the germ band. They subsequently migrate from both ends of the embryo towards its middle following four different routes.

At stage 15, hemocytes are evenly distributed throughout the embryo, and by the end of embryogenesis, approximately 90–95% of the hemocytes have become phagocytic as they have encountered apoptotic cells and are termed "macrophages" (Abrams et al., 1992, 1993; Tepass et al., 1994). In conditions of increased apoptosis, all hemocytes appear to be competent to become macrophages (Tepass et al., 1994).

There is no evidence that *Drosophila* macrophages are required for apoptosis. Indeed, mutants like *Bicaudal D* or *twist/snail* that lack all head mesoderm are unable to give rise to hemocytes and differentiated macrophages, yet proceed with apoptosis (Tepass et al., 1994). In this regard, *Drosophila* macrophages might behave differently than mammalian macrophages. Indeed, using a liposome-mediated macrophage elimination technique, Diez-Roux and Lang have reported that the programmed capillary regression in the mouse developing eye is dependent upon macrophage-induced apoptosis (Diez-Roux and Lang, 1997).

Drosophila macrophage proliferation is independent of cell death as the number of hemocytes does not increase in a context of induced PCD (Tepass et al., 1994). Conversely, hemocytes can be found even in the absence of PCD (White et al., 1994).

Embryonic Hemocyte Differentiation

Although the origin of embryonic hemocytes has been well documented, little was known about the mechanisms of differentiation of these cells. Recently, a gene, *serpent* (*srp*), which encodes a protein that belongs to the GATA family of transcription factors (previously described as ABF or dGATAb) (Abel et al., 1993), was shown to be essential for embryonic hematopoiesis in flies. Indeed, mutant embryos for *srp* are devoid of mature hemocytes (Rehorn et al., 1996). Interestingly, in vertebrates, transcriptional regulators of the GATA family have an essential function in differentiation and organ development. GATA-1 is required for primitive and definitive erythropoiesis (Pevny et al., 1991), GATA-2 for early hematopoiesis (Tsai et al., 1994), while GATA-3 is implicated in the differentiation of T lymphocytes (Ko et al., 1991). GATA-4, -5 and -6 are expressed during development of various organs of endodermal origin (Arceci et al., 1993; Laverriere et al., 1994; Soudais et al., 1995). *Srp* is also essential for endodermal development in flies (Abel et al., 1993; Sam et al., 1996), and therefore is considered as a functional homologue of several if not all members of the vertebrate GATA family. The role of *srp* in embryonic hematopoiesis suggests that some aspects of the molecular mechanisms that lead to blood cell development are early acquisitions of metazoan evolution and may be common to most higher organisms.

Recently, Glide/Gcm, a novel protein with DNA binding properties that was first reported as a glial promoting factor, was also shown to be a very early marker for the hemocyte lineage. *Glide/gcm* expression is dependent on the activity of *srp* and appears to be necessary and sufficient to induce hemocyte fate as ectopic expression of this gene induces some hemocyte differentiation (Bernardoni et al., 1997).

Embryonic Hemocytes Role in Production of Extracellular Matrix Proteins

Embryonic hemocytes are believed to play a key role in preserving tissue cohesion during development. Indeed, they are major producers of extracellular matrix (ECM) proteins, which accumulate in the intercellular spaces of the embryo and form basement membranes that underlie epithelia, surround muscle and fat body cells, enclose the central nervous system and form the boundary of the hemocoel (reviews in Fessler and Fessler, 1989; Fessler et al., 1994). Interestingly, a major signal for activation of phagocytes in vertebrates is their interaction with the extracellular matrix (review in Brown, 1994). A number of ECM proteins have been identified in *Drosophila* that are mainly produced by embryonic hemocytes, such as Peroxidasin, the two Collagen IV homologues, DCGL (Natzle et al., 1982) and Viking (Abrams et al., 1992; Yasothornsrikul et al., 1997), Tenascin (Baumgartner et al., 1994), Laminins, Glutactin, Tiggrin and the proteoglycan Papilin (review in Fessler et al., 1994). In their study, Tepass and his colleagues took advantage of this function of hemocytes to follow their pattern of development using an antibody directed against Peroxidasin (Tepass et al., 1994). Interestingly, Peroxidasin is an ECM protein that combines ECM motifs and a functional peroxidase domain also found in human oxidative defense proteins (Nelson et al., 1994). Peroxidasin is primarily expressed in hemocytes and in the fat body cells in later embryonic stages and may contribute to immune mechanisms in a similar manner to oxidative defense proteins. Most recently, another ECM protein, a macrophage derived proteoglycan, MDP-1, was characterized that is secreted exclusively by migratory embryonic macrophages (Horstch et al., 1998). Embryos that are deficient for the three regulators of PCD in *Drosophila*, *reaper*, *hid* and *grim* (White et al., 1994, 1996; Grether et al., 1995; Chen et al., 1996), show a greatly reduced amount of MDP-1 production, which suggests that apoptosis may have a role in the terminal differentiation of *Drosophila* hemocytes into migratory phagocytic hemocytes, the macrophages (Horstch et al., 1998).

Embryonic Hemocytes and Scavenger Receptor Activity

Embryonic macrophages are the key effectors of phagocytosis of apoptotic cells during development (Tepass et al., 1994). Interestingly, embryonic macrophages were shown to exhibit scavenger receptor-mediated endocytosis (Abrams et al., 1992). Scavenger receptors are defined by their ability to mediate the recognition and uptake of a wide range of polyanionic molecules, such as modified low density lipoproteins (LDL) which may play an important role in normal and pathological lipid metabolism (reviews in Pearson, 1996; Rigotti et al., 1997). Three classes of scavenger receptors have been found so far. Both *in vivo* and *in vitro* studies of the members of the first two classes of scavenger receptors identified in vertebrates indicated that they bind a wide variety of pathogens, suggesting that these receptors may serve as recognition and phagocytic receptors in innate immunity (review in Krieger, 1997). Other evidences support their role as phagocytic receptors for apoptotic cells (review in Savill, 1997b).

The scavenger receptor DSR-CI was the first hemocyte receptor described in *Drosophila* that defines the third class of scavenger receptor (Pearson et al., 1995). The *dSR-CI* gene was isolated from L2 Schneider cells, an embryonic cell line that exhibits a scavenger activity. DSR-CI, like SR-AI/II, the mammalian scavenger receptor class A type I and II, exhibits high-affinity binding to a broad array of polyanionic ligands, which includes β-glucan (review in Pearson, 1996). Interestingly, β-glucan has been described as a potent inducer of the *Drosophila* immune response (Hultmark et al., 1993). Among several other domains, the extracellular portion of DSR-CI contains two amino-terminal complement control protein domains (CCP) (Pearson et al., 1995). In mammals, CCP domains mediate the binding of complement receptors and complement regulatory proteins to the central component of the complement cascade, C3 (Reid et al., 1989). Therefore, like the other scavenger receptors, DSR-CI might be involved in immune response, and possibly in phagocytosis of apoptotic cells during development (review in Pearson, 1996).

Another protein, Malvolio (*mvl*), which was first characterized for its function in taste behavior of the fly, shares striking similarities to natural resistance-associated macrophage proteins (NRAMPs) found in mammals (Rodrigues et al., 1995). *Mvl* is also expressed in embryonic hemocytes. This protein has a consensus motif that is found in a number of ATP-coupled transporters that might participate in nitrite transport (Vidal et al., 1993). Recently, an ATP-binding cassette transporter, ABC1, was characterized that participates in phagocytosis of apoptotic cells in mice (Luciani et al., 1996). The characterization of both of these proteins suggests that invertebrates and vertebrates rely on similar types of proteins and probably similar mechanisms of recognition and phagocytosis, and that these mechanisms may require several signaling pathways acting together.

B. Larval and Pupal Stages

Larval Hematopoietic Tissues and Blood Cell Differentiation

It remains unclear at this point whether embryonic hemocytes persist to the larval stage and may be at the origin of larval hemocytes. However, it is known that a new complement of hemocytes is produced by the hematopoietic organs or so-called "lymph glands" (functional analogue of the bone marrow) at larval stage. These glands derive from the cardiogenic region in the lateral mesoderm (Shrestha and Gateff, 1982). When fully developed, they appear as five to seven paired lobes along the dorsal vessel next to the ring gland and brain. The role for the lymph glands in larval hematopoiesis was previously substantiated by the characterization of a mutant, *lethal (3) hematopoiesis missing* (*l(3)hem*) that is devoid of lymph glands and also devoid of circulating hemocytes (reviews in Gateff, 1994a,b). The gene affected by this mutation has been proposed to be involved in cell-type specific inhibition of cell division. Recently, in a phenotypic screen for mutations that affect blood cells, Braun and colleagues have characterized a novel mutation, *domino*, which also results in the depletion of circulating hemocytes by late larval stages, apparently due to the melanization of the lymph glands (Braun et al., 1997). *Domino* mutant larvae have prolonged third instars and die after puparium formation (as early pupae). All diploid tissues (imaginal structures, neuroblasts, germline cells) show overgrowth to various extents and therefore *domino* might also have a function in cell proliferation. Furthermore, mutations in the bithorax complex (*BXC*) transform posterior parasegments to parasegments with a more anterior identity and cause the development of additional lymph gland material in these parasegments. *Dorothy*, a gene specifically expressed in the lymph glands, is regulated by the *BXC* (Rodriguez et al., 1996). This supports previous observations that this complex is involved in regulating the identity of these cells.

The lineage of larval hemocytes has not yet been fully worked out. However, they can be tentatively divided into two cell lineages according to their morphologies and functions (Rizki, 1957a; Gateff, 1980; Rizki and Rizki, 1980): (1) the crystal cells, which are characterized by prominent cytoplasmic paracrystalline inclusions, account for 5–10% of the blood cell population; (2) the proplasmatocytes that differentiate into plasmatocytes, small spherical and nonadhesive cells, which account for 85–90% of the blood cell population. Plasmatocytes eventually give rise to lamellocytes, which appear as large discoidal and adhesive cells (5–10%). Just prior to the onset of apoptosis in the prepupal stage there is a new massive release of blood cells from the hematopoietic organs. This seems to coincide with a peak of 20-hydroxy-ecdysone levels occurring six hours before pupation. Ecdysone is a steroid hormone that initiates a variety of developmental changes throughout the animal (review in Thummel, 1996). After pupation, a second ecdysone peak initiates the lysis of most larval tissues (review in Baehrecke, 1996). It has been proposed that plasmatocytes are concurrently signaled to differentiate into lamellocytes at this stage (5–10% of the blood cell population) (Gateff, 1980; review in Rizki and Rizki, 1984). These blood cells are believed to persist until adulthood, and new complements of blood cells may well originate from the lymph glands in adult flies.

Blood Cell Proliferation and Differentiation: A Model for Cancer

Melanotic tumors are black masses of tissues that can be free floating in the hemocoel, or attached to various tissues. They are thought to result from abnormal development (class 1 melanotic tumors) (Rizki and Rizki, 1980; Watson et al., 1991), or the disregulation of blood cells function (class 2) (Watson et al., 1991). They are the result of a lamellocyte-mediated encapsulation directed against the organism itself. The lamellocytes form multilayers around the tissue. These capsules get melanized due to the activity of the crystal cells, which contain the enzymes and the substrate of the prophenoloxidase cascade and produce melanin (reviews in Rizki, 1957a, 1957b; Rizki and Rizki, 1980; Shestra and Gateff, 1982). Indeed, a mutation, *Black cells* (*Bc*) results in melanin-filled crystal cells that function abnormally (Rizki et al., 1980). Mutations in genes that activate the cellular immune system lead to the formation of melanotic tumors. Very often, melanotic tumor mutations are also associated with an overgrowth phenotype of the affected tissues. These mutations affect a number of tumor suppressor genes that have been reported in *Drosophila* (review in Bryant et al., 1993; Watson et al., 1994; Harrison et al., 1995; Luo et al., 1995; Gateff, 1996). In this regard, *Drosophila* has become a model system for cancer studies and the characterization of such mutants will undoubtedly provide insights into the mechanisms of cell proliferation.

So far, six lethal mutants of *Drosophila melanogaster* were characterized that cause malignant growth of the larval hematopoietic system, namely $hop^{Tum\text{-}l}$ (*hopscotch Tumorous lethal*) (see below), *air8* (*l(1)* aberrant immune response), *l(1)mbn*, *l(2)mbn* and *l(3)mbn* (*lethal malignant blood neoplasm* mutants found on the X, second and third chromosome), and *Toto* (Gateff, 1978a,b; Hanratty and Ryerse, 1981; Watson et al., 1992; Rodriguez et al., 1996). These mutants develop enlarged hematopoietic organs and show a considerable increase in the amount of free hemocytes in their hemolymph. The gain-of-function mutation of the *hopscotch* gene, $hop^{Tum\text{-}l}$, causes overproliferation of the plasmatocytes, and aggregation of these cells into large melanizing tumors in hematopoietic and gut tissues (Hanratty and Ryerse, 1981; Harrison et al., 1995; Luo et al., 1995). Loss-of-function of *hop* causes a variety of abnormalities, including defective embryonic pattern formation and larval lethality in *Drosophila* (Binari and Perrimon, 1994). *Hop* encodes a janus kinase (JAK), a class of non-receptor tyrosine kinases that activates *stat92E*, a *Drosophila* member of the signal transducers and activators of transcription family (STAT). *Stat92E* loss-of-function phenotype suppresses the gain-of-function mutation phenotype of $hop^{Tum\text{-}l}$ (Yan et al., 1996). *Stat92E* loss-of-function mutation has a similar phenotype to *hop* loss-of-function mutations (Hou et al., 1996). This suggests that both genes encode components of the same pathway. Similar to mammals, the components of the JAK/STAT pathway are critical for both cell proliferation and cell differentiation in *Drosophila* (reviews in Hou and Perrimon, 1997, Mathey-Prevot and Perrimon, 1998).

Mutations in *air8* lead to overproliferation and early differentiation of hemocytes (among other tissues) (Watson et al., 1992; review in Bryant et al., 1993). This gene encodes a homologue of human ribosomal protein S6. Ribosomal S6 protein expression and phosphorylation are associated with changes in cell proliferation in mammalian cells (Kanda et al., 1997) and the *air8* mutant includes proliferation defects in most tissues (Watson et al., 1992).

The genes affected by the *l(2)mbn* and *l(3)mbn* mutations are not known. However, the identification of these two mutations facilitated the establishment of two tumorous blood cell lines *in vitro*, named *l(2)mbn* and *l(3)mbn* (Gateff, 1980). These cells have the ability to distinguish between self and non-self, but have lost their ability to melanize or encapsulate foreign particles. The availability of these cell lines is of considerable interest as it allows *in vitro* studies of *Drosophila* hemocytes. More recently, in a genetic screen for mutations that affect one or several of the immune tissues (i.e., blood cells, lymph glands, fat body cells) or genes that are induced upon immune challenge, Rodriguez and colleagues identified two additional mutations, *wizard* and *dappled,* which also cause melanotic tumor formation (Rodriguez et al., 1996).

The idea that *Drosophila* and mammalian systems share common mechanisms of blood cell development is further substantiated by the finding that Lozenge (*Lz*), a transcription factor that contains a runt domain homologous to a portion of a human transcription factor, AML1, which is associated with acute myelogenous leukemia, is implicated in crystal cells formation as suggested by its genetic interaction with *Bc* (Daga et al., 1996; Chosa et al., 1997). Interestingly, AML1 was recently found to also be necessary for definitive hematopoiesis in mice (Orkin et al., 1996).

The Roles of Larval Blood Cells in Innate Immunity and Signal Transduction

Each blood cell subtype has been shown to be involved in various aspects of the innate immune response of *Drosophila* larvae and adults. Plasmatocytes participate, to some extent, in the synthesis of antimicrobial peptides during the humoral response (Samakovlis et al., 1990; Meister et al., 1994) and assume the function of phagocytosis of microorganisms (reviews in Gupta, 1979; Ratcliffe, 1993), while lamellocytes and crystal cells play respective roles in encapsulation and melanization of larger intruders (reviews in Rizki and Rizki, 1984; Nappi and Vass, 1993).

Studies of the humoral response by the fat body cells has demonstrated that *Drosophila* innate immunity exhibits strong similarities with the mammalian innate immune response. For example both utilize similar signaling pathways that activate transcription factors of the *Rel/NF-kB* family (reviews in Hoffmann and Reichhart, 1997; Hoffmann et al., 1999). The Toll signaling pathway, which also controls dorsal-ventral patterning during *Drosophila* embryogenesis (review in Belvin and Anderson, 1996), regulates the nuclear translocation of Dorsal, a NF-kB related protein, in the fly immune response (Lemaitre et al., 1996). This pathway is specifically involved in the regulation of antifungal peptide synthesis, and, to a lesser extent, affects the synthesis of certain antibacterial peptides. Indeed, in homozygous mutants for the *immune deficiency* gene, *imd*, the antifungal response remains fully inducible but the antibacterial activity is not properly up-regulated and flies are more susceptible to bacterial infection (Lemaitre et al., 1995).

In the *Drosophila* immune response, the transcription of *Toll* (*Tl*) and a related gene, *18-wheeler* (*18w*), both encoding molecules related to the interleukin-1 receptor (IL-1R) of mammals, are induced upon infection (Eldon et al., 1994). 18-Wheeler is present in all immune responsive tissues, including hemocytes. Its expression is induced upon infection and *18-Wheeler* mutant larvae are more susceptible to bacterial infection, which demonstrates that it plays an active role in signaling during the immune response (Williams et al., 1997). This suggests that the 18-Wheeler pathway may also be responsible for synthesis of antimicrobial peptides by hemocytes.

More recently, new genes have been identified that are required for the normal transcriptional induction of antibacterial peptides during the immune response and define new components of signaling pathways that also involve Rel-like factors (Wu and Anderson, 1998). At least three Rel-related factors have been characterized in the fly, Dorsal, Dif, and Relish, which may be recruited together or alone by one or several of the various identified pathways (Ip et al., 1993; Dushay et al., 1996). Recently, a homologue of the *Drosophila* Toll protein, huToll, now called Toll Receptor 4 (T1-R4) was identified and shown to be involved in the human immune response as well (Medzhitov et al., 1997). Indeed, ectopic expression of a constitutively active form of this receptor leads to the activation of NF-kB and the production of inflammatory cytokines in human cells.

Interestingly loss of function mutations in *cactus,* the *Drosophila I kappa B* homologue, and constitutively active form of a *Toll,* or constitutive expression of *dorsal,* induce lamellocyte differentiation and also cause the formation of melanotic tumors (Qiu et al., 1998). Cactus and Toll proteins are expressed in the nascent hemocytes of the larval lymph glands. Tube, a unique protein, and Pelle, a serine/threonine kinase, two of the downstream components of the Toll signaling cascade, are also expressed in these tissues (Qiu et al., 1998). The Toll/Tube/Pelle signal leads to the phosphorylation and degradation of Cactus and the concomitant nuclear translocation of Dorsal (review in Belvin and Anderson, 1996). In their study, Qiu and colleagues show that a primary phenotype of the zygotic loss-of-function of *cactus* is lethality, which correlates with the number of melanized capsules present in these animals. *Cact* mutant larvae have enlarged lymph glands that contain many more hemocyte precursors than wild-type larvae. Their lethality can be rescued by expression of wild-type Cactus protein in the lymph gland, which strongly argues that the normal function of zygotic Cactus is to regulate hemocyte formation. Loss-of-function mutations in *Tl*, *tub*, and *pll* suppress *cact* lethality, and therefore imply that Toll, Tube and Pelle proteins convey an intercellular signal to Cactus within hemocytes precursors, as is the case for their maternal counterparts in the dorso-ventral polarity signaling (Qiu et al., 1998). Signaling through Toll and Cactus proteins parallels signaling induced by Toll/IL-1R activation in mammalian cells where these genes mediate inflammation and other immune-related functions (Baldwin, 1996).

The immune response of the fly can be triggered by several stimuli, such as bacteria, fungi, parasites, and cell debris (review in Hultmark, 1993). Exposure to the endotoxic lipopolysaccharide (LPS), a component of the bacterial cell wall, initiates an immune response in *l(2)mbn* tumorous blood cells and leads to the activation of a mitogen activated protein (MAP) kinase, DJNK (Sluss et al., 1996). DJNK is a *Drosophila* homologue of the mammalian c-Jun amino-terminal kinase (JNK). It is expressed in fat body cells and *l(2)mbn* tumorous blood cells where it is activated after only five minutes of treatment with LPS. This result suggests that DJNK is a target in an early event of the response to the endotoxin. Mammalian JNK has been implicated in a variety of immune-related signaling pathways, including the response to pro-inflammatory cytokines, and the response to endotoxic LPS and T cell activation (Sluss et al., 1994; Su et al., 1994; Westwick et al., 1994; Raingaud et al., 1995; Whitmarsh et al., 1995). The characterization of DJNK illustrates once more the similarities between the innate immune response of the fly and mammals.

III. CROQUEMORT: A CD36-RELATED RECEPTOR THAT PLAYS A ROLE IN PHAGOCYTOSIS OF APOPTOTIC CELLS

A. Characterization of the *Croquemort* Gene (*crq*)

In an attempt to understand how *Drosophila* macrophages can distinguish between self and non-self during infection, or between self and abnormal self in the removal of apoptotic cells during development, we used a polymerase chain reaction approach to search for candidate genes that would encode macrophage receptors. We characterized a novel gene which shares similarity with the human cluster of differentiation 36 antigen, CD36 (review in Greenwalt et al., 1992) which we named *croquemort* (*crq*) (the choice of this name will be explained later in this chapter) (Franc et al., 1996). *Crq* is localized at position 21C4 on the polytene chromosomes and encodes a protein of 457 amino acids. It shares 23–28% identity with various members of the CD36 superfamily of receptors (an alignment of Croquemort with various members of the CD36 superfamily of receptors is shown in Figure 1). Of note is that the overall identity of the various members is not higher within one species than between members of different species.

Croquemort can be divided into four domains that are illustrated in Figure 2: (a) a short 12-residue N-terminal domain, which contains four grouped cysteine residues and a putative protein kinase C (PKC) phosphorylation site; (b) a hydrophobic domain of 24 residues from residues 13 to 36; (c) a C-terminal region, which begins with a 200-residue stretch containing six potential N-glycosylation sites

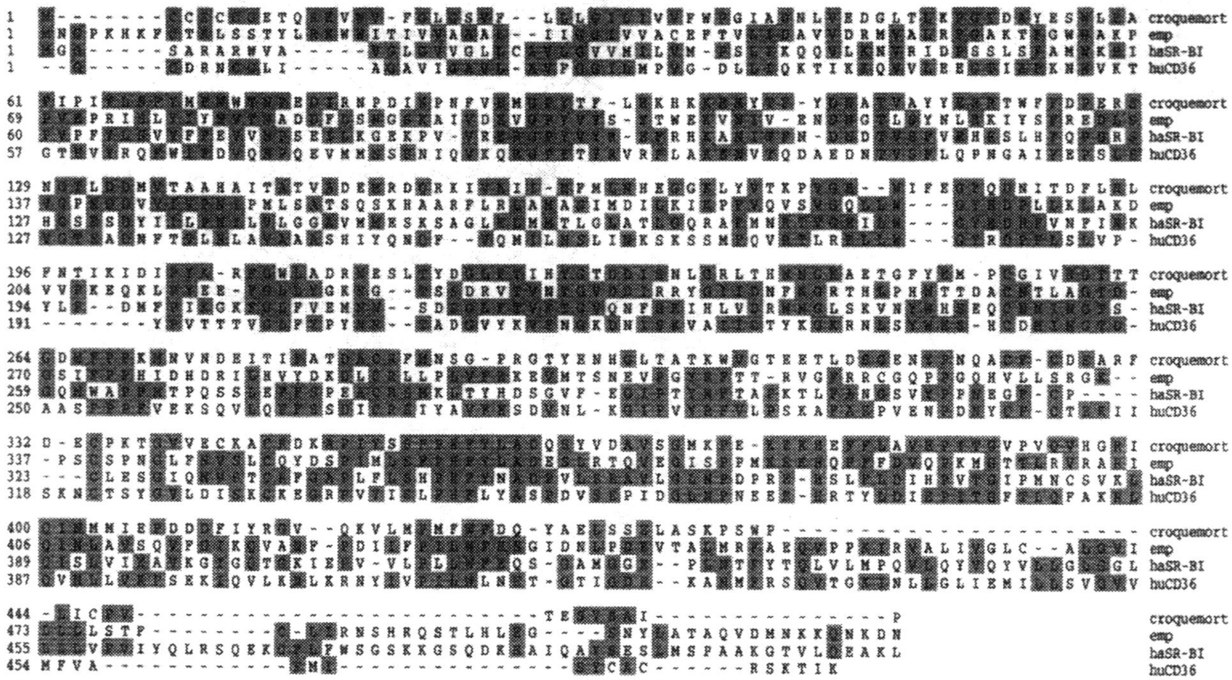

Figure 1. Croquemort is a member of the CD36 superfamily of receptor. An amino acid sequence alignment of the *Drosophila* Croquemort protein with other CD36-related receptors, the *Drosophila* emp (Hart and Wilcox, 1993), the hamster SR-BI (Acton et al., 1994), and the human CD36 (Tandon et al., 1989b). Conserved residues are shaded. The alignment was defined by computer analysis using Megalign-Lasergene DNASTAR, Macintosh.

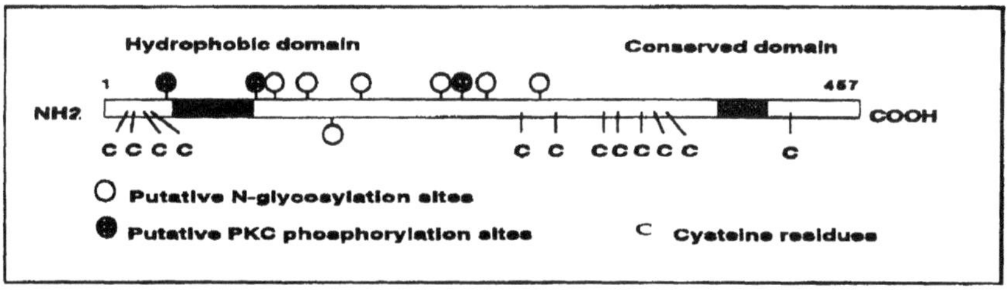

Figure 2. Schematic representation of the Croquemort protein. A schematic linear representation of Croquemort is shown where cysteines (C) are indicated throughout the amino acid sequence at their respective positions (2, 3, 5, 6, 256, 285, 324, 326, 334, 342, 345, 447). Putative PKC phosphorylation and N-glycosylation sites are represented respectively by shaded and white circles. A putative hydrophobic transmembrane region is black-boxed. A particularly well conserved region among all CD36-related receptors is represented by a shaded box although the functional significance of this domain remains unclear. Croquemort has the particularity to possess a single amino-terminal hydrophobic domain while all the other members of the CD36 superfamily possess an extra hydrophobic domain at their carboxyl-terminus.

and two additional putative PKC-phosphorylation sites, (d) followed by a stretch of residues with seven clustered cysteines and a seventh N-glycosylation site consensus sequence. Strikingly, a 13 amino acid motif is present in the C-terminal cysteine-rich domain of the molecule that is conserved among all the members of the CD36 superfamily (Figure 1). Therefore, this motif might have functional significance that remains to be defined and may be necessary for an ancestral function common to all of these CD36-related receptors.

Croquemort is distinguishable from the other members of the CD36 family as it does not possess a hydrophobic domain at its carboxy-terminus. Although the amino-acid sequence predicts a 54 kDa polypeptide, Western blot analyses reveal that Croquemort is a 68 kDa membrane protein with properties that are expected of an integral membrane protein. This difference in molecular weight might result from phosphorylation at the PKC sites or from posttranslational addition of O and/or N-linked glycans at the seven potential glycosylation sites. N-linked glycan consensus motifs are conserved within the CD36 family members and they have been shown to be occupied by complex oligosaccharides in human CD36 (review in Greenwalt et al., 1992).

B. The CD36 Superfamily of Receptors: Ligands and Functions

CD36 is a 88-kDa heavily (20%) glycosylated protein that has attracted considerable interest over the last 10 years. It was initially identified as an antigen associated with leukocyte differentiation using monoclonal antibodies (Knowles et al., 1984; Asch et al., 1987). In humans, CD36 is expressed in a wide range of cells, including monocytes, platelets, microvascular endothelial cells, and mammary gland epithelial cells (review in Greenwalt et al., 1992). CD36 is expressed in some melanoma cells and erythroid precursors and has been implicated in various myeloproliferative diseases (review in Daviet and McGregor, 1997). It has also been found in activated keratinocytes where it may play a role in the early events of wound healing in the mouse skin (Simon et al., 1996). CD36 participates in the initial phase of platelet binding to collagen (Tandon et al., 1989a), mediates platelet binding to thrombospondin, a protein of the extracellular matrix (Asch et al., 1987; Asch et al., 1992) and induces platelet activation and blood clotting (Ockenhouse et al., 1989; Tandon et al., 1989b). Interestingly, in platelets, CD36 is associated with the Fyn, Lyn, and Yes protein tyrosine kinases of the *src* gene family and might act as a signal transduction molecule (Huang et al., 1991). In all these cases, CD36 functions as an essential adhesion molecule during blood clotting, probably in association with other membrane proteins such as GP IIb-IIIa ($\alpha^{IIb}\beta_3$ integrin or CD41/CD61) (reviewed in Shattil and Brugge, 1991). CD36 mediates the adherence of *Plasmodium falciparum* infected erythrocytes to capillary endothelial cells, via its binding to sequestrin, a protein encoded by the parasite and targeted to the cell surface of the infected erythrocytes (Ockenhouse et al., 1989, 1991; Oquendo et al., 1989). This binding contributes to the lethality of patients with malarial disease. Certainly the most well-documented role for CD36 is its function in the recognition and phagocytosis of apoptotic cells, such as neutrophils and T lymphocytes, by macrophages (reviews in Savill et al., 1993; Savill, 1997a). In this regard, CD36 cooperates with the vitronectin receptor, an integrin ($\alpha_V\beta_3$ or CD51/CD61) via a bridge of thrombospondin in order to recognize and trigger the phagocytosis of the senescent cells (Savill et al., 1992; Ren et al., 1995). CD36 is also present in retinal pigment epithelium (RPE) where it plays a role in the phagocytosis of rod outer segments (ROSs), a function that is critical to the normal vision process (Ryeom et al., 1996b). In this regard, phosphatidyl serine (PS) and phosphatidyl inositol (PI) are believed to be the ligands of CD36 (Ryeom et al., 1996a).

Several new members of the CD36 family of receptors have been recently identified in various species which highlighted new functions for CD36. An adipocyte membrane glycoprotein, FAT, which participates in fatty acid transport is 85% similar to CD36 and is likely to be the rat homologue of CD36 (Abumrad et al., 1993). Human CD36 is also implicated in fatty acid binding and transport in foam cells in atherosclerosis, and acts as a scavenger receptor in mediating uptake of modified LDLs (Endemann et al., 1993).

In rat adipocytes, LIMPII, a lysosomal integral membrane protein, which is unrelated to the previously described classes (Classes A and B) of lysosomal proteins, has been identified that shares about 34% of identity with CD36 (Vega et al., 1991) and defines a new subclass of CD36-related molecules.

In hamster, another CD36-related molecule, HaSR-BI, has also been characterized that shares, respectively, 31% (muCD36), 32% (human CD36) and 33% (rat FAT and LIMPII) of identity with the murine, human and rat members of the CD36 superfamily (Acton et al., 1994). HaSR-BI, like CD36, acts as a scavenger receptor but sets itself apart as it is also a receptor for high density lipoproteins (HDL) and plays an extremely important role in mediating selective cholesterol uptake in liver cells and steroidogenic tissues (Acton et al., 1996; review in Rigotti et al., 1997). Both rat and mouse homologues of SR-BI have been characterized that have a similar function in cholesterol transport (Landschulz et al., 1996; review in Rigotti et al., 1997). A new human CD36-related receptor, CLA-1 (CD36/LIMP II analogue), was recently characterized that shares 80% of identity with murine and hamster SR-BI, and is functionally related to these molecules (Calvo et al., 1995, 1997). CLA-1 recognizes apoptotic thymocytes and acts as a docking receptor for HDL in connection with selective uptake of cholesterol esters (Murao et al., 1997). Therefore, SR-Bs and CLA-1 constitute a third class of CD36-related molecules. However, members of both SR-B and CD36 subclasses have been shown to be expressed in endothelial cells and macrophages (review in Greenwalt et al., 1992; Abumrad et al., 1993). They act as receptors for anionic phospholipids, such as phosphatidyl serine, and it is believed that this binding activity accounts for their role as phagocytic receptors of apoptotic cells (Rigotti et al., 1995, 1997; Fukasawa et al., 1996; reviews in Savill, 1997b).

It is also possible that some, if not all, of these CD36 members might be essential for various tissue development as well. In fact, in *Drosophila*, Emp, an

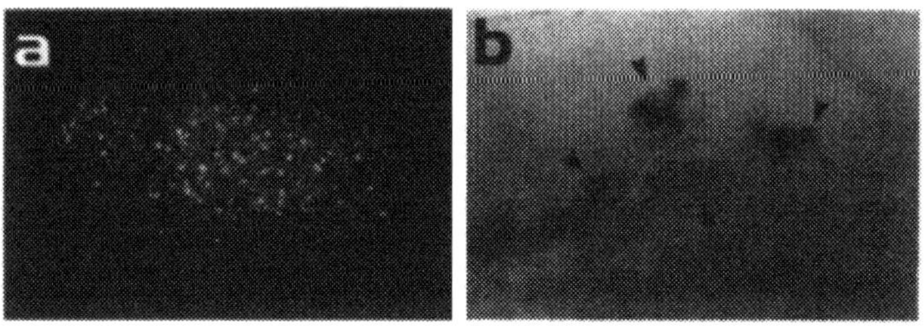

Figure 3. Croquemort is a macrophage-specific receptor during *Drosophila* embryogenesis. Whole-mounts of Oregon^R embryos were stained with the anti-Croquemort polyserum that was detected with a fluorescent FITC-coupled secondary antibody (**a**) or an alkaline phosphatase-coupled secondary antibody (**b**). (**a**) Is a confocal image of a 20x magnification of a dorsal view of a stage 14 embryo (11 hours after egg laying) where stained cells (staining appears white) are spread throughout the embryo in a classical pattern for hemocytes (Tepass et al., 1994). (**b**) Is a 90x magnification of macrophages within the brain region of a stage 13 embryo (10 hours after egg laying). Croquemort staining (black) is seen at the membrane level of subcellular vesicles which contain apoptotic corpses as assayed by TUNEL (not visible in figure). Arrowheads point to apoptolic corpses inside Croquemort-positive vesicles.

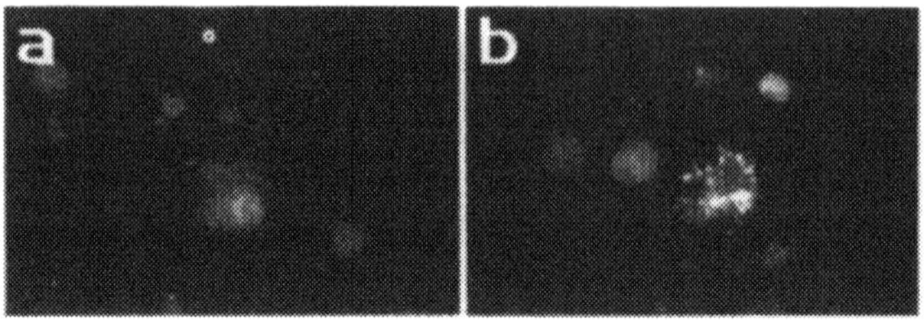

Figure 4. Croquemort is specifically expressed in a subset of the tumorous blood cell line *l(2)mbn*. *l(2)mbn* tumorous blood cells were fixed and stained with preimmune (**a**) and specific anti-Croquemort polysera (**b**). Both polysera were detected by a fluorescent FITC-coupled second antibody. (**a**) No staining is observed using the preimmune polyserum. (**b**) Is a fluorescent image of stained *l(2)mbn* tumorous blood cells where Croquemort is restricted to the larger cells, or so-called lamellocytes (phagocytic cells). In these cells, Croquemort is seen at the membrane level of subcellular vesicles around the nucleus, a distribution which suggests that these vesicles might be lysosomes.

epithelial membrane protein, has been characterized as a new member of the CD36 family of receptor, and is similar to CD36 and LIMPII to about the same degree (25% and 32%, respectively) (Hart and Wilcox, 1993). *Emp* is expressed in embryonic tissues that are derived from the ectoderm. During larval development, *emp* is found in epithelial cells of wing imaginal discs, and in adult, it is expressed in the precursor cells of epidermal structures. Its expression pattern suggests a role for Emp in the proper development or function of epithelial cells. However this remains to be further investigated.

C. Croquemort Is a Macrophage-Specific Receptor

Among the various member of the CD36 superfamily, Croquemort is most similar to SR-BI (33% of identity) which suggested to us that Croquemort might have a similar pattern of expression and similar functions in *Drosophila*. During embryogenesis, the pattern of Croquemort expression is almost identical to that observed for Peroxidasin and follows the classical distribution of macrophages as described by Tepass and colleagues (Figure 3a) (Tepass et al., 1994). However, there is no Croquemort expression in very early embryonic stages. In fact, Croquemort is first detected in late stage 11 embryos. This stage coincides with the first wave of apoptosis as defined by the work of Steller and co-workers (Abrams et al., 1993; White et al., 1994). Interestingly, Croquemort expression within macrophages is most prominently at the membrane surface of subcellular vesicles that contain apoptotic corpses (Figure 3b), and this specific expression within macrophages persists throughout embryogenesis. This result provides cir-

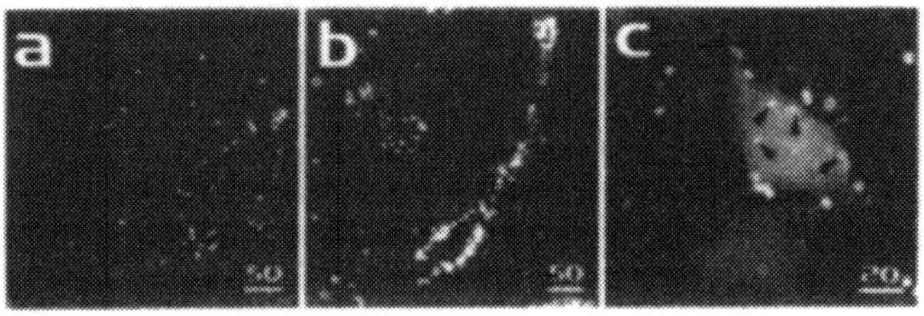

Figure 5. COS-7 cells transfected with the *croquemort* cDNA bind and engulf murine apoptotic thymocytes. (**a**) is a confocal image of a representative field of murine FITC-labeled thymocytes (small white cells) that failed to bind to COS-7 cells transfected with *croquemort* cDNA (large dark gray cells). (**b**) Is a confocal image of a representative field of murine FITC-labeled apoptotic thymocytes (small white or light blue cells) that bound to *croquemort* transfected COS-7 cells (large dark blue stained cells). (**c**) Is a confocal image of a representative field of *croquemort* transfected COS-7 cells that are in the process of engulfing or have engulfed (black arrowheads) bound FITC-labeled apoptotic murine thymocytes. Binding and phagocytosis of murine apoptotic thymocytes to these cells were both inhibited by the anti-Croquemort polyserum, but not preimmune polyserum demonstrating that Croquemort is specifically involved in these processes (Franc et al., 1996). Scale bars are indicated.

cumstantial evidence that Croquemort may mediate the uptake of apoptotic cells *in vivo* during *Drosophila* embryogenesis.

Interestingly, Croquemort immunostaining of *l(2)mbn* tumorous blood cells, a heterogeneous population of larval blood cells where large flattened cells with a typical morphology of lamellocytes are the most phagocytic cells (Gateff, 1980; Dimarcq et al., 1997), reveals that Croquemort is restricted to these larger cells (Figure 4). These cells are competent to express antimicrobial genes in response to immune challenge, and increase their phagocytic activity when treated with 20-hydroxy ecdysone (Dimarcq et al., 1997). Interestingly, the ecdysteroid hormone also induces massive programmed cell death during metamorphosis of the fly (review in Thummel, 1996; Jiang et al., 1997). Therefore, programmed cell death may well precede and signal the terminal differentiation of these cells, as well as that of hemocytes in larval and adult stages of the fly. Croquemort expression within these cells suggests that it might also be a marker for phagocytic hemocytes in larval and subsequent developmental stages. Of note is the perinuclear distribution of the vesicles where Croquemort is expressed within these cells which suggests that these subcellular organelles might be lysosomes. Interestingly, lysosomes have been reported to fuse with the plasma membrane during the formation of the phagosome (Silverstein et al., 1977), a fusion that allows the digestion of ingested particles by lysosomal enzymes.

D. Croquemort Mediates Phagocytosis of Cells Undergoing Apoptosis

The similarity in the pattern of expression of Croquemort with the pattern of apoptosis during embryogenesis, the typical macrophage morphology (round scattered cells with large vacuoles) of the cells expressing Croquemort, and its sequence identity with human and murine CD36 and in particular SR-BI suggested to us that this protein might be indeed a phagocytic receptor for apoptotic cells. In order to seek out more direct evidence for this role of Croquemort, we transfected the full-length cDNA encoding *croquemort* into Cos-7 cells. Although these cells are normally non-phagocytic, they became able to bind apoptotic murine thymocytes 72 hours after transfection (Figure 5). The binding was specific for apoptotic cells as non-apoptotic cells did not bind to the cells, and the cell association of apoptotic thymocytes was specifically inhibited by the anti-croquemort serum (Franc et al., 1996). These results suggest that Croquemort is sufficient to trigger the recognition of apoptotic cells by Cos-7 cells. Furthermore, the recognition process was sufficient to trigger the entire phagocytic machinery of these cells as apoptotic thymocytes were fully ingested after only four hours of incubation (Figure 5c). The feature of a "catcher of death" of Croquemort prompted us to name this receptor "croquemort" after the French word "croque-mort," which means pallbearer. The origin of this word, however, comes from an ancient French tradition in which the pallbearer was expected to bite the dead in order to confirm their death. Literally translated, croque-mort means "bite death" or "bite the dead."

We recently used *Drosophila* genetics to further investigate the potential role of *croquemort* in phagocytosis of apoptotic cells *in vivo* during embryogenesis (Franc et al., 1999). Two deficiencies, *Df(2L)al* (*aristales* deficiency) and *Df(2L)TE99(Z)XW88* (*W88* deficiency), which remove the *croquemort* locus, were tested for phagocytosis of apoptotic cells using an *in vitro* assay. The assay consisted in a triple staining with the anti-Peroxidasin antibody, the anti-Croquemort antibody, and a nuclear dye, 7-amino actinomycin D (7-AAD). In these experiments, the *croquemort*-deficient homozygous embryos were identified by lack of Croquemort immunostaining and/or by morphology. The Peroxidasin immunostaining allowed us to detect hemocytes particularly in *croquemort*-deficient embryos that lacked Croquemort. The nuclear dye stained all nuclei and allowed us to distinguish between regular nuclei and apoptotic nuclei. Indeed, due to nuclear condensation during the apoptotic program, apoptotic nuclei appear as extremely bright round corpses. Our results show that removing the *croquemort* locus results in a severe defect in phagocytosis of apoptotic corpses in the embryo. While, in the wild-type embryo, the phagocytic index is about 4.0 ingested apoptotic corpses per macrophage, it drops dramatically to about 0.3 injested corpses in *croquemort*-deficient embryos. As a consequence of their poor phagocytic activity, hemocytes in *croquemort*-deficient embryos appear very small as compared to wild-type macrophages that

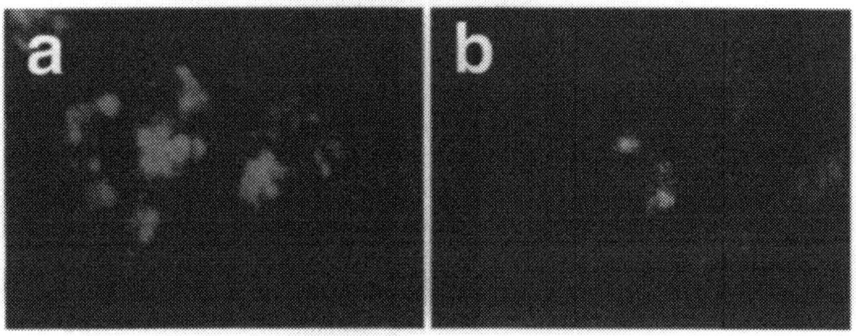

Figure 6. Macrophages in *croquemort*-deficient embryos are poorly phagocytic and remain small and round. (**a**) and (**b**) are confocal micrographs of macrophages that were detected by Peroxidasin immunostaining (the anti-Peroxidasin antibody was a gift of Dr. J. Fessler at UCLA). Both images are 400X magnification of macrophages in (**a**) a wild-type embryo and (**b**) an *aristaless*-deficient embryo. While wild-type macrophages that have engulfed apoptotic corpses appear as very large cells (**a**) macrophages in *aristaless*-deficient embryos that lack *croquemort* and are defective in phagocytosis of apoptotic cells remain small and round (**b**).

have engulfed apoptotic corpses (Figure 6). The phagocytosis defect can be rescued by ubiquitous expression of *croquemort* from a transgene in *aristaless*-deficient embryos. Therefore, *croquemort* is essential for phagocytosis of apoptotic cells in the fly embryo.

All other hemocyte functions, however, are not impaired in *croquemort*-deficient embryos. Hemocytes in *croquemort*-deficient embryos are properly differentiated and produce ECM proteins such as Peroxidasin and MDP-1. Moreover, their scavenger activity is not affected as *croquemort*-deficient hemocytes are still able to bind and to uptake acetylated LDLs. This is intriguing as the homology between Croquemort and SR-BI suggested that it might function as a scavenger receptor. However, our scavenger activity assay, which consisted of injecting acetylated LDLs into living embryos, is strictly qualitative. It is possible that embryonic hemocytes may utilize several redundant mechanisms of uptake of acetylated LDLs. We cannot, at this point, exclude that Croquemort may not act as a scavenger receptor, and this will be tested.

Most interestingly, our study also pointed to a new function for embryonic hemocytes (Franc et al., 1999). We had previously suggested that perhaps both phagocytosis of apoptotic cells during development and phagocytosis of pathogens in innate immunity may share similar mechanisms of recognition (Franc et al., 1996). Therefore, we also investigated the role of *croquemort* in innate immunity and in phagocytosis of various pathogens *in vivo* (Franc et al., 1999). Both Gram-negative and Gram-positive fluorescently labeled bacteria were injected into wild-type and *croquemort*-deficient living embryos. Engulfment was later monitored using confocal microscopy. Rapid phagocytosis of bacteria by wild-type embryonic macrophages was observed, demonstrating the ability of embryonic macrophages to sustain immune functions. Interestingly, phagocytosis of bacteria does not appear to be impaired in *croquemort*-deficient embryos. Our data define distinct pathways for the phagocytosis of apoptotic corpses and bacteria by *Drosophila* macrophages and promote *Drosophila* as being a useful model system to study and genetically dissect the signaling pathways that lead to phagocytosis.

Our recent data also show that *croquemort* expression is regulated by the amount of PCD (Franc et al., 1999). The amount of Croquemort protein in each macrophage is reduced by 75% in embryos that lack developmentally regulated PCD as compared to wild-type embryos. When PCD is induced by X-ray irradiation, Croquemort expression is increased threefold. MDP-1 levels were also shown to correlate with the amount of PCD (Hortsch et al., 1998). These results suggest that a signal from the apoptotic corpse is conveyed to the macrophage, which in turn upregulates and/or activates the components of its phagocytic machinery. Interestingly, CD36 is also upregulated upon exposure to oxidized LDLs, one type of its ligands (Nagy et al., 1998; Nakagawa et al., 1998). We are currently taking advantage of these findings to screen for mutants that fail to regulate *croquemort* in response to PCD. It is not yet known whether Croquemort such as CD36 requires other proteins in order to function in phagocytosis of apoptotic cells in the fly or what the Croquemort ligands might be at the surface of the dying cells, and we believe that this screen may lead to the identification of such molecules. We are also currently using both biochemical and genetic approaches to uncover putative partners or ligands of Croquemort.

IV. CONCLUSION

The recent and expanding interest in invertebrates innate immunity and in particular that of the fruitfly, *Drosophila*, has led to the molecular characterization of a number of antigens that are specifically expressed in hematopoietic tissues or the hemocytes of the flies. Interestingly, many of these molecules share similarities with mammalian molecules and are responsible for similar blood cells functions. This chapter was designed to summarize the current knowledge of hematopoiesis and blood cell functions in *Drosophila* and to illustrate the interest of using this model system to identify new molecules involved in these processes. It is likely that these molecules will be found to be conserved throughout evolution. Indeed, many blood cell functions are similar in vertebrate and invertebrate systems. Although a large number of functional studies of various mammalian blood cell types have been possible using *in vitro* culture systems, the complexity and the redundancy of the molecules involved in these functions have hampered their *in vivo* genetic studies. *Drosophila* represents a powerful system to study these functions *in vivo* as an ever-growing array of genetic tools has been generated over the past years.

Studies of innate immunity in the fly have highlighted its similarities with innate immunity of mammals, although it appears as a much simpler immune system. Among the conserved functions of blood cells in the fly is the phagocytosis of pathogens in innate immunity, as well as the phagocytosis of apoptotic cells during development. Although many phagocytic receptors have been characterized and studied in mammalian systems, many of the molecular mechanisms underlying each step of the phagocytic process remain unclear. We recently characterized a new *Drosophila* macrophage receptor, croquemort, which is a member of the CD36 superfamily. Croquemort acts as a phagocytic receptor of apoptotic cells during *Drosophila* development. This indicates that Croquemort behaves as a functional analogue of various mammalian CD36 family members, which also participate in similar processes. We believe that the study of this molecule in *Drosophila* will help to identify new molecules and unravel new pathways that are involved in both the terminal differentiation of macrophages and in phagocytosis. The identification of these molecules may reveal more similarities in the conserved mechanisms of hematopoiesis and phagocytosis in invertebrates and vertebrates.

ACKNOWLEDGMENTS

I wish to express my gratitude to R. Alan B. Ezekowitz and Kristin White for their patience and constant support. I am thankful to Istvan Ando, Claude Desplan, Alan Ezekowitz, François Huet, Kellie Watson and Kristin White for stimulating discussions, and for their comments and corrections of this manuscript. I also wish to thank members of Kristin White's laboratory and members of the Cutaneous Biology Research Center at Harvard who make the laboratory such an interactive and stimulating environment.

REFERENCES

Abel, T., Michelson, A.M., & Maniatis, T. (1993). A *Drosophila* GATA family member that binds to *Adh* regulatory sequences is expressed in the developing fat body. Development 119, 623–633.

Abrams, J.M., Lux, A., Steller, H., & Krieger, M. (1992). Macrophages in *Drosophila* embryos and L2 cells exhibit scavenger receptor-mediated endocytosis. Proc. Natl. Acad. Sci. USA 89, 10375–10379.

Abrams, J.M., White, K., Fessler, L.I., & Steller, H. (1993). Programmed cell death during *Drosophila* embryogenesis. Development 117, 29–43.

Abumrad, N.A., El-Mahrabi, M.R., Amri, E-Z., Lopez, E., & Grimalds, P.A. (1993). Cloning of a rat adipocyte membrane protein implicated in binding and transport of long-chain fatty acids that is induced during preadipocyte differentiation; homology with human CD36. J. Biol. Chem. 268, 17665–17668.

Acton, S.L., Rigotti, A., Landschulz, K.T., Xu, S., Hobbs, H.H. & Krieger, M. (1996). Identification of a scavenger receptor *SR-BI* as a high density lipoprotein receptor. Science 271, 518–520.

Acton, S.L., Scherer, P.E., Lodish, H.F., & Krieger, M. (1994). Expression cloning of *SR-BI*, a CD36-related class B scavenger receptor. J. Biol. Chem. 269, 21003–21009.

Akbar, A.N., Savill, J., Gombert, W., Bofill, M., Borthwick, N.J., Whitelaw, F., Grundy, J., Janossy, G., & Salmon, M. (1994). The specific recognition by macrophages of $CD8^+$, $CD45RO^+$ T cells undergoing apoptosis: a mechanism for T cell clearance during resolution of viral infections. J. Exp. Med. 180, 1943–1947.

Arceci, R.J., King, A.A.J., Simon, M.C., Orkin, S.H., & Wilson, D.B. (1993). Mouse GATA-4: a retinoic acid-inducible GATA-binding transcription factor expressed in endodermally derived tissues and heart. Mol. Cell. Biol. 13, 2235–2246.

Asch, A.S., Barnwell, J., Silverstein, R.L., & Nachman, R.L. (1987). Isolation of the thrombospondin membrane receptor. J. Clin. Invest. 79, 1054–1061.

Asch, A.S., Silbiger, S., Heimer, E., & Nachman, R.L. (1992). Thrombospondin sequence motif (CSVTCG) is responsible for CD36 Binding. Biochem. Biophys. Res. Commun. 182, 1208–1217.

Baehrecke, E.H. (1996). Ecdysone signaling cascade and regulation of *Drosophila* metamorphosis. Arch. Insect Biochem. Physiol. 33, 231–244.

Baldwin, A.S. Jr. (1996). The NF-kappa B and I kappa B proteins: new discoveries and insights. Ann. Rev. Immunol. 14, 649–683.

Baumgartner, S., Martin, D., Hagios, C., & Chiquet-Ehrismann, R. (1994). Tenm, a *Drosophila* gene related to tenascin, is a new pair-rule gene. EMBO J. 15, 3728–3740.

Belvin, M.P., & Anderson, K.V. (1996). A conserved signaling pathway: The *Drosophila* Toll/Dorsal pathway. Annu. Rev. Cell. Dev. Biol. 12, 393–416.

Bernardoni, R., Vivancos, V., & Giangrande, A. (1997). *glide/gcm* is expressed and required in the scavenger cell lineage. Devel. Biol. 191, 118–130.

Binari, R., & Perrimon, N. (1994). Stripe-specific regulation of pair-rule genes by *hopscotch*, a putative Jak family tyrosine kinase in *Drosophila*. Genes Dev. 8, 300–312.

Boman, H. G. (1995). Peptide antibiotics and their role in innate immunity. Annu. Rev. Immunol. 13, 61–92.

Braun, A., Lemaitre, B., Lanot, R., Zachary, D., & Meister, M. (1997). *Drosophila* immunity: analysis of larval hemocytes by P-element-mediated enhancer trap. Genetics 147, 623–634.

Brown, E.J. (1994). Phagocytosis. Bioessays 17, 109–117.

Bryant, P.J., Watson, K.L., Justice, R.W., & Woods, D.F. (1993). Tumor suppressor genes encoding proteins required for cell interactions and signal transduction in *Drosophila*. Dev. (Suppl.) 239–249.

Calvo, D., Dopazo, J., & Vega, M.A. (1995). The CD36, CLA-1 (CD36L1), and LIMPII (CD36L2) gene family: cellular distribution, chromosomal location, and genetic evolution. Genomics 25, 100–106.

Calvo, D., Gomez-Coronado, D., Lasuncion, M.A., & Vega, M.A. (1997). CLA-1 is an 85-kD plasma membrane glycoprotein that acts as a high affinity receptor for both native (HDL, LDL, and VLDL) and modified (OxLDL and AC LDL) lipoproteins. Arterioscler. Thromb. Vasc. Biol. 17, 2341–2349.

Chen, P., Nordstrom, W., Gish, B., & Abrams, J.M. (1996). *grim*, a novel cell death gene in *Drosophila*. Genes Dev. 15, 1773–1782.

Chosa, N., Fukumitsu, T., Fujimoto, K., & Ohnishi, E. (1997). Activation of prophenoloxidase A1 by an activating enzyme in *Drosophila melanogaster*. Insect Biochem. Molec. Biol. 27, 61–68.

Daga, A., Karlovich, C.A., Dumstrei, K., & Banerjee, U. (1996). Patterning of cells in the *Drosophila* eye by lozenge, which shares homologous domains with AML1. Genes Dev. 10, 1194–1205.

Daviet, L., & McGregor, J.L. (1997). Vascular biology of CD36: roles of this new adhesion molecule family in different disease states. Thromb. Haemost. 78, 65–69.

Diez-Roux, G., & Lang, R. (1997). Macrophages induce apoptosis in normal cells *in vivo*. Development. 124, 3633–3638.

Dimarcq, J.L., Imler, J.L., Lanot, R., Ezekowitz, R.A.B., Hoffmann J.A., Janeway, C.A., & Lagueux, M. (1997). Treatment of *l(2)mbn Drosophila* tumorous blood cells with the steroid hormone ecdysone amplifies the inducibility of antimicrobial peptide gene expression. Insect. Biochem. Mol. Biol. 27, 877–886.

Dushay, M.S., Asling, B., & Hultmark, D. (1996). Origins of immunity: Relish, a compound Rel-like gene in the antibacterial defense of *Drosophila*. Proc. Natl. Acad. Sci. USA 93, 10343–10347.

Eldon, E., Kooyer, S., D'Evelyn, D., Duman, M., Lawinger, P., Botas, J., & Bellen, H. (1994). The *Drosophila* 18 wheeler is required for morphogenesis and has striking similarities to Toll. Development 120, 885–99.

Endemann, G., Stanton, L.W., Madden, K.S., Bryant, C.M., White, R.T., & Protter, A.A. (1993). CD36 is a receptor for oxidized low density lipoprotein. Biol. Chem. 268, 11811–11816.

Ezekowitz, R.A.B. (1989). Macrophages—biological aspects. In: *Natural Immunity.* (Nelson, D.S., Ed.), pp. 15–38. Academic Press, Australia.

Fessler, J.H., & Fessler, L.I. (1989). *Drosophila* extracellular matrix. Annu. Rev. Cell Biol. 5, 309–339.

Fessler, L.I., Nelson, R.E., & Fessler, J.H. (1994). *Drosophila* extracellular matrix. Methods Enzymol. 245, 271–294.

Franc, N.C., Dimarcq, J.L., Lagueux, M., Hoffmann, J.A., & Ezekowitz, R.A.B. (1996). Croquemort, a novel *Drosophila* hemocyte/macrophage receptor that recognizes apoptotic cells. Immunity 4, 431–443.

Franc, N.C., Heitzler, P., Ezekowitz, R.A., & White, K. (1999). Requirement for croquemort in phagocytosis of apoptotic cells in *Drosophila*. Science 284, 1991–1994.

Fusakawa, M., Adachi, H., Hirota, K., Tsujimoto, M., Arai, H., & Inoue, K. (1996). *SR-BI*, a class B scavenger receptor recognizes both negatively charged liposomes and apoptotic cells. Exp. Cell Res. 222, 246–250.

Gateff, E. (1978a). Malignant neoplasms of genetic origin in *Drosophila melanogaster.* Science. 200, 1448–1459.

Gateff, E. (1978b). Malignant and benign neoplasms of *Drosophila melanogaster.* In: *Genetics and Biology of* Drosophila (Ashburner, M. & Wright, T.R.F., Eds.). Vol. 2b, pp. 181–272. Academic Press, New York.

Gateff, E. (1994a). Tumor suppressor and overgrowth suppressor genes of *Drosophila melanogaster.* Developmental aspects. Int. J. Dev. Biol. 38, 565–590.

Gateff, E. (1994b). Tumor-suppressor genes, hematopoietic malignancies and other hematopoietic disorders of *Drosophila melanogaster.* In: *Primordial Immunity: Foundation for the Vertebrate Immune System.* (Beck, G., Cooper, G.S., Habight, G.S., & Marchalonis, J.L., Eds.), Annals of the New York Academy of Science. Vol. 712, pp. 260–279. The New York Academy of Science, New York.

Gateff, E., Gissmann, L., Shrestha, R., Plus, N., Pfister, H., Schroder, J., & Zur Hausen, H. (1980). Characterization of two tumorous blood cell lines of *Drosophila melanogaster* and the viruses they contain. In: *Invertebrate Systems* In Vitro (Kurstak, E. Maramorosch, K., & Dubendorfer, A., Eds.), pp. 517–533. North Holland Biomedical Press, Elsevier, Amsterdam.

Gateff, E., Kurzik-Dumke, U., Wismar, J., Loffler, T., Habtemichael, N., Konrad, L., Dreschers, S., Kaiser, S., & Protin, U. (1996). *Drosophila* differentiation genes instrumental in tumor suppression. Int. J. Dev. Biol. 40, 149–156.

Greenwalt, B.D.E., Lipsky, R.H., Ockenhouse, C.F., Ikdea, H., Tandon, N.N., & Jamieson, G.A. (1992). Membrane glycoprotein CD36: a review of its roles in adherence, signal transduction, and transfusion medicine. Blood 80, 1105–1115.

Grether, M.E., Abrams, J.M., Agapite, J., White, K., & Steller, H. (1995). The head involution defective gene of *Drosophila melanogaster* functions in programmed cell death. Genes Dev. 9, 1694–1708.

Gupta, A.P. (1979). In: *Insect Hemocytes: Development, Forms, Functions and Techniques* (Gupta, A.P., Ed.). Cambridge University Press, Cambridge, England.

Hanratty, W.P., & Ryerse, J.S. (1981). A genetic melanotic neoplasm of *Drosophila melanogaster*. Dev. Biol. 83, 238–249.

Harrison, D.A., Binari, R., Nahreini, T., Gilman, M., & Perrimon, N. (1995). Activation of a *Drosophila Janus kinase* (JAK) causes hematopoietic neoplasia and developmental defects. EMBO J. 14, 2857–2865.

Hart, K., & Wilcox, M. (1993). A *Drosophila* gene encoding an epithelium membrane protein with homology to CD36/LIMP II. J. Mol. Biol. 234, 249–253.

Hoffmann, J.A., Kafatos, F.C., Janeway, C.A., & Ezekowitz, R.A. (1999). Phylogenetic perspectives in innate immunity. Science 284, 1313–1318.

Hoffmann, J.A., & Reichhart, J.M. (1997). *Drosophila* immunity. Trends Cell Biol. 7, 309–316.

Hoffmann, J.A., Reichhart, J.M., & Hetru, C. (1996). Innate immunity in higher insects. Curr. Opin. Immunol. 8, 8–13.

Hortsch, M., Olson, A., Fishman, S., Soneral, S.N., Marikar, Y., Dong, R., & Jacobs, J.R. (1998). The expression of MDP-1, a component of *Drosophila* embryonic basement membranes, is modulated by apoptotic cell death. Int. J. Dev. Biol. 42, 33–42.

Hou, X.S., & Perrimon, N. (1997). The JAK-STAT pathway in *Drosophila*. Trends Genet. 13, 105–110.

Hou, X.S., Melnick, M.B., & Perrimon, N. (1996). *marelle* acts downstream of the *Drosophila* HOP/JAK kinase and encodes a protein similar to the mammalian STATs. Cell 84, 411–419.

Huang, M.-M., Bolen, J.B., Barnwell, J.W., Shattil, S.F., & Brugge, J.S. (1991). Membrane glycoprotein IV (CD36) is physically associated with the Fyn, Lyn, and Yes protein-tyrosine kinases in human platelets. Proc. Natl. Acad. Sci. USA 88, 7844–7848.

Hultmark, D. (1993). Immune reactions in *Drosophila* and other insects: a model for innate immunity. Trends Genet. 9, 178–183.

Ip, Y.T., Reach, M., Engström, Y., Kadalayil, L., Cai, H., Gonzalez-Crespo, S., Tatei, K., & Levine, M. (1993). *Dif*, a dorsal-related gene that mediates an immune response in *Drosophila*. Cell 75, 753–763.

Jiang, C., Baehrecke, E.H., & Thummel, C.S. (1997). Steroid regulated programmed cell death during *Drosophila* metamorphosis. Development 124, 4673–4683.

Kanda, S., Hodgkin, M.N., Woodfield, R.J., Wakelam, M.J., Thomas, G., & Claesson-Welsh, L. (1997). Phosphatidylinositol 3′-kinase-independent p70 S6 kinase activation by fibroblast growth factor receptor-1 is important for proliferation but not differentiation of endothelial cells. J. Biol. Chem. 272, 23347–23353.

Knowles, D.M., Tollidjian, B., Marboe, C., D'agati, V., Grimes, M., & Chess, L. (1984). Monoclonal anti-human monocyte antibodies OKM1 and OKM5 possess distinctive tissue distributions including differential reactivity with vascular endothelium. J. Immunol. 132, 2170–2173.

Ko, L.J., Yamamoto, M., Leonard, M.W., George, K.M., Ting, P., & Engel, J.D. (1991). Murine and human T-lymphocyte GATA-3 factors mediate transcription through a cis-regulatory element within the human T-cell receptor δ gene enhancer. Mol. Cell. Biol. 11, 2778–2784.

Krieger, M. (1997). The other side of scavenger receptors: pattern recognition for host defense. Curr. Opin. Lipidol. 8, 275–280.

Landschulz, K.T., Pathak, R.K., Rigotti, A., Krieger, M., & Hobbs, H.H. (1996). Regulation of scavenger receptor, Class B, type I, a high density lipoprotein receptor, in liver and steroidogenic tissues of the rat. J. Clin. Invest. 98, 984–995.

Laverriere, A.C., MacNeill, C., Mueller, C., Poelmann, R.E., Burch, J.B.E., & Evans, T. (1994). GATA-4/5/6, a subfamily of three transcription factors transcribed in developing heart and gut. J. Biol. Chem. 269, 23177–23184.

Lemaitre, B., Kromer-Metzger, E., Michaut, L., Nicolas, E., Meister, M., Georgel, P., Reichhart, J.M., & Hoffmann, J.A. (1995). A novel mutation, immune deficiency, defines two distinct control pathways in the *Drosophila* host defense. Proc. Natl. Acad. Sci. USA 92, 9465–9469.

Lemaitre, B., Nicolas, E., Michaut, L., Reichhart, J.M., & Hoffmann, J.A. (1996). The dorsoventral regulatory gene cassette spätzle/Toll/cactus controls the potent antifungal response in *Drosophila* adults. Cell 86, 973–984.

Luciani, M.-F., & Chimini, G. (1996). The ATP binding cassette transporter ABC1, is required for the engulfment of corpses generated by apoptotic cell death. EMBO J. 15, 226–235.

Luo, H., Hanratty, P., & Dearolf, C.R. (1995). An amino-acid substitution in the *Drosophila* $hop^{Tum\text{-}l}$ Jak kinase causes leukemia-like hematopoietic defects. EMBO J. 14, 1412–1420.

Mathey-Prevot, B., & Perrimon, N. (1998). Mammalian and *Drosophila* blood: JAK and all trades. Cell 92, 697–700.

Medzhitoz, R., Preston-Hurlburt, P., & Janeway, C.A. Jr. (1997). A human homologue of *Drosophila* Toll protein signals activation of adaptive immunity. Nature. 388, 394–397.

Meister, M., Georgel, P., Lemaitre, B., Kappler, C., Lagueux, M., Reichhart, J.M., & Hoffmann, J.A. (1994). Immune gene expression in *Drosophila.* In: Phylogenetic Perspectives in Immunity: The Insect Host Defense (Hoffmann, J.A., Janeway, C.A. Jr., & Natori, S.R.G, Eds.), pp. 167–181. Landes Company, Austin.

Murao, K., Terpstra, V., Green, S.R., Kondratenko, N., Steinberg, D., & Quehenberger, O. (1997). Characterization of CLA-1, a human homologue of rodent scavenger receptor BI, as a receptor for high density lipoprotein and apoptotic thymocytes. J. Biol. Chem. 272, 17551–17557.

Nagy, L., Tontonoz, P., Alvarez, J.G., Chen, H., & Evans, R.M. (1998). Oxidized LDL regulates macrophage gene expression through ligand activation of PPARγ. Cell 93, 229–240.

Nakagawa, T., Nozaki, S., Nishida, M., Yakub, J.M., Tomiyama, Y., Nakata, A., Matsumoto, K., Funahashi, T., Kameda-Takemura, K., Kurata, Y., Yamashita, S., & Matsuzawa, Y. (1998). Oxidized LDL increases and interferon-γ decreases expression of CD36 in human monocyte-derived macrophages. Arterioscler. Thromb. Vasc. Biol. 18, 1350–1357.

Nappi, A.J., & Vass, E. (1993). Melanogenesis and the generation of cytotoxic molecules during insect cellular immune reactions. Pigment Cell Res. 6, 117–126.

Natzle, J.E., Monson, J.M., & McCarthy, B.J. (1982). Cytogenetic location and expression of a collagen-like gene in *Drosophila*. Nature 296, 368–371.

Nelson, R.E., Fessler, L.I., Takagi, Y., Blumberg, B., Keene, D.R., Olson, P.F., Parker, C.G., & Fessler, J.H. (1994). Peroxidasin: a novel enzyme-matrix protein of *Drosophila* development. EMBO J. 13, 3438–3447.

Ockenhouse, C.F., Klotz, F.W., Tandon, N.N., & Jamieson, G. (1991). Sequestrin, a CD36 recognition protein on *Plasmodium falciparum* malaria-infected erythrocytes identified by anti-idiotype antibodies. Proc. Natl. Acad. Sci. USA 88, 3175–3179.

Ockenhouse, C.F., Magowan, C., & Chulay, J.D. (1989). Activation of monocytes and platelets by monoclonal antibodies or malaria-infected erythrocytes binding to the CD36 surface receptor *in vitro*. J. Clin. Invest. 84, 468–475.

Oquendo, P., Hundt, E., Lawler, J., & Seed, B. (1989). CD36 directly mediates cytoadherence of *Plasmodium falciparum* parasitized erythrocytes. Cell. 58, 95–101.

Orkin, S.H. (1996). Development of the hematopoietic system. Curr. Opin. Genet. Dev. 6, 597–602.

Pearson, A.M. (1996). Scavenger receptors in immunity. Curr. Opin. Immunol. 8, 20–28.

Pearson, A.M., Lux, A., & Krieger, M. (1995). Expression cloning of dSR-CI, a class C macrophage-specific scavenger receptor from *Drosophila melanogaster*. Proc. Natl. Acad. Sci. USA 92, 4056–4060.

Pevny, L., Simon, M.C., Robertson, E., Klein, W.H., Tsai, S.-F., D'agati, V., Orkin, S.H., & Costantini, F. (1991). Erythroid differentiation in chimaeric mice blocked by a targeted mutation in the gene for transcription factor GATA-1. Nature 349, 257–260.

Qiu, P., Pan, P.C., & Govind S. (1998). A role for the *Drosophila* Toll/Cactus pathway in larval hematopoiesis. Development 125, 1909–1920.

Raff, M.C. (1992). Social controls on survival and cell death. Nature. 356, 397–400.

Raingeaud, J., Gupta, S., Rogers, J.S., Dickens, M., Han, J., Ulevitch, R.J., & Davis R.J. (1995). Proinflammatory cytokines and environmental stress cause p38 mitogen-activated protein kinase activation by dual phosphorylation on tyrosine and threonine. J. Biol. Chem. 270, 7420–7426.

Ratcliffe, N.A. (1993). Cellular defense responses in insects: unresolved problems. In: Parasites and Pathogens of Insects, Vol. 1 (Bechage, N.E., Thompson, S.N., & Federici, B.A., Eds.), pp. 579–604. Academic Press, San Diego.

Rehorn, K.-P., Thelen, H., Michelson, A., & Reuter, R. (1996). A molecular aspect of hematopoiesis and endoderm development common to vertebrates and *Drosophila*. Development 122, 4023–4031.

Reid, K.B.M., & Day, A.J. (1989). Structure-function relationships of the complement components. Immunol. Today. 6, 177–180.

Ren, Y., Silverstein, R.L., Allen, J., & Savill, J. (1995). CD36 gene transfer confers capacity for phagocytosis of cells undergoing apoptosis. J. Exp. Med. 181, 1857–1862.

Rigotti, A., Acton, S.L., & Krieger, M. (1995). The class B scavenger receptors *SR-BI* and CD36 are receptors for anionic phospholipids. J. Biol. Chem. 270, 16221–16224.

Rigotti, A., Trigatti, B., Babbit, J., Penman, M., Xu, S., & Krieger, M. (1997). Scavenger receptor BI: a cell surface receptor for high density lipoprotein. Curr. Opin. Lipidol. 8, 181–188.

Rizki, T.M. (1957a). Alterations in the hemocyte population of *Drosophila melanogaster*. J. Morphol. 100, 437–458.

Rizki, T.M. (1957b). Tumor formation in relation to metamorphosis in Drosophila melanogaster. J. Morphol. 100, 459–472.

Rizki, T.M., & Rizki, R.M. (1980). Developmental analysis of a temperature-sensitive melanotic tumor mutant in *Drosophila melanogaster*. Wilhelm Roux's Arch. Dev. Biol. 189, 197–206.

Rizki, T.M., & Rizki, R.M. (1984). The cellular defense system of *Drosophila melanogaster*. In: Insect Ultrastructure, Vol. 2 (King, R.C., & Akai, H., Eds.), pp. 579–604. Plenum, New York.

Rizki, T.M., Rizki, R.M., & Grell, H. (1980). A mutant affecting the crystal cells in *Drosophila melanogaster*. Wilhelm Roux's Arch. Dev. Biol. 188, 91–99.

Rodrigues, V., Cheah, P.Y., Ray, K., & Chia, W. (1995). *malvolio*, the *Drosophila* homologue of mouse NRAMP-1 (*Bcg*), is expressed in macrophages and in the nervous system and is required for normal taste behavior. EMBO J. 14, 3007–3020.

Rodriguez, A., Zhou, Z., Tang, M.L., Meller, S., Chen, J., Bellen, H., & Kimbrell, D.A. (1996). Identification of immune system and response genes, and novel mutations causing melanotic tumor formation in *Drosophila melanogaster*. Genetics. 143, 929–940.

Ryeom, S.W., Silverstein, R.L., Scotto, A., & Sparrow J.R. (1996a). Binding of anionic phospholipids to retinal pigment epithelium may be mediated by the scavenger receptor CD36. J. Biol. Chem. 271, 20536–20539.

Ryeom, S.W., Sparrow J.R., & Silverstein, R.L. (1996b). CD36 participates in the phagocytosis of rod outer segments by retinal pigment epithelium. J. Cell Sci. 109, 387–395.

Sam, S., Leise, W., & Hoshizaki, D.K. (1996). The *serpent* gene is necessary for progression through the early stages of fat-body development. Mech. Dev. 60, 197–205.

Samakovlis, C., Kimbrell, D.A., Kylsten, P., Engström, A., & Hultmark, D. (1990). The immune response in *Drosophila*: pattern of cecropin expression and biological activity. EMBO J. 9, 2969–2976.

Savill, J. (1997a). Apoptosis in resolution of inflammation. J. Leukoc. Biol. 61, 375–380.

Savill, J. (1997b). Recognition and phagocytosis of cells undergoing apoptosis. Br. Med. Bull. 53, 491–508.

Savill, J., Fadock, V., Henson, P., & Haslett, C. (1993). Phagocyte recognition of cells undergoing apoptosis. Immunol. Today 14, 131–136.

Savill, J., Hogg, N., Ren, Y., & Haslett, C. (1992). Thrombospondin cooperates with CD36 and the vitronectin receptor in macrophage recognition of neutrophils undergoing apoptosis. J. Clin. Invest. 90, 1513–1522.

Shattil, S., & Brugge, J. (1991). Protein tyrosine phosphorylation and the adhesive functions of platelets. Curr. Opin. Cell Biol. 3, 869–879.

Shrestha, R., & Gateff, E. (1982). Ultrastructure and cytochemistry of the cell types in the larval hematopoietic organs and hemolymph of *Drosophila melanogaster.* Dev. Growth Diff. 24, 65–82.

Silverstein, S.C., Steinman, R.M., & Cohn, Z.A. (1977). Endocytosis. Annu. Rev. Biochem. 46, 669–689.

Simon, M. Jr, Juhasz, I., Herlyn, M., & Hunyadi, J. (1996). Thrombospondin receptor (CD36) expression of human keratinocytes during wound healing in a SCID mouse/human skin repair model. J. Dermat. 23, 305–309.

Sluss, H.K., Barrett, T., Dérijard, B., & Davis, R.J. (1994). Signal transduction by tumor necrosis factor mediated by JNK protein kinases. Mol. Cell. Biol. 14, 8376–8384.

Sluss, H.K., Han, Z., Barett, T., Davis, R.T., & Ip, T. (1996). A JNK signal transduction pathway that mediates morphogenesis and an immune response in *Drosophila.* Genes Dev. 10, 2745–2758.

Soudais, C., Bielinska, M., Keikinheimo, M., MacArthur, C.A., Narita, N., Saffiz, J.E., Simon, M.C., Leiden, J.M., & Wilson, D.B. (1995). Targeted mutagenesis of the transcription factor GATA-4 gene in mouse embryonic stem cells disrupts visceral endoderm differentiation *in vitro.* Development 121, 3877–3888.

Su, B., Jacinto, E., Hibi, M., Kallunki, T., Karin, M., & Ben-Neriah, Y. (1994). JNK is involved in signal transduction during costimulation of T lymphocytes. Cell 77, 727–736.

Tandon, N.N., Kralisz, U., & Jamieson, G.A. (1989a). Identification of Glycoprotein IV (CD36) as a primary receptor for platelet-collagen adhesion. J. Biol. Chem. 264, 7576–7583.

Tandon, N.N., Lipsky, R.H., Burgess, W., & Jamieson, G.A. (1989b). Isolation and characterization of platelet Glycoprotein IV (CD36). J. Biol. Chem. 264, 7570–7575.

Tepass, U., Fessler, L.I., Aziz, A., & Hartenstein, V. (1994). Embryonic origin of hemocytes and their relationship to cell death in *Drosophila.* Development. 120, 1829–1837.

Thummel, C.S. (1996). Flies on steroids-*Drosophila* metamorphosis and the mechanisms of steroid hormone action. Trends Genet. 12, 306–310.

Tsai, F.Y., Keller, G., Kuo, F.C., Weiss, M., Chen, J.Z., Rosenblatt, M., Alt, F.W., & Orkin, S.H. (1994). An early hematopoietic defect in mice lacking the transcription factor GATA-2. Nature 371, 221–226.

Unanue, E.R. (1984). Antigen: presenting function of the macrophage. Annu. Rev. Immunol. 2, 395–428.

Vega, M.A., Segui-Real, B., Alcade Garcia, J., Calès, C., Rodriguez, F., Vanderkerckhove, J., & Sandoval, I.V. (1991). Cloning, sequencing, and expression of a cDNA encoding rat LIMP II, a novel 74kDa lysosomal membrane protein related to the surface adhesion protein CD36. J. Biol. Chem. 266, 16818–16824.

Vidal, S.M., Malo, D., Vogan, K., Skamene, E., & Gross, P. (1993). Natural resistance to infection with intracellular parasites: isolation of a candidate for *Bcg.* Cell 73, 469–485.

Watson, K.L., Johnson, T.K., & Denell, R.E. (1991). *Lethal (1) aberrant immune response* mutations leading to melanotic tumor formation in *Drosophila melanogaster.* Dev. Genet. 12, 173–187.

Watson, K.L., Justice, R.W., & Bryant, P.J. (1994). *Drosophila* in cancer research: the first fifty tumor suppressor genes. J. Cell Sci. 18, 19–33.

Watson, K.L., Konrad, K.D., Woods, D.F., & Bryant, P.J. (1992). The *Drosophila* homologue of the human S6 ribosomal protein is required for tumor suppression in the hematopoietic system. Proc. Natl. Acad. Sci. USA 89, 11302–11306.

Westwick, J.K., Weitzel, C., Minden, A., Karin, M., & Brenner, D.A. (1994). Tumor necrosis factor alpha stimulates AP-1 activity through prolonged activation of the c-Jun kinase. J. Biol. Chem. 269, 26396–26401.

White, K., Grether, M.E., Abrams, J.M., Young, L., Farrell, K., & Steller, H. (1994). Genetic control of programmed cell death in *Drosophila.* Science. 264, 677–683.

White, K., Tahaoglu, E., & Steller, H. (1996). Cell killing by the *Drosophila* gene *reaper*. Science. 271, 805–807.

Whitmarsh, A.J., Shor, P., Sharrocks, A.D., & Davis, R.J. (1995). Integration of MAP kinase signal transduction at the serum response element. Science. 269, 403–407.

Williams M.J., Rodriguez, A., Kimbrell, D.A., & Eldon, E.D. (1997). The *18 wheeler* mutation reveals complex antibacterial gene regulation in *Drosophila* host defense. EMBO J. 16, 6120–6130

Wu, L.P., & Anderson, K.V. (1998). Regulated nuclear import of Rel proteins in the *Drosophila* immune response. Nature 392, 93–97.

Wyllie, A.H., Kerr, J.F.R., & Lurrie, A.R. (1980). Cell death: the significance of apoptosis. Int. Rev. Cytol. 68, 251–306.

Yan, R., Small, S., Desplan, C., Dearolf, C.R., & Darnell, J.E. Jr. (1996). Identification of a STAT gene that functions in *Drosophila* development. Cell 84, 421–430.

Yasothornsrikul, S., Davis, W.J., Cramer, G., Kimbrell, D.A., & Dearolf, C.R. (1997). *viking*: identification and characterization of a second type IV collagen in *Drosophila*. Gene 198, 17–25.

PHAGOCYTOSIS BY NONPROFESSIONAL PHAGOCYTES

Debora Williams-Herman and Zena Werb

I. INTRODUCTION

In multicellular organisms a significant amount, perhaps even the bulk, of phagocytosis is carried out by nonprofessional phagocytes. These working class phagocytes are usually fibroblasts or epithelial cells whose phagocytic function facilitates general processes such as maintenance of the extracellular matrix,

Advances in Cell and Molecular Biology of Membranes and Organelles
Volume 5, pages 47-67.

ISBN: 1-55938-999-0

response to injury, development of inflammation, and regulation of immune responses. Phagocytosis by these working class cells is also essential to specific processes such as maintenance of vision, prevention of premature aging, clearance of apoptotic cells, presentation of certain antigens, and perhaps even thymic selection. Further, nonprofessional phagocytosis is an important vehicle through which pathogens gain access to the intracellular milieu. The term "non-professional phagocyte" was first used around 1970 when it was observed that antiserum enhanced macrophage, but inhibited fibroblast, uptake of gluteraldehyde-fixed red blood cells (RBCs). Both cell types were inherently phagocytic for RBC in the absence of antiserum (Rabinovitch, 1995). This chapter is designed to provide a broad overview of the roles played by nonprofessional phagocytes and to remind the reader that not only macrophages, granulocytes, and B cells, but also fibroblasts, epithelial cells, and others, are capable of receptor-mediated and nonspecific phagocytosis.

II. RENEWAL OF THE SENSORY RETINA AND THE NONPROFESSIONAL PHAGOCYTE

Debris is shed from the outer segments of the cells of the sensory retina, the rods and cones, daily. These discarded rod outer segments (ROSs) enter the extracellular milieu located apical to the retinal epithelium. They are then specifically bound and engulfed by cells of the retinal pigmented epithelium (RPE) (Bok, 1993) which target them along a pathway that sorts recyclable lipids from nonrecyclable material to be degraded (Bazan et al., 1992). In animals with circadian rhythms, shedding is synchronized and occurs once a day (LaVail, 1976). This remarkable process requires that each RPE cell phagocytose on the order of 25–30,000 segments per day and thus defines these specialized cells as the ultimate in working class phagocytes (Ershov et al., 1996).

RPE binding of ROSs appears to be mediated by the integrin $\alpha v\beta 1$, otherwise known as the vitronectin receptor (Finnemann et al., 1997). The actual process of phagocytosis, however, requires the assistance of CD36, a receptor for thrombospondin, collagen, and oxidized low-density lipoproteins (hence its name the "scavenger" receptor) (Sparrow et al., 1997). Most likely, co-receptors, such as the mannose receptor and those as yet unidentified, are also involved (Boyle et al., 1991). RPE tissue-specific absence of CD36 (Sparrow et al., 1997) has been observed in the Royal College of Surgeons (RCS) rat, an animal model for retinal degeneration leading to complete blindness. *In vitro* and *in vivo* studies of this model have demonstrated a decreased ability of RPE cells to phagocytose ROSs (Edwards and Szamier, 1977). Extracellular debris accumulates in the subretinal space, and blindness ensues.

While RPE cells from both normal and RCS rats can phagocytose a variety of particles in a relatively nonspecific manner (Seyfried-Williams et al., 1984), ingestion of ROSs by normal RPE triggers specific signal transduction cascades that have not yet been entirely defined. ROS phagocytosis by normal RPE is accompanied by altered protein phosphorylation patterns that are different than those observed in dystrophic RPE; these patterns also vary depending on whether the ingested materials are ROSs or nonspecific particles (heth and Schmidt, 1992). In normal RPE, specific ROS phagocytosis (unlike phagocytosis of latex beads) also induces a decrease in phosphatidylinositol bisphosphate (PIP2) accompanied by an increase in diacylglycerol (DAG) and inositol triphosphate (IP3) (Heth and Marescalchi, 1994). These changes do not occur in RPE from the RCS rat; attempts to restore phagocytosis in the RCS rat RPE by bypassing proximal signaling paths through the use of carbachol to increase IP3 levels have met with varied success (Heth et al., 1995; Hall et al., 1996). Basic fibroblast growth factor (FGF) is markedly reduced in the RCS rat RPE at the crucial developmental age when ROS phagocytosis begins. Treatment with basic FGF, a protein expressed in many professional phagocytes, restores phagocytic ability to RPE from the RCS rat by an undefined mechanism that involves ongoing transcription (McLaren and Inana, 1997).

Ironically, inhibition of protein kinase C (PKC) enhances phagocytosis of ROSs by RPE and may mediate the increased phagocytosis observed with transforming growth factor beta (TGF-β) treatment (Herron et al., 1974). Furthermore, activation of PKC and increases in intracellular calcium, both of which presumably follow increases in DAG and IP3, inhibit phagocytosis of ROSs by normal RPE (Hall et al., 1991). Nitric oxide and superoxide, which can be produced by RPE or by neighboring cells in the presence of inflammation *in vivo*, also inhibit phagocytosis (Becquet et al, 1994a,b). Inhibition of normal phagocytosis by a cAMP-independent isoproterenol-mediated mechanism and by activation of cAMP-dependent protein kinase I and II suggests there is additional complexity to the signaling story (Hall et al., 1993; Kuriyama et al., 1995). Overall, however, these studies suggest: (1) that binding of ROSs occurs first and then stimulates and/or downregulates one or more signaling pathways allowing phagocytosis to proceed, and (2) that these pathway(s) are deficient in the RCS rat.

In RPE cells from normal animals, a non-mitochondrial, intracellular respiratory burst accompanies the specific phagocytosis of ROS, but not the nonspecific phagocytosis of latex beads (Miceli et al., 1994). These studies implicate the involvement of phagosomal NAPDH oxidase and/or peroxisomal β-oxidation in the process of ROS phagocytosis and/or degradation. Downstream effects of binding/engulfment of ROSs by RPE include upregulation of the anti-oxidants catalase and metallothionein (Tate et al., 1995), which may serve to protect the eye from degenerative processes. Exogenous administration of H_2O_2 can also induce expression of these protective proteins (Tate et al., 1995). Specific transcription factors/early response genes induced by ROS phagocytosis include *zif*-268 (*egr*-1), c-*fos*, and *tis*-1 (*NGF1-B*); nuclear translocation of *zif*-268, *tis*-1, AP-1, and AP-2 (but not NFκB) also occurs. These events are absent in RPE from the

RCS rat (Ershov et al., 1996), and their downstream effects require further delineation.

III. REMODELING OF THE EXTRACELLULAR MATRIX AND THE NONPROFESSIONAL PHAGOCYTE

Although some tissue macrophages participate in the active process of collagen homeostasis, the majority of collagen phagocytosis is performed by fibroblasts (Everts et al., 1996). In fact, collagen turnover in the periodontium, gingiva, uterus, and during the process of wound healing is directly proportional to the amount of banded fibrils observed in fibroblast phagosomes (Dyer and Peppler, 1997; Svoboda et al., 1981; McGaw and Ten Cate, 1983). During postpartum uterine involution fibroblasts, along with smooth muscle cells and macrophages, work in concert with extracellular degradation and ingest enormous amounts of collagen over a brief period of time. Dysregulation of collagen phagocytosis underscores its importance. Too little phagocytosis can lead to fibrosis associated with accumulation of extracellular collagen stores (Hall and Squier, 1982; McGaw and Porter, 1988). Fibroblasts isolated from fibrotic tissues have decreased phagocytic capacity; further, agents associated with the development of fibrosis (phenytoin and nifedipine) decrease collagen phagocytosis by fibroblasts *in vitro* (McCulloch and Knowles, 1993). In both cases, overall decreased phagocytosis reflects decreased numbers of phagocytic cells rather than decreased phagocytic capacity of individual cells. Too much phagocytosis can lead to abnormal tissue resorption or decreased collagen stores: senescent fibroblasts are generally more phagocytic for collagen-coated beads and less discriminating about the material they ingest (Lee and McCulloch, 1997). Increased tissue resorption associated with decreased extracellular collagen stores occurs with aging of the gingiva, periodontal ligament, bone, skin and eye.

One of the most well-studied examples of the role of phagocytosis in tissue remodeling is the continual engulfment of collagen by the periodontal fibroblast. Shifting movement of teeth throughout the lifetime of an animal requires continual renewal and strengthening of the attachment to bone via their anchoring, the periodontal ligament (PDL) (Vignery and Baron, 1980; Sodek and Ferrier, 1988). Extracellular matrix (ECM) homeostasis is primarily maintained by a balance between collagen synthesis and degradation. Degradation can occur via two distinct but interrelated pathways: extracellular degradation (primarily via secreted metalloproteinases) and intracellular degradation involving lysosomal proteolysis of extracellular material, which has been internalized via phagocytosis (Werb, 1982; Murphy and Reynolds, 1984). Electron microscopy (EM) revealing intracellularly sequestered collagen fibrils that accumulate to a greater extent upon inhibition of lysosomal proteinases suggests that the intracellular degradation is the predominant physiologic pathway (Everts et al., 1996; Creemers et al., 1998).

However, some extracellular collagen breakdown may be required to cut intact collagen fibrils into fragments small enough to be engulfed by fibroblasts (Svoboda et al., 1979). Indeed, some fibers are engulfed end on so that the intracellular end is located within a phagosome, while the opposite end continues to project into the extracellular milieu (Yajima, 1996).

Although CD36 and transmembrane alkaline phosphatase can bind and mediate collagen phagocytosis, the prototypical collagen I receptor on fibroblasts is the integrin, α2β1. Antibodies against this receptor and peptides containing an arg-gly-asp (RGD) sequence downregulate collagen phagocytosis (Lee et al., 1996). The RGD integrin-binding sequence promotes adhesion and phagocytosis by fibroblasts and a variety of other cells via integrins, including α5β1, α2β1, and αvβ3. Interestingly, collagen (particularly when partly degraded) (Knowles et al., 1991) avidly binds fibronectin, suggesting a probable role for α5β1 in mediating collagen phagocytosis. In fact, EM studies of collagen phagocytosis show fibronectin localized to contact points linking cell processes and collagen (Pitaru et al., 1987). Engagement of α2β1 integrin by collagen I in substrate form triggers transient tyrosine phosphorylation of platelet-derived growth factor (PDGF) β-receptors (Sundberg and Rubin, 1996) and upregulates expression of the secreted matrix metalloproteinase, collagenase-1 (MMP1) (Langholz et al., 1995).

It has long been known that many cell types, including fibroblasts, keratinocytes, myocytes, endothelial cells and epithelial cells, phagocytose particles of different sizes (Grinnell, 1980; Takashima and Grinnell, 1984; Schwachula et al., 1994). The majority of early experiments were performed over many hours in the presence of serum or after coating the particles with fibronectin. The particles not coated with fibronectin acquire serum proteins or secreted fibronectin during the course of the experiment. Thus, it is likely that fibronectin promotes phagocytosis via the integrin α5β1. It is therefore not surprising that nonprofessional phagocytosis is associated with altered integrin expression and increased expression of proteins such as cytokines, MMPs, and ECM components (Werb and Reynolds, 1974; Brinckerhoff et al., 1982; Wigley et al., 1983; Havemose-Poulsen and Holmstrup, 1997; Lee and McCulloch, 1997).

In synovial fibroblasts, engagement of α5β1 by soluble anti-α5 antibody generates a respiratory burst and upregulates expression of interleukin-1 alpha (IL-1α) via a Rac-dependent mechanism (Kheradmand et al., 1998). IL-1α then acts as an autocrine factor to upregulate collagenase-1. Phagocytosis of beads coated with fibronectin or anti-α5 antibody (2.0–4.5 μm) also upregulates collagenase-1 secretion (Figure 1).

Cytokines play a key role in determining the fate of the extracellular matrix (reviewed in Schwachula et al., 1994; Bienkowski and Gotkin, 1995; see also Havemose-Poulsen and Holmstrup, 1997). As noted above, phagocytosis of fibronectin-coated particles increases fibroblast secretion of IL-1α. Not only is IL-1 an inflammatory cytokine that helps recruit and activate cells of the immune system, it also increases production of collagenase-1, which itself may be cleaved and

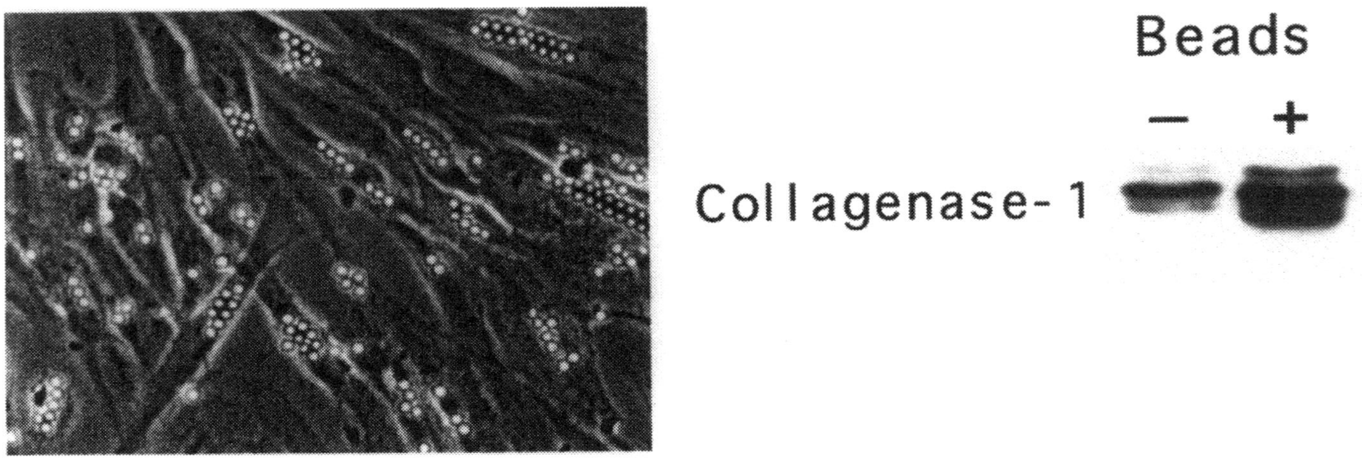

Figure 1. **A.** Synovial fibroblasts phagocytose 4.5 μm beads coated with anti-integrin antibody. **B.** Western blot for collagenase-1 24 hours after beads added.

activated by enzymes secreted by the inflammatory cells recruited. This cycle, if unchecked, can lead to massive tissue destruction as seen in disease states such as rheumatoid arthritis and periodontal disease/gingivitis. Because IL-1 can inhibit phagocytosis of collagen (van der Zee et al., 1995), it may also play a role in the generation of fibrosis following inhalation of particulate matter. Tumor necrosis factor alpha (TNF-α), another cytokine involved in the evolution of fibrosis, inactivates the collagen receptor, integrin $\alpha 2\beta 1$, and thereby inhibits the phagocytosis of collagen (Chou et al., 1996). TGF-β, alternatively, may enhance collagen phagocytosis, possibly by inhibiting PKC (van der Zee et al., 1995).

IV. PATHOGENS AND THE NONPROFESSIONAL PHAGOCYTE

A. Phagocytosis

The ECM proteins fibronectin and laminin also act to enhance uptake of microorganisms (Bohnsack et al., 1985; Brown, 1986; Hermann et al., 1990). While this pathway most likely promotes pathogen clearance by facilitating uptake into, and infection of, a variety of cells, many pathogens, such as *Yersinia pseudotuberculosis*, directly engage receptors on the surface of their target cells (Isberg and Tran Van Nhieu, 1995). They can therefore gain intracellular entry by receptor-mediated phagocytosis using receptors usually confined to adhesive functions. Some, such as *Salmonella* and *Shigella*, are even clever enough to induce phagocytic uptake via a "nonzipper" mechanism which involves triggering extensive small GTPase-mediated cell ruffling in nearby cells (reviewed in Finlay and Cossart, 1997). Further, somc, such as *Y. pseudotuberculosis* (the Yop proteins, see below) and *Escherichia coli* (cytotoxic necrotizing factor), secrete factors which can selectively inhibit or induce nonprofessional phagocytosis (Fiorentini et al., 1997a,b). In all cases, the triggering of signal transduction cascades leading to cytoskeletal rearrangement and cell shape change is likely to be a common theme. (Table 1).

Y. pseudotuberculosis first gains access to epithelial cells by expressing the protein, invasin, which binds to integrin $\alpha 5\beta 1$ (the fibronectin receptor) and mediates uptake of the bacterium into enterocytes overlying Peyer's patches in the gut (Leong et al., 1995). The pathogen also utilizes this protein to gain intracellular access to other cells such as fibroblasts. Invasin binding to $\alpha 5\beta 1$ results in protein tyrosine phosphorylation of many proteins. Tyrosine kinase inhibitors decrease phagocytosis of the organism (Rosenshine et al., 1992). YopH, a protein with tyrosine phosphatase activity, also inhibits uptake. Interestingly, once *Yersinia* has gained access to its preferable extracellular sites, it upregulates expression of YopH and similar proteins, presumably limiting its further phagocytosis (Persson et al., 1995; Sory et al., 1995).

Table 1. Overview of Strategies Used by Selected Organisms to Gain Intracellular Access to Nonprofessional Phagocytes via Phagocytosis

Organism	*Cell Type(s)*	*Host Receptor/Mechanism*
Aspergillus fumigata (Henwick et al., 1993; Bouchara et al., 1995; Paris et al., 1997)	Epithelium including alveolar type II cells, and endothelium	Unknown, phagocytosis in macrophages via CR1, CR3; laminin appears to be opsonin
Candida albicans (Filler et al., 1995; Zink et al., 1996)	Endothelium	Unknown, but actin cytoskeletal rearrangement and microtubule assembly involved
Chlamydia species (Byrne and Moulder, 1978; Hodinka and Wyrick, 1986; Wyrick et al., 1989; Majeed and Kihlström, 1991)	Epithelium including alveolar type II cells, fibroblasts	Unknown, but involves clathrin coat assembly and actin cytoskeleton rearrangement
Legionella Pneumophilia (Garduño et al., 1998; Maruta et al., 1998)	Epithelium including alveolar type I, II cells	Phagocytosis in macrophages via CR1, CR3; involves actin cytoskeleton rearrangement
Listeria monocytogenes (Theriot, 1995; Finlay and Cossart, 1997)	Fibroblasts, smooth muscle, and hepatocytes	Unknown, involves cytoskeletal rearrangement
Neisseria species (Grassmé et al., 1997; Jerse and Rest, 1997; Merz and So, 1997; Naumann et al., 1997; Dehio et al., 1998a,b)	Epithelium and fibroblasts	Involves the vitronectin receptor, CD66, and acidic sphingomyelinase as well as actin cytoskeletal rearrangement and NFκB and AP-1 translocation
Rickettsia species (Wisseman et al., 1976; Silverman and Bond, 1984; Feng et al., 1993)	Embryonal cells, endothelium, and perivascular myocytes	Unknown
Salmonella species (Chen et al., 1996; Clark et al., 1998a; Hardt et al., 1998; Jensen et al., 1998)	Epithelium including M cells and enterocytes	Involves actin cytoskeletal rearrangement and the small GTPase, CDC42
Shigella species (Clerc and Sansonetti, 1989; Perdomo et al., 1994a,b; Zychlinsky et al., 1994; Theriot, 1995; Watarai et al., 1996; Finlay and Cossart, 1997)	Epithelium including M cells and basal surface of enterocytes	Adhesion to $\alpha 5\beta 1$, Involves actin cytoskeletal rearrangement, the small GTPase, Rho, and possibly clathrin recruitment
Staphylococcus species (Vercellotti et al., 1985; Blumberg et al., 1998; Lowy et al., 1988; Almeida et al., 1996; Zavizion et al., 1997; Bayles et al., 1998)	Endothelium and epithelium	Invasion potentiated by bacterial ECM coating, acidic FGF, plasminogen; involves actin cytoskeletal rearrangement
Streptococcus species (Cundell et al., 1995; Calvinho and Oliver, 1998)	Endothelium, epithelium	Phagocytosis via platelet activating receptor

(continued)

Table 1. (Continued)

Organism	Cell Type(s)	Host Receptor/Mechanism
Toxoplasma gondii (Werk, 1985; Furtado et al., 1992; Sibley, 1995)	All cells with exception of mature erythrocytes	Rapid translocation, laminin receptor may be involved
Trypanosoma cruzii (McCabe et al., 1984; Andrews, 1995; Burleigh and Andrews, 1995)	Epithelium, fibroblasts, macrophages, and myocytes	Rapid translocation; involves microtubules assembly
Yersinia pseudotuberculosis (Isberg and Tran Van Nhieu, 1995; Leong et al., 1995; Clark et al., 1998b)	Epithelium including M cells and enterocytes	Phagocytosis via $\alpha5\beta1$; involves clathrin

Clathrin is essential to invasin-mediated phagocytosis. Not only are sheets containing clathrin and the adaptor protein, AP2, localized to the nascent phagocytic cup (consistent with early observation of phagocytosis of latex beads; Aggeler and Werb, 1982), but intracellular anti-clathrin antibodies prevent phagocytic uptake (Tran Van Nhieu et al., 1996). Although factors recruiting coat proteins to the phagocytic cup have yet to be identified, mutations in the $\beta1$ integrin cytoplasmic domain that inhibit $\beta1$ localization to adhesive focal complexes enhance its role in phagocytic activities. Interestingly, invasin-coated beads are ingested more effectively than fibronectin-coated beads: invasin has higher affinity for $\alpha5\beta1$ than does fibronectin, and it appears that both ligand-receptor affinity and $\alpha5\beta1$ receptor density determine the rate and extent of phagocytosis (Tran Van Nhieu et al., 1996).

Shigella flexneri also depends on the presence of $\alpha5\beta1$ for basolateral entry into epithelial cells (Dehio et al., 1995; Watarai et al., 1996); however, it appears to induce its cntry by stimulating exaggerated ruffles that eventually create a sealed-off phagosome surrounding the organism (Finlay and Cossart, 1997). This ruffling may be induced by a diffusable factor which mediates its signal via the small guanosine triphosphate (GTP)-binding protein, Rho. Src, a non-receptor tyrosine kinase is phosphorylated and localized to the site of entry, where it appears to phosphorylate the actin binding protein. Host pp125FAK and paxillin also become tyrosine phosphorylated. Salmonella utilizes a different member of the Rho family of GTPases, Cdc42, to induce ruffling and gain apical access to enterocytes (reviewed in Finlay and Cossart, 1997).

B. Antigen Presentation

Once an organism has invaded the nonprofessional phagocyte has it evaded host immunity? This question, of course, has no simple answer. As described elsewhere in this volume, pathogens have developed clever ways of avoiding detection by the host immune system. Many manipulate the cellular endocytic machinery for their own benefit: *Mycobacteria* prevent vacuolar acidification by excluding adenosine triphosphatase from its limiting membrane (Sturgill-Koszycki et al., 1994). Not only does this prevent lysosomal degradation of the organism, it also promotes sequestration away from the MHCII antigen presenting molecules. Other organisms can travel from cell to cell without leaving the intracellular environment: *Listeria*, by using actin polymerization to propel itself, can distend the plasma membrane of one epithelial cell to project into the plasma membrane of another (as long as adjacent cells are tethered by adhesion molecules). The membrane then pinches off to create a double-walled vesicle from which the organism escapes into the cytoplasm of the new cell (reviewed in Theriot, 1995). The working class phagocyte, however, does its share to promote clearance of organisms and antigen recognition.

One example of this process is that circulating fibronectin-coated organisms and particles are readily cleared not only by the phagocytic liver macrophages, or Kupffer cells, but also by high endothelial venules (HEV) and hepatocytes. These cells are capable of massive phagocytosis as demonstrated *in vivo,* when downregulation of the macrophage/polymorphonuclear complement 3 receptor (CR3) results in increased clearance of *Listeria* by hepatocytes and HEV of the spleen, both of which lack CR3 (Conlan and North, 1992). Also, although antigen presentation by nonprofessional phagocytes is just recently beginning to be explored, interferon gamma (IFNγ)-stimulated human dermal fibroblasts are capable of presenting tetanus toxoid antigen to a tetanus-specific T cell clone (Umetsu et al., 1985). T cells from a primed donor can also be specifically stimulated, but only in the presence of exogenous IL-2 (Umetsu et al., 1986). Furthermore, antigen-presenting fibroblasts are synergistic with small numbers of monocytes for stimulating resting T cells. This suggests that there must be underlying inflammation and/or T cell activation for fibroblasts to act as antigen presenting cells to T cells.

Antigen presentation by other nonprofessional cells including epithelium (Kaneda et al., 1997), myoblasts (Garlepp et al., 1995; Curnow et al., 1998), keratinocytes (de Bueger et al., 1993; Nickoloff and Turka, 1994), and tumor cells (Bellone et al., 1997; Marti et al., 1997) is also possible. As in the fibroblast studies above, effective antigen presentation by nonprofessional cells frequently requires stimulation with IFNγ to upregulate major histocompatibility antigen class II (MHCII); it also requires co-expression of accessory molecules on the presenting cell and/or the presence of exogenous stimulatory products. These findings support the current dogma that antigen presentation in the absence of accessory co-stimulatory molecules induces immunotolerance, and at least one study has addressed this issue in more detail (Roth and Dröge, 1994). Further, the processing of an antigen might also determine whether a strong stimulatory T cell response is elicited: unlike an entire protein exogenously administered, an endogenously expressed peptide derived from that protein can induce proliferation by a responsive T cell hybridoma when presented by fibroblasts (Moreno et al., 1991).

V. CLEARANCE OF APOPTOTIC CELLS AND THE NONPROFESSIONAL PHAGOCYTE

Although the major work of clearing apoptotic cells has been attributed to macrophages and monocytes, it is becoming increasingly clear that many nonprofessional phagocytes play an important role in this process. This is particularly true for tissues that undergo frequent remodeling, such as epithelial barriers, mammary gland, and the uterine cavity during trophoblast invasion. It is also true for tissues which have been damaged by trauma or other means. Unlike the majority of professional phagocytes, nonprofessional phagocytes are inherently "on the scene," thus allowing them the opportunity for quiet, immediate recognition and phagocytosis of cells undergoing programmed cell death. Apoptosis in combination with rapid clearance of the apoptotic body provides a means by which an organism can clear dead cells without inciting an inflammatory response directly and without injuring nearby cells secondary to released intracellular contents (reviewed in Savill, 1997; Platt et al., 1998). In this way, one could hypothesize that nonprofessional phagocytosis of apoptotic cells protects the immunocompetent multicellular organism from unneeded inflammation and the development of autoimmune disease.

Examples of this type of phagocytosis have been reported for different types of cells from numerous tissues. Following traumatic or recurrent bleeds, fibroblasts are capable of amassing large numbers of intracellular RBCs (Wakefield and Hicks, 1974; Wolter and Lichter, 1983). Mesothelial cells from serous fluids also show evidence of having ingested RBCs (Zaharopoulos et al., 1998). Phagocytosis by invading trophoblast allows for nutrient uptake and clearance of RBCs and apoptotic uterine epithelial cells (El-Shershaby and Hinchliffe, 1974, 1975; Poelmann, 1975). Alveolar cells phagocytose their apoptotic neighbors in involuting mammary glands (Walker et al., 1989). Olfactory epithelium plays a major role in phagocytosis of replacement neurons following bulbectomy in the mouse and may play a role in phagocytosis of apoptotic neurons resulting from the massive ongoing renewal of olfactory neurons in the normal adult (Suzuki et al., 1996). Hepatocytes and liver endothelial cells phagocytose apoptotic bodies (Dini et al., 1992, 1995). Sertoli cells phagocytose apoptotic spermatogenic cells (Shiratsuchi et al., 1997). Renal mesangial cells engulf apoptotic neutrophils in a process which may be defective in lupus glomerulonephritis (Hughes et al., 1997; Botto et al., 1998). Fibroblasts also phagocytose apoptotic neutrophils and tissue mast cells, a role that has the possibility of containing active inflammation (Atkins et al., 1985b; Hall et al., 1994). In the pathological condition amyloidosis, fibroblasts engulf melanosomes and accumulate pigment/amyloid (Ishii et al., 1984). Rather than promoting host inflammatory responses, tumor cells have the potential of postponing their own death by phagocytosing neighboring apoptotic tumor cells (Büchi et al., 1994; Li et al., 1997). Lastly, epithelial-like thymic nurse cells are known to harbor many apoptotic thymocytes, perhaps influencing self vs. non-self recognition by the surviving lymphocytes (Hiramine et al., 1996).

The events that follow phagocytosis of apoptotic cells appear to be dictated by a number of factors including the receptor(s) and potential opsonin(s) involved in the cell-cell interaction, the type of cell engulfing the apoptotic body, the downstream intracellular signaling events triggered during the process, and the tissue in which the process occurs. Some of the receptors involved in the process have been identified including CD14 and the scavenger receptor, CD36: CD14 is important for uptake of apoptotic B cells by professional macrophages, while CD36 has roles in the uptake of aging neutrophils by macrophages and in the uptake of ROSs by nonprofessional retinal epithelium (Savill, 1997; Devitt et al., 1998; and see above section on retinal epithelium). Other molecules, such as the vitronectin receptor, integrin αvβ3, are essential to phagocytosis of apoptotic cells by some cells (Hall et al., 1994; Hughes et al., 1997; Rubartelli et al., 1997) but may act as co-receptors for other cells types. Transfection of CD36 into non-phagocytic cells bearing αvβ3 confers phagocytic capability (Ren et al., 1995). Interestingly, while CD36 recognizes known ligands such as phosphatidylserine on apoptotic cells, a direct receptor-ligand interaction is not necessary for phagocytosis to proceed (reviewed in Platt et al., 1998). Many additional factors concerning the role of CD36 in apoptotic cell uptake have yet to be identified; however, it is known that in some cases thrombospondin is required and appears to act as an opsonin (Savill, 1997).

In addition to isolating potentially inflammatory material from the extracellular milieu, phagocytosis of apoptotic cells also appears to alter the balance between anti-inflammatory and pro-inflammatory cytokines. Although cytokine profiles following ingestion of apoptotic cells by nonprofessional phagocytes have yet to be generated, cytokine profiles of macrophages stimulated with both lipopolysaccharide (LPS) and apoptotic neutrophils show lower levels of inflammatory proteins and higher levels of the non-inflammatory cytokine, IL-10, than when macrophages are stimulated with LPS alone (Voll et al., 1997). Similarly, unlike challenging macrophages with LPS, which binds CD14 and upregulates expression of TNF-α, challenging macrophages with apoptotic B cells that bind CD14 does not increase TNF-α expression (Devitt et al., 1998). This reinforces the hypothesis that phagocytosis of apoptotic cells is, in general, a silent non-inflammatory process.

Analogous to the role of limiting inflammation via phagocytosis of the apoptotic cell, nonprofessional phagocytes also control the degree of inflammation and regulate the immune response by engulfing parts of cells and inflammatory products released from granulocytes during exocytosis. During Wallerian degeneration following peripheral nerve damage there is intense phagocytic uptake of myelin sheaths, first by the Schwann cells, then by surrounding fibroblasts and macrophages (Röyttä et al., 1987). Increased Schwann cell expression of *c-fos* is thought to correlate with their altered phagocytic phenotype; currently it is unclear whether inflammatory mechanisms are dampened or upregulated following the phenotypic change (Liu et al., 1995). In the involuting uterus, in addition to phagocytosis of collagen, fibroblasts also engulf cytoplasmic protrusions and isolated blebs of smooth muscle cells (Inouye et al., 1983).

A most fascinating finding is that at sites of mast cell exocytosis, fibroblasts, endothelial cells, and their professional colleagues act as scavengers to retrieve and phagocytose remnants of toxic granules thus preventing the inflammatory process from accelerating out of control (Atkins et al., 1985a,b; Takeda, 1985). Although some mast cells can be seen discharging their contents almost directly into neighboring fibroblasts, phagocytosis of granules is generally a cytochalasin-sensitive process occurring minutes to hours following degranulation. At this time, histamine has diffused away, and the remaining granules primarily contain neutral proteases, acid hydrolases, and oxidative enzymes bound to heparin. Phagocytosis of these remaining heparin-protein granules limits their extracellular biologic activity, and the granules are subsequently degraded by the phagocyte (Atkins and Metcalfe, 1983). In fibroblasts, phagocytosis of mast cell granules results in increased expression of collagenase and β-hexosaminidase (Subba Rao et al., 1983). This suggests that not only are fibroblasts capable of limiting or altering the potential biologic response elicited by mast cell degranulation, but that mast cell degranulation is capable of altering the process of tissue remodeling. By altering the pattern of enzymes and cytokines secreted by fibroblasts, phagocytosis of mast cell granules also participates in shaping the immune response.

VI. SUMMARY

This chapter begins to touch on the important phagocytic roles played by many of the working class phagocytes present in the multicellular organism. It seems the original distinctions drawn between professional and nonprofessional phagocytes have become blurred, but the roles are more distinct. The more we understand the phagocytic roles these nonprofessionals play, the more we will gain insight into disease processes such as infectious illness, atherosclerosis, autoimmunity, progressive blindness, and malignancy; the more we will understand processes such as inflammation, aging, nerve growth and regeneration; and the better we might become at the targeting and delivery of intracellular agents.

ACKNOWLEDGMENTS

This study was supported by funds from the NIH (K11 AI01283, AR20684 and CA72006) and by the Ruth and Milton Steinbach Fund.

REFERENCES

Aggeler, J., & Werb, Z. (1982). Initial events during phagocytosis by macrophages viewed from outside and inside the cell: membrane-particle interactions and clathrin. J. Cell Biol. 94, 613–623.

Almeida, R.A., Matthews, K.R., Cifrian, E., Guidry, A.J., & Oliver, S.P. (1996). *Staphylococcus aureus* invasion of bovine mammary epithelial cells. J. Dairy Sci. 79, 1021–1026.

Andrews, N.W. (1995). Lysosome recruitment during host cell invasion by Trypanosome cruzi. Trends Cell Biol. 5, 133–137.

Atkins, F.M., & Metcalfe, D.D. (1983). Degradation of the heparin matrix of mast cell granules by cultured fibroblasts. J. Immunol. 131, 1420–1425.

Atkins, F.M., Friedman, M.M., & Metcalfe, D.D. (1985a). Biochemical and microscopic evidence for the internalization and degradation of heparin-containing mast cell granules by bovine endothelial cells. Lab. Invest. 52, 278–286.

Atkins, F.M., Friedman, M.M., Subba Rao, P.V., & Metcalfe, D.D. (1985b). Interactions between mast cells, fibroblasts and connective tissue components. Int. Arch. Allergy Appl. Immunol. 77, 96–102.

Bayles, K.W., Wesson, C.A., Liou, L.E., Fox, L.K., Bohach, G.A., & Trumble, W.R. (1998). Intracellular *Staphylococcus aureus* escapes the endosome and induces apoptosis in epithelial cells. Infect. Immun. 66, 336–342.

Bazan, N.G., Gordon, W.C., & Rodriguez de Turco, E.B. (1992). Docosahexaenoic acid uptake and metabolism in photoreceptors: retinal conservation by an efficient retinal pigment epithelial cell-mediated recycling process. Adv. Exp. Med. Biol. 318, 295–306.

Becquet, F., Courtois, Y., & Goureau, O. (1994a). Nitric oxide decreases in vitro phagocytosis of photoreceptor outer segments by bovine retinal pigmented epithelial cells. J. Cell Physiol. 159, 256–262.

Becquet, F., Goureau, O., Soubrane, G., Coscas, G., Courtois, Y., & Hicks, D. (1994b). Superoxide inhibits proliferation and phagocytic internalization of photoreceptor outer segments by bovine retinal pigment epithelium in vitro. Exp. Cell. Res. 212, 374–382.

Bellone, M., Iezzi, G., Martin-Fontecha, A., Rivolta, L., Manfredi, A.A., Protti, M.P., Freschi, M., Dellabona, P., Casorati, G., & Rugarli, C. (1997). Rejection of a nonimmunogenic melanoma by vaccination with natural melanoma peptides on engineered antigen-presenting cells. J. Immunol. 158, 783–789.

Bienkowski, R.S., & Gotkin, M.G. (1995). Control of collagen deposition in mammalian lung. Proc. Soc. Exp. Biol. Med. 209, 118–140.

Blumberg, E.A., Hatcher, V.B., & Lowy, F.D. (1988). Acidic fibroblast growth factor modulates *Staphylococcus aureus* adherence to human endothelial cells. Infect. Immun. 56, 1470–1474.

Bohnsack, J.F., Kleinman, H.K., Takahashi, T., O'Shea, J.J., & Brown, E.J. (1985). Connective tissue proteins and phagocytic cell function. Laminin enhances complement and Fc-mediated phagocytosis by cultured human macrophages. J. Exp. Med. 161, 912–923.

Bok, D. (1993). The retinal pigment epithelium: a versatile partner in vision. J. Cell Sci. Suppl. 17, 189–195.

Botto, M., Dell'Agnola, C., Bygrave, A.E., Thompson, E.M., Cook, H.T., Petry, F., Loos, M., Pandolfi, P.P., & Walport, M.J. (1998). Homozygous C1q deficiency causes glomerulonephritis associated with multiple apoptotic bodies [see comments]. Nat. Genet. 19, 56–59.

Bouchara, J.P., Tronchin, G., Larcher, G., & Chabasse, D. (1995). The search for virulence determinants in *Aspergillus fumigatus*. Trends Microbiol. 3, 327–330.

Boyle, D., Tien, L.F., Cooper, N.G., Shepherd, V., & McLaughlin, B.J. (1991). A mannose receptor is involved in retinal phagocytosis. Invest. Ophthalmol. Vi.s Sci. 32, 1464–1470.

Brinckerhoff, C.E., Gross, R.H., Nagase, H., Sheldon, L., Jackson, R.C., & Harris, E.D., Jr. (1982). Increased level of translatable collagenase messenger ribonucleic acid in rabbit synovial fibroblasts treated with phorbol myristate acetate or crystals of monosodium urate monohydrate. Biochemistry 21, 2674–2679.

Brown, E.J. (1986). The role of extracellular matrix proteins in the control of phagocytosis. J. Leukoc. Biol. 39, 579–591.

Büchi, E.R., Bernauer, W., & Daicker, B. (1994). Cell death and disposal in retinoblastoma: an electron microscopic study. Graefes Arch. Clin. Exp. Ophthalmol. 232, 635–645.

Burleigh, B.A., & Andrews, N.W. (1995). The mechanisms of *Trypanosoma cruzi* invasion of mammalian cells. Annu. Rev. Microbiol. 49, 175–200.

Byrne, G.I., & Moulder, J.W. (1978). Parasite-specified phagocytosis of *Chlamydia psittaci* and Chlamydia trachomatis by L and HeLa cells. Infect. Immun. 19, 598–606.

Calvinho, L.F., & Oliver, S.P. (1998). Invasion and persistence of *Streptococcus dysgalactiae* within bovine mammary epithelial cells. J. Dairy Sci. 81, 678–686.

Chen, L.M., Hobbie, S., & Galán, J.E. (1996). Requirement of CDC42 for *Salmonella*-induced cytoskeletal and nuclear responses. Science 274, 2115–2118.

Chou, D.H., Lee, W., & McCulloch, C.A. (1996). TNF-alpha inactivation of collagen receptors: implications for fibroblast function and fibrosis. J. Immunol. 156, 4354–4362.

Clark, M.A., Hirst, B.H., & Jepson, M.A. (1998a). Inoculum composition and Salmonella pathogenicity island 1 regulate M-cell invasion and epithelial destruction by *Salmonella typhimurium.* Infect. Immun. 66, 724–731.

Clark, M.A., Hirst, B.H., & Jepson, M.A. (1998b). M-cell surface beta1 integrin expression and invasin-mediated targeting of Yersinia pseudotuberculosis to mouse Peyer's patch M cells. Infect. Immun. 66, 1237–1243.

Clerc, P.L., & Sansonetti, P.J. (1989). Evidence for clathrin mobilization during directed phagocytosis of *Shigella flexneri* by HEp2 cells. Microb. Pathog. 7, 329–336.

Conlan, J.W., & North, R.J. (1992). Monoclonal antibody NIMP-R10 directed against the CD11b chain of the type 3 complement receptor can substitute for monoclonal antibody 5C6 to exacerbate listeriosis by preventing the focusing of myelomonocytic cells at infectious foci in the liver. J. Leukoc. Biol. 52, 130–132.

Creemers, L.B., Hoeben, K.A., Jansen, D.C., Buttle, D.J., Beertsen, W., & Everts, V. (1998). Participation of intracellular cysteine proteinases, in particular cathepsin B, in degradation of collagen in periosteal tissue explants. Matrix Biol. 16, 575–584.

Cundell, D.R., Gerard, N.P., Gerard, C., Idanpaan-Heikkila, I., & Tuomanen, E.I. (1995). *Streptococcus pneumoniae* anchor to activated human cells by the receptor for platelet-activating factor. Nature 377, 435–438.

Curnow, S.J., Willcox, N., & Vincent, A. (1998). Induction of primary immune responses by allogeneic human myoblasts: dissection of the cell types required for proliferation, IFNγ secretion and cytotoxicity. J. Neuroimmunol. 86, 53–62.

de Bueger, M., Bakker, A., & Goulmy, E. (1993). Human keratinocytes activate primed major and minor histocompatibility antigen specific Th cells in vitro. Transpl. Immunol. 1, 52–59.

Dehio, C., Freissler, E., Lanz, C., Gómez-Duarte, O.G., David, G., & Meyer, T.F. (1998a). Ligation of cell surface heparan sulfate proteoglycans by antibody-coated beads stimulates phagocytic uptake into epithelial cells: a model for cellular invasion by *Neisseria gonorrhoeae*. Exp. Cell Res. 242, 528–539.

Dehio, C., Prévost, M.C., & Sansonetti, P.J. (1995). Invasion of epithelial cells by *Shigella flexneri* induces tyrosine phosphorylation of cortactin by a pp60c-src-mediated signalling pathway. EMBO J. 14, 2471–2482.

Dehio, M., Gómez-Duarte, O.G., Dehio, C., & Meyer, T.F. (1998b). Vitronectin-dependent invasion of epithelial cells by *Neisseria gonorrhoeae* involves alpha(v) integrin receptors. FEBS Lett. 424, 84–88.

Devitt, A., Moffatt, O.D., Raykundalia, C., Capra, J.D., Simmons, D.L., & Gregory, C.D. (1998). Human CD14 mediates recognition and phagocytosis of apoptotic cells [see comments]. Nature 392, 505–509.

Dini, L., Autuori, F., Lentini, A., Oliverio, S., & Piacentini, M. (1992). The clearance of apoptotic cells in the liver is mediated by the asialoglycoprotein receptor. FEBS Lett. 296, 174–178.

Dini, L., Lentini, A., Diez, G.D., Rocha, M., Falasca, L., Serafino, L., & Vidal-Vanaclocha, F. (1995). Phagocytosis of apoptotic bodies by liver endothelial cells. J. Cell Sci. 108, 967–973.

Dyer, R.F., & Peppler, R.D. (1977). Intracellular collagen in the nonpregnant and IUD-containing rat uterus. Anat. Rec. 187, 241–247.

Edwards, R.B., & Szamier, R.B. (1977). Defective phagocytosis of isolated rod outer segments by RCS rat retinal pigment epithelium in culture. Science 197, 1001–1003.

El-Shershaby, A.M., & Hinchliffe, J.R. (1974). Cell redundancy in the zona-intact preimplantation mouse blastocyst: a light and electron microscope study of dead cells and their fate. J. Embryol. Exp. Morphol. 31, 643–654.

El-Shershaby, A.M., & Hinchliffe, J.R. (1975). Epithelial autolysis during implantation of the mouse blastocyst: an ultrastructural study. J. Embryol. Exp. Morphol. 33, 1067–1080.

Ershov, A.V., Lukiw, W.J., & Bazan, N.G. (1996). Selective transcription factor induction in retinal pigment epithelial cells during photoreceptor phagocytosis. J. Biol. Chem. 271, 28458–28462.

Everts, V., van der Zee, E., Creemers, L., & Beertsen, W. (1996). Phagocytosis and intracellular digestion of collagen, its role in turnover and remodelling. Histochem. J. 28, 229–245.

Feng, H.M., Wen, J., & Walker, D.H. (1993). *Rickettsia australis* infection: A murine model of a highly invasive vasculopathic rickettsiosis. Am. J. Pathol. 142, 1471–1482.

Filler, S.G., Swerdloff, J.N., Hobbs, C., & Luckett, P.M. (1995). Penetration and damage of endothelial cells by *Candida albicans*. Infect. Immun. 63, 976–983.

Finlay, B.B., & Cossart, P. (1997). Exploitation of mammalian host cell functions by bacterial pathogens. Science 276, 718–725.

Finnemann, S.C., Bonilha, V.L., Marmorstein, A.D., & Rodriguez-Boulan, E. (1997). Phagocytosis of rod outer segments by retinal pigment epithelial cells requires alpha(v)beta5 integrin for binding but not for internalization. Proc. Natl. Acad. Sci. USA 94, 12932–12937.

Fiorentini, C., Fabbri, A., Flatau, G., Donelli, G., Matarrese, P., Lemichez, E., Falzano, L., & Boquet, P. (1997a). *Escherichia coli* cytotoxic necrotizing factor 1 (CNF1), a toxin that activates the Rho GTPase. J. Biol. Chem. 272, 19532–19537.

Fiorentini, C., Fabbri, A., Matarrese, P., Falzano, L., Boquet, P., & Malorni, W. (1997b). Hinderance of apoptosis and phagocytic behaviour induced by *Escherichia coli* cytotoxic necrotizing factor 1: two related activities in epithelial cells. Biochem. Biophys. Res. Commun. 241, 341–346.

Furtado, G.C., Cao, Y., & Joiner, K.A. (1992). Laminin on *Toxoplasma gondii* mediates parasite binding to the beta 1 integrin receptor alpha 6 beta 1 on human foreskin fibroblasts and Chinese hamster ovary cells. Infect. Immun. 60, 4925–4931.

Garduño, R.A., Quinn, F.D., & Hoffman, P.S. (1998). HeLa cells as a model to study the invasiveness and biology of *Legionella pneumophila*. Can. J. Microbiol. 44, 430–440.

Garlepp, M.J., Chen, W., Tabarias, H., Baines, M., Brooks, A., & McCluskey, J. (1995). Antigen processing and presentation by a murine myoblast cell line. Clin. Exp. Immunol. 102, 614–619.

Grassmé, H., Gulbins, E., Brenner, B., Ferlinz, K., Sandhoff, K., Harzer, K., Lang, F., & Meyer, T.F. (1997). Acidic sphingomyelinase mediates entry of *N. gonorrhoeae* into nonphagocytic cells. Cell 91, 605–615.

Grinnell, F. (1980). Fibroblast receptor for cell-substratum adhesion: studies on the interaction of baby hamster kidney cells with latex beads coated by cold insoluble globulin (plasma fibronectin). J. Cell Biol. 86, 104–112.

Hall, B.K., & Squier, C.A. (1982). Ultrastructural quantitation of connective tissue changes in phenytoin-induced gingival overgrowth in the ferret. J. Dent. Res. 61, 942–952.

Hall, M.O., Abrams, T.A., & Mittag, T.W. (1991). ROS ingestion by RPE cells is turned off by increased protein kinase C activity and by increased calcium. Exp. Eye Res. 52, 591–598.

Hall, M.O., Abrams, T.A., & Mittag, T.W. (1993). The phagocytosis of rod outer segments is inhibited by drugs linked to cyclic adenosine monophosphate production. Invest. Ophthalmol. Vis. Sci. 34, 2392–2401.

Hall, M.O., Burgess, B.L., Abrams, T.A., & Martinez, M.O. (1996). Carbachol does not correct the defect in the phagocytosis of outer segments by Royal College of Surgeons rat retinal pigment epithelial cells. Invest. Ophthalmol. Vis. Sci. 37, 1473–1477.

Hall, S.E., Savill, J.S., Henson, P.M., & Haslett, C. (1994). Apoptotic neutrophils are phagocytosed by fibroblasts with participation of the fibroblast vitronectin receptor and involvement of a mannose/fucose-specific lectin. J. Immunol. 153, 3218–3227.

Hardt, W.D., Chen, L.M., Schuebel, K.E., Bustelo, X.R., & Galán, J.E. (1998). *S. typhimurium* encodes an activator of Rho GTPases that induces membrane ruffling and nuclear responses in host cells. Cell 93, 815–826.

Havemose-Poulsen, A., & Holmstrup, P. (1997). Factors affecting IL-1-mediated collagen metabolism by fibroblasts and the pathogenesis of periodontal disease: a review of the literature. Crit. Rev. Oral Biol. Med. 8, 217–236.

Henwick, S., Hetherington, S.V., & Patrick, C.C. (1993). Complement binding to *Aspergillus conidia* correlates with pathogenicity. J. Lab. Clin. Med. 122, 27–35.

Hermann, M., Jaconi, M.E., Dahlgren, C., Waldvogel, F.A., Stendahl, O., & Lew, D.P. (1990). Neutrophil bactericidal activity against *Staphylococcus aureus* adherent on biological surfaces. Surface-bound extracellular matrix proteins activate intracellular killing by oxygen-dependent and -independent mechanisms. J. Clin. Invest. 86, 942–951.

Herron, W.L., Jr., Riegel, B.W., Brennan, E., & Rubin, M.L. (1974). Retinal dystrophy in the pigmented rat. Invest. Ophthalmol. 13, 87–94.

Heth, C.A., & Marescalchi, P.A. (1994). Inositol triphosphate generation in cultured rat retinal pigment epithelium. Invest. Ophthalmol. Vis. Sci. 35, 409–416.

Heth, C.A., & Schmidt, S.Y. (1992). Protein phosphorylation in retinal pigment epithelium of Long-Evans and Royal College of Surgeons rats. Invest. Ophthalmol. Vis. Sci. 33, 2839–2847.

Heth, C.A., Marescalchi, P.A., & Ye, L. (1995). IP3 generation increases rod outer segment phagocytosis by cultured Royal College of Surgeons retinal pigment epithelium. Invest. Ophthalmol. Vis. Sci. 36, 984–989.

Hiramine, C., Nakagawa, T., Miyauchi, A., & Hojo, K. (1996). Thymic nurse cells as the site of thymocyte apoptosis and apoptotic cell clearance in the thymus of cyclophosphamide-treated mice. Lab. Invest. 75, 185–201.

Hodinka, R.L., & Wyrick, P.B. (1986). Ultrastructural study of mode of entry of *Chlamydia psittaci* into L-929 cells. Infect. Immun. 54, 855–863.

Hughes, J., Liu, Y., Van Damme, J., & Savill, J. (1997). Human glomerular mesangial cell phagocytosis of apoptotic neutrophils: mediation by a novel CD36-independent vitronectin receptor/thrombospondin recognition mechanism that is uncoupled from chemokine secretion. J. Immunol. 158, 4389–4397.

Inouye, S., Iyama, K., Araki, C., & Usuku, G. (1983). Phagocytosis of cytoplasmic blebs derived from smooth muscle cells by fibroblast-like cells and macrophages in the post-partum rat uterus. Virchows Arch. B Cell Pathol. Incl. Mol. Pathol. 42, 235–242.

Isberg, R., & Tran Van Nhieu, G. (1995). The mechanism of phagocytic uptake promoted by invasin-integrin interaction. Trends Cell Biol. 5, 120–124.

Ishii, M., Terao, Y., Asai, Y., & Hamada, T. (1984). High melanosome engulfing activity of cutaneous fibroblasts in macular amyloidosis: an electron microscopic study. J. Cutan. Pathol. 11, 476–484.

Jensen, V.B., Harty, J.T., & Jones, B.D. (1998). Interactions of the invasive pathogens *Salmonella typhimurium*, Listeria monocytogenes, and *Shigella flexneri* with M cells and murine Peyer's patches. Infect. Immun. 66, 3758–3766.

Jerse, A.E., & Rest, R.F. (1997). Adhesion and invasion by the pathogenic neisseria. Trends Microbiol. 5, 217–221.

Kaneda, R., Iwabuchi, K., Kasai, M., Murakami, M., Uede, T., & Onoé, K. (1997). Selective antigen presenting activity of cortical thymic epithelial cells against CD4+ T cells associated with both lack of costimulatory molecules and inefficient presentation of MHC-peptide ligands. Cell Immunol. 181, 163–171.

Kheradmand, F., Werner, E., Tremble, P., Symons, M., & Werb, Z. (1998). Role of Rac1 and oxygen radicals in collagenase-1 expression induced by cell shape change. Science 280, 898–902.

Knowles, G.C., McKeown, M., Sodek, J., & McCulloch, C.A. (1991). Mechanism of collagen phagocytosis by human gingival fibroblasts: importance of collagen structure in cell recognition and internalization. J. Cell. Sci. 98, 551–558.

Kuriyama, S., Hall, M.O., Abrams, T.A., & Mittag, T.W. (1995). Isoproterenol inhibits rod outer segment phagocytosis by both cAMP-dependent and independent pathways. Invest. Ophthalmol. Vis. Sci. 36, 730–736.

Langholz, O., Röckel, D., Mauch, C., Kozlowska, E., Bank, I., Krieg, T., & Eckes, B. (1995). Collagen and collagenase gene expression in three-dimensional collagen lattices are differentially regulated by alpha 1 beta 1 and alpha 2 beta 1 integrins. J. Cell Biol. 131, 1903–1915.

LaVail, M.M. (1976). Rod outer segment disk shedding in rat retina: relationship to cyclic lighting. Science 194, 1071–1074.

Lee, W., & McCulloch, C.A. (1997). Deregulation of collagen phagocytosis in aging human fibroblasts: effects of integrin expression and cell cycle. Exp. Cell Res. 237, 383–393.

Lee, W., Sodek, J., & McCulloch, C.A. (1996). Role of integrins in regulation of collagen phagocytosis by human fibroblasts. J. Cell. Physiol. 168, 695–704.

Leong, J.M., Morrissey, P.E., Marra, A., & Isberg, R.R. (1995). An aspartate residue of the Yersinia pseudotuberculosis invasin protein that is critical for integrin binding. EMBO J. 14, 422–431.

Li, M., Carcangiu, M.L., & Rosai, J. (1997). Abnormal intracellular and extracellular distribution of basement membrane material in papillary carcinoma and hyalinizing trabecular tumors of the thyroid: implication for deregulation of secretory pathways. Hum. Pathol. 28, 1366–1372.

Liu, H.M., Yang, L.H., & Yang, Y.J. (1995). Schwann cell properties: 3. C-fos expression, bFGF production, phagocytosis and proliferation during Wallerian degeneration. J. Neuropathol. Exp. Neurol. 54, 487–496.

Lowy, F.D., Fant, J., Higgins, L.L., Ogawa, S.K., & Hatcher, V.B. (1988). Staphylococcus aureus—human endothelial cell interactions. J. Ultrastruct. Mol. Struct. Res. 98, 137–146.

Majeed, M., & Kihlström, E. (1991). Mobilization of F-actin and clathrin during redistribution of Chlamydia trachomatis to an intracellular site in eucaryotic cells. Infect. Immun. 59, 4465–4472.

Marti, W.R., Oertli, D., Meko, J.B., Norton, J.A., & Tsung, K. (1997). Induction of antigen-presenting capacity in tumor cells upon infection with non-replicating recombinant vaccinia virus encoding murine MHC class II and costimulatory molecules. J. Immunol. Methods 200, 191–198.

Maruta, K., Miyamoto, H., Hamada, T., Ogawa, M., Taniguchi, H., & Yoshida, S. (1998). Entry and intracellular growth of *Legionella dumoffii* in alveolar epithelial cells. Am. J. Respir. Crit. Care Med. 157, 1967–1974.

McCabe, R.E., Remington, J.S., & Araujo, F.G. (1984). Mechanisms of invasion and replication of the intracellular stage in *Trypanosoma cruzi*. Infect. Immun. 46, 372–376.

McCulloch, C.A., & Knowles, G.C. (1993). Deficiencies in collagen phagocytosis by human fibroblasts in vitro: A mechanism for fibrosis? J. Cell. Physiol. 155, 461–471.

McGaw, W.T., & Porter, H. (1988). Cyclosporine-induced gingival overgrowth: An ultrastructural stereologic study. Oral Surg. Oral Med. Oral Pathol. 65, 186–190.

McGaw, W.T., & Ten Cate, A.R. (1983). A role for collagen phagocytosis by fibroblasts in scar remodeling: an ultrastructural stereologic study. J. Invest. Dermatol. 81, 375–378.

McLaren, M.J., & Inana, G. (1997). Inherited retinal degeneration: basic FGF induces phagocytic competence in cultured RPE cells from RCS rats. FEBS Lett. 412, 21–29.

Merz, A.J., & So, M. (1997). Attachment of piliated, Opa- and Opc- gonococci and meningococci to epithelial cells elicits cortical actin rearrangements and clustering of tyrosine-phosphorylated proteins. Infect. Immun. 65, 4341–4349.

Miceli, M.V., Liles, M.R., & Newsome, D.A. (1994). Evaluation of oxidative processes in human pigment epithelial cells associated with retinal outer segment phagocytosis. Exp. Cell Res. 214, 242–249.

Moreno, J., Vignali, D.A., Nadimi, F., Fuchs, S., Adorini, L., & Hämmerling, G.J. (1991). Processing of an endogenous protein can generate MHC class II-restricted T cell determinants distinct from those derived from exogenous antigen. J. Immunol. 147, 3306–3313.

Murphy, G. & Reynolds, J.J. (1984). Current views of collagen degradation: Progress towards understanding the resorption of connective tissues. Bioessays 2, 55–60.

Naumann, M., Wessler, S., Bartsch, C., Wieland, B., & Meyer, T.F. (1997). *Neisseria gonorrhoeae* epithelial cell interaction leads to the activation of the transcription factors nuclear factor kappaB and activator protein 1 and the induction of inflammatory cytokines. J. Exp. Med. 186, 247–258.

Nickoloff, B.J., & Turka, L.A. (1994). Immunological functions of non-professional antigen-presenting cells: New insights from studies of T-cell interactions with keratinocytes. Immunol. Today 15, 464–469.

Paris, S., Boisvieux-Ulrich, E., Crestani, B., Houcine, O., Taramelli, D., Lombardi, L., & Latgé, J.P. (1997). Internalization of *Aspergillus fumigatus* conidia by epithelial and endothelial cells. Infect. Immun. 65, 1510–1514.

Perdomo, J.J., Gounon, P., & Sansonetti, P.J. (1994a). Polymorphonuclear leukocyte transmigration promotes invasion of colonic epithelial monolayer by *Shigella flexneri*. J. Clin. Invest. 93, 633–643.

Perdomo, O.J., Cavaillon, J.M., Huerre, M., Ohayon, H., Gounon, P., & Sansonetti, P.J. (1994b). Acute inflammation causes epithelial invasion and mucosal destruction in experimental shigellosis. J. Exp. Med. 180, 1307–1319.

Persson, C., Nordfelth, R., Holmström, A., Håkansson, S., Rosqvist, R., & Wolf-Watz, H. (1995). Cell-surface-bound Yersinia translocate the protein tyrosine phosphatase YopH by a polarized mechanism into the target cell. Mol. Microbiol. 18, 135–150.

Pitaru, S., Aubin, J.E., Bhargava, U., & Melcher, A.H. (1987). Immunoelectron microscopic studies on the distributions of fibronectin and actin in a cellular dense connective tissue: The periodontal ligament of the rat. J. Periodontal. Res. 22, 64–74.

Platt, N., da Silva, R.P., & Gordon, S. (1998). Recognizing death: The phagocytosis of apoptotic cells. Trends Cell Biol. 8, 365–372.

Poelmann, R.E. (1975). An ultrastructural study of implanting mouse blastocysts: Coated vesicles and epithelium formation. J. Anat. 119, 421–434.

Rabinovitch, M. (1995). Professional and non-professional phagocytes. Trends Cell Biol. 5, 85–87.

Ren, Y., Silverstein, R.L., Allen, J., & Savill, J. (1995). CD36 gene transfer confers capacity for phagocytosis of cells undergoing apoptosis. J. Exp. Med. 181, 1857–1862.

Rosenshine, I., Duronio, V., & Finlay, B.B. (1992). Tyrosine protein kinase inhibitors block invasin-promoted bacterial uptake by epithelial cells. Infect. Immun. 60, 2211–2217.

Roth, S., & Dröge, W. (1994). Glutathione reverses the inhibition of T cell responses by superoptimal numbers of "nonprofessional" antigen presenting cells. Cell Immunol. 155, 183–194.

Röyttä, M., Salonen, V., & Peltonen, J. (1987). Reversible endoneurial changes after nerve injury. Acta Neuropathol. (Berlin) 73, 323–329.

Rubartelli, A., Poggi, A., & Zocchi, M.R. (1997). The selective engulfment of apoptotic bodies by dendritic cells is mediated by the alpha(v)beta3 integrin and requires intracellular and extracellular calcium. Eur. J. Immunol. 27, 1893–1900.

Savill, J. (1997). Recognition and phagocytosis of cells undergoing apoptosis. Br. Med. Bull. 53, 491–508.

Schwachula, A., Riemann, D., Kehlen, A., & Langner, J. (1994). Characterization of the immunophenotype and functional properties of fibroblast-like synoviocytes in comparison to skin fibroblasts and umbilical vein endothelial cells. Immunobiology 190, 67–92.

Seyfried-Williams, R., McLaughlin, B.J., & Cooper, N.G. (1984). Phagocytosis of lectin-coated beads by dystrophic and normal retinal pigment epithelium. Exp. Cell Res. 154, 500–509.

Shiratsuchi, A., Umeda, M., Ohba, Y., & Nakanishi, Y. (1997). Recognition of phosphatidylserine on the surface of apoptotic spermatogenic cells and subsequent phagocytosis by Sertoli cells of the rat. J. Biol. Chem. 272, 2354–2358.

Sibley, L.D. (1995). Invasion of vertebrate cells by *Toxoplasma gondii.* Trends Cell Biol. 5, 129–132.

Silverman, D.J., & Bond, S.B. (1984). Infection of human vascular endothelial cells by *Rickettsia rickettsii.* J. Infect. Dis. 149, 201–206.

Sodek, J., & Ferrier, J.M. (1988). Collagen remodelling in rat periodontal tissues: compensation for precursor reutilization confirms rapid turnover of collagen. Coll Relat Res. 8, 11–21.

Sory, M.P., Boland, A., Lambermont, I., & Cornelis, G.R. (1995). Identification of the YopE and YopH domains required for secretion and internalization into the cytosol of macrophages, using the cyaA gene fusion approach. Proc. Natl. Acad. Sci. USA 92, 11998–12002.

Sparrow, J.R., Ryeom, S.W., Abumrad, N.A., Ibrahimi, A., & Silverstein, R.L. (1997). CD36 expression is altered in retinal pigment epithelial cells of the RCS rat. Exp Eye Res. 64, 45–56.

Sturgill-Koszycki, S., Schlesinger, P.H., Chakraborty, P., Haddix, P.L., Collins, H.L., Fok, A.K., Allen, R.D., Gluck, S.L., Heuser, J., & Russell, D.G. (1994). Lack of acidification in Mycobacterium phagosomes produced by exclusion of the vesicular proton-ATPase [see comments] [published erratum appears in Science (1994) 263, 1359]. Science 263, 678–681.

Subba Rao, P.V., Friedman, M.M., Atkins, F.M., & Metcalfe, D.D. (1983). Phagocytosis of mast cell granules by cultured fibroblasts. J. Immunol. 130, 341–349.

Sundberg, C., & Rubin, K. (1996). Stimulation of beta1 integrins on fibroblasts induces PDGF independent tyrosine phosphorylation of PDGF beta-receptors. J. Cell. Biol. 132, 741–752.

Suzuki, Y., Takeda, M., & Farbman, A.I. (1996). Supporting cells as phagocytes in the olfactory epithelium after bulbectomy. J. Comp. Neurol. 376, 509–517.

Svoboda, E.L., Brunette, D.M., & Melcher, A.H. (1979). *In vitro* phagocytosis of exogenous collagen by fibroblasts from the periodontal ligament: an electron microscopic study. J. Anat. 128, 301–314.

Svoboda, E.L., Shiga, A., & Deporter, D.A. (1981). A stereologic analysis of collagen phagocytosis by fibroblasts in three soft connective tissues with differing rates of collagen turnover. Anat. Rec. 199, 473–480.

Takashima, A., & Grinnell, F. (1984). Human keratinocyte adhesion and phagocytosis promoted by fibronectin. J. Invest. Dermatol. 83, 352–358.

Takeda, Y. (1985). Phagocytosis of mast cell granules by fibroblasts of the human gingiva. Virchows Arch. A. Pathol. Anat. Histopathol. 406, 197–201.

Tate, D.J., Jr., Miceli, M.V., & Newsome, D.A. (1995). Phagocytosis and H2O2 induce catalase and metallothionein gene expression in human retinal pigment epithelial cells. Invest. Ophthalmol. Vis. Sci. 36, 1271–1279.

Theriot, J.A. (1995). The cell biology of infection by intracellular bacterial pathogens. Ann. Rev. Cell. Dev. Biol. 11, 213–239.

Tran Van Nhieu, G., Krukonis, E.S., Reszka, A.A., Horwitz, A.F., & Isberg, R.R. (1996). Mutations in the cytoplasmic domain of the integrin beta1 chain indicate a role for endocytosis factors in bacterial internalization. J. Biol. Chem. 271, 7665–7672.

Umetsu, D.T., Katzen, D., Jabara, H.H., & Geha, R.S. (1986). Antigen presentation by human dermal fibroblasts: activation of resting T lymphocytes. J. Immunol. 136, 440–445.

Umetsu, D.T., Pober, J.S., Jabara, H.H., Fiers, W., Yunis, E.J., Burakoff, S.J., Reiss, C.S., & Geha, R.S. (1985). Human dermal fibroblasts present tetanus toxoid antigen to antigen-specific T cell clones. J. Clin. Invest. 76, 254–260.

van der Zee, E., Everts, V., Hoeben, K., & Beertsen, W. (1995). Cytokines modulate phagocytosis and intracellular digestion of collagen fibrils by fibroblasts in rabbit periosteal explants. Inverse effects on procollagenase production and collagen phagocytosis. J. Cell. Sci. 108, 3307–3315.

Vercellotti, G.M., McCarthy, J.B., Lindholm, P., Peterson, P.K., Jacob, H.S., & Furcht, L.T. (1985). Extracellular matrix proteins (fibronectin, laminin, and type IV collagen) bind and aggregate bacteria. Am. J. Pathol. 120, 13–21.

Vignery, A., & Baron, R. (1980). Dynamic histomorphometry of alveolar bone remodeling in the adult rat. Anat. Rec. 196, 191–200.

Voll, R.E., Herrmann, M., Roth, E.A., Stach, C., Kalden, J.R., & Girkontaite, I. (1997). Immunosuppressive effects of apoptotic cells [letter]. Nature 390, 350–351.

Wakefield, J.S., & Hicks, R.M. (1974). Erythrophagocytosis by the epithelial cells of the bladder. J. Cell. Sci. 15, 555–573.

Walker, N.I., Bennett, R.E., & Kerr, J.F. (1989). Cell death by apoptosis during involution of the lactating breast in mice and rats. Am. J. Anat. 185, 19–32.

Watarai, M., Funato, S., & Sasakawa, C. (1996). Interaction of Ipa proteins of *Shigella flexneri* with alpha5beta1 integrin promotes entry of the bacteria into mammalian cells. J. Exp. Med. 183, 991–999.

Werb, Z. (1982). Degradation of collagen. In: Collagen in Health and Disease (Jayson, M.I.V., & Weiss, J.B., Eds.), pp. 121–134. Churchill Livingstone, Edinburgh.

Werb, Z., & Reynolds, J.J. (1974). Stimulation by endocytosis of the secretion of collagenase and neutral proteinase from rabbit synovial fibroblasts. J. Exp. Med. 140, 1482–1497.

Werk, R. (1985). How does *Toxoplasma gondii* enter host cells? Rev. Infect. Dis. 7, 449–457.

Wigley, F.M., Fine, I.T., & Newcombe, D.S. (1983). The role of the human synovial fibroblast in monosodium urate crystal-induced synovitis. J. Rheumatol. 10, 602–611.

Wisseman, C.L., Jr., Edlinger, E.A., Waddell, A.D., & Jones, M.R. (1976). Infection cycle of *Rickettsia rickettsii* in chicken embryo and L-929 cells in culture. Infect. Immun. 14, 1052–1064.

Wolter, J.R., & Lichter, P.R. (1983). Fibroblast-like cells on intraocular lens implants: phagocytosing erythrocytes. Br. J. Ophthalmol. 67, 641–645.

Wyrick, P.B., Choong, J., Davis, C.H., Knight, S.T., Royal, M.O., Maslow, A.S., & Bagnell, C.R. (1989). Entry of genital Chlamydia trachomatis into polarized human epithelial cells. Infect. Immun. 57, 2378–2389.

Yajima, T. (1986). Acid phosphatase activity and intracellular collagen degradation by fibroblasts in vitro. Cell Tissue Res. 245, 253–260.

Zaharopoulos, P., Wen, J.W., & Wong, J. (1998). Membranous lamellar cytoplasmic inclusions in histiocytes and mesothelial cells of serous fluids. Their relationship to phagocytosis of red blood cells. Acta Cytol. 42, 607–613.

Zavizion, B., White, J.H., & Bramley, A.J. (1997). *Staphylococcus aureus* stimulates urokinase-type plasminogen activator expression by bovine mammary cells. J. Infect. Dis. 176, 1637–1640.

Zink, S., Nass, T., Rösen, P., & Ernst, J.F. (1996). Migration of the fungal pathogen *Candida albicans* across endothelial monolayers. Infect. Immun. 64, 5085–5091.

Zychlinsky, A., Perdomo, J.J., & Sansonetti, P.J. (1994). Molecular and cellular mechanisms of tissue invasion by *Shigella flexneri*. Ann. NY Acad. Sci. 730, 197–208.

SECTION II

RECEPTORS

SCAVENGER RECEPTORS AND PHAGOCYTOSIS OF BACTERIA AND APOPTOTIC CELLS

Nick Platt, Richard Haworth, Rosangela P. da Silva, and Siamon Gordon

Advances in Cell and Molecular Biology of Membranes and Organelles
Volume 5, pages 71-85.

ISBN: 1-55938-999-0

I. INTRODUCTION

Scavenger receptors (SRs) are membrane glycoproteins which are defined by their capacity to bind and internalize modified lipoprotein (Brown and Goldstein, 1983). Their ability to endocytose large quantities of chemically-modified low-density lipoprotein (LDL) underlies the mechanism through which macrophages (Mϕ) are converted into the foam cells that are characteristic of atherosclerotic plaques and vascular disease (Goldstein et al., 1979). Although this aspect of SR biology has been the major focus of research, there is an accumulating body of evidence that supports the hypothesis that SRs have a number of other important functions besides their role in atherogenesis (Krieger, et al., 1993; Krieger and Hertz, 1994; Pearson, 1996; Platt and Gordon, 1998). One of these, that of contributing to the phagocytic activity of cells, will be discussed in this review.

II. CLASSES AND STRUCTURES OF SCAVENGER RECEPTORS

Molecular biological techniques have enabled the recent identification of several new SRs (Figure 1) and facilitate their classification into at least five classes based on sequence homologies and binding properties (Pearson, 1996; Platt and Gordon, 1998). There have been reports of SR-like activities that are currently ill-defined, suggesting it is likely that other molecules remain to be identified (Van Berkel et al., 1998). All SRs are multi-domained plasma membrane proteins, but there is no single domain that is common to all SRs. Rather, receptors of a particular class may share a specific domain; for example the three receptors that at present comprise class A all have a collagenous domain that includes the binding site for acetylated low density lipoprotein (acLDL) (Figure 1) (Kodama et al., 1996). Studies of synthetic trimerized peptides have confirmed the predicted triple-helical organization of this domain and the position of the unpaired lysine residues that are available for binding negatively-charged ligands (Anachi et al., 1995; Mielewczyk et al., 1996; Tanaka et al., 1996). The five classes of SRs show that these molecules are structurally very diverse, yet they can bind similar and overlapping ligands, which implies that determination of their quaternary structures may reveal how different primary amino acid sequences form similar three-dimensional configurations

A. General Ligand-Binding Properties of Scavenger Receptors

Although SRs are defined by their ability to bind modified lipoprotein, experimental evidence has shown that they display broad, high-affinity ligand-binding activities (Figure 2) (Krieger et al., 1993; Krieger and Hertz, 1994). This property distinguishes them from the vast majority of receptors which can only interact

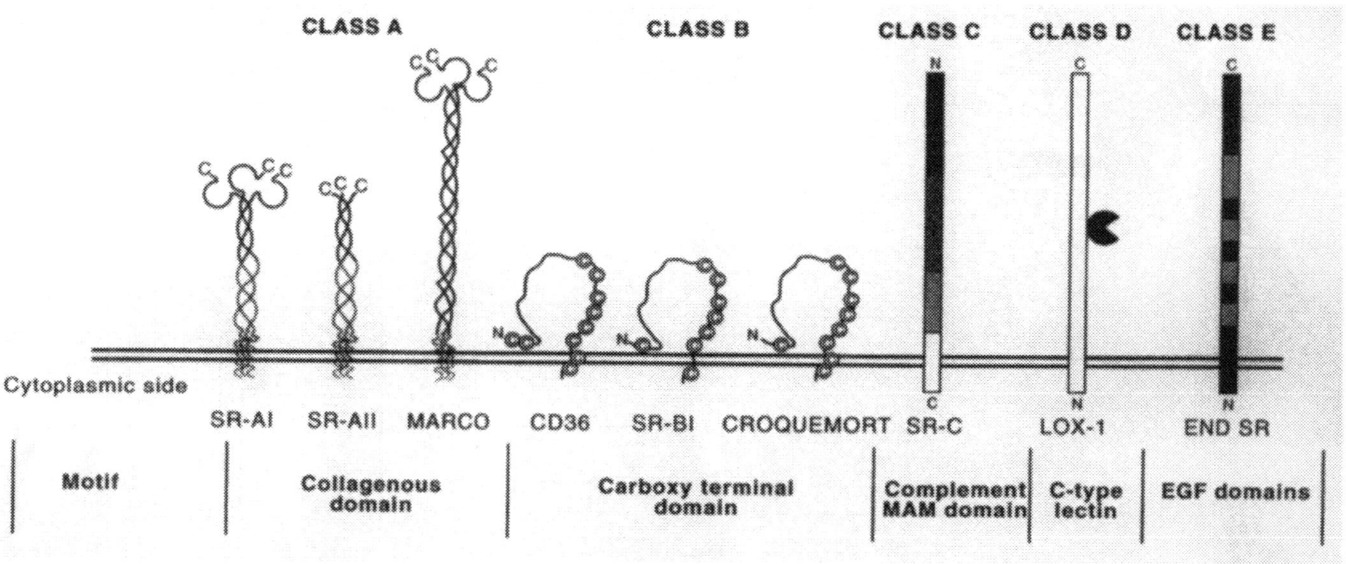

Figure 1. Structures of the five classes of scavenger receptors. Five classes of scavenger receptors have been recognized on the basis of structure and profiles of ligand-binding. Class A receptors are characterized by the presence of a collagenous domain (solid line). SR-AI and II are alternative transcripts of the same gene. MARCO is a distinct gene. Class B receptors have a single extracellular domain with conserved cysteine residues (enclosed c). Class C composes a single species isolated from *Drosophila* that consists of multiple domains, including homologues of complement control proteins. Class D and class E are each represented by a single species that have a C-type lectin domain or multiple epidermal growth factor repeats, respectively.

A	SR-A I
B	SR-A II
C	MARCO
D	CD36
E	Croquemort
F	SR-BI/CLA-1
G	SR-C1
H	Lox-1
I	End SR

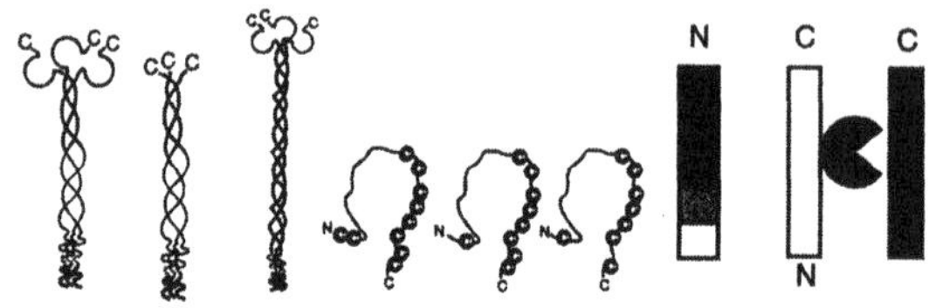

	A	B	C	D	E	F	G	H	I
acLDL	+	+		+/-		+	+	-	+
oxLDL	+	+		+		+		+	+
LDL	-	-		-		+	-	-	-
HDL						+			
m-BSA	+	+		+		+	+		
mal-BSA	+	+							+
BSA	-	-		-		-	-		
Fucoidan	+	+		-		-	+		
Dextran sulphate									
Chondroitan sulphate	-	-				-			
Heparin	-	-							
poly (g)/poly (I)	+	+		-		-	+	+	+
poly (c)/poly (A)	-	-		-		-	-		
Phosphatidylserine	-	-		+		+	-	+	
Phosphatidylcholine	-	-		-		-	-		
Apoptotic cells	+	+		+	+	+		+	

Other Ligands	
SR-AI	asbestos, silica, gram-negative bacteria. LPS, gram-positivebacteria, LTA
MARCO	*E. coli, S. aureus*
CD36	Collagen, thrombospondin, *Plasmodium*-infected erythrocytes
SR-CI	β-glucan, poly D-glutamic acid

Figure 2. Scavenger receptors and their ligands. The ligand binding profiles of representatives of each of the five classes of receptors are illustrated. (+) represents that the receptor can either recognize the structure directly or that it can inhibit binding of a recognized ligand. (–) indicates when a molecule is unable to bind or compete. Abbreviations: LDL, low-density liporotein; acLDL, acetylated low-density lipoprotein; oxLDL, oxidized low-density lipoprotein; HDL, high-density lipoprotein; poly (G), polyguanylic acid; poly (I), polyinosinic acid; poly (C), polycytidylic acid, poly (A), polyadenylic acid; m-BSA, maleylated bovine serum albumin; mal-BSA, malondialdehyde bovine serum albumin; LPS, lipopolysaccharide; LTA, lipoteichoic acid.

with a single, or very restricted, spectrum of ligands. The term "molecular flypaper" has been coined to describe their promiscuous binding properties (Krieger et al., 1993). Ligands for SRs include modified proteins and lipoproteins, polyribonucleotides and non-physiological compounds such as asbestos and silica. They also include sugars, phospholipids, lipopolysaccharide (LPS) and lipoteichoic acid (LTA), which may act as ligands for phagocytosis. While all these ligands are macromolecular and polyanionic, the precise requirements for binding are not yet fully understood and being polyanionic alone is insufficient to confer binding to the receptor(s). The list of ligands is extensive, but is unlikely to be complete. So far ligands have been identified by their ability to inhibit the endocytosis of modified lipoprotein and it is possible therefore that there are other molecules that bind to sites on the receptors distinct from that for lipoprotein.

B. Tissue Expression of Scavenger Receptors

The availability of specific probes for a number of SRs has permitted determination of their patterns of expression. Some SRs have a rather restricted distribution, and others are expressed on a variety of cell types. Class A SRs type I and II (SR-A I and II) are essentially restricted to cells of the macrophage lineage and have been shown on a number of tissue Mϕ populations in the mouse (Hughes et al., 1995). The other class A receptor, MARCO, is even more limited in its distribution and is normally present only on selected or subpopulations of Mϕ in the spleen, lymph node and peritoneum (Elomaa et al., 1995). In contrast, the class B receptor CD36 is more widely expressed and is found on monocytes, Mϕ, endothelia and platelets (Greenwalt et al., 1992). Clearly, the presence of a number of SRs on the so-called professional phagocytes (Mϕ), as well as on cells that are less phagocytically-active is consistent with potential roles in phagocytosis. However, descriptions of expression are not complete; the distribution of two of the more recently identified SRs (Lox-1 and End SR) has yet to be fully described.

III. SCAVENGER RECEPTORS AND PHAGOCYTOSIS

The presence of certain SRs on particular cells of the immune system implies contribution to host defence and perhaps the phagocytic activity of these cells. However, while the functioning of the cellular immune system is typically thought of in terms of dealing with foreign or non-self components, such as pathogens, it is sometimes forgotten that unwanted self or altered-self cells also have to be efficiently removed. Thus phagocytes need to be capable of distinguishing self from both non-self and altered self. The altered self consists of the large numbers of host cells that die by apoptosis and are rapidly cleared by phagocytes (typically Mϕ) shortly after induction of the death program (Wyllie et al., 1980). Although there is growing evidence that SRs are involved in internalizing both kinds of phagocytic particle (bacteria and senescent host cells) and may contribute to each of these in comparable ways, for clarity we shall consider them separately.

IV. CLASS A SCAVENGER RECEPTORS AS PATTERN RECOGNITION MOLECULES IN INNATE IMMUNE RESPONSES

A number of receptors have been identified that contribute to the recognition of pathogens by the innate immune system (Janeway, 1992). Mechanisms that lead to phagocytosis may be cellular based, through direct recognition of surface structures of the foreign cell or dependent upon the humoral system, which opsonizes the target prior to ingestion via a phagocytic receptor. Although an unidentified serum component is required for SR-A mediated Mϕ adherence to tissue culture plastic *in vitro* (Fraser et al., 1993), there is as yet no direct evidence that opsonization is necessary for phagocytosis by SRs. However, this remains a possibility. Rather, the peculiar ligand-binding properties that we have already outlined are consistent with SRs functioning as pattern-recognition receptors. Other receptors with these recognition properties are two lectins, the Mϕ mannose receptor and DEC 205 on dendritic cells (Jiang et al., 1995; Stahl and Ezekowitz, 1998), collectins (Hoppe and Reid, 1994), and CD14 (Pugin et al., 1994). Particularly relevant to the argument for a role for SR-A in innate responses is that the microbial surface component LPS of gram-negative and LTA of gram-positive bacteria are both ligands for the receptor (Figure 2) (Hampton et al., 1991; Dunne et al., 1994) . It is very probable that the same pattern-recognition mechanisms are exploited for the identification of altered self, which is seen as alterations of the surface of dying as opposed to healthy cells. This ability to bind microbial products is not limited to mammalian SRs: *Drosophila* SR-C1 can bind β-glucans, implying an immunological role in this relatively simple organism (Pearson et al., 1995).

A. Class A Scavenger Receptors and the Phagocytosis of Bacteria

The original evidence for SR-A involvement in the recognition of intact and specific components of bacteria came from a number of *in vitro* studies. Endotoxin, lipid A, ReLPS and its precursor lipidIV_Awere able to compete for endocytic uptake of acLDL (Ashkenas et al., 1993). In addition, *in vivo* clearance of LPS from the circulation by the liver was reduced by pretreatment with SR ligands. Dunne and co-workers (Dunne et al., 1994) demonstrated that SR-AI could recognize the surface of bacteria directly. They showed that a soluble form of bovine SR-A I bound a number of gram-positive bacteria, including *Streptococcus pyogenes*, *Staphylococcus aureus*, *Enterococcus hirae* and *Listeria monocytogenes* and the interaction could be competed for by several SR-A ligands and LTA, a cell wall component of these microbes. However, because of experimental design, binding, but not SR-A-mediated ingestion, was demonstrated.

Characterizing the spectrum of ligands for SRs expressed on microorganisms would be important in defining the role of the receptors in host defence and may help identify those pathogens which are likely to interact with the host in a SR-dependent manner. Greenberg et al. (1996) have extended the original demonstration of SR-A recognition of gram-positive organisms, by investigating the ability and relative affinity of various surface LTAs and lipoglycans to bind to the receptor. LTA of *S. aureus* displayed binding with the highest affinity; those of *Enterococcus faecalis* and *L. monocytogenes* bound to a lesser extent, while LTAs of *Streptococcus pneumonia* and *Clostridium innoculum*, which are substituted with positively-charged sugar residues, were unable to bind. However, as mentioned previously, in relation to the structure of other ligands, interaction with SR-A is not governed purely by the degree of negative charge. The lipoglycans of *Bifidobacterium bifidium* and *Micrococcus luteus* have comparable negative charges, but only the former was able to bind. The failure of *M. luteus* to bind to SR-A may be of significance for addressing the question of whether SR-A interacts with the human pathogens *Mycobacterium tuberculosis* and *M. leprae*. There is a report suggestive of such an interaction (Zimmerli et al., 1996) , but because the lipoglycans of the two mycobacteria structurally resemble that of *M. luteus*, this may require re-examination.

B. SR-A I and II and Host Defense—the SR-A-Deficient Mouse

The best evidence that SR-A is important in the innate response has come from studies of mice with targeted deletion of the SR-A gene (Suzuki et al., 1997) because they permit definitive analyses of the contribution of SR-A *in vitro* and *in vivo*. SR-A null animals were reported to be more susceptible to *L. monocytogenes* and herpes simplex virus (HSV-1) infections, because of a defect in either uptake or killing. Haworth and colleagues (Haworth et al., 1998) directly tested for SR-A involvement in phagocytosis by comparing the ingestion of heat killed *Escherichia coli* by activated peritoneal Mϕ from SR-A deficient mice with that by wild-type cells *in vitro*. Cells lacking SR-A took up only 60% compared to wild-type cells, the ingestion was independent of the presence of serum and the failure of less-specific SR ligands to enhance the difference implied that other SRs are not part of the SR-A-independent uptake (Figure 3). Whether the absence of SR-A results in diminished anti-microbial activity is not currently known.

As a consequence of gram-negative infections, primed Mϕ are stimulated by bacterial-derived endotoxin to secrete pro-inflammatory molecules that can induce toxic shock (Ulevitch and Tobias, 1995). A number of different receptors on the surface of Mϕ have been shown to bind LPS (Ulevich and Tobias, 1995), including SR-A (see above). Injection of endotoxin into deficient and control mice primed with viable bacillus Calmette Guerin (BCG) revealed that SR-A null mice

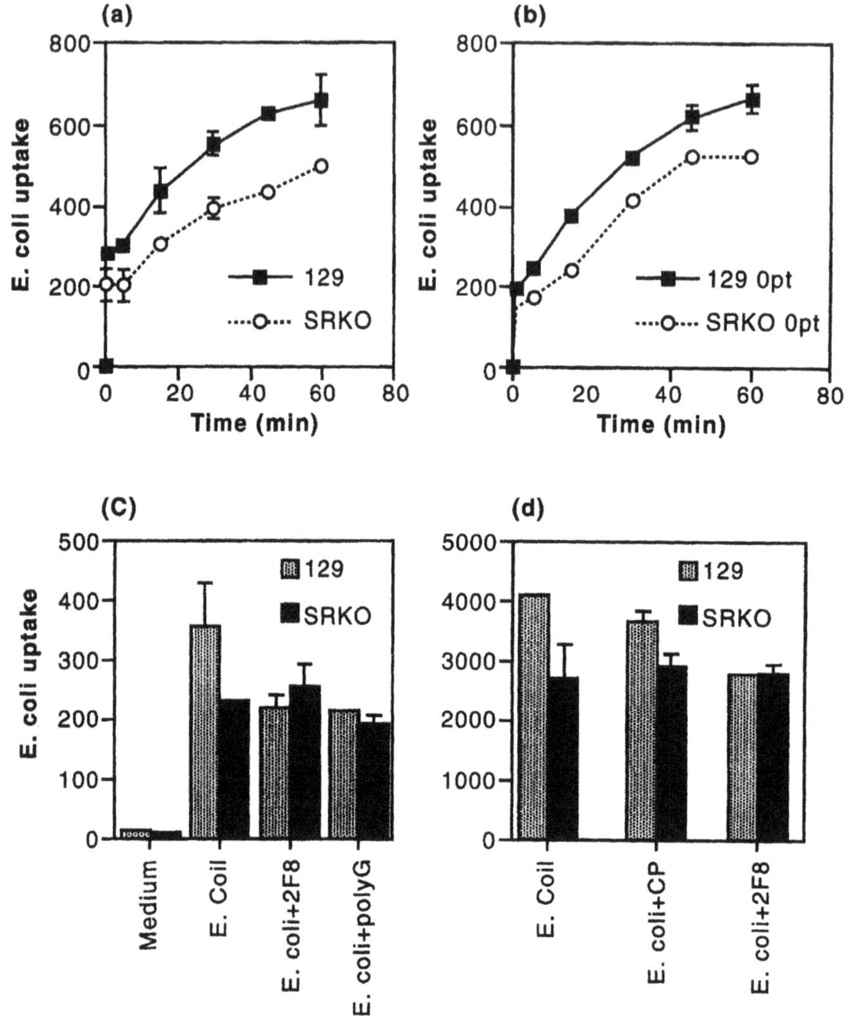

Figure 3. Phagocytosis of *E. coli* by 129 and SR-A deficient Mϕ. (**a**) and (**b**) Dynamics of *E. coli* uptake by 129 and SR-A deficient Mϕ. Uptake of fluorescent heat killed *E. coli* by 129 and SR-A deficient BCG-activated peritoneal Mϕ. X-axis represents median fluorescence determined by flow cytometric analysis. (**c**) and (**d**) Effect of SR inhibitors on *E. coli* uptake by 129 and SR-A-deficient Mϕ. Effect on uptake of *E. coli* by pretreatment of BCG-activated peritoneal Mϕ with anti-SR-A mAb 2F8 or control isotype matched mAb CP at 20ug/ml or poly (G) at 200ug/ml, in either presence (**c**) or absence (**d**) of serum. X-axis represents median fluorescence determined by flow cytometric analysis.

were hypersenstive to endotoxin shock, through enhanced production of TNF-α and possibly other cytokines such as interleukin (IL)-6 (Haworth et al., 1997). Therefore, although the precise mechanism(s) await elucidation, SR-A is protective against the adverse affects of LPS.

The third member of the class A SRs, MARCO, can bind *E. coli* and *S. aureus* (Elomaa et al., 1995), but a broader range of bacteria has not been reported. The binding properties of a series of MARCO truncation mutants have indicated the sequence RGRAEEVYYSGT is essential for bacterial binding, but interestingly this region is poorly conserved in SR-A I and II (Elomaa et al., 1998). In contrast, the basic residues at the end of the collagenous domain of SR-A I and II that mediate acLDL binding are present in MARCO. Testing of a larger number of putative ligands, together with more extensive mutagenesis, that could include the construction of chimeric receptors, may provide information of the MARCO-bacteria interactions and permit detailed comparisons with SR-A I and II. The development of mice with inactivated MARCO will further our understanding of the importance of this receptor in innate immunity.

The expression patterns of class A SRs are consistent with roles in host defence. SR-A is upregulated in mouse brain following LPS injection (Bell et al., 1994); MARCO is also increased in tissues from human subjects who died from septic meningitis and in mice with systemic *Klebsiella pneumoniae* infection (Elomaa et al., 1998).

V. SCAVENGER RECEPTORS AND THE CLEARANCE OF APOPTOTIC CELLS

The phagocytosis of apoptotic cells is now recognized as a vital and conserved stage in programmed cell death, that allows for the removal of unwanted cells without deleterious effects to the host (Savill et al., 1993; Platt et al., 1998). In particular, the failure to generate a pro-inflammatory immune response is characteristic of apoptotic cell uptake and is in direct contrast to what typically happens following ingestion of foreign cells (Savill, 1997). Most of the studies of apoptotic cell clearance have focused on the receptor(s) on the plasma membrane of the phagocyte that mediate recognition and rapid internalization, shortly after induction of the death program. Although no single receptor has been identified that is responsible for all apoptotic cell uptake, SRs have been implicated in many of the systems that have been investigated.

Evidence supporting the involvement of SRs in the phagocytic recognition of apoptotic cells has included investigations of non-mammalian species. Such studies have demonstrated not only the conservation of phagocytic mechanisms, but suggest that this may be the ancestral function of the receptors. In *Drosophila melanogaster*, the protein *Croquemort*, a homologue of the vertebrate class B receptor CD36, is expressed by hemocytes and can bind apoptotic thymocytes when transfected into COS cells (Franc et al., 1996). The fly expresses several other SRs, including SR-CI and II, which may also have this function. Although genetic studies of mutants of the nematode *Caenorhabitis elegans* that display defective phagocytosis have not yet implicated SRs (Ellis et al., 1991), the worm genome does encode a number of proteins structurally related to class B receptors (Crombie and Silverstein, 1998).

A. Studies of Apoptotic Cell Uptake by Scavenger Receptors

SRs have been identified as the receptors on the surface of phagocytes (particularly Mϕ) which bind and ingest apoptotic cells. This has been determined by blocking receptor activity *in vitro*. For example, specific monoclonal antibodies (mAb) against CD36 have inhibited the uptake of apoptotic neutrophils by human blood-derived Mϕ (Savill et al., 1992), the ingestion of apoptotic thymocytes by murine thymic and peritoneal Mϕ with a mAb to SR-A I and II (Platt et al., 1996), and binding of apoptotic cells and aged erythrocytes to endothelial cells with a mAb directed against Lox-1 (Oka et al., 1998). In all of these cases, inhibition with a single reagent is not complete (being typically in the order of 50%) and therefore it is likely that multiple receptors are engaged simultaneously to achieve full activity. Neutrophil recognition is achieved through a molecular complex involving CD36 and the $\alpha_v\beta_3$ integrin on the plasma membrane and soluble thrombospondin (Savill et al., 1992). Preliminary evidence using less-specific ligands suggests that multiple SRs may be required for ingestion of dying thymocytes by thymic Mϕ (Platt et al., 1996; Platt, unpublished observations).

Despite the accumulating evidence of the role of various SRs in the recognition of dying cells, at present there is little data that demonstrates *in vivo* activity. Thymic Mϕ prepared from SR-A knock-out mice have reduced uptake *in vitro* comparable to mAb blockade (Platt et al., 1996), but analyses of the thymus *in vivo* have failed to detect a phenotype consistent with inefficient clearance (Platt, in preparation). Receptor redundancy, for example, may indicate that it will be necessary to manipulate more than one recognition receptor. However, there is one experimental model that displays defective phagocytosis and highlights the potential pathological consequences of impaired removal of apoptotic cells. Mice deficient in the complement component C1q, a molecule structurally related to class A SRs but which cannot bind acLDL, accumulate apoptotic bodies in the kidney, an appearance that mimics the symptoms of the auto-immune condition systemic lupus erythematosus (Botto et al., 1998). A recent important demonstration has been that myeloid dendritic cells (DCs) can ingest virally-infected apoptotic cells and that viral peptides can enter the class I pathway and stimulate a cytotoxic T cell response (Albert et al., 1998). This has extensive implications for immunological function and as SR-A I and II are expressed on murine bone marrow-derived DCs (Haworth et al., 1998), they represent potential candidate receptors responsible for this activity.

The experimental approach that has been adopted to test if a particular receptor is involved in just the initial binding or tethering of the dying cell by the phagocyte, or is sufficient for internalization, has been the expression of SRs in relatively non-phagocytic cell types. Transfected cells may then have the ability to bind and/or take up apoptotic cells. COS cells expressing human CD36 were first used to demonstrate transfer of phagocytic ability with respect to apoptotic cells (Ren et al., 1995), and this has subsequently been shown for *Croquemort*, SR-B1, Lox-1, and SR-AI and II (Franc et al., 1996; Fukasawa et al., 1996; Oka et al., 1998; Platt, unpublished observations).

B. Apoptotic Ligands for Scavenger Receptors

An appropriate ligand on the surface of the dying cell is clearly a requirement for efficient binding and internalization by the phagocyte. Currently relatively little is known about alterations of the plasma membrane of a cell undergoing apoptosis, but alterations in glycosylation and charge have been suggested (Savill, 1997; Platt et al., 1998). The ability of SRs to bind secondary modifications on molecules, rather than primary sequences, makes them attractive candidates to recognize these changes. The best characterised of these is the exposure of phosphatidylserine (PtdSer) on the outer leaflet of the plasma membrane, which normally resides on the inner leaflet of healthy cells (Zwaal and Schroit, 1997). Experiments competing apoptotic cell ingestion with PtdSer-containing liposomes suggest some populations of phagocytes use the phospholipid as a ligand for uptake (Fadok et al., 1992). Although certain SRs, including CD36, SR-B1 and Lox-1 can bind PtdSer, the precise nature of the receptor remains elusive.

VI. INTRACELLULAR EVENTS ASSOCIATED WITH SCAVENGER RECEPTOR-MEDIATED PHAGOCYTOSIS

Extensive studies of a number of receptors, particularly Fc and complement, have illustrated a number of events characteristic of phagocytosis, as discussed elsewhere in this volume. As we have stated, internalization begins with the binding of specific receptor(s) on the phagocyte with ligand(s) on the particle. This initiates polymerization of actin beneath the site of binding, there then occurs actin-mediated uptake and subsequent formation and maturation of the phagosome (Allen and Aderem, 1996). Specific cellular signalling pathways coordinate these sequence of events (Greenberg, 1995). In contrast to Fc receptor-mediated ingestion, our understanding of the signals that follow SR-dependent phagocytosis is rather poor. This is probably because a number of the receptors have only recently been isolated, while for the others the focus has been on their endocytic activities. Re-organization of the cytoskeleton has been visualized, in the form of F-actin-rich phagocytic cups beneath sites of apoptotic neutrophil binding to SR-A transfected Chinese hamster ovary cells (Platt et al., 1998), but beyond this preliminary study, relatively little has been reported.

Conserved sequence motifs in the cytoplasmic tails of phagocytic receptors have been shown to be important for promoting phagocytosis. For example, deletion mutagenesis, together with detection of phosphorylation changes and kinase recruitment, have demonstrated that in response to ligation, the ITAM motif of human FcγRII undergoes tyrosine phosphorylation by src kinases, which results in the recruitment and activation of syk kinase, initiation of cytoskeletal rearrangements, particle internalization and subsequent transcriptional activity (Ravetch, 1997). The cytoplasmic tail of SR-A I and II, which is approximately 50 residues long, does not contain any of these motifs found in Fc and complement receptors, so it is likely that the details of signal transduction may well be different. However, it does possess a domain for protein kinase C (PKC) activation (Ashkenas et al., 1993), as happens in FcγR-mediated phagocytosis (Indik et al., 1995). Other events downstream of SR-A I and II, including phosphorylation, calcium fluxes and specific gene transcription have been recorded (Hsu et al., 1998). However, these studies have examined changes that accompany endocytic uptake and those that are initiated by phagocytic mechanisms may be different. Secondly, these experiments have typically used rather non-specific ligands, combined with cell types that have more than one SR, which does not allow for the discrimination of signals originating from a single receptor. Receptor co-immunoprecipitation found that lyn kinase is physically associated with the intracellular domain of SR-A I and II (Miki et al., 1996). Application of techniques such as the yeast two-hybrid system might identify other cytosolic proteins that can interact with SR intracellular domains.

VII. CONCLUDING REMARKS

SRs have only relatively recently been identified as phagocytic receptors and therefore the extent of our knowledge is currently limited. Some SRs have yet to be evaluated for their phagocytic capability. Interest in this group of receptors stems from their unusual binding properties, which mean they have the ability to interact with ligands that are not bound by other receptors, including Fc and complement. They therefore occupy a specialized niche within the total phagocytic capacity of many cells, especially professional phagocytes.

This pattern of ligand binding facilitates contributions to several immune functions; the innate system through the recognition of both infectious and damaged host cells and perhaps to acquired immunity through the presentation of antigen (Haworth et al., 1998). It will be of interest to investigate how these receptors function in discrete biological situations and also to understand how engagement of the same receptor can generate different cellular outcomes. As an example, we have already discussed that SR-A can be involved in the phagocytosis of both microbes and dying host cells, but the cellular responses of the phagocytes (pro-inflammatory as against non-inflammatory) are very different. Important areas of future research will include fully characterizing SR-ligand interactions and the downstream cellular consequences.

REFERENCES

Albert, M.L., Sauter, B., & Bhardwaj, N. (1998). Dendritic cells acquire antigen from apoptotic cells and induce class I-restricted CTLs. Nature 392, 86–89.

Allen, L.A.H., & Aderem, A.A. (1996). Mechanisms of phagocytosis. Curr. Opin. Immunol. 8, 36–40.

Anachi, R.B., Siegel, D.L., Baum, J., & Brodsky, B. (1995). Acid destabilization of a triple-helical peptide model of the macrophage scavenger receptor. FEBS Lett. 368, 551–555.

Ashkenas, J., Penman, M., Vasile, E., Acton, S., Freeman, M., & Krieger, M. (1993). Structures and high and low affinity ligand binding properties of murine type I and type II macrophage scavenger receptors. J. Lipid Res. 34, 983–1000.

Bell, M.D., Lopez Gonzalez, R., Lawson, L., Hughes, D., Fraser, I., Gordon, S., & Perry, V.H. (1994). Upregulation of the macrophage scavenger receptor in response to different forms of injury in the CNS. J. Neurocytol. 23, 605–613.

Botto, M., Dell'Agnola, C., Bygrave, A.E., Thompson, E.M., Cook, H.T., Petry, F., Loos, M., Pandolfi, P.P., & Walport, M.J. (1998). Homozygous C1q deficiency causes glomerulonephritis associated with multiple apoptotic bodies. Nat. Genet. 19, 56–59.

Brown, M.S., &. Goldstein, J.L. (1983). Lipoprotein metabolism in the macrophage; implications for cholesterol deposition in atherosclerosis. Ann. Rev. Biochem. 52, 223–261.

Crombie, R., &. Silverstein, R.L. (1998). Lysosomal integral membrane protein II binds thrombospondin-1. J. Biol. Chem. 273, 15381–15385.

Dunne, D.W., Resnick, D., Greenberg, J., Krieger, M., & Joiner, K.A. (1994). The type I macrophage scavenger receptor binds to gram-positive bacteria and recognizes lipoteichoic acid. Proc. Natl. Acad. Sci. USA 91, 1863–1867.

Ellis, R.E., Yuan, J., & Horvitz, H.R. (1991). Cell death in the nematodc, *C. elegans*. Ann. Rev. Cell Biol. 7, 663–698.

Elomaa, O., Kangas, M., Sahlberg, C., Tuukkanen, J., Sormunen, R., Liakka, A., Thesleff, I., Kraal, G., & Tryggvason, K. (1995). Cloning of a novel bacteria-binding receptor structurally related to scavenger receptors and expressed in a subset of macrophages. Cell 80, 603–619.

Elomaa, O., Sankala, M., Pikkarainen, T., Bergmann, U., Tuuttila, A., Raatikainen-Aholas, A., Sariola, H., & Tryggvason, K. (1998). Structure of the human macrophage MARCO receptor and characterisation of its bacteria-binding region. J. Biol. Chem. 273, 4530–4538.

Fadok, V.A., Savill, J.S., Haslett, C., Bratton, D.L., Doherty, D.E., Campbell, P.A., & Henson, P.M. (1992). Different populations of macrophages use either the vitronectin receptor or the phosphatidylserine receptor to recognize and remove apoptotic cells. J. Immunol. 149, 4029–4035.

Franc, N.C., Dimarcq, J-L., Langeux, M., Hoffman, J., & Ezekovitz, R.A.B. (1996). Croquemort, a novel Drosophila hemocyte/macrophage receptor that recognises apoptotic cells. Immunity 4, 431–443.

Fraser, I., Hughes, D., & Gordon, S. (1993). Divalent cation-independent macrophage adhesion inhibited by monoclonal antibody to murine scavenger receptor. Nature 364, 343–346.

Fukasawa, M., Adachi, H., Hirota, K., Tsujimoto, M., Arai, H., & Inoue, K. (1996). SRB1, a class B scavenger receptor, recognizes both negatively charged liposomes and apoptotic cells. Exp. Cell. Res. 222, 246–250.

Goldstein, J.L., Ho, Y.K., Basu, S.K., & Brown, M.S. (1979). Binding site on macrophages that mediates uptake and degradation of acetylated low density lipoprotein, producing massive cholesterol deposition. Proc. Natl. Acad. Sci. USA 76, 333–337.

Greenberg, J.W., Fischer, W., & Joiner, K.A. (1996). Influence of lipoteichoic acid structure on recognition by the macrophage scavenger receptor. Infect. Immun. 64, 3318–3325.

Greenberg, S. (1995). Signal transduction of phagocytosis. Trends Cell Biol. 5, 93–99.

Greenwalt, D.E., Lipsky, R.H., Ockenhouse, C.F., Ikeda, H., Tandon, N.N., & Jamieson, G.A. (1992). Membrane glycoprotein CD36: a review of its roles in adherence, signal transduction and transfusion medicine. Blood 80, 1105–1115.

Hampton, R.Y., Golenbock, D.T., Penman, M., Krieger, M., & Raetz, C.R. (1991). Recognition and plasma clearance of endotoxin by scavenger receptors. Nature 352, 342–344.

Haworth, R., Platt, N., Keshav, S., Hughes, D., Darley, E., Suzuki, H., Kurihara, Y., Kodama, T., & Gordon, S. (1997). The macrophage scavenger receptor type A is expressed by activated macrophages and protects the host against lethal endotoxic shock. J. Exp. Med. 186, 1431–1439.

Haworth, R., Platt, N., Suzuki, H., Kodama, T., & Gordon, S. (1998). Gene ablation in mice establishes that SR-A mediates scavenger-like uptake of ligand by dendritic cells and phagocytosis of gram negative bacteria by activated peritoneal macrophages. J. Immunol. In Press.

Hoppe, H.J., & Reid, K.B. (1994). Collectins—soluble proteins containing collagenous regions and lectin domains—and their roles in innate immunity. Protein Sci. 3, 1143–1158.

Hsu, H.Y., Hajjar, D.P., Khan, K.M., & Falcone, D.J. (1998). Ligand binding to macrophage scavenger receptor-A induces urokinase-type plasminogen activator expression by protein kinase-dependent signaling pathway. J. Biol. Chem. 273, 1240–1246.

Hughes, D.A., Fraser, I.P., & Gordon, S. (1995). Murine macrophage scavenger receptor: *in vivo* expression and function as receptor for macrophage adhesion in lymphoid and non-lymphoid organs. Eur. J. Immunol. 25, 466–473.

Indik, Z.K., Park, J.G., Hunter, S., & Schreiber, A.D. (1995). The molecular dissection of gamma receptor mediated phagocytosis. Blood 86, 4389–4399.

Janeway, C.A. (1992). The immune system evolved to discriminate infectious nonself from noninfectious self. Immunol. Today 13, 11–16.

Jiang, W., Swiggard, W.J., Heufler, C., Peng, M., Mirza, A., Steinman, R.M., & Nussenzweig, M.C. (1995). The receptor DEC-205 expressed by dendritic cells and thymic epithelial cells is involved in antigen presentation. Nature 375, 151–155.

Kodama, T., Doi, T., Suzuki, H., Takahashi, K., Wada, Y., & Gordon, S. (1996). Collagenous macrophage scavenger receptors. Curr. Opin. Lipidol. 7, 287–291.

Krieger, M., & Herz, J. (1994). Structures and functions of multiligand lipoprotein receptors: macrophage scavenger receptors and LDL receptor-related protein (LRP). Ann. Rev. Biochem. 63, 601–637.

Krieger, M., Acton, S., Ashkenas, J., Pearson, A., Penman, M., & Resnick, D. (1993). Molecular flypaper, host defense, and atherosclerosis. Structure, binding properties, and functions of macrophage scavenger receptors. J. Biol. Chem. 268, 4569–4572.

Miki, S., Tsukada, S., Nakamura, Y., Aimoto, S., Hojo, H., Sato, B., Yamamoto, M., & Miki, Y. (1996). Functional and possible physical association of scavenger receptor with cytoplasmic tyrosine kinase Lyn in monocytic THP-1-derived macrophages. FEBS Lett. 399, 241–244.

Oka, K., Sawamura, T., Kikuta, K., Itokawa, S., Kume, N., Kita, T., & Masaki, T. (1998). Lectin-like oxidized low-density lipoprotein receptor 1 mediates phagocytosis of aged/apoptotic cells in endothelial cells. Proc. Natl. Acad. Sci. USA 95, 9535–9540.

Pearson, A., Lux, A., & Krieger, M. (1995). Expression cloning of dSR-CI, a class C macrophage-specific scavenger receptor from *Drosophila melanogaster*. Proc. Natl. Acad. Sci. USA 92, 4056–4060.

Pearson, A.M. (1996). Scavenger receptors in innate immunity. Curr. Opin. Immunol. 8, 20–28.

Platt, N., &. Gordon, S. (1998). Scavenger receptors: diverse activities and promiscuous binding of polyanionic ligands. Chem. Biol. 5, 193–203.

Platt, N., da Silva, R.P., & Gordon, S. (1998). Recognising death: the phagocytosis of apoptotic cells. Trends Cell Biol. 8, 365–372.

Platt, N., Suzuki, H., Kurihara, Y., Kodama, T., & Gordon, S. (1996). Role for the class A macrophage scavenger receptor in the phagocytosis of apoptotic thymocytes *in vitro*. Proc. Natl. Acad. Sci. USA 93, 12456–12460.

Pugin, J., Heumann, D., Tomasz, A., Kravchenko, V.V., Akamatsu, R., Nishijima, M., Glauser, M.P., Tobias, P.S., & Ulevitch, R.J. (1994). CD14 is a pattern recognition molecule. Immunity 1, 509–516.

Ravetch, J.V. (1997). Fc receptors. Curr. Opin. Immunol. 9, 121–125.

Ren, Y., Silverstein, R.L., Allen, J., & Savill, J. (1995). CD36 gene transfer confers capacity for phagocytosis of cells undergoing apoptosis. J. Exp. Med. 181, 1857–1862.

Savill, J. (1997). Recognition and phagocytosis of cells undergoing apoptosis. Br. Med. Bull. 53, 491–508.

Savill, J., Fadok, V., Henson, P., & Haslett, C. (1993). Phagocyte recognition of cells undergoing apoptosis. Immunol. Today 14, 131–136.

Savill, J., Hogg, N., Ren, Y., & Haslett, C. (1992). Thrombospondin cooperates with CD36 and the vitronectin receptor in macrophage recognition of neutrophils undergoing apoptosis. J. Clin. Invest. 90, 1513–1522.

Stahl, P.D., &. Ezekowitz, R.A.B. (1998). The mannose receptor is a pattern recognition receptor involved in host defense. Curr. Opin. Immunol. 10, 50–56.

Suzuki, H., Kurihara, Y., Takeya, M., Kamada, N., Kataoka, M., Jishage, K., Ueda, O., Sakaguchi, H., Higashi, T., Suzuki, T., Takashima, Y., Kawabe, Y., Cynshi, O., Wada, Y., Honda, M., Kurihara, H., Aburatani, H., Doi, T., Matsumoto, A., Azuma, S., Noda, T., Toyoda, Y., Itakura, H., Yazaki, Y., & Kodama, T. (1997). A role for macrophage scavenger receptors in atherosclerosis and susceptibility to infection. Nature 386, 292–296.

Tanaka, T., Nishikawa, A., Tanaka, Y., Nakamura, H., Kodama, T., Imanishi, T., & Doi, T. (1996). Synthetic collagen-like domain derived from the macrophage scavenger receptor binds acetylated low-density lipoprotein in vitro. Protein Eng. 9, 307–313.

Ulevich, R.J., &. Tobias, P.S. (1995). Receptor-dependent mechanisms of cell stimulation by bacterial endotoxin. Ann. Rev. Immunol. 13, 437–457.

Van Berkel, T.J.C., Van Velzen, A., Kruijt, J.K., Suzuki, H., & Kodama, T. (1998). Uptake and catabolism of modified LDL in scavenger receptor class A type I/II knock-out mice. Biochem. J. 331, 29–35.

Wyllie, A.H., Kerr, J.F.R., & Currie, A.R. (1980). Cell death: the significance of apoptosis. Int. Rev. Cytol. 68, 251–306.

Zimmerli, S., Edwards, S., & Ernst, J.D. (1996). Selective receptor blockade during phagocytosis does not alter the survival and growth of Mycobacterium tuberculosis in human macrophages. Am. J. Respir. Cell. Mol. Biol. 15, 760–70.

Zwaal, R.F., & Schroit, A.J. (1997). Pathophysiologic implications of membrane phospholipid asymmetry in blood cells. Blood 89, 1121–1132.

MANNOSE RECEPTOR AND PHAGOCYTOSIS

Iain P. Fraser and R. Alan. B. Ezekowitz

Advances in Cell and Molecular Biology of Membranes and Organelles
Volume 5, pages 87-101.

ISBN: 1-55938-999-0

I. INTRODUCTION

Plasma membrane receptors provide cells with a molecular link to their extracellular environment. These receptors may transduce signals from the outside of the cell, or assist in the internalization of extracellular molecules and particles. Macrophages, by virtue of their anatomic localization, cell surface receptor repertoire and cellular machinery, play an important role in host defense. In this capacity, they are required to function both as initial discriminators of self from non-self, as well as effector cells in the destruction of invading pathogens.

The immune system has evolved a number of molecular mechanisms whereby pathogens may be recognized, internalized and destroyed. These may be ordered into a hierachy of recognition. The opsonic receptors for the Fc portions of immunoglobulin (reviewed in Ravetch, 1994), and for cleavage products of complement components (reviewed in Carroll and Fischer, 1997) are the most sophisticated and best-described phagocytic receptors. Other opsonic receptors, such as those for the collectins, are now becoming increasingly recognized and studied (Tenner et al., 1995; Nepomuceno et al., 1997).

Macrophages and other cells are also able to recognize and internalize pathogens directly via non-opsonic receptors. Pathogen-associated molecular patterns (PAMPs) are invariant motifs shared by many different groups of microorganisms. These PAMPs are in turn recognized by host-encoded pattern recognition receptors (PRRs), which couple recognition to effector functions (Medzhitov and Janeway, 1997). It is likely that the primary function of these PRRs in primitive organisms was initially in development, and that they subsequently evolved to play a role in host defense. Glycoconjugates on the surfaces of pathogenic microorganisms serve as a class of PAMP that has been selected for by the innate immune system. The lectin-like PRRs that recognize these PAMPs are exemplified by the mannose receptor, the subject of this review.

II. THE MANNOSE RECEPTOR

A. Primary Structure

The mannose receptor (MR) is a ≈175 kDa transmembrane receptor expressed on resident macrophage subpopulations, dendritic cells, and on subsets of endothelial cells (Takahashi et al., 1998). Molecular cloning of the cDNA for this receptor has revealed the following predicted primary structure of a type I transmembrane molecule (Figure 1): an N-terminal, cysteine-rich domain, followed by a fibronectin type II-like domain, eight C-type lectin carbohydrate recognition domains (CRDs) in tandem, a single transmembrane domain, and a 45 residue cytoplasmic tail at the C-terminus (Ezekowitz et al., 1990; Taylor et al., 1990; Harris et al., 1992). A single murine gene encoding the MR and consisting

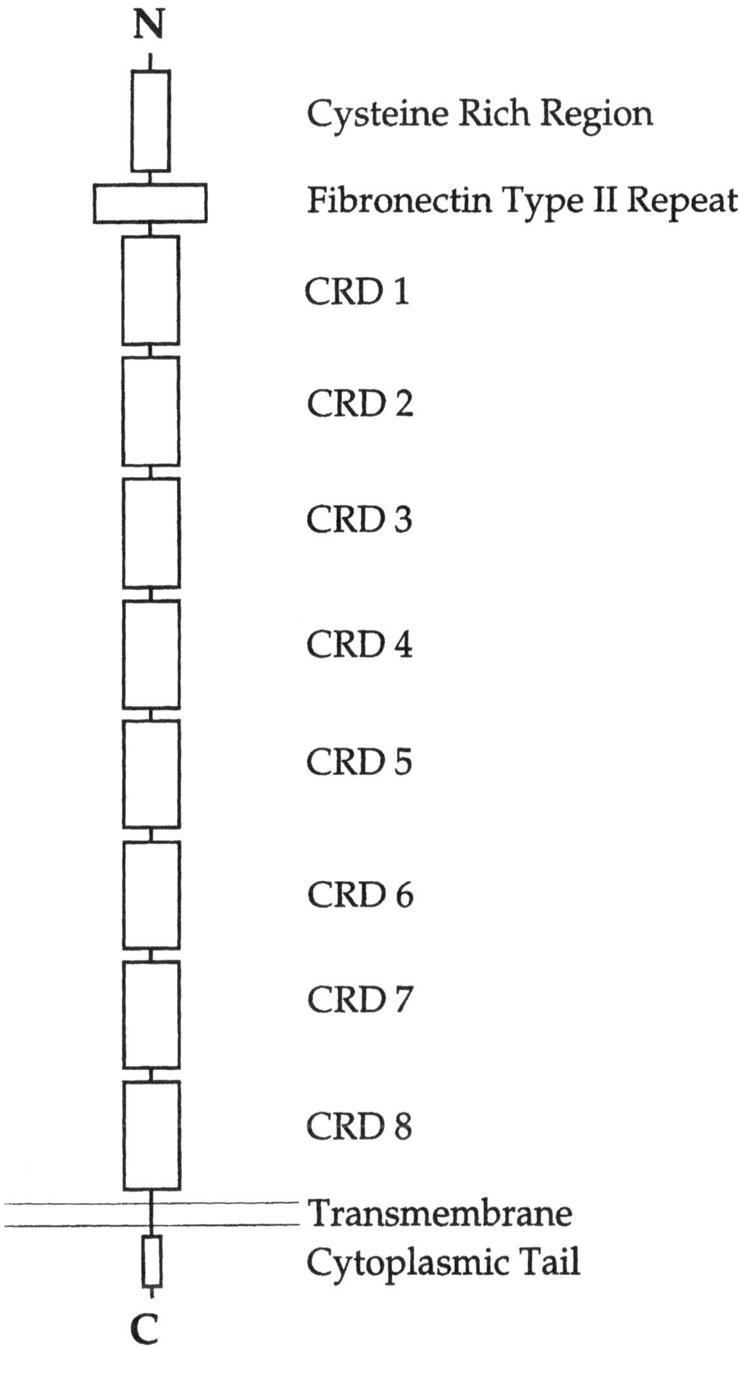

Figure 1. Mannose receptor: predicted primary structure.

Table 1. Mouse Mannose Receptor—Gene Structure and Domain Organization

Exon Number	*Exon Size (bp)*	*Protein Domain*	*Domain Size (aa)*
1	160	5′ UTR	
2	401	Cysteine rich region	134
3	173	Fibronectin type II repeat	58
4	164		
5	113	CRD 1	142
6	146		
7	185		
8	157	CRD 2	151
9	110		
10	115		
11	148	CRD 3	154
12	199		
13	127		
14	87	CRD 4	120
15	144		
16	41	Stalk	14
17	163		
18	67	CRD 5	158
19	100		
20	142		
21	114		
22	166	CRD 6	128
23	102		
24	232	CRD 7	133
25	165		
26	149		
27	113	CRD 8	143
28	164		
29	41	Stalk	14
30	868	Stalk;	13
		transmembrane;	27
		cytoplasmic tail;	45
		(3′ UTR)	

Note: Abbreviations: bp, base pairs; aa, amino acids; UTR, untranslated region; CRD, carbohydrate recognition domain.

of 30 exons spanning approximately 70 kb on chromosome 2 has been identified (Harris et al., 1994). The contribution of individual exons to specific MR protein domains is illustrated in Table 1.

B. Carbohydrate Recognition Domains and Ligand Binding

The MR was originally described as a mannose/*N*-acetylglucosamine-specific receptor on the surface of alveolar macrophages that was responsible for the binding and endocytic uptake of lysosomal glycosidases (Stahl et al., 1978). The calcium-dependent carbohydrate specificity of the MR (fucose = mannose > *N*-acetylglucosamine ≈ glucose (Shepherd et al., 1981)) resides in its CRDs (Ezekowitz et al., 1990; Taylor et al., 1990).

The individual CRDs in each MR polypeptide chain contribute differently to the binding activity of the intact receptor. The first three CRDs have little affinity for carbohydrate ligands, whereas CRDs 4, 5 and 7 are required for high affinity binding of glycoconjugates (Taylor et al., 1992). While CRD-4 demonstrates both high affinity binding, and multispecificity for various monosaccharides, it cannot in isolation account for the affinity of the intact MR for complex carbohydrate ligands (Taylor et al., 1992). These data, together with those indicating that the MR exists in monomeric form (Taylor and Drickamer, 1993), suggest that the high affinity binding of the MR is achieved through clustering of CRDs in a single polypeptide chain. This is in contrast to the soluble mannose binding protein (MBP), which achieves high affinity carbohydrate binding by multimerization of several polypeptide chains, each containing a single CRD (Sheriff et al., 1994; Weis and Drickamer, 1994).

The structural basis for carbohydrate recognition by CRDs has been best described for the rat MBP-A (Weis et al., 1991, 1992). In the MBP-A CRD, a calcium ion in the binding site coordinates with equatorial 3-OH and 4-OH groups in the terminal residue of the carbohydrate ligand. In the MR, CRD-4 is the only CRD whose ligand-binding properties have been studied extensively in isolation. As is the case with the soluble MBP, CRD-4 of the MR requires two calcium ions for sugar binding (Mullin et al., 1994). One of these calcium binding sites is conserved with that of MBP, while the second, auxillary site appears to be unique. Furthermore, the MR CRD-4 utilizes a unique stacking interaction between a tyrosine residue and the ligated sugar (Mullin et al., 1997). The differences in ligand binding properties of the MR CRD-4, and the CRDs of other C-type lectins that have been studied, appear to be related to the fact that the MR does not require multimerization to confer optimal ligand binding.

CRD binding specificity confers some differentiation of self from non-self, as cognate carbohydrate ligands are rarely found at the termini of mammalian glycoconjugates (notable exceptions being lysosomal hydrolases (Stahl et al., 1980) and tissue plasminogen activator (Narita et al., 1995), both of which are cleared physiologically from the circulation by the MR). In contrast, mannose and *N*-acetyl-

Table 2. Microbial Ligands for the Mannose Receptor

Microorganism	*Reference*
Klebsiella pneumoniae	Athamana et al., 1991; Kabha et al., 1995
Pseudomonas aeruginosa	Speert et al., 1988
Mycobacterium tuberculosis	Schlesinger, 1993; Schlesinger et al., 1994
Atypical mycobacteria	Bermudez et al., 1991; Kruskal et al., 1994
HIV-1 (gp120)	Larkin et al., 1989
Yeast	Sung et al., 1983
Candida albicans	Marodi et al., 1991
Cryptococcus neoformans	Bolanos and Mitchell, 1989
Aspergillus fumigatus	Kan and Bennett, 1988
Pneumocystis carinii	Ezekowitz et al., 1991
Leishmania donovani	Blackwell et al., 1985
Leishmania mexicana	Peters et al., 1995
Trypanosoma cruzi	Kahn et al., 1995

glucosamine residues frequently decorate the surfaces of a variety of pathogenic microorganisms including bacteria, mycobacteria, yeasts, and parasites, rendering them capable of binding to the MR (Table 2).

While the MR was the first protein demonstrated to contain multiple CRDs in a single polypeptide chain, subsequent efforts have identified a family of related molecules with a similar primary structure. The endocytic mannose receptor C-type lectins are type I transmembrane proteins, containing a cysteine-rich domain, fibronectin type II-like domain, and either 8 or 10 tandem CRD repeats. Aside from the MR, members include the DEC-205 molecule (Jiang et al., 1995), phospholipase A_2 receptor (Ishizaki et al., 1994; Ancian et al., 1995), and an unnamed novel lectin (Wu et al., 1996). The ligand binding profiles and physiological functions of these molecules remain to be elucidated in detail.

C. The N-Terminal Domains

Interestingly, the N-terminal cysteine-rich and fibronectin type II domains are among the most conserved regions between mouse and human MR. In addition, Harris and co-workers found that the cysteine-rich region has homology to the galactose-binding B chain of the plant lectin ricin (Harris et al., 1994). The presumed function for these two N-terminal domains of the MR is now the focus of much attention. It had been determined previously that these two domains play no role in the calcium-dependent binding of carbohydrates to the MR (Taylor et al., 1992). More recent studies have suggested that the cysteine-rich domain may interact with cell-associated ligands in murine spleen and lymph nodes (Martinez-Pomares et al., 1996). These putative ligands were upregulated in spleen and lymph node following immunization, suggesting a role for the MR in the generation of immune responses. These authors therefore postulated that this domain may serve to target MR-bound carbohydrate antigens to lymphocytes in the spleen and lymph node.

Other studies have indicated that the MR is also able to bind oligosaccharides terminating in Gal*N*Ac-4-SO_4 at a site independent from that at which mannosylated ligands are bound (Fiete and Baenziger, 1997; Fiete et al., 1997). This sulfated carbohydrate motif is present in glycoprotein hormones such as lutropin and thyrotropin. Receptor-mediated hepatic clearance is thought to modulate hormone serum half-life. Recent studies with MR deletion mutants have shown that this calcium-independent Gal*N*Ac-4-SO_4 binding is mediated by the MR cysteine-rich domain, implicating the MR in hormonal homeostasis (Fiete et al., 1998). We believe that evidence is accumulating to suggest that the cysteine-rich domain of the MR plays a role as a distinct pattern recognition receptor. The repertoire of MR ligands may therefore be even broader than previously anticipated.

III. PHAGOCYTOSIS AND THE MANNOSE RECEPTOR

A. Phagocytosis

Phagocytosis is the process whereby cells ingest and clear particles > 0.5 μm in diameter (reviewed in Allen and Aderem, 1996). This process is initiated by specific interactions between phagocyte receptors and particle-associated ligands. These interactions lead to localized actin polymerization, and internalization of the particle. In this manner microorganisms, apoptotic cells, and extracellular debris may be cleared. Participation of the actin cytoskeleton in internalization is a key feature differentiating phagocytosis from receptor mediated endocytosis. Much of what is known about phagocytosis in macrophages derives from studies of the Fc receptors (reviewed in Ravetch, 1994), while the mechanisms of MR-mediated phagocytosis remain less well elucidated. The demonstration of phagocytosis by a mannose-specific receptor in the amoeba *Acanthamoeba castellanii* (Allen and Dawidowicz, 1990), suggests that lectinophagocytosis is a highly conserved process, and that it must have predated the evolution of opsonic phagocytic receptors.

B. Demonstration that the MR Is a Phagocytic Receptor

The earliest reports describing the macrophage MR focused their attention on receptor mediated endocytosis. Indeed, the MR has been extensively studied as a classic recycling endocytic receptor. The first study to suggest that the MR could mediate phagocytosis demonstrated that the binding of intact yeast particles to alveolar macrophages could be inhibited by known MR endocytic inhibitors (Warr, 1980). It was subsequently shown that soluble yeast mannans could inhibit the uptake of intact zymosan particles by mouse peritoneal macrophages (Sung et al., 1983). In addition, plating these macrophages on mannose-containing substrates could prevent subsequent phagocytosis of zymosan, presumably by sequestering MR from the zymosan-accessible apical cell membrane. An additional, novel phagocytic function was ascribed to the MR when retinal pigment epithelial cells were noted to phagocytose photoreceptor outer segments by an MR-dependent mechanism (Boyle et al., 1991).

Formal proof that the MR could indeed function as an opsonin-independent phagocytic receptor was obtained after cloning of the MR cDNA. Transfection of simian Cos cells with the human MR cDNA conferred on these cells the ability to bind and ingest unopsonized yeast particles (Ezekowitz et al., 1990). Approximately 10% of transfected cells ingested between 1 and 5 yeast particles, while 7% of transfectants ingested >5 particles. Similarly, transfection of the murine MR cDNA into Cos cells conferred on them the ability to phagocytose unopsonized zymosan particles (Harris et al., 1992). Transfection of human MR cDNA into Cos cells resulted in the binding and uptake of *Pneumocystis carinii* organisms (Ezekowitz et al., 1991). This binding could be inhibited by mannan or mannosylated BSA, known MR ligands. These experiments provided important evidence for a role for the MR in first-line host defense, particularly in situations where opsonins may be absent or present in limiting concentrations.

Transfection of tailless MR mutant constructs into Cos cells enabled these cells to bind, but not phagocytose yeast particles (Ezekowitz et al., 1990). Interestingly, these cells demonstrated detectable, although markedly reduced MR-mediated endocytic activity. In an attempt to define further the MR domains required for optimal phagocytosis, chimeric molecules coupling the ectodomain of FcγRI with combinations of MR and FcγRI transmembrane and cytoplasmic domains were constructed. Transfection of these constructs into Cos cells showed that the MR phagocytic signal resides in both the transmembrane and cytoplasmic tail of the MR, and that these domains act synergistically in signaling phagocytosis (Kruskal et al., 1992). In these studies, site-directed mutagenesis of the cytoplasmic tyrosine residue reduced both phagocytosis and endocytosis by approximately 50%, but did not abolish either process.

At present, it is not clear whether MR-mediated endocytosis and phagocytosis are mediated by different receptor domains, or indeed if there is a requirement for additional accessory chain(s). In the case of the integrin β3 subunit, it appears that different domains of the cytoplasmic tail independently mediate processes such as phagocytosis, spreading, and targeting to adhesion plaques (Ylanne et al., 1995). While human FcγRII transfected into Cos cells can mediate phagocytosis, transfection with FcγRI confers binding and endocytosis, but not phagocytosis of ligands (Indik et al., 1991). Phagocytic activity of FcγRI in Cos cells can be restored by co-transfection with the γ subunit of FcεRI, indicating that an accessory chain may be used to transduce a phagocytic signal, and that independent mechanisms may mediate endocytosis and phagocytosis (Davis et al., 1995).

The existence of separate MR signal transduction pathways has been suggested by a recent study indicating that MR-mediated phagocytosis, but not endocytosis of the same carbohydrate ligands, triggers macrophage interleukin (IL)-12 production (Shibata et al., 1997).

C. Consequences of MR Ligation

A number of downstream events triggered by MR binding and/or internalization have been described. These include the release of reactive oxygen intermediates (Berton and Gordon, 1983; Ezekowitz et al., 1991), arachidonic acid metabolites (Rouzer et al., 1982), and cytokines such as IL-1, IL-6, granulocyte macrophage colony-stimulating factor (Yamamoto et al., 1997), tumor necrosis factor-α (Stein and Gordon, 1991; Garner et al., 1994), and IL-12 (Shibata et al., 1997). In addition, MR ligation has been implicated in upregulation of macrophage FcγR activity (Murai et al., 1995, 1996), as well as in enhancement of macrophage-mediated bactericidal (Lefkowitz et al., 1997) and candidacidal (Marodi et al., 1991, 1993) activities. While one would expect that the upregulation of MR levels would be a prerequisite for enhanced uptake and killing of an MR microbial ligand, this does not appear to be the case. Interferon-γ downregulates MR expression and subsequent endocytic activity at the level of transcription (Harris et al., 1992), but paradoxically enhances the MR-mediated phagocytosis and killing of *Candida albicans* (Marodi et al., 1993). These observations were extended by Raveh and colleagues who showed that interferon-γ inhibited IL-4-induced MR-mediated endocytosis, but that cells treated with both of these cytokines displayed enhanced MR phagocytic activity (Raveh et al., 1998). These results suggest that T_h1 and T_h2-type cytokines act on macrophages in concert to stimulate MR-mediated phagocytosis, but have antagonistic effects on MR-mediated endocytosis.

Taken together, the above studies suggest that the coupling of MR to intracellular endocytic and phagocytic machinery may be separate and independently regulated processes. While the signal transduction pathways coupled to phagocytic Fcγ receptors are becoming increasingly well-defined (Cox et al., 1997; Crowley et al., 1997; Hackam et al., 1997), the same can not be said for the MR. While both intracellular calcium fluxes (Marodi et al., 1993) and tyrosine kinase activation (Murai et al., 1996) have been implicated in MR signal transduction, more specific data are lacking. Given that FcγR-mediated phagocytosis, but not phagocytosis of latex beads or unopsonized bacteria, is defective in macrophages lacking the Syk tyrosine kinase (Crowley et al., 1997), it would appear that not all macrophage phagocytic events are coupled to the same signaling pathways. A number of investigators are now actively pursuing MR-coupled signal transduction pathways, particularly those coupled to phagocytosis.

IV. THE MANNOSE RECEPTOR AND THE IMMUNE RESPONSE

A. A Link Between the Innate and Clonal Immune Systems

Until recently, most studies of MR-mediated physiological functions have concentrated on its role in host defense, and in particular on its role as an effector mechanism in clearance and destruction of pathogens. A growing body of evidence supports the concept of the MR as an important molecule in the induction of clonal immune responses. In this way, it is postulated that the broad non-self specific recognition pattern of the MR may be coupled to the fine-tuned specificity of the clonal immune system.

A role for the MR in antigen presentation and immune stimulation was first proposed by Lorenz and colleagues in 1990 (Lorenz et al., 1990). In their *in vitro* studies of antigen processing and presentation, the coupling of mannose residues to an antigen resulted in a 10-fold increase in T lymphocyte stimulation. This process was inhibited by mannan, implicating the MR in antigen uptake for presentation. These authors concluded that receptor-mediated uptake is a critical step in the generation of immune responses, and that this uptake step is an important initial discriminator favoring the presentation of non-self over self antigens.

B. The MR and Dendritic Cells

Expression of the MR on dendritic cells was suggested in 1993 (Reis e Sousa et al., 1993), and formally demonstrated in 1995 (Sallusto et al., 1995). Immature dendritic cells were shown to use the MR to capture and deliver efficiently antigens to an intracellular major histocompatibility complex (MHC) Class II-containing compartment (Sallusto et al., 1995). This concept of the MR as an efficient capture and delivery system received further support when it was demonstrated that the MR mediated the uptake of mycobacterial lipoarabinomannan (LAM), and its intracellular delivery to CD1b antigen presenting molecules (Prigozy et al., 1997). In keeping with this concept, MR antagonists inhibited the presentation of this antigen to LAM-reactive T lymphocytes (Prigozy et al., 1997). Experiments with peptide and protein antigens revealed that mannosylation of antigens increased their ability to stimulate MHC-restricted T lymphocytes by 10^2- (Engering et al., 1997) to 10^5- (Tan et al., 1997) fold.

While the antigen presentation experiments mentioned above were performed with soluble antigen, in reality antigens are more likely to be encountered initially in a more complex, particle-associated form. MR-mediated phagocytosis by macrophages is well-established (see above), but few data attest to MR-mediated phagocytosis by dendritic cells. Freshly isolated Langerhans cells (representing immature dendritic cells) have been shown to be capable of phagocytosing zymosan, yeast, bacteria, and latex beads (Reis e Sousa et al., 1993). Other studies have demonstrated the ability of dendritic cells to phagocytose intact microorganisms, and to process their antigens for presentation to T lymphocytes (Inaba et al., 1993; Filgueria et al., 1996; Svensson et al., 1997). The involvement of specific phagocytic receptors in this process have not been described. The inhibition of dendritic cell zymosan phagocytosis by mannan strongly implicated the MR (Reis e Sousa et al., 1993), but a role for other receptors such as the DEC 205 molecule (Jiang et al., 1995) in this process cannot be excluded. It is likely that the nature of the specific receptors involved in this process will only be resolved once MR and related knockout mice become available.

V. CONCLUSION

The mannose receptor mediates the uptake of ligands bearing appropriately configured carbohydrate residues. Its ligand binding profile confers on it a broad capacity to recognize non-self over self molecules, in keeping with its designation as a pattern recognition receptor. Ligation of the receptor is followed by internalization of the bound ligand, and is coupled to a variety of cellular processes directed towards destruction of pathogens, participation in the inflammatory response, or stimulation of the immune system. While this receptor is capable of both endocytic and phagocytic uptake, the signaling pathways utilized remain to be determined.

REFERENCES

Allen, L.-A.H., & Aderem, A. (1996). Mechanisms of phagocytosis. Curr. Opin. Immunol. 8, 36–40.

Allen, P.G., & Dawidowicz, E.A. (1990). Phagocytosis in Acanthamoeba: I. A mannose receptor is responsible for the binding and phagocytosis of yeast. J. Cell Physiol. 145, 508–513.

Ancian, P., Lambeau, G., Mattei, M.-G., & Lazdunski, M. (1995). The human 180-kDa receptor for secretory phospholipases A2: molecular cloning, identification of a secreted soluble form, expression, and chromosomal localization. J. Biol. Chem. 270, 8963–8970.

Athamana, A., Ofek, I., Keisari, Y., Markowitz, S., Dutton, G.G.S., & Sharon, N. (1991). Lectinophagocytosis of encapsulated Klebsiella pneumoniae mediated by surface lectins of guinea pig alveolar macrophages and human monocyte-derived macrophages. Infect. Immun. 59, 1673–1682.

Bermudez, L.E., Young, L.S., & Enkel, H. (1991). Interaction of Mycobacterium avium complex with human macrophages: roles of membrane receptors and serum proteins. Infect. Immun. 59, 1697–1702.

Berton, G., & Gordon, S. (1983). Modulation of macrophage mannosyl-specific receptors by cultivation on immobilised zymosan. Effects on superoxide-anion release and phagocytosis. Immunol. 49, 705–715.

Blackwell, J.M., Ezekowitz, R.A., Roberts, M.B., Channon, J.Y., Sim, R.B. & Gordon, S. (1985). Macrophage complement and lectin-like receptors bind Leishmania in the absence of serum. J. Exp. Med. 162, 324–331.

Bolanos, B., & Mitchell, T.G. (1989). Phagocytosis and killing of Cryptococcus neoformans by rat alveolar macrophages in the absence of serum. J. Leuk. Biol. 46, 521–528.

Boyle, D., Tien, L.F., Cooper, N.G., Shepherd, V., & McLaughlin, B.J. (1991). A mannose receptor is involved in retinal phagocytosis. Invest. Ophthalmol. Vis. Sci. *32*, 1464–1470.

Carroll, M.C., & Fischer, M.B. (1997). Complement and the immune response. Curr. Opin. Immunol. 9, 64–69.

Cox, D., Chang, P., Zhang, Q., Reddy, P.G., Bokoch, G.M., & Greenberg, S. (1997). Requirements for both Rac1 and Cdc42 in membrane ruffling and phagocytosis in leukocytes. J. Exp. Med. 186, 1487–1494.

Crowley, M.T., Costello, P.S., Fitzer-Attas, C.J., Turner, M., Meng, F., Lowell, C., Tybulewicz, V.L.J., & DeFranco, A.L. (1997). A critical role for Syk in signal transduction and phagocytosis mediated by Fcγ receptors on macrophages. J. Exp. Med. 186, 1027–1039.

Davis, W., Harrison, P.T., Hutchinson, M.J., & Allen, J.M. (1995). Two distinct regions of FcγRI initiate separate signalling pathways involved in endocytosis and phagocytosis. EMBO J. 14, 432–441.

Engering, A.J., Cella, M., Fluitsma, D., Brockhaus, M., Hoefsmit, E.C., Lanzavecchia, A., & Pieters, J. (1997). The mannose receptor functions as a high capacity and broad specificity antigen receptor in human dendritic cells. Eur. J. Immunol. 27, 2417–2425.

Ezekowitz, R.A.B., Sastry, K., Bailly, P., & Warner, A. (1990). Molecular characterization of the human macrophage mannose receptor: demonstration of multiple carbohydrate recognition-like domains and phagocytosis of yeasts in Cos-1 cells. J. Exp. Med. 172, 1785–1794.

Ezekowitz, R.A.B., Williams, D.J., Koziel, H., Armstrong, M.Y.K., Warner, A., Richards, F.F., & Rose, R.M. (1991). Uptake of *Pneumocystis carinii* mediated by the macrophage mannose receptor. Nature 351, 155–158.

Fiete, D., & Baenziger, J.U. (1997). Isolation of the SO_4-4-GalNAcβ1, 4GlcNAcβ1,2Manα-specific receptor from rat liver. J. Biol. Chem. 272, 14629–14637.

Fiete, D., Beranek, M.C., & Baenzinger, J.U. (1997). The macrophage/endothelial cell mannose receptor cDNA encodes a protein that binds oligosaccharides terminating with SO_4-4-GalNAcβ1, 4GlcNAcβ or Man at independent sites. Proc. Natl. Acad. Sci. USA 94, 11256–11261.

Fiete, D.J., Beranek, M.C., & Baenziger, J.U. (1998). The cysteine-rich domain of the "mannose" receptor mediates GalNAc-4-SO_4 binding. Proc. Natl. Acad. Sci. USA 95, 2089–2093.

Filgueria, L., Nestle, F.O., Rittig, M., Joller, H.I., & Groscurth, P. (1996). Human dendritic cells phagocytose and process Borrelia burgdorferi. J. Immunol. 157, 2998–3005.

Garner, R.E., Rubanowice, K., Sawyer, R.T., & Hudson, J.A. (1994). Secretion of TNF-alpha by alveolar macrophages in response to *Candida albicans* mannan. J. Leuk. Biol. 55, 161–168.

Hackam, D.J., Rotstein, O.D., Schreiber, A., Zhang, W.-J., & Grinstein, S. (1997). Rho is required for the initiation of calcium signalling and phagocytosis by Fcγ receptors in macrophages. J. Exp. Med. 186, 955–966.

Harris, N., Peters, L.L., Eicher, E.M., Rits, M., Raspberry, D., Eichbaum, Q.G., Super, M., & Ezekowitz, R.A.B. (1994). The exon-intron structure and chromosomal localization of the mouse macrophage mannose receptor gene *Mrc1*: identification of a ricin-like domain at the N-terminus of the receptor. Biochem. Biophys. Res. Comm. 198, 682–692.

Harris, N., Super, M., Rits, M., Chang, G., & Ezekowitz, R.A.B. (1992). Characterization of the murine macrophage mannose receptor: demonstration that the downregulation of receptor expression mediated by Interferon-γ occurs at the level of transcription. Blood 80, 2363–2373.

Inaba, K., Inaba, M., Naito, M., & Steinman, R.M. (1993). Dendritic cell progenitors phagocytose particulates, including Bacillus Calmette-Guerin organisms, and sensitize mice to mycobacterial Ags in vivo. J. Exp. Med. 178, 479–488.

Indik, Z., Kelly, C., Chien, P., Levinson, A.I., & Schreiber, A.D. (1991). Human FcγRII in the absence of other Fcγ receptors, mediates a phagocytic signal. J. Clin. Invest. 88, 1766–1771.

Ishizaki, J., Hanasaki, K., Higashino, K.-I., Kishino, J., Kikuchi, N., Ohara, O., & Arita, H. (1994). Molecular cloning of pancreatic group I phospholipase A2 receptor. J. Biol. Chem. 269, 5897–5904.

Jiang, W., Swiggard, W.J., Heufler, C., Peng, M., Mirza, A., Steinman, R.M., & Nussenzweig, M.C. (1995). The receptor DEC-205 expressed by dendritic cells and thymic epithelial cells is involved in antigen processing. Nature 375, 151–155.

Kabha, K., Nissimov, L., Athamna, A., Keisari, Y., Parolis, L.A., Grue, R.M., Schlepper-Schafer, J., Ezekowitz, R.A., Ohman, D.E., & Ofek, I. (1995). Relationships among capsular structure, structure, phagocytosis, and mouse virulence in Klebsiella pneumoniae. Infect. Immun. 63, 847–852.

Kahn, S., Wleklinski, M., Aruffo, A., Farr, A., Coder, D., & Kahn, M. (1995). *Trypanosoma cruzi* amastigote adhesion to macrophages is facilitated by the mannose receptor. J. Exp. Med. 182, 1243–1258.

Kan, V.L., & Bennett, J.E. (1988). Lectin-like attachment sites on murine pulmonary alveolar macrophages bind Aspergillus fumigatus conidia. J. Infect. Dis. 158, 407–414.

Kruskal, B.A., Sastry, K., Warner, A.B., Mathieu, C.E., & Ezekowitz, R.A.B. (1992). Phagocytic chimeric receptors require both transmembrane and cytoplasmic domains from the mannose receptor. J. Exp. Med. 176, 1673–1680.

Kruskal, B.A., Super, M., Rogers, R., & Ezekowitz, R.A.B. (1994). The macrophage mannose receptor can mediate the uptake of Mycobacterium avium-intracellulare. Ped. Res. 35, 184A.

Larkin, M., Childs, R.A., Matthews, T.J., Thiel, S., Mizuochi, T., Lawson, A.M., Savill, J.S., Haslett, C., Diaz, R., & Feizi, T. (1989). Oligosaccharide-mediated interactions of the envelope glycoprotein gp120 of HIV-1 that are independent of CD4 recognition. AIDS 3, 793–798.

Lefkowitz, D.L., Lincoln, J.A., Lefkowitz, S.S., Bollen, A., & Moguilevsky, N. (1997). Enhancement of macrophage-mediated bactericidal activity by macrophage mannose receptor-ligand interaction. Immunol. Cell Biol. 75, 136–141.

Lorenz, R.G., Blum, J.S., & Allen, P.M. (1990). Constitutive competition by self proteins for antigen presentation can be overcome by receptor-enhanced uptake. J. Immunol. 144, 1600–1606.

Marodi, L., Korchak, H.M., & Johnson, R.B.J. (1991). Mechanisms of host defense against Candida species. I. Phagocytosis by monocytes and monocyte-derived macrophages. J. Immunol. 146, 2783–2789.

Marodi, L., Schreiber, S., Anderson, D.C., MacDermott, R.P., Korchak, H.M., & Johnston, R.B.J. (1993). Enhancement of macrophage candidacidal activity by interferon-gamma. Increased phagocytosis, killing, and calcium signal mediated by a decreased number of mannose receptors. J. Clin. Invest. 91, 2596–2601.

Martinez-Pomares, L., Kosco-Vilbois, M., Darley, E., Tree, P., Herren, S., Bonnefoy, J.-Y., & Gordon, S. (1996). Fc chimeric protein containing the cysteine-rich domain of the murine mannose receptor binds to macrophages from splenic marginal zone and lymph node subcapsular sinus and to germinal centers. J. Exp. Med. 184, 1927–1937.

Medzhitov, R., & Janeway, C.A.J. (1997). Innate immunity: the virtues of a nonclonal system of recognition. Cell 91, 295–298.

Mullin, N.P., Hall, K.T., & Taylor, M.E. (1994). Characterization of ligand binding to a carbohydrate-recognition domain of the macrophage mannose receptor. J. Biol. Chem. 269, 28405–28413.

Mullin, N.P., Hitchen, P.G., & Taylor, M.E. (1997). Mechanism of Ca^{2+} and monosaccharide binding to a C-type carbohydrate-recognition domain of the macrophage mannose receptor. J.Biol. Chem. 272, 5668–5681.

Murai, M., Aramaki, Y., & Tsuchiya, S. (1995). Contribution of mannose receptor to signal transduction in Fc gamma receptor-mediated phagocytosis of mouse peritoneal macrophages induced by liposomes. J. Leuk. Biol. 57, 687–691.

Murai, M., Aramaki, Y., & Tsuchiya, S. (1996). Alpha 2-macroglobulin stimulation of protein tyrosine phosphorylation in macrophages via the mannose receptor for Fc gamma receptor-mediated phagocytosis activation. Immunol. 89, 436–441.

Narita, M., Bu, G., Herz, J., & Schwartz, A.L. (1995). Two receptor systems are involved in the plasma clearance of tissue-type plasminogen activator (t-PA) in vivo. J. Clin. Invest. 96, 1164–1168.

Nepomuceno, R.R., Henschen-Edman, A.H., Burgess, W.H., & Tenner, A.J. (1997). cDNA cloning and primary structure analysis of C1qR(P), the human C1q/MBL/SPA receptor that mediates enhanced phagocytosis in vitro. Immunity 6, 119–129.

Peters, C., Aebischer, T., Stierhof, Y.D., Fuchs, M., & Overath, P. (1995). The role of macrophage receptors in adhesion and uptake of Leishmania mexicana amastigotes. J. Cell Sci. 108, 3715–3724.

Prigozy, T.I., Sieling, P.A., Clemens, D., Stewart, P.L., Behar, S.M., Porcelli, S.A., Brenner, M.B., Modlin, R.L., & Kronenberg, M. (1997). The mannose receptor delivers lipoglycan antigens to endosomes for presentation to T cells by CD1b molecules. Immunity 6, 187–197.

Raveh, D., Kruskal, B.A., Farland, J., & Ezekowitz, R.A.B. (1998). Th1 and Th2 cytokines cooperate to stimulate mannose receptor-mediated phagocytosis. J. Leuk. Biol. 64, 108–113.

Ravetch, J.V. (1994). Fc receptors: Rubor redux. Cell 78, 553–560.

Reis e Sousa, C., Stahl, P.D., & Austyn, J.M. (1993). Phagocytosis of antigens by Langerhans cells in vitro. J. Exp. Med. 178, 509–519.

Rouzer, C.A., Scott, W.A., Hamill, A.L., & Cohn, Z.A. (1982). Synthesis of leukotriene C and other arachidonic acid metabolites by mouse pulmonary macrophages. J. Exp. Med. 155, 720–733.

Sallusto, F., Cella, M., Danieli, C., & Lanzavecchia, A. (1995). Dendritic cells use macropinocytosis and the mannose receptor to concentrate macromolecules in the major histocompatibility complex class II compartment: downregulation by cytokines and bacterial products. J. Exp. Med. 182, 389–400.

Schlesinger, L.S. (1993). Macrophage phagocytosis of virulent but not attenuated strains of Mycobacterium tuberculosis is mediated by mannose receptors in addition to complement receptors. J. Immunol. 150, 2920–2930.

Schlesinger, L.S., Hull, S.R., & Kaufman, T.M. (1994). Binding of the terminal mannosyl units of lipoarabinomannan from a virulent strain of Mycobacterium tuberculosis to human macrophages. J. Immunol. 152, 4070–4079.

Shepherd, V.L., Lee, Y.C., Schlesinger, P.H., & Stahl, P.D. (1981). L-Fucose terminated glycoconjugates are recognized by pinocytosis receptors on macrophages. Proc. Natl. Acad. Sci. USA 78, 1019–1022.

Sheriff, S., Chang, C.Y., & Ezekowitz, R.A.B. (1994). Human mannose-binding protein carbohydrate recognition domain trimerizes through a triple alpha-helical coiled-coil. Nature Struct. Biol. 1, 789–794.

Shibata, Y., Metzger, W.J., & Myrvik, Q.N. (1997). Chitin particle-induced cell-mediated immunity is inhibited by soluble mannan: mannose receptor-mediated phagocytosis initiates IL-12 production. J. Immunol. 159, 2462–2467.

Speert, D.P., Wright, S.D., Silverstein, S.C., & Mah, B. (1988). Functional characterization of macrophage receptors for in vitro phagocytosis of unopsonized Pseudomonas aeruginosa. J. Clin. Invest. 82, 872–879.

Stahl, P., Schlesinger, P.H., Sigardson, E., Rodman, J.S., & Lee, Y.C. (1980). Receptor-mediated pinocytosis of mannose glycoconjugates by macrophages: characterization and evidence for receptor recycling. Cell 19, 207–215.

Stahl, P.D., Rodman, J.S., Miller, M.J. & Schlesinger, P.H. (1978). Evidence for receptor-mediated binding of glycoproteins, glycoconjugates, and lysosomal glycosidases by alveolar macrophages. Proc. Natl. Acad. Sci. USA 75, 1399–1403.

Stein, M.L., & Gordon, S. (1991). Regulation of tumour necrosis factor release by murine peritoneal macrophages: role of cell stimulation and specific phagocytic plasma membrane receptors. Eur. J. Immunol. 21, 431–437.

Sung, S.-S.J., Nelson, R.S., & Silverstein, S.C. (1983). Yeast mannans inhibit binding and phagocytosis of zymosan by mouse peritoneal macrophages. J. Cell Biol. 96, 160–166.

Svensson, M., Stockinger, B., & Wick, M.J. (1997). Bone marrow-derived dendritic cells can process bacteria for MHC-I and MHC-II presentation to T cells. J. Immunol. 158, 4229–4236.

Takahashi, K., Donovan, M.J., Rogers, R.A., & Ezekowitz, R.A.B. (1998). Distribution of murine mannose receptor expression from early embryogenesis through to adulthood. Cell Tiss. Res. 292, 311–323.

Tan, M.C., Mommaas, A.M., Drijfhout, J.W., Jordens, R., Onderwater, J.J., Verwoerd, D., Mulder, A.A., van der Heiden, A.N., Scheidegger, D., Oomen, L.C., Ottenhoff, T.H., Tulp, A., Neefjes, J.J., & Koning, F. (1997). Mannose receptor-mediated uptake of antigens strongly enhances HLA class II-restricted antigen presentation by cultured dendritic cells. Eur. J. Immunol. 27, 2426–2435.

Taylor, M.E., & Drickamer, K. (1993). Structural requirements for high affinity binding of complex ligands by the macrophage mannose receptor. J. Biol. Chem. 268, 399–404.

Taylor, M.E., Bezouska, K., & Drickamer, K. (1992). Contribution to ligand binding by multiple carbohydrate-recognition domains in the macrophage mannose receptor. J. Biol. Chem. 267, 1719–1726.

Taylor, M.E., Conary, J.T., Lennartz, M.R., Stahl, P.D., & Drickamer, K. (1990). Primary structure of the mannose receptor contains multiple motifs resembling carbohydrate recognition domains. J. Biol. Chem. 265, 12156–12162.

Tenner, A.J., Robinson, S.L., & Ezekowitz, R.A.B. (1995). Mannose binding protein (MBP) enhances mononuclear phagocyte function via a receptor that contains the 126,000 M(r) component of the C1q receptor. Immunity 3, 485–493.

Warr, G.A. (1980). A macrophage receptor for (mannose/glucosamine)-glycoproteins of potential importance in phagocytic activity. Bichem. Biophys. Res. Commun. 93, 737–745.

Weis, W.I., & Drickamer, K. (1994). Trimeric structure of a C-type mannose binding protein. Structure 2, 1227–1240.

Weis, W.I., Drickamer, K., & Hendrickson, W.A. (1992). Structure of a C-type mannose-binding protein complexed with an oligosaccharide. Nature 360, 127–134.

Weis, W.I., Kahn, R., Fourme, R., Drickamer, K., & Hendrickson, W.A. (1991). Structure of the calcium-dependent lectin domain from a rat mannose-binding protein determined by MAD phasing. Science 254, 1608–1615.

Wu, K., Yuan, J., & Lasky, L.A. (1996). Characterization of a novel member of the macrophage mannose receptor type C lectin family. J. Biol. Chem. 271, 21323–21330.

Yamamoto, Y., Klein, T.W. & Friedman, H. (1997). Involvement of mannose receptor in cytokine interleukin-1 beta (IL-1 beta), IL-6, and granulocyte-macrophage colony-stimulating factor responses, but not in chemokine macrophage inflammatory protein 1 beta (MIP-1 beta), MIP-2, and KC responses, caused by attachment of Candida albicans to macrophages. Infect. Immun. 65, 1077–1082.

Ylanne, J., Huuskonen, J., O'Toole, T.E., Ginsberg, M.H., Virtanen, I., & Gahmberg, C.G. (1995). Mutation of the cytoplasmic domain of the integrin β3 subunit: differential effects on cell spreading, recruitment to adhesion plaques, endocytosis and phagocytosis. J. Biol. Chem. 270, 9550–9557.

INTEGRIN RECEPTORS OF PHAGOCYTES

Scott D. Blystone and Eric J. Brown

Advances in Cell and Molecular Biology of Membranes and Organelles
Volume 5, pages 103-147.

ISBN: 1-55938-999-0

I. INTRODUCTION

In recent years, there has been a rapid expansion in information about the mechanisms involved in cell adhesions, both to other cells or pathogens and to the extracellular matrix in which these cells exist. This increase in understanding has arisen primarily from more detailed knowledge of the structure of receptors for extracellular matrix and related proteins and the signaling events initiated upon their ligation. These advances have been summarized in numerous recent reviews (Hynes, 1987; Hemler, 1988; Albelda and Buck, 1990; Springer, 1990; Ruoslahti, 1991; Hynes, 1992; Brown and Lindberg, 1993; Damsky et al., 1993; Juliano and Haskill, 1993; Luscinskas and Lawler, 1994; Varner et al., 1995). This information has underscored the important influence of cell-cell adhesion and cell-matrix adhesion on cell phenotype during development, normal homeostasis, metastasis, tissue repair, and inflammation. Myeloid cells present a particularly intriguing set of problems in this regard. During the normal maturation of myeloid cells, they reside in the bone marrow, an environment rich in extracellular matrix molecules and tightly packed with neighboring cells, in which adhesive phenomena are an important part of cell development. Upon maturation, myeloid cells move into the bloodstream, where significant contact with other cells is minimal and in which, under normal circumstances, there is no exposure to extracellular matrix. However, a key role for monocytes is to replenish the supply of tissue macrophages. Thus, these cells move back into an area rich in extracellular matrix ligands. Another essential function of both monocytes and neutrophils is to move to areas of inflammation or infection to provide essential host defense and tissue repair functions. In these processes of emigration from the bloodstream to damaged or infected extravascular tissues, the recognition of endothelium, extracellular matrix, and pathogens has a critical role. Thus, leukocytes possess mechanisms for precise modulation of expression and function of their adhesion receptors during development, while in the circulation, during emigration into solid tissues, and in the resolution of inflammatory events. The focus of this chapter is the role of integrin receptors in myeloid cell function with emphasis on host-pathogen interactions.

II. INTEGRINS

A. Integrin Structure

Receptors for extracellular matrix and plasma proteins fall into several gene families, the best studied of which is the integrin superfamily (Brown, 1986; Hemler et al., 1990; Virtanen et al., 1990; Phillips et al., 1991; Hynes, 1992). Integrins are heterodimeric proteins consisting of noncovalently associated α and β subunits generally of ~150 and ~100 kDa, respectively (Figure 1, Table 1). At present, 14 α- and 8 β-chains have been characterized molecularly and at least 22 different

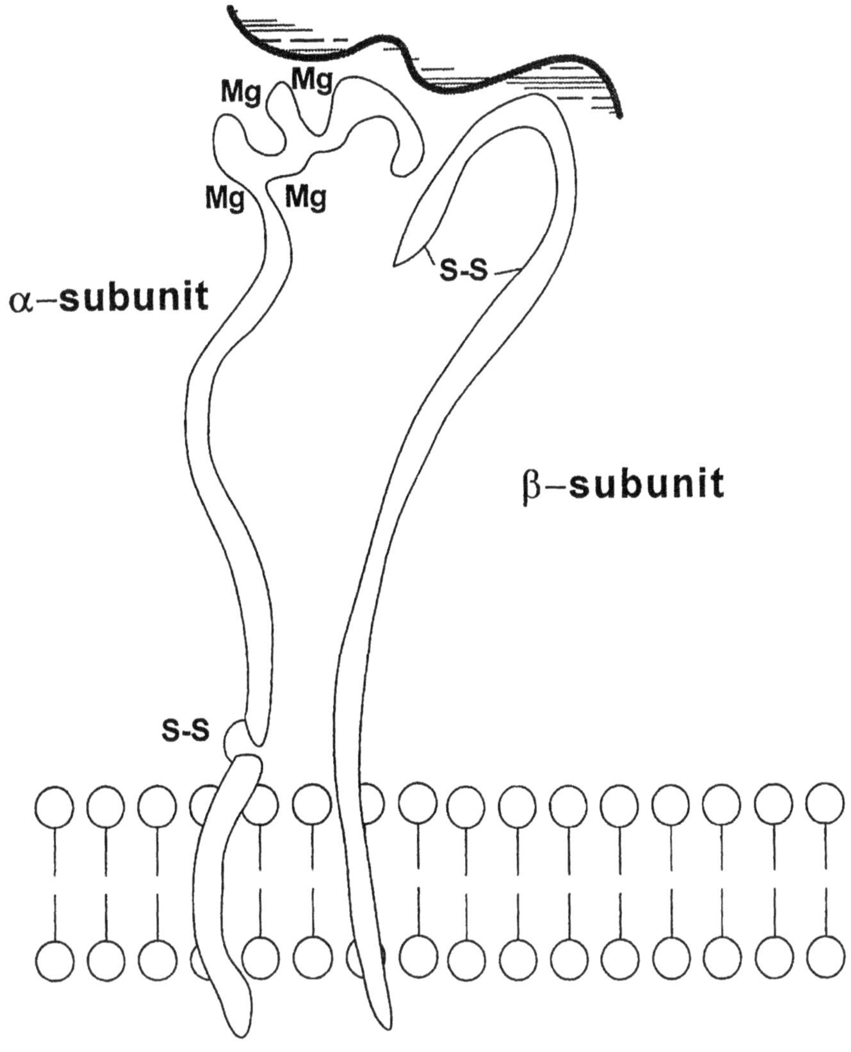

Figure 1. Integrin receptor structure. This diagram depicts the interaction of integrin α- and β-subunits with each other, a representational ligand, and the cell membrane. Mg represents divalent cation pockets, S–S represents internal disulfide bridges.

heterodimers have been found. The integrin superfamily can be subdivided according to β-chain usage into several subfamilies (Hynes, 1987). While there are several integrin receptors with unique β-chains which do not fall into one of these families, the majority of myeloid integrin receptors belong to either the β_1 (VLA) family, the β_2 (Leu-CAM) family, or the β_3 (Cytoadhesin) family. Myeloid cell populations can express at least 12 of these integrins, sometimes simultaneously displaying members of all three major subfamilies (Kishimoto et al., 1987b, 1990; Brown and Goodwin, 1988; Hemler, 1988, 1990a; Sanchez-Madrid and Corbi, 1992; Brown and Lindberg, 1993; Erle et al., 1994).

Both α and β subunits are transmembrane glycoproteins, and the cytoplasmic domains of several integrins can interact indirectly with the actin cytoskeleton via the proteins talin, vinculin and α-actinin (Singer, 1982; Horwitz et al., 1986; Tamkun et al., 1986; Pavalko et al., 1991; Miyamoto et al., 1995; Schaller et al., 1995; Ezzell et al., 1997; Vignoud et al., 1997; Retta et al., 1998). Each subunit exhibits a large extracellular domain, a single type I transmembrane domain and a short cytoplasmic tail averaging 35 amino acids for α-chains and 53 amino acids for β-chains (Corbi et al., 1987; Hynes, 1987; Suzuki et al., 1987; Takada et al., 1989; Hemler, 1990a; Larson and Springer, 1990). A notable exception is the cytoplasmic tail of the β_4 subunit which exceeds 1,000 residues and interacts with intermediate filaments rather than microfilaments (Hogervorst et al., 1990; Stepp et al., 1990; Marchisio et al., 1991; Kennel et al., 1993; Niessen et al., 1994). However, this integrin, which is an important component of hemidesmosomes, is not expressed in leukocytes (Schaller et al., 1992). At the amino-terminus of each α subunit are seven homologous tandem repeats (I though VII) (Larson et al., 1989a; Stanley et al., 1994). The last three or four tandem repeats are thought to contain EF-hand type domains which bind divalent cations essential for integrin function (Dransfield and Hogg, 1989; Kirchhofer et al., 1990; van Kooyk et al., 1991; Dransfield et al., 1992; Masumoto and Hemler, 1993a). Some α subunits exhibit an inserted domain or I-domain, located between repeat domains II and III (Larson et al., 1989a; Takada and Hemler, 1989; Diamond et al., 1993; Masumoto and Hemler, 1993a; Kamata and Takada, 1994; Kern et al., 1994; Randi and Hogg, 1994; Puzon-McLaughlin and Takada, 1996). The extracellular regions of other α subunits are cleaved post translationally to create heavy and light chains joined by a single disulfide bond (Table 1) (Teixid et al., 1992). The extracellular portions of integrin β-chains are heavily disulfide bonded, with each β-chain containing 48–56 cysteines (Hynes, 1987; Hogervorst et al., 1990; Ramaswamy and Hemler, 1990; Sheppard et al., 1990; Suzuki et al., 1990; Giunta et al., 1991; Yuan et al., 1991). Presumably, the role for the disulfide bonding is to create a structure with considerable rigidity. This would allow the ligand binding site in the integrin heterodimer to project some distance from the plasma membrane, potentially allowing more efficient interaction with ligand.

I domains are very important for ligand recognition by all β_2 integrins and by $\alpha_2\beta_1$. Recombinant I domains can bind ligands autonomously. The atomic structures of the I-domains of both α_L and α_M have revealed a similar unique cation binding site in both I domains that has been given the acronym MIDAS, for metal ion dependent activation site (Fairbanks et al., 1995; Lee et al., 1995; Rieu et al., 1996a; Emsley et al., 1997; Qu and Leahy, 1998). The hypothesis that the MIDAS motif is involved in ligand binding is supported by numerous mutagenesis studies (Randi and Hogg, 1994; Bajt et al.,1995; McGuire and Bajt, 1995; Goodman and

Table 1. Biochemical Characteristics of Integrin Subunits

Subunit Name	*CD Classification (Human)*	*Size (kDa) (Nonreduced/ Reduced)*	*Alternative Splicing (Cytoplasmic Domain*	*I Domain*	*SS Bridge*
α_1	CD49a	200/210		+	
α_2	CD49b	160/165		+	
α_3	CD49c	150/135; 30	+		+
α_4	CD49d	140/150; 40			+
		180/80; 70			
α_5	CD49e	155/135; 20			+
α_6	CD49f	150/130; 30 (31)	+		+
α_7		120/100; 30	+		+
α_8 (chicken)		160/140; 25			
α_9		140/15			
α_L	CD11a	/180		+	
α_M	CD11b	/170		+	
α_X	CD11c	/150		+	
α_{IIb}	CD41	145/120; 25			+
α_v	CD51	150; 125; 25			+
β_1	CD29	120/130	+		
β_2	CD18	90/95			
β_3	CD61	95/115	+		
β_4	CD104	200/205	+ (two types)		
β_5		97/110			
β_6		110/			
β_7		105/120			
β_8		95/97			

Note: Known integrin subunits are listed with CD classification numbers, approximate molecular mass, and special features of subunit structure.

Bajt, 1996; Puzon-McLaughlin and Takada, 1996). The I-domain can exist in two distinct conformations determined by the nature of the cation occupying the MIDAS (Fairbanks et al., 1995; McGuire and Bajt, 1995). This is of interest because it suggests a possible mechanism for the well-known ability of leukocytes and platelets to regulate integrin-mediated ligand binding (Altieri et al., 1988; D'Sousa et al., 1990; O'Toole et al., 1990; Smith and Cheresh, 1991; Dransfield et al., 1992; Kunicki et al., 1996; Rieu et al., 1996b; Tozer et al., 1996, 1998). Based on sequence homologies, integrin β-chains may also contain a MIDAS motif (Lin et al., 1997). This suggests the possibility that MIDAS-mediated conformational changes and regulation of ligand avidity may be more broadly applicable than for just I domain-containing integrins. However, this hypothesis remains controversial and direct cation binding by integrin β-chains has not been demonstrated. Thus the MIDAS may provide insights into the capacity of integrins to modulate their affinity for ligand in response to changes in divalent cations.

For integrins lacking an I domain, ligand recognition by integrins is a shared function of both the α and β subunits (Cheresh and Harper, 1987; Ginsberg et al., 1990; Sonnenberg et al., 1990; D'Souza et al., 1991). Chemical crosslinking studies have demonstrated required elements from both chains (Calvete et al., 1992, 1994). Ligand specificity of integrins appears to depend primarily on the α-chain (Kawaguchi and Hemler, 1993). For example, $\alpha_5\beta_1$ is a receptor for fibronectin (Pytela et al., 1985a; Argraves et al., 1987), while $\alpha_6\beta_1$ is a receptor for laminin (Shaw et al., 1990), and $\alpha_2\beta_1$ is a receptor for collagen (Staatz et al., 1989; Kawaguchi and Hemler, 1993; Kamata and Takada, 1994). The α chain repeats have been modeled as a β propeller, similar in overall structure to the WD40 domains of G_β subunits (Springer, 1997). This domain, with seven-fold symmetry, is thought to extend from the plasma membrane on a stalk composed of the rest of the α-chain, suggesting an ideal location to encounter ligand. In I domain-containing integrins, the I-domain projects from this β-propeller structure away from the plasma membrane toward a potential ligand-bearing surface (Kern et al., 1994; Fairbanks et al., 1995). Integrin α-chains may also regulate heterodimer expression (O'Toole et al., 1989; Krissansen et al., 1990; Shimizu et al., 1990b; Kim and Yamada, 1997). In many cells there is retention of an intracellular pool of immature β_1 integrin chains associated with calnexin, a chaperone protein (Lenter and Vestweber, 1994). Surface expression of mature receptor is regulated by the synthesis, assembly, and transport of the α-chain. Synthesis of integrin α-chains can be enhanced by molecules such as inflammatory cytokines which in turn increase cell surface expression of the regulated integrin (Clark et al., 1989; Hickstein et al., 1989; Ballard et al., 1991; Milam et al., 1991). In contrast, in some cells, α_v seems to be stored in the endoplasmic reticulum and the rate limiting step in integrin expression is β chain synthesis (Krissansen et al., 1990; Defilippi et al., 1991; De Nichilo and Burns, 1993).

B. Integrin Function

Known ligands for integrin receptors include proteins of the extracellular matrix, cell surface proteins of the immunoglobulin superfamily, plasma proteins, and a variety of pathogens. Included among integrin ligands are fibronectin (Bevilacqua et al., 1981; Cardarelli and Pierschbacher, 1987), vitronectin (Smith et al., 1990b; Pytela et al., 1985b; Bodary and McLean, 1990), collagen (Takada and Hemler, 1989; Dedhar and Gray, 1990; Staatz et al., 1990; Elices et al., 1991), and laminin (Horwitz et al.,1985; Kramer et al., 1990; Sonnenberg et al., 1990) in the matrix; fibrinogen (Altieri et al., 1988; Bennett et al., 1988; Wright et al., 1988), fibrin (Blystone et al., 1991; Katagiri et al., 1995), thrombin (Bar-Shavit et al., 1991), and von Willebrand Factor (Cheresh, 1987; Cook et al., 1992) in provisional clot matrices; intercellular adhesion molecules (ICAMs) (Fawcett et al., 1992; de Fougerolles et al., 1994), vascular cell adhesian molecule-1 (VCAM-1) (Lobb et al., 1991 Masumoto and Hemler, 1993b), and mucosal addressin cell adhesion molecule-1 (MadCAM-1) (Erle et al., 1994; Tidswell et al., 1997) on other cell surfaces; fibrinogen (Altieri et al., 1988) and fibronectin (Blystone and Kaplan, 1992) in plasma; and pathogens such as *Bordetella spp.* (Relman et al., 1990; Register et al., 1994), *Mycobacteria spp.* (Schlesinger and Horwitz, 1991), *Escherichia spp.* (Frankel et al., 1996), foot-and-mouth disease virus (Neff et al., 1998), and *Borrelia burgdorferi* (Lyme disease) (Cinco et al., 1997), among many others (Tables 2–5).

In most cells, integrins function as adhesive receptors. As adhesion molecules, they are involved in many biologic processes including embryonic development (Sorokin et al., 1990; Zusman et al., 1990), maintenance of tissue integrity (DeLuca et al., 1990; Pober and Cotran, 1991), hemostasis (Shattil, 1995), and leukocyte homing (Holzmann et al., 1989; Springer, 1990). Their structures provide a direct linkage from extracellular matrix to the actin cytoskeleton (Burridge et al., 1990; Pavalko et al., 1991; Ruoslahti, 1991). Thus integrin ligation can be a mechanism for cells to monitor forces exerted upon tissue, a means for cells to determine their location within tissues, and a link to organization of cells into tissues. As integrins are expressed on virtually all cells in higher organisms (with the exception of erythrocytes) and are a predominant receptor for extracellular matrix ligands, it is likely that these heterodimers are a primary means of attachment to tissue for all cells in the body. A unique feature of leukocyte integrins is the ability to regulate their adhesive function (Kishimoto et al., 1990; Hogg et al., 1993; van Kooyk and Figdor, 1993). This crucial feature of leukocyte integrins is discussed in detail below. For cells of non-hematopoietic origin, integrins are thought to be always in a state capable of mediating adhesion (Hughes et al., 1996). However, growth factors and other environmental cues may affect integrin-mediated adhesion even in these cells.

In addition to their adhesive function, integrins also provide an efficient means of traction during the migration of cells (Aota et al., 1991; Pasqualini and Hemler, 1994). Integrin-mediated migration is a required component of development (Krotoski and Bronner Fraser, 1990) and wound healing (Hourihan et al., 1993) involving many cell types, including leukocytes (Parkos et al., 1991). For example, following vascular injury, macrophages initiate repair by migration to the site of injury, and fibroblasts, smooth muscle cells, and endothelial cells must migrate through provisional matrices to repopulate injured tissue and restore vascular integrity (Clyman et al., 1992; Leavesley et al., 1993). These migration events are mediated by integrin receptors whose expression and function are regulated by growth factors present at a site of injury. As mentioned above, migration through tissue is also a major component of leukocyte-mediated host defense. Integrin mediated migration in leukocytes is discussed below.

Certainly, the related functions of adhesion and migration are the best understood and most prominent roles for integrins. However, integrins also are required for the production and remodeling of extracellular matrices. Specifically, both $\alpha_5\beta_1$ and $\alpha_v\beta_3$ can bind and initiate polymerization of fibronectin into an insoluble matrix which can then support cell adhesion and differentiation (Wu et al., 1995, 1996). It has recently been appreciated that integrin ligation also initiates numerous signaling cascades with a multitude of effects on cell behavior. Integrin-mediated adhesion induces activation of many signaling enzymes including serine/threonine and tyrosine kinases (FAK, Src, Fgr, Csk, Syk, protein kinase C (PKC), MAP kinase, ILK) and induces the assembly at adhesion sites of SH2–SH3 containing signaling molecules (Crk, Grb2, PI-3K, IRS-1, SHC, PLC), and small GTPases (Ras, Rho, mSOS1, C3G, RasGAP). Generation of phospholipid mediators (PIP-5K, PI-3K, cPLA2, arachidonic acid, 5-lipoxygenase) also results from integrin ligation. Interested readers should see Dedhar and Hannigan (1996) for a review of integrin signaling and downstream effects of activation of these signaling pathways. Activation of these signaling pathways presumably mediates most integrin ligation-dependent effects on cells. As integrins exhibit no intrinsic enzymatic activity, initiation of signaling cascades likely results from mobilization of adaptor and signaling molecules into a functionally stable complex through tethering either to ligand-clustered integrins or to cytoskeletal or membrane elements which redistribute upon receptor ligation.

III. LEUKOCYTE INTEGRINS

A. β_1 (VLA) Integrin Family

The β_1 or very late antigen (VLA) family of integrins consists of heterodimers in which 10 distinct α-chains ($\alpha_1..\alpha_9$, α_v) can combine with the β_1- (CD29) chain (Table 2) (Hemler et al., 1987; Hemler, 1990b; Shimizu and Shaw, 1991). In the immune system, the β_1 family was first identified immunologically on lympho-

Table 2. The VLA Family of Integrins

Receptor Name	*Synonyms*	*Ligands*	*Leukocyte Distribution*
$\alpha_1\beta_1$	VLA-1	Collagen (I, IV), Ln	Monocytes
$\alpha_2\beta_1$	VLA-2, GPIa-IIa	Collagen (I, IV), Ln	Monocytes, macrophages
$\alpha_3\beta_1$	VLA-3	Collagen I, Ln, Ėn, Ep, Fn	None
$\alpha_4\beta_1$	VLA-4, LPAM-2	Fn, VCAM-1	Monocytes, macrophages
$\alpha_5\beta_1$	VLA-5, FnR, GPIc-IIa	Fn	Monocytes, macrophages
$\alpha_6\beta_1$	VLA-6	Ln	Monocytes, macrophages, PMN
$\alpha_7\beta_1$	VLA-7	Ln	None
$\alpha_8\beta_1$		?	None
$\alpha_9\beta_1$		?	PMN
$\alpha_v\beta_1$		Fn, Vn	None

Note: Heterodimers of the VLA integrin family are listed with alternate names, known ligands, and leukocyte distribution.

cytes (Hemler et al., 1985) and ultimately shown to be identical to adhesion protein families identified on monocytes, myoblasts, and sarcoma cells (Hemler et al., 1987; Hemler and Jacobson, 1987; Takada et al., 1987a,b, 1988). The VLA and cluster of differentiation (CD) nomenclature, once popular for identifying these integrins, has fallen into disuse, in favor of a more descriptive designation in which each integrin is identified by its component α- and β-chains, e.g., $\alpha_5\beta_1$. Multiple β_1 integrins are expressed on monocytes, macrophages, and lymphocytes, with $\alpha_4\beta_1$ and $\alpha_5\beta_1$ among the most highly expressed (Hemler, 1990a). Polymorphonuclear leukocytes (PMN) express β_1 integrins at much lower levels, but $\alpha_3\beta_1$, $\alpha_4\beta_1$, $\alpha_5\beta_1$, and $\alpha_6\beta_1$ have been identified unequivocally on these cells by biochemical and/or functional assays (Gresham et al., 1989; Bohnsack et al.,1990). Singer et al. (1989) showed that there was a significant pool of these receptors in PMN secretory granules.

$\alpha_4\beta_1$ (VLA-4) binds both a cell ligand, VCAM-1, expressed on activated endothelial cells, bone marrow stromal cells, and developing myocytes, and an extracellular ligand, fibronectin (Elices et al., 1990; Masumoto and Hemler, 1993a). The fibronectin domain recognized by $\alpha_4\beta_1$ is distinct from that recognized by another integrin fibronectin receptor, $\alpha_5\beta_1$, suggesting that these two integrins may interact independently and possibly synergistically with fibronectin in the extracellular matrix (Elices et al., 1990; Masumoto and Hemler, 1993b). In the adult animal, $\alpha_4\beta_1$ is expressed almost exclusively on leukocytes, and its interaction with VCAM-1 is critically important in migration of monocytes and lymphocytes through endothelium at sites of inflammation (Kohn et al., 1991; Walsh et al., 1991; Hourihan et al., 1993). $\alpha_4\beta_1$ also has an essential role in B cell development in the bone marrow and in homing of lymphoid precursors to the thymus in postnatal life (Shimizu et al., 1990a; Miyake et al., 1991; Williams et al., 1991; Kincade, 1992; Arroyo et al., 1996). $\alpha_4\beta_1$ has no role in PMN migration through endothelium, perhaps because it is minimally surface-expressed on these cells until they have migrated out of the vasculature (Walsh et al., 1991; Johnston et al., 1996). However, $\alpha_4\beta_1$ can play a role in PMN interaction with fibronectin once the cells enter the extravascular extracellular matrix. The α_4 chain can be expressed on cell surfaces in either a 180 or a 150 kDa form; the latter can also appear as a cleaved configuration with 80 and 70 kDa chain fragments (Hemler et al., 1990). Different functions for these variant forms are unknown but may be important in differential activation of this integrin by cellular signals.

$\alpha_5\beta_1$ is a fibronectin receptor which is not known to recognize cell surface ligands. Ligand binding by $\alpha_5\beta_1$ is regulated differently from $\alpha_4\beta_1$-mediated interactions with fibronectin, since the tyrosine phosphatase CD45 affects adhesion through the former but not the latter fibronectin receptor. Since $\alpha_5\beta_1$, but not $\alpha_4\beta_1$ can mediate focal contact formation (Ylänne et al., 1993) and clot retraction (Tanoue et al., 1996), while $\alpha_4\beta_1$ is the superior integrin for migration, differential regulation of the function of these two receptors may have significant consequences for cell function (Chan et al., 1991a; Hourihan et al.,1993; Porter and Hogg, 1997).

The functions of other β_1 integrins on myeloid cells are less thoroughly understood. $\alpha_3\beta_1$ mediates PMN binding to the basement membrane protein entactin (Gresham et al., 1996), and $\alpha_6\beta_1$ mediates a Mac-1 independent binding of PMN to laminin (Bohnsack,1992). $\alpha_2\beta_1$ ligation by collagen leads to macrophage activation when the cells are embedded in a three-dimensional matrix (Schiro et al.,1991; Goldman et al., 1992).

B. β_2 (Leu-CAM, CD18) Integrin Family

The β_2 (leukocyte cell adhesion molecule (Leu-CAM)) integrin family contains the most highly expressed integrins on leukocytes. All four members of the Leu-CAM family, which share a common β_2 subunit, are expressed exclusively on leukocytes (Table 3). Expression of Leu-CAM integrins begins during the early promyelocytic stage of leukocyte development and continues to increase during differentiation (Hickstein et al., 1987, 1988; Back et al., 1991). On mature human monocytes and neutrophils, $\alpha_M\beta_2$ is the most highly expressed, with over 500,000 copies per cell. During monocyte differentiation to a mature macrophage, the expression of $\alpha_x\beta_2$ increases markedly, sometimes in coordination with a decrease in the expression of $\alpha_M\beta_2$ (Myones et al., 1988).

The β_2 integrins have a major role in cell-cell adhesion events with other leukocytes, endothelial cells, and the cells making up solid organs, including fibroblasts and epithelial cells. Several cell surface ligands for β_2 integrins have been identified (Table 3). Many of these are members of the immunoglobulin superfamily, as is VCAM-1, the ligand for $\alpha_4\beta_1$ (Dobrina et al., 1991; Masumoto and Hemler, 1993b). ICAM-1 is a ligand for both $\alpha_L\beta_2$ and $\alpha_M\beta_2$ (Makgoba et al., 1988; Diamond et al., 1990), whose expression on many cell types is markedly increased by inflammatory stimuli, including lipopolysacharides, interleukin 1, and interferon

Table 3. The Leu-CAM Family of Integrins

Receptor Name	*Synonyms*	*Ligands*	*Leukocyte Distribution*
$\alpha_L\beta_2$	LFA-1	ICAM-1, ICAM-2, ICAM-3	All leukocytes
$\alpha_M\beta_2$	Mac-1, Mo-1	iC3b, Fg, Factor X, ICAM-1	Monocytes, macrophages, PMN, NK
$\alpha_x\beta_2$	p150, 95	iC3b, Fg	Monocytes, macrophages, PMN
$\alpha_d\beta_2$		ICAM-3	Macrophage subpopulations

Note: Heterodimers of the LeuCAM family of integrins are listed with alternate names, known ligands, and leukocyte distribution.

(Dustin et al., 1986; Fawcett et al., 1992). Interaction between leukocyte $\alpha_L\beta_2$ and newly expressed ICAM-1 is thought to be a major mechanism whereby leukocytes, especially PMN, adhere to endothelium overlying sites of inflammation (Gahmberg et al., 1990; Randi and Hogg, 1994).

$\alpha_M\beta_2$ and $\alpha_x\beta_2$ integrins also can recognize several ligands in the extracellular matrix or at sites of inflammation, including fibrin(ogen), coagulation Factor X, and complement component iC3b (Altieri and Edgington, 1988; Altieri et al.,1988; Wright et al., 1983b, 1988). $\alpha_M\beta_2$ also has been implicated in nonopsonic recognition of pathogens such as Histoplasma, Leishmania, and gram-negative bacteria (Wyler et al., 1985; Russell and Wright, 1988; Talamas Rohana et al., 1990; Drevets et al., 1992). One puzzling aspect of $\alpha_M\beta_2$ is its apparent recognition of multiple diverse particulate ligands. This has led us and others to suggest that a major role for $\alpha_M\beta_2$ is as a fundamental mediator for connection of multiple low affinity receptors to cytoskeleton, rather than as a receptor for each member of this very diverse set of particulate or adherent ligands (Graham et al., 1989; Brown, 1991; Gresham et al., 1991; Mosser et al., 1992).

The fourth β_2 integrin, $\alpha_D\beta_2$, is the most recently discovered, and little information concerning its function is yet available (Van der Vieren et al., 1995; Wong et al., 1996). Its known ligand is ICAM-3,a leukocyte-restricted member of the immunoglobulin superfamily (Van der Vieren et al., 1995; de Fougerolles et al., 1994). This suggests that the physiologic role for $\alpha_D\beta_2$ may be in regulated cognate interaction among cells of the immune system (El-Gabalawy et al., 1996).

The importance of β_2 integrins to host defense is well documented and best illustrated by a genetic disease termed leukocyte adhesion deficiency type I (LAD-1) (Springer et al., 1984; Todd and Freyer, 1988). This disease has been reported in humans, dogs, and cattle (Kehrli et al., 1990, 1992; Yoder et al., 1990; Mazzone and Ricevuti, 1995). Individuals of each species afflicted with LAD-1 have spontaneously arising genetically determined defects in synthesis of the common β_2 subunit, resulting in varying levels of expression and function of all four β_2 integrins (Kishimoto et al., 1987a, 1989; Arnaout et al., 1990; Sligh et al., 1992). These patients generally come to medical attention due to repeated and severe bacterial infections, persistent leukocytosis, and commonly, a delayed separation of the umbilical cord after birth (Etzioni et al., 1992; Mazzone and Ricevuti, 1995). Sites of infection in LAD patients lack neutrophil accumulation, while monocyte migration is normal or only slightly decreased, presumably because monocyte $\alpha_4\beta_1$ initiates an alternative mechanism for transendothelial migration (Carlos et al., 1991; Vonandrian et al., 1993; Lipnick et al., 1996). The severity of disease in LAD patients correlates with the level of β_2 integrin expression. The most severely affected patients have little or no Leu-CAM expression and often do not survive childhood without bone marrow transplantation, while heterozygotes who express 50% of normal β_2 integrin levels are clinically normal.

A fundamental difference between $\alpha_M\beta_2$ and $\alpha_L\beta_2$ has emerged from studies of mice deficient in one or all of the β_2 integrins and from *in vivo* use of blocking antibodies directed at single β_2 integrins (Vedder et al., 1988; Tuomanen et al., 1989; Morisaki et al., 1991; Von Andrian et al., 1991; Mizgerd et al., 1997). While $\alpha_L\beta_2$ seems to play a more important role than $\alpha_M\beta_2$ in transendothelial migration of PMN, $\alpha_M\beta_2$ appears to be extremely important in PMN activation at sites of inflammation (Furie et al., 1991; Parkos et al., 1991; Simpson et al., 1992; Lu et al., 1997; Tang et al., 1997). Thus animals deficient in $\alpha_M\beta_2$ function have markedly impaired PMN-mediated inflammatory responses.

C. β_3 (Cytoadhesin) Integrin Family

Integrins of the β_3 (Cytoadhesin) family share a common β_3 subunit and exhibit highly promiscuous ligand binding based on the recognition of a tripeptide sequence Arg-Gly-Asp (RGD) present in many extracellular matrix and inflammatory proteins (Table 4) (Cheresh and Spiro, 1987; Smith and Cheresh, 1990; Smith et al., 1990a; Kieffer et al., 1991). Only two α chains have been demonstrated to associate with β_3. $\alpha_{IIb}\beta_3$ is expressed only on platelets and some carcinomas (Bray et al., 1987; Chen et al., 1992), while the closely associated integrin, $\alpha_v\beta_3$, is more widely expressed (Varner et al.,1995). Circulating human monocytes express very little β_3 integrin on their plasma membranes but expression increases markedly during the process of differentiation into macrophages (Savill et al., 1990). $\alpha_v\beta_3$ expression on macrophages is markedly enhanced by granulocyte macrophage colony-stimulating factor (GM-CSF), while expression of the closely related $\alpha_v\beta_5$ integrin appears to be stimulated by exposure to M-CSF (De Nichilo and Burns, 1993). A subpopulation of lymphocytes, including activated cytotoxic cells, also expresses $\alpha_v\beta_3$ (Maxfield et al., 1989; Shevach, 1992). Lymphocyte and monocyte $\alpha_v\beta_3$ may be involved in transendothelial migration, perhaps through recognition of the endothelial Ig superfamily member platelet-endothelial cell adhesion molecule (PECAM-1) (Buckley et al., 1996; Weerasinghe et al., 1998).

PMN also express $\alpha_v\beta_3$, but, like β_1 integrins, this integrin is expressed at very low levels on these cells (Brown and Goodwin, 1988; Gresham et al., 1989; Gresham et al., 1992). Nonetheless, this $\alpha_v\beta_3$ is important during migration of PMN

Table 4. The Cytoadhesin Family of Integrins

Receptor Name	*Synonyms*	*Ligands*	*Leukocyte Distribution*
$\alpha_{IIb}\beta_3$	GPIIb-IIIa	Fg, Fn, vWF, Vn	None
$\alpha_v\beta_3$	VnR	Fg, Fn, vWF, Vn, TSP, Osp, Bsp	Macrophages, PMN?, lymphocytes
$\alpha_?\beta_3$	Leukocyte response integrin (LRI)	Fn, en, Vn, Fg	PMN

Note: Heterodimers of the Cytoadhesin integrin family are listed with alternate names, known ligands, and leukocyte distribution.

on vitronectin-coated surfaces (Hendey et al., 1996). Adhesion of PMN to Arg-Gly-Asp containing proteins which are $\alpha_v\beta_3$ ligands also modulates their phagocytic and respiratory burst potential (Gresham et al., 1989; Gresham and Brown, 1990; Senior et al., 1992). Some monoclonal antibodies reactive with either α_v or β_3 can inhibit PMN responses to Arg-Gly-Asp containing matrix proteins. The integrin which mediates Arg-Gly-Asp-dependent PMN activation has a binding specificity distinct from $\alpha_v\beta_3$ or $\alpha_{IIb}\beta_3$ on other cells since this β_3 integrin will bind the peptide sequence Lys-Gly-Ala-Gly-Asp-Val (KGAGDV) (Gresham et al., 1992). Recently we have determined that expression of cDNAs encoding $\alpha_v\beta_3$ in a cell of hematopoietic origin (K562) confers on these cells the ability to recognize KGAGDV, while the same cDNAs expressed in a fibroblast yield an $\alpha_v\beta_3$ integrin with "classical" binding properties that does not recognize this peptide sequence (Blystone and Brown, unpublished observations). Thus, the most likely explanation for the difference in binding specificity between $\alpha_v\beta_3$ integrins on PMN and other cells is that ligand binding by this integrin is dependent upon the cell in which it is expressed. This has also been shown to be true for $\alpha_2\beta_1$ which binds laminin only when expressed in certain cell types (Elices and Hemler, 1989; Chan et al., 1991; Mecham, 1991; Delwel et al., 1993; Kawaguchi and Hemler, 1993). However it is also possible that a unique, as yet uncharacterized α-chain associates with PMN β_3.

D. Additional Leukocyte Integrins

Several integrins in addition to those decribed above are expressed on leukocytes (Table 5). One such integrin present on mature macrophages is $\alpha_v\beta_5$. This receptor, which is more ubiquitously expressed than $\alpha_v\beta_3$ (Pasqualini et al., 1993), also recognizes Arg-Gly-Asp peptide and vitronectin but has lower affinity for fibronectin or fibrinogen (Smith et al., 1990; Vogel et al., 1993). Data in fibroblast cell systems suggest that vitronectin ligation of $\alpha_v\beta_3$ and $\alpha_v\beta_5$ has different consequences for cell adhesion and cytoskeletal conformation (Wayner et al., 1991;

Table 5. Additional Integrin Family Members

Receptor Name	*Synonyms*	*Ligands*	*Leukocyte Distribution*
$\alpha_E\beta_7$	M290 IEL, HML-1	E-cadherin	Lymphocytes
$\alpha_4\beta_7$	LPAM-1	Fn(CS-1), VCAM-1, MadCAM	Lymphocytes
$\alpha_6\beta_4$	TSP-180, A9, Ea-1	Ln	None
$\alpha_v\beta_5$		Vn	Macrophages, lymphocytes?
$\alpha_v\beta_6$		Fn, Vn	None
$\alpha_v\beta_8$		?	None

Note: Heterodimers not of the VLA, Leu-CAM, or Cytoadhesin families are listed with alternate names, known ligands, and leukocyte distribution.

Delannet et al., 1994; Klemke et al., 1994). M-CSF and GM-CSF have distinct effects on the expression of these two α_v receptors since the former increases $\alpha_v\beta_5$ expression and the latter $\alpha_v\beta_3$ (De Nichilo and Burns, 1993). However, many macrophages express both these receptors. The significance of the co-expression of these two closely related receptors on mature macrophages and whether ligation of these receptors have differing effects on macrophage phenotype are unknown. Currently, phagocytes are thought not to express the β_4, β_6, β_7 or β_8 integrin chains.

IV. LEUKOCYTE INTEGRIN FUNCTION

A. Integrin Ligand Recognition

A unique feature of leukocyte integrins is that, despite expression at the membrane, they exhibit little ligand affinity on circulating cells (Lotz et al., 1989; Tozer et al., 1996). Teleologically, this is a necessary property for leukocyte adhesion molecules; otherwise, some leukocyte integrins, such as $\alpha_4\beta_1$ and $\alpha_5\beta_1$, would be continually occupied by ligands such as fibronectin and vitronectin which circulate with cells in the blood. Other integrins might mediate adhesion and transendothelial migration at inappropriate sites in the body. As integrins represent a signaling system by which phagocytes move out of the bloodstream and into sites of inflammation, receptor occupancy in the circulation would result in deleterious unwarranted leukocyte activation and, potentially, uncontrolled inflammation. Thus, integrin function is suppressed in circulating leukocytes, in contrast to the constitutively active integrin phenotype exhibited by cells residing within solid tissues (Kucik et al., 1996).

Activation of leukocytes via any of a variety of molecules found at sites of inflammation leads to enhancement of integrin-mediated adhesion. ADP, LTB4, fMLP, C5a, and immune complexes are among the inflammatory mediators shown to increase integrin avidity in leukocytes (Michl et al., 1979; Marder et al., 1985; Altieri and Edgington, 1988; Goldman et al., 1991; Gresham et al., 1991; Monk and Banks, 1991; Graham et al., 1993). Two signaling pathways for activation have been described, although neither is understood in great detail. Integrin activation can occur either through G-protein coupled seven transmembrane receptors or through a tyrosine kinase mediated mechanism which requires phosphoinositol 3-kinase (Garratt and Humphries, 1995; Zhang et al., 1996a,b; Hughes et al., 1997). Both pathways seem to utilize PKC, and PMA, which is a direct activator of protein kinase C, can enhance integrin-mediated adhesion in leukocytes independent of tyrosine kinases, phosphoinositol 3-kinase, or receptor-coupled G-proteins (Rothlein and Springer, 1986; Wright and Meyer, 1986; Kolanus and Seed, 1997). An important characteristic of the integrin activation induced by physiologic stimuli is that it is reversible. The cycling of integrins through low and high affinity states may have an important role in motility and phagcytosis (Hendey and Maxfield, 1993; Kucik et al., 1996; Mandeville and Maxfield, 1997).

What are the molecular mechanisms by which cell activation promotes enhanced integrin-mediated adhesion? Currently, there are four major non-mutually exclusive hypotheses: affinity modulation, integrin expression levels, integrin diffusion in the plasma membrane, and direct integrin interaction with cytoplasmic components. For some integrins, this latter mechanism may depend upon tyrosine phosphorylation of the integrin β-chain. The relative importance of each of these mechanisms may vary depending upon the integrin and the cell in which it is expressed.

Affinity modulation as a result of cell activation occurs when the affinity of an individual, isolated integrin for its ligand is increased, presumably as the result of a conformational change in the integrin. In general, individual unactivated integrins exhibit low affinity for ligand, estimated at 1–10 x 10^6 M. Thus the binding of monovalent ligand to integrins on circulating leukocytes is minimal. When leukocytes exit the vasculature, they can encounter multivalent insoluble ligands in the extracellular matrix, where the net effect of multiple low affinity interactions would be a greater total cell avidity even without integrin activation. Affinity for monovalent ligand can be increased by cell activation. This is clearly proven for the platelet integrin $\alpha_{IIb}\beta_3$ and the myeloid integrin $\alpha_M\beta_2$ (Plow and Ginsberg, 1989; O'Toole et al., 1990; Du et al., 1991; Garratt and Humphries, 1995; Jones et al., 1998). Affinity of activated $\alpha_M\beta_2$ for its complement ligand iC3b has been measured to be as high as 3.0 nM (Cai et al.,1995). A mechanism for alteration in affinity of $\alpha_M\beta_2$ has been proposed which involves conformational change of the I-domain and potentially in the interaction between the I-domain and the β propeller structure created by the seven tandem repeats in the α chain (Lee et al., 1995; Oxvig and Springer, 1998). The significance of affinity modulation for other integrins, including other I-domain containing integrins, is controversial.

A second method used by leukocytes to regulate integrin avidity for ligand is rapid increase of the number of receptors expressed at the plasma membrane. For example, upon PMN activation by a variety of inflammatory stimuli, the expression of $\alpha_M\beta_2$ at the plasma membrane is enhanced 10-fold due to release from intracellular storage granules (Berger et al., 1984, 1991; Bainton et al., 1987; Jones et al., 1990; Miller et al., 1987). This process results from regulated secretion, is dependent on increases in $[Ca^{2+}]_i$, and is activated by PKC, but does not require mRNA transcription or new protein synthesis (Berger et al., 1985; O'Shea et al., 1985; Bainton et al., 1987; Miller et al., 1987). Several approaches have demonstrated that the rapid increase in PMN $\alpha_M\beta_2$-mediated adhesion which accompanies cell activation is essentially independent of this upregulation of receptor number (Diamond and Springer, 1994; Jones et al., 1998). Moreover, there is almost no intracellular pool of $\alpha_L\beta_2$ which undergoes equally dramatic upregulation of function after cell activation. It may be that increase in receptor number is an effect on the ability of the cell to sustain integrin-mediated adhesion, rather than on the initial activation of adhesion.

A fundamental question concerning adhesion regulation is whether the default state of a leukocyte integrin is adhesive or nonadhesive. Often it is assumed that the default state is nonadherent. However, there is some experimental evidence that this is not the case. In resting lymphocytes, integrin diffusion in the plasma membrane is restricted by cytoskeletal barriers. Phorbol ester treatment of these cells increases the rate of integrin diffusion in the cell membrane (Kucik et al., 1996). The ability to mediate adhesion has also been correlated with increased diffusion for $\alpha_4\beta_1$ and $\alpha_L\beta_2$ in several experimental systems (Kucik et al., 1996; Edidin et al., 1994; Yauch et al., 1997). There are several implications of these results. First, these data suggest that there is a diffusion-limited step required for cell adhesion. The theoretical basis for this is sound. Second, these data suggest that resting lymphocytes actively restrict integrin diffusion to limit adhesion and that one effect of cell activation is to release these cellular barriers to diffusion. These barriers have been shown in several different experimental systems to require the cortical cytoskeleton, although the molecular mechanism is unclear (Gaidano et al., 1990; Coates et al., 1991; Crawford and Eggleton, 1991; Zhou et al., 1991; Gregorio et al., 1992; Ross et al., 1992). One experimentally testable implication of the hypothesis is that cytochalasin, by interfering with microfilament assembly, should release integrins to diffuse more freely and activate adhesion. This prediction has been verified in three different experimental systems (Chan et al., 1991b; Edidin et al., 1994; Peter and O'Toole, 1995; Kucik et al., 1996; Yauch et al., 1997). This observation initially appears to contradict the fundamental role of integrins to associate cell adhesion with cytoskeleton, which predicts that cell activation would enhance integrin association with cytoskeleton. However, an effect of cell activation which allows transient integrin redistribution is compatible with a model in which integrins can associate with cytoskeleton both prior to activation and during the process of firm adhesion (Kucik et al., 1996). The implication of this model is that association of unligated integrin with cytoskeleton in a resting cell is quite distinct from the much better understood association during stable adhesion. The molecular differences between these modes of integrin-cytoskeleton interaction are unknown, although Pavalko has suggested that α-actinin association induces tight interaction between integrins and cytoskeleton, as would occur with firm adhesion, but not in the resting lymphocyte with restricted integrin diffusion (Pavalko et al., 1991; Otey et al., 1993; Pavalko and Laroche, 1993).

Recent data suggest that phosphorylation of the actin-bundling protein l-plastin may be a key event in release of integrins from cytoskeletal contraint (Jones et al., 1998). Protease activity, specifically calpain activated by the increase in $[Ca^{2+}]_i$ which occurs on ligation of most activating receptors, may also play an important role in the release of unactivated integrin from cytoskeletal constraint (Du et al., 1995; Huttenlocher et al., 1997; Schoenwaelder et al., 1997; Yuan et al., 1997; Stewart et al., 1998). However, calpain cleavage of integrin-cytoskeleton connections cannot explain all integrin activation, since phorbol esters, which are potent stimuli for integrin-mediated adhesion, do not increase intracytoplasmic $[Ca^{2+}]_i$ (Rosales and Brown, 1992).

Posttranslational modification of integrin cytoplasmic tails also has been suggested as a mechanism for activation of adhesion. Serine and/or threonine phosphorylation of the α_6 and of the β_2 chain has been correlated with increased adhesion, and tyrosine phosphorylation of β_1 in Rous-Sarcoma Virus (RSV)-transformed fibroblasts has been correlated with loss of adhesion (Chatila et al., 1989; Buyon et al., 1990; Horvath et al., 1990; Shaw et al., 1990; Aneskievich et al.,1991; Valmu et al., 1991). However, for these integrins, it has been difficult to show a direct role for phosphorylation in regulation of adhesion. The best and most detailed study of the role of phosphorylation in modulation of integrin avidity has been $\alpha_v\beta_3$ expressed in the myelogenous leukemia cell line K562 (Blystone et al., 1996, 1997). In this cell as in macrophages, firm adhesion via $\alpha_v\beta_3$ requires cell activation. Firm adhesion is accompanied by tyrosine phosphorylation of the β_3 cytoplasmic tail. The tyrosine at position 747 of the β_3 cytoplasmic tail is required both for stimulated adhesion and for tyrosine phosphorylation. Vitronectin binding to $\alpha_v\beta_3$ induces phosphorylation of Tyr747 even in the absence of cell activation. Thus phosphorylation of Tyr747precedes stimulated adhesion and inhibition of phosphorylation prevents adhesion, suggesting that this biochemical modification is required for activation of leukocyte $\alpha_v\beta_3$. Neither tyrosine phosphorylation nor cell activation are required for the $\alpha_v\beta_3$ conformational changes detected by antibodies to neoepitopes in the integrin which accompany high affinity ligand binding. Interestingly, the requirement for tyrosine phosphorylation of β_3 for firm adhesion appears restricted to leukocytes and, therefore, Tyr747 is likely involved in the process by which cell activation induces firm adhesion rather than in the adhesion event itself (Blystone et al., 1997). The β_3 motif containing Tyr747 is conserved in β_1, β_5, β_6, and β_7, and several of these integrin chains have been shown to be tyrosine phosphorylated. Thus tyrosine phosphorylation of the β-subunit maybe a general feature of leukocyte integrin activation. Presumably, the phosphotyrosine in the β-chain cytoplasmic tail represents a docking site for additional cytoplasmic molecules required for stable adhesion. While Shc, Grb2, and myosin have all been shown to bind to a phosphorylated β_3 cytoplasmic tail peptide, no relevant *in vivo* phosphorylation-dependent associations have been shown (Law et al., 1996). In contrast, the β_2 chain has phenylalanine at the position analagous to β_3 Tyr747. Thus, there are likely to be mechanisms for activation of β_2 integrins which are distinct from those requiring tyrosine phosphorylation. One of these may be the physical association between the β_2 cytoplasmic tail and cytohesin 1 or 2 which has been shown to increase adhesion mediated by β_2, but not β_1 integrins (Kolanus et al., 1996). Cytohesins are guanine nucleotide exchange factors for low molecular weight G-proteins of the arf family and thus may provide a link between integrins and rac- or rho-mediated regulation of cytoskeletal assembly and cell morphology

(Colombo et al., 1995; Kolanus et al., 1996; Meacci et al., 1997; Van Aelst and D'Souza-Schorey, 1997).

Finally, membrane lipid changes which occur upon cell activation may affect integrin avidity. Lysophosphatidic acid activates β_3 integrins (Smyth et al., 1992). While this may occur through a G-protein-mediated mechanism, direct interaction of lysophosphatidic acid with $\alpha_{IIb}\beta_3$ seems to be sufficient for its activation *in vitro* (Smyth et al., 1992; Smyth and Parise, 1993). Similarly, a lipid generated by phorbol ester treatment of PMN can activate $\alpha_M\beta_2$ directly (Hermanowski-Vosatka et al., 1992), and arachidonic acid is required for integrin-mediated adhesion and spreading in macrophages (Lefkowith et al., 1991). Moreover, lipid environment has been shown to affect $\alpha_v\beta_3$-mediated adhesion (Conforti et al., 1995). Presumably these changes in lipid environment affect adhesion by modifying integrin conformation and affecting affinity.

Thus, there are multiple mechanisms which may influence integrin mediated adhesion in leukocytes. An interaction with cytoskeleton which restricts diffusion appears to be necessary for maintenance of the nonadherent state in unactivated cells. Cell activation is accompanied by integrin diffusion, changes in clustering, affinity, phosphorylation, association with intracytoplasmic effectors and lipid environment, all of which may contribute to the more highly adherent state of integrins in activated phagocytes.

B. Integrin Adhesion

Adhesion to Endothelium

A complete discussion of leukocyte interaction with the endothelium is beyond the scope of this chapter, and reviews of this topic are available elsewhere (Butcher, 1992; Springer, 1994). However, a brief review of the role of integrins in leukocyte-endothelium interaction is included to illustrate this critical aspect of leukocyte integrin physiology. The site of extravasation of leukocytes during inflammation is the postcapillary venule (Kansas et al., 1993). Normally, the endothelial surface is nonadhesive for leukocytes. At sites of inflammation and in lymph nodes, endothelia express a variety of molecules which interact with integrins and other leukocyte adhesion molecules (Cotran et al., 1986; Pober et al., 1986; Bevilacqua et al., 1987; Jalkanen et al., 1988; Ley et al., 1991; Spertini et al., 1991; Kansas et al., 1993; Mayadas et al., 1993). When these molecules are expressed on endothelium, leukocyte movement through the venule initially is slowed by interactions which occur with rapid on- and off-rates (Lawrence and Springer, 1991; Von Andrian et al., 1991). The result of repetitive interactions of this sort in the presence of the shear stress imparted by blood flow is a characteristic rolling of leukocytes along the venular wall. In most cases, initial leukocyte-endothelium interactions are mediated by carbohydrates and selectins rather than integrins (Kuijpers et al., 1991; McEver, 1991; Polley et al., 1991; Springer and

Lasky, 1991). However, both $\alpha_4\beta_1$ and $\alpha_4\beta_7$ are capable of mediating rolling interactions (Erle et al., 1994; Johnston et al., 1996).

Following these initial low affinity but rapid interactions, integrin activation occurs. Activation most often depends on endothelial expression of chemokines, which are chemoattractant cytokines. Chemokines are ligands for seven-transmembrane receptors expressed on leukocytes. Chemokine interaction with their receptors is likely to be a major mechanism for generation of specificity in cell emigration through venular endothelium. Different combinations of chemokine receptors are expressed on each of the major leukocyte types, and chemokines selectively chemoattractant for PMN, monocytes, lymphocytes, and eosinophils have already been defined. For example, genetic deletion of CCR2, a major receptor for MCP-1, leads to a failure of monocyte migration through inflammatory endothelium (Carr et al., 1996; Weber et al., 1996).

Actual migration of phagocytes through interendothelial junctions is the least understood step of the transmigration process. Indeed, it has been suggested recently that PMN may migrate directly through endothelial cells in some cases, rather than though interendothelial junctions. Antibodies to two molecules present at endothelial junctions, PECAM and Integrin-Associated Protein (IAP), block specifically at this step (Cooper et al., 1995; Piali et al., 1995). In addition, it is clear that β_2 integrins are involved in the transmigration process and it is likely that $\alpha_L\beta_2$ is the most important family member in transendothelial migration (Doerschuk et al., 1990; Furie et al., 1991; Kavanaugh et al., 1991). However, the β_2 ligand in this step is not yet defined.

Adhesion to Extracellular Matrix

Recognition of extracellular matrix by phagocyte integrins is a major signal that the cell is out of the vasculature and in an environment in which its proinflammatory and host defense effector mechanisms are required. Ligation of phagocyte integrins by matrix proteins can in some circumstances directly activate effector functions such as phagocytosis and respiratory burst (Wright et al., 1983a; Gresham et al., 1988, 1989; Blystone et al., 1994), while in other cases, integrin ligation primes cell responses to other effectors such as tumor necrosis factor-α (TNF-α) (Nathan et al., 1989; Nathan and Sanchez, 1990). Multiple integrins are involved in recognition of extracellular matrix, and several of them can participate in this activation. Ligation of the β_2 integrin $\alpha_M\beta_2$, for example, is critical for PMN responses to TNF-α (Nathan et al., 1989). In the absence of $\alpha_M\beta_2$ ligation, TNF-α cannot induce a PMN respiratory burst or activate tyrosine phosphorylation. Ligation of $\alpha_L\beta_2$ and $\alpha_x\beta_2$ induces assembly of the NADPH oxidase in PMN (Berton et al., 1992). Ligation of the $\alpha_2\beta_1$ collagen receptor causes enhanced macrophage phagocytosis (Newman and Tucci, 1990). Despite their low abundance, β_3 integrins are critical for activation responses to extracellular matrix. Ligation of β_3 integrins on PMN leads to enhanced IgG mediated phagcytosis, cell adhesion via $\alpha_M\beta_2$, migration, and assembly of the NADPH oxidase (Gresham et al., 1989; Gresham and Brown, 1990; Marks et al., 1991; Senior et al., 1992).

In signaling phagocyte activation, the β_3 integrins act in association with a second plasma membrane molecule as a signaling complex. This second molecule, called CD47 or integrin-associated protein (IAP) is an ubiquitously expressed immunoglobulin superfamily member (Rosales et al., 1992; Lindberg et al., 1993, 1994; Reinhold et al., 1995). IAP physically associates with $\alpha_v\beta_3$ through its Ig domain, extending the paradigm of interactions between integrins and Ig superfamily members from interactions leading to cell-cell adhesion to lateral associations between molecules on a single plasma membrane (Gao et al., 1996a). Recent analysis has suggested that any mechanism of anchoring IAP to the plasma membrane is sufficent to induce its interaction with $\alpha_v\beta_3$ and that for signaling leukocyte activation, the β_3 cytoplasmic tail is both necessary and sufficient (Blystone et al., 1995; Lindberg et al., 1996). Thus, a model has emerged in which the role for IAP in signal transduction by the IAP/β_3 complex is to influence the avidity of the β_3 integrin for its ligands rather than to particpitate directly in signal transduction. However, this model may be too simplistic, since IAP can itself bind thrombospondin, and IAP interaction with this ligand can drastically alter the function of $\alpha_v\beta_3$ (Gao et al., 1996a,b). The mechanisms by which the β_3 cytoplasmic tail signals leukocyte activation is incompletely understood. However, it seems to be independent of the well-described tyrosine kinase cascades which can be activated by integrin-mediated adhesion, but it depends instead on serine/threonine kinases and perhaps trimeric G-proteins (Blystone et al., 1995).

C. Migration

With our currently incomplete understanding of integrins in leukocyte adhesion, it is difficult to provide definitive information on the role of leukocyte integrins in migration. Combining studies from numerous investigators working in varied systems, it is evident that most, if not all, leukocyte integrins are capable of providing the traction required for migration upon their respective ligands. As mentioned earlier, $\alpha_4\beta_1$ blockade prevents leukocyte emigration from the vasculature (Chuluyan and Issekutz, 1993; Chuluyan et al., 1995; Issekutz et al., 1995). Whether VLA-4 is mediating an adhesive or migration event, or both, is unknown (Hakkert et al., 1991). Additionally, antibody blockade of $\alpha_L\beta_2$ can also block transendothelial migration of leukocytes *in vitro* (Furie et al., 1991), as can antibody blockade of CD31 (an $\alpha_v\beta_3$ ligand) or CD47 (Cooper et al., 1995; Piali et al., 1995). In a transfection system of integrins expressed on a hematopoietic cell background, we can demonstrate significant migratory potential by $\alpha_v\beta_3$, $\alpha_v\beta_5$, $\alpha_5\beta_1$, $\alpha_4\beta_1$, and $\alpha_M\beta_2$ on ligands specific for each receptor (Blystone and Brown, unpublished observations).

The mechanism of migration of PMN on extracellular matrix has been studied in detail by Maxfield's group. They showed that PMN migration on the extracellular matrix proteins fibronectin and vitronectin requires transient increases in cytosolic calcium concentrations for uropod release (Hendey and Maxfield, 1993). If cytosolic calcium changes are prevented by experimental buffering, stimulated leukocytes adherent to these surfaces fail to migrate because of failure to release from adhesive contacts (Mandeville and Maxfield, 1997). In PMN adherent to vitronectin, the effector mechanism activated by the increase in cytosolic calcium involves the serine/threonine phosphatase calcineurin which is required for the release of integrin attachments (Hendey et al., 1992). However, this seems not to be the mechanism for the involvement of increased $[Ca^{2+}]_i$ in PMN migration on fibronectin. Whether calcineurin activation leads to alterations in integrin affinity, alterations in integrin association with cytoskeleton, or another mechanism for regulation of adhesion is unknown. The integrin released from its matrix ligand moves through the cell from the cell rear (uropod) to the cell front (lamellipod) where it is re-expressed on the cell surface, and reactivated to begin the adhesive cycles again (Kucik et al., 1991). Buffering $[Ca^{2+}]_i$ seems to prevent this integrin recycling as well. The effect of this integrin trafficking is release of integrin contacts at the cell rear and establishment of contacts at the leading edge, with a net result of forward movement of the cell (Kucik et al., 1989; Schmidt et al., 1993).

D. Phagocytosis

Phagocytosis, as an essential component of host defense, wound healing, and connective tissue repair, is a major function of leukocytes. In retrospect, it is clear that a role for integrin receptors in leukocyte phagocytosis was demonstrated prior to their biochemical characterization. More than 30 years ago, Saba and Diluzio described a nonspecific opsonin of plasma which enhanced leukocyte phagocytosis of particulate ligands (Saba and Di Luzio, 1965). This opsonin, then termed $\alpha2$ opsonic glycoprotein, was later found to be antigenically related to the LETS protein (large external transformation sensitive) of tumor fibroblasts and both were renamed fibronectins, the prototypical Arg-Gly-Asp containing matrix ligand for integrin receptors. At least six fibronectin binding integrins can be expressed by mature leukocytes (Brown and Goodwin, 1988; Blystone and Kaplan, 1993). The most likely candidates for mediating phagocytosis of fibronectin opsonized particles are $\alpha_5\beta_1$ and $\alpha_v\beta_3$ (Blystone and Kaplan, 1993; Blystone et al., 1994). As nonspecific serum opsonins, both vitronectin and fibronectin can act as molecular bridges between tissue debris or pathogens and leukocytes (Blystone et al., 1994). Several pathogenic bacteria, including *Staphylococcus aureus*, *Streptococcus pyogenes*, and *Mycobacteria* all can infect cells using fibronectin as an opsonin (Stanislawski et al., 1985; McGavin et al., 1991; Schlesinger and Horwitz, 1991; Kuroda et al., 1993). Internalization via integrins often does not lead to a PMN respiratory burst or metabolism of arachidonic acid (Berton et al., 1992; Gudewicz et al., 1982). Thus pathogens may take advantage of an internalization pathway deficient in these bactericidal functions to gain host entry via integrin-mediated phagocytosis.

As a receptor which recognizes complement, $\alpha_M\beta_2$ is particularly important in leukocyte phagcytosis. Indeed, $\alpha_M\beta_2$ has been implicated through monoclonal antibody inhibition as a means of internalization for several pathogens, including *Histoplasma spp.*, *Bordetella spp.*, *Escherichia coli*, and *Leishmania spp.*, but these interactions have not been defined at a molecular level, and the actual receptor-ligand interaction remains unknown (Brown, 1991). It is possible, at least for monocytes, that the cells can synthesize and secrete complement component C3, the precursor to the iC3b ligand for $\alpha_M\beta_2$, and that iC3b opsonizes the target (Mosser et al., 1992). While PMN are thought not to have the capacity to synthesize this complement opsonin, serum components are stored in their secretory granules, and this may provide an alternative mechanism for opsonization of phagocytic targets via C3 (Borregaard et al., 1992a,b; Botto et al., 1992).

In circulating leukocytes, neither the β_2 nor the β_1 integrins are capable of direct ingestion. However, following activation from a variety of signals, these leukocyte receptors attain this ability. The complete details of this activation process remain undetermined, but likely are closely related to the direct activation of integrin receptors for adhesion described above. It is clear from the work of Griffin and Silverstein that the complement receptors must be mobile in the plasma membrane to mediate ingestion, which parallels the requirement for increased integrin diffusion to mediate adhesion (Griffin and Silverstein, 1974; Griffin et al., 1975; Griffin and Mullinax, 1981; Kucik et al., 1996). In addition, cell differentiation can affect integrin-mediated phagocytosis. For example, purified circulating monocytes will not ingest fibronectin opsonized targets, while purified resident peritoneal macrophages will, probably because tissue macrophages, like other tissue resident cells, have consitutively active integrins (Bianco et al., 1975; Gudewicz et al., 1980). However, peritoneal macrophages elicited by inflammatory stimuli exhibit an even greater propensity for integrin mediated phagocytosis. Thus this mechanism of internalization is likely to be prominent in inflammatory leukocytes. In our transfection model of leukocyte integrins, we have demonstrated direct phagocytosis via $\alpha_M\beta_2$, $\alpha_5\beta_1$, $\alpha_4\beta_1$, $\alpha_v\beta_3$, and $\alpha_v\beta_5$; efficient phagocytosis via these integrins requires activation of the cell and, at least in the case of $\alpha_5\beta_1$, an increase in the affinity of the integrin for ligand (Blystone et al., 1994).

V. INTEGRIN REGULATION OF PHAGOCYTE PHENOTYPE

Another critical role of integrins on leukocytes is their ability to modulate the functions of other membrane receptors. More than 20 years ago Ehlenberger et al. demonstrated that ligation of $\alpha_M\beta_2$, known then only as a receptor for complement component C3,dramatically enhanced the phagocytic potential of receptors for Ig and Fc (Fcγ receptors) on PMN and monocytes (Ehlenberger and Nussenzweig, 1977). This effect of integrin ligation on Fc receptor-mediated phagocytosis was confirmed genetically in patients with LAD whose Fc receptors bind IgG opsonized particles normally, but fail to internalize these targets efficiently in the absence of $\alpha_M\beta_2$ (Arnaout et al., 1983; Gresham et al., 1991). These studies demonstrate that cooperation between $\alpha_M\beta_2$ and Fcγ receptors is a necessary component of the phagocytic process. Ligation of $\alpha_M\beta_2$ is also required for the additional PMN Fcγ receptor functions of sustained adhesion and spreading, arachidonate metabolism, and LTB_4 synthesis (Gresham et al., 1991; Graham et al., 1993).

The molecular mechanisms of cooperation between $\alpha_M\beta_2$ and Fc receptors have been studied by several laboratories and have been reviewed recently. A physical association between $\alpha_M\beta_2$ and FcγRIII bin transfected cells has been demonstrated by coimmunoprecipitation, resonance energy transfer, and restriction of diffusion (Zhou et al.,1991, 1993). FcγRIIIb associates with the plasma membrane via a glycan phosphoinositol (gpi) anchor, and thus is not a transmembrane protein. It has been proposed that signaling through FcγRIIIb is mediated via its association with $\alpha_M\beta_2$. Furthermore, $\alpha_M\beta_2$ can physically associate with several other gpi-linked proteins, so that it is possible that interaction with this integrin represents a general mechanism for signaling through receptors which do not span the lipid bilayer. In addition, FcγRIIa, which is a transmembrane Fcγ receptor, and $\alpha_M\beta_2$ are in physical proximity and functionally cooperate in PMN and monocytes (Zhou and Brown, 1994; Graham et al., 1994). The likely mechanism of cooperation is that ligation of FcγRIIa stimulates tyrosine kinase cascades which lead to activation of $\alpha_M\beta_2$-mediated adhesion (Brown, 1993). Integrin-mediated adhesion in turn leads to signaling which enhances the function of multiple effectors of inflammation and host defense. The control of leukocyte activation, particularly phagocytic potential, may explain why $\alpha_M\beta_2$ has a role in the internalization of pathogens which may not be direct ligands for this receptor.

Integrins of the VLA family also regulate the leukocyte inflammatory phenotype. Leukocyte activation as a result of adhesion to the extracellular matrix proteins collagen and entactin is dependent upon ligation of $\alpha_2\beta_1$ and $\alpha_3\beta_1$ respectively (Dedhar et al., 1992; Goldman et al., 1992; Senior et al., 1992; Xing et al., 1992). Similarly, adhesion to laminin and the subsequent leukocyte activation requires ligation of $\alpha_6\beta_1$ (Shaw et al., 1990). Leukocyte adhesion via these VLA integrins facilitates phagocytosis via Fc receptors, suggesting that integrin-mediated adhesion is a general mechanism for enhancing IgG-mediated phagocytosis.

In PMN, ligation of $\alpha_v\beta_3$ promotes Fc receptor phagocytosis as well as respiratory burst activity (Gresham et al., 1989; Gresham and Brown, 1990; Senior et al., 1992). For this reason, this integrin has been termed the leukocyte response integrin (LRI). LRI activation of PMN following adhesion to Arg-Gly-Asp containing matrix proteins is dependent upon activation of phospholipase C and PKC via a pertussis sensitive G-protein (Brown et al., 1987; Gresham et al., 1987). The signals transducing this matrix-induced activation are also dependent upon IAP and an intact actin cytoskeleton. Thus activation of leukocytes via attachment to extracellular matrix likely requires a complex arrangement of integrin, associated signaling molecules, and elements of the cortical cytoskeleton for efficient communication to leukocyte phagocytosis receptors or activation machinery.

VI. INTEGRIN CROSSTALK

The extracellular matrix is a complex gel with multiple integrin ligands. A major question in integrin biology is how simultaneous ligation of several integrins, with different ligands and distinct but overlapping functions, can be sensed by a single cell. Our laboratory has studied this question in monocyte-derived macrophages and in a transfection model of leukocyte integrins (Blystone et al., 1994, 1995). We have examined how signals from these multiply ligated integrins are coordinated into a net effect on cell phenotype. In both macrophages and a K562 cell transfection model of macrophage integrins, $\alpha_5\beta_1$ is the preferred integrin for attachment to fibronectin coated surfaces, migration on fibronectin, and for phagocytosis of fibronectin opsonized particles. This is true despite the fact that $\alpha_v\beta_3$ is expressed at equal levels to $\alpha_5\beta_1$ on these cells and can also recognize fibronectin. We have found however, that when both integrins are ligated, the net effect is a decrease in fibronectin-mediated phagocytosis and migration. Ligation of $\alpha_v\beta_3$ does not alter $\alpha_5\beta_1$-mediated adhesion to fibronectin coated surfaces. The differing effects of $\alpha_v\beta_3$ ligation upon these three $\alpha_5\beta_1$ functions is explained by the requirement for $\alpha_5\beta_1$ to achieve its high affinity state in order to mediate migration and phagocytosis, but not to mediate adhesion. Thus simultaneous $\alpha_v\beta_3$ ligation prevents $\alpha_5\beta_1$ high affinity function. This blockade occurs subsequent to ligand binding. The $\alpha_5\beta_1$ high affinity functions of migration and phagocytosis require an elevation in the activity of the calcium and calmodulin dependent protein kinase II (CamKII). Ligation of $\alpha_v\beta_3$ on the same cell prevents the increase in CamKII activity, thus blocking high affinity $\alpha_5\beta_1$ functions (Blystone and Brown, unpublished observations). The target of CamKII activity in $\alpha_5\beta_1$ function remains undetermined, but may be myosin light chain kinase. We have termed the regulation of one integrin by ligation of another on the same cell "integrin crosstalk."

Numerous examples of integrin crosstalk have been discovered, including $\alpha_v\beta_3$ downregulation of β_1-dependent smooth muscle cell migration (Bilato et al., 1997), $\alpha_5\beta_1$ stimulation of $\alpha_2\beta_1$-mediated adhesion (Pacifici et al., 1994), $\alpha_v\beta_3$ inhibition of $\alpha_5\beta_1$-dependent migration in kidney cells (Simon et al., 1997), and $\alpha_L\beta_2$ inhibition of $\alpha_4\beta_1$-mediated adhesion in T cells (Porter and Hogg, 1997). In the case of leukocyte activation by extracellular matrix, this type of regulatory interaction allows for combinatorial modulation of responses to the necessarily complex interaction of a cell with extracellular matrix, and may explain why cells express multiple integrin receptors for the same ligand. Upon emigration from the vasculature, multiple leukocyte integrins are engaged. Some, such as $\alpha_M\beta_2$ and $\alpha_6\beta_1$ may directly participate in the activation of the leukocyte. Additional integrins may communicate via crosstalk mechanisms to ensure that cell behavior is appropriate, coordinated and reflective of the extracellular environment. $\alpha_v\beta_3$ expression is regulated during monocyte differentiation and is subject to cytokine stimulation (De Nichilo and Burns, 1993). Its expression only on cells which have exited the vasculature may prevent excessive $\alpha_5\beta_1$-mediated migration mechanisms. The downregulation of $\alpha_5\beta_1$ phagocytosis upon $\alpha_v\beta_3$ ligation may result in internalization of fibronectin opsonized pathogens via other pathways more favorable for pathogen destruction. Few integrin crosstalk mechanisms have been defined at the molecular level. A more complete definition of the roles of individual leukocyte integrins in crosstalk will improve the understanding of how simultaneous ligation of multiple integrins regulates leukocyte functions *in vivo.*

REFERENCES

Albelda, S.M., & Buck, C.A. (1990). Integrins and other cell adhesion molecules. FASEB J. 4, 2868–2880.

Altieri, D.C., & Edgington, T.S. (1988). The saturable high affinity association of factor X to ADP-stimulated monocytes defines a novel function of the Mac-1 receptor. J. Biol. Chem. 263, 7007–7015.

Altieri, D.C., Bader, R., Mannucci, P.M., & Edgington, T.S. (1988). Oligospecificity of the cellular adhesion receptor Mac-1 encompasses an inducible recognition specificity for fibrinogen. J. Cell Biol. 107, 1893–1900.

Aneskievich, B.J., Haimovich, B., & Boettiger, D. (1991). Phosphorylation of integrin in differentiating ts-Rous sarcoma virus-infected myogenic cells. Oncogene 6, 1381–1390.

Aota, S., Nagai, T., Olden, K., Akiyama, S.K., & Yamada, K.M. (1991). Fibronectin and integrins in cell adhesion and migration. Biochem. Soc. Trans. 19, 830–835.

Argraves, W.S., Suzuki, S., Arai, H., Thompson, K., Pierschbacher, M.D., & Ruoslahti, E. (1987). Amino acid sequence of the human fibronectin receptor. J. Cell Biol. 105, 1183–1190.

Arnaout, M.A., Dana, N., Gupta, S.K., Tenen, D.G., & Fathallah, D.M. (1990). Point mutations impairing cell surface expression of the common beta subunit (CD18) in a patient with leukocyte adhesion molecule (Leu-CAM) deficiency. J. Clin. Invest. 85, 977–981.

Arnaout, M.A., Todd, R.F., III, Dana, N., Melamed, J., Schlossman, S.F., & Colten, H.R. (1983). Inhibition of phagocytosis of complement C3- or immunoglobulin G- coated particles and of C3bi binding by monoclonal antibodies to a monocyte granulocyte membrane glycoprotein (Mo1). J. Clin. Invest. 72, 171–179.

Arroyo, A.G., Yang, J.T., Rayburn, H., & Hynes, R.O. (1996). Differential requirements for alpha4 integrins during fetal and adult hematopoiesis. Cell 85, 997–1008.

Back, A.L., Gollahon, K.A., & Hickstein, D.D. (1991). Regulation of expression of the leukocyte integrin CD11a (LFA-1) molecule during differentiation of HL-60 cells along the monocyte/macrophage pathway. J. Immunol. 148, 710–714.

Bainton, D.F., Miller, L.J., Kishimoto, T.K., & Springer, T.A. (1987). Leukocyte adhesion receptors are stored in peroxidase-negative granules of human neutrophils. J. Exp. Med. 166, 1641–1653.

Bajt, M.L., Goodman, T., & McGuire, S.L. (1995). β_2 (CD18) mutations abolish ligand recognition by I domain integrins LFA-1 ($\alpha_L\beta_2$, CD11a/CD18) and MAC-1 ($\alpha_M\beta_2$, CD11b/CD18). J. Biol. Chem. 270, 94–98.

Ballard, L.L., Brown, E.J., & Holers, V.M. (1991). Expression of the fibronectin receptor VLA-5 is regulated during human B cell differentiation and activation. Clin. Exp. Immunol. 84, 336–346.

Bar-Shavit, R., Sabbah, V., Lampugnani, M.G., Marchisio, P.C., Fenton, J.W., Vlodavsky, I., & Dejana, E. (1991). An Arg-Gly-Asp sequence within thrombin promotes endothelial cell adhesion. J. Cell Biol. 112, 335–344.

Bennett, J.S., Shattil, S.J., Power, J.W., & Gartner, T.K. (1988). Interaction of fibrinogen with its platelet receptor: differential effects of alpha and gamma chain fibrinogen peptides on the glycoprotein IIb-IIIa complex. J. Biol. Chem. 263, 12948–12953.

Berger, M., Birx, D.L., Wetzler, E.M., O'Shea, J.J., Brown, E.J., & Cross, A.S. (1985). Calcium requirements for optimal complement receptor expresson during neutrophil activation. J. Immunol. 135, 1342–1348.

Berger, M., O'Shea, J.J., Cross, A.S., Folks, T.M., Chused, T.M., Brown, E.J., & Frank, M.M. (1984). Human neutrophils increase expression of C3bi as well as C3b receptors upon activation. J. Clin. Invest. 74, 1566–1571.

Berger, M., Wetzler, E.M., Welter, E., Turner, J.R., & Tartakoff, A.M. (1991). Intracellular sites for storage and recycling of C3b receptors in human neutrophils. Proc. Natl. Acad. Sci. USA 88, 3019–3023.

Berton, G., Laudanna, C., Sorio, C., & Rossi, F. (1992). Generation of signals activating neutrophil functions by leukocyte integrins: LFA-1 and gp150/95, but not CR3, are able to stimulate the respiratory burst of human neutrophils. J. Cell Biol. 116, 1007–1017.

Bevilacqua, M.P., Amrani, D.L., Mosesson, M.W., & Bianco, C. (1981). Receptors for cold-insoluble globulin (plasma fibronectin) on human monocytes. J. Exp. Med. 153, 42–60.

Bevilacqua, M.P., Pober, J.S., Mendrick, D.L., Cotran, R.S., & Gimbrone, M.A., Jr. (1987). Identification of an inducible endothelial-leukocyte adhesion molecule. Proc. Natl. Acad. Sci. USA 84, 9238–9242.

Bianco, C., Griffin, F.M., Jr., & Silverstein, S.C. (1975). Studies of the macrphage complement receptor. Alteration of receptor function upon macrophage activation. J. Exp. Med. 141, 1278–1291.

Bilato, C., Curto, K., Monticone, R.E., Pauly, R.R., White, A.J., & Crow, M.T. (1997). The inhibition of vascular smooth muscle cell migration by peptide and antibody agonists of the alphaV/beta3 integrin complex is reversed by activated Calcium/Calmodulin-dependent Protein Kinase II. J. Clin. Invest. 100, 693–704.

Blystone, S.D. & Kaplan, J.E. (1992). Isolation of an amino-terminal fibronectin-binding protein on human U937 cells and rat peritoneal macrophages. J. Biol. Chem. 267, 3968–3975.

Blystone, S.D. & Kaplan, J.E. (1993). The role of fibronectin in macrophage fibrin binding: a potential mechanism for high affinity, high capacity clearance of circulating fibrin. [Review] [131 refs]. Blood Coagul. Fibrinolysis 4, 769–781.

Blystone, S.D., Graham, I.L., Lindberg, F.P., & Brown, E.J. (1994). Integrin alpha-v beta-3 differentially regulates adhesive and phagocytic functions of the fibronectin receptor alpha-5 beta-1. J. Cell Biol. 127, 1129–1137.

Blystone, S.D., Lindberg, F.P., LaFlamme, S.E., & Brown, E.J. (1995). Integrin beta3 cytoplasmic tail is necessary and sufficient for regulation of alpha5/beta1 phagocytosis by alphaV/beta3 and integrin associated protein. J. Cell Biol. 130, 745–754.

Blystone, S.D., Lindberg, F.P., Williams, M.P., McHugh, K.P., & Brown, E.J. (1996). Inducible tyrosine phosphorylation of the beta-3 integrin requires the alpha-V cytoplasmic tail. J. Biol. Chem. 271, 31458–31462.

Blystone, S.D., Weston, L.K., & Kaplan, J.E. (1991). Fibronectin dependent macrophage fibrin binding. Blood 78, 2900–2907.

Blystone, S.D., Williams, M.P., Slater, S.E., & Brown, E.J. (1997). Requirement of integrin beta3 tyrosine 747 for beta3 tyrosine phosphorylation and regulation of alphav/beta3 avidity. J. Biol. Chem. 272, 28757–28761.

Bodary, S.C., & McLean, J.W. (1990). The integrin β1 subunit associates with the vitronectin receptor αv subunit to form a novel vitronectin receptor in a human embryonic kidney cell line. J. Biol. Chem. 265, 5938–5941.

Bohnsack, J.F. (1992). CD11/CD18-independent neutrophil adherence to laminin is mediated by the integrin VLA-6. Blood 79, 1545–1552.

Bohnsack, J.F., Akiyama, S.K., Damsky, C.H., Knape, W.A., & Zimmerman, G.A. (1990). Human neutrophil adherence to laminin in vitro. Evidence for a distinct neutrophil integrin receptor for laminin. J. Exp. Med. 171, 1221–1237.

Borregaard, N., Kjeldsen, L., Lollike, K., & Sengelov, H. (1992a). Ca^{2+}-dependent translocation of cytosolic proteins to isolated granule subpopulations and plasma membrane from human neutrophils. FEBS Lett. 304, 195–197.

Borregaard, N., Kjeldsen, L., Rygaard, K., Bastholm, L., Nielsen, M.H., Sengelov, H., Bjerrum, O.W., & Johnsen, A.H. (1992b). Stimulus-dependent secretion of plasma proteins from human neutrophils. J. Clin. Invest. 90, 86–96.

Botto, M., Lissandrini, D., Sorio, C., & Walport, M.J. (1992). Biosynthesis and secretion of complement component (C3) by activated human polymorphonuclear leukocytes. J. Immunol. 149, 1348–1355.

Bray, P.F., Rosa, J.-P., Johnston, G.I., Shiu, D.T., Cook, R.G., Lau, C., Kan, Y.W., McEver, R.P., & Shuman, M.A. (1987). Platelet glycoprotein IIb: chromosomal localization and tissue expression. J. Clin. Invest. 80, 1812–1817.

Brown, D. (1993). The tyrosine kinase connection: how GPI-anchored proteins activate T cells. Curr. Opin. Immunol. 5, 349–354.

Brown, E.J., & Goodwin, J.L. (1988). Fibronectin receptors of phagocytes: characterization of the Arg-Gly-Asp binding proteins of human monocytes and polymorphonuclear leukocytes. J. Exp. Med. 167, 777–793.

Brown, E.J., & Lindberg, F.P. (1993). Matrix receptors of myeloid cells. In: Blood Cell Biochemistry, Volume 5: Macrophages and Related Cells (Horton, M.A., Ed.), pp. 279–306. Plenum Press, New York.

Brown, E.J. (1986). The interaction of connective tissue proteins with phagocytic cells. J. Leuk. Biol. 39, 579–591.

Brown, E.J. (1991). Complement receptors and phagocytosis. Curr. Opin. Immunol. 3, 76–82.

Brown, E.J., Newell, A.M., & Gresham, H.D. (1987). Molecular regulation of phagocyte function: evidence for involvement of a GTP binding protein in opsonin mediated phagocytosis by monocytes. J. Immunol. 139, 3777–3782.

Buckley, C.D., Doyonnas, R., Newton, J.P., Blystone, S.D., Brown, E.J., & Watt, S.M. (1996). Identification of alphav/beta3 as a heterotypic ligand for CD31/PECAM. J. Cell Sci. 109, 437–445.

Burridge, K., Nuckolls, G., Otey, C., Pavalko, F., Simon, K., & Turner, C. (1990). Actin-membrane interaction in focal adhesions. Cell Differ. Dev. 32, 337–342.

Butcher, E.C. (1992). Leukocyte-endothelial cell adhesion as an active, multi-step process: A combinatorial mechanism for specificity and diversity in leukocyte targeting. Adv. Exp. Med. Biol. 323, 181–194.

Buyon, J.P., Slade, S.G., Reibman, J., Abramson, S.B., Philips, M.R., Weissmann, G., & Winchester, R. (1990). Constitutive and induced phosphorylation of the alpha- and beta- chains of the CD11/CD18 leukocyte integrin family. Relationship to adhesion-dependent functions. J. Immunol. 144, 191–197.

Calvete, J.J., McLane, M.A., Stewart, G.J., & Niewiarowski, S. (1994). characterization of the cross-linking site of disintegrins albolabrin, bitistatin, echistatin, and eristostatin on isolated human platelet integrin GPIIb/IIIa. Biochem. Biophys. Res. Commun. 202, 135–140.

Calvete, J.J., Schäfer, W., Mann, K., Henschen, A., & González-Rodrguez, J. (1992). Localization of the cross-linking sites of RGD and KQAGDV peptides to the isolated fibrinogen receptor, the human platelet integrin glicoprotein IIb/IIIa—Influence of peptide length. Eur. J. Biochem. 206, 759–765.

Cardarelli, P.M., & Pierschbacher, M.D. (1987). Identification of fibronectin receptors on T lymphocytes. J. Cell Biol. 105, 499–506.

Carlos, T., Kovach, N., Schwartz, B., Rosa, M., Newman, B., Wayner, E., Benjamin, C., Osborn, L., Lobb, R., & Harlan, J. (1991). Human monocytes bind to two cytokine-induced adhesive ligands on cultured human endothelial cells: Endothelial-leukocyte adhesion molecule-1 and vascular cell adhesion molecule-1. Blood 77, 2266–2271.

Carr, M.W., Alon, R., & Springer, T.A. (1996). The C-C chemokine MCP-1 differentially modulates the avidity of beta 1 and beta 2 integrins on T lymphocytes. Immunity 4, 179–187.

Chan, B.M.C., Matsuura, N., Takada, Y., Zetter, B.R., & Hemler, M.E. (1991a). In vitro and in vivo consequences of VLA-2 expression on rhabdomyosarcoma cells. Science 251, 1600–1602.

Chan, P.Y., Lawrence, M.B., Dustin, M.L., Ferguson, L.M., Golan, D.E., & Springer, T.A. (1991b). Influence of receptor lateral mobility on adhesion strengthening between membranes containing LFA-3 and CD2. J. Cell Biol. 115, 245–255.

Chatila, T.A., Geha, R.S., & Arnaout, M.A. (1989). Constitutive and stimulus-induced phosphorylation of CD11/CD18 leukocyte adhesion molecules. J. Cell Biol. 109, 3435–3444.

Chen, Y.Q., Gao, X., Timar, J., Tang, D., Grossi, I.M., Chelladurai, M., Kunicki, T.J., Fligiel, S.E., Taylor, J.D., & Honn, K.V. (1992). Identification of the alpha IIb beta 3 integrin in murine tumor cells. J. Biol. Chem. 267, 17314–17320.

Cheresh, D.A. (1987). Human endothelial cells synthesize and express an Arg-Gly-Asp-directed adhesion receptor involved in attachment to fibrinogen and von Willebrand factor. Proc. Natl. Acad. Sci. USA 84, 6471–6475.

Cheresh, D.A., & Harper, J.R. (1987). Arg-Gly-Asp recognition by a cell adhesion receptor requires its 130-kDa alpha subunit. J. Biol. Chem. 262, 1434–1437.

Cheresh, D.A., & Spiro, R.C. (1987). Biosynthetic and functional properties of an Arg-Gly-Asp-directed receptor involved in human melanoma cell attachment to vitronectin,fibrinogen,and von Willebrand factor. J. Biol. Chem. 262, 17703–17711.

Chuluyan, H.E., & Issekutz, A.C. (1993). VLA-4 integrin can mediate CD11/CD18-independent transendothelial migration of human monocytes. J. Clin. Invest. 92, 2768–2777.

Chuluyan, H.E., Schall, T.J., Yoshimura, T., & Issekutz, A.C. (1995). IL-1 activation of endothelium supports VLA-4 (CD49d/CD29)-mediated monocyte transendothelial migration to C5a, MIP-1 alpha, RANTES, and PAF but inhibits migration to MCP-1: a regulatory role for endothelium-derived MCP-1. J. Leukoc. Biol. 58, 71–79.

Cinco, M., Murgia, R., Presani, G., & Perticarari, S. (1997). Integrin CR3 mediates the binding of nonspecifically opsonized Borrelia burgdorferi to human phagocytes and mammalian cells. Infect. Immun. 65, 4784–4789.

Clark, J.G., Dedon, T.F., Wayner, E.A., & Carter, W.G. (1989). Effects of interferon-gamma on expression of cell surface receptors for collagen and deposition of newly synthesized collagen by cultured human lung fibroblasts. J. Clin. Invest. 83, 1505–1511.

Clyman, R.I., Mauray, F., & Kramer, R.H. (1992). β_1 and β_3 integrins have different roles in the adhesion and migration of vascular smooth muscle cells on extracellular matrix. Exp. Cell Res. 200, 272–284.

Coates, T.D., Torkildson, J.C., Torres, M., Church, J.A., & Howard, T.H. (1991). An inherited defect of neutrophil motility and microfilamentous cytoskeleton associated with abnormalities in 47-Kd and 89-Kd proteins. Blood 78, 1338–1346.

Colombo, M.I., Inglese, J., D'Souza-Schorey, C., Beron, W., & Stahl, P.D. (1995). Heterotrimeric G proteins interact with the small GTPase ARF. Possibilities for the regulation of vesicular traffic. J. Biol. Chem. 270, 24564–24571.

Cook, J.J., Trybulec, M., Lasz, E.C., Khan, S., & Niewiarowski, S. (1992). Binding of glycoprotein IIIa-derived peptide 217-231 to fibrinogen and von Willebrand factors and its inhibition by platelet glycoprotein IIb/IIIa complex. Biochim. Biophys. Acta Protein Struct. Mol. Enzymol. 1119, 312–321.

Cooper, D., Lindberg, F.P., Gamble, J.R., Brown, E.J., & Vadas, M.A. (1995). The transendothelial migration of neutrophils involves integrin associated protein (CD47). Proc. Natl. Acad. Sci. USA 92, 3978–3982.

Corbi, A.L., Miller, L.J., O'Connor, K., Lee, A., Larson, R.S., & Springer, T.A. (1987). cDNA cloning and complete primary structure of the alpha subunit of a leukocyte adhesion glycoprotein p150,95. EMBO J. 6, 4023–4028.

Cotran, R.S., Gimbrone, M.A.,Jr., Bevilacqua, M.P., Mendrick, D.L., & Pober, J.S. (1986). Induction and detection of a human endothelial activation antigen in vivo. J. Exp. Med. 164, 661–666.

Crawford, N., & Eggleton, P. (1991). Dynamic changes in neutrophil cytoskeleton during priming and subsequent surface stimulated functions. Biochem. Soc. Trans. 19, 1048–1055.

D'Sousa, S.E., Ginsberg, M.H., Burke, T.A., & Plow, E.F. (1990). The ligand binding site of the platelet integrin receptor gpIIb- IIIa is proximal to the second calcium binding domain of its α subunit. J. Biol. Chem. 265, 3440–3446.

D'Souza, S.E., Ginsberg, M.H., Matsueda, G.R., & Plow, E.F. (1991). A discrete sequence in a platelet integrin is involved in ligand recognition. Nature 350, 66–68.

Damsky, C., Sutherland, A., & Fisher, S. (1993). Extracellular matrix 5: Adhesive interactions in early mammalian embryogenesis, implantation, and placentation. FASEB J. 7, 1320–1329.

de Fougerolles, A.R., Qin, X., & Springer, T.A. (1994). Characterization of the function of intercellular adhesion molecule (ICAM)-3 and comparison with ICAM-1 and ICAM-2 in immune responses. J. Exp. Med. 179, 619–629.

De Nichilo, M.O. & Burns, G.F. (1993). Granulocyte-macrophage and macrophage colony-stimulating factors differentially regulate alpha-v integrin expression on cultured human macrophages. Proc. Natl. Acad. Sci. USA 90, 2517–2521.

Dedhar, S., & Gray, V. (1990). Isolation of a novel integrin receptor mediating Arg-Gly-Asp-directed cell adhesion to fibronectin and type I collagen from human neuroblastoma cells. Association of a novel β1–related subunit with αv. J. Cell Biol. 110, 2185–2193.

Dedhar, S., & Hannigan, G.E. (1996). Integrin cytoplasmic interactions and bidirectional transmembrane signalling. [Review] [86 refs]. Curr. Opin. Cell Biol. 8, 657–669.

Dedhar, S., Jewell, K., Rojiani, M., & Gray, V. (1992). The receptor for the basement membrane glycoprotein entactin is the integrin alpha3/beta1. J. Biol. Chem. 267, 18908–18914.

Defilippi, P., Truffa, G., Stefanuto, G., Altruda, F., Silengo, L., & Tarone, G. (1991). Tumor necrosis factor α and interferon gamma modulate the expression of the vitronectin receptor (integrin β_3) in human endothelial cells. J. Biol. Chem. 266, 7638–7645.

Delannet, M., Martin, F., Bossy, B., Cheresh, D.A., Reichardt, L.F., & Duband, J.-L. (1994). Specific roles of the αVβ1, αVβ3 and αVβ5 integrins in avian neural crest cell adhesion and migration on vitronectin. Development 120, 2687–2702.

DeLuca, M., Tamura, R.N., Kajiji, S., Bondanza, S., Rossino, P., Cancedda, R., Marchisio, P.C., & Quaranta, V. (1990). Polarized integrin mediates human keratinocyte adhesion to basal lamina. Proc. Natl. Acad. Sci. USA 87, 6888–6892.

Delwel, G.O., Hogervorst, F., Kuikman, I., Paulsson, M., Timpl, R., & Sonnenberg, A. (1993). Expression and function of the cytoplasmic variants of the integrin α6 subunit in transfected K562 cells. Activation-dependent adhesion and interaction with isoforms of laminin. J. Biol. Chem. 268, 25865–25875.

Diamond, M.S., & Springer, T.A. (1994). The dynamic regulation of integrin adhesiveness. [Review] [118 refs]. Curr. Biol. 4, 506–517.

Diamond, M.S., Garcia-Aguilar, J., Bickford, J.K., Corbi, A.L., & Springer, T.A. (1993). The I domain is a major recognition site on the leukocyte integrin Mac-1 (CD11b/CD18) for four distinct adhesion ligands. J. Cell Biol. 120, 1031–1043.

Diamond, M.S., Staunton, D.E., de Fougerolles, A.R., Stacker, S.A., Garcia-Aguilar, J., Hibbs, M.L., & Springer, T.A. (1990). ICAM-1 (CD54): a counter-receptor for Mac-1 (CD11b/CD18). J. Cell Biol. 111, 3129–3139.

Dobrina, A., Menegazzi, R., Carlos, T.M., Nardon, E., Cramer, R., Zacchi, T., Harlan, J.M., & Patriarca, P. (1991). Mechanisms of eosinophil adherence to cultured vascular endothelial cells. Eosinophils bind to the cytokine-induced endothelial ligand vascular cell adhesion molecule-1 via the very late activation antigen-4 integrin receptor. J. Clin. Invest. 88, 20–26.

Doerschuk, C.M., Winn, R.K., Coxson, H.O., & Harlan, J.M. (1990). CD18-dependent and -independent mechanisms of neutrophil emigration in the pulmonary and systemic microcirculation of rabbits. J. Immunol. 144, 2327–2333.

Dransfield, I., & Hogg, N. (1989). Regulated expression of Mg2+ binding epitope on leukocyte integrin alpha subunits. EMBO J. 8, 3759–3765.

Dransfield, I., Cabañas, C., Craig, A., & Hogg, N. (1992). Divalent cation regulation of the function of the leukocyte integrin LFA-1. J. Cell Biol. 116, 219–226.

Drevets, D.A., Canono, B.P., & Campbell, P.A. (1992). Listericidal and nonlistericidal mouse macrophages differ in complement receptor type 3-mediated phagocytosis of *L. monocytogenes* and in preventing escape of the bacteria into the cytoplasm. J. Leukoc. Biol. 52, 70–79.

Du, X., Plow, E.F., Frelinger, A.L., III, O'Toole, T.E., Loftus, J.C., & Ginsberg, M.H. (1991). Ligands "activate" integrin $\alpha_{IIb}\beta_3$ (platelet GPIIb-IIIa). Cell 65, 409–416.

Du, X., Saido, T.C., Tsubuki, S., Indig, F.E., Williams, M.J., & Ginsberg, M.H. (1995). Calpain cleavage of the cytoplasmic domain of the integrin beta 3 subunit. J. Biol. Chem. 270, 26146–26151.

Dustin, M.L., Rothlein, R., Bhan, A.K., Dinarello, C.A., & Springer, T.A. (1986). Induction by IL-1 and interferon, tissue distribution, biochemistry, and function of a natural adherence molecule (ICAM-1). J. Immunol. 137, 245–254.

Edidin, M., Zuniga, M.C., & Sheetz, M.P. (1994). Truncation mutants define and locate cytoplasmic barriers to lateral mobility of membrane glycoproteins. Proc. Natl. Acad. Sci. USA 91, 3378–3382.

Ehlenberger, A.G. & Nussenzweig, V. (1977). The role of memberane receptors for C3b and C3d in phagocytosis. J. Exp. Med. 145, 357–371.

El-Gabalawy, H., Canvin, J., Ma, G.M., Van der Vieren, M., Hoffman, P., Gallatin, M., & Wilkins, J. (1996). Synovial distribution of alpha d/CD18, a novel leukointegrin. Comparison with other integrins and their ligands. Arthritis Rheum. 39, 1913–1921.

Elices, M.J., & Hemler, M.E. (1989). The human integrin VLA-2 is a collagen receptor on some cells and a collagen/laminin receptor on others. Proc. Natl. Acad. Sci. USA 86, 9906–9910.

Elices, M.J., Osborn, L., Takada, Y., Crouse, C., Luhowskyj, S., Hemler, M.E., & Lobb, R.R. (1990). VCAM-1 on activated endothelium interacts with the leukocyte integrin VLA-4 at a site distinct from the VLA-4/fibronectin binding site. Cell 60, 577–584.

Elices, M.J., Urry, L.A., & Hemler, M.E. (1991). Receptor functions for the integrin VLA-3: fibronectin, collagen, and laminin binding are differentially influenced by ARG-GLY-ASP peptide and by divalent cations. J. Cell Biol. 112, 169–181.

Emsley, J., King, S.L., Bergelson, J.M., & Liddington, R.C. (1997). Crystal structure of the I domain from integrin alpha2beta1. J. Biol. Chem. 272, 28512–28517.

Erle, D.J., Briskin, M.J., Butcher, E.C., Garcia-Pardo, A., Lazarovits, A.I., & Tidswell, M. (1994). Expression and function of the MAdCAM-1 receptor, integrin $\alpha 4\beta 7$, on human leukocytes. J. Immunol. 153, 517–528.

Etzioni, A., Frydman, M., Pollack, S., Avidor, I., Phillips, M.L., Paulson, J.C., & Gershoni-Baruch, R. (1992). Brief report: recurrent severe infections caused by a novel leukocyte adhesion deficiency. N. Engl. J. Med. 327, 1789–1792.

Ezzell, R.M., Goldmann, W.H., Wang, N., Parasharama, N., & Ingber, D.E. (1997). Vinculin promotes cell spreading by mechanically coupling integrins to the cytoskeleton. Exp. Cell Res. 231, 14–26.

Fairbanks, M.B., Pollock, J.R., Prairie, M.D., Scahill, T.A., Baczynskyj, L., Heinrikson, R.L., & Stockman, B.J. (1995). Purification and structural characterization of the CD11b/CD18 integrin alpha subunit I domain reveals a folded conformation in solution. FEBS Lett. 369, 197–201.

Fawcett, J., Holness, C.L.L., Needham, L.A., Turley, H., Gatter, K.C., Mason, D.Y., & Simmons, D.L. (1992). Molecular cloning of ICAM-3, a third ligand for LFA-1, constitutively expressed on resting leukocytes. Nature 360, 481–484.

Frankel, G., Lider, O., Hershkoviz, R., Mould, A.P., Kachalsky, S.G., Candy, D.C.A., Cahalon, L., Humphries, M.J., & Dougan, G. (1996). The cell-binding domain of intimin from enteropathogenic Escherichia coli binds to beta1 integrins. J. Biol. Chem. 271, 20359–20364.

Furie, M.B., Tancinco, M.C.A., & Smith, C.W. (1991). Monoclonal antibodies to leukocyte integrins CD11a/CD18 and CD11b/CD18 or intercellular adhesion molecule-1 inhibit chemoattractant-stimulated neutrophil transendothelial migration in vitro. Blood 78, 2089–2097.

Gahmberg, C.G., Nortamo, P., Kantor, C., Autero, M., Kotovuori, P., Hemiö, L., Salcedo, R., & Patarroyo, M. (1990). The pivotal role of the Leu-CAM and ICAM molecules in human leukocyte adhesion. Cell Differ. Dev. 32, 239–246.

Gaidano, G., Bergui, L., Schena, M., Gaboli, M., Cremona, O., Marchisio, P.C., & Caligaris Cappio, F. (1990). Integrin distribution and cytoskeleton organization in normal and malignant monocytes. Leukemia 4, 682–687.

Gao, A.G., Lindberg, F.P., Dimitry, J.M., Brown, E.J., & Frazier, W.A. (1996a). Thrombospondin modulates alpha v beta 3 function through integrin-associated protein. J. Cell Biol. 135, 533–544.

Gao, A.G., Lindberg, F.P., Finn, M.B., Blystone, S.D., Brown, E.J., & Frazier, W.A. (1996b). Integrin-associated protein is a receptor for the C-terminal domain of thrombospondin. J. Biol. Chem. 271, 21–24.

Garratt, A.N. & Humphries, M.J. (1995). Recent insights into ligand binding, activation and signalling by integrin adhesion receptors. [Review] [112 refs]. Acta Anatomica 154, 34–45.

Ginsberg, M.H., Loftus, J.C., D'Souza, S., & Plow, E.F. (1990). Ligand binding to integrins: common and ligand specific recognition mechanisms. Cell Differ. Dev. 32, 203–214.

Giunta, M., Favre, A., Ramarli, D., Grossi, C.E., & Corte, G. (1991). A novel integrin involved in thymocyte-thymic epithelial cell interactions. J. Exp. Med. 173, 1537–1548.

Goldman, G., Welbourn, R., Valeri, C.R., Shepro, D., & Hechtman, H.B. (1991). Thromboxane A_2 induces leukotriene B_4 synthesis that in turn mediates neutrophil diapedesis via CD 18 activation. Microvasc. Res. 41, 367–375.

Goldman, R., Harvey, J., & Hogg, N. (1992). VLA-2 is the integrin used as a collagen receptor by leukocytes. Eur. J. Immunol. 22, 1109–1114.

Goodman, T.G., & Bajt, M.L. (1996). Identifying the putative metal ion-dependent adhesion site in the beta2 (CD18) subunit required for alphaLbeta2 and alphaMbeta2 ligand interactions. J. Biol. Chem. 271, 23729–23736.

Graham, I.L., Anderson, D.C., Holers, V.M., & Brown, E.J. (1994). Complement receptor 3 (CR3, Mac-1, Integrin alpha-M beta-2, CD11b/CD18) is required for tyrosine phosporylation of paxillin in adherent and nonadherent neutrophils. J. Cell Biol. 127, 1139–1147.

Graham, I.L., Gresham, H.D., & Brown, E.J. (1989). An immobile subset of plasma membrane CD11b/CD18 (Mac-1) is involved in phagocytosis of targets recognized by multiple receptors. J. Immunol. 142, 2352–2358.

Graham, I.L., Lefkowith, J.B., Anderson, D.C., & Brown, E.J. (1993). Immune complex-stimulated neutrophil LTB4 production is dependent on beta2 integrins. J. Cell Biol. 120, 1509–1517.

Gregorio, C.C., Kubo, R.T., Bankert, R.B., & Repasky, E.A. (1992). Translocation of spectrin and protein kinase C to a cytoplasmic aggregate upon lymphocyte activation. Proc. Natl. Acad. Sci. USA 89, 4947–4951.

Gresham, H.D., & Brown, E.J. (1990). Fibrinogen gamma chaim peptide KQAGDV stimulates neutrophil phagocytosis. Clin. Res. 38, 480a. Abstract.

Gresham, H.D., Adams, S.P., & Brown, E.J. (1992). Ligand binding specificity of the leukocyte response integrin expressed by human neutrophils. J. Biol. Chem. 267, 13895–13902.

Gresham, H.D., Clement, L.D., Volanakis, J.E., & Brown, E.J. (1987). Cholera toxin and pertussis toxin regulate the Fc- receptor mediated phagocytosis response of human neutrophls in a manner analagous to regulation by the monoclonal antibody 1C2. J. Immunol. 139, 4159–4166.

Gresham, H.D., Goodwin, J.L., Anderson, D.C., & Brown, E.J. (1989). A novel member of the integrin receptor family mediates Arg-Gly-Asp-stimulated neutrophil phagocytosis. J. Cell Biol. 108, 1935–1943.

Gresham, H.D., Graham, I.L., Anderson, D.C., & Brown, E.J. (1991). Leukocyte adhesion deficient (LAD) neutrophils fail to amplify phagocytic function in response to stimulation: evidence for CD11b/CD18-dependent and -independent mechanisms of phagocytosis. J. Clin. Invest. 88, 588–597.

Gresham, H.D., McGarr, J.A., Shackelford, P.G., & Brown, E.J. (1988). Studies on the molecular mechanisms of receptor-mediated phagocytosis: amplification of ingestion is dependent on the generation of reactive oxygen metabolites and is deficient in PMN from patients with chronic granulomatous disease. J. Clin. Invest. 82, 1192–1201.

Griffin, F.M., Jr., & Mullinax, P.J. (1981). Augmentation of macrophage complement receptor function in vitro. III. C3b receptors that promote phagocytosis migrate within the plane of the macrophage plasma membrane. J. Exp. Med. 154, 291.

Griffin, F.M., Jr., & Silverstein, S.C. (1974). Segmental response of the macrophage plasma membrane to a phagocytic stimulus. J. Exp. Med. 139, 323–336.

Griffin, F.M., Jr., Griffin, J.A., Leider, J.E., & Silverstein, S.C. (1975). Studies on the mechanism of phagocytosis: I. Requirements for circumferential attachment of particle-bound ligands to specific receptors on the macrophage plasma membrane. J. Exp. Med. 142, 1263–1282.

Gudewicz, P.W., Beezhold, D.H., van Alten, P., & Molnar, J. (1982). Lack of stimulation of post-phagocytic metabolic activities of polymorphonuclear leukocytes by fibronectin opsonized particles. J. Reticuloendothelial Soc. 32, 143–154.

Gudewicz, P.W., Molnar, J., Lai, M.Z., Beezhold, D.W., Siefring, G.E., Jr., Credo, R.B., & Lorand, L. (1980). Fibronectin-mediated uptake of gelatin-coated latex particles by peritoneal macrophages. J. Cell Biol. 87, 427–433.

Hakkert, B.C., Kuijpers, T.W., Leeuwenberg, J.F., van Mourik, J.A., & Roos, D. (1991). Neutrophil and monocyte adherence to and migration across monolayers of cytokine-activated endothelial cells: the contribution of CD18, ELAM-1, and VLA-4. Blood 78, 2721–2726.

Hemler, M.E. (1988). Adhesive protein receptors on hematopoietic cells. Immunol. Today 9, 109–113.

Hemler, M.E. (1990). VLA proteins in the integrin family: structures, functions, and their role on leukocytes. Annu. Rev. Immunol. 8, 365–400.

Hemler, M.E., & Jacobson, J.G. (1987). Cell matrix adhesion-related proteins VLA-1 and VLA 2: regulation of expression on T cells. J. Immunol 138, 2941–2948.

Hemler, M.E., Elices, M.J., Parker, C., & Takada, Y. (1990). Structure of the integrin VLA-4 and its cell-cell and cell-matrix adhesion functions. Immunol Rev. 114, 45–65.

Hemler, M.E., Huang, C., & Schwarz, L. (1987). The VLA protein family: characterization of five distinct cell surface heterodimers each with a common 130,000 molecular weight beta subunit. J. Biol. Chem. 262, 3300–3309.

Hemler, M.E., Jacobson, J.G., & Strominger, J.L. (1985). Biochemical characterization of VLA-1 and VLA-2. Cell surface heterodimers on activated T cells. J. Biol. Chem. 260, 15246–15252.

Hendey, B., & Maxfield, F.R. (1993). Regulation of neutrophil motility and adhesion by intracellular calcium transients. Blood Cells 19, 143–164.

Hendey, B., Klee, C.B., & Maxfield, F.R. (1992). Inhibition of neutrophil chemkinesis on vitronectin by inhibitors of calcineurin. Science 258, 296–299.

Hermanowski-Vosatka, A., Van Strijp, J.A., Swiggard, W.J., & Wright, S.D. (1992). Integrin modulating factor-1: a lipid that alters the function of leukocyte integrins. Cell 68, 341–352.

Hickstein, D.D., Back, A.L., & Collins, S.J. (1989). Regulation of expression of the CD11b and CD18 subunits of neutrophil adherence receptor during human myeloid differentiation. J. Biol. Chem. 264, 21812–21817.

Hickstein, D.D., Howard, M., Meuller, L., Hickey, M.J., & Collins, S.J. (1988). Expression of the beta-subunit of the human leukocyte adherence receptor depends upon cell type and stage of differentiation. J. Immunol. 141, 4313–4317.

Hickstein, D.D., Smith, A., Fisher, W., Beatty, P.G., Schwartz, B.R., Harlan, J.M., Root, R.K., & Locksley, R.M. (1987). Expression of leukocyte adherence-related glycoproteins during differentiation of HL-60 promyelocytic leukemia cells. J. Immunol. 138, 513–519.

Hogervorst, F., Kuikman, I., Von dem Borne, A.E.G.K., & Sonnenberg, A. (1990). Cloning and sequence analysis of beta-4 cDNA: an integrin subunit that contains a unique 118kd cytoplasmic domain. EMBO J. 9, 765–770.

Hogg, N., Harvey, J., Cabanas, C., & Landis, R.C. (1993). Control of leukocyte integrin activation. [Review]. Am. Rev. Respir. Dis. 148, S55–S59.

Holzmann, B., McIntyre, B.W., & Weissman, I.L. (1989). Identification of a murine Peyer's patch-specific lymphocyte homing receptor as an integrin molecule with an α chain homologous to human VLA-4. Cell 56, 37–46.

Horvath, A.R., Elmore, M.A., & Kellie, S. (1990). Differential tyrosine-specific phosphorylation of integrin in Rous sarcoma virus transformed cells with differing transformed phenotypes. Oncogene 5, 1349–1357.

Horwitz, A., Duggan, K., Buck, C., Beckerle, M.C., & Burridge, K. (1986). Interaction of plasma membrane fibronectin receptor with talin-a transmembrane linkage. Nature 320, 531–533.

Horwitz, A., Duggan, K., Greggs, R., Decker, C., & Buck, C. (1985). The cell substrate attachment (CSAT) antigen has properties of a receptor for laminin and fibronectin. J. Cell Biol. 101, 2134–2144

Hourihan, H., Allen, T.D., & Ager, A. (1993). Lymphocyte migration across high endothelium is associated with increases in $\alpha_4\beta_1$ integrin (VLA-4) affinity. J. Cell Sci. 104, 1049–1059.

Hughes, P.E., Diaz-Gonzalez, F., Leong, L., Wu, C., McDonald, J.A., Shattil, S.J., & Ginsberg, M.H. (1996). Breaking the integrin hinge. A defined structural constraint regulates integrin signaling. J. Biol. Chem. 271, 6571–6574.

Hughes, P.E., Renshaw, M.W., Pfaff, M., Forsyth, J., Keivens, V.M., Schwartz, M.A., & Ginsberg, M.H. (1997). Suppression of integrin activation: a novel function of a Ras/Raf-initiated MAP kinase pathway. Cell 88, 521–530.

Huttenlocher, A., Palecek, S.P., Lu, Q., Zhang, W., Mellgren, R.L., Lauffenburger, D.A., Ginsberg, M.H., & Horwitz, A.F. (1997). Regulation of cell migration by the calcium-dependent protease calpain. J. Biol. Chem. 272, 32719–32722.

Hynes, R.O. (1987). Integrins: a family of cell surface receptors. Cell 48, 549–554.

Hynes, R.O. (1992). Integrins: versatility, modulation, and signaling in cell adhesion. Cell 69, 11–25.

Issekutz, A.C., Chuluyan, H.E., & Lopes, N. (1995). CD11/CD18-independent transendothelial migration of human polymorphonuclear leukocytes and monocytes: involvement of distinct and unique mechanisms. J. Leukoc. Biol. 57, 553–561.

Jalkanen, S., Jalkanen, M., Bargatze, R.F., Tammi, M., & Butcher, E.C. (1988). Biochemical properties of glycoproteins involved in lymphocyte recognition of high endothelial venules in man. J. Immunol. 141, 1615–1623.

Johnston, B., Issekutz, T.B., & Kubes, P. (1996). The alpha 4-integrin supports leukocyte rolling and adhesion in chronically inflamed postcapillary venules in vivo. J. Exp. Med. 183, 1995–2006.

Jones, D.H., Schmalstieg, F.C., Hawkins, H.K., Burr, B.L., Rudloff, H.E., Krater, S., Smith, C.W., & Anderson, D.C. (1990). Characterization of a new mobilizable Mac-1 (CD11b/CD18) pool that co-localizes with gelatinase in human neutrophils. In: Leukocyte Adhesion Molecules: Structure, Function, and Regulation (Springer, T.A., Anderson, D.C., Rosenthal, A.S., & Rothlein, R., Eds.), pp. 106–124. Springer-Verlag, New York.

Jones, S.L., Knaus, U.G., Bokoch, G.M., & Brown, E.J. (1998). Two signaling mechanisms for activation of alphaM beta2 avidity in polymorphonuclear neutrophils. J. Biol. Chem. 273, 10556–10566.

Juliano, R.L., & Haskill, S. (1993). Signal transduction from the extracellular matrix. J. Cell Biol. 120, 577–585.

Kamata, T. & Takada, Y. (1994). Direct binding of collagen to the I domain of integrin $\alpha2\beta1$ (VLA-2, CD49b/CD29) in a divalent cation-independent manner. J. Biol. Chem. 269, 26006–26010.

Kansas, G.S., Ley, K., Munro, J.M., & Tedder, T.F. (1993). Regulation of leukocyte rolling and adhesion to high endothelial venules through the cytoplasmic domain of L-selectin. J. Exp. Med. 177, 833–838.

Katagiri, Y., Hiroyama, T., Akamatsu, N., Suzuki, H., Yamazaki, H., & Tanoue, K. (1995). Involvement of alpha v beta 3 integrin in mediating fibrin gel retraction. J. Biol. Chem. 270, 1785–1790.

Kavanaugh, A.F., Lightfoot, E., Lipsky, P.E., & Oppenheimer-Marks, N. (1991). Role of CD11/CD18 in adhesion and transendothelial migration of T cells: analysis utilizing CD18-deficient T cell clones. J. Immunol. 146, 4149–4156.

Kawaguchi, S., & Hemler, M.E. (1993). Role of the subunit cytoplasmic domain in regulation of adhesive activity mediated by the integrin VLA-2. J. Biol. Chem. 268, 16279–16285.

Kehrli, M.E., Jr., Ackermann, M.R., Shuster, D.E., Van der Maaten, M.J., Schmalstieg, F.C., Anderson, D.C., & Hughes, B.J. (1992). Bovine leukocyte adhesion deficiency: β_2 integrin deficiency in young Holstein cattle. Am. J. Pathol. 140, 1489–1492.

Kehrli, M.E., Jr., Schmalstieg, F.C., Anderson, D.C., Van der Maaten, M.J., Hughes, B.J., Ackermann, M.R., Wilhelmsen, C.L., Brown, G.B., Stevens, M.G., & Whetstone, C.A. (1990). Molecular definition of the bovine granulocytopathy syndrome: identification of deficiency of the Mac-1 (CD11b/CD18) glycoprotein. Am. J. Vet. Res. 51, 1826–1836.

Kennel, S.J., Foote, L.J., Cimino, L., Rizzo, M.G., Chang, L.-Y., & Sacchi, A. (1993). Sequence of a cDNA encoding the β_4 subunit of murine integrin. Gene 130, 209–216.

Kern, A., Briesewitz, R., Bank, I., & Marcantonio, E.E. (1994). The role of the I domain in ligand binding of the human integrin $\alpha_1\beta_1$. J. Biol. Chem. 269, 22811–22816.

Kieffer, N., Fitzgerald, L.A., Wolf, D., Cheresh, D.A., & Phillips, D.R. (1991). Adhesive properties of the β_3 integrins: comparison of GP IIb-IIIa and the vitronectin receptor individually expressed in human melanoma cells. J. Cell Biol. 113, 451–461.

Kim, L.T., & Yamada, K.M. (1997). The regulation of expression of integrin receptors. [Review] [118 refs]. Proc. Soc. Exp. Biol. Med. 214, 123–131.

Kincade, P.W. (1992). Cell interaction molecules and cytokines which participate in B lymphopoiesis. [Review] [171 refs]. Baillieres Clin. Haematol. 5, 575–598.

Kirchhofer, D., Gailit, J., Ruoslahti, E., Grzesiak, J., & Pierschbacher, M.D. (1990). Cation-dependent changes in the binding specificity of the platelet receptor GPIIb/IIIa. J. Biol. Chem. 265, 18525–18530.

Kishimoto, T.K., Hollander, N., Roberts, T.M., Anderson, D.C., & Springer, T.A. (1987a). Heterogeneous mutations in the β subunit common to the LFA-1, Mac-1, and p150,95 glycoproteins cause leukocyte adhesion deficiency. Cell 50, 192–202.

Kishimoto, T.K., Larson, R.S., Corbi, A.L., Dustin, M.L., Staunton, D.E., & Springer, T.A. (1990). Leukocyte integrins. In: Leukocyte Adhesion Molecules: Structure, Function, and Regulation (Springer, T.A., Anderson, D.C., Rosenthal, A.S., & Rothlein, R., Eds), pp. 7–43. Springer-Verlag, New York.

Kishimoto, T.K., O'Connor, K., & Springer, T.A. (1989). Leukocyte adhesion deficiency: Aberrant splicing of a conserved integrin sequence causes a moderate deficiency phenotype. J. Biol. Chem. 264, 3588–3595.

Kishimoto, T.K., O'Connor, K., Lee, A., Roberts, T.M., & Springer, T.A. (1987b). Cloning of the β subunit of the leukocyte adhesion proteins: homology to an extracellular matrix receptor defines a novel supergene family. Cell 48, 681–690.

Klemke, R.L., Yebra, M., Bayna, E.M., & Cheresh, D.A. (1994). Receptor tyrosine kinase signaling required for integrin $\alpha v\beta 5$-directed cell motility but not adhesion on vitronectin. J. Cell Biol. 127, 859–866.

Kohn, F.R., Grigg, M.E., & Klingemann, H.-G. (1991). Fibronectin receptor subunit (α^5, α^4 and β_1) mRNA and cell surface expression in human peripheral blood B lymphocytes. Immunol. Lett. 28, 27–30.

Kolanus, W., & Seed, B. (1997). Integrins and inside-out signal transduction: converging signals from PKC and PIP3. [Review] [77 refs]. Curr. Opin. Cell Biol. 9, 725–731.

Kolanus, W., Nagel, W., Schiller, B., Zeitlmann, L., Godar, S., Stockinger, H., & Seed, B. (1996). Alpha L beta 2 integrin/LFA-1 binding to ICAM-1 induced by cytohesin-1, a cytoplasmic regulatory molecule. Cell 86, 233–242.

Kramer, R.H., Cheng, Y.F., & Clyman, R. (1990). Human microvascular endothelial cells use beta 1 and beta 3 integrin receptor complexes to attach to laminin. J. Cell Biol. 111, 1233–1243.

Krissansen, G.W., Elliott, M.J., Lucas, C.M., Stomski, F.C., Berndt, M.C., Cheresh, D.A., Lopez, A.F., & Burns, G.F. (1990). Identification of a novel integrin β subunit expressed on cultured monocytes (macrophages): evidence that one α subunit can associate with multiple β subunits. J. Biol. Chem. 265, 823–830.

Krotoski, D., & Bronner Fraser, M. (1990). Distribution of integrins and their ligands in the trunk of Xenopus laevis during neural crest cell migration. J. Exp. Zool. 253, 139–150.

Kucik, D.F., Dustin, M.L., Miller, J.M., & Brown, E.J. (1996). Adhesion-activating phorbol ester increases the mobility of leukocyte integrin LFA-1 in cultured lymphocytes. J. Clin. Invest. 97, 2139–2144.

Kucik, D.F., Elson, E.L., & Sheetz, M.P. (1989). Forward transport of glycoproteins on leading lamellipodia in locomoting cells [see comments]. Nature 340, 315–317.

Kucik, D.F., Kuo, S.C., Elson, E.L., & Sheetz, M.P. (1991). Preferential attachment of membrane glycoproteins to the cytoskeleton at the leading edge of lamella. J. Cell Biol. 114, 1029–1036.

Kuijpers, T.W., Hakkert, B.C., Hoogerwerf, M., Leeuwenberg, J.F.M., & Roos, D. (1991). Role of endothelial leukocyte adhesion molecule-1 and platelet-activating factor in neutrophil adherence to IL-1-prestimulated endothelial cells: endothelial leukocyte adhesion molecule-1-mediated CD18 activation. J. Immunol. 147, 1369–1376.

Kunicki, T.J., Annis, D.S., Deng, Y.J., Loftus, J.C., & Shattil, S.J. (1996). A molecular basis for affinity modulation of Fab ligand binding to integrin alphaIIb beta3. J. Biol. Chem. 271, 20315–20321.

Kuroda, K., Brown, E.J., Telle, W.B., Russell, D.G., & Ratliff, T.L. (1993). Characterization of the internalization of Bacillus Calmette-Guerin (BCG) by human bladder tumor cells. J. Clin. Invest. 91, 69–76.

Larson, R.S., & Springer, T.A. (1990). Structure and function of leukocyte integrins. Immunol. Rev. 114, 181–217.

Larson, R.S., Corbi, A.L., Berman, L., & Springer, T. (1989a). Primary structure of the leukocyte function-associated molecule-1 α subunit: an integrin with an embedded domain defining a protein superfamily. J. Cell Biol. 108, 703–712.

Larson, R.S., Corbi, A.L., Berman, L., & Springer, T. (1989b). Primary structure of the leukocyte function-associated molecule-1 alpha subunit: an integrin with an embedded domain defining a protein superfamily. J. Cell Biol. 108, 703–712.

Law, D.A., Nannizzi-Alaimo, L., & Phillips, D.R. (1996). Outside-in signal transduction. Alpha-IIb/beta-3 (GP IIb-IIIa)-tyrosine phosphorylation induced by platelet aggregation. J. Biol. Chem. 271, 10811–10815.

Lawrence, M.B., & Springer, T.A. (1991). Leukocytes roll on a selectin at physiologic flow rates: distinction from and prerequisite for adhesion through integrins. Cell 65, 859–873.

Leavesley, D.I., Schwartz, M.A., Rosenfeld, M., & Cheresh, D.A. (1993). Integrin $\beta 1$- and $\beta 3$-mediated endothelial cell migration is triggered through distinct signaling mechanisms. J. Cell Biol. 121, 163–170.

Lee, J., Rieu, P., Arnaout, M., & Liddington, R. (1995). Crystal structure of the A domain from the alpha subunit of integrin CR3 (CD11b/CD18). Cell 80, 631–638.

Lefkowith, J.B., Rogers, M., Lennartz, M.R., & Brown, E.J. (1991). Essential fatty acid deficiency impairs macrophage spreading and adherence: role of arachidonate in cell adhesion. J. Biol. Chem. 266, 1071–1076.

Lenter, M., & Vestweber, D. (1994). The integrin chains β_1 and α_6 associate with the chaperone calnexin prior to integrin assembly. J. Biol. Chem. 269, 12263–12268.

Ley, K., Gaehtgens, P., Fennie, C., Singer, M.S., Lasky, L.A., & Rosen, S.D. (1991). Lectin-like cell adhesion molecule 1 mediates leukocyte rolling in mesenteric venules in vivo. Blood 77, 2553–2555.

Lin, E.C.K., Ratnikov, B.I., Tsai, P.M., Gonzalez, E.R., McDonald, S., Pelletier, A.J., & Smith, J.W. (1997). Evidence that the integrin beta3 and beta5 subunits contain a metal ion-dependent adhesion site-like motif but lack an I domain. J. Biol. Chem. 272, 14236–14243.

Lindberg, F.P., Gresham, H.D., Reinhold, M.I., & Brown, E.J. (1996). Integrin-associated protein immunoglobulin domain is necessary for efficient vitronectin bead binding. J. Cell Biol. 134, 1313–1322.

Lindberg, F.P., Gresham, H.D., Schwarz, E., & Brown, E.J. (1993). Molecular cloning of integrin-associated protein: an immunoglobulin family member with multiple membrane spanning domains implicated in alpha-v, beta-3-dependent ligand binding. J. Cell Biol. 123, 485–496.

Lindberg, F.P., Lublin, D.M., Telen, M.J., Veile, R.A., Miller, Y.E., Donis-Keller, H., & Brown, E.J. (1994). Rh-related antigen CD47 is the signal-transducer integrin associated protein. J. Biol. Chem. 269, 1567–1570.

Lipnick, R., Iliopoulos, A., Salata, K., Hershey, J., Melnick, D., & Tsokos, G. (1996). Leukocyte adhesion deficiency: report of a case and review. Clin. Exp. Rheumatol. 74, 33–39.

Lobb, R., Chi-Rosso, G., Leone, D., Rosa, M., Newman, B., Luhowskyj, S., Osborn, L., Schiffer, S., Benjamin, C., Dougas, I., Hession, C., & Chow, P. (1991). Expression and functional characterization of a soluble form of vascular cell adhesion molecule 1. Biochem. Biophys. Res. Commun. 178, 1498–1504.

Lotz, M.M., Burdsal, C.A., Erickson, H.P., & McClay, D.R. (1989). Cell adhesion to fibronectin and tenascin: quantitative measurements of initial binding and subsequent strengthening response. J. Cell Biol. 109, 1795–1805.

Luscinskas, F.W., & Lawler, J. (1994). Integrins as dynamic regulators of vascular function. FASEB J. 8, 929–938.

Makgoba, M.W., Sanders, M.E., Ginther Luce, G.E., Dustin, M.L., Springer, T.A., Clark, E.A., Mannoni, P., & Shaw, S. (1988). ICAM-1: a ligand for LFA-1-dependent adhesion of B, T and myeloid cells. Nature 331, 86–88.

Mandeville, J.T., & Maxfield, F.R. (1997). Effects of buffering intracellular free calcium on neutrophil migration through three-dimensional matrices. J. Cell. Physiol.171, 168–178.

Marchisio, P.C., Bondanza, S., Cremona, O., Cancedda, R., & De Luca, M. (1991). Polarized expression of integrin receptors (alpha 6 beta 4, alpha 2 beta 1, alpha 3 beta 1, and alpha v beta 5) and their relationship with the cytoskeleton and basement membrane matrix in cultured human keratinocytes. J. Cell Biol. 112, 761–773.

Marder, S.R., Chenoweth, D.E., Goldstein, I.M., & Perez, H.D. (1985). Chemotactic responses of human peripheral blood monocytes to the complement derived peptides C5a and C5a des Arg. J. Immunol 134, 746–757.

Marks, P.W., Hendey, B., & Maxfield, F.R. (1991). Attachment to fibronectin or vitronectin makes human neutrophil migration sensitive to alterations in cytosolic free calcium concentration. J. Cell Biol. 112, 149–158.

Masumoto, A., & Hemler, M.E. (1993a). Mutation of putative divalent cation sites in the α^4 subunit of the integrin VLA-4: Distinct effects on adhesion to CS1/fibronectin, VCAM-1, and invasin. J. Cell Biol. 123, 245–253.

Masumoto, A., & Hemler, M.E. (1993b). Multiple activation states of VLA-4. Mechanistic differences between adhesion to CS1/fibronectin and to vascular cell adhesion molecule-1. J. Biol. Chem. 268, 228–234.

Maxfield, S.R., Moulder, K., Koning, F., Elbe, A., Stingl, G., Coligan, J.E., Shevach, E.M., & Yokoyama, W.M. (1989). Murine T cells express a cell surface receptor for multiple extracellular matrix proteins. J. Exp. Med. 169, 2173–2190.

Mayadas, T.N., Johnson, R.C., Rayburn, H., Hynes, R.O., & Wagner, D.D. (1993). Leukocyte rolling and extravasation are severely compromised in P selectin-deficient mice. Cell 74, 541–554.

Mazzone, A., & Ricevuti, G. (1995). Leukocyte CD11/CD18 integrins: biological and clinical relevance. [Review] [135 refs]. Haematologica 80, 161–175.

McEver, R.P. (1991). Selectins: novel receptors that mediate leukocyte adhesion during inflammation. Thromb. Haemost. 65, 223–228.

McGavin, M.J., Raucci, G., Gurusiddappa, S., & Höök, M. (1991). Fibronectin binding determinants of the *Staphylococcus aureus* fibronectin receptor. J. Biol. Chem. 266, 8343–8347.

McGuire, S.L. & Bajt, M.L. (1995). Distinct ligand binding sites in the I domain of integrin alpha M beta 2 that differentially affect a divalent cation-dependent conformation. J. Biol. Chem. 270, 25866–25871.

Meacci, E., Tsai, S.C., Adamik, R., Moss, J., & Vaughan, M. (1997). Cytohesin-1, a cytosolic guanine nucleotide-exchange protein for ADP-ribosylation factor. Proc. Natl. Acad. Sci. USA 94, 1745–1748.

Mecham, R.P. (1991). Receptors for laminin on mammalian cells. FASEB J. 5, 2538–2546.

Michl, J., Pieczonka, M.M., Unkeless, J.C., & Silverstein, S.C. (1979). Effects of immobilized immune complexes on Fc- and complement-receptor function in resident and thioglycollate-elicited mouse peritoneal macrophages. J. Exp. Med. 150, 607–621.

Milam, S.B., Magnuson, V.L., Steffensen, B., Chen, D., & Klebe, R.J. (1991). IL-1β and prostaglandins regulate integrin mRNA expression. J. Cell. Physiol. 149, 173–183.

Miller, L.J., Bainton, D.F., Borregaard, N., & Springer, T.A. (1987). Stimulated mobilization of monocyte Mac-1 and p150,95 adhesion proteins from an intracellular vesicular compartment to the cell surface. J. Clin. Invest. 80, 535–544.

Miyake, K., Weissman, I.L., Greenberger, J.S., & Kincade, P.W. (1991). Evidence for a role of the integrin VLA-4 in lympho-hemopoiesis. J. Exp. Med. 173, 599–607.

Miyamoto, S., Teramoto, H., Coso, O.A., Gutkind, J.S., Burbelo, P.D., Akiyama, S.K., & Yamada, K.M. (1995). Integrin function: molecular hierarchies of cytoskeletal and signaling molecules. J. Cell Biol. 131, 791–805.

Monk, P.N., & Banks, P. (1991). The role of protein kinase C activation and inositol phosphate production in the regulation of cell-surface expression of Mac-1 by complement fragment C5a. Biochim. Biophys. Acta Mol. Cell Res. 1092, 251–255.

Morisaki, T., Goya, T., Toh, H., Nishihara, K., & Torisu, M. (1991). The anti Mac-1 monoclonal antibody inhibits neutrophil sequestration in lung and liver in a septic murine model. Clin. Immunol. Immunopathol. 61, 365–375.

Mosser, D.M., Springer, T.A., & Diamond, M.S. (1992). Leishmania promastigotes require opsonic complement to bind to the human leukocyte integrin Mac-1 (CD11b/CD18). J. Cell Biol. 116, 511–520.

Myones, B.L., Dalzell, J.G., Hogg, N., & Ross, G.D. (1988). Neutrophil and monocyte cell surface p150,95 has iC3b receptor (CR4) activity. J. Clin. Invest. 82, 640–651.

Nathan, C., & Sanchez, E. (1990). Tumor necrosis factor and CD11/CD18 (β2) integrins act synergistically to lower cAMP in human neutrophils. J. Cell Biol. 111, 2171–2181.

Nathan, C., Srimal, S., Farber, C., Sanchez, E., Kabbash, L., Asch, A., Gailit, J., & Wright, S.D. (1989). Cytokine-induced respiratory burst of human neutrophils: dependence on extracellular matrix protens and CD11/CD18 integrins. J. Cell Biol. 109, 1341–1349.

Neff, S., Sa-Carvalho, D., Rieder, E., Mason, P.W., Blystone, S.D., Brown, E.J., & Baxt, B. (1998). Foot-and-mouth disease virus virulent for cattle utilizes the integrin alpha(v)beta3 as its receptor. J. Virol. 72, 3587–3594.

Newman, S.L., & Tucci, M.A. (1990). Regulation of human monocyte/macrophage function by extracellular matrix. Adherence of monocytes to collagen matrices enhances phagocytosis of opsonized bacteria by activation of complement receptors and enhancement of Fc receptor function. J. Clin. Invest. 86, 703–714.

Niessen, C.M., Hogervorst, F., Jaspars, L.H., De Melker, A.A., Delwel, G.O., Hulsman, E.H.M., Kuikman, I., & Sonnenberg, A. (1994). The $\alpha6\beta4$ integrin is a receptor for both laminin and kalinin. Exp. Cell Res. 211, 360–367.

O'Shea, J.J., Brown, E.J., Seligmann, B.E., Metcalf, J.E., Frank, M.M., & Gallin, J.I. (1985). Evidence for distinct intracellular pools for CR1 and CR3 in human neutrophils. J. Immunol. 134, 2580–2587.

O'Toole, T.E., Loftus, J.C., Du, X., Glass, A.A., Ruggeri, Z.M., Shattil, S.J., Plow, E.F., & Ginsberg, M.H. (1990). Affinity modulation of the $\alpha_{IIb}\beta_3$ integrin (platelet GPIIb-IIIa) is an intrinsic property of the receptor. Cell Regulation 1, 883–893.

O'Toole, T.E., Loftus, J.C., Plow, E.F., Glass, A.A., Harper, J.R., & Ginsberg, M.H. (1989). Efficient surface expression of platelet GPIIb-IIIa requires both subunits. Blood 74, 14–18.

Otey, C.A., Vasquez, G.B., Burridge, K., & Erickson, B.W. (1993). Mapping of the α-actinin binding site within the β_1 integrin cytoplasmic domain. J. Biol. Chem. 268, 21193–21197.

Oxvig, C., & Springer, T.A. (1998). Experimental support for a beta-propeller domain in integrin alpha-subunits and a calcium binding site on its lower surface. Proc. Natl. Acad. Sci. USA 95, 4870–4875.

Pacifici, R., Roman, J., Kimble, R., Civitelli, R., Brownfield, C.M., & Bizzarri, C. (1994). Ligand binding to monocyte alpha 5 beta 1 integrin activates the alpha 2 beta 1 receptor via the alpha 5 subunit cytoplasmic domain and protein kinase C. J. Immunol. 153, 2222–2233.

Parkos, C.A., Delp, C., Arnaout, M.A., & Madara, J.L. (1991). Neutrophil migration across a cultured intestinal epithelium. Dependence on a CD11b/CD18-mediated event and enhanced efficiency in physiological direction. J. Clin. Invest. 88, 1605–1612.

Pasqualini, R., & Hemler, M.E. (1994). Contrasting roles for integrin β_1 and β_5 cytoplasmic domains in subcellular localization, cell proliferation, and cell migration. J. Cell Biol. 125, 447–460.

Pasqualini, R., Bodorova, J., Ye, S., & Hemler, M.E. (1993). A study of the structure, function and distribution of β_5 integrins using novel anti-β_5 monoclonal antibodies. J. Cell Sci. 105, 101–111.

Pavalko, F.M., & Laroche, S.M. (1993). Activation of human neutrophils induces an interaction between the integrin β_2-subunit (CD18) and the actin binding protein α-actinin. J. Immunol. 151, 3795–3807.

Pavalko, F.M., Otey, C.A., Simon, K.O., & Burridge, K. (1991). α-Actinin: A direct link between actin and integrins. Biochem. Soc. Trans. 19, 1065–1069.

Peter, K., & O'Toole, T.E. (1995). Modulation of cell adhesion by changes in alpha(L)beta(2) (LFA-1, CD11a/CD18) cytoplasmic domain/cytoskeleton interaction. J. Exp. Med. 181, 315–326.

Phillips, D.R., Charo, I.F., & Scarborough, R.M. (1991). GPIIb-IIIa: the responsive integrin. Cell 65, 359–362.

Piali, L., Hammel, P., Uherek, C., Bachmann, F., Gisler, R.H., Dunon, D., & Imhof, B.A. (1995). CD31/PECAM is a ligand for alpha-v/beta-3 integrin involved in adhesion of leukocytes to endothelium. J. Cell Biol. 130, 451–460.

Plow, E.F., & Ginsberg, M.H. (1989). Cellular adhesion: GPIIb-IIIa as a prototypic adhesion receptor. Prog. Hemost. Thromb. 9, 117–156.

Pober, J.S., & Cotran, R.S. (1991). What can be learned from the expression of endothelial adhesion molecules in tissues. Lab. Invest. 64, 301–305.

Pober, J.S., Gimbrone, M.A., Jr., Lapierre, L.A., Mendrick, D.L., Fiers, W., Rothlein, R., & Springer, T.A. (1986). Overlapping patterns of activation of human endothelial cells by interleukin 1, tumor necrosis factor, and immune interferon. J. Immunol. 137, 1893–1896.

Polley, M.J., Phillips, M.L., Wayner, E., Nudelman, E., Singhal, A.K., Hakomori, S., & Paulson, J.C. (1991). CD62 and endothelial cell-leukocyte adhesion molecule 1 (ELAM-1) recognize the same carbohydrate ligand, sialyl-Lewis x. Proc. Natl. Acad. Sci. USA 88, 6224–6228.

Porter, J.C. & Hogg, N. (1997). Integrin cross talk: activation of lymphocyte function-associated antigen-1 on human T cells alters alpha4/beta1- and alpha5/beta1-mediated function. J. Cell Biol. 138, 1437–1447.

Puzon-McLaughlin, W., & Takada, Y. (1996). Critical residues for ligand binding in an I domain-like structure of the integrin beta1 subunit. J. Biol. Chem. 271, 20438–20443.

Pytela, R., Pierschbacher, M.D., & Ruoslahti, E. (1985a). Identification and isolation of a 140kd cell surface glycoprotein with properties expected of a fibronectin receptor. Cell 40, 191–196.

Pytela, R., Pierschbacher, M.D., & Ruoslahti, E. (1985b). A 125/115-kDa cell surface receptor specific for vitronectin interacts with the arginine-glycine-aspartic acid adhesion sequence derived from fibronectin. Proc. Natl. Acad. Sci. USA 82, 5766–5770.

Qu, A., & Leahy, D. (1995). Crystal structure of the I-domain from the CD11a/CD18 (LFA-1, alpha L beta. Proc. Nat. Acad. Sci. USA 92, 10277–10281.

Ramaswamy, H., & Hemler, M.E. (1990). Cloning, primary structure and properties of a novel human integrin β subunit. EMBO J. 9, 1561–1568.

Randi, A.M., & Hogg, N. (1994). I domain of β2 integrin lymphocyte function-associated antigen-1 contains a binding site for ligand intercellular adhesion molecule-1. J. Biol. Chem. 269, 12395–12398.

Register, K.B., Ackermann, M.R., & Kehrli, M.E.,Jr. (1994). Non-opsonic attachment of Bordetella bronchiseptica mediated by CD11/CD18 and cell surface carbohydrates. Microbial Pathogenesis 17, 375–385.

Reinhold, M.I., Lindberg, F.P., Plas, D., Reynolds, S., Peters, M.G., & Brown, E.J. (1995). In vivo expression of alternateively spliced forms of integrin-associated protein (CD47). J. Cell Sci. 108, 3419–3425.

Relman, D., Tuomanen, E., Falkow, S., Golenbock, D.T., Saukkonen, K., & Wright, S.D. (1990). Recognition of a bacterial adhesin by an integrin: macrophage CR3 (αMβ2, CD11b/CD18) binds filamentous hemagglutinin of bordetella pertussis. Cell 61, 1375–1382.

Retta, S.F., Balzac, F., Ferraris, P., Belkin, A.M., Fassler, R., Humphries, M.J., De Leo, G., Silengo, L., & Tarone, G. (1998). beta1-integrin cytoplasmic subdomains involved in dominant negative function. Mol. Biol. Cell 9, 715–731.

Rieu, P., Sugimori, T., Griffith, D.L., & Arnaout, M.A. (1996b). Solvent-accessible residues on the metal ion-dependent adhesion site face of integrin CR3 mediate its binding to the neutrophil inhibitory factor. J. Biol. Chem. 271, 15858–15861.

Rosales, C., & Brown, E.J. (1992). Calcium channel blockers nifedipine and diltiazem inhibit Ca^{+2} release from intracellular stores in neutrophils. J. Biol. Chem. 267, 1443–1448.

Rosales, C., Gresham, H.D., & Brown, E.J. (1992). Expression of the 50kD integrin associated protein on myeloid cells and erythrocytes. J. Immunol. 149, 2759–2764.

Ross, G.D., Reed, W., Dalzell, J.G., Becker, S.E., & Hogg, N. (1992). Macrophage cytoskeleton association with CR3 and CR4 regulates receptor mobility and phagocytosis of iC3b-opsonized erythrocytes. J. Leukocyte Biol. 51, 109–117.

Rothlein, R., & Springer, T.A. (1986). The requirement for lymphocyte function-associated antigen 1 in homotypic leukocyte adhesion stimulated by phorbol ester. J. Exp. Med. 163, 1132–1149.

Ruoslahti, E. (1991). Integrins. J. Clin. Invest. 87, 1–5.

Russell, D.G., & Wright, S.D. (1988). Complement receptor type 3 (CR3) binds to an Arg-Gly-Asp-containing region of the major surface glycoprotein, gp63, of Leishmania promastigotes. J. Exp. Med. 168, 279–292.

Saba, T.M., & Di Luzio, N.R. (1965). Kupffer cell phagocytosis and metabolism of a variety of particles as a function of opsonization. J. Reticuloendothelial Soci. 2, 437–453.

Sanchez-Madrid, F., & Corbi, A.L. (1992). Leukocyte integrins: structure, function and regulation of their activity. [Review]. Semin. Cell Biol. 3, 199–210.

Savill, J., Dransfield, I., Hogg, N., & Haslett, C. (1990). Vitronectin receptor-mediated phagocytosis of cells undergoing apoptosis. Nature 343, 170–173.

Schaller, M.D., Borgman, C.A., Cobb, B.S., Vines, R.R., Reynolds, A.B., & Parsons, J.T. (1992). pp125FAK a structurally distinctive protein-tyrosine kinase associated with focal adhesions. Proc. Natl. Acad. Sci. USA 89, 5192–5196.

Schaller, M.D., Otey, C.A., Hildebrand, J.D., & Parsons, J.T. (1995). Focal adhesion kinase and paxillin bind to peptides mimicking beta integrin cytoplasmic domains. J. Cell Biol. 130, 1181–1187.

Schiro, J.A., Chan, B.M.C., Roswit, W.T., Kassner, P.D., Pentland, A.P., Hemler, M.E., Eisen, A.Z., & Kupper, T.S. (1991). Integrin $\alpha^2\beta_1$ (VLA-2) mediates reorganization and contraction of collagen matrices by human cells. Cell 67, 403–410.

Schlesinger, L.S., & Horwitz, M.A. (1991). Phagocytosis of *Mycobacterium leprae* by human monocyte-derived macrophages is mediated by complement receptors CR1 (CD35), CR3 (CD11b/CD18), and CR4 (CD11c/CD18) and IFN-gamma activation inhibits complement receptor function and phagocytosis of this bacterium. J. Immunol. 147, 1983–1994.

Schmidt, C.E., Horwitz, A.F., Lauffenburger, D.A., & Sheetz, M.P. (1993). Integrin-cytoskeletal interactions in migrating fibroblasts are dynamic, asymmetric, and regulated. J. Cell Biol. 123, 977–991.

Schoenwaelder, S.M., Yuan, Y., Cooray, P., Salem, H.H., & Jackson, S.P. (1997). Calpain cleavage of focal adhesion proteins regulates the cytoskeletal attachment of integrin alphaIIbbeta3 (platelet glycoprotein IIb/IIIa) and the cellular retraction of fibrin clots. J. Biol. Chem. 272, 1694–1702.

Senior, R.M., Gresham, H.D., Griffin, G.L., Brown, E.J., & Chung, A.E. (1992). Entactin stimulates neutrophil adhesion and chemotaxis through interactions between its Arg-Gly-Asp (RGD) domain and the leukocyte response integrin (LRI). J. Clin. Invest. 90, 2251–2257.

Shattil, S.J. (1995). Function and regulation of the beta 3 integrins in hemostasis and vascular biology. [Review] [98 refs]. Thrombosis & Haemostasis 74, 149–155.

Shaw, L.M., Messier, J.M., & Mercurio, A.M. (1990). The activation dependent adhesion of macrophages to laminin involves cytoskeletal anchoring and phosphorylation of the $\alpha6\beta1$ integrin. J. Cell Biol. 110, 2167–2174.

Sheppard, D., Rozzo, C., Starr, L., Quaranta, V., Erle, D.J., & Pytela, R. (1990). Complete amino acid sequence of a novel integrin β subunit ($\beta6$) identified in epithelial cells using the polymerase chain reaction. J. Biol. Chem. 265, 11502–11507.

Shevach, E.M. (1992). Integrins, gammaδ T cells, and autoimmunity. Adv. Exp. Med. Biol. 323, 49–55.

Shimizu, Y., & Shaw, S. (1991). Lymphocyte interactions with extracellular matrix. FASEB J. 5, 2292–2299.

Shimizu, Y., Van Seventer, G.A., Horgan, K.J., & Shaw, S. (1990a). Regulated expression and binding of three VLA (beta 1) integrin receptors on T cells. Nature 345, 250–253.

Shimizu, Y., Van Seventer, G.A., Horgan, K.J., & Shaw, S. (1990b). Roles of adhesion molecules in T-cell recognition: fundamental similarities between four integrins on resting human T cells (LFA- 1, VLA-4, VLA-5, VLA-6) in expression, binding, and costimulation. Immunol. Rev. 114, 109–143.

Simon, K.O., Nutt, E.M., Abraham, D.G., Rodan, G.A., & Duong, L.T. (1997). The alphaV/beta3 integrin regulates alpha5/beta1-mediated cell migration toward fibronectin. J. Biol. Chem. 272, 29380–29389.

Simpson, P.J., Todd, R.F., III, Fantone, J.C., Mickelson, J.K., Griffin, J.D., & Lucchesi, B.R. (1988). Reduction of experimental canine myocardial reperfusion injury by a monoclonal antibody (Anti-Mo-1, anti-CD11b) that inhibits leukocyte adhesion. J. Clin. Invest. 81, 624–629.

Singer, I.I. (1982). Association of fibronectin and vinculin with focal contacts and stress fibers in staionary hamster fibroblasts. J. Cell Biol. 92, 398–408.

Singer, I.I., Scott, S., Kawka, D.W., & Kazazis, D.M. (1989). Adhesomes: specific granules containing receptors for laminin, C3bi/fibrinogen, fibronectin, and vitronectin in human polymorphonuclear leukocytes and monocytes. J. Cell Biol. 109, 3169–3182.

Sligh, J.E., Jr., Hurwitz, M.Y., Zhu, C., Anderson, D.C., & Beaudet, A.L. (1992). An initiation codon mutation in CD18 in association with the moderate phenotype of leukocyte adhesion deficiency. J. Biol. Chem. 267, 714–718.

Smith, J.W., & Cheresh, D.A. (1990). Integrin (αvβ3)-ligand interaction: identification of a heterodimeric RGD binding site on the vitronectin receptor. J. Biol. Chem. 265, 2168–2172.

Smith, J.W., & Cheresh, D.A. (1991). Labeling of integrin $\alpha_v\beta_3$ with ^{58}Co(III). Evidence of metal ion coordination sphere involvement in ligand binding. J. Biol. Chem. 266, 11429–11432.

Smith, J.W., Ruggeri, Z.M., Kunicki, T.J., & Cheresh, D.A. (1990a). Interaction of integrins αvβ3 and glycoprotein IIb-IIIa with fibrinogen: differential peptide recognition accounts for distinct binding sites. J. Biol. Chem. 265, 12267–12271.

Smith, J.W., Vestal, D.J., Irwin, S.V., Burke, T.A., & Cheresh, D.A. (1990b). Purification and functional characterization of integrin αvβ5: an adhesion receptor for vitronectin. J. Biol. Chem. 265, 11008–11013.

Smyth, S.S., & Parise, L.V. (1993). Regulation of ligand binding to glycoprotein IIb-IIIa (integrin alpha IIb beta 3) in isolated platelet membranes. Biochem. J. 292, 749–758.

Smyth, S.S., Hillery, C.A., & Parise, L.V. (1992). Fibrinogen binding to purified platelet glycoprotein IIb-IIIa (integrin $\alpha_{IIb}\beta_3$) is modulated by lipids. J. Biol. Chem. 267, 15568–15577.

Sonnenberg, A., Linders, C.J.T., Modderman, P.W., Damsky, C.H., Aumailley, M., & Timpl, R. (1990). Integrin recognition of different cell-binding fragments of laminin (P1, E3, E8) and evidence that α6β1 but not α6β4 functions as a major receptor for fragment E8. J. Cell Biol. 110, 2145–2155.

Sorokin, L., Sonnenberg, A., Aumailley, M., Timpl, R., & Ekblom, P. (1990). Recognition of the laminin E8 cell-binding site by an integrin possessing the alpha 6 subunit is essential for epithelial polarization in developing kidney tubules. J. Cell Biol. 111, 1265–1273.

Spertini, O., Kansas, G.S., Munro, J.M., Griffin, J.D., & Tedder, T.F. (1991). Regulation of leukocyte migration by activation of the leukocyte adhesion molecule-1 (LAM-1) selectin. Nature 349, 691–694.

Springer, T.A. (1990). Adhesion receptors of the immune system. Nature 346, 425–433.

Springer, T.A. (1994). Traffic signals for lymphocyte recirculation and leukocyte emigration: the multistep paradigm. [Review]. Cell 76, 301–314.

Springer, T.A. (1997). Folding of the N-terminal, ligand-binding region of integrin alpha-subunits into a beta-propeller domain. Proc. Natl. Acad. Sci. USA 94, 65–72.

Springer, T.A., & Lasky, L.A. (1991). Cell adhesion: Sticky sugars for selectins. Nature 349, 196–197.

Springer, T.A., Thompson, W.S., Miller, L.J., Schmalstieg, F.C., & Anderson, D.C. (1984). Inherited deficiency of the Mac-1, LFA-1, p150,95 glycoprotein family and its molecular basis. J. Exp. Med. 160, 1901–1918.

Staatz, W.D., Rajpara, S.M., Wayner, E.A., Carter, W.G., & Santoro, S.A. (1989). The membrane glycoprotein Ia-IIa (VLA-2) complex mediates the Mg^{++}-dependent adhesion of platelets to collagen. J. Cell Biol. 108, 1917–1924.

Staatz, W.D., Walsh, J.J., Pexton, T., & Santoro, S.A. (1990). The alpha 2 beta 1 integrin cell surface collagen receptor binds to the alpha 1 (I)-CB3 peptide of collagen. J. Biol. Chem. 265, 4778–4781.

Stanislawski, L., Simpson, W.A., Hasty, D., Sharon, N., Beachey, E.H., & Ofek, I. (1985). Role of fibronectin in attachment of *Streptococcus pyogenes* and *Escherichia coli* to human cell lines and isolated oral eithelial cells. Infect. Immun. 48, 257–259.

Stanley, P., Bates, P.A., Harvey, J., Bennett, R.I., & Hogg, N. (1994). Integrin LFA-1 α subunit contains an ICAM-1 binding site in domains V and VI. EMBO J. 13, 1790–1798.

Stepp, M.A., Spurr-Michaud, S., Tisdale, A., Elwell, J., & Gipson, I.K. (1990). Alpha 6 beta 4 integrin heterodimer is a component of hemidesmosomes. Proc. Natl. Acad. Sci. USA 87, 8970–8974.

Stewart, M.P., McDowall, A., & Hogg, N. (1998). LFA-1-mediated adhesion is regulated by cytoskeletal restraint and by a Ca2+-dependent protease, calpain. J. Cell Biol. 140, 699–707.

Suzuki, S., Argraves, W.S., Arai, H., Languino, L.R., Pierschbacher, M.D., & Ruoslahti, E. (1987). Amino acid sequence of the vitronectin receptor a subunit and comparative expression of adhesion receptor mRNAs. J. Biol. Chem. 262, 14800–14805.

Suzuki, S., Huang, Z.S., & Tanihara, H. (1990). Cloning of an integrin beta subunit exhibiting high homology with integrin beta 3 subunit. Proc. Natl. Acad. Sci. USA 87, 5354–5358.

Takada, Y., & Hemler, M.E. (1989). The primary structure of the VLA-2/collagen receptor $\alpha 2$ subunit (platelet GPIa): homology to other integrins and the presence of a possible collagen- binding domain. J. Cell Biol. 109, 397–407.

Takada, Y., Elices, M.J., Crouse, C., & Hemler, M.E. (1989). The primary structure of the alpha 4 subunit of VLA 4: homology to other integrins and a possible cell-cell adhesion function. EMBO J. 8, 1361–1368.

Takada, Y., Huang, C., & Hemler, M.E. (1987a). Fibronectin receptor structures in the VLA family of heterodimers. Nature 326, 607–609.

Takada, Y., Strominger, J.L., & Hemler, M.E. (1987b). The very late antigen family of heterodimers is part of a superfamily of molecules involved in adhesion and embryogenesis. Proc. Natl. Acad. Sci. USA 84, 3239–3243.

Takada, Y., Wayner, E.A., Carter, W.G., & Hemler, M.E. (1988). Extracellular matrix receptors, ECM-RII and ECMRI, for collagen and fibronectin correspond to VLA-2 and VLA 3 in the VLA family of heterodimers. J. Cell Biochem. 37, 385–393.

Talamas Rohana, P., Wright, S.D., Lennartz, M.R., & Russell, D.G. (1990). Lipophosphoglycan from Leishmania mexicana promastigotes binds to members of the CR3, p150,95 and LFA-1 family of leukocyte integrins. J. Immunol. 144, 4817–4824.

Tamkun, J.W., Desimone, D.W., Fonda, D., Patel, R.S., Buck, C., Horowitz, A.F., & Hynes, R.O. (1986). Structure of integrin, a glycoprotein involved in the transmembrane linkage between fibronectin and actin. Cell 46, 271–282.

Tanoue, K., Katagiri, Y., & Suzuki, H. (1996). Molecular mechanisms of fibrin gel clot retraction by platelets and cultured tumor cells. Polish Journal of Pharmacology 48, 341–343.

Teixidó, J., Parker, C.M., Kassner, P.D., & Hemler, M.E. (1992). Functional and structural analysis of VLA-4 integrin α^4 subunit cleavage. J. Biol. Chem. 267, 1786–1791.

Tidswell, M., Pachynski, R., Wu, S.W., Qiu, S.Q., Dunham, E., Cochran, N., Briskin, M.J., Kilshaw, P.J., Lazarovits, A.I., Andrew, D.P., Butcher, E.C., Yednock, T.A., & Erle, D.J. (1997). Structure-function analysis of the integrin beta 7 subunit: identification of domains involved in adhesion to MAdCAM-1. J. Immun. 159, 1497–1505.

Todd, R.F., & Freyer, D.R. (1988). The CD11/CD18 leukocyte glycoprotein deficiency. Hematol. Oncol. Clin. North Am. 2, 13–31.

Tozer, E., Hughes, P., & Loftus, J. (1996). Ligand binding and affinity modulation of integrins. Biochem. Cell Biol. 74, 785–798.

Tuomanen, E.I., Saukkonen, K., Sande, S., Cioffe, C., & Wright, S.D. (1989). Reduction of inflammation, tissue damage, and mortality in bacterial meningitis in rabbits treated with monoclonal antibodies against adhesion-promoting receptors of leukocytes. J. Exp. Med. 170, 959–969.

Valmu, L., Autero, M., Siljander, P., Patarroyo, M., & Gahmberg, C.G. (1991). Phosphorylation of the β-subunit of CD11/CD18 integrins by protein kinase C correlates with leukocyte adhesion. Eur. J. Immunol. 21, 2857–2862.

Van Aelst, L., & D'Souza-Schorey, C. (1997). Rho GTPases and signaling networks. [Review] [295 refs]. Genes Dev. 11, 2295–2322.

Van der Vieren, M., Le Trong, H., Wood, C.L., Moore, P.F., St. John, T., Staunton, D.E., & Gallatin, W.M. (1995). A novel leukointegrin, alpha d beta 2, binds preferentially to ICAM-3. Immunity 3, 683–690.

van Kooyk, Y., & Figdor, C.G. (1993). Activation and inactivation of adhesion molecules. [Review]. Curr. Top. Microbiol. Immunol. 184, 235–256.

van Kooyk, Y., Weder, P., Hogervorst, F., Verhoeven, A.J., van Seventer, G., te Velde, A.A., Borst, J., Keizer, G.D., & Figdor, C.G. (1991). Activation of LFA-1 through a Ca^{+2}-dependent epitope stimulates lymphocyte adhesion. J. Cell Biol. 112, 345–354.

Varner, J.A., Brooks, P.C., & Cheresh, D.A. (1995). Review: the integrin alpha V beta 3: angiogenesis and apoptosis. [Review] [60 refs]. Cell Adhes. Commun. 3, 367–374.

Vedder, N.B., Winn, R.K., Rice, C.L., Chi, E.Y., Arfors, K.-E., & Harlan, J.M. (1988). A monoclonal antibody to the adherence-promoting leukocyte glycoprotein, CD18, reduces organ injury and improves survival from hemorrhagic shock and resuscitation in rabbits. J. Clin. Invest. 81, 939–944.

Vignoud, L., Albiges-Rizo, C., Frachet, P., & Block, M.R. (1997). NPXY motifs control the recruitment of the alpha5/beta1 integrin in focal adhesions independently of the association of talin with the beta1 chain. J. Cell Sci. 110, 1421–1430.

Virtanen, I., Korhonen, M., Kariniemi, A.L., Gould, V.E., Laitinen, L., & Ylänne, J. (1990). Integrins in human cells and tumors. Cell Differ. Dev. 32, 215–227.

Vogel, B.E., Lee, S.-J., Hildebrand, A., Craig, W., Pierschbacher, M.D., Wong-Staal, F., & Ruoslahti, E. (1993). A novel integrin specificity exemplified by binding of the $\alpha_v\beta_5$ integrin to the basic domain of the HIV Tat protein and vitronectin. J. Cell Biol. 121, 461–468.

Von Andrian, U.H., Berger, E.M., Ramezani, L., Chambers, J.D., Ochs, H.D., Harlan, J.M., Paulson, J.C., Etzioni, A., & Arfors, K.E. (1993). In vivo behavior of neutrophils from 2 patients with distinct inherited leukocyte adhesion deficiency syndromes. J. Clin. Inv. 17, 2893–2897.

Von Andrian, U.H., Chambers, J.D., McEvoy, L.M., Bargatze, R.F., Arfors, K.-E., & Butcher, E.C. (1991). Two-step model of leukocyte-endothelial cell interaction in inflammation: Distinct roles for LECAM-1 and the leukocyte β_2 integrins *in vivo*. Proc. Natl. Acad. Sci. USA 88, 7538–7542.

Walsh, G.M., Mermod, J.-J., Hartnell, A., Kay, A.B., & Wardlaw, A.J. (1991). Human eosinophil, but not neutrophil, adherence to IL-1-stimulated human umbilical vascular endothelial cells is $\alpha_4\beta_1$ (very late antigen-4) dependent. J. Immunol. 146, 3419–3423.

Wayner, E.A., Orlando, R.A., & Cheresh, D.A. (1991). Integrins αvβ3 and αvβ5 contribute to cell attachment to vitronectin but differentially distribute on the cell surface. J. Cell Biol. 113, 919–929.

Weber, C., Alon, R., Moser, B., & Springer, T.A. (1996). Sequential regulation of alpha 4 beta 1 and alpha 5 beta 1 integrin avidity by CC chemokines in monocytes: implications for transendothelial chemotaxis. J. Cell Biol. 134, 1063–1073.

Williams, D.A., Rios, M., Stephens, C., & Patel, V.P. (1991). Fibronectin and VLA-4 in haematopoietic stem cell-microenvironment interactions. Nature 352, 438–441.

Wong, D.A., Davis, E.M., LeBeau, M., & Springer, T.A. (1996). Cloning and chromosomal localization of a novel gene-encoding a human beta 2-integrin alpha subunit. Gene 171, 291–294.

Wright, S.D., & Meyer, B.C. (1986). Phorbol esters cause sequential activation and deactivation of complement receptors on polymorphonuclear leukocytes. J. Immunol. 136, 1758–1764.

Wright, S.D., Craigmyle, L.S., & Silverstein, S.C. (1983a). Fibronectin and serum amyloid P-component stimulate C3b- and C3bi-mediated phagocytosis in cultured human monocytes. J. Exp. Med. 158, 1338–1342.

Wright, S.D., Rao, P.E., Wesley, C., van Voorhis, W.C., Craigmyle, L.S., Iida, K., Talle, M.A., Westberg, E.F., Goldstein, G., & Silverstein, S.C. (1983b). Identification of the C3bi receptor of human monocytes and macrophages by using monoclonal antibodies. Proc. Natl. Acad. Sci. USA 80, 5699–5703.

Wright, S.D., Weitz, J.I., Huang, A.J., Levin, S.M., Silverstein, S.C., & Loike, J.D. (1988). Complement receptor type three (CD11b/CD18) of human polymorphonuclear leukocytes recognizes fibrinogen. Proc. Natl. Acad. Sci. USA 85, 7734–7738.

Wu, C., Hughes, P.E., Ginsberg, M.H., & McDonald, J.A. (1996). Identification of a new biological function for the integrin alpha v beta 3: initiation of fibronectin matrix assembly. Cell Adhesion & Communication 4, 149–158.

Wu, C., Keivens, V.M., O'Toole, T.E., McDonald, J.A., & Ginsberg, M.H. (1995). Integrin activation and cytoskeletal interaction are essential for the assembly of a fibronectin matrix. Cell 83, 715–724.

Wyler, D.J., Sypek, J.P., & McDonald, J.A. (1985). In vitro parasite-monocyte interactions in human leishmaniasis: possible role of fibronectin in parasite attachment. Infect. Immun. 49, 305–311.

Xing, Z., Jordana, M., & Gauldie, J. (1992). IL-1β and IL-6 gene expression in alveolar macrophages: Modulation by extracellular matrices. Am. J. Physiol. Lung Cell. Mol. Physiol. 262, L600–L605.

Yauch, R.L., Felsenfeld, D.P., Kraeft, S.K., Chen, L.B., Sheetz, M.P., & Hemler, M.E. (1997). Mutational evidence for control of cell adhesion through integrin diffusion/clustering, independent of ligand binding. J. Exp. Med. 186, 1347–1355.

Ylänne, J., Chen, Y., O'Toole, T.E., Loftus, J.C., Takada, Y., & Ginsberg, M.H. (1993). Distinct functions of integrin α and subunit β cytoplasmic domains in cell spreading and formation of focal adhesions. J. Cell Biol. 122, 223–233.

Yoder, M.C., Checkley, L.L., Giger, U., Hanson, W.L., Kirk, K.R., Capen, R.L., & Wagner, W.W.,Jr. (1990). Pulmonary microcirculatory kinetics of neutrophils deficient in leukocyte adhesion-promoting glycoproteins. J. Appl. Physiol. 69, 207–213.

Yuan, Q., Jiang, W., Hollander, D., Leung, E., Watson, J.D., & Krissansen, G.W. (1991). Identity between the novel integrin β_7 subunit and an antigen found highly expressed on intraepithelial lymphocytes in the small intestine. Biochem. Biophys. Res. Commun. 176, 1443–1449.

Yuan, Y., Dopheide, S.M., Ivanidis, C., Salem, H.H., & Jackson, S.P. (1997). Calpain regulation of cytoskeletal signaling complexes in von Willebrand factor-stimulated platelets. Distinct roles for glycoprotein Ib-V-IX and glycoprotein IIb-IIIa (integrin alphaIIbbeta3) in von Willebrand factor-induced signal transduction. J. Biol. Chem. 272, 21847–21854.

Zhang, J., Shattil, S.J., Cunningham, M.C., & Rittenhouse, S.E. (1996a). Phosphoinositide 3-kinase gamma and p85/phosphoinositide 3- kinase in platelets. Relative activation by thrombin receptor or beta-phorbol myristate acetate and roles in promoting the ligand- binding function of alphaIIbbeta3 integrin. J. Biol. Chem. 271, 6265–6272.

Zhang, Z., Vuori, K., Wang, H., Reed, J.C., & Ruoslahti, E. (1996b). Integrin activation by R-ras. Cell 85, 61–69.

Zhou, M., Todd, R.F., III, Van de Winkel, J.G.J., & Petty, H.R. (1993). Cocapping of the leukoadhesin molecules complement receptor type 3 and lymphocyte function-associated antigen-1 with Fc_{gamma} receptor III on human neutrophils: possible role of lectin-like interactions. J. Immunol. 150, 3030–3041.

Zhou, M.-J., & Brown, E.J. (1994). CR3 (Mac-1, $\alpha_M\beta_2$, CD11b/CD18) and Fc(gamma)RIII cooperate in generation of a neutrophil respiratory burst: requirement for Fc(gamma)RII and tyrosine phosphorylation. J. Cell Biol. 125, 1407–1416.

Zhou, M.-J., Todd, R.F., III, & Petty, H.R. (1991). Detection of transmembrane linkages between immunoglobulin or complement receptors and the neutrophil's cortical microfilaments by resonance energy transfer microscopy. J. Mol. Biol. 218, 263–268.

Zusman, S., Patel King, R.S., Ffrench Constant, C., & Hynes, R.O. (1990). Requirements for integrins during Drosophila development. Development 108, 391–402.

Fc RECEPTOR–MEDIATED PHAGOCYTOSIS

Steven Greenberg

Advances in Cell and Molecular Biology of Membranes and Organelles
Volume 5, pages 149-191.

ISBN: 1-55938-999-0

I. INTRODUCTION

Phagocytosis is a phylogenetically ancient response of motile cells to particulate stimuli. This response is utilized by specialized cells of the immune system to aid in host defenses. The striking resemblance between phagocytosis mediated by simple organisms, such as *Dictyostelium*, and of higher-order eukaryotes, implies that the fundamental mechanisms of pseudopod extension and particle engulfment are conserved. The adaptation by components of the acquired immune system to the host phagocytic machinery suggests that Fc receptors function by utilizing a shared program of signaling cascades and cytoskeletal rearrangements.

II. STRUCTURE AND FUNCTION OF THE FAMILY OF FC RECEPTORS

A. Fc_γ Receptors

There are three principal types of human receptors which recognize the Fc domain of immunoglobulin G (reviewed in Hulett and Hogarth, 1994; Ravetch, 1994; Indik et al., 1995; Unkeless et al., 1995) (Figure 1). All three types are members of the Ig superfamily and all are capable of independently triggering phagocytosis (Anderson et al., 1990).

Fc_γ Receptor I (CD64)

In humans, $Fc_\gamma RI$ is expressed on monocytes, macrophages and IFN-γ-treated neutrophils and eosinophils (reviewed in van de Winkel et al., 1991) (Table 1). Three genes for $Fc_\gamma RI$ have been identified (Ernst et al., 1992); however, only one ($Fc_\gamma RIA$) encodes a transmembrane protein. A distinguishing feature of $Fc_\gamma RI$ is its relatively high affinity for IgG (K_a of 10^8 to 10^9 M^{-1} in the human (Lubeck et al., 1985) and 10^7 to 10^8 M^{-1} in the mouse (Unkeless and Eisen, 1975)). The structural basis for the ability to bind Ig with high affinity resides, in part, in a third binding domain within the extracellular region of the receptor

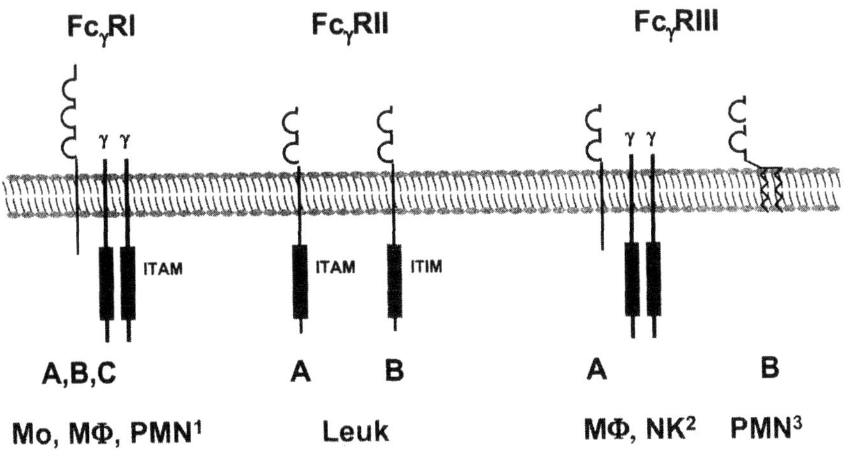

Figure 1. Domain structure of Fc$_\gamma$Rs. A,B,C refer to transcripts of individual α (i.e., ligand binding domain) Fc$_\gamma$R subunits. Mo, monocytes; Mϕ, macrophages; Leuk, leukocytes; PMN, polymorphonuclear leukocytes.[1] Fc$_\gamma$RI in neutrophils is expressed only upon cytokine stimulation (e.g., IFN-γ). [2]In NK cells, the a subunit of Fc$_\gamma$RIII is associated with a ζ subunit homodimer, instead of a γ subunit homodimer. [3]Fc$_\gamma$RIIIB, the isoform expressed in neutrophils, is GPI-linked.

that is not present in Fc$_\gamma$RII and III (Hulett et al., 1991). A recent study implicates the associated γ chain in the acquisition of the receptor's high affinity for IgG. This property endows the receptor with the ability to bind monomeric immunoglobulin, similar to the high-affinity receptor for IgE. Freshly harvested monocytes express approximately Fc$_\gamma$RI 10–40,000 molecules on their surfaces (Perussia et al., 1987); this expression is increased by several agents, including IFN-γ (Perussia et al., 1987), G-CSF, granulocyte macrophage colony-stimulating factor (GM-CSF) (Buckle and Hogg, 1989), interleukin (IL)-10 (te Velde et al., 1992), glucocorticoids (Girard et al., 1987), and the complement component C5a (Yancey et al., 1985). Not surprisingly, neutrophils isolated from patients with acute pyogenic infections express greater quantities of Fc$_\gamma$RI (but not Fc$_\gamma$RII or III) on their surfaces (Simms et al., 1989). Using erythrocytes coated with $F(ab')_2$ fragments of mAb 32.2 (which recognize human Fc$_\gamma$RI), Anderson and colleagues confirmed the phagocytic capacity of this type of Fc receptor (Anderson et al., 1990). Fc$_\gamma$RI is also capable of mediating production of superoxide anion (Anderson et al., 1986) and tumor necrosis factor (TNF)-α (Debets et al., 1988, 1990) production. The identification of four related healthy individuals whose monocytes lack surface expression of Fc$_\gamma$RI (Ceuppens et al., 1988) suggests that expression of Fc$_\gamma$RI in humans is not essential for immunity to infections.

Table 1. Cellular Distribution of Fc Receptors

Cell Types	*Receptor*	*Target*	*Ligand*	*References*
Mo[a], Mϕ, PMN[b]	$Fc_\gamma RI$ (CD64)	IgG-opsonized bacteria	Fc portion of IgG	Anderson et al., 1990
Leuk, Plts	$Fc_\gamma RIIA$ (CD32)	IgG-opsonized bacteria	Fc portion of IgG	Looney et al., 1986; Anderson et al., 1990
Mϕ, Mo[c]	$Fc_\gamma RIIIA$ (CD16)	IgG-opsonized bacteria	Fc portion of IgG	Huizinga et al., 1989; Anderson et al., 1990
PMN	$Fc_\gamma RIIIB$ (CD16)[d]	IgG-opsonized bacteria	Fc portion of IgG	Salmon et al., 1987
Mast, Eo	$Fc_\varepsilon RI$	IgE-opsonized particles	Fc portion of IgE	Daeron et al., 1994; Pierini et al., 1996
PMN, Mo	$Fc_\alpha R$ (CD89)	IgA-opsonized particles	Fc portion of IgA	Weisbart et al., 1988; Shen et al., 1989;

Notes: [a] Mo, monocytes; Mϕ, macrophages; Leuk, leukocytes; PMN, polymorphonuclear leukocytes; Plts, platelets; Eo, eosinophils; Mast, mast cells.
[b] $Fc_\gamma RI$ is present only on IFN-γ-stimulated PMNs.
[c] A subset of circulating monocytes (~13%) express $Fc_\gamma RIIIA$ (Passlick et al., 1989).
[d] $Fc_\gamma RIIIB$ on PMNs is GPI-linked.

Fc_γ Receptor II (CD32)

Isoforms of $Fc_\gamma RII$ are expressed on nearly all hematopoetic cells, including platelets (reviewed in Hulett and Hogarth, 1994) (Table 1). In the human, three distinct genes, termed IIA, IIB and IIC, each encoding one or more transcript, have been identified (Brooks et al., 1989). There is no murine homologue of hFcRIIA. Transcripts of all three genes are expressed in monocytes and macrophages, whereas human neutrophils express predominantly $Fc_\gamma RIIA$, with lower levels of expression of $Fc_\gamma RIIB$ and C (Brooks et al., 1989). Two well-described allelic variants of $Fc_\gamma RIIA$ have been identified, termed high-responder (HR) and low-responder (LR) (corresponding to strong or weak interaction with murine IgG1). This polymorphism has been mapped to amino acid residue 131 within the second extracellular domain (Warmerdam et al., 1990; Clark et al., 1991). In contrast to results using murine Ig, the presence of arginine at this position confers weak binding to human IgG2, whereas the presence of histidine confers strong binding to this human Ig isotype (Warmerdam et al., 1991; Parren et al., 1992). Since $Fc_\gamma RIIA$ is the only Fc receptor that binds hIgG2 appreciably, the expression of $Fc_\gamma RIIA$-R131 (i.e., the HR allele) would be expected to lead to a generalized defect in the binding of hIgG2 by phagocytic leukocytes. Studies of individuals expressing the HR allele of $Fc_\gamma RIIA$ have verified this (Salmon et al., 1992), and linkage studies have established a correlation between the expression of $Fc_\gamma RIIA$-R131 and severity of immune complex disease in patients with lupus (Duits et al.,

1995; Salmon et al., 1996). Whether disease susceptibility is due to decreased clearance of IgG2-containing immune complexes remains to be established.

The surface expression of Fc_{γ}RII varies from approximately 30,000 receptors on the human neutrophil (Selvaraj et al., 1988) to about 260,000 receptors on the human alveolar macrophage (Rossman et al., 1989). IL-4 decreases the surface expression of Fc_{γ}RII on cultured human monocytes (te Veld et al., 1990).

Fc_{γ}RIIA mediates phagocytosis of IgG-coated particles by human neutrophils and mononuclear cells (Anderson et al., 1990). Studies using COS cells (Indik et al., 1991), fibroblasts (Tuijnman et al., 1992), and macrophages (Odin et al., 1991) transfected with cDNAs for either human Fc_{γ}RIIA and IIB indicate that the former, and not the latter, mediates phagocytosis. Distinct residues within the cytosolic domains of Fc_{γ}RIIA are required for this function. These residues, which lie within the immunoreceptor tyrosine activation motif (ITAM) (see below), confer phagocytic signaling. In contrast, Fc_{γ}RIIB, which is also expressed on myeloid cells, and is the exclusive Fc_{γ}RII isoform expressed on B cells (Brooks et al., 1989), contains an immunoreceptor tyrosine inhibition motif (ITIM) (see below). Fc_{γ}RII also mediates ADCC (Walker et al., 1991) and superoxide anion (Huizinga et al., 1989) and TNF-α (Debets et al., 1990) secretion.

Fc_{γ} Receptor III (CD16)

Fc_{γ}RIII is expressed on macrophages, subpopulations of monocytes, neutrophils, eosinophils, NK cells, and γδ T cells (reviewed in Hulett and Hogarth, 1994) (Table 1). Two genes encoding Fc_{γ}RIII have been identified that encode structurally distinct cell-specific proteins. Fc_{γ}RIIIA is expressed as a transmembrane protein in macrophages, NK cells, and γδ T cells, whereas Fc_{γ}RIIIB is expressed as a GPI-linked protein in neutrophils and eosinophils (Huizinga et al., 1988; Hibbs et al., 1989). Like Fc_{γ}RI, surface expression of the transmembrane form of Fc_{γ}RIII is dependent on the co-expression of either the γ subunit or the ζ subunit of the T cell antigen receptor/CD3 complex (Kurosaki and Ravetch, 1989; Lanier et al., 1989; Anderson et al., 1990). Two allelic forms of FcγRIIIb have been identified (NA-1 and NA-2) that differ in four amino acids which encode two N-linked glycosylation sites. Expression of the NA-2 allele is associated with decreased phagocytic capacity of this Fc_{γ}R (Salmon et al., 1990). Other polymorphisms have been detected, including one (pheylalanine to valine substitution at amino acid position 176) that is correlated with greater binding of hIgG1 and 3 (Wu et al., 1997). Transforming growth factor (TGF)-β, but not a variety of other cytokines, increases the surface expression of Fc_{γ}RIII on human monocytes (Welch et al., 1990), whereas IFN-γ, G-CSF, and GM-CSF increase surface expression of Fc_{γ}RIII on human neutrophils (Buckle and Hogg, 1989).

Fc_{γ}RIIIA on human macrophages mediates phagocytosis (Anderson et al., 1990). Despite its non-transmembrane topology, Fc_{γ}RIIIB on human neutrophils is capable of mediating transmembrane signaling, including increases in cytosolic free calcium concentration ($[Ca^{2+}]_i$); (Kimberly et al., 1990) and F-actin content (Salmon et al., 1991). While Fc_γRIIIB demonstrates some independent capacity to mediate phagocytosis, the function of Fc_γRIIIB *in vivo* may be to enhance binding of IgG-bound ligands and act in concert with Fc_γRII to promote phagocytosis in human neutrophils. Salmon and co-workers found that mAb 3G8 (directed against Fc_γRIII) inhibited ingestion, but not binding, of Con A-coated erythrocytes and unopsonized *Escherichia coli* by human neutrophils. They suggested that clustering of Fc_γRIII via its mannose-containing oligosaccharides generated a phagocytic signal (Salmon et al., 1987). Thus, Fc_γRIIIB, like CR3 (Ross et al., 1985; Altieri and Edgington, 1988; Wright et al., 1988, 1989; Relman et al., 1990; Diamond et al., 1995), has more than one functional ligand binding domain.

While the exact mechanism of signaling by GPI-linked proteins is in doubt, an emerging consensus is that they occupy microdomains on the cell surface, possibly physically associating with other proteins bearing enzymatic activities (reviewed in Simons and Ikonen, 1997). This mechanism of signaling seems likely in light of recent data indicating that a variety of GPI-linked proteins have the capacity to augment Ca^{2+} signaling by Fc_γRIIA, when co-clustered with this Fc_γR (Green et al., 1997). Another plausible mechanism for signaling by Fc_γRIIIB is through its physical association with other phagocytosis-promoting receptors, such as CR3 (Zhou et al., 1993; Poo et al., 1995).

Other functions of Fc_γRIII include ADCC of tumor cells by NK cells and neutrophils (Perussia et al., 1984), and clearance of immune complexes from the bloodstream (Clarkson et al., 1986).

B. Fc_ε Receptors

There are two recognized receptors for the Fc portion of IgE. Both are capable of mediating cytoskeletal rearrangements and phagocytosis (Pfeiffer et al., 1985; Yokota et al., 1992; Pierini et al., 1996). It is not clear to what extent their phagocytic function is important in host defenses.

Fc_εRI

Fc_εRI is expressed on mast cells, basophils, eosinophils, dendritic cells, activated monocytes, and platelets (reviewed in Adamczewski and Kinet, 1994). In the human, one gene encoding one transcript has been identified for the α subunit. This subunit binds the Fc portion of IgE with high-affinity (K_a of 10^{10} M^{-1}). The α subunit, containing two Ig-like domains, associates non-covalently with a β subunit and a γ subunit homodimer. This subunit is identical to the γ subunit of Fc_γRI and III (Ra et al., 1989). Hydropathicity plots predict that the β subunit has four transmembrane domains. Surface expression of the α subunit requires expression of the γ subunit, suggesting that inter-subunit association is required for plasma membrane targeting and/or stability of the complex (Miller et al., 1989). There are no recognized polymorphisms of the α subunit in the human. Recent studies have identified polymorphisms in the gene for the β subunit that are associated with atopy and/or bronchial hyperresponsiveness (Hill et al., 1995; Hill and Cookson, 1996; Palmer et al., 1997). While a functional role for these polymorphisms has not been established, recent studies by Kinet and colleagues indicate that the β subunit may function as an amplifier of effector responses (Lin et al., 1996) (see below); it is possible that polymorphisms which affect the coding region of the β subunit influence this function. Cytokines that up-regulate expression of $Fc_{\varepsilon}RI$ include stem cell factor, IL-3 (Rottem et al., 1994; Lantz and Huff, 1995), IL-4 (Xia et al., 1997), and IL-9 (Louahed et al., 1995).

The best-characterized function of $Fc_{\varepsilon}RI$ is to mediate degranulation and secretion of various inflammatory mediators, such as arachidonate and its metabolites, histamine, and neutral proteases, including tryptase (reviewed in Beaven and Metzger, 1993; Gilfillan, 1997; Hamawy and Swaim, 1997). Ligation of $Fc_{\varepsilon}RI$ mediates membrane ruffling (Pfeiffer et al., 1985) and phagocytosis of antigen-derivatized beads (Pierini et al., 1996). Aggregation of $Fc_{\varepsilon}RI$ also leads to gene expression (e.g., TNF-α, IL-6, and GM-CSF) (Richards et al., 1988; Gordon and Galli, 1990; Gurish et al., 1991).

$Fc_{\varepsilon}RII$ (CD23)

$Fc_{\varepsilon}RII$ (CD23) is expressed on monocytes, macrophages, eosinophils, lymphocytes and platelets (reviewed in Bonnefoy et al., 1995, 1996). It belongs to the C-lectin family of type II membrane proteins and binds IgE (with low-affinity), L-fucose, CD21, CD11b, and CD11c. In the human $Fc_{\varepsilon}RII$ is encoded by one gene; two mRNA species, termed $Fc_{\varepsilon}RIIa$ and $Fc_{\varepsilon}RIIb$, are generated by utilizing different transcription initiation sites. $Fc_{\varepsilon}RIIa$ is expressed constitutively on B cells while $Fc_{\varepsilon}RIIb$ is expressed on IL-4- and IL-13-stimulated B cells and monocytes (McKenzie et al., 1993). IL-4-stimulated expression of CD23 is antagonized by TNF-α (Hashimoto et al., 1995). Expression of CD23 is enhanced by IL-7 in human T cells (Carini and Fratazzi, 1996) and by ligation of CD40 on B cells (Gordon et al., 1991; Maliszewski et al., 1993). Studies in CD23-knockout mice indicate that a major function of this Fc receptor is to mediate downregulation of IgE synthesis following immunization with thymus-dependent antigens (Yu et al., 1994). $Fc_{\varepsilon}RIIb$ expressed on macrophages mediates phagocytosis of IgE-coated particles, while $Fc_{\varepsilon}RIIa$ does not. Asn and Pro residues in the cytosolic domain of $Fc_{\varepsilon}RIIb$ were shown to be critical for conferring phagocytosis by this receptor (Yokota et al., 1992). The observation that ligation of CD23 on human monocytes leads to production of iNOS mRNA and nitric oxide (Vouldoukis et al., 1995) suggests that it may play a more general role in inflammation during a T_H2-type lymphocyte response, or when IL-4 is generated locally. The interaction of soluble CD23 with CD11b and CD11c on human monocytes results in expression of pro-inflammatory cytokines, such as IL-1β, IL-6, and TNF-α (Lecoanet-Henchoz et al., 1995).

C. Fc_{α}Receptor

$Fc_{\alpha}R$ is expressed on monocytes, macrophages, neutrophils and eosinophils (reviewed in Morton et al., 1996). Its gene has been cloned and its sequence identifies it as a member of the Ig superfamily (Maliszewski et al., 1990). Three distinct transcripts have been identified in monocytes and macrophages, although only two have been shown to be expressed at the cell surface. Alveolar macrophages preferentially express $Fc_{\alpha}Ra.2$, which lacks 22 amino acids otherwise found in the extracellular domain of $Fc_{\alpha}Ra.1$ (Patry et al., 1996). $Fc_{\alpha}R$ binds the Fc portion of monomeric and polymeric IgA1 and IgA2, and associates non-covalently with a γ subunit homodimer common to $Fc_{\varepsilon}RI$, $Fc_{\gamma}RI$, and $Fc_{\gamma}RIIIA$ (Pfefferkorn and Yeaman, 1994). However, unlike the latter Fc receptors, expression of $Fc_{\alpha}RI$ does not require the presence of the γ subunit for cell surface expression (Maliszewski et al., 1990). Cell surface expression of $Fc_{\alpha}R$ is rapidly increased following the addition of chemoattractants, suggesting that there are latent intracellular pools of this receptor (Hostoffer et al., 1993). Surface $Fc_{\alpha}R$ expression is increased in cells derived from bronchoalveolar lavage of patients with cystic fibrosis and following addition of TNF-α (Hostoffer et al., 1994). In contrast, TGF-β1 downregulates expression of $Fc_{\alpha}R$ on human monocytes (Reterink et al., 1996). Addition of GM-CSF to human neutrophils led to enhanced affinity of an IgA receptor and capacitated the neutrophils for IgA-dependent phagocytosis of IgA (Weisbart et al., 1988). It is not known whether this reflected activation of CD89 or another receptor recognizing IgA on these cells, although evidence suggests that CD89 is the principal IgA receptor on human neutrophils (Stewart et al., 1994). Studies using My 43, a mAb of the IgM isotype, demonstrate that $Fc_{\alpha}R$ is capable of mediating phagocytosis and superoxide anion production (Shen et al., 1989).

D. Other Fc Receptors

The polymeric IgA/IgM receptor, expressed on the basolateral surfaces of glandular epithelial cells, is responsible for the transcytosis of these immunoglobulins from the blood into the lumen (reviewed in Mostov et al., 1995). There is no evidence that this receptor mediates phagocytosis. A receptor for the Fc portion of IgM ($Fc_{\mu}R$) has been identified on human natural killer cells that associates with the γ and ζ subunit of Fc receptors (Rabinowich et al., 1996). FcRn is a receptor on intestinal epithelial cells that mediates the transfer of maternal Ig from milk to the bloodstream of newborn rodents. A human homolog exists. β_2-microglobulin-deficient mice, which also lack expression of FcRn, display increased clearance of IgG from the serum, corresponding to enhanced lysosomal delivery of endocytosed IgG (Israel et al., 1996; Junghans and Anderson, 1996). FcRn thus plays a major role in the regulation of the concentration of IgG in the serum.

III. ACTIVATION AND DEACTIVATION SIGNALS FOLLOWING FC RECEPTOR LIGATION

A. The Immunoreceptor Tyrosine Activation Motif (ITAM)

First recognized by Reth (Reth, 1989), the ITAM consensus sequence, $YxxLx_{5-12}Yx_{2-3}L/I$, is present in the ζ, ν, δ, and ε subunits of the T cell antigen receptor complex, the B cell antigen receptor subunits Igα and Igβ, and in the γ subunit and other subunits of Fc receptors (reviewed in Cambier, 1995; Chan and Shaw, 1996; Isakov, 1997). Using various transfected cell models, mutation of either of the tyrosine residues within this sequence markedly impairs receptor signaling, including phagocytosis (Bonnerot et al., 1992; Park et al., 1993; Daeron et al., 1994; Mitchell et al., 1994; Paolini et al., 1995; Park and Schreiber, 1995). A study of γ subunit knock-out mice indicates that $Fc_{\gamma}R$-mediated phagocytosis in the mouse requires expression of the γ subunit (Takai et al., 1994). Thus, the ITAM is an indispensable component of the phagocytic signaling cascade of Fc receptors.

Tyrosine Kinases and ITAM Phosphorylation

Many aspects of FcR signaling are inhibited by a variety of tyrosine kinase inhibitors (Greenberg et al., 1993; Ghazizadeh and Fleit, 1994; Davis et al., 1995; Fallman et al., 1995; Allen and Aderem, 1996). Upon oligomerization of the ligand binding subunits of Fc γ, ε, or α receptors by clustered Fc residues, the tyrosine residues within the associated γ subunit ITAMs become phosphorylated (Paolini et al., 1991; Duchemin et al., 1994; Greenberg et al., 1994; Pfefferkorn and Yeaman, 1994). Once phosphorylated, these domains serve as high-affinity binding sites for members of the Syk family of tyrosine kinases (Agarwal et al., 1993; Benhamou et al., 1993; Kiener et al., 1993; Law et al., 1993). This family, of which only two members are known, contain tandem SH2 domains, each of which interact with tyrosine phosphorylated ITAMs (reviewed in Yanagi et al., 1995).

The identity of the "initiating" kinase which phosphorylates tyrosine residues within the ITAM is unknown. There is evidence in T cells that the Src family members Lck and Fyn promote this function (reviewed in Qian and Weiss, 1997). Indeed, several workers have coprecipitated multiple Src family members and various Fc_{γ} and ε receptors (Hamada et al., 1993; Salcedo et al., 1993; Ghazizadeh et al., 1994; Sarmay et al., 1994; Wang et al., 1994; Durden et al., 1995; Duchemin and Anderson, 1997). In a study using macrophages derived from knock-out mice that lack expression of Lyn, Fgr, and Hck, the Src family members normally expressed in these cells, $Fc_{\gamma}R$-mediated phagocytosis occurred at a reduced rate (Crowley et al., 1997). An earlier study suggested that Hck was the predominant Src family member that was responsible for optimal $Fc_{\gamma}R$-mediated phagocytosis (Lowell et al., 1994). These studies did not address whether $Fc_{\gamma}R$ ligation in macrophages lacking expression of Src family members led to tyrosine phosphorylation of the γ subunit. A role for Src family tyrosine kinases in phosphorylation and activation of Syk has been demonstrated in cells transfected with subunits of $Fc_{\varepsilon}RI$ (Jouvin et al., 1994), and macrophages derived from Src family knock-out mice show markedly decreased activation of Syk (Crowley et al., 1997). While Lyn is capable of a direct association with Syk (Jouvin et al., 1994), experiments in P815 mastocytoma cells transfected with $Fc_{\varepsilon}RI$ suggest that the β subunit of $Fc_{\varepsilon}R$ is the primary binding partner for Lyn. These data led Kinet and co-workers to suggest a model for $Fc_{\varepsilon}RI$ signaling in which the unphosphorylated β subunit of $Fc_{\varepsilon}R$ serves to recruit Lyn, which acts to amplify the activation of Syk that is bound to phosphorylated γ subunits (Lin et al., 1996). However, this model may not be applicable to $Fc_{\gamma}Rs$, which are expressed in cells that lack the β subunit. It is possible that other ITAM-bearing subunits substitute for the β subunit in phagocytic leukocytes, or that other Lyn-binding proteins are recruited to ligated $Fc_{\gamma}Rs$. Collectively, these studies demonstrate that one or more Src family tyrosine kinases are not required for $Fc_{\gamma}R$-mediated phagocytosis, but rather serve to amplify the rate or extent of phagocytosis, probably by contributing to activation of Syk.

ITAM-Dependent Signaling and Syk

The constitutive association of Syk with subunits of the antigen receptor in resting B cells (Law et al., 1993) and with $Fc_{\gamma}RIIA$ in THP-1 cells, a human monocyte-like cell line (Ghazizadeh et al., 1995), suggests that Syk may directly phosphorylate the γ subunit. In this capacity, Syk would fulfill the role of the initiating kinase in ITAM-dependent phagocytosis. According to this view, a small percentage of Syk molecules promotes tyrosine phosphorylation of ITAM-bearing subunits in the absence of receptor ligation. This creates a limited number of Syk/ITAM complexes in resting cells. Following engagement of Fc_{γ} receptors by IgG, further recruitment of unphosphorylated ITAMs occurs; these become phosphorylated by nearby Syk/ITAM complexes, leading to further Syk recruitment from the cytosolic pool (Figure 2). Syk activation may result from multiple interrelated events: (1) binding to phosphorylated ITAMs and localized submembranous recruitment; (2) activation of intrinsic Syk kinase activity by binding phosphorylated ITAMs (Rowley et al., 1995; Shiue et al., 1995; Kimura et al., 1996;); (3) Syk phosphorylation, both by neighboring Syk molecules ("autophosphorylation") (Kurosaki et al., 1995) and by other recruited kinases, including Src family members (Kurosaki et al., 1994; Sidorenko et al., 1995; Ting et al., 1995; Amoui et al., 1997); (4) activation by increased cytosolic free calcium concentration

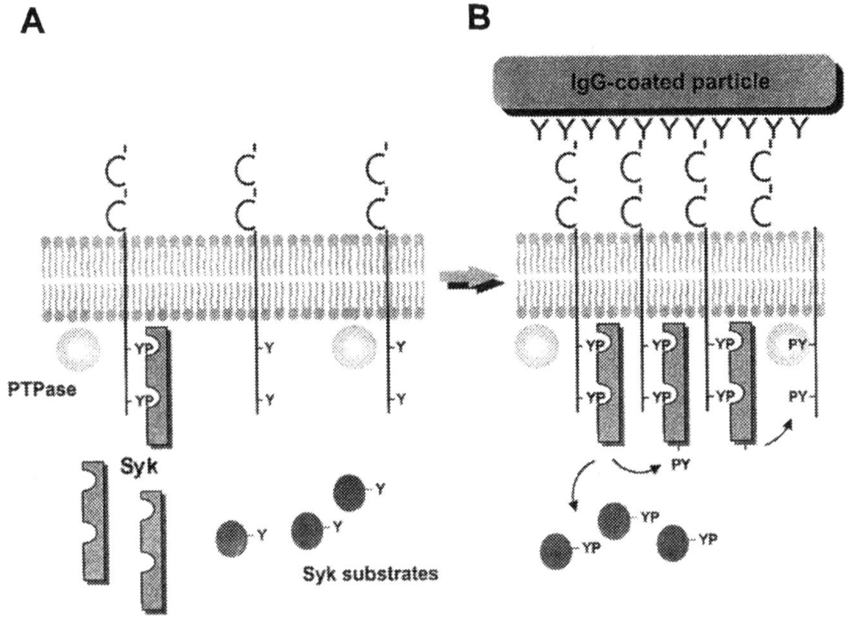

Figure 2. Model for early signal transduction events during phagocytosis by Fc$_{\gamma}$RIIA. An analogous model applies for both Fc$_{\gamma}$RI and Fc$_{\gamma}$RIIIA. (**A**) Fc$_{\gamma}$Rs are freely diffusible within the plane of the plasma membrane. Constitutively associated with the plasma membrane are various protein tyrosine phosphatases (PTPase) which dephosporylate phosphorylated substrates of plasma membrane-associated tyrosine kinases, creating a dynamic balance between the two. Syk is present predominantly in the cytosol; a small fraction is constitutively associated with freely mobile Fc$_{\gamma}$RIIA molecules via SH2 domain-phosphotyrosine interactions. (**B**) Following engagement of Fc$_{\gamma}$Rs with IgG-opsonized targets, Fc$_{\gamma}$Rs aggregate and conform to the distribution of Fc ligands bound to the surface of the particle. This causes close apposition of Fc$_{\gamma}$Rs, some of which are already associated with Syk. This allows Syk to phosphorylate ITAMs of neighboring Fc$_{\gamma}$RIIA serving to recruit additional Syk molecules to phosphorylated ITAMs. A localized imbalance between tyrosine kinase and tyrosine phosphatase activities ensues favoring the former. This results in enhanced tyrosine phosphorylation of Syk (autophosphorylation) and Syk substrate (arrows). Not depicted are additional tyrosine kinases, such as Lyn, which are recruited to Fc$_{\gamma}$Rs and which phosphorylate and further activate Syk.

(Wang et al., 1994) and reactive oxygen species (Schieven et al., 1993; Qin et al., 1995). It is not clear whether Ca^{2+} or reactive oxygen species directly enhance Syk tyrosine kinase activity. Interestingly, addition of *N*-acetyl-L-cysteine inhibited both Syk activation and effector functions following ligation of Fc$_{\varepsilon}$RI in mast cells (Valle and Kinet, 1995), suggesting that reactive oxygen species are important activators of Syk *in vivo*. It is possible that generation of oxidants and calcium fluxes following ligation of $Fc_{\varepsilon}RI$ serve to amplify Syk activation by other mechanisms.

The evidence that Syk is required for ITAM-dependent signaling in macrophages and mast cells is compelling. Pharmacological blockade of Syk kinase activity using a Syk-selective inhibitor inhibited multiple activities following ligation of $Fc_{\varepsilon}RI$ in mast cells (Oliver et al., 1994). Downregulation of Syk expression in monocytes using antisense oligonucleotides abrogated phagocytosis (Matsuda et al., 1996), and expression of a Syk construct lacking the kinase domain inhibited release of arachidonate following antigen stimulation in RBL cells (Hirasawa et al., 1995). Studies using either Syk-deficient lymphocytes (Cox et al., 1996), mast cells (Zhang et al., 1996), or macrophages (Crowley et al., 1997), indicate an essential role for Syk in mediator release, phagocytosis, and cytoskeletal assembly. In Syk-deficient mast cells, phosphorylation of the γ subunit upon receptor aggregation was greatly impaired, indicating that Syk plays an important role in the phosphorylation of the γ subunit. However, the ability of both the β and γ subunits to undergo some degree of enhanced tyrosine phosphorylation upon receptor aggregation in the absence of Syk (Zhang et al., 1996) suggests that other kinases may compensate for the loss of Syk in these cells.

ITAM-Independent Signaling by Clustered Fc Receptors

While the tyrosine-phosphorylated ITAMs are essential for triggering many aspects of FcR signaling, other residues in the cytosolic domains of $Fc_{\gamma}Rs$ and their subunits may influence ITAM-dependent signaling. For example, transfection of WT or truncation mutants of the ligand binding subunit of $Fc_{\gamma}RIIIA$ in a Jurkat cell line that expressed the ζ subunit demonstrated a role for the membrane-proximal amino acids of $Fc_{\gamma}RIIIA$ in calcium signaling (Hou et al., 1996). Since the magnitude of the calcium response did not correlate with the extent of association of $Fc_{\gamma}RIIIA$ and the ζ subunit, these results imply that ITAM function may be influenced by neighboring amino acid residues, either directly or indirectly.

Other residues in ITAM-bearing subunits besides tyrosine are phosphorylated upon receptor engagement (reviewed in Rivera, 1997). Ser^{229} within the β subunit ITAM becomes phosphorylated upon $Fc_{\varepsilon}RI$ clustering (Pribluda et al., 1997). This residue, which is conserved in rodent and human β subunits, is a potential site for phosphorylation by casein kinase II; however, it is not known whether this enzyme is activated following ligation of $Fc_{\varepsilon}RI$. In contrast, phosphorylation of Thr^{52} is the likely site of non-tyrosine phosphorylation within the γ subunit ITAM (Pribluda et al., 1997). This residue, which is conserved among several species, is flanked by weakly basic sites and may serve as a consensus sequence for phosphorylation by protein kinase C (PKC)-δ. This isoform of PKC can be coprecipitated with $Fc_{\varepsilon}RI$, and receptor ligation leads to its enhanced phosphorylated on tyrosine residues (Haleem-Smith et al., 1995). Recombinant PKC-δ was capable of phosphorylating the γ chain cytoplasmic domain *in vitro* (Germano et al., 1994). The functional consequences of either serine phosphorylation of the β subunit or threonine phosphorylation of the γ subunit are not known.

B. The Immunoreceptor Tyrosine Inhibitory Motif (ITIM)

Coligation of $Fc_{\gamma}RIIB$ with the antigen receptor in B cells (BCR) leads to decreased cellular activation (Phillips and Parker, 1983). The structural basis for this observation was only apparent in 1992, when Amigorena and co-workers demonstrated that a 13 amino acid sequence in $Fc_{\gamma}RIIB1$ is required for inhibition of B-cell activation (Amigorena et al., 1992). Since then, similar motifs, now termed ITIMs, have been identified in several other receptors in hematopoetic cells, including CD22 and the so-called killer cell inhibitory receptors (KIRs) (reviewed in Vivier and Daeron, 1997). While the minimal sequence requirements for the ITIM are in doubt, I/VxYxxL/V has been proposed as a provisional consensus motif (Daeron, 1996). Phosphorylation of the tyrosine residue within this motif is required for delivering the inhibitory signal (Muta et al., 1994). The capacity of receptors bearing this motif to inhibit cellular activation correlates with the recruitment of either the tyrosine phosphatases SHP-1/2 (Olcese et al., 1996) or an inositol 5′-phosphatase, SHIP (Lioubin et al., 1996; Ono et al., 1996; Osborne et al., 1996). SHIP dephosphorylates either phosphatidylinositol-3,4,5-trisphosphate (PIP_3) to yield phosphatidylinsotiol-3,4-bisphosphate, or inositol (1,3,4,5) tetrakisphosphate to yield inositol-(1,3,4)-trisphosphate (Lioubin et al., 1996; Ono et al., 1996; Osborne et al., 1996). Specific receptors preferentially recruit either SHP (e.g., KIR) or SHIP (e.g., $Fc_{\gamma}RIIB$) (Gupta et al., 1997; Ono et al., 1997; Vely et al., 1997). Inhibition of antigen receptor signaling by $Fc_{\gamma}RIIB$ is correlated with premature termination of BCR-induced elevations in inositol trisphosphate (IP_3) (Bijsterbosch and Klaus, 1985) and a predominant reduction in BCR-mediated calcium influx, rather than impaired release of calcium from intracellular stores fluxes (Bolland et al., 1998). This activity requires catalytic activity of SHIP (Ono et al., 1997). A recent study provides strong evidence that SHIP contributes to the recruitment of BTK, a PH domain-containing kinase, possibly by inhibiting the accumulation of PIP_3 (Bolland et al., 1998). PIP_3 would be expected to participate in recruitment of BTK to the membrane, via its PH domain (Salim et al., 1996). Furthermore, BTK has been shown to activate a plasma membrane calcium influx pathway (Fluckiger et al., 1998) and is required for optimal activation of phospholipase C (PLC)-γ in reponse to antigen receptor ligation in B cells (Takata and Kurosaki, 1996); see below.

C. Tyrosine Phosphatases and Other Modulators of ITAM Function

Several tyrosine phosphatases have been identified which modulate ITAM-mediated responses. These include the membrane-bound tyrosine phosphatase CD45 and SHP-1/SHP-2. CD45 is required for signaling via the T (Pingel and

Thomas, 1989; Koretzky et al., 1990; Koretzky et al., 1991) and B cell antigen receptors (Justement et al., 1991) and for degranulation mediated by $Fc_{\varepsilon}RI$ in mast cells (Berger et al., 1994). The requirement for this phosphatase is most likely due to its ability to dephosphorylate a conserved tyrosine residue in the C-terminus of Src family members that is required for their activation (reviewed in Brown and Cooper, 1996; Neel, 1997). The role of CD45 in $Fc_{\gamma}R$-mediated signaling is uncertain. Both SHP-1 and SHP-2 are associated with $Fc_{\varepsilon}RI$ (Kimura et al., 1997) and SHP-1 associates with ZAP-70, and negatively regulates signaling by the T cell antigen receptor (Plas et al., 1996). The role of these phosphatases in signaling via $Fc_{\gamma}Rs$ in phagocytic cells is not known, although it is likely that either SHP-1 or SHP-2 negatively regulate $Fc_{\gamma}R$ activation.

Other proteins have the potential to negatively regulate ITAM-mediated signaling. For example, Cbl is a 120 kDa protein that undergoes enhanced tyrosine phosphorylation following ligation of $Fc_{\gamma}Rs$ (Marcilla et al., 1995; Tanaka et al., 1995; Matsuo et al., 1996). Its overexpression leads to inhibition of Syk tyrosine kinase activity and association of Syk with the γ subunit of $Fc_{\varepsilon}RI$ (Ota and Samelson, 1997).

IV. EARLY DOWNSTREAM SIGNALS FOLLOWING ITAM CLUSTERING

A. Phosphatidylinositol 3-Kinase(s)

Syk is required for ITAM-dependent actin assembly in a transfected cell model (Cox et al., 1996), for $Fc_{\gamma}R$-mediated phagocytosis in human monocytes (Matsuda et al., 1996), and for $Fc_{\gamma}R$-mediated phagocytosis in mouse macrophages (Crowley et al., 1997). What are the key events that are triggered by Syk recruitment that result in phagocytosis? One likely candidate is the activation of one or more members of the phosphatidylinositol 3-kinase (PI 3-kinase) family (reviewed in Carpenter and Cantley, 1996). The p85/p110 isoform of PI 3-kinase is a heterodimeric enzyme which phosphorylates the 3′ hydroxyl group within the *myo*-inositol ring of phosphatidylinositol and its phosphorylated derivatives. Activation of $Fc_{\gamma}RIIA$ in platelets induces its association with PI 3-kinase (Chacko et al., 1996) and $Fc_{\gamma}R$ activation in myeloid cells is accompanied by enhanced PI 3-kinase activity associated with phosphotyrosine residues (Kanakaraj et al., 1994; Ninomiya et al., 1994). Aggregation of $Fc_{\gamma}RI$ in IFN-γ-primed U937 cells results in accumulation of PIP_3 due, in part, to activation of a p85-binding isoform of PI 3-kinase (Melendez et al., 1998). Furthermore, addition of inhibitors of PI 3-kinases to macrophages and neutrophils leads to impaired $Fc_{\gamma}R$-mediated phagocytosis (Ninomiya et al., 1994; Araki et al., 1997; Cox et al., 1999). In contrast to growth factor-mediated signaling, PI 3-kinase is not responsible for triggered actin assembly during $Fc_{\gamma}R$-mediated phagocytosis (Cox et al., 1999). Since PI 3-kinases have been implicated in vesicular trafficking (Shepherd et al., 1996), one possible role for isoforms of PI 3-kinase is in the regulation of membrane traficking necessary to support phagocytosis. Inhibition of PI 3-kinase results in a concurrent decrease in pseudopod extension and exocytosis, consistent with a model of pseudopod extension that requires localized membrane insertion from a latent intracellular vesicular pool (Cox et al., 1999). PI 3-kinase may also be required for phagosomal closure (Araki et al., 1997).

Enhanced phosphorylation of other tyrosine kinase substrates following $Fc_{\gamma}R$ ligation has been reported. These include PLCγ-1 and PLCγ-2 (Ting et al., 1991; Liao et al., 1993; Shen et al., 1994) (see below); paxillin (Greenberg et al., 1994), HS1 (Zeng et al., 1995), Raf-1 (Park et al., 1996), Vav (Darby et al., 1994; Xu and Chong, 1996) (see below); the adapter proteins Shc (Shen et al., 1994; Park et al., 1996) and Nck (Li et al., 1992; Park and Rhee, 1992), and Cbl (Marcilla et al., 1995; Tanaka et al., 1995; Matsuo et al., 1996) (see above). The role of these tyrosine kinase substrates in phagocytosis is unknown.

B. Activation of Phospholipases

$Fc_{\gamma}R$ ligation leads to the activation of one or more isoforms of phospholipase A_2 (PLA_2); (Suzuki et al., 1982; Aderem et al., 1985; Lennartz and Brown, 1991). In human monocytes, a variety of PLA_2 inhibitors inhibit phagocytosis; the ability of exogenous arachidonate to restore phagocytosis in the presence of PLA_2 inhibitors strongly supports a role for PLA_2 in phagocytosis (Lennartz and Brown, 1991). Initial characterization of the $Fc_{\gamma}R$-activated PLA_2 activity revealed it be Ca^{2+}-insensitive, utilizing phosphatidylethanolamine as a preferred substrate (Lennartz et al., 1993). The requirement for PLA_2 in regulating endosomal trafficking (Mayorga et al., 1993) raises the possibility that PLA_2, like PI 3-kinase, may be required for upregulating a vesicular compartment necessary for phagocytosis to occur. Recent ultrastructural studies support this concept and implicate the plasma membrane as a proximate source for this compartment (Lennartz et al., 1997). Treatment of macrophages and HeLa cells with inhibitors of PLA_2 inhibited cell spreading (Lefkowith et al., 1991; Auer and Jacobson, 1995), and macrophages derived from mice with essential fatty acid deficiency displayed defects in cell spreading (Lefkowith et al., 1991), suggesting that PLA_2 may be required for adhesive (e.g., binding to matrix) or protrusive (e.g., extension of pseudopods) events in general.

Other phospholipases are activated following $Fc_{\gamma}R$ ligation, including phospholipase C (PLC); (Della Bianca et al., 1990; Scholl et al., 1992; Ting et al., 1992; Della Bianca et al., 1993; Dusi et al., 1994; Shen et al., 1994) and phospholipase D (PLD); (Gewirtz and Simons, 1997; Melendez et al., 1998). Recent data indicate that PLD is required for $Fc_{\gamma}R$-mediated phagocytosis (Kusner et al., 1999). Interestingly, overexpression of PLDs, an isoform that localizes to the plasma membrane, leads to cytoskeletal alterations, suggesting involvement of Rho family GTPases (Colley et al., 1997). However, inhibition of PLD activity does not affect Rac1- or Cdc42-mediated cytoskeletal alterations in macrophages and PLD activation may be downstream of these or other (ARF) GTPases (Olson and Lambeth, 1996). PLD activity is required for $Fc_{\gamma}RI$-directed sphingosine kinase, which activates calcium mobilization and degradation of internalized immune complexes (Melendez et al., 1998). $Fc_{\gamma}R$-mediated sphingomyelin hydrolysis acts as a negative regulator of $Fc_{\gamma}R$-mediated phagocytosis (Glick and Barenholz, 1996; Suchard et al., 1997).

C. Alterations in $[Ca^{2+}]_i$ and Other Cation Fluxes

Many investigators have shown that $Fc_{\gamma}R$ ligation triggers elevations in $[Ca^{2+}]_i$ in a variety of leukocytes (Young et al., 1984; Lew et al., 1985; Sawyer et al., 1985; Di Virgilio et al., 1988; Hishikawa et al., 1991). Peak $[Ca^{2+}]_i$ are observed in periphagosomal regions (Sawyer et al., 1985) and correspond to a redistribution of several markers of intracellular Ca^{2+} stores (Stendahl et al., 1994). Are alterations in $[Ca^{2+}]_i$ required for phagocytosis? Use of intracellular Ca^{2+} chelators has produced disparate results, depending on the cell type, the nature of the Ca^{2+} chelator, and the type of particle used (Lew et al., 1985; McNeil et al., 1986; Di Virgilio et al., 1988; Della Bianca et al., 1990; Greenberg et al., 1991; Hishikawa et al., 1991; Odin et al., 1991; Rosales and Brown, 1991). The ability of mouse peritoneal macrophages to undergo $Fc_{\gamma}R$-mediated phagocytosis and actin polymerization without detectable alterations in $[Ca^{2+}]_i$ indicates that $[Ca^{2+}]_i$ fluxes are unlikely to play a central role in the ingestion process in these cells (McNeil et al., 1986; Greenberg et al., 1991). Since $Fc_{\gamma}R$-mediated phagocytosis in murine macrophages is dependent on expression of the γ subunit (Takai et al., 1994), these findings suggest that γ subunit-dependent cytoskeletal alterations are independent of fluxes of $[Ca^{2+}]_i$. In contrast, mutations in the cytosolic domain of human $Fc_{\gamma}RIIA$ that inhibited $[Ca^{2+}]_i$ fluxes also inhibited phagocytosis (Edberg et al., 1995), suggesting that alterations in $[Ca^{2+}]_i$ may play a role in γ subunit-independent phagocytosis. It is also possible that these residues within $Fc_{\gamma}RIIA$ are important for recruitment of other signaling modules. Cytosolic transients may be involved in other aspects of phagocytic function, such as chemotactic peptide enhancement of $Fc_{\gamma}R$-mediated phagocytosis in neutrophils (Rosales and Brown, 1991), arachidonic acid production (Aderem et al., 1986), and phagosome-lysosome fusion (Jaconi et al., 1990). However, here too, $Fc_{\gamma}R$-mediated $[Ca^{2+}]_i$-independent enhanced PLA_2 activity (Lennartz et al., 1993) and $[Ca^{2+}]_i$-independent phagosome-lysosome fusion have been reported (Zimmerli et al., 1996).

Other cation fluxes that accompany $Fc_{\gamma}R$ ligation include those due to activation of plasma membranes channels (Nelson et al., 1985; Ince et al., 1988) and a Na^+/H^+ exchanger (Fukushima et al., 1996). Alterations in cation fluxes are unlikely to play a role in promoting phagocytosis since $Fc_{\gamma}R$-mediated phagocytosis is insensitive to the cationic species of the extracellular medium (Pfefferkorn, 1984).

D. Regulation of $[Ca^{2+}]_i$ Fluxes by Multiple Classes of Kinases

Activation of PLC isofoms by many classes of cell surface receptors, including Fc receptors, is a familiar pathway leading to $[Ca^{2+}]_i$ fluxes. However, IP_3-independent Ca^{2+} mobilization has been described, and has been attributed to the activation of a sphingosine kinase, leading to the production of shingosine-1-phosphate (reviewed in Beaven, 1996). This pathway is likely to play a role in Fc receptor signaling, since activation of $Fc_\varepsilon RI$ in RBL-2H3 cells leads to the enhanced production of sphingosine-1-phosphate, and inhibitors of sphingosine kinase inhibit $Fc_\varepsilon RI$-mediated alterations in $[Ca^{2+}]_i$ (Choi et al., 1996).

Recent evidence provides a complex picture of PLC-γ and $[Ca^{2+}]_i$ regulation by ITAM-bearing receptors. PLC-γl is activated by phosphorylation of Tyr^{771}, Tyr^{783}, and Tyr^{1254} (reviewed in Rhee and Bae, 1997). A role for both Syk (Takata et al., 1994; Law et al., 1996) and BTK (Takata and Kurosaki, 1996) in the activation of PLC-γ has been described. Syk and BTK may regulate PLC-γ by direct phosphorylation. Activation of BTK *in vivo* may depend on PI 3-kinase activity, probably via production of PIP_3 and its binding to the PH domain of BTK (Scharenberg et al., 1998). BTK is required for sustained Ca^{2+} entry across the plasma membrane following BCR ligation (Fluckiger et al., 1998). Since this event requires the expression of IP_3 receptors (Sugawara et al., 1997), BTK may enhance Ca^{2+} entry by production of IP_3 and activation of an IP_3-dependent plasma membrane channel, or by activating other types of plasma membrane Ca^{2+} channels (Fluckiger et al., 1998).

The enzymatic activity of PLC-γ itself is increased by phospholipids, including PIP_3, which binds directly to the PH domain of the enzyme (Bae et al., 1998). This interaction probably facilitates binding of PLC-γ to membranes (Falasca et al., 1998), and may explain why addition of wortmannin to RBL-2H3 cells inhibits plasma membrane translocation of PLC-γ following addition of antigen (Barker et al., 1998). PI 3-kinase-generated PIP_3, and subsequent activation of PLC-γ and/or BTK may also be required for $[Ca^{2+}]_i$ signaling following $Fc_\gamma R$ ligation in human neutrophils (Vossebeld et al., 1997).

In summary, FcR-directed $[Ca^{2+}]_i$ signaling pathways, which may be required for secretion, phagosome-lysosome fusion, and gene expression, depend on the concerted activities of several classes of enzymes, including PLC-γ, multiple tyrosine kinases including Syk and BTK, one or more members of the PI 3-kinase family, and a sphingosine kinase.

E. Serine/Threonine Kinases and Phagocytosis

Activation of many protein serine/threonine kinases has been reported following $Fc_\gamma R$ ligation. These include PKC (Brozna et al., 1988; Zheleznyak and Brown, 1992), protein kinase A (Smolen et al., 1980; Nitta and Suzuki, 1982), casein kinase II (Hirata and Suzuki, 1987), calcium/calmodulin-dependent protein kinase II (Liang and Huang, 1995), Akt (Tilton et al., 1997), histone H4 protein kinase (Liang and Huang, 1995) and multiple isoforms of mitogen-activated protein (MAP) kinases (Durden et al., 1995; Liang and Huang, 1995; Park et al., 1996; Trotta et al., 1996; Rose et al., 1997). The specific roles of these kinases in $Fc_{\gamma}R$-mediated signaling pathways are largely unknown. There is evidence that PKC activity is required for $Fc_{\gamma}R$-mediated phagocytosis (Zheleznyak and Brown, 1992; Allen and Aderem, 1996), but a requirement for this family of enzymes in phagocytosis may not be universal (Newman et al., 1991; Greenberg et al., 1993). Further work is needed to identify which isoforms of PKC are activated during phagocytosis, and to define the specific stage in the phagocytic process that may be regulated by PKC.

V. RECEPTOR COOPERATIVITY AND PHAGOCYTOSIS

While many studies of phagocytosis involve the use of artificial particles, such as erythrocytes opsonized with mono-specific ligands, it is apparent that bacterial and fungal pathogens express multiple adhesins on their surfaces (reviewed in Hauschildt and Kleine, 1995; Ofek et al., 1995). For example, phagocytosis of unopsonized *Pseudomonas aeruginosa* by human macrophages is mediated by multiple receptors including the macrophage mannose receptor, CR1 and CR3, and $Fc_{\gamma}R$ (Speert et al., 1988). Phagocytic receptors may function independently to promote particle ingestion, or may cooperate to generate a phagocytic signal. Cooperativity may involve activation of the phagocytic capacity one receptor by another, including those which bind secreted cytokines and other substances (Table 2), or may reflect the physical association of two phagocytic receptors. Additionally, the production of common signaling intermediates, such as activated tyrosine kinases or Rho family GTPases (see below), could be amplified by the simultaneous activation of different classes of phagocyte surface receptors.

A. Activation of One Receptor Enhances Phagocytic Activity of Another

Examples of this include activation of $Fc_{\gamma}Rs$ on human neutrophils by the chemotactic peptide fMet-Leu-Phe and by RGD-containing proteins. Work by Brown and colleagues has shown that the RGD-dependent activation of both $Fc_{\gamma}R$ and complement receptor-mediated phagocytosis occurs via ligation of a "leukocyte response integrin" which is immunologically related to a β_3 integrin. This integrin was found to be associated with a 50 kDa associated protein (IAP) (Brown et al., 1990), later shown to be identical to CD47 (Lindberg et al., 1994). IAP is a 50 kDA protein with three or five membrane-spanning segments whose precise function is unknown (Lindberg et al., 1993). Mice rendered deficient in IAP demonstrated defects in phagocytosis and increased susceptibility to bacterial infection (Lindberg et al., 1996). Another interesting example of $Fc_{\gamma}R$ activation is the

Table 2. Secreted Proteins That Enhance the Efficiency of $Fc_{\gamma}R$-Mediated Phagocytosis

Protein	*Reference*
Surfactant protein A (SP-A)	Tenner et al., 1989
Mannan binding protein (MBP)	Kuhlman et al., 1989; Tenner et al., 1995
C1q	Bobak et al., 1987, 1988
Laminin	Bohnsack et al., 1985
Fibronectin	Pommier et al., 1983; Wright et al., 1983
Serum amyloid P	Wright et al., 1983
IL-10*	Capsoni et al., 1995
IL-1	Moxey-Mims et al., 1991; Simms et al., 1991
IL-4	Sampson et al., 1991
GM-CSF	Capsoni et al., 1991; Collins and Bancroft, 1992
M-CSF	Sampson et al., 1991
TNF-α	Klebanoff et al., 1986; Gresham et al., 1990

Note: * The effect of IL-10 and other cytokines on phagocytosis requires a prolonged incubation with the cells, probably reflecting a requirement for protein synthesis.

ability of Fc_{γ}RIIIB to enhance the activity of Fc_{γ}RIIA in human neutrophils. Ligation of the former induced an increase in Fc_{γ}RIIA-mediated phagocytosis that was inhibited by scavengers of reactive oxygen radicals, suggesting that Fc_{γ}RIIA activation by Fc_{γ}RIIIB requires the participation of reactive oxygen species (Salmon et al., 1995).

Cooperation between receptors during phagocytosis is likely to be important in the alveolar lining of the lung. The lung is the major portal for inhaled pathogenic microorganisms and noxious particles. Surfactant protein A (SP-A) is the most abundant protein constituent of surfactant, the surface lining material of the lung. SP-A was shown to enhance $Fc_{\gamma}R$-mediated phagocytosis (Tenner et al., 1989). SP-A is a member of a family of proteins containing collagen-like regions contiguous with globular domains. Other members of the family include the complement protein C1q and mannose-binding lectin. The cDNA cloning of a receptor for this family of proteins was recently reported (Nepomuceno et al., 1997).

B. Phagocytic Receptors Associate in Multi-Subunit Complexes

Addition of mAbs to CR3 inhibit not only CR3-dependent phagocytosis, but $Fc_{\gamma}R$-mediated phagocytosis, as well (Arnaout et al., 1983). Some mAbs which recognize CR3 also block binding of IgG-coated targets (Brown et al., 1988), raising the possibility that $Fc_{\gamma}Rs$ and CR3 are physically associated with each other. We now know that many cell surface receptors, including CR3 and Fc_{γ}RIIIB (Poo et al., 1995), occupy microdomains on cell surfaces. Since Fc_{γ}RIIIB is a GPI-linked protein, the association between these proteins is most likely between their extracellular domains and involves lectin-like interactions (Zhou et al., 1993).

Indeed, soluble Fc_{γ}RIIIB interacts with both CD11b/CD18 and CD11c/CD18 and triggers release of IL-6 and IL-8 (Galon et al., 1996). The ability of soluble Fc_{γ}RI-IIB to trigger cell responses may explain how this receptor retains its signaling function when shed from cell surfaces.

VI. THE CYTOSKELETON AND FcγR-MEDIATED PHAGOCYTOSIS

Cells undergoing phagocytosis exhibit certain characteristic ultrastructural features. For example, during Fc_{γ}R-mediated phagocytosis, F-actin-rich pseudopods extend from the perimeter of the cells and conform to the circumference of the test particles. The classic observation that ligand needed to be present around the entire circumference of the phagocytic particle for phagocytosis to be completed helped define the "zipper hypothesis" (Griffin et al., 1975; Griffin et al., 1976). Recently, alternative mechanisms of phagocytosis have been demonstrated. For example, *Salmonella* stimulates localized membrane ruffling resembling the process of macropinocytosis (Francis et al., 1993). Recent elegant studies by Galán and colleagues have identified the likely molecular basis for "triggered" phagocytosis. They have shown that SopE, a substrate of a type III secretion system utilized by *Salmonella typhimurium*, induces nucleotide exchange in several Rho GTPases *in vitro* and stimulates membrane ruffling that resembles that seen during invasion of host cells by *Salmonella* (Hardt et al., 1998). One or more members of the Rho family are required for ingestion of *Salmonella* (Chen et al., 1996) and other forms of phagocytosis (see below).

A. Rho Family GTPases and Phagocytosis

Both Rac1 and Cdc42 are required for Fc_{γ}R-mediated phagocytosis and its underlying cytoskeletal alterations (Cox et al., 1997). While structurally homologous, Rac1 and Cdc42 play distinct roles in modifying the actin-based cytoskeleton. GTPase-deficient alleles of Rac1 induce membrane ruffling in a variety of cells types while GTPase-deficient versions of Cdc42 induce filopodia/microspike formation (Van Aelst and D'Souza-Schorey, 1997; Hall, 1998). What is the molecular basis for these differences? Both GTPases are capable of translocating from the cytosol to either the membrane (Abo et al., 1994; Bokoch et al., 1994; Dusi et al., 1996) or the cytoskeleton (Dash et al., 1995) upon activation. Both are capable of interacting with IQGAP (Hart et al., 1996; Erickson et al., 1997), an F-actin cross-linking protein (Bashour et al., 1997) which contains a calponin-homology domain and which accumulates in F-actin-rich ruffles during cellular stimulation (Kuroda et al., 1996). However, only Cdc42 was capable of stimulating actin assembly in a cell-free system (Zigmond et al., 1997). Furthermore, Cdc42, but not Rac, binds N-WASP, a protein which induces microspike formation and nucleates actin assembly in a cell-free system (Miki et al., 1998). Given these differences, it is tempting to speculate that Cdc42 regulates actin nucleation per se while Rac coordinates alterations in the cytoskeleton with large membrane-protrusive structures such as membrane ruffles. Indeed, Rac-dependent actin filament assembly in permeabilized fibroblasts required the presence of a member of the ERM family of F-actin-binding proteins; these proteins also contain membrane-targeting domains (Mackay et al., 1997). Rac may also promote uncapping of gelsolin from actin filaments, thereby generating free barbed ends for filament growth (Hartwig et al., 1995; Arcaro, 1998).

A requirement for RhoA in $Fc_{\gamma}R$-mediated phagocytosis is uncertain. RhoA has been implicated in $Fc_{\gamma}R$-mediated phagocytosis and early events that accompany $Fc_{\gamma}R$ ligation, including efficient binding of IgG ligand and $Fc_{\gamma}R$ capping (Hackam et al., 1997). These data are reminiscent of studies of CHO cells interacting with *Shigella*, in which addition of C3 exotoxin, an inhibitor of RhoA function, blocked early signaling events associated with phagocytosis, including enhanced protein tyrosine phosphorylation and bacterial association with the host cells (Watarai et al., 1997). Together, these studies suggest that intact RhoA function is required for maintaining the affinity and/or mobility of several classes of cell surface receptors. These findings are consistent with an earlier study showing that RhoA is required for cell adhesion via β_1 and β_2 integrins (Laudanna et al., 1996). In contrast, in a recent study of Rho GTPases and phagocytosis, Caron and Hall found that RhoA is required for CR3-mediated phagocytosis, but not for $Fc_{\gamma}R$-mediated phagocytosis (Caron and Hall, 1998). While the effect of RhoA inhibition on CR3-mediated phagocytosis is consistent with alterations in CR3 affinity and/or mobility, it is unclear why, in this most recent study, there was a lack of effect of RhoA inhibition on $Fc_{\gamma}R$-mediated phagocytosis. It is possible that the inhibition of RhoA function was incomplete and/or that integrin-mediated phagocytosis is more sensitive to RhoA inhibition.

Activators and Effectors of Rho Family Members In Vivo

The mechanisms by which Rac1 and Cdc42 are activated *in vivo* is the subject of current investigation. Activation of Rho family members is likely to involve one or more guanine nucleotide exchange factors. One of the best studied of these is Vav, a 95 kDa tyrosine kinase substrate that contains several modular domains, including a Dbl-homology domain and a potential actin binding domain (reviewed in Bustelo, 1996). Its guanine nucleotide exchange activity is enhanced by tyrosine phosphorylation (Crespo et al., 1997), and Vav undergoes enhanced tyrosine phosphorylation upon $Fc_{\gamma}R$ ligation (Darby et al., 1994; Xu and Chong, 1996). In addition, it binds directly to phosphorylated tyrosyl residues within the linker region of Syk (Deckert et al., 1996). However, macrophages derived from Vav knock-out mice ingest IgG-coated erythrocytes (Cox, Tybulewicz, Greenberg, unpublished results), arguing against an essential role for this protein in $Fc_{\gamma}R$-

mediated phagocytosis. It is possible that Vav2, a homologue of Vav, can substitute for Vav as a guanine nucleotide exchange factor during phagocytosis, or that other proteins or substances fulfill this role.

Evidence for the regulation of phospholipid-modifying enzymes by Rho family GTPases has stimulated several models of cell motility. Rho is capable of activating a phosphatidylinositol 4-phosphate 5-kinase in mammalian cells (Chong et al., 1994), and addition of an activated form of Rac1 induced the accumulation of PIP_2 in permeabilized platelets which correlated with the uncapping of actin filaments in these cells (Hartwig et al., 1995). Furthermore, addition of a polyphosphoinositide-binding peptide inhibited the Rac-mediated creation of free barbed ends of actin filaments (Hartwig et al., 1995). The existence of PIP_2-independent pathways of actin assembly in cells must also exist, however, since a substantial degree of actin assembly stimulated by thrombin receptor ligation in platelets preceded a rise in PIP_2 levels (Hartwig et al., 1995), and Cdc42-stimulated actin assembly in a cell-free system did not correlate with PIP_2 synthesis or availability (Zigmond et al., 1997). Furthermore, activation of most receptors that result in PLC activation results in decreases, rather than increases, in PIP_2 synthesis. This result was confirmed in a study employing a GFP-tagged PH domain derived from phospholipase C δ1 (GFP-PH) which preferentially binds $PI(4,5)P_2$. This study demonstrated that addition of PAF to RBL cells expressing this construct resulted in a rapid and global decrease in the plasma membrane localization of GFP-PH (Stauffer et al., 1998), consistent with a decline in plasma membrane PIP_2 content.

In addition to promoting uncapping of actin filaments, Rac and Cdc42 also participate in *de novo* nucleation of acting filaments. Recent studies implicate actin-related proteins (Arps) in actin filament nucleation (reviewed in Machesky and Gould, 1999). Rac and Cdc42 interact with members of the WASP family, which stimulates actin nucleation with the Arp2/3 complex (Yarar et al., 1999).

B. The Role of Other GTPases and $Fc_\gamma R$-Mediated Cytoskeletal Alterations—ARF6

Recent studies have provided evidence that ARF6 is required for $Fc_\gamma R$-mediated phagocytosis. ARF6 is a member of a family of GTPases which have been implicated in numerous membrane trafficking events in eukaryotic cells (reviewed in Donaldson et al., 1995; Moss and Vaughan, 1995). For example, GTP binding to ARF1 results in a conformational change of the protein and enhanced affinity for membranes. Once membrane bound, ARF1 participates in recruitment of cytosolic coat proteins, including COPI, to Golgi membranes. The regulated binding of these coat proteins is required for budding and fission of membrane vesicles, and maintenance of the structure and function of the GA (Donaldson et al., 1992; Robinson and Kreis, 1992; Palmer et al., 1993; Traub et al., 1993). Activation of most ARF proteins is blocked by brefeldin A (BFA) (Donaldson et al., 1992; Helms and Rothman, 1992; Randazzo et al., 1993; Morinaga et al., 1996; Peyroche et al.,

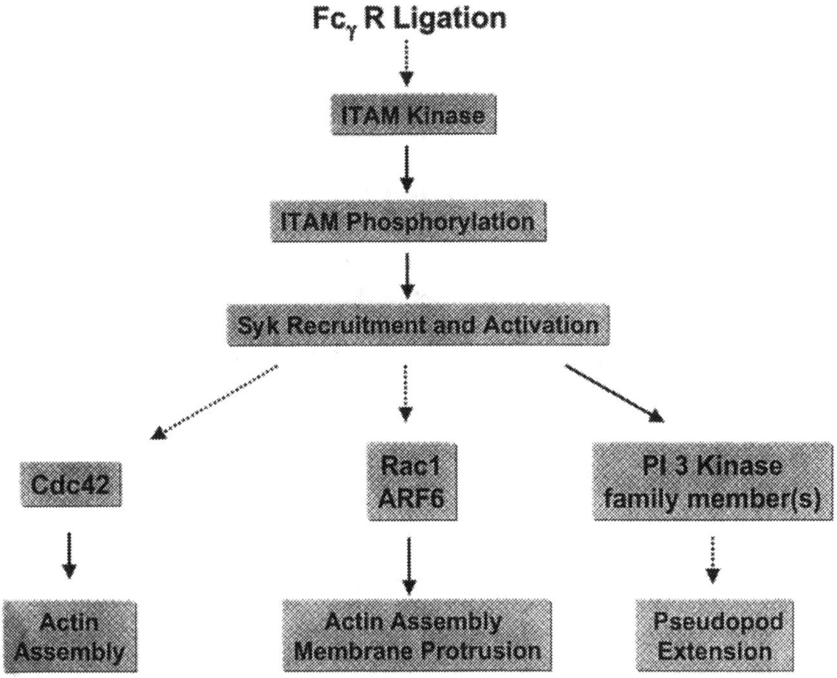

Figure 3. Sequence of events that occur following $Fc_\gamma R$ ligation that culminate in phagocytosis. Dashed lines refer to events that are likely to occur, but have not yet been firmly established. For details, see text.

1996). However, ARF6 is unique among the ARFs in that its association with membranes is not sensitive to BFA (Peters et al., 1995; Cavenagh et al., 1996; Radhakrishna et al., 1996). ARF6 resides primarily on the plasma membrane, probably in its GTP-bound form, while GDP-ARF6 resides in an internal tubulovesicular compartment (Peters et al., 1995; Radhakrishna and Donaldson, 1997; D'Souza-Schorey et al., 1998). Expression of the Q67 ARF6 (a GTPase-deficient allele) results in F-actin-rich plasma membrane protrusions in HeLa cells (Radhakrishna et al., 1996) and accumulation of F-actin at the cell periphery in CHO cells (D'Souza-Schorey et al., 1997). In macrophages expressing a GTP binding-deficient allele of ARF6 (ARF6 T27N), phagocytosis, but not binding, of EIgG was markedly impaired. Expression of ARF6 T27N resulted in a decrease in the formation of F-actin-rich phagocytic cups (Zhang et al., 1998). While the exact nature of the requirement for ARF6 in phagocytosis is not clear, the ability of ARF6 and Rac1 to interact with a common protein, POR-1 (which may be required for membrane ruffling) (Van Aelst et al., 1996) suggests that ARF6 may be required for Rac1-based motile events in general. A model for the participation of GTPases in $Fc_\gamma R$-mediated phagocytosis is presented in Figure 3.

VII. MEMBRANE REMODELING AND PHAGOCYTOSIS

The phagocytic capacity of macrophages is prodigous. Single adherent leukocytes can ingest many scores of bacteria, yet the leukocytes themselves do not exhibit an obvious decline in membrane surface area. This is indicative of ongoing recruitment and remodeling of plasma membrane. However, there is no clear picture as to the source of this membrane. Earlier studies of macrophages spreading on immune complex-coated substrates showed that the Golgi apparatus redistributed to areas beneath the plasma membrane adjacent to the substrate (Bainton et al., 1987). However, BFA, a compound which disrupts Golgi integriy, does not inhibit phagocytosis (Zhang et al., 1998). More recent studies have indicated that a plasma membrane-derived endosomal compartment is recruited locally during phagocytosis. Inhibitors of PLA_2 blocked ingestion of IgG-coated particles by human monocytes and induced the accumulation of plasma membrane-derived electron-lucent vesicles beneath the particles (Lennartz et al., 1997). This raises the possibility that $Fc_{\gamma}R$-directed PLA_2 activation precedes, and is required for, the insertion of endosomal membrane into the nascent phagosome. However, this does not address the proximate source of additional membrane needed for replenishment of the pool that is internalized during the engulfment process. Perhaps an endosomal pool that normally recycles at remote areas of the plasma membrane is recruited to forming phagosomes, and the endosomal pool itself is replenished from other intracellular stores. An alternative view is that one or more compartments comprising a recycling pathway participate in membrane renewal during phagocytosis. This may include an ARF6-containing recycling compartment (see above) and a study using bacterial toxins indicates that one or more v-SNAREs are required for optimal $Fc_{\gamma}R$-mediated phagocytosis (Hackam et al., 1998). The demonstration that transferrin receptors are localized to areas of membrane ruffling in cells expressing activated alleles of Rac1 (Bretscher and Aguado-Velasco, 1998) underscores the close association of membrane recycling pathways and areas of membrane protrusion. Perhaps ligated $Fc_{\gamma}Rs$, like growth factor receptors, utilize localized membrane insertion in a Rac and/or ARF6-dependent fashion. Further studies are needed to define the precise molecular architecture and signaling components that underly pseudopod extension and phagosome formation.

ACKNOWLEDGMENTS

Work from this laboratory was supported by grant HL04164 from the National Institutes of Health and grant CB-170 from the American Cancer Society. S.G. is an Established Investigator of the American Heart Association.

REFERENCES

Abo, A., Webb, M.R., Grogan, A., & Segal, A.W. (1994). Activation of NADPH oxidase involves the dissociation of p21^{rac} from its inhibitory GDP/GTP exchange protein (rhoGDI) followed by its translocation to the plasma membrane. Biochem. J. 3, 585–591.

Adamczewski, M., & Kinet, J.P. (1994). The high-affinity receptor for immunoglobulin E. Chem. Immunol. 59, 173–190.

Aderem, A.A., Scott, W.A., & Cohn, Z.A. (1986). Evidence for sequential signals in the induction of the arachidonic acid cascade in macrophages. J. Exp. Med. 163, 139–154.

Aderem, A.A., Wright, S.D., Silverstein, S.C., & Cohn, Z.A. (1985). Ligated complement receptors do not activate the arachidonic acid cascade in resident peritoneal macrophages. J. Exp. Med. 161, 617–622.

Agarwal, A., Salem, P., & Robbins, K.C. (1993). Involvement of p72^{syk}, a protein-tyrosine kinase, in Fc_{γ} receptor signaling. J. Biol. Chem. 268, 15900–15905.

Allen, L.A.H., & Aderem, A. (1996). Molecular definition of distinct cytoskeletal structures involved in complement- and Fc receptor-mediated phagocytosis in macrophages. J. Exp. Med. 184, 627–637.

Altieri, D.C., & Edgington, T.S. (1988). The saturable high affinity association of factor X to ADP-stimulated monocytes defines a novel function of the Mac-1 receptor. J. Biol. Chem. 263, 7007–7015.

Amigorena, S., Bonnerot, C., Drake, J.R., Choquet, D., Hunziker, W., Guillet, J.G., Webster, P., Sautes, C., Mellman, I., & Fridman, W.H. (1992). Cytoplasmic domain heterogeneity and functions of IgG Fc receptors in B lymphocytes. Science 256, 1808–1812.

Amoui, M., Draberova, L., Tolar, P., & Draber, P. (1997). Direct interaction of syk and lyn protein tyrosine kinases in rat basophilic leukemia cells activated via type I Fcε receptors. Eur. J. Immunol. 27, 321–328.

Anderson, C.L., Guyre, P.M., Whitin, J.C., Ryan, D.H., Looney , R.J., & Fanger, M.W. (1986). Soluble circulating Fcγ receptors on human mononuclear phagocytes. Antibody characterization and induction of superoxide production in a monocyte cell line. J. Biol. Chem. 261, 12856–12864.

Anderson, C.L., Shen, L., Eicher, D.M., Wewers, M.D., & Gill, J.K. (1990a). Phagocytosis mediated by three distinct Fc_{γ} receptor classes on human leukocytes. J. Exp. Med. 171, 1333–1345.

Anderson, P., Caliguiri, M., O'Brien, C., Manley, T., Ritz, J., & Schlossman, S.F. (1990b). Fcγ receptor type III (CD16) is included in the ζ NK receptor complex expressed by human natural killer cells. Proc. Natl. Acad. Sci. USA 87, 2274–2278.

Araki, N., Johnson, M.T., & Swanson, J.A. (1997). A role for phosphoinositide 3-kinase in the completion of macropinocytosis and phagocytosis by macrophages. J. Cell Biol. 135, 1249–1260.

Arcaro, A. (1998). The small GTP-binding protein Rac promotes the dissociation of gelsolin from actin filaments in neutrophils. J. Biol. Chem. 273, 805–813.

Arnaout, M.A., Todd, R.F., Dana, N., Melamed, J., Schlossman, S.F., & Colten, H.R. (1983). Inhibition of phagocytosis of complement C3 or immunoglobulin G coated particles and of C3bi binding by monoclonal antibodies to the monocyte-granulocyte membrane glycoprotein (Mo1). J. Clin. Invest. 72, 171–179.

Auer, K.L., & Jacobson, B.S. (1995). β_1 integrins signal lipid second messengers required during cell adhesion. Mol. Biol. Cell 6, 1305–1313.

Bae, Y.S., Cantley, L.G., Chen, C.-C., Kim, S.-R., Kwon, K.-S., & Rhee, S.G. (1998). Activation of phospholipase C-γ by phosphatidylinositol 3,4,5-trisphosphate. J. Biol. Chem. 273, 4465–4469.

Bainton, D.F., Takemura, R., Stenberg, P.E., & Werb, Z. (1987). Rapid fragmentation and reorganization of Golgi membranes during frustrated phagocytosis of immobile immune complexes by macrophages. Am. J. Pathol. 134, 15–26.

Barker, S.A., Caldwell, K.K., Pfeiffer, J.R., & Wilson, B.S. (1998). Wortmannin-sensitive phosphorylation, translocation, and activation of PLCγ1, but not PLCγ2, in antigen-stimulated RBL-2H3 mast cells. Mol. Biol. Cell 9, 483–496.

Bashour, A.M., Fullerton, A.T., Hart, M.J., & Bloom, G.S. (1997). IQGAP1, a Rac- and Cdc42-binding protein, directly binds and cross-links microfilaments. J. Cell Biol. 137, 1555–1566.

Beaven, M.A. (1996). Calcium signalling: sphingosine kinase versus phospholipase C? Curr. Biol. 6, 798–801.

Beaven, M.A., & Metzger, H. (1993). Signal transduction by Fc receptors: the Fc epsilon RI case. Immunol. Today 14, 222–226.

Benhamou, M., Ryba, N.J., Kihara, H., Nishikata, H., & Siraganian, R.P. (1993). Protein-tyrosine kinase $p72^{syk}$ in high affinity IgE receptor signaling. Identification as a component of pp72 and association with the receptor γ chain after receptor aggregation. J. Biol. Chem. 268, 23318–23324.

Berger, S.A., Mak, T.W., & Paige, C.J. (1994). Leukocyte common antigen (CD45) is required for immunoglobulin E-mediated degranulation of mast cells. J. Exp. Med. 180, 471–476.

Bijsterbosch, M.K., & Klaus, G.G. (1985). Crosslinking of surface immunoglobulin and Fc receptors on B lymphocytes inhibits stimulation of inositol phospholipid breakdown via the antigen receptors. J. Exp. Med. 162, 1825–1836.

Bobak, D.A., Frank, M.M., & Tenner, A.J. (1988). C1q acts synergistically with phorbol dibutyrate to activate CR1-mediated phagocytosis by human mononuclear phagocytes. Eur. J. Immunol. 18, 2001–2007.

Bobak, D.A., Gaither, T.A., Frank, M.M., & Tenner, A.J. (1987). Modulation of FcR function by complement: subcomponent C1q enhances the phagocytosis of IgG-opsonized targets by human monocytes and culture-derived macrophages. J. Immunol. 138, 1150–1156.

Bohnsack, J.F., Kleinman, H.K., Takahashi, T., Oshea, J.J., & Brown, E.J. (1985). Connective tissue proteins and phagocytic cell function. Laminin enhances complement and Fc-mediated phagocytosis by cultured human phagocytes. J. Exp. Med. 161, 912–923.

Bokoch, G.M., Bohl, B.P., & Chuang, T.-H. (1994). Guanine nucleotide exchange regulated membrane translocation of Rac/Rho GTP-binding proteins. J. Biol. Chem. 269, 31674–31679.

Bolland, S., Pearse, R.N., Kurosaki, T., & Ravetch, J.V. (1998). SHIP modulates immune receptor responses by regulating membrane association of BTK. Immunity 8, 509–516.

Bonnefoy, J.Y., Lecoanet-Henchoz, S., Aubry, J.P., Gauchat, J.F., & Graber, P. (1995). CD23 and B-cell activation. Curr. Opin. Immunol. 7, 355–359.

Bonnefoy, J.Y., Plater-Zyberk, C., Lecoanet-Henchoz, S., Gauchat, J.F., Aubry, J.P., & Graber, P. (1996). A new role for CD23 in inflammation. Immunol. Today 17, 418–420.

Bonnerot, C., Amigorena, S., Choquet, D., Pavlovich, R., Choukroun, V., & Fridman, W.H. (1992). Role of γ-chain in tyrosine kinase activation via murine $Fc_{\gamma}RIII$. EMBO J. 11, 2747–2757.

Bretscher, M.S., & Aguado-Velasco, C. (1998). EGF induces recycling membrane to form ruffles. Curr. Biol. 8, 721–724.

Brooks, D.G., Qui, W.Q., Luster, A.D., & Ravetch, J.V. (1989). Structure and expression of human IgG FcRII (CD32) Functional heterogeneity is encoded by the alternatively spliced products of multiple genes. J. Exp. Med. 170, 1369–1385.

Brown, E., Hooper, L., Ho, T., & Gresham, H. (1990). Integrin-associated protein: a 50 kD plasma membrane antigen physically and functionally associated with integrins. J. Cell Biol. 111, 2785–2794.

Brown, E.J., Bohnsack, J.F., & Gresham, H.D. (1988). Mechanism of inhibition of immunoglobulin G-mediated phagocytosis by monoclonal antibodies that recognizes the Mac-1 antigen. J. Clin. Invest. 81, 365–375.

Brown, M.T., & Cooper, J.A. (1996). Regulation, substrates and functions of src. Biochim. Biophys. Acta 1287, 121–149.

Brozna, J.P., Hauff, N.F., Phillips, W.A., & Johnston, J., R.B. (1988). Activation of the respiratory burst in macrophages. Phosphorylation specifically associated with Fc receptormediated stimulation. J. Immunol. 141, 1642–1647.

Buckle, A.M., & Hogg, N. (1989). The effect of IFN-γ and colony-stimulating factors on the expression of neutrophil cell membrane receptors. J. Immunol. 143, 2295–2301.

Bustelo, X.R. (1996). The VAV family of signal transduction molecules. Crit. Rev. Oncogenesis 7, 65–88.

Cambier, J.C. (1995). Antigen and Fc receptor signaling: The awesome power of the immunoreceptor tyrosine-based activation motif (ITAM). J. Immunol. 155, 3281–3285.

Capsoni, F., Bonara, P., Minonzio, F., Ongari, A.M., Colombo, G., Rizzardi, G.P., & Zanussi, C. (1991). The effect of cytokines on human neutrophil Fc receptor-mediated phagocytosis. J. Clin. Lab. Immunol. 34, 115–124.

Capsoni, F., Minonzio, F., Ongari, A.M., Carbonelli, V., Galli, A., & Zanussi, C. (1995). IL-10 up-regulates human monocyte phagocytosis in the presence of IL-4 and IFN-γ. J. Leuk. Biol. 58, 351–358.

Carini, C., & Fratazzi, C. (1996). CD23 expression in activated human T cells is enhanced by interleukin-7. Int. Arch. Allergy. Immunol. 110, 23–30.

Caron, E., & Hall, A. (1998). Identification of two distinct mechanisms of phagocytosis controlled by different Rho GTPases. Science 282, 1717–1721.

Carpenter, C.L., & Cantley, L.C. (1996). Phosphoinositide kinases. Curr. Opin. Cell Biol. 8, 153–158.

Cavenagh, M.M., Whitney, J.A., Carroll, K., Zhang, C.J., Boman, A.L., Rosenwald, A.G., Mellman, I., & Kahn, R.A. (1996). Intracellular distribution of Arf proteins in mammalian cells. Arf6 is uniquely localized to the plasma membrane. J. Biol. Chem. 271, 21767–21774.

Ceuppens, J.L., Baroja, M.L., Van Vaeck, F., & Anderson, C.L. (1988). Defect in the membrane expression of high affinity 72-kD Fcγ receptors on phagocytic cells in four healthy subjects. J. Clin. Invest. 82, 571–578.

Chacko, G.W., Brandt, J.T., Coggeshall, K.M., & Anderson, C.L. (1996). Phosphoinositide 3-kinase and $p72^{syk}$ noncovalently associate with the low affinity Fcγ receptor on human platelets through an immunoreceptor tyrosine-based activation motif—Reconstitution with synthetic phosphopeptides. J. Biol. Chem. 271, 10775–10781.

Chan, A.C., & Shaw, A.S. (1996). Regulation of antigen receptor signal transduction by protein tyrosine kinases. Curr. Opin. Immunol. 8, 394–401.

Chen, L.-M., Hobbie, S., & Galan, J.E. (1996). Requirement of CDC42 for *Salmonella*-induced cytoskeletal and nuclear responses. Nature 274, 2115–2118.

Choi, O.H., Kim, J.H., & Kinet, J.P. (1996). Calcium mobilization via sphingosine kinase in signalling by the FcεRI antigen receptor. Nature 380, 634–636.

Chong, L.D., Traynor-Kaplan, A., Bokoch, G.M., & Schwartz, M.A. (1994). The small GTP-binding protein Rho regulates a phosphatidylinositol 4-phosphate 5-kinase in mammalian cells. Cell 79, 507–513.

Clark, M.R., Stuart, S.G., Kimberly, R.P., Ory, P.A., & Goldstein, I.M. (1991). A single amino acid distinguishes the high-responder from the low-responder form of Fc receptor II on human monocytes. Eur. J. Immunol. 21, 1911–1916.

Clarkson, S.B., Kimberly, R.P., Valinsky, J.E., Witmer, M.D., Bussel, J.B., Nachman, R.L., & Unkeless, J.C. (1986). Blockade of clearance of immune complexes by an anti-Fcγ receptor monoclonal antibody. J. Exp. Med. 164, 474–489.

Colley, W.C., Sung, T.C., Roll, R., Jenco, J., Hammond, S.M. Altshuller, Y., Bar-Sagi, D., Morris, A.J., & Frohman, M.A. (1997). Phospholipase D2, a distinct phospholipase D isoform with novel regulatory properties that provokes cytoskeletal reorganization. Curr. Biol. 7, 191–201.

Collins, H.L., & Bancroft, G.J. (1992). Cytokine enhancement of complement-dependent phagocytosis by macrophages: synergy of tumor necrosis factor-α and granulocyte-macrophage colony-stimulating factor for phagocytosis of *Cryptococcus neoformans*. Eur. J. Immunol. 22, 1447–1454.

Cox, D., Chang, P., Kurosaki, T., & Greenberg, S. (1996). Syk tyrosine kinase is required for immunoreceptor tyrosine activation motif-dependent actin assembly. J. Biol. Chem. 271, 16597–16602.

Cox, D., Chang, P., Zhang, Q., Reddy, P.G., Bokoch, G.M., & Greenberg, S. (1997). Requirements for both Rac1 and Cdc42 in membrane ruffling and phagocytosis in leukocytes. J. Exp. Med. 186, 1487–1494.

Cox, D., Tseng, C-C., Bjekic, G., & Greenberg, S. (1999). A requirement for phosphatidylinositol 3-kinase in pseudopod extension. J. Biol. Chem. 274, 1240–1247.

Crespo, P., Schuebel, K.E., Ostrom, A.A., Gutkind, J.S., & Bustelo, X.R. (1997). Phosphotyrosine-dependent activation of Rac-1 GDP/GTP exchange by the *vav* proto-oncogene product. Nature 385, 169–172.

Crowley, M.T., Costello, P.S., Fitzer-Attas, C.J., Turner, M., Meng, F., Lowell, C., Tybulewicz, V.L.J., & DeFranco, A.L. (1997). A critical role for Syk in signal transduction and phagocytosis mediated by Fc_γ receptors on macrophages. J. Exp. Med. 186, 1027–1039.

Daeron, M. (1996). Building up the family of ITIM-bearing negative coreceptors. Immunol. Lett. 54, 73–76.

Daeron, M., Malbec, O., Bonnerot, C., Latour, S., Segal, D.M., & Fridman, W.H. (1994). Tyrosine-containing activation motif-dependent phagocytosis in mast cells. J. Immunol. 152, 783–792.

Darby, C., Geahlen, R.L., & Schreiber, A.D. (1994). Stimulation of macrophage FcγRIIIA activates the receptor-associated protein tyrosine kinase Syk and induces phosphorylation of multiple proteins including p95Vav and p62/GAP-associated protein. J. Immunol. 152, 5429–5437.

Dash, D., Aepfelbacher, M., & Siess, W. (1995). Integrin $\alpha_{IIb}\beta_3$-mediated translocation of CDC42Hs to the cytoskeleton in stimulated human platelets. J. Biol. Chem. 270, 17321–17326.

Davis, W., Harrison, P.T., Hutchinson, M.J., & Allen, J.M. (1995). Two distinct regions of Fc_γ RI initiate separate signalling pathways involved in endocytosis and phagocytosis. EMBO J. 14, 432–441.

Debets, J.M., van de Winkel, J.G., Ceuppens, J.L., Dieteren, I.E., & Buurman, W.A. (1990). Cross-linking of both FcγRI and FcγRII induces secretion of tumor necrosis factor by human monocytes, requiring high affinity Fc-FcγR interactions. Functional activation of FcγRII by treatment with proteases or neuraminidase. J. Immunol. 144, 1304–1310.

Debets, J.M., Van der Linden, C.J., Dieteren, I.E., Leeuwenberg, J.F., & Buurman, W.A. (1988). Fc-receptor cross-linking induces rapid secretion of tumor necrosis factor (cachectin) by human peripheral blood monocytes. J. Immunol. 141, 1197–1201.

Deckert, M., Tartare-Deckert, S., Couture, C., Mustelin, T., & Altman, A. (1996). Functional and physical interactions of Syk family kinases with the Vav proto-oncogene product. Immunity 5, 591–604.

Della Bianca, V., Grzeskowiak, M., & Rossi, F. (1990). Studies on molecular regulation of phagocytosis and activation of the NADPH oxidase in nuetrophils-IgG- and C3b-mediated ingestion and associated respiratory burst independent of phospholipid turnover and Ca^{2+} transients. J. Immunol. 144, 1411–1417.

Della Bianca, V., Grzeskowiak, M., Dusi, S., & Rossi, F. (1993). Formation of inositol (1,4,5) trisphophate and increase of cytosolic Ca^{2+} mediated by Fc receptors in human neutrophils. Biochem. Biophys. Res. Comm. 196, 1233–1239.

Di Virgilio, F., Meyer, B.C., Greenberg, S., & Silverstein, S.C. (1988). Fc receptor-mediated phagocytosis occurs in macrophages at exceedingly low cytosolic Ca^{2+} levels. J. Cell Biol. 106, 657–666.

Diamond, M.S., Alon, R., Parkos, C.A., Quinn, M.T., & Springer, T.A. (1995). Heparin is an adhesive ligand for the leukocyte integrin Mac-1 (CD11b/CD18). J. Cell Biol. 130, 1473–1482.

Donaldson, J.G., Cassel, D., Kahn, R.A., & Klausner, R.D. (1992a). ADP-ribosylation factor, a small GTP-binding protein, is required for binding of the coatomer protein β-COP to Golgi membranes. Proc. Natl. Acad. Sci. USA 89, 6408–6412.

Donaldson, J.G., Finazzi, D., & Klausner, R.D. (1992b). Brefeldin A inhibits Golgi membrane-catalysed exchange of guanine nucleotide onto ARF protein. Nature 360, 350–352.

Donaldson, J.G., Radhakrishna, H., & Peters, P.J. (1995). The ARF GTPases: defining roles in membrane traffic and organelle structure. Cold Spring Harb. Symp. Quant. Biol. 60, 229–234.

D'Souza-Schorey, C., Boshans, R.L., McDonough, M., Stahl, P.D. & Van Aelst, L. (1997). A role for POR1, a Rac1-interacting protein, in ARF6-mediated cytoskeletal rearrangements. EMBO J. 16, 5445–5454.

D'Souza-Schorey, C., van Donselaar, E., Hsu, V.W., Yang, C., Stahl, P.D., & Peters, J. P. (1998). ARF6 targets recycling vesicles to the plasma membrane: insights from an ultrastructural investigation. J. Cell Biol. 140, 603–616.

Duchemin, A.-M., & Anderson, C.L. (1997). Association of non-receptor protein tyrosine kinases with the FcγRI/γ-chain complex in monocytic cells. J. Immunol. 158, 865–871.

Duchemin, A.M., Ernst, L.K., & Anderson, C.L. (1994). Clustering of the high affinity Fc receptor for immunoglobulin G (FcγRI) results in phosphorylation of its associated γ-chain. J. Biol. Chem. 269, 12111–12117.

Duits, A.J., Bootsma, H., Derksen, R.H., Spronk, P.E., Kater, L., Kallenberg, C.G., Capel, P.J., Westerdaal, N.A., Spierenburg, G.T., et al. (1995). Skewed distribution of IgG Fc receptor IIa (CD32) polymorphism is associated with renal disease in systemic lupus erythematosus patients. Arthritis Rheum. 38, 1832–1836.

Durden, D.L., Kim, H.M., Calore, B., & Liu, Y.B. (1995). The FcγRI receptor signals through the activation of *hck* and MAP kinase. J. Immunol. 154, 4039–4047.

Dusi, S., Donini, M., & Rossi, F. (1996). Mechanisms of NADPH oxidase activation: translocation of $p40_{phox}$, Rac1 and Rac2 from the cytosol to the membranes in human neutrophils lacking $p47_{phox}$ or $p67_{phox}$. Biochem. J. 314, 409–412.

Dusi, S., Donini, M., Della, B.V., & Rossi, F. (1994). Tyrosine phosphorylation of phospholipase C-γ2 is involved in the activation of phosphoinositide hydrolysis by Fc receptors in human neutrophils. Biochim. Biophys. Acta 201, 1100–1108.

Edberg, J.C., Lin, C.T., Lau, D., Unkeless, J.C., & Kimberly, R.P. (1995). The Ca2+ dependence of human Fc gamma receptor-initiated phagocytosis. J. Biol. Chem. 270, 22301–22307.

Erickson, J.W., Cerione, R.A., & Hart, M.J. (1997). Identification of an actin cytoskeletal complex that includes IQGAP and the Cdc42 GTPase. J. Biol. Chem. 272, 24443–24447.

Ernst, L.K., van de Winkel, J.G.J., Chiu, I.-M., & Anderson, C.L. (1992). Three genes for the human high affinity Fc receptor for IgG (FcγRI) encode four distinct transcription products. J. Biol. Chem. 267, 15692–15700.

Falasca, M., Logan, S.K., Lehto, V.P., Baccante, G., Lemmon, M.A., & Schlessinger, J. (1998). Activation of phospholipase Cγ by PI 3-kinase-induced PH domain-mediated membrane targeting. EMBO J. 17, 414–422.

Fallman, M., Andersson, K., Hakansson, S., Magnusson, K.-E., Stendahl, O., & Wolf-Watz, H. (1995). *Yersinia pseudotuberculosis* inhibits Fc receptor-mediated phagocytosis in J774 cells. Infect. Immun. 63, 3117–3124.

Fluckiger, A.-C., Li, Z., Kato, R.M., Wahl, M.I., Ochs, H.D., Longnecker, R., Kinet, J.-P., Witte, O.N., Scharenberg, A.M., & Rawlings, D.J. (1998). Btk/Tec kinase regulate sustained increases in intracellular Ca^{2+} following B-cell receptor activation. EMBO J. 17, 1973–1985.

Francis, C.L., Ryan, T.A., Jones, B.D., Smith, S.J., & Falkow, S. (1993). Ruffles induced by *Salmonella* and other stimuli direct macropinocytosis of bacteria. Nature 364, 639–642.

Fukushima, T., Waddell, T.K., Grinstein, S., Goss, G.G., Orlowski, J., & Downey, G.P. (1996). Na^+/H^+ exchange activity during phagocytosis in human neutrophils: role of Fcγ receptors and tyrosine kinases. J. Cell Biol. 132, 1037–1052.

Galon, J., Gauchat, J.F., Mazieres, N., Spagnoli, R., Storkus, W., Lotze, M., Bonnefoy, J.Y., Fridman, W.H., & Sautes, C. (1996). Soluble Fcγ receptor type III (FcγRIII, CD16) triggers cell activation through interaction with complement receptors. J. Immunol. 157, 1184–1192.

Germano, P., Gomez, J., Kazanietz, M.G., Blumberg, P.M., & Rivera, J. (1994). Phosphorylation of the γ chain of the high affinity receptor for immunoglobulin E by receptor-associated protein kinase C-δ. J. Biol. Chem. 269, 23102–23107.

Gewirtz, A.T., & Simons, E.R. (1997). Phospholipase D mediates Fcγ receptor activation of neutrophils and provides specificity between high-valency immune complexes and fMLP signaling pathways. J. Leuk. Biol. 61, 522–528.

Ghazizadeh, S., & Fleit, H.B. (1994). Tyrosine phosphorylation provides an obligatory early signal for Fcγ RII-mediated endocytosis in the monocytic cell line THP-1. J. Immunol. 152, 30–41.

Ghazizadeh, S., Bolen, J.B., & Fleit, H.B. (1994). Physical and functional association of Src-related protein tyrosine kinases with Fcγ RII in monocytic THP-1 cells. J. Biol. Chem. 269, 8878–8884.

Ghazizadeh, S., Bolen, J.B., & Fleit, H.B. (1995). Tyrosine phosphorylation and association of Syk with Fcγ RII in monocytic THP-1 cells. Biochem. J. 305, 669–674.

Gilfillan, A.M. (1997). Signal transduction pathways regulating arachidonic acid metabolite generation following FcεRI aggregation. In: IgE Receptor (FcεRI) Function in Mast Cells and Basophils (Hamawy, M.M., Ed.), pp. 181–208. R.G. Landes Company.

Girard, M.T., Hjaltadottir, S., Fejes-Toth, A.N., & Guyre, P.M. (1987). Glucocorticoids enhance γ-interferon augmentation of human monocyte immunoglobulin G Fc receptor expression. J. Immunol. 138, 3235–3241.

Glick, D., & Barenholz, Y. (1996). IgG immunoglobulins induce activation of the sphingomyelin cycle in HL-60 cells. FEBS Lett. 394, 237–240.

Gordon, J., Katira, A., Strain, A.J., & Gillis, S. (1991). Inhibition of interleukin 4-promoted CD23 production in human B lymphocytes by transforming growth factor-β, interferons or anti-CD19 antibody is overriden on engaging CD40. Eur. J. Immunol. 21, 1917–1922.

Gordon, J.R., & Galli, S.J. (1990). Mast cells as a source of both preformed and immunologically inducible TNF-α/cachectin. Nature 346, 274–276.

Green, J.M., Schreiber, A.D., & Brown, E.J. (1997). Role for a glycan phosphoinositol anchor in Fcγ receptor synergy. J. Cell Biol. 139, 1209–1217.

Greenberg, S., Chang, P., & Silverstein, S.C. (1993). Tyrosine phosphorylation is required for Fc receptor-mediated phagocytosis in mouse macrophages. J. Exp. Med. 177, 529–534.

Greenberg, S., Chang, P., & Silverstein, S.C. (1994). Tyrosine phosphorylation of the γ subunit of Fc_{γ} receptors, $p72^{syk}$, and paxillin during Fc receptor-mediated phagocytosis in macrophages. J. Biol. Chem. 269, 3897–3902.

Greenberg, S., El Khoury, J., Di Virgilio, F., Kaplan, E.M., & Silverstein, S.C. (1991). Ca^{2+}-independent F-actin assembly and disassembly during Fc receptor-mediated phagocytosis in mouse macrophages. J. Cell Biol. 113, 757–767.

Gresham, H.D., Zheleznyak, A., Mormol, J.S., & Brown, E.J. (1990). Studies on the molecular mechanisms of human neutrophil Fc receptor-mediated phagocytosis. Evidence that a distinct pathway for activation of the respiratory burst results in reactive oxygen metabolite-dependent amplification of ingestion. J. Biol. Chem. 265, 7819–7826.

Griffin, F.M.J., Griffin, J.A., & Silverstein, S.C. (1976). Studies on the mechanism of phagocytosis. II. The interaction of macrophages with anti-immunoglobulin IgG-coated bone marrow-derived lymphocytes. J. Exp. Med. 144, 788–809.

Griffin, F.M.J., Griffin, J.A., Leider, J.E., & Silverstein, S.C. (1975). Studies on the mechanism of phagocytosis. I. Requirements for circumferential attachment of particle-bound ligands to specific receptors on the macrophage plasma membrane. J. Exp. Med. 142, 1263–1282.

Gupta, N., Scharenberg, A.M., Burshtyn, D.N., Wagtmann, N., Lioubin, M.N., Rohrschneider, L.R., Kinet, J.P., & Long, E.O. (1997). Negative signaling pathways of the killer cell inhibitory receptor and Fc gamma RIIb1 require distinct phosphatases. J. Exp. Med. 186, 473–478.

Gurish, M.F., Ghildyal, N., Arm, J., Austen, K.F., Avraham, S., Reynolds, D., & Stevens, R.L. (1991). Cytokine mRNA are preferentially increased relative to secretory granule protein mRNA in mouse bone marrow-derived mast cells that have undergone IgE-mediated activation and degranulation. J. Immunol. 146, 1527–1533.

Hackam, D.J., Rotstein, O.D., Schreiber, A., Zhang, W.J., & Grinstein, S. (1997). Rho is required for the initiation of calcium signaling and phagocytosis by Fcγ receptors in macrophages. J. Exp. Med. 186, 955–966.

Hackam, D.J., Rotstein, O.D., Sjolin, C., Schreiber, A.D., Trimble, W.S., & Grinstein, S. (1998). v-SNARE-dependent secretion is required for phagocytosis. Proc. Natl. Acad. Sci. USA 95, 11691–11696.

Haleem-Smith, H., Chang, E.-Y., Szallasi, Z., Blumberg, P.M., & Rivera, J. (1995). Tyrosine phosphorylation of protein kinase C-δ in response to the activation of the high-affinity receptor for immunoglobulin E modifies its substrate recognition. Proc. Natl. Acad. Sci. USA 92, 9112–9116.

Hall, A. (1998). Rho GTPases and the actin cytoskeleton. Science 279, 509–513.

Hamada, F., Aoki, M., Akiyama, T., & Toyoshima, K. (1993). Association of immunoglobulin G Fc receptor II with Src-like protein-tyrosine kinase Fgr in neutrophils. Proc. Natl. Acad. Sci. USA 90, 6305–6309.

Hamawy, M.M., & Swaim, W.D. (1997). FcεRI-mediated cell degranulation, proliferation and adhesion. In: IgE Receptor (FcεRI) Function in Mast Cells and Basophils (Hamawy, M.M., Ed.), pp. 174–180. R.G. Landes.

Hardt, W.-D., Chen, L.-M., Schuebel, K.E., Bustelo, S.R., & Galan, J.E. (1998). *S. typimurium* encodes an activator of Rho GTPases that induces membrane ruffling and nuclear responses in host cells. Cell 93, 815–826.

Hart, M.J., Callow, M.G., Souza, B., & Polakis, P. (1996). IQGAP1, a calmodulin-binding protein with a rasGAP-related domain, is a potential effector for cdc42Hs. EMBO J. 15, 2997–3005.

Hartwig, J.H., Bokoch, G.M., Carpenter, C.L., Janmey, P.A., Taylor, L.A., Toker, A., & Stossel, T.P. (1995). Thrombin receptor ligation and activated Rac uncap actin filament barbed ends through phosphoinositide synthesis in permeabilized human platelets. Cell 82, 643–653.

Hashimoto, S., Koh, K., Tomita, Y., Amemiya, E., Sawada, S., Yodoi, J., & Horie, T. (1995). TNF-α regulates IL-4-induced FcεRII/CD23 gene expression and soluble FcεRII release by human monocytes. Internat. Immunol. 7, 705–713.

Hauschildt, S., & Kleine, B. (1995). Bacterial stimulators of macrophages. In: International Review of Cytology—A Survey of Cell Biology, Vol. 161, pp. 263–331.

Helms, J.B., & Rothman, J.E. (1992). Inhibition by brefeldin A of a Golgi mebrane enzyme that catalyses exchange of guanine nucleotide bound to ARF. Nature 360, 352–354.

Hibbs, M.L., Selvaraj, P., Carpen, O., Springer, T.A., Kuster, H., Jouvin, M.E., & Kinet, J. (1989). Mechanisms for regulating expression of membrane isoforms of FcγRIII (CD16). Science 246, 1608–1611.

Hill, M.R., & Cookson, W.O. (1996). A new variant of the beta subunit of the high-affinity receptor for immunoglobulin E (FcεRI-beta E237G): associations with measures of atopy and bronchial hyper-responsiveness. Hum. Mol. Genet. 5, 959–962.

Hill, M.R., James, A.L., Faux, J.A., Ryan, G., Hopkin, J.M., le Souef, P., Musk, A.W., & Cookson, W.O. (1995). FcεRI-beta polymorphism and risk of atopy in a general population sample. Brit. Med. J. 311, 776–779.

Hirasawa, N., Scharenberg, A., Yamamura, H., Beaven, M.A., & Kinet, J.P. (1995). A requirement for Syk in the activation of the microtubule-associated protein kinase phospholipase A_2 pathway by Fc_ε R1 is not shared by a G protein-coupled receptor. J. Biol. Chem. 270, 10960–10967.

Hirata, Y., & Suzuki, T. (1987). Protein kinase activity associated with Fcγ_{2a} receptor of a murine macrophage like cell line, P388D_1. Biochem. 26, 8189–8195.

Hishikawa, T., Cheung, J.Y., Yelamarty, R.V., & Knutson, D.W. (1991). Calcium transients during Fc receptor-mediated and nonspecific phagocytosis by murine peritoneal macrophages. J. Cell Biol. 115, 59–66.

Hostoffer, R.W., Krukovets, I., & Berger, M. (1993). Increased FcαR expression and IgA-mediated funcion on neutrophils induced by chemoattractants. J. Immunol. 150, 4532–4540.

Hostoffer, R.W., Krukovets, I., & Berger, M. (1994). Enhancement by tumor necrosis factor-α of Fc_α receptor expression and IgA-mediated superoxide generation and killing of *Pseudomonas aeruginosa* by polymorphonuclear leukocytes. J. Infect. Dis. 170, 82–87.

Hou, X.H., Dietrich, J., Odum, N., & Geisler, C. (1996). The cytoplasmic tail of FcγRIIIAα is involved in signaling by the low affinity receptor for Immunoglobulin G. J. Biol. Chem. 271, 22815–22822.

Huizinga, T.W.J., van der Schoot, C.E., Jost, C., Klaassen, R., Kleijer, M., von dem Borne, A.E.G.K., Roos, D., & Tetteroo, P.A.T. (1988). The PI-linked receptor FcRIII is released on stimulation of neutrophils. Nature 333, 667–669.

Huizinga, T.W.J., van Kemenade, F., Koenderman, L., Dolman, K.M., Von den Borne, A.E.G., Tetteroo, P.A.T., & Roos, R. (1989). The 40-kDa Fc_γ receptor (FcRII) on human neutrophils is essential for the IgG-induced respiratory burst and IgG-induced phagocytosis. J. Immunol. 142, 2365–2369.

Hulett, M.D., & Hogarth, P.M. (1994). Molecular basis of Fc receptor function. In: Advances in Immunology, Vol. 57, (Dixon, F.J., Ed.), pp. 1–127. Academic Press.

Hulett, M.D., Osman, N., McKenzie, I.F., & Hogarth, P.M. (1991). Chimeric Fc receptors identify functional domains of the murine high affinity receptor for IgG. J. Immunol. 147, 1863–1868.

Ince, C., Coremans, J.M.C.C., Ypey, D.L., Leijh, P.C.J., Verveen, A.A., & van Furth, R. (1988). Phagocytosis by human macrophages is accompanied by changes in ionic channel currents. J. Cell Biol. 106, 1873–1878.

Indik, Z., Kelly, C., Chien, P., Levinson, A.I., & Schreiber, A.D. (1991). Human FcγRII, in the absence of other Fcγ receptors, mediates a phagocytic signal. J. Clin. Invest. 88, 1766–1771.

Indik, Z.K., Park, J.-G., Hunter, S., & Schreiber, A.D. (1995). Structure/function relationships of Fcγ receptors in phagoctyosis. Sem. Immunol. 7, 45–54.

Isakov, N. (1997). Immunoreceptor tyrosine-based activation motif (ITAM), a unique module linking antigen and Fc receptors to their signaling cascades. J. Leukoc. Biol. 61, 6–16.

Israel, E.J., Wilsker, D.F., Hayes, K.C., Schoenfield, D., & Simister, N.E. (1996). Increased clearance of IgG in mice that lack β2-microglobulin: Possible protective role of FcRn. Immunol. 89, 573–578.

Jaconi, M.E.E., Lew, D.P., Carpentier, J.-L., Magnusson, K.E., Sjogren, M., & Stendahl, O. (1990). Cytosolic free calcium elevation mediates the phagosome-lysosome fusion during phagocytosis in human neutrophils. J. Cell Biol. 110, 1555–1564.

Jouvin, M.H., Adamczewski, M., Numerof, R., Letourneur, O., Valle, A., & Kinet, J.P. (1994). Differential control of the tyrosine kinases Lyn and Syk by the two signaling chains of the high affinity immunoglobulin E receptor. J. Biol. Chem. 269, 5918–5925.

Junghans, R.P., & Anderson, C.L. (1996). The protection receptor for IgG catabolism is the β2-microglobulin-containing neonatal intestinal transport receptor. Proc. Natl. Acad. Sci. U.S.A. 93, 5512–5516.

Justement, L.B., Campbell, K.S., Chien, N.C., & Cambier, J.C. (1991). Regulation of B cell antigen receptor signal transduction and phosphorylation by CD45. Science 252, 1839–1842.

Kanakaraj, P., Duckworth, B., Azzoni, L., Kamoun, M., Cantley, L.C., & Perussia, B. (1994). Phosphatidylinsoitol-3 kinase activation induced upon FcγRIIIA-ligand interaction. J. Exp. Med. 179, 551–558.

Kiener, P.A., Rankin, B.M., Burkhardt, A.L., Schieven, G.L., Gilliland, L.K., Rowley, R.B., Bolen, J.B., & Ledbetter, J.A. (1993). Cross-linking of Fcγ receptor I (Fcγ RI) and receptor II (Fcγ RII) on monocytic cells activates a signal transduction pathway common to both Fc receptors that involves the stimulation of $p72^{Syk}$ protein tyrosine kinase. J. Biol. Chem. 268, 24442–24448.

Kimberly, R.P., Ahlstrom, J.W., Click, M.E., & Edberg, J.C. (1990). The glycosyl phosphatidylinositol-linked $Fc\gamma RIII_{PMN}$ mediates transmembrane signaling events distinct from FcγRII. J. Exp. Med. 171, 1239–1255.

Kimura, T., Sakamoto, H., Appella, E., & Siraganian, R.P. (1996). Conformational changes induced in the protein tyrosine kinase $p72^{syk}$ by tyrosine phosphorylation or by binding of phosphorylated immunoreceptor tyrosine-based activation motif peptides. Mol. Cell Biol. 16, 1471–1478.

Kimura, T., Zhang, J., Sagawa, K., Sakaguchi, K., Appella, E., & Siraganian, R.P. (1997). Syk-independent tyrosine phosphorylation and association of the protein tyrosine phosphatases SHP-1 and SHP-2 with the high affinity IgE receptor. J. Immunol. 159, 4426–4434.

Klebanoff, S.J., Vadas, M.A., Harlan, J.M., Sparks, L.H., Gamble, J.R., Agosti, J.M., & Waltersdorph, A.M. (1986). Stimulation of neutrophils by tumor necrosis factor. J. Immunol. 136, 4220–4225.

Koretzky, G.A., Picus, J., Schultz, T., & Weiss, A. (1991). Tyrosine phosphatase CD45 is required for T-cell antigen receptor and CD2-mediated activation of a protein tyrosine kinase and interleukin 2 production. Proc. Natl. Acad. Sci. USA 88, 2037–2041.

Koretzky, G.A., Picus, J., Thomas, M.L., & Weiss, A. (1990). Tyrosine phosphatase CD45 is essential for coupling T-cell antigen receptor to the phosphatidyl inositol pathway. Nature 346, 66–68.

Kuhlman, M., Joiner, K., & Ezekowitz, R.A.B. (1989). The human mannose-binding protein functions as an opsonin. J. Exp. Med. 169, 1733–1745.

Kuroda, S., Fukata, M., Kobayashi, K., Nakafuku, M., Nomura, N., Iwamatsu, A., & Kaibuchi, K. (1996). Identification of IQGAP as a putative target for the small GTPases, Cdc42 and Rac1. J. Biol. Chem. 271, 23363–23367.

Kurosaki, T., & Ravetch, J.V. (1989). A single amino acid in the glycosyl phosphatidyl inositol attachment domain determines the membrane topology of FcγRIII. Nature 342, 805–807.

Kurosaki, T., Johnson, S.A., Pao, L., Sada, K., Yamamura, H., & Cambier, J.C. (1995). Role of the Syk autophosphorylation site and SH2 domains in B cell antigen receptor signaling. J. Exp. Med. 182, 1815–1823.

Kurosaki, T., Takata, M., Yamanashi, Y., Inazu, T., Taniguchi, T., Yamamoto, T., & Yamamura, H. (1994). Syk activation by the Src-family tyrosine kinase in the B cell receptor signaling. J. Exp. Med. 179, 1725–1729.

Kusner, D.J., Hall, C.F., & Jackson, S. (1999). Fcγ receptor-mediated activation of phospholipase D. regulates macrophage phagocytosis of IgG-opsomized particles. J. Immunol. 162, 2266–2274.

Lanier, L.L., Cwirla, S., Yu, G., Testi, R., & Phillips, J.H. (1989). Membrane anchoring of a human IgG Fc receptor (CD16) determined by a single amino acid. Science 246, 1611–1613.

Lantz, C.S., & Huff, T.F. (1995). Murine KIT+ lineage-bone marrow progenitors express Fcγ-RII but do not express Fcε-RI until mast cell granule formation. J. Immunol. 154, 355–362.

Laudanna, C., Campbell, J.J., & Butcher, E.C. (1996). Role of Rho in chemoattractant-activated leukocyte adhesion through integrins. Science 271, 981–983.

Law, C.L., Chandran, K.A., Sidorenko, S.P., & Clark, E.A. (1996). Phospholipase C-γ1 interacts with conserved phosphotyrosyl residues in the linker region of Syk and is a substrate for Syk. Mol. Cell Biol. 16, 1305–1315.

Law, D.A., Chan, V.W.F., Datta, S.K., & DeFranco, A.L. (1993). B-cell antigen receptor motifs have redundant signalling capabilities and bind the tyrosine kinases PTK72, Lyn and Fyn. Curr. Biol. 3, 645–657.

Lecoanet-Henchoz, S., Gauchat, J.F., Aubry, J.P., Graber, P., Life, P., Paul-Eugene, N., Ferrua, B., Corbi, A.L., Dugas, B., & Plater-Zyberk, C. (1995). CD23 regulates monocyte activation through a novel interaction with the adhesion molecules CD11b-CD18 and CD11c-CD18. Immunity 3, 119–125.

Lefkowith, J.B., Rogers, M., Lennartz, M.R., & Brown, E.J. (1991). Essential fatty acid deficiency impairs macrophage spreading and adherence. Role of arachidonate in cell adhesion. J. Biol. Chem. 266, 1071–1076.

Lennartz, M.R., & Brown, E.J. (1991). Arachidonic acid is essential for IgG Fc receptor-mediated phagocytosis by human monocytes. J. Immunol. 147, 621–626.

Lennartz, M.R., Lefkowith, J.B., Bromley, F.A., & Brown, E.J. (1993). Immunoglobulin G-mediated phagocytosis activates a calcium-independent, phosphatidylethanolamine-specific phospholipase. J. Leuk. Biol. 54, 389–398.

Lennartz, M.R., Yuen, A.F.C., Masi, S.M., Russell, D.G., Buttle, K.F., & Smith, J.J. (1997). Phospholipase A_2 inhibition results in sequestration of plasma membrane into electron-lucent vesicles during IgG-mediated phagocytosis. J. Cell Sci. In Press.

Lew, D.P., Andersson, T., Hed, J., DiVirgilio, F., Pozzan, T., & Stendahl, O. (1985). Ca^{2+}-dependent and Ca^{2+}-independent phagocytosis in human neutrophils. Nature 315, 509–511.

Li, W., Hu, P., Skolnik, E.Y., Ullrich, A., & Schlessinger, J. (1992). The SH2 and SH3 domain-containing Nck protein is oncogenic and a common target for phosphorylation by different surface receptors. Mol. Cell Biol. 12, 5824–5833.

Liang, L., & Huang, C.-K. (1995). Activation of multiple proterin kinase induced by cross-linking of FcγRII in human neutrophils. J. Leuk. Biol. 57, 326–331.

Liao, F., Shin, H.S., & Rhee, S.G. (1993). Cross-linkng of FcγRIIIA on natural killer cells results in tyrosine phosphorylation of PLC-γ1 and PLC-γ2. J. Immunol., 2668–2674.

Lin, S.Q., Cicala, C., Scharenberg, A.M., & Kinet, J.P. (1996). The FcεRI β subunit functions as an amplifier of FcεRI γ-mediated cell activation signals. Cell 85, 985–995.

Lindberg, F.P., Bullard, D.C., Caver, T.E., Gresham, H.D., Beaudet, A.L., & Brown, E.J. (1996). Decreased resistance to bacterial infection and granulocyte defects in IAP-deficient mice. Science 274, 795–798.

Lindberg, F.P., Gresham, H.D., Schwarz, E., & Brown, E.J. (1993). Molecular cloning of integrin-associated protein: an immunoglobulin family member with multiple membrane-spanning domains implicated in $\alpha_v\beta_3$-dependent ligand binding. J. Cell Biol. 123, 485–496.

Lindberg, F.P., Lublin, D.M., Telen, M.J., Veile, R.A., Miller, Y.E., Donis, K.H., & Brown, E.J. (1994). Rh-related antigen CD47 is the signal-transducer integrin-associated protein. J. Biol. Chem. 269, 1567–1570.

Lioubin, M.N., Algate, P.A., Tsai, S., Carlberg, K., Aebersold, A., & Rohrschneider, L.R. (1996). $p150^{Ship}$, a signal transduction molecule with inositol polyphosphate-5-phosphatase activity. Genes Devel. 10, 1084–1095.

Looney, R.J., Ryan, D.H., Takahashi, K., Fleit, H.B., Cohen, H.J., Abraham, G.N., & Anderson, C.L. (1986). Identification of a second class of IgG Fc receptors on human neutrophils. J. Exp. Med. 163, 826–836.

Louahed, J., Kermouni, A., Van Snick, J., & Renauld, J.C. (1995). IL-9 induces expression of granzymes and high-affinity IgE receptor in murine T helper clones. J. Immunol. 154, 5061–5070.

Lowell, C.A., Soriano, P., & Varmus, H.E. (1994). Functional overlap in the src gene family: inactivation of *hck* and *fgr* impairs natural immunity. Genes Develop. 8, 387–398.

Lubeck, M.D., Steplewski, Z., Baglia, F., Klein, M.H., Dorrington, K.D., & Koprowski, H. (1985). The interaction of murine IgG subclass proteins with human monocyte Fc receptors. J. Immunol. 135, 1299–1304.

Ludger, J., & Galli, T. (1998). Exocytosis: SNAREs drum up! Eur. J. Neurosci. 10, 415–422.

Machesky, L.M., & Gould, K.L. (1999). The Arp2/3 complex: A multifunctional actin organizer. Curr. Opin. Cell Biol. 11, 117–121.

Mackay, D.J.G., Esch, F., Furthmayr, H., Hall, A. (1997) Rho- and Rac-dependent assembly of focal adhesion complexes and actin filaments in permeabilized fibroblasts: An essential role for ezrin/radixin/moesin proteins. J. Cell Biol. 138, 927–938.

Maliszewski, C.R., Grabstein, K., Fanslow, W.C., Armitage, R., Spriggs, M.K., & Sato, T.A. (1993). Recombinant CD40 ligand stimulation of murine B cell growth and differentiation: cooperative effects of cytokines. Eur. J. Immunol. 23, 1044–1049.

Maliszewski, C.R., March, C.J., Schoenborn, M.A., Gimpel, S., & Shen, L. (1990). Expression cloning of a human Fc receptor for IgA. J. Exp. Med. 172, 1665–1672.

Marcilla, A., Riverolezcano, O.M., Agarwal, A., & Robbins, K.C. (1995). Identification of the major tyrosine kinase substrate in signaling complexes formed after engagement of Fcγ receptors. J. Biol. Chem. 270, 9115–9120.

Matsuda, M., Park, J.G., Wang, D.C., Hunter, S., Chien, P., & Schreiber, A.D. (1996). Abrogation of the Fcγ receptor IIA-mediated phagocytic signal by stem-loop Syk antisense oligonucleotides. Mol. Biol. Cell 7, 1095–1106.

Matsuo, T., Hazeki, K., Hazeki, O., Katada, T., & Ui, M. (1996). Specific association of phosphatidylinositol 3-kinase with the protooncogene product Cbl in Fcγ receptor signaling. FEBS Lett. 382, 11–14.

Mayorga, L.S., Colombo, M.I., Lennartz, M., Brown, E.J., Rahman, K.H., Weiss, R., Lennon, P.J., & Stahl, P.D. (1993). Inhibition of endosome fusion by phospholipase A_2 (PLA_2) inhibitors points to a role for PLA_2 in endocytosis. Proc. Natl. Acad. Sci. USA 90, 10255–10259.

McKenzie, A.N., Culpepper, J.A., de Waal Malefyt, R., Briere, F., Punnonen, J., Aversa, G., Sato, A., Dang, W., Cocks, B.G., & Menon, S. (1993). Interleukin 13, a T-cell-derived cytokine that regulates human monocyte and B-cell function. Proc. Natl. Acad. Sci. USA 90, 3735–3739.

McNeil, P.L., Swanson, J.A., Wright, S.D., Silverstein, S.C., & Taylor, D.L. (1986). Fc-receptor-mediated phagocytosis occurs in macrophages without an increase in average $[Ca^{++}]_i$. J. Cell Biol. 102, 1586–1592.

Melendez, A., Floto, R.A., Gillooly, D.J., Harnett, M.M., & Allen, J.M. (1998a). FcγRI coupling to phospholipase D initiates sphingosine kinase-mediated calcium mobilization and vesicular trafficking. J. Biol. Chem. 273, 9393–9402.

Melendez, A.J., Gillooly, D.J., Harnett, M.M., & Allen, J.M. (1998b). Aggregation of the human high affinity immunoglobulin G receptor (FcγRI) activates both tyrosine kinase and G protein-coupled phosphinositide 3-kinase isoforms. Proc. Natl. Acad. Sci. USA 95, 2169–2174.

Miki, H., Sasaki, T., Takai, Y., & Takenawa, T. (1998). Induction of filopodium formation by a WASP-related actin-depolymerizing protein N-WASP. Nature 391, 93–96.

Miller, L., Blank, U., Metzger, H., & Kinet, J.-P. (1989). Expression of high affinity binding of human IgE by transfected cells. Science 244, 334–337.

Mitchell, M.A., Huang, M.M., Chien, P., Indik, Z.K., Pan, X.Q., & Schreiber, A.D. (1994). Substitutions and deletions in the cytoplasmic domain of the phagocytic receptor FcγRIIA: effect on receptor tyrosine phosphorylation and phagocytosis. Blood 84, 1753–1759.

Morinaga, N., Tsai, S.-C., Moss, J., & Vaughan, M. (1996). Isolation of a brefeldin A-inhibited guanine nucleotide-exchange protein for ADP ribosylation factor (ARF) 1 and ARF3 that contains a Sec7-like domain. Proc. Natl. Acad. Sci. USA 93, 12856–12860.

Morton, H.C., van Egmond, M., & van de Winkel, J.G. (1996). Structure and function of human IgA Fc receptors (FcαR). Crit. Rev. Immunol. 16, 423–440.

Moss, J., & Vaughan, M. (1995). Structure and function of ARf proteins: activators of cholera toxin and critical components of intracellular vesicular transport processes. J. Biol. Chem. 270, 12327–12330.

Mostov, K.E., Altschuler, Y., Chapin, S.J., Enrich, C., Low, S.H., Luton, F., Richman-Eisenstat, J., Singer, K.L., Tang, K., & Weimbs, T. (1995). Regulation of protein traffic in polarized epithelial cells: the polymeric immunoglobulin receptor model. Cold Spring Harb. Symp. Quant. Biol. 60, 775–781.

Moxey-Mims, M.M., Simms, H.H., Frnak, M.M., Lin, E.Y., & Gaither, T.A. (1991). The effects of IL-1, IL-2, and tumor necrosis factor on polymorphonuclear leukocyte Fcγ receptor-mediated phagocytosis. J. Immunol. 147, 1823–1830.

Muta, T., Kurosaki, T., Misulovin, Z., Sanchez, M., Nussenzweig, M.C., & Ravetch, J.V. (1994). A 13-amino-acid motif in the cytoplasmic domain of $Fc_\gamma RIIB$ modulates B-cell receptor signalling. Nature 368, 70–73.

Neel, B.G. (1997). Role of phosphatases in lymphocyte activation. Curr. Opin. Immunol. 9, 405–420.

Nelson, D.J., Jacobs, E.R., Tang, J.M., Zeller, J.M., & Bone, R.C. (1985). Immunoglobulin-G-induced single ionic channels in human alveolar macrophage membranes. J. Clin. Invest. 76, 500–507.

Nepomuceno, R.R., Henschen-Edman, A.H., Burgess, W.H., & Tenner, A.J. (1997). cDNA cloning and primary structure analysis of $C1qR_P$, the human C1q/MBL/SPA receptor that mediates enhanced phagocytosis in vitro. Immunity 6, 119–129.

Newman, S.L., Mikus, L.K., & Tucci, M.A. (1991). Differential requirements for cellular cytoskeleton in human macrophage complement receptor- and Fc receptor-mediated phagocytosis. J. Immunol. 146, 967–974.

Ninomiya, N., Hazeki, K., Fukui, Y., Seya, T., Okada, T., Hazeki, O., & Ui, M. (1994). Involvement of phosphatidylinositol 3-kinase in Fcγ receptor signaling. J. Biol. Chem. 269, 22732–22737.

Nitta, T., & Suzuki, T. (1982). $Fc_\gamma 2b$ receptor-mediated prostaglandin synthesis by a murine macrophage cell line $P388D_1$. J. Immunol. 128, 2527–2532.

Odin, J.A., Edberg, J.C., Painter, C.J., Kimberly, R.P., & Unkeless, J.C. (1991). Regulation of phagocytosis and $[Ca^{2+}]_i$ flux by distinct regions of an Fc receptor. Science 254, 1785–1788.

Ofek, I., Goldhar, J., Keisari, Y., & Sharon, N. (1995). Nonopsonic phagocytosis of microorganisms. Annu. Rev. Microbiol. 49, 239–276.

Olcese, L., Lang, P., Vely, F., Cambiaggi, A., Marguet, D., Blery, M., Hippen, K.L., Biassoni, R., Moretta, A., Moretta, L., Cambier, J.C., & Vivier, E. (1996). Human and mouse killer-cell inhibitory receptors recruit PTP1C and PTP1D protein tyrosine phosphatases. J. Immunol. 156, 4531–4534.

Oliver, J.M., Burg, D.L., Wilson, B.S., Mclaughlin, J.L., & Geahlen, R.L. (1994). Inhibition of mast cell $Fc_\varepsilon R1$-mediated signaling and effector function by the Syk-selective inhibitor, piceatannol. J. Biol. Chem. 269, 29697–29703.

Olson, S.C., & Lambeth, J.D. (1996). Biochemistry and cell biology of phospholipase D in human neutrophils. Chem. Phys. Lipids 80, 3–19.

Ono, M., Bolland, S., Tempst, P., & Ravetch, J.V. (1996). Role of the inositol phosphatase SHIP in negative regulation of the immune system by the receptor FcγRIIB. Nature 383, 263–266.

Ono, M., Okada, H., Bolland, S., Yanagi, S., Kurosaki, T., & Ravetch, J.V. (1997). Deletion of SHIP or SHP-1 reveals two distinct pathways for inhibitory signaling. Cell 90, 293–301.

Osborne, M.A., Zenner, G., Lubinus, M., Zhang, X.L., Songyang, Z., Cantley, L.C., Majerus, P., Burn, P., & Kochan, J.P. (1996). The inositol 5′-phosphatase SHIP binds to immunoreceptor signaling motifs and responds to high affinity IgE receptor aggregation. J. Biol. Chem. 271, 29271–29278.

Ota, Y., & Samelson, L.E. (1997). The product of the proto-oncogene c-cbl: A negative regulator of the Syk tyrosine kinase. Science 276, 418–420.

Palmer, D.J., Helms, J.B., Beckers, C.J.M., Orci, L., & Rothman, J.E. (1993). Binding of coatomer to Golgi emebranes requires ADP-ribosylation factor. J. Biol. Chem. 268, 12083–12089.

Palmer, L.J., Pare, P.D., Faux, J.A., Moffatt, M.F., Daniels, S.E., LeSouef, P.N., Bremner, P.R., Mockford, E., Gracey, M., Spargo, R., Musk, A.W., & Cookson, W.O. (1997). Fc epsilon R1-β polymorphism and total serum IgE levels in endemically parasitized Australian aborigines. Am. J. Hum. Genet. 61, 182–188.

Paolini, R., Jouvin, M.H., & Kinet, J.P. (1991). Phosphorylation and dephosphorylation of the high-affinity receptor for immunoglobulin E immediately after receptor engagement and disengagement. Nature 353, 855–858.

Paolini, R., Renard, V., Vivier, E., Ochiai, K., Jouvin, M.H., Malissen, B., & Kinet, J.P. (1995). Different roles for the FcεRI γ chain as a function of the receptor context. J. Exp. Med. 181, 247–255.

Park, D., & Rhee, S.G. (1992). Phosphorylation of Nck in response to a variety of receptors, phorbol myristate acetate, and cyclic AMP. Mol. Cell. Biol. 12, 5816–5823.

Park, J.G., & Schreiber, A.D. (1995). Determinants of the phagocytic signal mediated by the type IIIA Fc gamma receptor, FcγRIIIA: sequence requirements and interaction with protein-tyrosine kinases. Proc. Natl. Acad. Sci. USA 92, 7381–7385.

Park, J.G., Murray, R.K., Chien, P., Darby, C., & Schreiber, A.D. (1993). Conserved cytoplasmic tyrosine residues of the γ subunit are required for a phagocytic signal mediated by Fc_{γ}RIIIA. J. Clin. Invest. 92, 2073–2079.

Park, R.K., Liu, Y.B., & Durden, D.L. (1996). A role for Shc, Grb2, and Raf-1 in FcγRI signal relay. J. Biol. Chem. 271, 13342–13348.

Parren, P.W., Warmerdam, P.A., Boeije, L.C., Arts, J., Westerdaal, N.A., Vlug, A., Capel, P.J., Aarden, L.A., & van de Winkel, J.G. (1992). On the interaction of IgG subclasses with the low affinity FcγRIIa (CD32) on human monocytes, neutrophils, and platelets. Analysis of a functional polymorphism to human IgG2. J. Clin. Invest. 90, 1537–1546.

Passlick, B., Flieger, D., & Ziegler-Heitbrock, H.W.L. (1989). Identification and characterization of a novel monocyte subpopulation in human peripheral blood. Blood 74, 2527–2534.

Patry, C., Sibille, Y., Lehuen, A., & Monteiro, R.C. (1996). Identification of Fcα receptor (CD89) isoforms generated by altyernative splicing that are differentially expressed between blood monocytes and alveolar macrophages. J. Immunol. 156, 4442–4448.

Perussia, B., Kobayashi, M., Rossi, M.E., Anegon, I., & Trinchieri, G. (1987). Immune interferon enhances functional properties of human granulocytes: role of Fc receptors and effect of lymphotoxin, tumor necrosis factor, and granulocyte-macrophage colony-stimulating factor. J. Immunol. 138, 765–774.

Perussia, B., Trinchieri, G., Jackson, A., Warner, N.L., Faust, J., Rumpold, H., Kraft, D., & Lanier, L.L. (1984). The Fc receptor for IgG on human natural killer cells: phenotypic, functional, and comparative studies with monoclonal antibodies. J. Immunol. 133, 180–189.

Peters, P.J., Hsu, V.W., Ooi, C.E., Finazzi, D., Teal, S.B., Oorschot, V., Donaldson, J.G., & Klausner, R.D. (1995). Overexpression of wild-type and mutant ARF1 and ARF6: distinct perturbations of nonoverlapping membrane compartments. J. Cell Biol. 128, 1003–1017.

Peyroche, P., Paris, S., & Jackson, C.L. (1996). Nucleotide exchange on ARF mediated by yeast Gea1 protein. Nature 384, 479–481.

Pfefferkorn, L.C. (1984). Transmembrane signaling: Anion flux-independent model for signal transduction by complexed Fc receptors. J. Cell Biol. 99, 2231–2240.

Pfefferkorn, L.C., & Yeaman, G.R. (1994). Association of IgA-Fc receptors (FcαR) with FcεRI γ2 subunits in U937 cells—Aggregation induces the tyrosine phosphorylation of γ2. J. Immunol. 153, 3228–3236.

Pfeiffer, J.R., Seagrave, J.C., Davis, B.H., Deanin, G.G., & Oliver, J.M. (1985). Membrane and cytoskeletal changes associated with IgE-mediated serotonin release from rat basophilic leukemia cells. J. Cell Biol. 101, 2145–2155.

Phillips, N.E., & Parker, D.C. (1983). Cross-linking of B lymphocyte Fc gamma receptors and membrane immunoblobulin inhibits anti-immunoglobulin-induced blastogenesis. J. Immunol. 132, 627–632.

Pierini, L., Holowka, D., & Baird, B. (1996). FcεRI-mediated association of 6-μm beads with RBL-2H3 mast cells results in exclusion of signaling proteins from the forming phagosome and abrogation of normal downstream signaling. J. Cell Biol. 134, 1427–1439.

Pingel, J.T., & Thomas, M.L. (1989). Evidence that the leukocyte common antigen is required for antigen-induced T lymphocyte proliferation. Cell 58, 1055–1061.

Plas, D.R., Johnson, R., Pingel, J.T., Matthews, R.J., Dalton, M., Roy, G., Chan, A.C., & Thomas, M.L. (1996). Direct regulation of ZAP-70 by SHP-1 in T cell antigen receptor signaling. Science 272, 1173–1176.

Pommier, C.G., Inada, S., Fries, L.F., Takahashi, T., Frank, M.M., & Brown, E.J. (1983). Plasma fibronectin enhances phagocytosis of opsonized particles by human peripheral blood monocytes. J. Exp. Med. 157, 1844–1854.

Poo, H., Krauss, J.C., Mayo, B.L., Todd, R.r., & Petty, H.R. (1995). Interaction of Fc_γ receptor type IIIB with complement receptor type 3 in fibroblast transfectants: evidence from lateral diffusion and resonance energy transfer studies. J. Molec. Biol. 247, 597–603.

Pribluda, V.S., Pribluda, C., & Metzger, H. (1997). Biochemical evidence that the phosphorylated tyrosines, serines, and threonines on the aggregated high affinity receptor for IgE are in the immunoreceptor tyrosine-based activation motifs. J. Biol. Chem. 272, 11185–11192.

Qian, D.P., & Weiss, A. (1997). T cell antigen receptor signal transduction. Curr. Opin. Cell Biol. 9, 205–212.

Qin, S., Inazu, T., & Yamamura, H. (1995). Activation and tyrosine phosphorylation of $p72^{syk}$ as well as calcium mobilization after hydrogen peroxide stimulation in peripheral blood lymphocytes. Biochem. J. 308, 347–352.

Ra, C., Jouvin, M.E., Blank, U., & Kinet, J. (1989). A macrophage Fcγ receptor and the mast cell receptor for IgE share an identical subunit. Nature 341, 752–754.

Rabinowich, H., Manciulea, M., Metes, D., Sulica, A., Herberman, R.B., Corey, S.J., & Whiteside, T.L. (1996). Physical and functional association of Fcμ receptor on human natural killer cells with the ζ- and FcεRI γ-chains and with src family protein tyrosine kinases. J. Immunol. 157, 1485–1491.

Radhakrishna, H. & Donaldson, J.G. (1997). ADP-ribosylation factor 6 regulates a novel plasma membrane recycling pathway. J. Cell Biol. 139, 49–61.

Radhakrishna, H., Klausner, R.D., & Donaldson, J.G. (1996). Aluminum fluoride stimulates surface protrusions in cells overexpressing the ARF6 GTPase. J. Cell Biol. 134, 935–947.

Randazzo, P.A., Yang, Y.C., Rulka, C., & Kahn, R.A. (1993). Activation of ADP-ribosylation factor by Golgi membranes: evidence for a brefeldin A and protease sensitive activating factor on Golgi membranes. J. Biol. Chem. 268, 9555–9563.

Ravetch, J.V. (1994). Fc receptors: rubor redux. Cell 78, 553–560.

Relman, D., E., T., Falkow, S., Golenbock, D.T., Saukkonen, K., & Wright, S.D. (1990). Recognition of a bacterial adhesion by an integrin: macrophage CR3($\alpha_M\beta_2$, CD11b/CD18) binds filamentous hemagglutinin of Bordetella pertussis. Cell 61, 1375–1382.

Reterink, T.J., Levarht, E.W., Klar, M.N., Van Es, L.A., & Daha, M.R. (1996). Transforming growth factor-β1 (TGF-β1) down-regulates IgA Fc-receptor (CD89) expression on human monocytes. Clin. Exp. Immunol. 103, 161–166.

Reth, M. (1989). Antigen receptor tail clue. Nature 338, 384–384.

Rhee, S.G., & Bae, Y.S. (1997). Regulation of phosphoinositide-specific phospholipase C isozymes. J. Biol. Chem. 272, 15045–15048.

Richards, A.L., Okuno, T., Takagaki, Y., & Djeu, J.Y. (1988). Natural cytotoxic cell-specific cytotoxic factor produced by IL-3-dependent basophilic/mast cells. Relationship to TNF. J. Immunol. 141, 3061–3066.

Rivera, J. (1997). Phosphorylation of serines and threonines in FcεRI-mediated responses in mast cells. In: IgE Recetpor (FcεRI) Function in Mast Cells and Basophils (Hamawy, M.M., Ed.), pp. 107–126. R.G. Landes Company.

Robinson, M.S., & Kreis, T.E. (1992). Recruitment of coat proteins onto Golgi membranes in intact and permealbilzed cells: effects of brefeldin A and G protein activators. Cell 69, 129–138.

Rosales, C., & Brown, E.J. (1991). Two mechanisms for IgG Fc-receptor-mediated phagocytosis by human neutrophils. J. Immunol. 146, 3937–3944.

Rose, D.M., Winston, B.W., Chan, E.D., Riches, D.W.H., Gerwins, P., Johnson, G.L., & Henson, P.M. (1997). Fcγ receptor cross-linking activates p42, p38, and JNK/SAPK mitogen-activated protein kinases in murine macrophages—Role for $p42^{MAPK}$ in Fcγ receptor-stimulated TNF-α synthesis. J. Immunol. 158, 3433–3438.

Ross, G.D., Cain, J.A., & Lachmann, P.J. (1985). Membrane complement receptor type three (CR_3) has lectin-like properties analogous to bovine conglutinin and functions as a receptor for zymosan and rabbit erythrocytes as well as a receptor for iC3b. J. Immunol. 134, 3307–3315.

Rossman, M.D., Chen, E., Chien, P., Rotten, M., Cprek, A., & Schreiber, A.D. (1989). Fcγ receptor recognition of IgG ligand by human monocytes and macrophages. Am. J. Respir. Cell Mol. Biol. 1, 211–220.

Rottem, M., Hull, G., & Metcalfe, D.D. (1994). Demonstration of differential effects of cytokines on mast cells derived from murine bone marrow and peripheral blood mononuclear cells. Exp. Hematol. 22, 1147–1155.

Rowley, R.B., Burkhardt, A.L., Chao, H.G., Matsueda, G.R., & Bolen, J.B. (1995). Syk protein-tyrosine kinase is regulated by tyrosine-phosphorylated Igα/Igβ immunoreceptor tyrosine activation motif binding and autophosphorylation. J. Biol. Chem. 270, 11590–11594.

Salcedo, T.W., Kurosaki, T., Kanakaraj, P., Ravetch, J.V., & Perussia, B. (1993). Physical and functional association of $p56^{lck}$ with Fc_γ RIIIA (CD16) in natural killer cells. J. Exp. Med. 177, 1475–1480.

Salim, K., Bottomley, M.J., Querfurth, E., Zvelebil, M.J., Gout, I., Scaife, R., Margolis, R.L., Gigg, R., Smith, C.I., Driscoll, P.C., Waterfield, M.D., & Panayotou, G. (1996). Distinct specificity in the recognition of phosphoinositides by the pleckstrin homology domains of dynamin and Bruton's tyrosine kinase. EMBO J. 15, 6241–6250.

Salmon, J.E., Brogle, N.L., Edberg, J.C., & Kimberly, R.P. (1991). Fc_γ receptor III induces actin polymerization in human neutrophils and primes phagocytosis mediated by Fc_γ receptor II. J. Immunol. 146, 997–1004.

Salmon, J.E., Edberg, J.C., & Kimberly, R.P. (1990). Fcγ receptor III on human neutrophils. Allelic variants have functionally distinct capacities. J. Clin. Invest. 85, 1287–1295.

Salmon, J.E., Edberg, J.C., Brogle, N.L., & Kimberly, R.P. (1992). Allelic polymorphisms of human Fcγ receptor IIA and Fcγ receptor IIIB. Independent mechanisms for differences in human phagocyte function. J. Clin. Invest. 89, 1274–1281.

Salmon, J.E., Kapur, S., & Kimberly, R.P. (1987). Opsonin-independent ligation of Fcγ receptors. The 3G8-bearing receptors on neutrophils mediate the phagocytosis of concanavalin A-treated erythrocytes and nonopsonized *Escherichia coli*. J. Exp. Med. 166, 1798–1813.

Salmon, J.E., Millard, S., Schachter, L.A., Arnett, F.C., Ginzler, E.M., Gourley, M.F., Ramsey Goldman, R., Peterson, M.G., & Kimberly, R.P. (1996). FcγRIIA alleles are heritable risk factors for lupus nephritis in African Americans. J. Clin. Invest. 97, 1348–1354.

Salmon, J.E., Millard, S.S., Brogle, N.L., & Kimberly, R.P. (1995). Fcγ receptor IIIb enhances Fcγ receptor IIa function in an oxidant-dependent and allele-sensitive manner. J. Clin. Invest. 95, 2877–2885.

Sampson, L.L., Heuser, J., & Brown, E.J. (1991). Cytokine regulation of complement receptor-mediated ingestion by mouse peritoneal macrophages. M-CSF and IL-4 activate phagocytosis by a common mechanism requiring autostimulation by IFN-β. J. Immunol. 146, 1005–1013.

Sarmay, G., Pecht, I., & Gergely, J. (1994). Protein-tyrosine kinase activity tightly associated with human type II Fcγ receptors. Proc. Natl. Acad. Sci. USA 91, 4140–4144.

Sawyer, D.W., Sullivan, J.A., & Mandell, G.L. (1985). Intracellular free calcium localization in neutrophils during phagocytosis. Science 230, 663–666.

Scharenberg, A.M., El-Hillal, O., Fruman, D.A., Beitz, L.O., Li, Z., Lin, S., Gout, I., Cantley, L.C., Rawlings, D.J., & Kinet, J.-P. (1998). Phosphatidylinositol-3,4,5-trisphosphate (PtdIns-3,4,5-P_3)/Tec kinase-dependent calcium signaling pathway: a target for SHIP-mediated inhibitory signals. EMBO J. 17, 1961–1972.

Schieven, G.L., Kirihara, J.M., Burg, D.L., Geahlen, R.L., & Ledbetter, J.A. (1993). $p72^{syk}$ tyrosine kinase is activated by oxidizing conditions that induce lymphocyte tyrosine phosphorylation and Ca^{2+} signals. J. Biol. Chem. 268, 16688–16692.

Scholl, P.R., Ahern, D., & Geha, R.S. (1992). Protein tyrosine phosphorylation induced via the IgG receptors FcγRI and FcγRII in the human monocytic cell line THP-1. J. Immunol. 149, 1751–1757.

Selvaraj, P., Rosse, W.F., Silber, R., & Springer, T.A. (1988). The major Fc receptor in blood has a phosphatidylinositol anchor and is deficient in paroxysmal nocturnal haemoglobinuria. Nature 333, 565–567.

Shen, L., Lasser, R., & Fanger, M.W. (1989). My43, a monoclonal antibody that reacts with human myeloid cells inhibits monocyte IgA binding and triggers function. J. Immunol. 143, 4117–4122.

Shen, Z., Lin, C.T., & Unkeless, J.C. (1994). Correlations among tyrosine phosphorylation of Shc, $p72^{syk}$, PLCγ1, and $[Ca^{2+}]_i$ flux in Fcγ RIIA signaling. J. Immunol. 152, 3017–3023.

Shepherd, P.R., Reaves, B.J., & Davidson, H.W. (1996). Phosphoinositide 3-kinases and membrane traffic. Trends Cell Biol. 6, 92–97.

Shiue, L., Zoller, M.J., & Brugge, J.S. (1995). Syk is activated by phosphotyrosine-containing peptides representing the tyrosine-based activation motifs of the high affinity receptor for IgE. J. Biol. Chem. 270, 10498–10502.

Sidorenko, S.P., Law, C.L., Chandran, K.A., & Clark, E.A. (1995). Human spleen tyrosine kinase p72Syk associates with the Src-family kinase p53/56Lyn and a 120-kDa phosphoprotein. Proc. Natl. Acad. Sci. USA 92, 359–363.

Simms, H.H., Frank, M.M., Quinn, T.C., Holland, S., & Gaither, T.A. (1989). Studies on phagocytosis in patients with acute bacterial infections. J. Clin. Invest. 83, 252–260.

Simms, H.H., Gaither, T.A., Fries, L.F., & Frank, M.M. (1991). Monokines released during short-term Fcγ receptor phagocytosis up-regulate polymorphonuclear leukocytes and monocyte-phagocytic function. J. Immunol. 147, 265–272.

Simons, K., & Ikonen, E. (1997). Functional rafts in cell membranes. Nature 387, 569–572.

Smolen, J.E., Korchak, H.M., & Weissmann, G. (1980). Increased levels of cyclic adenosine-3′, 5′-monophosphate in human polymorphonuclear leukocytes after surface stimulation. J. Clin. Invest. 65, 1077–1085.

Speert, D.P., Wright, S.D., Silverstein, S.C., & Mah, B. (1988). Functional characterization of macrophage receptors for in vitro phagocytosis of unopsonized *Pseudomonas aeruginosa.* J. Clin. Invest. 82, 872–879.

Stauffer, T.P., Ahn, S., & Meyer, T. (1998). Receptor-induced transient reduction in plasma membrane PtdIns(4,5)P_2 concentration monitored in living cells. Curr. Biol. 8, 343–346.

Stendahl, O., Krause, K.-H., Krischer, J., Jerstrom, P., Theler, J.-M., Clark, R.A., Carpentier, J.-L., & Lew, D.P. (1994). Redistribution of intracellular Ca^{2+} stores during phagocytosis in human neutrophils. Science 265, 1439–1441.

Stewart, W.W., Mazengera, R.L., Shen, L., & Kerr, M.A. (1994). Unaggregated serum IgA binds to neutrophil FcαR at physiological concentrations and is endocytosed but cross-linking is necessary to elicit a respiratory burst. J. Leuk. Biol. 56, 481–487.

Suchard, S.J., Hinkovska-Galcheva, V., Mansfield, P.J., Boxer, L.A., & Shayman, J.A. (1997). Ceramide inhibits IgG-dependent phagocytosis in human polymorphonuclear leukocytes. Blood 89, 2139–2147.

Sugawara, H., Kurosaki, M., Takata, M., & Kurosaki, T. (1997). Genetic evidence for involvement of type 1, type 2 and type 3 inositol 1,4,5-trisphosphate receptors in signal transduction through the B-cell antigen receptor. EMBO J. 16, 3078–3088.

Suzuki, T., Saito-Taki, T., Sadasivan, R., & Nitta, T. (1982). Biochemical signal transmitted by Fcγ receptors: phospholipase A_2 activity of Fcγ2b receptor of murine macrophage cell line $P388D_1$. Proc. Natl. Acad. Sci. USA 79, 591–595.

Takai, T., Li, M., Sylvestre, D., Clynes, R., & Ravetch, J.V. (1994). FcR γ chain deletion results in pleiotropic effector cell defects. Cell 76, 519–529.

Takata, M., & Kurosaki, T. (1996). A role for Bruton's tyrosine kinase in B cell antigen receptor-mediated activation of phospholipase C-γ2. J. Exp. Med. 184, 31–40.

Takata, M., Sabe, H., Hata, A., Inazu, T., Homma, Y., Nukada, T., Yamamura, H., & Kurosaki, T. (1994). Tyrosine kinases Lyn and Syk regulate B cell receptor-coupled Ca^{2+} mobilization through distinct pathways. EMBO J. 13, 1341–1349.

Tanaka, S., Neff, L., Baron, R., & Levy, J.B. (1995). Tyrosine phosphorylation and translocation of the c-Cbl protein after activation of tyrosine kinase signaling pathways. J. Biol. Chem. 270, 14347–14351.

te Velde, A.A., de Waal Malefijt, R., Huijbens, R.J., de Vries, J.E., & Figdor, C.G. (1992). IL-10 stimulates monocyte FcγR surface expression and cytotoxic activity. Distinct regulation of antibody-dependent cellular cytotoxicity by IFN-γ, IL-4, and IL-10. J. Immunol. 149, 4048–4052.

te Veld, A.A., Huijbens, R.J., de Vries, J.E., & Figdor, C.G. (1990). IL-4 decreases FcγR membrane expression and FcγR-mediated cytotoxic activity of human monocytes. J. Immunol. 144, 3046–3051.

Tenner, A.J., Robinson, S.L., & Ezekowitz, R.A. (1995). Mannose binding protein (MBP) enhances mononuclear phagocyte function via a receptor that contains the 126,000 M_r component of the C1q receptor. Immunity 3, 485–493.

Tenner, A.J., Robinson, S.L., Borchelt, J., & Wright, J.R. (1989). Human pulmonary surfactant protein (SP-A), a protein structurally homologous to C1q, can enhance FcR- and CR1-mediated phagocytosis. J. Biol. Chem. 264, 13923–13928.

Tilton, B., Andjelkovic, Didichenko, S.A., Hemmings, B.A., & Thelen, M. (1997). G-protein-coupled receptors and Fcγ-receptors mediate activation of Akt/Protein kinase B in human phagocytes. J. Biol. Chem. 272, 28096–28101.

Ting, A.T., Dick, C.J., Schoon, R.A., Karnitz, L.R., Abraham, R.T., & Leibson, P.J. (1995). Interaction between lck and syk family tyrosine kinases in Fcγ receptor-initiated activation of natural killer cells. J. Biol. Chem. 270, 16415–16421.

Ting, A.T., Einspahr, K.J., Abraham, R.T., & Leibson, P.J. (1991). Fcγ receptor signal transduction in natural killer cells-coupling to phospholipase C via a G protein-independent, but tyrosine kinase-dependent pathway. J. Immunol. 147, 3122–3127.

Ting, A.T., Karnitz, L.M., Schoon, R.A., Abraham, R.T., & Leibson, P.J. (1992). Fcγ receptor activation induces the tyrosine phosphyrlation of both phospholipase C (PLC)-γ1 and PLC-γ2 in natural killer cells. J. Exp. Med. 176, 1751–1755.

Traub, L.M., Ostrom, J.A., & Kornfeld, S. (1993). Biochemical disection of AP-1 recruitment onto Golgi membranes. J. Cell Biol. 123, 561–573.

Trotta, R., Kanakaraj, P., & Perussia, B. (1996). FcγR-dependent mitogen-activated protein kinase activation in leukocytes: a common signal transduction event necessary for expression of TNF-α and early activation genes. J. Exp. Med. 184, 1027–1035.

Tuijnman, W.B., Capel, J.A., & van de Winkel, J.G.J. (1992). Human low-affinity IgG receptor FcγRIIa (CD32) introduced into mouse fibroblasts mediates phagocytosis of sensitized erythrocytes. Blood 79, 1651–1656.

Unkeless, J.C., & Eisen, H.N. (1975). Binding of monomeric immunoglobulin G to Fc receptors of mouse macrophages. J. Exp. Med. 142, 1520–1533.

Unkeless, J.C., Shen, Z., Lin, C.-W., & DeBeus, E. (1995). Function of human FcγRIIA and FcγRIIIB. Seminars Immunol. 7, 37–44.

Valle, A., & Kinet, J.P. (1995). *N*-acetyl-L-cysteine inhibits antigen-mediated Syk, but not Lyn tyrosine kinase activation in mast cells. FEBS Lett. 357, 41–44.

Van Aelst, L., & D'Souza-Schorey, C. (1997). Rho GTPases and signaling networks. Genes Develop. 11, 2295–2322.

Van Aelst, L., Joneson, T., & Bar-Sagi, D. (1996). Identification of a novel Rac1-interacting protein involved in membrane ruffling. EMBO J. 15, 3778–3786.

van de Winkel, J.G.J., Ernst, L.K., Anderson, C.L., & Chiu, I.-M. (1991). Gene organization of the human high affinity receptor for IgG, FcγRI (CD64): characterization and evidence for a second gene. J. Biol. Chem. 266, 13449–13455.

Vely, F., Olivero, S., Olcese, L., Moretta, A., Damen, J.E., Liu, L., Krystal, G., Cambier, J.C., Daeron, M., & Vivier, E. (1997). Differential association of phosphatases with hematopoietic co-receptors bearing immunoreceptor tyrosine-based inhibition motifs. Eur. J. Immunol. 27, 1994–2000.

Vivier, E., & Daeron, M. (1997). Immunoreceptor tyrosine-based inhibition motifs. Immunol. Today 286.

Vossebeld, P.J., Homburg, C.H., Schweizer, R.C., Ibarrola, I., Kessler, J., Koenderman, L., Roos, D., & Verhoeven, A.J. (1997). Tyrosine phosphorylation-dependent activation of phosphatidylinositide 3-kinase occurs upstream of Ca2+-signalling induced by Fcγ receptor cross-linking in human neutrophils. Biochem. J. 323, 87–94.

Vouldoukis, I., Riverosmoreno, V., Dugas, B., Ouaaz, F., Becherel, P., Debre, P., Moncada, S., & Mossalayi, M.D. (1995). The killing of Leishmania major by human macrophages is mediated by nitric oxide induced after ligation of the FcεRII/CD23 surface antigen. Proc. Natl. Acad. Sci. USA 92, 7804–7808.

Walker, B.A.M., Hagenlocker, B.E., Stubbs, E.B.J., Sandborg, R.R., Agranoff, B.W., & Ward, P.A. (1991). Signal transduction events and FcγR engagement in human neutrophils stimulated with immune complexes. J. Immunol. 146, 735–741.

Wang, A.V.T., Scholl, P.R., & Geha, R.S. (1994a). Physical and functional association of the high afinity immunoglobulin G receptor (FcγRI) with the kinases hck and lyn. J. Exp. Med. 180, 1165–1170.

Wang, X.Y., Sada, K., Yanagi, S., Yang, C., Rezaul, K., & Yamamura, H. (1994b). Intracellular calcium dependent activation of $p72^{syk}$ in platelets. J. Biochem. 116, 858–861.

Warmerdam, P.A., van de Winkel, J.G., Vlug, A., Westerdaal, N.A., & Capel, P.J. (1991). A single amino acid in the second Ig-like domain of the human Fcγ receptor II is critical for human IgG2 binding. J. Immunol. 147, 1338–1343.

Warmerdam, P.A.M., van de Winkel, J.G.J., Gosselin, E.J., & Capel, P.J.A. (1990). Molecular basis for a polymorphism of human Fcγ receptor II (CD32). J. Exp. Med. 172, 10–25.

Watarai, M., Kamata, Y., Kozaki, S., & Sasakawa, C. (1997). rho, a small GTP-binding protein, is essential for *Shigella* invasion of epithelial cells. J. Exp. Med. 185, 281–292.

Weisbart, R.H., Kacena, A., Schuh, A., & Golde, D.W. (1988). GM-CSF induces human neutrophil IgA-mediated phagocytosis by and IgA Fc receptor activation mechanism. Nature 332, 647–648.

Welch, G.R., Wong, H.L., & Wahl, S.M. (1990). Selective induction of FcγRIII on human monocytes by transforming growth factor-β. J. Immunol. 144, 3444–3448.

Wright, S.D., Craigmyle, L.S., & Silverstein, S.C. (1983). Fibronectin and serum amyloid P component stimulate C3b- and C3bi-mediated phagocytosis in cultured human monocytes. J. Exp. Med. 158, 1338–1343.

Wright, S.D., Levin, S.M., Jong, M.T.C., Chad, Z., & Kabbash, L.G. (1989). CR3 (CD11b/CD18) expresses one binding site for Arg-Gly-Asp-containing peptides and a second site for bacterial lipopolysaccharide. J. Exp. Med. 169, 175–183.

Wright, S.D., Weitz, J.I., Huang, A.J., Levin, S.M., Silverstein, S.C., & Loike, J.D. (1988). Complement receptor type three (CD11b/CD18) of human polymorphonuclear leukocytes recognizes fibrinogen. Proc. Natl. Acad. Sci. USA 85, 7734–7738.

Wu, J.M., Edberg, J.C., Redecha, P.B., Bansal, V., Guyre, P.M., Coleman, K., Salmon, J.E., & Kimberly, R.P. (1997). A novel polymorphism of FcγRIIIa (CD16) alters receptor function and predisposes to autoimmune disease. J. Clin. Invest. 100, 1059–1070.

Xia, H.Z., Du, Z., Craig, S., Klisch, G., Noben-Trauth, N., Kochan, J.P., Huff, T.H., Irani, A.M., & Schwartz, L.B. (1997). Effect of recombinant human IL-4 on tryptase, chymase, and Fcε receptor type I expression in recombinant human stem cell factor-dependent fetal liver-derived human mast cells. J. Immunol. 159, 2911–2921.

Xu, X.L., & Chong, S.F. (1996). Vav in natural killer cells is tyrosine phosphorylated upon cross-linking of FcγRIIIA and is constitutively associated with a serine/threonine kinase. Biochem. J. 318, 527–532.

Yanagi, S., Kurosaki, T., & Yamamura, H. (1995). The structure and function of nonreceptor tyrosine kinase $p72^{SYK}$ expressed in hematopoietic cells. Cellular Signal. 7, 185–193.

Yancey, K.B., O'Shea, J., Chused, T., Brown, E., Takahashi, T., Frank, M.M., & Lawley, T.J. (1985). Human C5a modulates monocyte Fc and C3 receptor expression. J. Immunol. 135, 465–470.

Yarar, D., To, W., Abo, A., & Welch, M.D. (1999). The Wiskott-Aldrich syndrome protein directs actin-based motility by stimulating actin nucleation with the Arp2/3 complex. Curr. Biol. 9, 555–558.

Yokota, A., Yukawa, K., Yamamoto, A., Sugiyama, K., Suemura, M., Tashiro, Y., Kishimoto, T., & Kikutani, H. (1992). Two forms of the low-affinity Fc receptor for IgE differentially mediate endocytosis and phagocytosis: identification of the critical cytoplasmic domains. Proc. Natl. Acad. Sci. USA 89, 5030–5034.

Young, J.D., Ko, S.S., & Cohn, Z.A. (1984). The increase in intracellular free calcium associated with IgG γ2b/γ1 Fc receptor-ligand interactions: role in phagocytosis. Proc. Natl. Acad. Sci. USA 81, 5430–5434.

Yu, P., Kosco-Vilbois, M., Richards, M., Kohler, G., & Lamers, M.C. (1994). Negative feedback regulation of IgE synthesis by murine CD23. Nature 369, 753–756.

Zeng, H., Yoshida, T., Kurosaki, T., Yamamura, H., Oshima, A., Kitamura, D., Watanabe, T., & Morikawa, M. (1995). Phosphorylation of HS1, GAP-associated p190 and a novel GAP-associated p60 protein by cross-linking of FcγRIIIA. J Biochem Tokyo 118, 1166–1174.

Zhang, J., Berenstein, E.H., Evans, R.L., & Siraganian, R.P. (1996). Transfection of Syk protein tyrosine kinase reconstitutes high affinity IgE receptor-mediated degranulation in a Syk-negative variant of rat basophilic leukemia RBL-2H3 cells. J. Exp. Med. 184, 71–79.

Zhang, Q., Cox, D., Tseng, C.-C., Donaldson, J.G., & Greenberg, S. (1998). A requirement for ARF6 in Fcγ receptor-mediated phagocytosis in macrophages. J. Biol. Chem. 273, 19977–19981.

Zheleznyak, A., & Brown, E.J. (1992). Immunoglobulin-mediated phagocytosis by human monocytes requires protein kinase C activation. Evidence for protein kinase C translocation to phagosomes. J. Biol. Chem. 267, 12042–12048.

Zhou, M.-J., Todd, R.F.d., van de Winkel, J.G.J., & Petty, H.R. (1993). Cocapping of the leukoadhesin molecules complement receptor type 3 and lymphocyte function-associated antigen-1 with Fc_{γ} receptor III on human neutrophils. Possible role of lectin-like interactions. J. Immunol. 150, 3030–3041.

Zigmond, S.H., Joyce, M., Borleis, J., Bokoch, G.M., & Devreotes, P.N. (1997). Regulation of actin polymerization in cell-free systems by GTPγS and Cdc42. J. Cell Biol. 138, 363–374.

Zimmerli, S., Majeed, M., Gustavsson, M., Stendahl, O., Sanan, D.A., & Ernst, J.D. (1996). Phagosome-lysosome fusion is a calcium-independent event in macrophages. J. Cell Biol. 132, 49–61.

SECTION III

SIGNALING

HETEROGENEITY IN MACROPHAGE PHAGOCYTOSIS

Alan Aderem and David M. Underhill

I. INTRODUCTION

Cells have evolved a variety of strategies to internalize particles and solutes, including pinocytosis, receptor-mediated endocytosis, and phagocytosis

Advances in Cell and Molecular Biology of Membranes and Organelles
Volume 5, pages 195-213.

ISBN: 1-55938-999-0

(reviewed in Rabinovitch, 1995; Silverstein, 1995; Swanson and Baer, 1995; Allen and Aderem, 1996). Pinocytosis usually refers to the uptake of fluid and solutes, and is closely related to receptor-mediated endocytosis, the specific process through which macromolecules, viruses and small particles enter cells. Pinocytosis and receptor-mediated endocytosis share a clathrin-based mechanism and usually occur independently of actin polymerization. By contrast, phagocytosis, the uptake of large particles (> 0.5 μm) into cells, occurs by an actin-dependent mechanism and is usually independent of clathrin. While lower organisms use phagocytosis primarily for the acquisition of nutrients, phagocytosis in Metazoa occurs primarily in specialized phagocytic cells such as macrophages and neutrophils, and has evolved into an extraordinarily complex process underlying a variety of critical biological phenomena. Thus phagocytosis by macrophages is critical for the uptake and degradation of infectious agents and senescent cells, and participates in development, tissue remodeling, the immune response, and inflammation. Monocytes/macrophages and neutrophils have been referred to as professional phagocytes and are very efficient at internalizing particles. On the other hand, most cells have some phagocytic capacity; for example thyroid and bladder epithelial cells phagocytose erythrocytes *in vivo,* and numerous cell types have been induced to phagocytose particles in culture. A group of cells termed paraprofessional phagocytes by Rabinovitch (who also coined the terms professional and non-professional phagocytes) have intermediate phagocytic ability (Rabinovitch, 1995). These include retinal epithelial cells that internalize the effete end of retinal rods . The major difference with respect to phagocytic capacity and efficiency of professional and non-professional phagocytes can probably be ascribed to the presence of an array of dedicated phagocytic receptors that increase particle range and phagocytic rate. Transfection of fibroblasts and epithelial cells with cDNA encoding Fc receptors dramatically increases the phagocytic rate (and obviously particle range), and this system has been used to dissect signaling pathways leading to particle internalization. However, it is clear that many other differences between professional and non-professional phagocytes exist that lead to the enhancement of both rate and efficiency of particle internalization.

The study of phagocytosis requires insight into the mechanisms of signal transduction, actin-based motility, membrane trafficking and infectious disease. While a basic description of phagocytosis has been available since the seminal studies of Metchnikoff (1905), investigations conducted over the last decade have begun to unravel the molecular basis of this process. In this brief overview, we will limit our focus to phagocytic mechanisms in macrophages. Other phagocyctic cells such as neutrophils certainly use similar mechanims, but important differences exist that may be important to the role each cell type plays in the immune response. In particular, we will discuss the initial steps of phagocytosis including particle binding and membrane engulfment. In this review we will not attempt to cover the ensuing maturation of the phagosome into a phagolysome. The reader is referred to a number of recent reviews for a more comprehensive understanding of the process

(Beron et al., Brown, 1995; Greenberg, 1995a; Rabinovitch, 1995; Russell, 1995; Silverstein, 1995; Swanson and Baer, 1995).

It has become clear that phagocytosis is extremely complex and that no single model can fully account for the diverse structures and outcomes associated with particle internalization. This complexity is in part due to the diversity of receptors capable of stimulating phagocytosis and in part due to the capacity of a variety of microbes to influence their fate as they are internalized. The fact that most particles are recognized by more than one receptor, and that these receptors are capable of cross-talk and synergy, further complicates our understanding. In addition, many phagocytic receptors have dual functions, often mediating both adhesion and particle internalization, and a complex relationship exists between these two related processes. Adhesion receptors and phagocytic receptors can both activate and inhibit each other's function. For example, ligation of the fibronectin receptor ($\alpha_5\beta_1$ integrin) at the substrate-adherent surface of a monocyte establishes preconditions within the cell that permits the otherwise inactive complement receptor, CR3 ($\alpha_M\beta_2$ integrin), to mediate phagocytosis (Pommier et al., 1983; Wright et al., 1983). On the other hand, adherent cells often round up during phagocytosis implying that there is competition for cytoskeletal and membrane components necessary for phagocytosis and adhesion. This notion is reinforced by the observation that many of the cytoskeletal components known to participate in adhesion are also enriched in the phagocytic cup. These include paxillin, talin, vinculin, α-actinin, protein kinase C (PKC) α, MARCKS and MacMARCKS (Allen and Aderem, 1995, 1996b).

Despite the complexity associated with different phagocytic mechanisms, a number of shared features exist: particle internalization is initiated by the interaction of specific receptors on the surface of the phagocyte with ligands on the surface of the particle. This leads to the polymerization of actin at the site of ingestion, and the internalization of the particle via an actin-based mechanism. After internalization actin is shed from the phagosome, and the phagosome matures by a series of fusion and fission events with components of the endocytic pathway, culminating in the formation of the mature phagolysosome. Since endosome-lysosome trafficking occurs primarily in association with microtubules, phagosome maturation requires the coordinated interaction of the actin and tubulin based cytoskeletons.

II. RECEPTORS: MECHANISMS OF RECOGNITION

A primary challenge to the innate immune system is the discrimination of a large number of potential pathogens from self, utilizing a restricted number of phagocytic receptors. This problem is compounded by the propensity of pathogens to mutate. This challenge has been met by the evolution of a variety of receptors that recognize conserved motifs on pathogens that are not found in higher eukaryotes.

These motifs have essential roles in the biology of the invading agents, and are therefore not subject to high mutation rates. Janeway (1992) has proposed calling the receptors pattern-recognition receptors (PRRs) and the targets for these receptors pathogen-associated molecular patterns (PAMPs). Pathogen-associated motifs include mannans in the yeast cell wall, formylated peptides in bacteria, and lipopolysaccharides and techoic acids on the surface of Gram-negative and Gram-positive bacteria. The recognition mechanisms leading to phagocytosis occurs either cellularly or humorally. Cellular receptors that recognize these patterns include the mannose-fucose receptor and DEC 205 that recognize mannans, and integrins (for example CD11b/CD18) and scavenger receptors that recognize surface components on bacteria including LPS (Sastry and Ezekowitz, 1993; Stahl and Ezekowitz, 1998). Humoral components which first opsonize the infectious agent before being recognized by a phagocytic receptor include the mannose binding protein, which binds mannans, and which is recognized by the C1q receptor; and surfactant protein A, which binds carbohydrates, and is recognized by a transmembrane receptor, SPR210 (Tenner et al., 1995; Epstein et al., 1996). Antibodies represent an intersection between adaptive and innate immunity: they recognize their cognate ligands on infectious agents with exquisite specificity, but are bound and internalized through their generic Fc domains by the Fc family of receptors (Ravetch, 1997). The complement system lies somewhere in between: the C3bi receptor binds to the C3bi fragment that is fixed nonspecifically to the carbohydrate surface of pathogens via the alternative pathway (Sengelov, 1995). Alternatively, complement is fixed to IgM that specifically recognizes epitopes on the surface of the pathogen.

A. Fc Receptor-Mediated Phagocytosis

Most of our understanding of the signaling pathways leading to phagocytosis in macrophages comes from studies of the Fc receptor (reviewed in Ravetch, 1997). Fc receptors fall into two general classes—those involved in effector functions and those that transport immunoglobulins across epithelial barriers. There are two major classes of Fcγ receptors: receptors that activate effector functions and receptors that inhibit these functions (Ravetch, 1997). Fc receptors that mediate phagocytosis in human macrophages fall within the activation class, and include FcγRI, FcγRIIA, and FcγRIII (Figure 1) (Ravetch, 1997). The human FcγRIIA is a single chain protein with an extracellular Fc binding domain, a transmembrane domain, and a cytoplasmic tail containing two YXXL ITAM motifs (for immunoglobulin gene family tyrosine activation motif) similar to those found in T cell and B cell receptors (Ravetch, 1994). In contrast to its human counterpart, murine FcγRII does not contain ITAM motifs, and is thought not to participate in phagocytosis (Ravetch, 1994). Ligand binding results in receptor cross-linking, and this causes tyrosine phosphorylation of the ITAMs (see below). FcγRI and FcγRIIIA have extracellular Fc binding domains similar to the FcγRIIA, but lack ITAMs on their

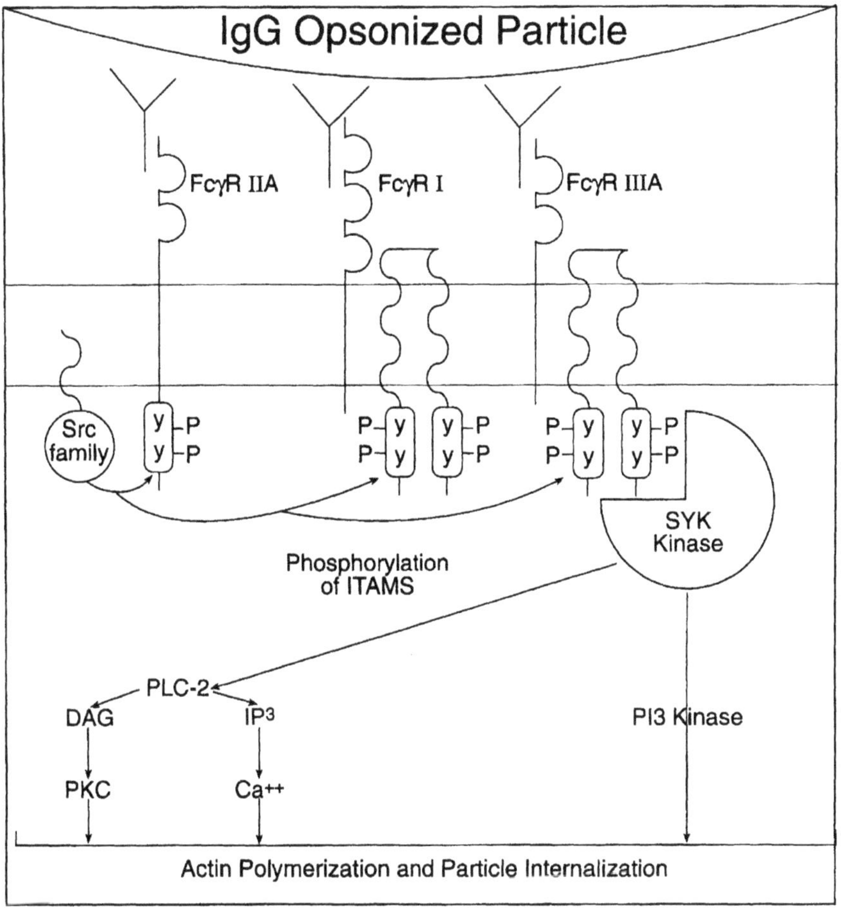

Figure 1. Fcγ receptors signal phagocytosis via their phosphorylated ITAM domains. Receptor cross-linking stimulates src family kinases to phosphorylate tyrosine (y) residues within the ITAM domain of FcγRIIA or within the dimerized γ subunits of FcγRI or FcγRIIIA. The tyrosine kinase Syk is then recruited to the phosphorylated ITAM domain, and upon its activation, is thought to mediate particle internalization by activating PI3 kinase and phospholipase C.

cytoplasmic tails (reviewed in Ravetch, 1994). For proper expression and signaling these receptors must interact with a dimer of γ subunits (FcγRI and FcγRIIIA), or ζ subunits (FcγRIIIA), small transmembrane proteins that contain the ITAMs needed for signal transduction (Figure 1). Ligation of Fcγ receptors I or III result in their cross-linking and in the tyrosine phosphorylation of the ITAM domains of their γ subunits (Ravetch, 1994). Deletion of the gene encoding the γ subunit of

Fcγ receptor in mice results in macrophages that are unable to express FcγR I or III, since these receptors are not transported to the surface of cells in the absence of their signaling subunit, and macrophages from these mice are unable to phagocytose IgG-coated particles (Takai et al., 1994).

The role of the ITAM motifs of the γ subunit of the Fcγ receptor have also been analyzed in COS cells; all three members of the FcγR family are capable of promoting phagocytosis when transfected into COS cells. Since COS cells aren't professional phagocytes, and all cells have some capacity to phagocytose (as discussed above), it is likely that much more than the presence or absence of Fc receptors is responsible for efficient phagocytosis by professional phagocytes. Indeed, COS cells expressing Fc receptors phagocytose IgG-opsonized particles much less efficiently than macophages. However, many of the early signaling events may be reconstituted in COS cells. In transfected COS-1 cells, FcγRIIIA or I mediates phagocytosis of IgG-opsonized particles, but only when coexpressed with the γ chain, and the ITAM motif of γ is required for a competent phagocytic signal (reviewed in Indik et al., 1995). The ζ chain of the T cell receptor contains sequences homologous to the γ chain including the conserved YXXL motifs, and can sbstitute for the γ chain in FcγRIIIA-dependent signaling of phagocytosis. However, the ζ chain is considerably less efficient in mediating FcγRIIIA-dependent phagocytosis than is the γ chain, and mutational analysis demonstrates that the functional differences between the γ and ζ subunits are due to the internal amino acids of the YXXL (Indik et al., 1995). Cross-linking of FcγRIIIA results in tyrosine phosphorylation of the γ subunit, and mutation of either tyrosine of the two YXXL motifs of the γ subunit ITAM eliminates both tyrosine phosphorylation and phagocytosis (Indik et al., 1995).

The protein tyrosine kinase responsible for this initial phosphorylation is thought to be a member of the src family (Greenberg, 1995a). Subsequently, a second protein tyrosine kinase, p72Syk, is recruited to the phosphorylated ITAM domains (Ravetch, 1994; Greenberg, 1995a). This results in the activation of the Syk kinase, which in turn triggers a plethora of pathways leading to transcriptional activation, cytoskeletal rearrangement and the release of inflammatory mediators. This model is at least partially correct for FcR-mediated signals leading to phagocytosis.

This model is supported by the observation that a chimera containing the extracellular domain of CD16 (FcγRIII), fused to the transmembrane stalk of CD7, and containing p72Syk intracellularly, is capable of signaling phagocytosis of IgG opsonized particles in transfected COS cells (Greenberg et al., 1995b). A competent phagocytic stimulus was independent of Syk SH2 domains, but required an active Syk kinase (Greenberg et al., 1995b). The related tyrosine kinase, ZAP70, could substitute for Syk in this system, whereas members of the src family of tyrosine kinases could not (Greenberg et al., 1995b). These studies have been extended in DT40 lymphocytes, a chicken cell line that has been valuable in dissecting signaling pathways because it undergoes a high rate of homologous recombination, and therefore permits gene deletions at high frequency. DT40 cells, expressing a fusion protein consisting of the extracellular domain of human FcγRIIIA and the ITAM-containing γ subunit of the Fc receptor, are capable of localized actin polymerization when the chimeric receptors are clustered (Cox et al., 1996). Actin assembly is dependent upon an intact ITAM, absent in cells lacking Syk, and exacerbated in cells overexpressing Syk (Cox et al., 1996), suggesting an absolute requirement for the Syk tyrosine kinase in ITAM-dependent actin assembly in DT40 cells. The requirement for Syk can probably be extended to all hematopoetic cells, since Fc receptor-mediated actin assembly and phagocytosis is abrogated in macrophages derived from the fetal livers of Syk null mice (Crowley et al., 1997). Further evidence for the involvement of Syk in phagocytosis is also derived from the COS cell system. Upon cross-linking of FcγRIIIA/γ and FcγRI/γ, Syk is phosphorylated and enhances the phagocytosis of IgG-opsonized erythrocytes, and this activity is dependent on the γ chain (Indik et al., 1995). Both SH2 domains of Syk are necessary for functional association with the γ subunit, and Syk is unable to induce either FcγRI or FcγRIIIA mediated phagocytosis by γ chain mutants in which YXXL tyrosine is replaced by phenylalanine (Indik et al., 1995). How the Syk tyrosine kinase stimulates actin assembly is unknown, although it is likely that PI 3-kinase is involved (see below).

There are clearly a number of problems with this model. First, it only applies to FγR mediated phagocytosis. Thus, while macrophages from Syk mice are incapable of FcγR-mediated phagocytosis, phagocytosis of latex particles, yeast and *Escherichia coli* is unimpaired (Crowley et al., 1997). Second, macrophages derived from mice deficient in the three members of the src-family kinases known to be expressed in these cells, Hck, Fgr, and Lyn, exhibit poor Syk activation when the FγR is ligated but are still capable of FcγR-mediated phagocytosis, albeit at a slightly slower rate (Crowley et al., 1997). Either a small activation of Syk is sufficient to support FcγR-mediated phagocytosis, or Syk participates by another means, perhaps by serving as an adapter. An answer to this conundrum might be found in the interesting observation that the c-fgr tyrosine kinase actually suppresses phagocytosis in macrophages (Gresham and Willman, unpublished data). These investigators found that while a fgr-negative murine macrophage line phagocytosed normally, FcγRI, FcγRII/FcγRIII, and C3bi-mediated phagocytosis was suppressed when the cells were transfected with wild-type c-fgr. While actin rearrangement and phagocytosis is suppressed, c-fgr has no effect on receptor expression or on attachment of the opsonized particle. The suppressive effect of c-fgr is independent of its kinase activity, implying that inhibition of phagocytosis may be mediated through an adapter function.

Downstream Effectors of Fc Receptor-Mediated Phagocytosis

The mechanism by which FcγRs stimulate the polymerization of actin, and induce the formation of phagosome is not known, although PI-3 kinase, the Rho family of GTPases, PKC, and motor proteins appear to participate.

PI-3 kinase. Recent evidence suggests that PI 3-kinase participates in the signaling cascade of phagocytic receptors. PI 3-kinase catalyzes phosphorylation at the D-3 position of the inositol ring of phosphatidylinositol (PI), PI(4)P and PI(4,5)P2, and is activated by many tyrosine kinase receptors which trigger the polymerization of actin (Toker and Cantley, 1997). In addition, there is compelling evidence in yeast that PI 3-kinase participates in membrane trafficking (De Camilli et al., 1996). Cross-linking of Fc_{γ}RI and RII increases PI 3-kinase activity, and FcR-mediated phagocytosis is prevented by wortmannin or LY294002, specific inhibitors of PI 3-kinase (Ninomiya et al., 1994; Araki et al., 1996). Elegant studies by Swanson and colleagues indicate that wortmannin and LY294002 PI do not inhibit actin-dependent formation of the phagocytic cup, but instead prevent the phagosome from sealing behind the particle (Araki et al., 1996). De Franco and co-workers confirmed these data, and further demonstrated that macrophages from Syk null mice are similarly capable of polymerizing actin beneath the FcγR-induced phagocytic cup, but unable to complete internalization (Crowley et al., 1997). This suggests that PI 3-kinase may participate in a Syk-dependent signaling pathway critical for FcγR-mediated phagocytosis.

GTPases. Members of the Rho family of GTPases have been shown to regulate the actin cytoskeleton in response to a variety of extracellular signals (Hall, 1998). In 3T3 cells, various members of the Rho family have been shown to act hierarchically during cell spreading: Cdc42 partipates in the formation of filopodia and in the activation of Rac, Rac stimulates membrane ruffling and activates Rho, and Rho stimulates the formation of focal adhesions and stress fibers (Hall, 1998). Recent evidence demonstrates that the Rho family also participates in phagocytosis. Microinjection of the J774 mouse macrophage cell line with the Rho-specific inhibitor, C3 exotoxin, inhibits FcγR-mediated phagocytosis by preventing receptor clustering, a prerequisite for efficient particle binding and internalization (Hackam et al., 1997). By contrast, inhibition of Rac1 and Cdc42, by expression of their dominant negative forms in the RAW mouse macrophage cell line does not affect particle binding to FcRs, but inhibits phagocytosis by preventing the accumulation of F-actin in the phagocytic cup (Cox et al., 1997). The precise mechanism by which these GTPases regulate F-actin structure has not yet been defined, but a variety of cytoskeletal regulators including PIP 5-kinase and myosin II have been implicated (Hall, 1998).

Members of the ARF family of GTPases have a role in most membrane trafficking events. ARF6 has been implicated in endocytosis, membrane recycling and regulated exocytosis (Roth and Sternweis, 1997). Expression of a mutant form of ARF6 that is incapable of hydrolyzing GTP causes profound rearrangement of F-actin in HeLa cells (Radhakrishna et al., 1996), and inhibits FcγR-mediated phagocytosis in a macrophage cell line (Zhang et al., 1998).

Protein kinase C. PKC also appears to have a role in phagocytosis (Zheleznyak and Brown, 1992; Allen and Aderem, 1995, 1996a); previous studies demonstrated that PKC is activated upon ligation of the FγR in human monocytes (Zheleznyak and Brown, 1992) and localized the α isozyme of PKC to nascent phagosomes in macrophages (Allen and Aderem, 1995). The involvement of PKC in phagocytosis is tantalizing since its major substrate, MARCKS, is known to regulate actin structure at the membrane (Aderem, 1992). MARCKS is rapidly phosphorylated during particle uptake, and MARCKS and PKCα are recruited to the forming zymosan phagosome with similar kinetics to F-actin (Allen and Aderem, 1995). MARCKS cross-links F-actin, and this activity is prevented by PKC-dependent phosphorylation and by calcium/calmodulin (Hartwig et al., 1992). Since the association of MARCKS with membranes is also regulated by PKC-dependent phosphorylation, it is an ideal candidate to regulate actin structure on the phagosome in response to signals from both PKC and calcium/calmodulin. This is supported by the observation that inhibitors of PKC prevent phagocytosis and block the accumulation of PKCα, MARCKS, F-actin and a number of other cytoskeletal proteins beneath bound zymosan (Allen and Aderem, 1995). MacMARCKS, another member of the MARCKS family, also associates with zymosan phagosomes (Li and Aderem, 1992; Zhu et al., 1995), and a mutant form of MacMARCKS appears to block phagocytosis when expressed in a macrophage cell line (Zhu et al., 1995). The significance of this observation is unclear since macrophages derived from MacMARCKS null mice phagocytose zymosan normally (Underhill et al., 1998).

Motor proteins. It is not clear whether actin polymerization alone is sufficient to drive pseudopod extension and particle internalization, or whether this also requires molecular motors. It has long been known that myosin II accumulates on the phagocytic cups of macrophages and neutrophils ingesting yeast, implying that it might act as a mechanical motor during particle internalization (Stendahl et al., 1980). Myosin I also colocalizes with F-actin on forming phagosomes, suggesting that it too might facilitate ingestion (Allen and Aderem, 1995). Despite these colocalization studies, and the observation that the broad spectrum myosin inhibitor BDM blocks phagocytosis (Underhill, unpublished observations), there is as yet no information on the specific roles of myosin isoforms in phagocytosis.

B. Complement Receptor-Mediated Phagocytosis

Complement proteins, present in serum, opsonize bacteria for phagocytosis by the C3b or C3bi receptors (CRs) on macrophages. Several receptors participating in phagocytosis of complement-opsonized particles are expressed on macrophages including CR1, CR3, and CR4 (Sengelov, 1995). CR1 is a single chain transmembrane protein consisting of a large extracellular lectin-like complement-binding domain and a short 43 amino acid cytosolic domain. CR1 binds

C3b, C4b and C3bi, and is thought to participate in particle binding; CR1 by itself is not a phagocytic receptor. CR3 and CR4 are integrin family members made up of heterodimers of different α chains (α_m for CR3 and α_x for CR4) and a shared β chain (β_2) (Sengelov, 1995). These two receptors bind specifically to C3bi and are responsible for particle internalization.

While FcRs are constitutively active for phagocytosis (Ravetch, 1994), the CRs of resident peritoneal macrophages bind but do not internalize particles in the absence of additional stimuli (Pommier et al., 1983; Wright et al., 1983). Particle ingestion by CRs can be induced by PKC activators such as PMA, as well as by tumor necrosis factor (TNF)-α, granulocyte/macrophage colony-stimulating factor (GM-CSF), or attachment to laminin- or fibronectin-coated substrata (Pommier et al., 1983; Wright et al., 1983; Wright and Griffin, 1985). Although all types of phagocytosis require actin polymerization at the site of ingestion (Allen and Aderem, 1996a), results of electron microscopy (EM) studies demonstrate that IgG- and complement-opsonized particles are internalized differently by macrophages (Kaplan, 1977; Allen and Aderem, 1996b). During FcγR-mediated phagocytosis, veils of membrane rise above the cell surface and tightly surround the particle before drawing it into the body of the macrophage (Kaplan, 1977; Allen and Aderem, 1996b) (Figure 2A). Silverstein and colleagues have demonstrated that ingestion occurs by a zippering process, in which FγRs in the macrophage plasma membrane interact sequentially with IgG molecules distributed over the surface of the ingested particle (Silverstein, 1995). On the other hand, EM data indicate that CR-mediated phagocytosis is a relatively passive process that occurs by a variation of the classic zipper model; complement-opsonized particles appear to sink into the cell with elaboration of small, if any, pseudopodia (Kaplan, 1977; Allen and Aderem, 1996b) (Figure 2B). Moreover, the phagosome membrane is less tightly opposed to complement-opsonized particles, with point-like contact areas separating regions of looser membrane. These point-like contact areas are enriched with a variety of cytoskeletal proteins including F-actin, vinculin, α-actinin, paxillin, and phosphotyrosine containing proteins, and their formation is blocked by inhibitors of PKC, but not by inhibitors of protein tyrosine kinases (although tyrosine phosphorylation increases the efficiency of phagocytosis) (Allen and Aderem, 1996b) (Figure 3E and F, and data not shown). By contrast, all of these protens are diffusely distributed on phagosomes containing IgG-coated particles (Figure 3C and D), and FcγR-mediated phagocytosis is blocked by both PKC and tyrosine kinase inhibitors (Allen and Aderem, 1996b). Thus, the signals required for particle ingestion, and the arrangement of cytoskeletal proteins on the phagosome surface, vary depending upon which phagocytic receptor is engaged. Moreover, complement receptor (CR)-mediated internalization requires intact microtubules and is accompanied by the accumulation of vesicles beneath the forming phagosome (Figure 2B, arrows), suggesting that membrane trafficking plays a key role in CR-mediated phagocytosis (Allen and Aderem, 1996b).

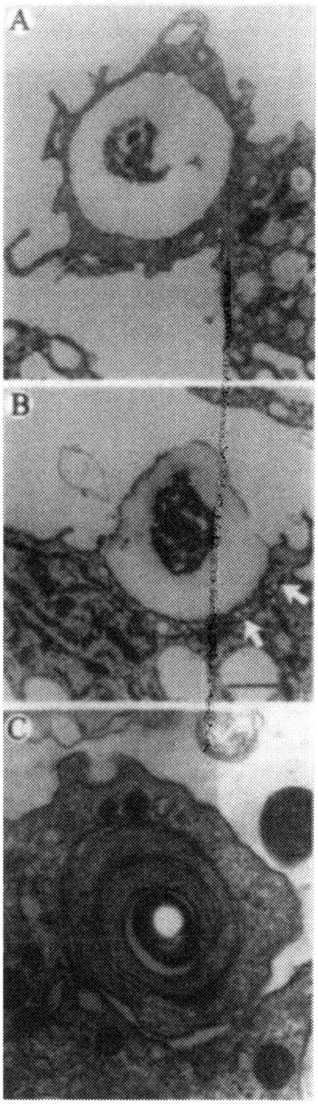

Figure 2. Different particles are internalized by different mechanisms. Cryo-EM sections of peritoneal macrophages that are in the process of ingesting IgG-opsonized particles (**A**) or complement-opsonized particles (**B**). Note that pseudopodia protrude from the macrophage surface to engulf the IgG-opsonized particle, whereas the complement-coated particle sinks directly into the cell. Arrows in **B** indicate vesicles directly beneath the forming complement phagosome that are absent beneath the FcR phagosome in **A**. Reprinted with permission from *The Journal of Experimental Medicine* (Allen, L.A.H., & Aderem, A. (1996). *J. Exp. Med.* 184, 627–637). (**C**) *pneumophila* is internalized into human monocytes by coiling phagocytosis. Reprinted with permission from *Cell* (Horwitz, M.A. (1984). *Cell* 36, 27–33).

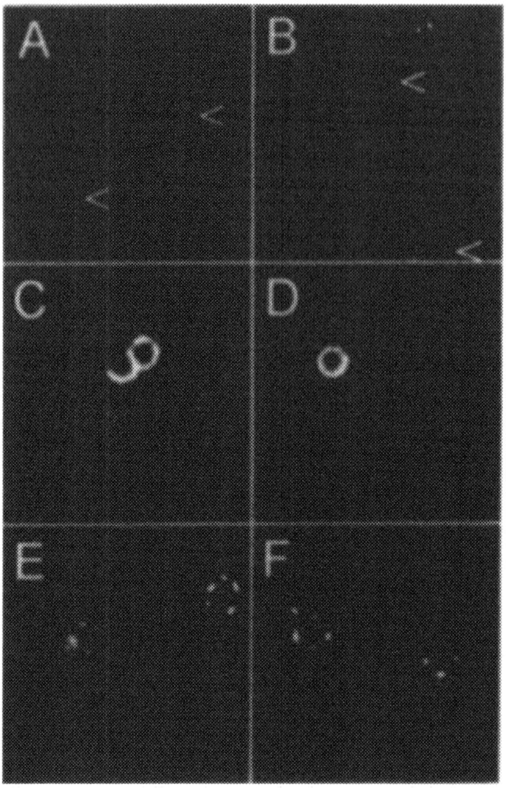

Figure 3. Vinculin and paxillin associate differently with different types of phagosomes. Zymosan containing phagosomes (**A** and **B**) are not stained with antibodies to vinculin (**A**) or paxillin (**B**). The arrowheads indicate the position of the phagosomes. FcR-mediated phagosomes (**C** and **D**) are enriched with vinculin (**C**) and paxillin (**D**) in a uniform pattern, while complement receptor-mediated phagosomes (**E** and **F**) are coated with vinculin (**E**) and paxillin (**F**) in discrete foci. Reprinted with permission from *The Journal of Experimental Medicine* (Allen, L.A.H., & Aderem, A. (1996). *J. Exp. Med.* 184, 627–637).

C. Mannose Receptor-Mediated Phagocytosis

The mannose receptor (MR) on macrophages recognizes mannose and fucose on the surfaces of pathogens and mediates phagocytosis of the organisms (Stahl and Ezekowitz, 1998). The high affinity of this receptor for branched mannose and fucose oligosacharrides, prototypic PAMPs as described above, makes the MR a phagocytic receptor with broad pathogen specificity.

The MR is a single chain receptor with a short cytoplasmic tail and an extracellular domain including eight lectin-like carbohydrate-binding domains. The lectin-like carbohydrate-binding domains share homology with other C-type lectins including the mannose-binding protein, collectins, DEC 205, and the phospholipase A2 receptor (Stahl and Ezekowitz, 1998). The cytoplasmic tail is crucial to both the endocytic and phagocytic functions of the receptor, but little is known about the signals that lead to phagocytosis (Ezekowitz et al., 1990; Stahl and Ezekowitz, 1998).

During MR-mediated phagocytosis of zymosan, the actin cytoskeleton is mobilized around the nascent phagosomes, and proteins such as F-actin, talin, PKCα, MARCKS, and myosin 1 are recruited (Allen and Aderem, 1996a). However, in contrast to FcR- and CR-mediated phagocytosis, vinculin and paxillin are not recruited to MR phagosomes (Figure 3A and B), reinforcing the notion that different phagocytic receptors send different signals to the actin cytoskeleton and initiate different mechanisms of internalization (Allen and Aderem, 1996a).

III. PHAGOCYTOSIS OF PATHOGENS

Even though one of the major functions of phagocytosis is to mediate the ingestion and sterilization of infectious agents, many pathogens such as *Salmonella typhimurium*, *Legionella pneumophila*, and *Mycobacterium tuberculosis* have evolved mechanisms for survival and even growth inside macrophage vacuoles. As with the other particles, we will only focus on the initial stages of pathogen ingestion, in order to extend our thesis that phagocytosis involves a large variety of heterogenous mechanisms.

A. *Salmonella typhimurium* Internalization Involves Macropinocytosis

After binding to the surface of a macrophage, virulent *S. typhimurium* induces profound membrane ruffling which results in the internalization of the bacterium into a compartment resembling a macropinosome (Alpuche-Aranda et al., 1994). The nascent vacuole is enormous relative to the size of the bacterium and has been called a "spacious phagosome" (Alpuche-Aranda et al., 1994). The macrophage receptors that bind *S. typhimurium* are not known, although internalization appears to be associated with an actin-dependent mechanism. By contrast, nonvirulent mutant strains of *S. typhimurium* are bound to the surface of the macrophage and do not induce membrane ruffling or macropinocytosis (Alpuche-Aranda et al., 1995). Rather, the bacteria are phagocytosed into a vacuole with a tightly opposed membrane.

B. *Legionella pneumophila* is Internalized by Coiling Phagocytosis

L. pneumophila is a facultative intracellular pathogen that invades and replicates in macrophages (Shuman and Horwitz, 1996; Vogel et al., 1998) . A bacterial surface protein, MOMP (major outer-membrane protein), fixes complement component C3 to the surface of the parasite thereby facilitating binding to the macrophage surface through complement receptors (Bellinger and Horwitz, 1990). Removing complement from the media, or blocking complement receptor CR3 with a specific antibody prevents bacterial adhesion to the macrophage surface. After binding, the parasite induces the formation of an extended host cell pseudopod that spirals around the bacterium forming a structure coined a "coiling phagosome" (Figure 2C) (Horwitz, 1984). Although CR3 is enriched in coiling phagosomes, there is no evidence that these receptors signal coiling phagocytosis. Since the structure of this phagosome is very different to that induced by CR3 (discussed above), it is likely that an additional, as yet uncharacterized, *L. pneumophila* signal must be required for inducing the coiling phagosome. Since coiling phagocytosis occurs with heat-killed and glutaraldehyde-killed *L. pneumophila*, this signal is likely to be due to a stable ligand on the parasite surface (Shuman and Horwitz, 1996).

C. *Mycobacterium tuberculosis* is Internalized via a Variety of Receptors

A plethora of macrophage receptors have been implicated in binding and internalization *of M. tuberculosis* (Ernst, 1998). As with other bacteria, complement fixes to the surface of the organism through the alternate pathway allowing deposition of complement proteins C3b and C3bi which are recognized by complement receptors CR1 and CR3 (Schlesinger et al., 1990). Alternately, in the absence of factors required for activation of the alternate pathway, a surface component of *M. tuberculosis* resembling complement component C4b can bind directly to C2b and form a C3 convertase analogous to the one formed in the classical complement cascade (Schorey et al., 1997) . This C3 convertase catalyzes the deposition of C3b onto the surface of the organism and facilitates binding to CR1. Blocking complement receptors drastically reduces binding and invasion of *M. tuberculosis,* but does not abolish it, suggesting that other receptors participate in their uptake. Consistent with this, *M. tuberculosis* has also been demonstrated to bind to the mannose receptor and the scavenger receptor (Schlesinger, 1993). In addition, surfactant protein A enhances macrophage binding and uptake of *M. tuberculosis,* probably by the surfactant protein A receptors (Zimmerli et al., 1996).

Phagocytosis of either Erdman or H37Ra *M. tuberculosis* in the presence of autologous non-immune serum is associated with an increase in phospholipase D activity in human monocyte-derived macrophages, and inhibition of phospholipase D prevents the uptake of the bacterium (Kusner et al., 1996). *M. tuberculosis* uptake is also associated with the tyrosine phosphorylation of multiple macrophage proteins, and tyrosine kinase inhibitors suppress the phagocytosis of the bacterium (Kusner et al., 1996). When murine bone marrow-derived macrophages ingest *M. smegmatis*, *M. tuberulosis* H37Rv, or its attenuated counterpart *M. tuberculosis* H37Ra, PKCα, and its substrates MARCKS and MacMARCKS, are recruited to the nascent mycobacterial phagosome concurrently with F-actin (Veis et al., submitted). Talin, paxillin, vinculin and phosphotyrosine-containing proteins are also components of the forming mycobacterial phagosome. In contrast to vacuoles containing zymosan or IgG opsonized particles, that retain MARCKS and PKCα, these proteins dissociate simultaneously with F-actin, talin, paxillin, vinculin, and phosphotyrosine-containing proteins from the maturing mycobacterial phagosome (Veis et al., submitted). Both MARCKS and MacMARCKS are rapidly and transiently phosphorylated by PKC upon ingestion of mycobacteria, and inhibitors of PKC prevent the internalization of the bacteria (Veis et al., submitted). Thus early signals leading to the internalization of *M. tuberculosis* into macrophages differ from those utilized in the internalization of other bacteria such as *S. typhimurium* and *L. pneumophila*, as well as the signals leading to phagocytosis by FcγR, CR and MR. It should be noted that we are only focusing on the early events in the uptake of *M. tuberculosis*. Elegant work from a number of laboratories demonstrating that this organism modifies vacuole maturation is beyond the scope of this review (Russell et al., 1997).

IV. CONCLUSIONS

It is clear that phagocytosis in macrophages is a diverse process; the signals leading to actin polymerization and particle internalization depend on the specific receptors that mediate the process, and on additional modifying signals that can be generated by complex particles. Complex particles, such as bacteria can activate multiple receptors whose signaling pathways may interact in intricate and unpredictable ways. In addition, living bacteria have the capacity to modify signaling pathways within eukaryotic cells. For example, *S. typhimurium* can introduce an GTP:GDP exchange factor into cells which modifies the way that Rho family of GTPases signals the actin cytoskeleton (Hardt et al., 1998), while *Yersinia* species introduce a broad spectrum tyrosine phosphatase (Finlay and Cossart, 1997).

In addition to the complexity relating to the different receptor systems and the capacity of microbes to modify phagocytosis discussed above, it is important to recognize that the formation of the phagosome can be a heterogenous process even when a single cell ingests two identical particles. While all phagocytosis involves actin remodeling around the phagocytic cup, we have observed that certain cytoskeletal proteins that decorate a particular phagosome are absent from otherwise identical phagosomes in the same cell. These differences are not temporal, but are likely to arise from stochastic, or even chaotic, processes. For example, actin cross-linking can be achieved using any of the diverse number of actin cross-linking proteins that are expressed in the cell. Thus if a specific actin cross-linking protein happens to have been enriched in a specific region of the cell, for example the leading edge of a motile cell, it may be used during phagocytosis at the leading edge while another protein may serve the same function on a phagosome formed at the trailing edge. It is clear that the molecular dissection of phagocytosis represents a daunting task; more than 100 actin-binding proteins have been identified, and many of these are expressed in the same cell. Two common approaches that have been used to establish the role of specific gene products in a particular biological phenomenon have been to express dominant negative forms of the protein, or to delete the gene encoding the protein. However, the interpretation of the results are complicated by the observation that these two approaches do not always yield the same phenotype. Ultimately, phagocytosis, like most other problems in biology, will have to be analyzed as a complex system rather than a linear series of isolated enzymatic reactions. In this guise, phagocytosis provides a window into the coordinate functioning of the actin and tubulin based cytoskeletons, and could serve as a model system for analyzing diverse biological phenomena including synaptic transmission, mitogenesis and morphogenesis.

ACKNOWLEDGMENTS

This work was supported by grants AI25032 and AI32972 from the National Institutes of Health. D.U. is an Irvington Institute Postdoctoral Fellow. We thank Drs. Hattie Gresham, Cheryl Willman and Steven Greenberg for communicating results to us prior to publication. Dr. Marcus Horwitz kindly provided the electron micrograph shown in Figure 2C.

REFERENCES

Aderem, A. (1992). The MARCKS brothers: a family of protein kinase C substrates. Cell 71, 713–716.

Allen, L.-A.H., & Aderem, A. (1995). A role for MARCKS, the α isozyme of protein kinase C, and myosin I in zymosan phagocytosis by macrophages. J. Exp. Med. 82, 829–840.

Allen, L.-A.H., & Aderem, A. (1996a). Mechanisms of phagocytosis. Curr. Opin. Immunol. 8, 36–40.

Allen, L.-A.H., & Aderem, A. (1996b). Molecular definition of distinct cytoskeletal structures involved in complement- and Fc receptor-mediated phagocytosis in macrophages. J. Exp. Med. 184, 627–637.

Alpuche-Aranda, C.M., Berthiaume, E.P., Mock, B., Swanson, J.A., & Miller, S.I. (1995). Spacious phagosome formation within mouse macrophages correlates with Salmonella serotype pathogenicity and host susceptibility. Infect. Immun. 63, 4456–4462.

Alpuche-Aranda, C.M., Raccoosin, E.L., Swanson, J.A., & Miller, S.I. (1994). *Salmonella* stimulate macrophage macropinocytosis and persist within spacious phagosomes. J. Exp. Med. 179, 601–608.

Araki, N., Johnson, M.T., & Swanson, J.A. (1996). A role for phosphoinositide 3-kinase in the completion of macropinocytosis and phagocytosis by macrophages. J. Cell Biol. 135, 1249–1260.

Bellinger, K.C., & Horwitz, M.A. (1990). Complement component C3 fixes selectively to the major outer membrane protein (MOMP) of Legionella pneumophila and mediates phagocytosis of liposome-MOMP complexes by human monocytes. J. Exp. Med. 172, 1201–1210.

Beron, W., Alvarez-Dominguez, C., Mayorga, L., & Stahl, P.D. (1995). Membrane trafficking along the phagocytic pathway. Trends Cell Biol. 5, 100–104.

Brown, E.J. (1995). Phagocytosis. BioEssays 17, 109–117.

Cox, D., Chang, P., Kurosaki, T., & Greenberg, S. (1996). Syk tyrosine kinase is required for immunoreceptor tyrosine activation motif-dependent actin assembly. J. Biol. Chem. 271, 16597–16602.

Cox, D., Chang, P., Zhang, Q., Reddy, P.G., Bokoch, G.M., & Greenberg, S. (1997). Requirements for both Rac1 and Cdc42 in membrane ruffling and phagocytosis in leukocytes. J. Exp. Med. 186, 1487–1494.

Crowley, M.T., Costello, P.S., Fitzer-Attas, C.J., Turner, M., Meng, F., Lowell, C., Tybulewicz, V.L., & DeFranco, A.L. (1997). A critical role for Syk in signal transduction and phagocytosis mediated by Fcgamma receptors on macrophages. J. Exp. Med. 186, 1027–1039.

De Camilli, P., Emr, S.D., McPherson, P.S., & Novick, P. (1996). Phosphoinositides as regulators in membrane traffic. Science 271, 1533–1539.

Epstein, J., Eichbaum, Q., Sheriff, S., & Ezekowitz, R.A. (1996). The collectins in innate immunity. Curr. Opin. Immunol. 8, 29–35.

Ernst, J.D. (1998). Macrophage receptors for Mycobacterium tuberculosis. Infect. Immun. 66, 1277–1281.

Ezekowitz, R.A., Sastry, K., Bailly, P., & Warner, A. (1990). Molecular characterization of the human macrophage mannose receptor: demonstration of multiple carbohydrate recognition-like domains and phagocytosis of yeasts in Cos-1 cells. J. Exp. Med. 172, 1785–1794.

Finlay, B.B., & Cossart, P. (1997). Exploitation of mammalian host cell functions by bacterial pathogens. Science 276, 718–725.

Greenberg, S. (1995). Signal transduction of phagocytosis. Trends Cell Biol. 5, 93–99.

Greenberg, S., Chang, P., Wang, D., Xavier, R., & Seed, B. (1996). Clustered syk tyrosine kinase domains trigger phagocytosis. Proc. Natl. Acad. Sci. USA 93,1103–1107.

Hackam, D.J., Rotstein, O.D., Schreiber, A., Zhang, W., & Grinstein, S. (1997). Rho is required for the initiation of calcium signaling and phagocytosis by Fcgamma receptors in macrophages. J. Exp. Med. 186, 955–966.

Hall, A. (1998). Rho GTPases and the actin cytoskeleton. Science 279, 509–514.

Hardt, W.-D., Chen, L.-M., Schuebel, K.E., Bustelo, X.R., & Galan, J.E. (1998). *S. typhimurium* encodes an activator of Rho GTPases that induces membrane ruffling and nuclear responses in host cells. Cell 93, 815–826.

Hartwig, J.H., Thelen, M., Rosen, A., Janmey, P.A., Nairn, A.C., & Aderem, A. (1992). MARCKS is an actin filament crosslinking protein regulated by protein kinase C and calcium-calmodulin. Nature 356, 618–622.

Horwitz, M.A. (1984). Phagocytosis of the Legionnaires' disease bacterium (Legionella pneumophila) occurs by a novel mechanism: engulfment within a pseudopod coil. Cell 36, 27–33.

Indik, Z.K., Park, J.G., Hunter, S., & Schreiber, A.D. (1995). The molecular dissection of Fcgamma receptor mediated phagocytosis. Blood 86, 4389–4399.

Janeway, C.A.J. (1992). The immune system evolved to discriminate infectious nonself from noninfectious self. Immunol. Today 13, 11–16.

Kaplan, G. (1977). Differences in the mode of phagocytosis with Fc and C3 receptors in macrophages. Scand. J. Immunol. 6, 797–807.

Kusner, D.J., Hall, C.F., & Schlesinger, L.S. (1996). Activation of phospholipase D is tightly coupled to the phagocytosis of Mycobacterium tuberculosis or opsonized zymosan by human macrophages. J. Exp. Med. 184, 585–595.

Li, J., & Aderem, A. (1992). MacMARCKS, a novel member of the MARCKS family of protein kinase C substrates. Cell 70, 791–801.

Metchnikoff, E. (1905). In: Immunity in Infective Diseases. Cambridge Press, London.

Ninomiya, N., Hazeki, K., Fukui, Y., Seya, T., Okada, T., Hazeki, O., & Ui, M. (1994). Involvement of phosphatidyl inositol 3-kinase in Fcgamma receptor signaling. J. Biol. Chem. 269, 22732–22737.

Pommier, C.G., Inada, S., Fries, L.F., Takahashi, T., Frank, M.M., & Brown, E.J. (1983). Plasma fibronectin enhances phagocytosis of opsonized particles by human peripheral blood monocytes. J. Exp. Med. 157, 1844–1854.

Rabinovitch, M. (1995). Profesional and non-profesional phagocytes: an introduction. Trends Cell Biol. 5, 85–87.

Radhakrishna, H., Klausner, R.D., & Donaldson, J.G. (1996). Aluminum fluoride stimulates surface protrusions in cells overexpressing the ARF6 GTPase. J. Cell Biol. 134, 935–947.

Ravetch, J.V. (1994). Fc receptors: rubor redux. Cell 78, 553–560.

Ravetch, J.V. (1997). Fc receptors. Curr. Opin. Immunol. 9, 121–125.

Roth, M.G., & Sternweis, P.C. (1997). The role of lipid signaling in constitutive membrane traffic. Curr. Opin. Cell Biol. 9, 519–526.

Russell, D.G. (1995). *Mycobacterium* and *Leishmania*: stowaways in the endosomal network. Trends Cell Biol. 5, 125–128.

Russell, D.G., Sturgill, K.S., Vanheyningen, T., Collins, H., & Schaible, U.E. (1997). Why intracellular parasitism need not be a degrading experience for Mycobacterium. Philos. Trans. R. Soc. Lond. B. Biol. Sci. 352, 1303–1310.

Sastry, K., & Ezekowitz, R.A. (1993). Collectins: pattern recognition molecules involved in first line host defense [erratum appears in Curr. Opin. Immunol. 1993 5, 566]. [Review]. Curr. Opin. Immunol. 5, 59–66.

Schlesinger, L.S. (1993). Macrophage phagocytosis of virulent but not attenuated strains of Mycobacterium tuberculosis is mediated by mannose receptors in addition to complement receptors. J Immunol. 150, 2920–2930.

Schlesinger, L.S., Bellinger-Kawahara, C.G., Payne, N.R., & Horwitz, M.A. (1990). Phagocytosis of Mycobacterium tuberculosis is mediated by human monocyte complement receptors and complement component C3. J. Immunol. 144, 2771–2780.

Schorey, J.S., Carroll, M.C., & Brown, E.J. (1997). A macrophage invasion mechanism of pathogenic mycobacteria. Science 277, 1091–1093.

Sengelov, H. (1995). Complement receptors in neutrophils. Crit. Rev. Immunol. 15, 107–131.

Shuman, H.A., & Horwitz, M.A. (1996). Legionella pneumophila invasion of mononuclear phagocytes. Curr. Top .Microbiol. Immunol. 209, 99–112.

Silverstein, S.C. (1995). Phagocytosis of microbes: insights and prospects. Trends Cell Biol. 5, 141–142.

Stahl, P.D., & Ezekowitz, R.A. (1998). The mannose receptor is a pattern recognition receptor involved in host defense. Curr. Opin. Immunol. 10, 50–55.

Stendahl, O.I., Hartwig, J.H., Brotschi, E.A., & Stossel, T.P. (1980). Distribution of actin-binding protein and myosin in macrophages during spreading and phagocytosis. J. Cell Biol. 84, 215–224.

Swanson, J.A., & Baer, S.C. (1995). Phagocytosis by zippers and triggers. Trends Cell Biol. 5, 89–93.

Takai, T., Li, M., Sylvestre, D., Clynes, R., & Ravetch, J.V. (1994). FcR gamma chain deletion results in pleiotrophic effector cell defects. Cell 76, 519–529.

Tenner, A.J., Robinson, S.L., & Ezekowitz, R.A. (1995). Mannose binding protein (MBP) enhances mononuclear phagocyte function via a receptor that contains the 126,000 M(r) component of the C1q receptor. Immunity 3, 485–493.

Toker, A., & Cantley, L.C. (1997). Signalling through the lipid products of phosphoinositide-3-OH kinase. Nature 387, 673–676.

Underhill, D.M., Chen, J., Allen, L.-A.H., & Aderem, A. (1998). MacMARCKS is not essential for phagocytosis in macrophages, J. Biol. Chem. 273, 33619–33623.

Vogel, J.P., Andrews, H.L., Wong, S.K., & Isberg, R.R. (1998). Conjugative transfer by the virulence system of Legionella pneumophila. Science 279, 873–876.

Wright, S.D. & Griffin, F.M.Jr. (1985). Activation of phagocytic cells' C3 receptors for phagocytosis. J. Leukoc. Biol. 38, 327–339.

Wright, S.D., Craigmyle, L.S., & Silverstein, S.C. (1983). Fibronectin and serum amyloid P component stimulate C3b- and C3bi-mediated phagocytosis in cultured human monocytes. J. Exp. Med. 158, 1338–1343.

Zhang, Q., Cox, D., Tseng-C-C, Donaldson, J.G., & Greenberg, S. (1998). A requirement for ARF6 in Fcγ receptor-mediated phagocytosis in macrophages. J. Biol. Chem. 273, 19977–19981.

Zheleznyak, A., & Brown, E.J. (1992). Immunoglobulin-mediated phagocytosis by human monocytes requires protein kinase c activation. J. Biol. Chem. 267, 12042–12048.

Zhu, Z., Bao, Z., & Li, J. (1995). MacMARCKS mutation blocks macrophage phagocytosis of zymosan. J. Biol. Chem. 270, 17652–17655.

Zimmerli, S., Edwards, S., & Ernst, J.D. (1996). Selective receptor blockade during phagocytosis does not alter the survival and growth of *Mycobacterium tuberculosis* in human macrophages. Am. J. Respir. Cell Mol. Biol. 15, 760–770.

SIGNALING THROUGH RHO GTPASES IN PHAGOCYTES

Arie E. Abo

Advances in Cell and Molecular Biology of Membranes and Organelles
Volume 5, pages 215-231.

ISBN: 1-55938-999-0

I. INTRODUCTION

Ras related GTPases act as molecular switches by cycling between the GDP to the GTP-bound form. In resting cells the GTPases are maintained in the inactive GDP-bound form. When cells are stimulated by an extracellular ligand such as platelet-derived growth factor (PDGF), a guanine nucleotide exchange factor (GEF) triggers the release of GDP and, subsequently, the binding of GTP (Bogouski and McCormick, 1993; Hall, 1998). In the active GTP-bound form the GTPases rapidly interact with an effector molecule to initiate biological output. As is implied from their name, GTPases are enzymes that catalyze the hydrolysis of the gamma phosphate on GTP to convert it to GDP. However, the intrinsic GTPase hydrolysis is rather slow and often is catalyzed by a class of enzymes known as GTPase activating proteins (GAPs). By rapid hydrolysis of GTP to GDP the GTPase is brought back to the inactive state and the cycle is complete. The GEFs, GAPs and effectors are specific and selective proteins for each GTPase (Bogouski and McCormick, 1993) (Figure 1).

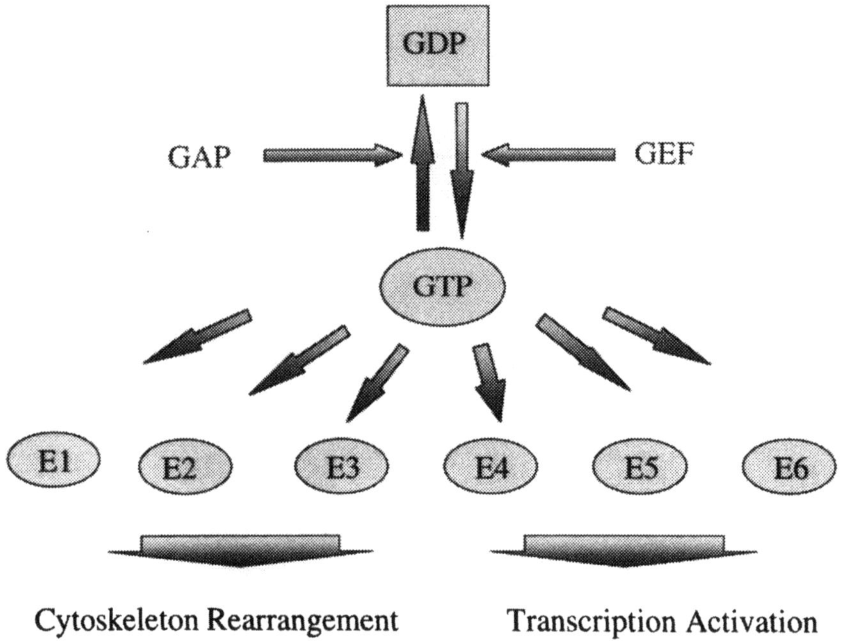

Figure 1. Multiple effectors (E) interact with the activated GTP form of a GTPase to regulate the reorganization of the actin cytoskeleton and to activate gene transcription. The cycle between the GTP- and the GDP-bound state is regulated by GEF and GAP. GEF catalyzes the nucleotide exchange to release GDP and, subsequently, binds GTP. GAP catalyzes the hydrolysis of GTP bound to the GTPase and consequently inactivates the GTPases.

More than 50 different GTPases are known as part of the Ras superfamily. The Ras superfamily is comprised of several subgroups including the Rho GTPases. Rho GTPases include Rho A, B, C, D and E, Rac1, 2, and 3, Cdc42, Chp, and TC10 (Hall, 1998). The sequence identity between the members of the subgroup is approximately 50%, whereas, Rac1 and 2 exhibit more than 90% sequence identity and ~70% sequence homology to Cdc42. The Rho GTPases are ubiquitously expressed except Rac2 which is found only in the hematopoietic cells.

II. REGULATORS FOR RHO GTPASES

A. Upstream from Rho GTPases-GEFs

More than 15 different GEFs for Rho GTPases were identified, however, it is not clear how these GEFs maintain specificity and selectivity (Van Aelst et al., 1998). For example, some of these GEFs can catalyze the nucleotide exchange activity of all of the Rho members *in vitro*, whereas some were reported to work on one member including Lbc for Rho (Zheng et al., 1995) and Vav for Rac (Crespo et al., 1997). All GEFs contain a dbl homolgy (DH) domain adjacent to a plekstrin homology (PH) domain (Cerione and Zheng, 1996). The DH/PH domains are responsible for the nucleotide exchange activity of the GEFs; deletion of this domain abrogates the exchange activity of all GEFs (Cerione and Zheng, 1996).

Since GEFs are the activators of GTPases, it is important to understand how GEFs are regulated in the cells. Various growth factors coupled to tyrosine kinase receptors and hetrotrimeric G-proteins stimulate the activation of Rho GTPases. In phagocytes, the chemoattractant receptor formyl-methionyl-leucyl-phenylalanine (fMLP) and other G-protein coupled receptors, including interleukin (IL)-8 and C5a, initiate signal transduction that is under the control of Rho GTPases. For example, activation of the NADPH oxidase by the fMLP receptor requires the Rac GTPase (Segal and Abo, 1993), whereas reorganization of the actin cytoskeleton is involved in all members of Rho GTPases (Hall, 1998). It was proposed by several groups that PI3 kinase is directly implicated in the activation of exchange factors for Rho GTPases (Hawkins et al., 1995) . Several reports demonstrated that inhibitors for PI3 kinase by wortmannin blocked the activation of Rho signal transduction. Stimulating cells with PDGF caused a dramtic rearrangment of the actin cytoskeleton (Kotani et al., 1994; Wennstrom et al., 1994; Nobes and Hall, 1995). This effect was completely blocked by wortmannin, whereas, cell-injected with activated Rac mutant were insensitive to the inhibitor. The data strongly suggest that Rac is downstream of PI3 kinase and proposed a role for the activation of Rho GEFs by PI3 kinase. Perhaps the most compelling evidence is the inhibition of the NADPH oxidase by PI3 kinase inhibitor wortmannin. Rac activation is indispensable for the activation of the NADPH oxidase (Stephens et al., 1994). It was shown that the total pool of Rac in phagocytes is complexed to a protein known as GDP dissociation inhibitor (GDI) and activation of the oxidase leads to the dissociation of Rac from GDI and subsequently the translocation of Rac to the plasma membrane (Abo et al., 1993). It is possible that the generated lipid product PIP3 by PI3 kinase plays a role in the dissociation of Rac from GDI and activation of specific GEF that can recognize the Rac GDI complex. Recently, a new PI3 kinase (PI3 kinaseγ) isoform was cloned and was shown to be specifically expressed in myloid cells (Hawkins et al., 1997). PI3 kinase γ is comprised of two subunits: the catalytic p110 and regulatory p101. The activity of this enzyme is directly stimulated by the hetrotrimeric G-protein subunits $\beta\gamma$ (Hawkins et al., 1997). Ligating the G-protein couple receptors leads to the dissociation of the $\beta\gamma$ complex from the α subunit and subsequently to the interaction with PI3 kinase γ. An alternative signaling pathway could be transmitted through the α subunit. Upon dissociation from the $\beta\gamma$ complex, the α subunit rapidly exchanges GDP to GTP and subsequently interact with putative effectors. Interestingly, it was demonstrated recently that the Gα12 and Gα13 subunits can bind directly to the Rho GEF p115 to stimulate the activation of Rho signaling (Hart et al., 1998; Kozasa et al., 1998). Similarly, it is possible that the Gαi2, the α subunit that is coupled to most of the serpentine receptors in phagocytes will directly activate a specific Rho GEF (Figure 2).

It is currently unknown which GEFs are implicated in the activation of Rho members in phagocytes. A good candidate for the activation of Rac is the GEF Vav. Vav expression is restricted to the hematopoietic cells and is regulated in T cells by tyrosine phosphorylation and PIP3 (Collins et al., 1997); however no direct connection between Gαi and Vav was reported for the stimulation of Rho GTPasses in phagocytes. Another level of specificity could be mediated by the GEFs. For example, it is conceivable that GEFs will couple GTPase to their targets. In such a scenario, each GEF will participate in forming an activation complex of a GTPase and an effector.

Tyrosine kinase receptors also activate Rho GTPases (Hall, 1998; Van Aelst et al., 1998). The most studied system is phagocytosis where it was shown that Rho GTPases play a central role. During phagocytosis the tyrosine kinase receptor Fcγ is activated to initiate a series of events associated with Rho GTPases (Cox et al., 1996). How tyrosine kinase receptors link a GEF to Rho signaling is poorly understood. Perhaps, the most clearly elucidated system is the activation of Ras by the GEF and PDGF receptors. Activation of these receptors leads to the recruitment of the adapter protein GRB2 complexed to the Ras GEF Sos to stimulate the nucleotide exchange activity of the membrane- bound Ras. It was recently shown that the DH domain found on Sos is responsible for Rac activation. An alternative pathway—the activation of PI3 kinase by Ras—may serve to activate a Rho GEF (Nimnul et al., 1998). A similar mechanism may be conserved in phagocytes.

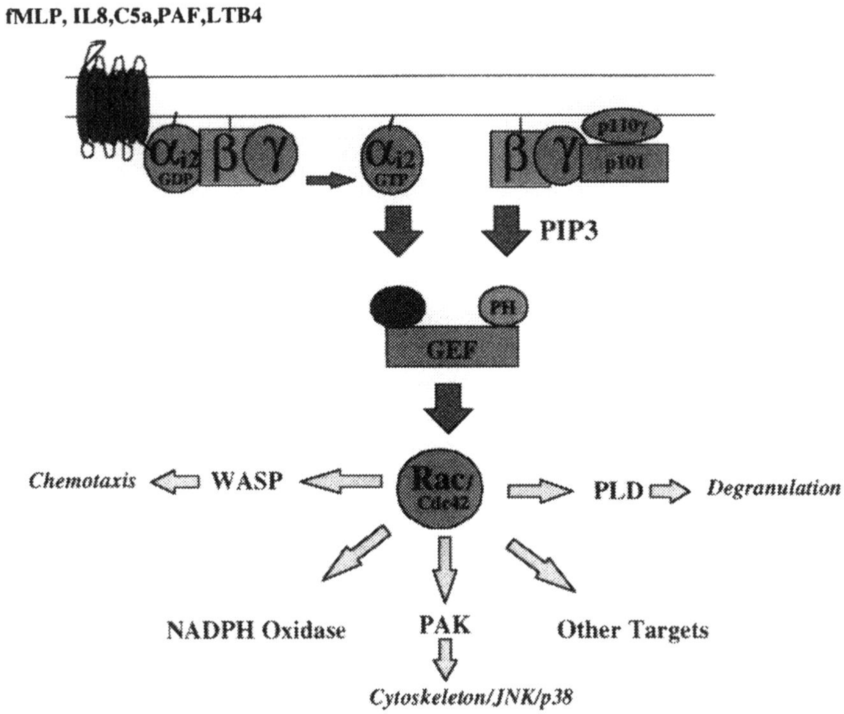

Figure 2. Activation of Rho GTPases and downstream effectors in phagocytes by chemoattractant receptors. According to this model a chemoattractant such as FMLP stimulates the dissociation of the hetrotrimeric Gαi from the βγ subunits to activate downstream cascades. The βγ subunits interact directly with the p101/p110 subunits of the PI3 kinaseγ and consequently activate it to produce PIP3. Subsequently, Rho GTPases are activated either by a direct stimulation of GEF by PIP3, or by Gαi subunits.

B. Mind the GAPs

GAPs for Rho GTPases are proteins that contain the bcr domain that is implicated in the activation of the GTP hydrolysis of the GTP bound to a GTPase (Diekmann et al., 1991). The rate of hydrolysis will determine the length of the signal output from the GTPases. Activated mutants that are deficient in the GTPase hydrolysis and are GAP insensitive generate a sustained signal output. Like GEFs, several GAPs were identified, however the specificity, selectivity, and regulation is poorly understood. The most exciting findings resulted from studies using the bcr knockout mice, a protein which exhibits Rac GAP activity and whose expression is restricted to the hematopoietic lineage. Mice lacking the bcr gene exhibited sustained activation of the NADPH oxidase (Voncken et al., 1995). This data strongly suggest that bcr is the GAP that is responsible for downregulating the NADPH oxidase by activating the GTP hydrolysis of Rac. Similar to GEFs, it is most likely that each GAP is coupled to a GTPase and a specific effector. The bcr knockout data support this model.

III. DOWNSTREAM EFFECTORS—PLENTY OF PARTNERS

How do these GTPases regulate diverse bilogical reponses? The most intuitive way to answer this question was to search for molecular partners that could interact with the activated form of the GTPases with a hope to pull down an effector protein that would directly link the GTPases to a biological output. More than 20 targets were recently identified, including protein and lipid kinases and adapter proteins. However, the link from Rho GTPases via the identified putative targets to a biolgical response including the actin cytoskeleton and transcription activation was not clear-cut. Conflicting findings were reported suggesting that a particular identified effector is or is not involved in the regulation of the actin cytoskeleton and gene regulation. A summary of the known effectors is presented in Figure 3, however, the discussion will be limited to effectors that were shown to be important in phagocyte biology.

A. NADPH Oxidase

The NADPH oxidase is an unique multimeric enzyme expressed in phagocytes and is rapidly activated by several agonists. When activated, the enzyme is assembled on the membrane-bound cytochrome b to catalyze the production of superoxide anion. The superoxide anion is a precursor for many toxic reactive oxygen metabolites that are involved in antimicrobial activity. Three cytosolic components were shown to be indispensable for NADPH oxidase activity including the phagocyte specific components p47-phox, p67-phox and the GTPase Rac (Segal and Abo, 1993). In the GTP-bound state Rac interacts with the N-terminus of p67-phox (Deikmann et al., 1995) and together with p47-phox form a complex with the phagocytes cytochrome b558. In resting cells, Rac is complexed to Rho GDI which keeps the GTPase away from the plasma membrane. Activation of neutrophils with fMLP and phorbol myristate acetate (PMA) caused the dissociation of Rac from GDI followed by its translocation to the plasma membrane (Abo et al., 1993). The other cytosolic components p47-phox and p67-phox contain two SH3 domains and a pxxp motif for binding an SH3 domain. Upon activation the series of SH3 domains serve to form a bridge between the cytochrome b and p47-phox and between p47-phox and p67-phox (De Mendez and Leto, 1996).

It was believed that phagocyte oxidase is an unique enzyme for phagocytic cells. Interestingly, it was recently reported that Rac activates the production of the superoxide in non-phagocytic cells that control mitogenesis (Irani et al., 1997;

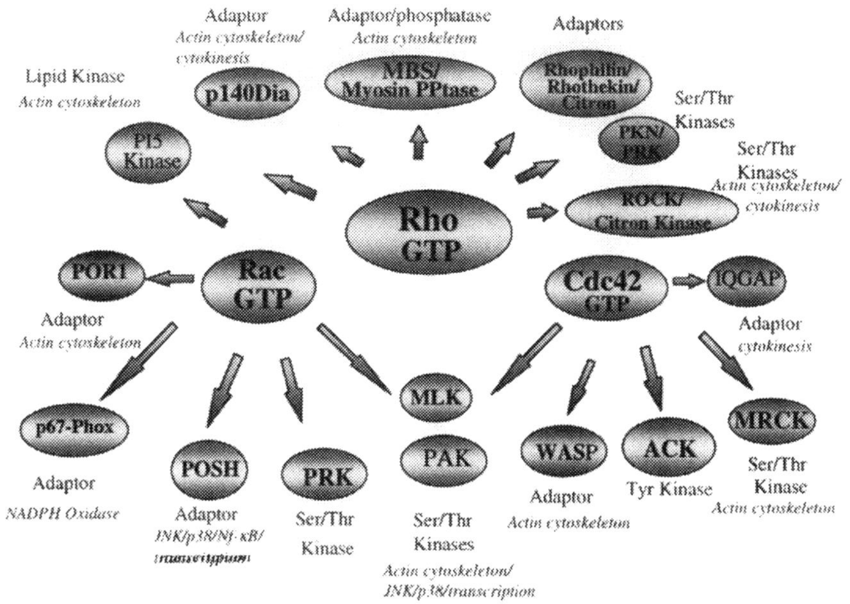

Figure 3. Multiple effectors for Rho GTPases. More than 20 different molecules were shown to interact with Rho GTPases including protein and lipid kinases, phosphatases, and adapter proteins.

Joneson and Bar Sagi, 1998). It remains to be seen if this enzyme will have similarity to the phagocytic oxidase.

B. Wiskott-Aldrich Syndrome Protein

Wiskott-Aldrich Syndrome (WAS) is an X-linked recessive disorder described as a clinical triad of immunodeficiency, eczema, and thrombocytopenia. Patients affected with this syndrome typically suffer from rashes, pyrogenic and opportunistic infections, and bleeding problems. The immunodeficiency in WAS is primarily associated with T and B lymphocyte dysfunction. Defective immune responses in WAS patients include defects in transmembrane signaling of T and B cells (Abo, 1998). In addition, phagocytes of WAS patients exhibit a chemotaxic defect. Recent studies demonstrated impairment in monocyte and macrophage motility (Zicha et al., 1998). The gene encoding the protein that is defective in WAS was recently cloned and was shown to interact with the Rho-like GTPase CDC42Hs through its GTPase binding domain (GBD). WAS proteins (WASPs) interact only with the active GTP-bound form of CDC42Hs and not with other Rho-like proteins (Asperstrom et al., 1996; Kolluri et al., 1996; Symons et al.,

1996). Actin polymerization was induced by the overexpression of WASP and was regulated by CDC42Hs (Symons et al., 1996). The exact role of WASP in the reorganization of the actin cytoskeleton is not clear. Recent studies with mice lacking WASP demonstrated a role for WASP in receptor capping in T cells (Snapper et al., 1998). It is possible that WASP is implicated in building actin structures that are important in transmembrane and receptor signaling. WASP may also play a different role in phagocytes than in T and B cells and may regulate phagocyte motiltity.

C. Phospholipase D

Phospholipase D (PLD) is an enzyme that catalyzes the hydrolysis phosphotidylcholine to generate phosphatidic acid and diacyle glycerol (Morris et al., 1997). Two isozymes were recently cloned and were shown to be ubiquitously expressed (Hammond et al., 1996). PLD was reported to play a central role in membrane trafficking and were shown to regulate exocytosis, degranulation and activation of the NADPH oxidase in neutrophils (Fensome et al., 1998). *In vitro*, PLD1 is activated by the Ras related GTPases Arf and the activity is further enhanced by Rho GTPases Rac, Rho and Cdc42 (Morris et al., 1997).

The precise mechanism through which PLD mediates changes in secretion remains unclear. Arf dependent PLD1 activity was reported to be associated with the exocytosis in neutrophils.

D. p21 Activated Kinase

p21 activated kinases (Paks) are protein kinases that are related to the yeast STE20 and are activated directly by the GTPases Rac and Cdc42. The Pak family consists of four isoforms that all contain the interacting domain for Rac and Cdc42 (known as CRIB/GBD) at the N-terminus and a kinase domain at the C-terminus (Sells and Chernoff, 1997). The kinase domains of Pak1, 2 and 3 are almost identical and the regulatory domains exhibit ~73% sequence homology. Recently a new isoform, Pak4, with a 64% homology to the rest of the Paks, was identified and was shown to selectively interact with Cdc42 (Abo et al., 1998). No homology to other genes was found in the Pak4 regulatory domain outside the CRIB/GBD.

All of the isoforms of Pak were shown to be rapidly activated in neutrophils by a variety of chemoattractants (Knaus et al., 1996; Huans et al., 1998) and can phosphorylate p47-phox (Knaus et al., 1996). However, the role of Paks in neutrophil signaling is currently unknown. A possible role for Pak is in the reorganization of the actin cytoskeleton and chemotaxis.

A role for Paks in the regulation of the actin cytoskeleton was proposed by several groups. Neutrophil endogenous Pak1 and microinjected fibroblast Pak was shown to localize at the lamellipodia (Dharmawardhane et al., 1997; Sells et al., 1997) and at the focal complexes (Manser et al., 1997; Sells et al., 1997). Microinjection of Pak1 protein into Swiss 3T3 cells rapidly induced the formation of lamellipodia and filopodia (Sells et al., 1997). What is puzzling about this data is the fact that kinase inactive Pak and a Pak mutant that lacks the CRIB/GBD domain was also able to induce the cytoskeletal changes. This data suggests that the interaction of Pak1 with Rac and Cdc42 and its kinase activity is not essential for the reorganization of the actin cytoskeleton (Manser et al., 1997; Sells et al., 1997). In contrast, Pak1 kinase activity was shown to be necessary in the regulation of focal complexes turnover (Manser et al., 1997). Constitutively active Pak caused the loss of stress fibers and focal adhesion which is similar to the effect observed by activated Cdc42.

Recent work identified a unique pxxp sequence that is conserved on the Pak1, 2 and 3 regulatory domains and that is responsible for targeting Pak to the focal adhesion complexes (Manser et al., 1998). Exciting work by Manser and colleagues (Manser et al., 1998) recently identified a novel Pak associated protein Pix that contains a SH3 domain and a DH domain that is found in Rho GEFs. Pix interacts through its SH3 domain with the pxxp motif found on the Pak regulatory domain to recruit Pak to the focal adhesion complexes. The Pak/Pix complex is implicated in the stimulation of the nucleotide exchange activity of Rac and consequently leads to the induction of lamellipodia formation. In this scenario, Pak is placed upstream of Rac but downstream of Cdc42 (Manser et al., 1998; Obermeier et al., 1998). If this model is correct it provides an elegant explanation for how Cdc42 and Rac dependent events are intimately coordinated. The first event in the cascade is the activation of Cdc42 which leads to the formation of the peripheral focal complexes, followed by Pix and Pak recruitment and activation of Rac and lamellipodia formation. The cycle ends by activation of Pak kinase activity by .

A good candidate is Pak4 whose role in filopodia formation requires its kinase activity and the interaction with Cdc42 (Abo et al., 1998). Interestingly, Pak4 can bind to CdcL61C40 mutants (a mutant that is implicated in filopodia formation), whereas the other Paks do not. This is in contrast to Pak4 and is consistent with the observation that Pak1 localization to the focal complexes and activation of the Rac pathway are independent of the interaction with Cdc42.

Finally, recent evidence from studies of the budding yeast *Saccharomyces cerevisiae* strongly suggests that Pak kinases regulate the actin cytoskeleton. Yeast lacking the Pak homologues STE20 and CLA4 exhibit cytoskeletal defects that are similar to Cdc42 mutants. The defect was corrected by active STE20 in permeabilized Cdc42 mutant cells (Eby et al., 1998).

E. Chemotaxis Rho GTPases and the Actin Cytoskeleton

Pioneer work by Ridley and Hall in the early 1990s (Ridley and Hall, 1992; Ridley et al., 1992) demonstrated dramatic changes on the actin cytoskeleton in Swiss 3T3 fibroblast cells that were micro-injected with various mutants of Rho GTPases. The activated allele of Rac-V12 induced the assembly of actin filaments at the cell periphery leading to the formation of membrane ruffles and lamellipodia. In contrast, the activated form of Rho stimulated the formation of stress fibers and in more recent studies it was shown that Cdc42 promotes the formation of the actin-rich hairlike protrusions known as filopodia (Kozma et al., 1995; Nobes and Hall, 1995). By expressing the dominant negative mutants of these GTPases it was shown that a heirarchical relationship exists between Cdc42, Rac and Rho (Ridley et al., 1992). Sustained activation of Cdc42 leads to the induction of Rac dependent lamellipodia formation; however, in contrast, within the first few minutes of expression of Cdc42 a rapid induction of filopodia formation is observed (Kozma et al., 1995; Nobes and Hall, 1995). In addition, for the organization of the actin cytoskeleton, Rho GTPases are implicated in the assembly of focal adhesion complexes at the cell surface that are associated with the filopodia and lamellipodia (Nobes and Hall, 1995). It is not clear if these focal complexes serve as nucleating sites for actin that is directly involved in the formation of the protrusion of the cell surface.

The link between Rho GTPases and the regulation of the actin cytoskeleton suggests a potential role for Rho GTPases in the regulation of cell motility. Early work by Vignais and colleagues demonstrated an inhibitory effect on chemoattractant-induced motility in neutrophils treated with the Rho specific inhibitor C3 exoenzyme (Stasia et al., 1991). Expression of Cdc42 mutants in monocytic cells showed that CC chemokines regulate the reorganization of the actin cytoskeleton and are implicated in CC chemokine-induced monocyte migration (Weber et al., 1998). In Bac1 macrophages it was recently shown that Rho regulates cell contraction whereas Rac and Cdc42 regulate the formation of lamellipodia and filopodia. Both Rho and Rac were required for colony-stimulating factor (CSF)-1 induced migration. In contrast, cells expressing the dominant negative Cdc42 mutant were able to migrate but failed to polarize towards the CSF gradient. It was proposed that Rac and Rho are indispensable for cell motility, whereas Cdc42 is important in determining the direction of cell migration (Allen et al., 1998).

The precise mechanism for the regulation of cell motility by Rho GTPases is poorly understood. However, several effector molecules were shown to play a role in reorganization of the actin cytoskeleton.

F. Effectors for Rac and Cdc42 that Are Implicated in the Reorganization of the Actin Cytoskeleton

Since Rac and Cdc42 are closely related (~73% sequence identity) it is not surprising that some of the targets can bind to both Rac and Cdc42. Although several targets were identified to interact with Rac and Cdc42, only some appeared to regulate the actin cytoskeleton. This includes the STE20 related protein kinases, Pak kinases (Pak1, 2, 3 and 4) (Sells and Chernoff, 1997; Abo et al., 1998), IQGAP (a protein that was shown to be implicated in cytokinesis) (Machesky, 1988), myotonic dystrophy kinase (MRCK) (Leung et al., 1998), and the protein that is defec-

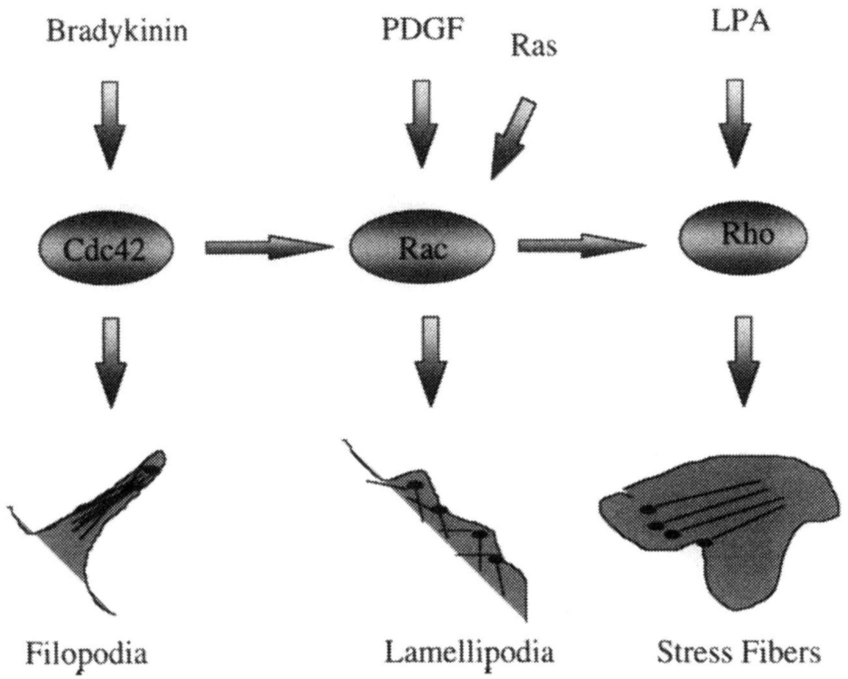

Figure 4. The Rho GTPase cascade and the actin cytoskeleton. In this model, Cdc42 is upstream of Rac and Rho. Activation of Cdc42 by a growth factor such as bradykinin leads to the rapid formation of protrusion structures known as filopodia. Activation of Rac either by a growth factor (PDGF), Ras or directly by Cdc42 leads to the polymerization of actin at the plasma membrane to produce lamellipodia. Rho is directly activated by the growth factor LPA or by Rac to stimulate the assembly of stress fibers. In addition, these GTPases are involved in regulating the assembly of focal complexes (shown as the dark circle that is associated with actin filaments (lines)).

tive in WAS (Abo, 1998). WASP, its relative N-WASP, and their yeast homologue Bee1/Las17 were shown to be implicated in the control of actin polymerization (Abo, 1998). Although N-WASP was recently suggested to induce the formation of filopodia (Miki et al., 1998), several studies suggest a role for WASP in transmembrane signaling and receptor capping by connecting receptors to the actin cytoskeleton (Abo, 1988).

MRCK is a protein kinase that was shown to bind activated Cdc42 to mediate cytoskeletal changes. Coexpression of the dominant negative form of MRCK and Cdc42 blocked the morphological changes associated with Cdc42 (Leung et al., 1998).

Another effector is the lipid kinase PIP5 that is implicated in the phosphorylation of PIP to generate PIP2. Activation of PIP2 production by Rac was shown to be important in uncapping the barbed end of the actin filament (Hartwig et al., 1995).

Several groups used a reconstitution system to identify Cdc42 targets. In a cell-free system, Cdc42 was shown to regulate actin polymerization when it was added to a cell lysate (Zigmond et al., 1997). What remains to be seen is if the use of this system will lead to the isolation and identification of a molecular target for Cdc42 that is responsible for actin polymerization.

G. Rho Effectors and the Actin Cytoskeleton

Several proteins were shown to interact with the activated form of Rho including lipid and protein kinases. Most of the attention was given to p160ROCK, a Ser/Thr kinase (also known as Rok kinase) that can induce stress fibers formation (Leung et al., 1996; Matsuiet et al., 1996). P160ROCK phosphorylates the regulatory subunit of the myosin light-chain phosphatase known as myosin binding subunit (MBS). Phosphorylation of MBS inactivates the phosphatase leading to an increase of phosphorylation of myosin by upstream kinases which results in an increase in actin myosin bundling and thereby the formation of stress fibers (Figure 4) (Kimura et al., 1996).

Other targets include the p140 mDia (Watanbe et al., 1997) and citron kinase which were shown to be involved in cytokinesis (Madaule et al., 1998). Studying the chain of events leading to the reorganization of the actin cytoskeleton will contribute to our understanding of the molecular mechanism of cell locomotion.

IV. A ROLE FOR RHO GTPASES IN PHAGOCYTOSIS

The interaction of opsonized particles with the Fc receptor on the surface of the phagocytes triggers a series of events leading to the engulfment of the particles. The process of phagocytosis is an actin-driven process which requires the reorganization of the actin cytoskeleton to complete the invagination and the budding of the phagosome from the plasma membrane. The link to the actin cytoskeleton strongly suggests that this process is regulated by Rho GTPases. Recent studies demonstrated that Rac and Cdc42 are required for the reorganization of the actin cytoskeleton in macrophages to form a-phagocytic cups which are essential for particle internalization during FcR mediated phagocytosis (Cox et al., 1997). In addition, inactivating Rho by C3 exoenzyme blocked Fcγ receptor-mediated phagocytosis (Hackam et al., 1997). A most recent study reported a two distinct mechanisms of phagocytosis: type I, mediated by the immunoglobilin receptor is controlled by Cdc42 and Rac, and type II, mediated by the complement receptor is regulated by Rho (Caron and Hall, 1998). The molecular effectors that are involved in this process are currently unknown.

V. REGULATION OF GENE TRANSCRIPTION BY RHO GTPASES

The GTPase Ras controls the activation of the MAP kinase cascade by direct stimulation of Raf kinase (Stokoe and McCormick, 1997). Activation of the MAP kinase cascade leads to the activation of the fos transcription factor and subsequently to the induction early genes expression. Similarly, the Rho GTPases were shown to regulate the JNK/stress activated and p38 MAP kinase cascades. Various mutants of Rac and Cdc42 were shown to play a key role in the activatoion of the JNK/p38 pathway by several groups (Coso et al. 1995; Minden et al. 1995). The JNK/p38 pathways are known to control gene transcription in response to cellular stresses including ultra violet light irradiation, osmotic shock, and inflammatory cytokines such as tumor recrosis factor α and IL-1 (Karin, 1998). The exact mechanism by which Rho GTPases activate the JNK/p38 cascade is currently not understood. Pak was proposed by several groups to link Rac and Cdc42 to the JNK/p38 pathway. Constitutively active Pak was shown to activate the JNK/p38 pathway (Bagrodia et al., 1995; Zhang et al., 1995), however, phosphorylation of the downstream kinase MEKK by Pak has not been demonstrated. Recently, it was reported that in neutrophils that C5a induced a sustained activation of all of the known Pak isoforms with no detected activation of the JNK/p38 pathway. This data suggests that Paks are not directly involved in the activation of JNK/p38 pathway. An attractive candidate is the MLK3 kinase that was shown to bind to the active form of Rac and Cdc42 and to stimulate the JNK/p38 pathway.

Rho GTPases were also were shown to be implicated in the activation of the translocation of the transcription factor NFκB (Perona et al., 1996; Sulciner et al., 1996) It is possible that Rho GTPases mediate the phosphorylation of the inhibitory subunit IκB which facilitate its dissociation and subsequently its degradation. Interestingly, it was demonstrated that Rac regulates the activation of the NFκB by cytokines through the induction of superoxide production. The control of these pathways in phagocytes is important during inflammation. For example, p38 was initially discovered as a protein that is rapidly activated during endotoxin shock. These pathways are activated by inflammatory cytokines and play a central role in inflammation.

REFERENCES

Abo, A. (1998). Understanding the molecular basis of Wiskott-Aldrich syndrome. CMLS 54, 1144–1152.

Abo, A., Qu J., Cammarano, M.S., Dan, C., Fritsch, A., Baud, V., Belisle, B., & Minden, A. (1998). PAK4, a novel effector for Cdc42Hs, is implicated in the reorganization of the actin cytoskeleton and in the formation of filopodia EMBO J. 17, 6527–6540.

Abo, A., Webb, M.R., Grogan, A., & Segal, A.W. (1994). Activation of NADPH oxidase involves the dissociation of p21rac from its inhibitory GDP/GTP exchange protein (rhoGDI) followed by its translocation to the plasma membrane. Biochem. J. 15, 585–591.

Allen, W.E., Zicha, D., Ridley, A.J., & Jones, G.E. (1998). A role for Cdc42 in macrophage chemotaxis. J. Cell Biol. 141,1147–1157.

Aspenstrom, P., Lindberg, U., & Hall, A. (1996). Two GTPases, Cdc42 and Rac, bind directly to a proteinimplicated in the immunodeficiency disorder Wiskott-Aldrich syndrome. Curr. Biol. 6, 70–75.

Bagrodia, S., Derijard, B., Davis, R.J., & Cerione, R.A. (1995). Cdc42 and PAK-mediated signaling leads to Jun kinase and p38 mitogen-activated protein kinase activation. J. Biol. Chem. 270, 27995–27998.

Boguski, M.S., & McCormick, F. (1993). Proteins regulating Ras and its relatives. Nature 16, 643–654.

Caron, E., & Hall, A. (1998). Identification of two distinct mechanisms of phagocytosis controlled by different Rho GTPases. Science 282, 1717–1721.

Cerione, R.A., & Zheng, Y. (1996). The Dbl family of oncogenes. Curr. Opin. Cell Biol. 8, 216–222.

Collins, T.L., Deckert, M., & Altman, A. (1997). Views on vav. Immunol. Today 18, 221–225.

Coso, O.A., Chiariello, M., Yu, J.C., Teramoto, H., Crespo, P., Xu, N., Miki, T., & Gutkind, J.S. (1995). The small GTP-binding proteins Rac1 and Cdc42 regulate the activity of the JNK/SAPK signaling pathway. Cell 81, 1137–1146.

Cox, D., Chang, P., Zhang, Q., Reddy, P.G., Bokoch, G.M., & Greenberg, S. (1997). Requirements for both Rac1 and Cdc42 in membrane ruffling and phagocytosis in leukocytes. J. Exp. Med. 186, 487–494.

de Mendez, I., Adams, A.G., Sokolic, R.A., Malech, H.L., & Leto, T.L. (1996). Multiple SH3 domain interactions regulate NADPH oxidase assembly in whole cells. EMBO J. 15, 1211–1220.

Dharmawardhane, S., Sanders, L.C., Martin, S.S., Daniels, R.H., & Bokoch, G.M. (1997). Localization of p21-activated kinase 1 (PAK1) to pinocytic vesicles and cortical actin structures in stimulated cells. J. Cell Biol. 138, 1265–1278.

Diekmann, D., Abo, A., Johnston, C., Segal, A.W., & Hall, A. (1994). Interaction of Rac with p67phox and regulation of phagocytic NADPH oxidase activity. Science 265, 531–533.

Diekmann, D., Brill, S., Garrett, M.D., Totty, N., Hsuan, J., Monfries, C., Hall, C., Lim, L., & Hall, A. (1991). Bcr encodes a GTPase-activating protein for p21rac. 351, 400–402.

Eby, J.J., Holly, S.P., van Drogen, F., Grishin, A.V., Peter, M., Drubin, D.G., & Blumer, K.J. (1998). Actin cytoskeleton organization regulated by the PAK family of protein kinases. Curr. Biol. 27, 967–970.

Fensome, A., Whatmore, J., Morgan, C., Jones, D., & Cockcroft, S. (1998). ADP-ribosylation factor and Rho proteins mediate fMLP-dependent activation of phospholipase D in humanneutrophils. J. Biol. Chem. 273, 13157–13164.

Hackam, D.J., Rotstein, O.D., Schreiber, A., Zhang, W.j., & Grinstein, S. (1997). Rho is required for the initiation of calcium signaling andphagocytosis by Fcgamma receptors in macrophages. J. Exp. Med. 186, 955–966.

Hall, A. (1998). Rho GTPases and the actin cytoskeleton. Science 279, 509–514.

Hammond, S.M., Altshuller, Y.M., Sung, T.C., Rudge, S.A., Rose, K., Engebrecht, J., Morris, A.J., & Frohman, M.A. (1995). Human ADP-ribosylation factor-activated phosphatidylcholine-specific phospholipase D defines a new and highly conserved gene family. J. Biol. Chem. 15, 29640–29643.

Hart, M.J., Jiang, X., Kozasa, T., Roscoe, W., Singer, W.D., Gilman, A.G., Sternweis, P.C., & Bollag, G. (1998). Direct stimulation of the guanine nucleotide exchange activity of p115 RhoGEF by Galpha13. Science 26, 2112–2124.

Hartwig, J.H., Bokoch, G.M., Carpenter, C.L., Janmey, P.A., Taylor, L.A., Toker, A., & Stossel, T.P. (1995). Thrombin receptor ligation and activated Rac uncap actin filament barbed ends through phosphoinositide synthesis in permeabilized human platelets. Cell 82, 643–653.

Hawkins, P.T., Eguinoa, A., Qiu, R.G., Stokoe, D., Cooke, F.T., Walters, R., Wennstrom, S., Claesson-Welsh, L., Evans, T., Symons, M., & Stephenes, L. (1995). PDGF stimulates an increase in GTP-Rac via activation of phosphoinositide 3-kinase. Curr. Biol. 5, 393–403.

Huang, R., Lian, J.P., Robinson, D., & Badwey, J.A. (1998). Neutrophils stimulated with a variety of chemoattractants exhibit rapid activation of p21-activated kinases (Paks): separate signals are required for activation and inactivation of Paks. Mol. Cell. Biol. 18, 7130–7138.

Irani, K., Xia, Y., Zweier, J.L., Sollott, S.J., Der, C.J., Fearon, E.R., Sundaresan, M., Finkel, T., & Goldschmidt-Clermont, P.J. (1997). Mitogenic signaling mediated by oxidants in Ras-transformed fibroblasts. Science 275, 1649–1652.

Joneson, T., & Bar-Sagi, D. (1998). A Rac1 effector site controlling mitogenesis through superoxide production. J. Biol. Chem. 273, 17991–17994.

Joneson, T., McDonough, M., Bar-Sagi, D., & Van Aelst, L. (1996). Stimulation of membrane ruffling and MAP kinase activation by distinct effectors of RAS. Science 274, 1374–1376.

Karin, M. (1998). Mitogen-activated protein kinase cascades as regulators of stress responses. Ann. NY Acad. Sci. 851, 139–146.

Kimura, K., Ito, M., Amano, M., Chihara, K., Fukata, Y., Nakafuku, M., Yamamori, B., Feng, J., Nakano, T., Okawa, K., Iwamatsu, A., & Kaibuchi, K. (1996). Regulation of myosin phosphatase by Rho and Rho-associated kinase (Rho-kinase). Science 273, 245–248.

Knaus, U.G., Morris, S., Dong, H.J., Chernoff, J., & Bokoch, G.M. (1995). Regulation of human leukocyte p21-activated kinases through G protein-coupled receptors. Science 269, 221–223.

Kolluri, R., Tolias, K.F., Carpenter, C.L., Rosen, F.S., & Kirchhausen, T. (1996). Direct interaction of the Wiskott-Aldrich syndrome protein with the GTPase Cdc42. Proc. Natl. Acad. Sci. USA 28, 5615–5618.

Kotani, K., Yonezawa, K., Hara, K., Ueda, H., Kitamura, Y., Sakaue, H., Ando, A, Chavanieu, A., Calas, B., Grigorescu, F., et al. (1994). Involvement of phosphoinositide 3-kinase in insulin- or IGF-1-induced membrane ruffling. EMBO J. 15, 2313–2321.

Kozasa, T., Jiang, X., Hart, M.J., Sternweis, P.M., Singer, W.D., Gilman, A.G., Bollag, G., & Sternweis, P.C. (1998). p115 RhoGEF, a GTPase activating protein for Galpha12 and Galpha13. Science 26, 2109–2111.

Kozma, R., Ahmed, S., Best, A., & Lim, L. (1996). The Ras-related protein Cdc42Hs and bradykinin promote formation of peripheral actin microspikes and filopodia in Swiss 3T3 fibroblasts. Mol. Cell Biol. 16, 5069–5080.

Leung, T., Chen, X.Q., Manser, E., & Lim, L. (1996). The p160 RhoA-binding kinase ROK alpha is a member of a kinase family and is involved in the reorganization of the cytoskeleton. Mol. Cell Biol. 16, 5313–5327.

Leung, T., Chen, X.Q., Tan, I., Manser, E., & Lim, L. (1998). Myotonic dystrophy kinase-related Cdc42-binding kinase acts as a Cdc42 effector in promoting cytoskeletal reorganization. Mol. Cell Biol. 18, 130–140.

Machesky, L. (1998). Cytokinesis: IQGAPs find a function. Curr. Biol. 8, R202–R205.

Madaule, P., Eda, M., Watanabe, N., Fujisawa, K., Matsuoka, T., Bito, H., Ishizaki, T., & Narumiya, S. (1998). Role of citron kinase as a target of the small GTPase Rho in cytokinesis. Nature 394, 491–494.

Manser, E., Huang, H.Y., Loo, T.H., Chen, X.Q., Dong, J.M., Leung, T., & Lim, L. (1997). Expression of constitutively active alpha-PAK reveals effects of the kinase on actin and focal complexes. Mol. Cell Biol. 17, 1129–1143.

Manser, E., Loo, T.H., Koh, C.G., Zhao, Z.S., Chen, X.Q., Tan, L., Tan, I., Leung, T., & Lim, L. (1998). PAK kinases are directly coupled to the PIX family of nucleotide exchange factors. Mol. Cell 1, 183–192.

Miki., H., Sasaki, T., Takai, Y., & Takenawa, T. (1988). Induction of filopodium formation by a WASP-related actin-depolymerizing protein N-WASP. Nature 391, 93–96.

Minden, A., Lin, A., Claret, F.X., Abo, A., & Karin, M. (1995). Selective activation of the JNK signaling cascade and c-Jun transcriptional activity by the small GTPases Rac and Cdc42Hs. Cell 81, 1147–1157.

Morris, A.J., Hammond, S.M., Colley, C., Sung, T.C., Jenco, J.M., Sciorra, V.A., Rudge, S.A., & Frohman, M.A. (1997). Regulation and functions of phospholipase D. Biochem. Soc. Trans. 25, 1151–1157.

Nimnual, A.S., Yatsula, B.A., & Bar-Sagi, D. (1998). Coupling of Ras and Rac guanosine triphosphatases through the Ras exchanger Sos. Science 279, 560–563.

Nobes, C.D., & Hall, A. (1995). Rho, rac, and cdc42 GTPases regulate the assembly of multimolecular focal complexes associated with actin stress fibers, lamellipodia, and filopodia. Cell 81, 53–62.

Nobes, C.D., Hawkins, P., Stephens, L., & Hall, A. Activation of the small GTP-binding proteins rho and rac by growth factor receptors. J. Cell. Sci. 108, 225–233.

Obermeier, A., Ahmed, S., Manser, E, Yen, S.C., Hall, C., & Lim, L. (1998). PAK promotes morphological changes by acting upstream of Rac. EMBO J. 17, 4328–4339.

Perona, R., Montaner, S., Saniger, L., Sanchez-Perez, I., Bravo, R., & Lacal, J.C. (1997). Activation of the nuclear factor-kappaB by Rho, CDC42, and Rac-1 proteins. Genes Dev. 11, 463–475.

Ridley, A.J., & Hall, A. (1992). The small GTP-binding protein rho regulates the assembly of focal adhesions and actin stress fibers in response to growth factors. Cell 70, 389–399.

Ridley, A.J., Paterson, H.F., Johnston, C.L., Diekmann, D., & Hall, A. (1992). The small GTP-binding protein rac regulates growth factor-induced membrane ruffling. Cell 70, 401–410.

Segal A.W., & Abo, A. (1993). The biochemical basis of the NADPH oxidase of phagocytes. Trends Biochem. Sci. 18, 43–47.

Sells, M.A., & Chernoff, J. (1997). Emerging from the Pak: the p21 activated kinase family. Trends Cell Biol. 7, 162–167.

Sells, M.A., Knaus, U.G., Bagrodia, S., Ambrose, D.M., Bokoch, G.M., & Chernoff, J. (1997). Human p21-activated kinase (Pak1) regulates actin organization in mammalian cells. Curr. Biol. 7, 202–205.

Snapper, S.B., Rosen, F.S., Mizoguchi, E., Cohen, P., Khan, W., Liu, C.H., Hagemann, T.L., Kwan, S.P., Ferrini, R., Davidson, L., Bhan, A.K., & Alt, F.W. (1998). Wiskott-Aldrich syndrome protein-deficient mice reveal a role for WASP in T but not B cell activation. Immunity 9, 81–91.

Stasia, M.J., Jouan, A., Bourmeyster, N., Boquet, P., & Vignais, P.V. (1991). ADP-ribosylation of a small size GTP-binding protein in bovine neutrophils by the C3 exoenzyme of Clostridium botulinum and effect on the cell motility. Biochem. Biophys. Res. Commun. 180, 615–622.

Stephens, L., Smrcka, A., Cooke, F.T. Jackson, T.R., Sternweis, P.C., & Hawkins, P.T. (1994). A novel phosphoinositide 3 kinase activity in myeloid-derived cells is activated by G protein beta gamma subunits. Cell 8, 83–93.

Stephens., L.R., Eguinoa, A., Erdjument-Bromage, H., Lui, M., Cooke, F., Coadwell, J., Smrcka, A.S., Thelen, M., Cadwallader, K., Tempst, P., & Hawkins, P.T. (1997). The G beta gamma sensitivity of a PI3K is dependent upon a tightly associated adaptor, p101. Cell 4, 105–114.

Stokoe, D., & McCormick, F. (1997). Activation of c-Raf-1 by Ras and Src through different mechanisms: activation in vivo and in vitro. EMBO J. 16, 2384–2396.

Sulciner, D.J., Irani, K., Yu, Z.X., Ferrans, V.J., Goldschmidt-Clermont, P., & Finkel, T. (1996). rac1 regulates a cytokine-stimulated, redox-dependent pathway necessary for NF-kappaB activation. Mol. Cell. Biol. 16, 7115–7121.

Symons, M., Derry, J.M., Karlak, B., Jiang, S., Lemahieu, V., Mccormick, F., Francke, U., & Abo, A. (1996). Wiskott-Aldrich syndrome protein, a novel effector for the GTPase CDC42Hs, is implicated in actin polymerization. Cell 84, 723–734.

Van Aelst, L., & D'Souza-Schorey, C. (1997). Rho GTPases and signaling network. Genes Dev. 15, 2295–2322.

Voncken, J.W., van Schaick, H., Kaartinen, V., Deemer, K., Coates, T., Landing, B., Pattengale, P., Dorseuil, O., Bokoch, G.M., Groffen, J., et al. (1995). Increased neutrophil respiratory burst in bcr-null mutants. Cell 80, 719–728.

Watanabe, N., Madaule, P., Reid, T., Ishizaki, T., Watanabe, G., Kakizuka, A., Saito, Y., Nakao, K., Jockusch, B.M., & Narumiya, S. (1997). p140mDia, a mammalian homolog of Drosophila

diaphanous, is a target protein for Rho small GTPase and is a ligand for profilin. EMBO J. 16, 3044–3056.
Weber, K.S., Klickstein, L.B., Weber, P.C., & Weber, C. (1998). Chemokine-induced monocyte transmigration requires cdc42-mediated cytoskeletal changes. Eur. J. Immunol. 28, 2245–2251.
Wennstrom, S., Hawkins, P., Cooke, F., Hara, K., Yonezawa, K., Kasuga, M., Jackson, T., Claesson-Welsh, L., & Stephens, L. (1994). Activation of phosphoinositide 3-kinase is required for PDGF-stimulated membrane ruffling. Curr. Biol. 4, 385–393.
Yang, N., Higuchi, O., Ohashi, K., Nagata, K., Wada, A., Kangawa, K., Nishida, E., & Mizuno, K. (1998). Nature 393, 809–812.
Zhang, S., Han, J., Sells, M.A., Chernoff, J., Knaus, U.G., Ulevitch, R.J., & Bokoch, G.M. (1995). Rho family GTPases regulate p38 mitogen-activated protein kinase through the downstream mediator Pak1. J. Biol. Chem. 270, 23934–23936.
Zheng, Y., Zangrilli, D., Cerione, R.A., & Eva, A. (1996). The pleckstrin homology domain mediates transformation by oncogenic dbl through specific intracellular targeting. J. Biol. Chem. 9, 19017–19020.
Zicha, D., Allen, W.E., Brickell, P.M., Kinnon, C., Dunn, G.A., Jones, G.E., & Thrasher, A.J. (1998). Chemotaxis of macrophages is abolished in the Wiskott-Aldrich syndrome. Br. J. Haematol. 101, 659–665.
Zigmond, S.H., Joyce, M., Borleis, J., Bokoch, G.M., & Devreotes, P.N. (1997). Regulation of actin polymerization in cell-free systems by GTPgammaS and Cdc42. J. Cell Biol. 138, 363–374.

REGULATORY ROLES OF PHOSPHATIDYLINOSITOL (4,5) BISPHOSPHATE IN CELL SIGNALING, MEMBRANE TRAFFIC, AND THE CYTOSKELETON

Shamshad Cockcroft

Advances in Cell and Molecular Biology of Membranes and Organelles
Volume 5, pages 233-263.

ISBN: 1-55938-999-0

I. INTRODUCTION

Phosphatidylinositol (4,5) bisphosphate (PIP_2) is a key player in diverse aspects of cell function although it is a minor component of the total cellular lipids (~0.4%). It is a phosphorylated product of phosphatidylinositol (PI) which represents 5–8% of the total cellular phospholipid pool. The first role for PIP_2 was identified in the early 1980s when it was established that hydrolysis of PIP_2 by phospholipase C (PLC) simultaneously generated two second messengers, inositol (1,4,5) trisphosphate (IP_3) and diacylglycerol (DG) (Figures 1 and 2) (Berridge, 1993). IP_3 mobilizes intracellular calcium by opening calcium channels at the endoplasmic reticulum and DG activates many members of the protein kinase C (PKC) family. In the late 1980s, evidence was obtained showing that PIP_2 could be phosphorylated at the D-3 position of the inositol ring to produce phosphatidylinositol (3,4,5) trisphosphate (PIP_3) (see Figures 1 and 2). The signaling functions of the 3-phosphorylated inositol lipids are just beginning to be unraveled, and have emerged as important in recruiting proteins to membrane compartments via specific protein-lipid interactions (Toker and Cantley, 1997). This is due to the presence of the pleckstrin homology (PH) domain found in over 100 signaling proteins which can interact with high specificity with either PIP_2/PIP_3 (Shaw, 1996; Lemmon et al., 1997).

Further roles for inositol lipids have emerged from studies where PIP_2 itself was identified as a cofactor for phospholipase D (PLD), a modulator of many cytoskeletal proteins and as a participant in the exocytotic process (Figure 1). This review focuses on the regulation of PIP_2 metabolism by the PLCs and phosphoinositide 3-kinases and on the more recent studies which identify the intact lipid as a signaling molecule. A major emphasis in this review is given to how cells can coordinate the availability of PIP_2 for the different cellular functions which include consumption by PLCs and phosphoinositide 3-kinases on one hand and a requirement for an increase on the other (PLD, exocytosis and cytoskeletal dynamics). The involvement of phosphoinositides in many aspects of phagocytosis can be predicted based on what we know about the function of these lipids.

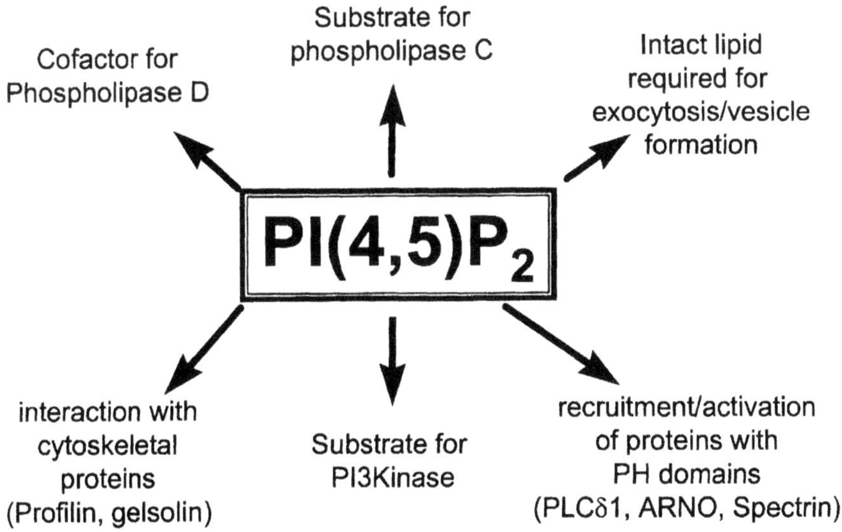

Figure 1. Multiple functions of PIP_2. PIP_2 is a substrate for generating second messengers and is also required as an intact lipid.

Phagocytosis can be defined as the interaction of particles with cell surface receptors that initiates signaling pathways that ultimately lead to pseudopod extension, and the engulfment of the particle. This is followed by fusion of secretory granules for the eventual digestion of the ingested particle. Phosphoinositides are likely to play major roles at several steps in the phagocytic processes, not only as intact lipid signaling molecules but as substrates for PLC and phosphoinositides 3-kinase.

The phagocytic particle delivers signals that mediate the assembly of filamentous actin. These signals are mediated via the Syk family of tyrosine kinases (Greenberg et al., 1993, 1996), which mediate the downstream activation of PLCγ1 and phosphoinositide 3-kinase. Activation of phosphoinositide 3-kinase and PLCγ1 will generate several second messengers important for intracellular signaling and will result in a drop in PIP_2 levels. On the other hand, phagocytosis requires the coordinated organization of the cytoskeleton and membrane fusion, processes which are dependent on phosphoinositides. The key to this puzzle must lie in spatial and temporal availability of these lipids for them to participate in the cytoskeletal reorganization and in the exocytic process, and at the same time be the sources of several second messengers.

Another lipid signaling pathway that is tightly coupled to phagocytosis is the activation of PLD (della Bianca et al., 1991; Fallman et al., 1992; Kusner et al., 1996; Serrander et al., 1996). Activation of tyrosine kinases is required for the downstream stimulation of PLD and PLD constitutes an early signal prior to actin reorganization (Kusner et al., 1996). Tyrosine kinases are also upstream of the activation of small GTPases such as Rac, Rho, Cdc42 and ARF. These GTPases are essential requirements for phagocytosis (Cox et al., 1997; Hackam et al., 1997). These GTPases could coordinate the reorganization of the cytoskeleton and fusion of membranes by an interplay of PIP_2, PIP_3 and phosphatidic acid (PA), in concert with other downstream effectors of these GTPases (Hall, 1998). The activation of these GTPases is dependent on exchange factors and it is notable that in all guanine nucleotide exchange factors (GEFs), the Sec7 or the Dbl domain is immediately followed by a PH domain. On the other hand, these small GTPases can regulate the synthesis of PIP_2. It thus appears that inositol lipids will operate upstream of the GTPases as well as downstream of the GTPases.

II. PHOSPHOLIPASE C

The first enzyme that was found to utilize PIP_2 as substrate was PLC (Figure 2). PLC is now firmly established as a lipid signaling transduction pathway coupled to many cell-surface receptors. These include G-protein-coupled receptors and receptors that are tyrosine kinases or that regulate tyrosine kinases. Regardless of the kind of receptor that is stimulated, the output is always the production of the second messengers, IP_3 and DG. By definition, this signaling event occurs at the plasma membrane where the receptor initially encounters its ligand. The possibility that the nucleus is an additional site for PLC signaling has recently emerged where the physical presence of PLC enzymes has been identified as well as the machinery to synthesize the substrate for the PLC (Martelli et al., 1992; Divecha and Irvine, 1995). Nuclear signaling remains a tantalizing puzzle since the molecular mechanism of how signals transmitted at the plasma membrane cross the cell interior remains ill-defined. The best studied agonist that is known to stimulate signaling at the nucleus is insulin-like growth factor (IGF-I) (Martelli et al., 1992).

A. Structural Relationships among Specific Phospholipase Enzymes

The existence of multiple forms of PLC enzymes became abundantly clear from purification of the first mammalian enzymes and subsequently molecular cloning. Currently, the PLC family has 10 mammalian members which are sub-divided into three families, designated PLCβ (1–4), PLCγ (1 and 2), and PLCδ (1–4) (Rhee and Bae, 1997). The general relationship of these three families to one another are presented schematically in Figure 3. The two most highly conserved domains have been designated X and Y and form the catalytic core (Essen et al., 1996). Other shared regions of homology are the N-terminal PH domain, 4 EF hands and a C2 domain. The PLCδ enzymes are the smallest and are composed essentially of the common regions, with the N-terminal PH domain, 4 EF hands, the catalytic domain and a C2 domain. Homologues are present in lower eukaryotes including

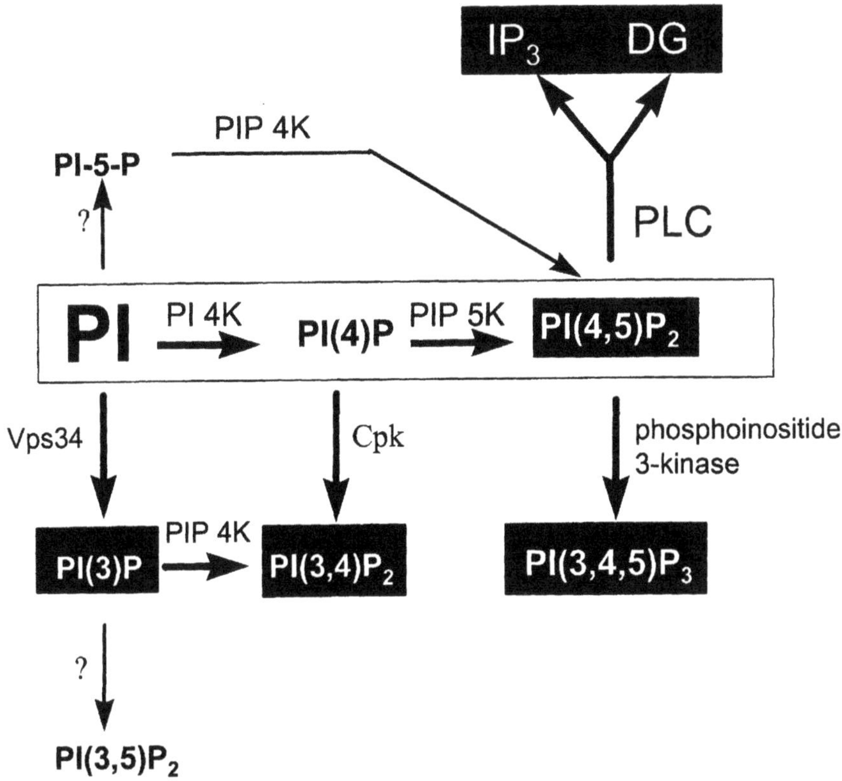

Figure 2. Metabolic pathways involving inositol lipids. The enzymes responsible for phosphorylation of the inositol ring are indicated. The molecules in boxes have been identified as having signaling functions.

yeasts and *Dictyostelium discoideum* as well as plants (Drayer and van Haastert, 1992; Yoko-O et al., 1993; Hirayama et al., 1995). The evolution of PLCs has preceded the evolution of IP_3 receptors since many of the lower organisms have a PLCδ like activity but no IP_3 receptors. The function of PLCδ-like enzymes in lower organisms indicates a role in sensing the environment and may be related to the use of inositides as an intact molecule for signaling purposes. Lower organisms have PIP_2-dependent PLD activity as well as all the elements of the cytoskeletal machinery. In addition, disruption in the genes encoding lipid kinases and phosphatases lead to many membrane trafficking defects indicating a central role for PIP_2 as an intact lipid in membrane trafficking (Garcia-Bustos et al., 1994; de Camilli et al., 1996; Srinivasan et al., 1997).

PLCβ and γ isoforms have arisen by acquisition of additional domains that impart unique regulatory properties. The PLCβ family are regulated by G-proteins

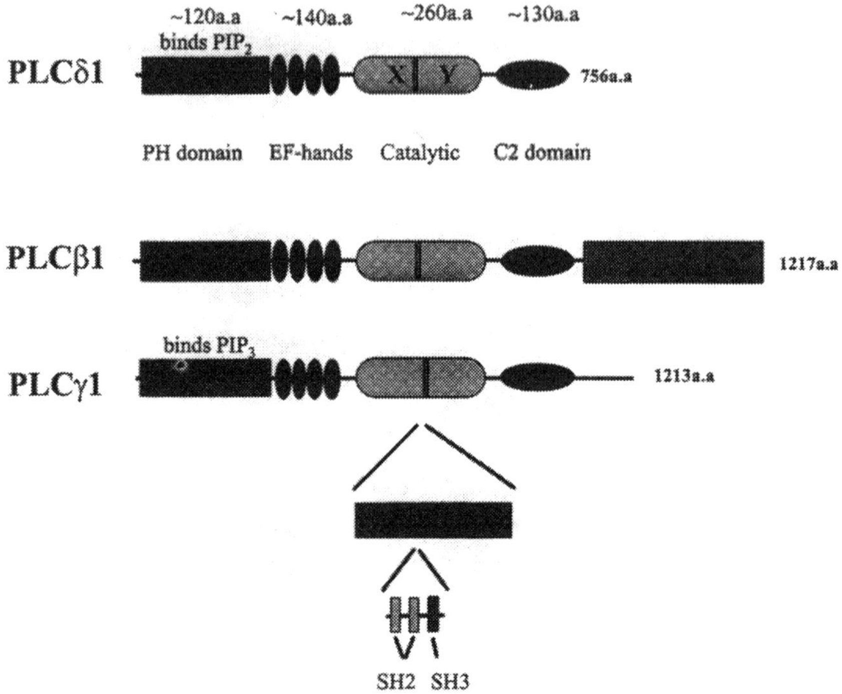

Figure 3. Domain structure of the PLC family. All members contain the highly homologous X and Y domains which constitute the catalytic domain, an N-terminal PH domain, four EF hands and a C2 domain. PLCγ1 also contains two SH2 and an SH3 domain inserted within a second PH domain.

and contain an extended carboxy terminus of about 500 amino acids that contain regions for membrane association and regulation by G-proteins. In contrast, the PLCγ enzymes have a large region inserted between the X and Y domains. The insert contains two SH2 domains and one SH3 domain and a split PH domain. SH2 domains bind to phosphotyrosine residues in a sequence-specific context while the SH3 domains bind to proline-rich residues.

B. Regulation of Mammalian Phospholipase C Enzymes

A role for G-proteins in the regulation of PIP_2 hydrolysis was anticipated when early studies revealed that guanine nucleotides could activate PLC activity (Cockcroft, 1987). In addition, PLC activation by some hormone receptors could be inhibited by pertussis toxin pretreatment, a toxin that was known to ADP-ribosylate Gi proteins and thus uncouple the G-proteins from their receptors. PLCβ enzymes are now known to be targets of heterotrimeric G-proteins. The GTP-bound α-subunit of the Gq family directly interacts with the carboxyl-terminal region of PLCβ (Park et al., 1993; Wu et al., 1993). In contrast Gβγ subunits derived from the Gi family of G-proteins interact with the amino-terminal region that also contains the PH domain preceding the X region (Lee et al., 1993).

The sensitivity of the PLCβ enzymes to Gαq and Gβγ subunits is different. Whereas Gαq subunits activate PLCβ isozymes according to the hierarchy of PLCβ1>PLCβ3>PLCβ4>PLCβ2, the sensitivity to Gβγ subunits decreases in the order of PLCβ3>PLCβ2>PLCβ1 (Rhee and Bae, 1997; Singer et al., 1997). These results suggest that cell-specific expression of PLC isozymes and G-protein subunits contributes to the diversity of the type and magnitude of the observed enzyme-mediated responses.

Regulation of PLCγ1 enzymes is controlled by receptors that regulate intrinsic or extrinsic tyrosine kinases. Polypeptide growth factor receptors such as epidermal growth factor (EGF), platelet derived growth factor (PDGF), and NGF are known to stimulate the hydrolysis of PIP_2 by activating PLCγ1 in a wide variety of cells. Dimerization of the receptor upon activation activates the intrinsic tyrosine kinase activity and leads to phosphorylation of target proteins including PLCγ1 and phosphoinositide 3-kinase and the receptor itself. The autophosphorylated receptor forms a docking site for the SH2 domain of PLCγ1 and so recruits the enzyme into a signaling complex. Phosphorylation of PLCγ1 by all growth factor receptors occurs at identical sites, tyrosine residues 771, 783 and 1254. Of these, phosphorylation of Tyr-783 was found to be essential for PLCγ1 activation (Kim et al., 1991).

Non-receptor protein tyrosine kinases also phosphorylate the same tyrosine residues as the growth factor receptor tyrosine kinase and activate PLCγ enzymes in response to the ligation of certain cell surface receptors. This includes the T cell antigen receptor, the membrane IgM receptor, the high affinity IgE receptor, and the IgG receptor (Rhee and Bae, 1997). These receptors, which comprise multiple polypeptide chains, are not themselves protein tyrosine kinases, but they activate a wide variety of non-receptor protein tyrosine kinases, such as the members of the src and Syk families. Some of these kinases have been known to associate with PLCγ1 and PLCγ2. PLCγ2 is mainly expressed in hematopoetic cells and is also phosphorylated at multiple tyrosine residues by non-receptor protein tyrosine kinases of the src family (Bianchini et al., 1993; Liao et al., 1993; Dusi et al., 1994; Atkinson and Yang, 1996; Kawakami et al., 1996).

The regulation of PLCδ1 by cell surface receptors has proved to be difficult to resolve. A novel p122 Rho-GAP (Homma and Emori, 1995) and an atypical G-protein, Gh which is a 69 kDa transglutaminase II, have been reported to activate this PLC (Feng et al., 1996). Further studies will be necessary to underpin the physiological significance of these observations. Another regulator of PLCδ1 that has emerged recently is Ca^{2+} (Allen et al., 1997). One of the paradoxes is that in many excitable cells, an increase in cytosol Ca^{2+} can lead to the hydrolysis of PIP_2 (Cockcroft and Thomas, 1992). The possibility that PLCδ1 may be responsible for this has been examined in permeabilized cells. In cytosol-depleted cells, when cytosol Ca^{2+} is titrated between 100 nM–10 μM, addition of PLCγ1 and PLCβ1 alone have no effect but this increase is sufficient for PLCδ1 to be active (Allen et al., 1997). From the structural analysis of PLCδ1, the C2 domain of PLCδ1 is the likely sensor for Ca^{2+} (Williams and Katan, 1996; Katan and Williams, 1997). Although the C2 domain is conserved in PLCγ1 and PLCβ1, they appear to require additional inputs for activation; Ca^{2+} only enhances their activity subsequently.

C. PH Domains

All PLCs have PH domains and the function of these PH domains in PLCs reveal novel features concerning PLC recruitment to membranes. Phosphoinositides can bind specifically to proteins, and a module that fulfils this function is the PH domain (Musacchio et al., 1993; Harlan et al., 1994; Lemmon et al., 1997). PH domains were first identified as protein modules found in pleckstrin, the major PKC substrate in platelets. They contain approximately 120 amino acids and are found in many proteins involved in signaling including PLCs, protein kinases and proteins that act as exchange factors for small G-proteins (see Table 2). Despite their relatively low sequence homology, the three-dimensional structure of the PH domains are conserved (Zhou et al., 1995).

Different PH domains have specificity for binding to particular phosphoinositides. PH domains can be broadly divided into two major categories: those that bind specifically to PIP_3 with high affinity and specificity and those that have only slight or no selectivity for PIP_3 over PIP_2. An example of a PH domain that binds PIP_3 with high affinity is that of PLCγ1 (Falasca et al., 1998) and a PH domain that bind PIP_2/PIP_3 with no selectivity is PLCδ1. Since PIP_2 is more abundant than PIP_3 *in vivo* (Carpenter and Cantley, 1996), only PH-domain-containing proteins with high affinity for PIP_3 will be regulated by products of phosphoinositide 3-kinase *in vivo* and others will be regulated by PIP_2.

D. The Function of PH Domains in Phospholipase C

The N-terminus of all PLCs have PH domains. The PH domain of PLCδ1 binds PIP_2 and is necessary for the binding of the enzyme to the plasma membrane (Ferguson et al., 1995; Paterson et al., 1995). The removal of the PH domain of PLCδ1 leads to no loss of catalytic activity when substrate is provided in mixed micelles but renders the enzyme inactive for hydrolysis of endogenous substrate in the cellular membrane (Allen et al., 1997). Thus, anchoring of PLCδ to the membranes via its PH domain is crucial for subsequent activation with Ca^{2+}. The attachment of PLCδ1 to membranes must be dynamic as permeabilization of cells does lead to loss of PLCδ1 in the extracellular medium (Allen et al., 1997).

In contrast to PLCδ1, the PH domain of PLCγ1 binds to $PI(3,4,5)P_3$ (Falasca et al., 1998). It is noteworthy that signaling via receptor tyrosine kinases frequently results in the concomitant activation of PLCγ and phosphoinositide 3-kinase. Thus activation of PLCγ can be coordinated to the activation of phosphoinositide 3-kinase in many cases but not all. There are examples where PLCγ activation is not accompanied by phosphoinositide 3-kinase activation and vice versa. EGF stimulates PLCγ activity but not phosphoinositide 3-kinase activity in A431 cells. Likewise CSF stimulates phosphoinositide 3-kinase activity but not PLCγ1 activity. The product of the phosphoinositide 3-kinase, $PI(3,4,5)P_3$ is apparently required for targeting PLCγ1 to the plasma membrane (Falasca et al., 1998). Both targeting to membranes and tyrosine phosphorylation of PLCγ1 appears to be required for increase in PIP_2 hydrolysis since mutations in the PH domain or interference in phosphoinositide 3-kinase activation led to inhibition of IP_3 production. There is a PH domain in PLCβ but there is no data indicating what its ligand is. In addition to phosphoinositides, PH domains can also interact with Gβγ (Touhara et al., 1994; Pitcher et al., 1995), and further studies will be required to address the issue of specificity for Gβγ and/or phosphoinositides.

E. Substrate Provision to Phospholipases Reveals a Requirement for a Cytosolic Protein, PITP

Several lines of evidence suggest that the synthesis of PIP_2 is tightly coupled to the receptor-regulated activity of PLC and that regulation of synthesis may be of equal importance for the generation of second messengers. Such a tight coupling may be of paramount importance in view of the multiple regulatory roles that have been identified for PIP_2. One protein that has been identified as a central component in substrate provision for the phospholipases is the cytosolic protein, phosphatidylinositol transfer protein (PITP) (Cockcroft, 1998).

Two related isoforms, PITPα and PITPβ, which are 94% homologous, have been described in mammalian cells. Its requirement in PLC signaling was uncovered when it was observed that activation of PLCβ by the G-protein-coupled receptor agonist, formylmethionylleucylphenylalanine (fMLP) was disrupted in cytosol-depleted HL60 cells. Cytosol-depleted cells retained all the known signaling components; the receptor, the relevant G-protein and the PLC. Addition of cytosol fully restored the ability of fMLP to stimulate the generation of IP_3 and PITP was identified as the major component in the cytosol responsible for reconstituting activity (Thomas et al., 1993; Cunningham et al., 1995). A second activity was identified as PLCβ. PLCβ and PITP cooperate in the production of IP_3 in fMLP-stimulated cells (Cunningham et al., 1995). The function of PITP in PLC signaling was suggested to be due to the ability of the protein to transfer PI from its intracellular site of synthesis, the endoplasmic reticulum, to the plasma membrane, the site of PLC-mediated hydrolysis of PIP_2 (Thomas et al., 1993). In addition, further studies revealed that the conversion of PI to PIP_2 was also dependent on PITP and the concept that PITP provided the PI directly to PI 4-kinase for phosphorylation at signaling sites was put forward (Cunningham et al., 1995).

Experimental evidence to support this concept came from two separate sources. Firstly, PITP was non-essential when GTPγS is used as an activator of G-proteins. GTPγS directly activates G-proteins escaping the control exerted by exchange factors, in this case the receptor. However, PITP substantially enhanced GTPγS-stimulated PLC hydrolysis suggesting that PITP was increasing the availability of the precursor PI to sites of PLC activation. When FMLP was used as an activator, PITP was an absolute requirement for IP_3 production. An interpretation of these results was that FMLP-stimulated PLC hydrolysis was critically dependent on the substrate supplied by PITP in the vicinity of the activated receptor and that this could be due to a coordinated recruitment of PITP to signaling sites.

Further evidence to support this concept came from studies with the EGF receptor. PITP was found to be essential for the activated EGF receptor on A431 cells to produce IP_3 (Kauffmann-Zeh et al., 1995). EGF regulates the activity of PLCγ1. Occupation of the EGF receptor increased the activity of PI 4-kinase and the consequent phosphorylation of PI to PIP was critically dependent on PITP. Additionally a multimeric complex which included the EGF receptor, PI 4-kinase, PLCγ and PITP could be induced following activation of the EGF receptor. All these data suggest that PITP is not a passive transporter of PI to plasma membranes but is recruited to the signaling sites upon stimulation by cell surface receptors. The most compelling evidence was obtained using deletion mutants of PITP. We noted that proteolytic cleavage of PITP could remove 24 amino acids from the C-terminus thereby inactivating the protein, both for *in vitro* lipid transfer and for PLC reconstitution (Prosser et al., 1997). We therefore engineered deletion mutants whereby we progressively removed 5 amino acid residues at a time (Hara et al., 1997). Removal of 5 amino acids was sufficient to inactivate the PLC reconstituting activity but retain transfer activity. In addition, this deletion mutant could inhibit the function of the wild-type protein indicating that they block the recruitment of PITP to their docking sites.

PITP, by its ability to bind and transport PI from one membrane compartment to another is ideally suited to provide the substrate, PI to the lipid kinases for phosphorylation in the vicinity of the activated receptor. PI synthesis occurs at the endoplasmic reticulum and this is where the concentration of PI is greatest. In principle, transport of PI to other membranes can take one of two forms, vesicular transport or transfer via lipid transport proteins. PITPα and PITPβ can bind one molecule of phospholipid and its binding site can accommodate either PI or PC. This dual specificity allows PITP to act as a supplier of PI to sites of PIP_2 consumption such as in PLC hydrolysis. The supply of PI to signaling domains for PIP_2 synthesis by the lipid kinases must be coordinately regulated to match the receptor-mediated hydrolysis of PIP_2. In a stimulated cell, the increase in IP_3 can exceed the resting pool of PIP_2 several times (Cunningham et al., 1995; Willars et al., 1998) and the combination of PITP and the lipid kinases work together to ensure that the cellular levels of PIP_2 are not depleted despite ongoing hydrolysis. In most cells, acute stimulation by an agonist leads to a very transient drop in PIP_2 levels but within seconds the levels of PIP_2 recover despite ongoing hydrolysis.

How PITP is recruited to these signaling sites has not yet been elucidated but the prediction is that the lipid kinases may engage PITP only under stimulated conditions. It has been found that antigen stimulation of RBL-mast cells leads to the active translocation of PITP from the cytosol to a membrane compartment (Way and Cockcroft, 1998). Additional regulatory steps exist in the synthesis of PIP_2. The lipid kinases belong to a multigene family and some members of the PIP 5-kinase family are regulated enzymes.

F. Regulation of the Synthesis of PIP_2

In the unstimulated cell, the proportion of PI that is in the phosphorylated state is ~10–12%. Approximately half is present as PIP and the remainder as PIP_2. What defines the upper limit of PIP_2 levels in resting cells is not known but the presence of the lipid kinases and phosphatases in the appropriate membrane compartments will dictate the phosphorylation status of PI. It is interesting to note that in the human erythrocyte where the only membrane is the plasma membrane, the majority of the inositol lipids are present in the phosphorylated state. The red cell also contains a PLC activity that hydrolyzes PIP_2 (and PIP). Only 50% of the PIP_2 (and PIP) is available for hydrolysis. The remainder is not accessible suggesting that it is sequestered by PIP_2 binding proteins, most likely cytoskeletal proteins. Below we review what is known about the lipid kinases that synthesize PIP_2.

III. LIPID KINASES INVOLVED IN THE PHOSPHORYLATION OF THE INOSITOL HEADGROUP

The first step in the synthesis of PIP_2 is the phosphorylation of PI by phosphatidylinositol 4-kinases (PI 4-kinase) (see Figure 2). PI kinases that phosphorylate phosphatidylinositol have been classified as Type I, II and III; Type I phosphorylates specifically at the D-3 position of the inositol ring and are known as phosphoinositide 3-kinases and they contribute to a different signal transduction pathway. Types II and III phosphorylate specifically at the D-4 position of the inositol ring. The 3-kinase and 4-kinase enzymes are highly homologous enzymes (Carpenter and Cantley, 1996). The second step in PIP_2 synthesis is the phosphorylation of PI(4)P to PI(4,5)P_2 and this is catalyzed by 5-kinases that form a distinct family with homologues in yeast. The different members of the lipid kinase family are summarized in Table 1.

Table 1. Mammalian Lipid Kinases

		References
	Phosphoinositide 3-Kinases (Type I)	
p110α, p110β and p110δ	Requires p85 as adapter; responsible for agonist-sensitive PIP_3 production by receptors that act via tyrosine kinases	Zvelebil et al., 1996; Toker and Cantley, 1997; Vanhaesebroek et al., 1997
p110γ	Associated with p101; responsible for agonist-sensitive PIP_3 production by G-protein-coupled receptors	Vanhaesebroek et al., 1997
Mammalian homologue of yeast Vps34p,	Forms a complex with Vps15p, a ser/thr kinase; only phosphorylates PI	Panaretou et al., 1997; Vanhaesebroek et al., 1997
Cpk	C2-domain containing phosphoinositide 3-kinase; Only phosphorylates PI and PI(4)P but not PI $(4,5)P_2$	Toker and Cantley, 1997; Vanhaesebroek et al., 1997
	PI 4-Kinases (Type II and III)	
Type II PI 4-kinase, 55 kDa, not cloned	Localized at the plasma membrane	Kauffmann-Zeh et al., 1994
Type IIIα PI 4-Kinase, 92 kDa and 230 kDa	Sensitive to wortmannin, localized at the endoplasmic reticulum	Wong et al., 1997
Type IIIβ, 110 kDa	Wortmannin-sensitive, localized at the Golgi and vacuoles	Meyers and Cantley, 1997; Wong et al., 1997
	PIP 5-Kinase	
Type I PI(4)P 5-kinase 1α and 1β, 68 kDa; α and β are 83% homologous	Regulated by PA and Rac and Rho; a 90 kDa form also found in brain	Jenkins et al., 1994; Ishihara et al., 1996; Loijens and Anderson, 1996; Ren et al., 1996; Tolias et al., 1998
Type II PIP 4-kinase, 53 kDa	Originally considered to be a 5-kinase but now identified as a 4-kinase that can phosphorylate PI(3)P or PI(5)P	Rameh et al., 1997

A. PI 4-Kinases

Based on size and sensitivity to various pharmacologic compounds, several PI 4-kinase activities were identified and were classified into two types, Type II and Type III (Endemann et al., 1987, 1991). The Type II enzyme was initially characterized as a membrane-associated 55 kDa protein whose lipid kinase activity was inhibited by adenosine and the monoclonal antibody, 4C5G, and was highly stimulated by nonionic detergents. The Type III enzyme has a much larger apparent molecular mass of >200 kDa and was membrane-associated as well. It was resistant to inhibition by adenosine and 4C5G antibody, and was less stimulated by nonionic detergent.

The Type III family of enzymes have been cloned and contain two family members; PI 4-kinase α and PI 4-kinase β (see Table 1). The first mammalian kinase to be cloned was a 92 kDa enzyme (PI 4-kinaseα) (Wong and Cantley, 1994). Subsequently, a 230 kDa PI 4-kinase was cloned and the 98% identity between these enzymes suggest that 92 kDa PI 4-kinaseα is a truncated form of 230 kDa protein. The full-length 230 kDa enzyme contains identifiable domains including a PH domain, proline-rich domain and a SH3 domain (Nakagawa et al., 1996; Balla et al., 1997). PI 4-kinaseα is sensitive to inhibition by wortmannin (Balla et al., 1997). The yeast homologue of this enzyme is STT4 and is also a target for wortmannin (Cutler et al., 1997). In yeast, STT4 is only essential when grown in standard yeast media. Osmotic support with sorbitol rescues *stt4* mutants. Genetic data suggest that the function of the kinase is to maintain a critical pool of PIP_2. The second Type III kinase that has been cloned is referred to as PI 4-kinaseβ (Balla et al., 1997; Meyers and Cantley, 1997). It is a 110 kDa protein and is also inhibited by 150 nM wortmannin. This PI 4-kinase is homologous to PIK1, an essential gene in yeast.

A wortmannin-sensitive PI 4 kinase has been previously suggested as responsible for regulating the synthesis of agonist-sensitive pools of polyphosphoinositides (Nakanishi et al., 1995; Downing et al., 1996; Balla et al., 1997), but none of these enzymes have been found at the plasma membrane (Wong et al., 1997). PI 4-kinaseα has been localized to the endoplasmic reticulum whereas PI 4-kinaseβ is in the cytosol and also present in the Golgi region (Nakagawa et al., 1996; Wong et al., 1997). Since recent studies indicate the importance of PIP_3 in membrane recruitment of PLCs, the inhibition by wortmannin of IP_3 production could be caused by lack of PIP_3 rather than inhibition of Type III PI 4kinases.

The activity found at the plasma membrane is mainly accounted for by the 55 kDa Type II, PI 4-kinase as judged by inhibition using the 4C5G antibody. Studies with the EGF receptor indicate that the 55 kDa Type II PI 4-kinase associates with the activated receptor and its activity may be regulated by phosphorylation (Kauffmann-Zeh et al., 1994). The Type II PI 4-kinase has evaded purification and has not been cloned.

In summary, PI 4-kinases comprise a family of several proteins and are present in many intracellular compartments in addition to the plasma membrane. The presence of PI 4-kinase activities at multiple sites strongly argues that compartment-specific synthesis of inositol lipids will be a key determinant in the localized availability of PIP_2. PI 4-kinase activity has been found at many locations including the Golgi, lysosomes, secretory granules, nucleus and the plasma membranes (Cockcroft et al., 1985; Del Vecchio and Pilch, 1991; Wiedemann et al., 1996). The Type III kinases are the likely candidates for the intracellular-localized PI 4-kinase activities while the plasma membrane activity is accounted for by the Type

II enzymes. This enzyme is the likely candidate for providing substrate to the PLCs since it is located at the appropriate membrane compartment.

B. PIP 5-Kinases

PI(4)P is phosphorylated to PI(4,5)P_2 by phosphatidylinositol 4-phosphate 5-kinases (PI(4)P 5-kinase). These enzymes originally consisted of Type I and Type II, based on size and *in vitro* characteristics. The Type I family has three members: Type 1α and 1β, which are both 68 kDa and are 83% identical, and a 90 kDa protein (Jenkins et al., 1994; Hay et al., 1995; Ishihara et al., 1996; Loijens and Anderson, 1996). The Type 1 enzymes can all be activated by phosphatidic acid (Jenkins et al., 1994; Ishihara et al., 1996). In addition, there is evidence suggesting that the two GTPases of the Rho family, Rac and RhoA, can also activate these enzymes (Chong et al., 1994; Tolias et al., 1995, 1998; Ren et al., 1996; Zigmond et al., 1997). The kinase associates with both the GDP-form and the GTP-form of Rac *in vitro* and the C-terminus of Rac is required for this interaction (Tolias et al., 1998). For activation of PIP_2 production *in vitro*, however, Rac has to be loaded with GTPγS (Zigmond et al., 1997).

The Type II kinase is a 53 kDa protein which has limited homology to Type I enzymes. The homology resides almost exclusively within the kinase domain core. The Type II enzyme which was originally purified as a PIP 5-kinase has had to be renamed to PIP 4-kinase because this enzyme has been recently found to phosphorylate the D-4 position on the inositol ring using either PI(3)P or PI(5)P (and not PI(4)P) as substrate as previously thought (see Figure 2) (Rameh et al., 1997). For the purification, PI(4)P was originally used as substrate and was obtained commercially. However, the PI(4)P was found to be contaminated with PI(5)P (Rameh et al., 1997). The enzyme PIP 4-kinase thus catalyzes 4-phosphorylation on the inositol ring and the enzyme can use either PI(5)P or PI(3)P. Both substrates are minor components (less than 5%) of the total PIP pool in cells. This kinase has been shown to translocate to the cytoskeleton in thrombin-stimulated platelets and is associated with an increase in PIP_2 levels with the cytoskeleton (Hinchcliffe et al., 1996).

From the above it is clear that the enzymology for the phosphorylation of PI to the higher phosphorylated derivatives involves several different enzymes and much needs to be learned about their regulation (and substrate specificity). In resting cells, the majority of the PIP_2 is localized at the plasma membrane and this is based on several kinds of studies including subcellular fractionation (Whatmore et al., 1996). A recent study has used the GFP-tagged PH-domain of PLCδ which is known to specifically bind PI(4,5)P_2 to localize PIP_2 in living cells and have confirmed that the majority of the lipid is in the plasma membrane (Stauffer et al., 1998).

C. Phosphoinositide 3 Kinases

Phosphoinositide 3-kinases belong to a multigene family and the field has been extensively reviewed (Stephens et al., 1993; Kapeller and Cantley, 1994; Toker and Cantley, 1997; Vanhaesebroek et al., 1997). The first enzyme to be identified was the heterodimer p85/p110 that associates with activated protein tyrosine kinases. p85 is the adaptor which has SH2 domains and can therefore be recruited by phosphotyrosine residues and p110 is the catalytic subunit which can phosphorylate $PI(4,5)P_2$ on the 3 position (see Figure 2 and Table 1). Some G-protein-coupled receptors also activate phosphoinositide 3-kinase notably in neutrophils and this is due to the presence of a $G\beta\gamma$-activated phosphoinositide 3-kinase γ isoform (Stoyanov et al., 1995). A phosphoinositide 3-kinase enzyme that only phosphorylates PI to PI(3)P has been identified in mammalian cells. The yeast homologue, Vps34 is required for trafficking from the Golgi to the vacuole (the mammalian equivalent being the lysosome). The major function of the 3-phosphorylated lipids is the recruitment of signaling proteins to the plasma membrane; this is summarized in Table 2 (Toker and Cantley, 1997).

IV. PHOSPHOLIPASE D

Phospholipase D hydrolyzes phosphatidylcholine (PC) to PA and free choline and is stimulated in nearly all cells by interaction of cell surface receptors with their appropriate agonists (Cockcroft, 1997; Exton, 1997a,b; Singer et al., 1997). In the presence of ethanol, the product of PLD activity is the corresponding phosphatidylethanol and this product is a hallmark of PLD activity. PLD has been cloned from yeast, plants and mammalian sources and they all share regions of similarity which most likely represents the catalytic domain. The first reported mammalian PLD (hPLD1a) was found to have 1072 amino acids and a molecular mass of 124 kDa. A shorter slice variant which lacks 38 amino acid residues has also been identified (hPLD1b) but no differences in regulation of hPLD1a and b have been reported (Hammond et al., 1995, 1997). A second PLD isoform, PLD2 has been cloned and has 932 amino acids and 51% amino acid identity to hPLD1a (Colley et al., 1997; Kodaki and Yamashita, 1997).

A. Enzymology and Regulation by PIP_2, ARF, RHO, and PKC

PLD1 and PLD2 both require $PI(4,5)P_2$ for activity. $PI(3, 4,5)P_3$ is also effective, but other acidic lipids including $PI(3,4)P_2$, PI(4)P and PI are not (Pertile et al., 1995; Hammond et al., 1997). Currently it is not clear whether physiological changes in PIP_2 or PIP_3 control the enzyme activity *in vivo* or whether PIP_2 is present in sufficient amounts in resting cells to fulfil its function as a cofactor for PLD activity. The activity of hPLD1 is directly controlled by inputs from three

Table 2. PIP_2/PIP_3 Binding Proteins

Proteins	*Comments*	*References*
	Cytoskeletal Proteins	
Profilin	Profilin sequesters actin monomers which can be dissociated by $PI(4,5)P_2$ micelles *in vitro*; profilin has high affinity for $PI(4,5)P_2$ binding at a proline-rich sequence	Lassing and Lindberg, 1985
Gelsolin, villin	Proteins that cap, sever and nucleate actin filaments; PI4P and $PI(4,5)P_2$ cause dissociation from monomeric and polymeric actin and inactivates their actin filament severing activity	Janmey and Stossel, 1987, 1989; Janmey et al., 1987, 1992
CapG	Actin capping protein; dissociation from filament ends regulated by $PI(4,5)P_2$; overexpression results in enhanced IP_3 generation	Sun et al., 1995a, 1997
Cofilin and destrin (also known as ADF)	Actin monomer binding protein whose interaction can be inhibited by phosphoinositides (PI, PIP and PIP_2); can also be regulated by dephosphorylation	Yonezawa et al., 1990; Suzuki et al., 1995
Vinculin	Protein folded by head-tail interactions which shields the binding site for F-actin and talin; acidic lipids including PS, PIP and PIP_2 all dissociate vinculin's head-tail interaction	Fukumi et al., 1994; Gilmore and Burridge, 1996; Weekes et al., 1996
α-actinin	Abundant PIP_2 binding protein; PLCγ1 hydrolyzes PIP_2 bound to α-actinin;	Fukumi et al., 1994
	Proteins With PH Domains	
PLCδ1	PH domain binds PIP_2 or $I(1,4,5)P_3$ and is essential for hydrolytic activity when substrate is presented in native membranes; recruits PLCδ1 to membranes	Paterson et al., 1995; Allen et al., 1997
PLCγ1	PH domain binds $PI(3,4,5)P_3$ and is recruited to plasma membranes; tyrosine phosphorylation is essential for stimulation of its activity	Falasca et al., 1998
PKB (Ser/Thr kinase)	$PI(3,4,5)P_3$ regulates enzyme activity by membrane recruitment; PKB requires phosphorylation for activation	Hemmings, 1997
Centaurin-α	PH domain binds $PI(3,4,5)P_3$ and $I(1,3,4,5)P_3$; homology to ARF-GAPs	Hammonds-Odie et al., 1996
SOS	Exchange factor of Ras and Rac; PH domain in tandem with DBL domain	Nimnual et al., 1998
Cytohesin, Grp1 and ARNO	Contains Sec7 domain that functions as an exchange factor for ARFs; PH domain of cytohesin and Grp1 binds $PI(3,4,5)P_3$ and that of ARNO binds $PI(4,5)P_2$	Chardin et al., 1996; Kolanus et al., 1996; Klarlund et al., 1997, 1998

(continued)

Table 2. Continued

Proteins	*Comments*	*References*
	Proteins With PH Domains (continued)	
Dynamin	GTPase with PH domain binds $PI(4.5)P_2$; forms rings around the neck of vesicles for release	Zheng et al., 1996
β-spectrin	PH domain binds $PI(4.5)P_2$ and is targeted to the plasma membranes *in vivo*	Wang et al., 1996
Vav	contains DBL domain that functions as an exchange factor for Rho proteins; PH domain binds $PI(3,4,5)P_3$ and requires tyrosine phosphorylation as well for activation	Han et al., 1998
	Miscellaneous	
SH2 domain of PI 3kinase and $pp60^{c\text{-}src}$	SH2 domains interacts with $PI(3,4,5)P_3$ and competes with binding to tyrosine-phosphorylated proteins	Rameh et al., 1995
Phospholipase A_2	PIP_2 enhances PLA_2 activity *in vitro* suggesting that there may be a potential PH domain responsible for membrane recruitment	Mosier et al., 1998
Phospholipase D1 and D2	Activity requires the presence of PIP_2; presence of a potential PH domain	Colley et al., 1997; Hammond et al., 1997
Synaptotagmin	Contain two C2 domains and C2A can bind to PIP_2 and PIP_3 in a Ca^{2+}-dependent manner	Schiavo et al., 1996

separate activators; two GTPases, ARF and Rho proteins and the Ca^{2+} and phospholipid-dependent isoforms of PKC (α and β) (Figure 4). PLD2 was discovered by cloning and no positive regulators of this enzyme has yet been identified although a 20 kDa negative regulator has been identified from bovine brain (Colley et al., 1997).

The most potent regulator of PLD1 is ARF (Cockcroft et al., 1994). In the majority of studies ARF1 has been used but it appears that all ARF proteins (ARF1, 3, 5, and 6) are activators (Massenburg et al., 1994). All the members of the Rho family (RhoA, Rac and Cdc42) activate PLD1 and a consistent finding is the synergistic activation when ARF and Rho proteins are present together (Hammond et al., 1997; Fensome et al., 1998). ARF and RhoA are indirectly activated by cell-surface receptors and this provides a link to PLD activation in intact cells (Laudanna et al., 1996; Fensome et al., 1998). The other link could be provided by the observation that PKC (α and β) can also regulate PLD1. From *in vitro* studies it appears that protein phosphorylation by PKC is not the main mechanism but is by direct protein-protein interaction. This does not preclude additional phosphorylation-dependent mechanisms that operate in intact cells (Marcil et al., 1998). Synergistic activation of PLD1 by the presence of all three regulators has been shown.

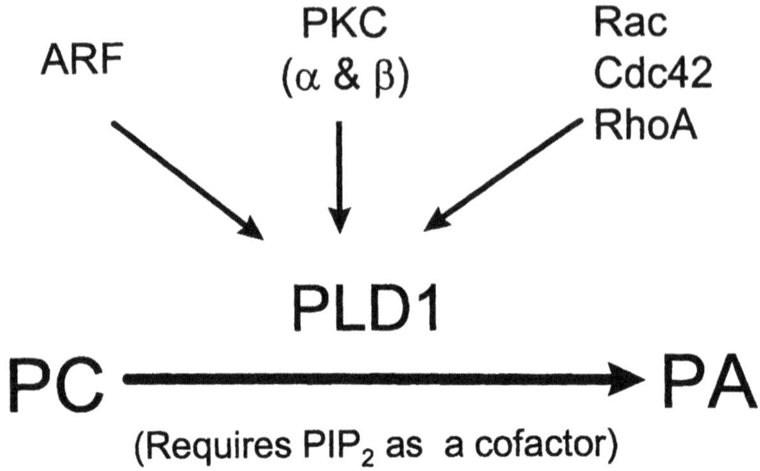

Figure 4. Regulation of PLD1 by the GTPases, ARF and Rho proteins and by PKC.

In all cells examined to date phorbol myristate acetate (PMA) has been found to be an activator of PLD. Recently, PMA was found to translocate ARF and RhoA to membranes (Fensome et al., 1998), indicating activation by PMA in intact cells is not entirely due to PKC only but is very likely due to synergistic activation by all the three activators. The mechanism of ARF and Rho activation by PMA is not clear but presumably the appropriate exchange factors are involved. The exchange factors for ARF belong to a large family of proteins (as defined by the presence of the Sec7 domain) which can be sub-divided into brefeldin A-sensitive or insensitive forms. ARNO, cytohesin, and Grp1 proteins are composed of Sec7 and PH domains (see Table 2) and are insensitive to brefeldin A (Chardin et al., 1996; Kolanus et al., 1996; Klarlund et al., 1997, 1998; Frank et al., 1998). Stimulation of PLD activity by the fMLP receptor is insensitive to brefeldin A indicating that the relevant exchange factors are likely to be similar to ARNO, cytohesin and GRP1 (Guillemain and Exton, 1997; Fensome et al., 1998).

B. Biological Function of Phospholipase D

The function of ARF proteins in vesicle formation is well-established (Rothman, 1994) but the evidence that PLD activity mediates this aspect of ARF function is controversial (Ktistakis et al., 1996; Kun et al., 1997; West et al., 1997; Jones et al., 1998). PLD activity is stimulated by activation of cell-surface receptors and the immediate product of PLD hydrolysis is PA. PLD could exert its biological effects by several mechanisms; this includes changes in the property of cellular membranes or by PA having a signaling role as a lipid. PA would remain in membranes but could interact with proteins in the membrane or cytosol. Downstream targets of PA have been identified from several *in vitro* studies and these include several kinases such as Type I PIP 5-kinase, an identified Ser/Thr kinase, PKCζ and Raf1-kinase (Cockcroft, 1997). As discussed previously, the Type I PIP 5-kinase has been cloned and is potently stimulated by PA *in vitro*. In permeabilized HL60 cells, ARF has been shown to activate PLD and so increase PA levels. ARF was also found to increase PIP_2 levels at the expense of PIP indicating the activation of a PIP 5-kinase (Fensome et al., 1996). Thus ARF-regulated PLD activity can drive PIP_2 synthesis in a physiological setting.

Another possibility is that a metabolite of PA is the candidate with the biological function. PA can be metabolized to either lyso-PA (a potent mitogen) or to DG (an activator of PKC). These possibilities are not mutually exclusive. Another possibility that deserves consideration is that *in vivo*, the cellular DG could serve as an acceptor of the phosphatidyl moiety from the phosphatidyl-PLD intermediate. In this case, the product of the reaction would be bisPA and in one study identified it as such (van Blitterswijk and Hilkmann, 1993). Deacylation would yield lyso-bisPA (LBPA). This lipid is specifically present in the intraluminal membrane of late endosomes and lysosomes. An antibody to this lipid perturbs the structure and function of the late endosomes, suggesting a critical role for LBPA in membrane trafficking events (Kobayashi et al., 1998).

Many roles for PLD activity in a biological setting have been proposed and these include a potential role in exocytosis (Stutchfield and Cockcroft, 1993; Fensome et al., 1996), vesicle trafficking by ARF in Golgi (Ktistakis et al., 1996; Kun et al., 1997), and control of actin polymerization (Ha and Exton, 1993; Cross et al., 1996). PA as a regulator of PIP_2 synthesis is potentially the most exciting possibility since this would explain the widespread occurrence of PLD activation by appropriate agonists acting at cell surface receptors. PIP_2 has many cellular second messenger functions and can be compared to Ca^{2+} or cAMP which have ubiquitous roles in controlling many aspects of cell function from exocytosis to gene transcription.

V. PIP_2 AS AN INTACT LIPID REQUIRED FOR SIGNALING

Evidence that the intact PIP_2 has signaling roles has accumulated from a variety of sources. In two cases, the identification of PITP as a regulator of exocytosis and vesicle formation provided the initial clues.

A. A Requirement for PIP_2 in Regulated Exocytosis and Vesicle Formation

The use of permeabilized cells revealed an unexpected role for PITP in regulated exocytosis. In this case, examination of Ca^{2+}-regulated release of noradrenaline from dense core granules in PC12 cells was under investigation. MgATP and cytosol are required for "priming" prior to Ca^{2+}-triggered exocytosis. A reconstitution assay was established to identify these priming factors in the cytosol. Three separate priming factors were detected and two of them were identified as PITP and the Type I PIP 5-kinase (Hay and Martin, 1993; Hay et al., 1995). These two proteins function together with PI 4-kinase that is membrane-localized in the conversion of PI to PIP_2.

The requirement for PITP in exocytosis has also been shown for HL60 cells and RBL cells (Fensome et al., 1996; Way and Cockcroft, 1998). HL60 cells (a cell-line related to the neutrophil) and RBL cells (a cell-line related to mast cells) can be induced to secrete specialized lysosomal granules from permeabilized cells by micromolar Ca^{2+} in the presence of GTPγS. Exocytosis becomes refractory in cytosol-depleted cells and this can be restored upon re-addition of PITP or ARF (Fensome et al., 1996). Secretion is MgATP dependent indicating that phosphorylation events need to take place, most likely PI phosphorylation to PIP_2.

The addition of PLCδ leads to inhibition of exocytosis indicating that intact PIP_2 is required (Hay et al., 1995; Way and Cockcroft, 1998). The target for PIP_2 in exocytosis has not been identified but the potential targets include synaptotagmin, cytoskeletal and actin-binding proteins (see Table 2). In addition to exocytosis, PITP has also been found to support budding of secretory vesicles from the Golgi membranes in PC12 cells (Ohashi et al., 1995) and again evidence has been presented to suggest that PITP functions by increasing PIP_2 levels (Tuscher et al., 1997).

B. PIP_2 Binding Proteins that Affect Cytoskeletal Assembly Reveal Novel Functions for These Lipids

A direct effect of PIP_2 on protein function is its ability to modulate the activities of components of the cytoskeleton. Table 2 summarizes some of the best studied proteins and further details can be found in recent reviews (Stossel, 1993; Janmey, 1994; Sun et al., 1995b). PI(4)P and PI(4,5)P_2 specifically interact with several actin-binding proteins, altering their affinity for actin. The first interaction of this kind to be identified was that of PIP_2 with profilin, a small ubiquitous soluble protein of 12–15 kDa that sequesters actin monomers (Lassing and Lindberg, 1985). PIP_2 decreases the affinity of profilin for actin and may thereby promote actin polymerization. It was later found that gelsolin, which severs and caps actin filaments and thus breaks up actin networks, is inactivated by PIP_2. Gelsolin binds extremely tightly to actin filaments and PIP_2 can dissociate these complexes (Janmey and Stossel, 1987). Since these early studies, PIP_2 is now known to interact with several actin binding proteins (Janmey, 1994) but these observations have been mainly made *in vitro* using purified proteins. The ability of gelsolin to bind PIP_2 is dependent on clusters of PIP_2. Between 5–10 inositol lipid headgroups must form to stabilize the protein lipid complex (Janmey and Stossel, 1989). *In vitro*, gelsolin can modulate the activity of PLC and PLD possibly through an effect on their interactions with PIP_2 (Banno et al., 1992; Steed et al., 1996; Sun et al., 1997). A PLD-gelsolin complex was purified on a GTP column and its activity was increased by PIP_2 (Steed et al., 1996). In contrast, overexpression of gelsolin depressed PLC activity and this could be relieved if gelsolin was allowed to leak out of the cells (Sun et al., 1997).

Vinculin is a prominent cytoskeletal protein found at focal adhesions and interacts with talin and actin. An intramolecular interaction between the head and tail domain of vinculin masks the binding sites for both talin and actin. Acidic lipids including PIP_2 can effectively relieve the intramolecular association and expose the actin and talin-binding site which would permit the recruitment of vinculin at focal adhesions (Gilmore and Burridge, 1996; Weekes et al., 1996).

Recent studies have begun to identify a role for phosphoinositides in cytoskeletal reorganization in the cellular context. For polyphosphoinositides to upregulate actin assembly, an accumulation of appropriate lipids is required. However, during receptor activation when actin assembly would be regulated, the PLC and phosphoinositide 3-kinase will actually consume these lipids leading to a drop in their concentration. This contradictory requirement can be easily met by segregation of the cellular pools of PIP_2 such that there are local increases in PIP_2 while there is on-going consumption of PIP_2 at the vicinity of the receptor. PITP is ideally suited to provide a flux of substrate for the kinases and the phospholipases thus sparing the pool of PIP_2 used for cytoskeletal reorganization.

The level of phosphoinositide synthesis has been correlated with the amount of actin assembled in RBL mast cells (Apgar, 1995). In this study, PMA, an activator of PKC was found to activate PI 4-kinase and PI(4)P 5-kinase activities leading to net accumulation of PIP_2 levels. This increase in PIP_2 correlated with actin polymerization. Although antigen-stimulation also caused an increase in actin polymerization, and this was associated with increases in lipid kinase activities, there was actually a net decrease in PIP_2 levels due to hydrolysis by PLC. Thus measurements of global levels can be misleading and the ability to visualize local changes in PIP_2 may resolve this issue. A recent study has examined the localization of PIP_2 in RBL mast cells using the PH-domain of PLCδ1 tagged with green flourescent protein (GFP). GFP-PH was highly enriched at the plasma membrane and uniformly distributed along the membrane. Upon stimulation with platelet activating factor (PAF), that induces the hydrolysis of PIP_2, a dissociation of the GFP-PH probe from the plasma membrane was observed within 60 seconds (Stauffer et al., 1998). Relocalization to the plasma membrane was observed in 3–8 minutes. In a separate study, a PIP_2 antibody was used and its distribution was mostly limited to the plasma membrane (Bubb et al., 1998). This distribution matches that of profilin II, the isoform that has the highest affinity for PIP_2.

If one assumes that these PH domains truly bind to PIP_2 *in vivo*, then an important point concerning PIP_2 distribution emerges. When the full-length PLCδ1, β-spectrin and the PH-domain of PLCδ1 were used in different studies, in all cases the probes were found to decorate the plasma membrane only (Paterson et al.,

1995; Wang et al., 1996; Stauffer et al., 1998). This would indicate that the majority of PIP_2 is present at the plasma membrane and very little is present in intracellular membrane compartments. However, PIP_2 which is sequestered by binding proteins would be invisible with this technique, underestimating the true distribution of this lipid.

In addition to the phosphoinositides, actin assembly and its organization is also regulated by the GTP binding proteins of the Rho family (Hall, 1998). This family includes Rac, Rho and Cdc42 and the proteins cycle between the active GTP-bound and inactive GDP-bound states. Guanine nucleotide exchange factors (GEFs) are responsible for converting the inactive G-protein to the active state and consist of a large family of proteins with a Dbl domain which is the exchange domain (Cerione and Zheng, 1996). These GEFs also have PH domains making them ideal targets for recruitment by PIP_2/PIP_3. From studies in fibroblasts, these three GTPases have been found to produce distinct patterns of actin assembly. Rho controls the assembly of contractile actin-myosin filaments, known as stress fibers; Rac assembles a meshwork of actin filaments at the cell periphery to produce lamellipodia and membrane ruffles; and Cdc42 has been shown to produce actin-rich protrusions called filopodia (Hall, 1998). Whether the GTPases control actin assembly via activation of downstream lipid metabolizing enzymes such as PLD and PIP 5-kinases is a matter of debate. Many downstream targets of these GTPases have now been identified and coordinated stimulation of different effectors may integrate the cellular response, such as phagocytosis.

To obtain direct evidence for phosphoinositide-mediated changes in actin assembly, studies have been conducted in permeabilized platelets (Hartwig et al., 1995) and in cell lysates (Zigmond et al., 1997; Ma et al., 1998). In permeabilized platelets, PI(4)P, PI(4,5)P2, PI(3,4)P_2 and PI(3,4,5)P_3 can all uncap F-actin in resting cells. Such uncapping has also been demonstrated with neutrophil lysates (DiNubile and Huang, 1997). GTPγS plus agonist also uncap F-actin and this could be mimicked by an active form of mutant Rac but not Rho. GTPγS plus agonist and the active form of mutant Rac both stimulated PIP_2 synthesis and they concluded that actin assembly involves Rac which increases phosphoinositides for mediating F-actin uncapping. Rac can potentially increase PIP_2 levels by directly activating PIP 5-kinase or via activation of PLD. The inability of Rho to mediate F-actin uncapping is surprising in view of previous studies where Rho has been shown to activate PIP 5-kinase activity (Chong et al., 1994; Ren et al., 1996).

Using lysates from neutrophils, it was found that GTPγS induced formation of filamentous actin. The G-protein responsible was identified as Cdc42. Only GTPγS-loaded Cdc42, but not Rho and Rac, induced actin-polymerization and this could be demonstrated in a high speed supernatant where membrane lipids would not be present. Actin polymerization did not correlate with PIP_2 synthesis although acidic lipids were required for exchange of GDP to GTP on Cdc42 (Zigmond et al., 1997). Most interestingly, Rac, which was able to stimulate PIP_2 synthesis, did not stimulate actin polymerization in neutrophil lysates (Zigmond et al., 1997), unlike the study in platelet lysates (Hartwig et al., 1995). This difference could be due to the cell preparation. The neutrophil study used lysates while the platelet study used permeabilized cells and lysis may have disrupted coupling of Rac to actin polymerization. Examination of actin assembly in cellular extracts of *Xenopus* egg indicated that Cdc42 was required in addition to membrane PIP_2. The conclusion of this study was that phosphoinositides were required to facilitate GTP exchange on Cdc42, as well as to anchor Cdc42 and actin nucleation activities (Ma et al., 1998).

It is clear that actin assembly is influenced by phosphoinositides at least two steps. The activation of the small GTPases can be dependent on phosphoinositides, acting in concert with the GEFs. The exchange factors all contain PH domains and the phosphoinositides play an important role in recruitment of the appropriate GEFs to the membrane, and here PIP_2 and PIP_3 are most important. Phosphoinositides are also directly able to uncap actin filaments by interacting with proteins such as gelsolin and by dissociating actin-profilin complexes to make available free actin for polymerization. The elucidation of the precise routes taken within a cell remains to be established but the application of permeabilized cell systems where the individual steps can be independently controlled will provide a fertile ground for such analysis.

VI. CONCLUDING REMARKS

Phosphoinositides are likely to play major roles at several steps in the phagocytic processes, as substrates for PLCγ and phosphoinositide 3-kinases and as intact lipid signaling molecules in membrane fusion and in cytoskeletal reorganization. It is to be predicted that the signaling machinery that will replenish the PIP_2 pools will get activated to enable localized increases of PIP_2. The key must lie in the spatial and temporal availability of these lipids for them to participate in the cytoskeletal reorganization as well as in the exocytic process. Future studies will need to address the machinery leading to the control of PIP_2 levels but will include the Rho family of GTPases for the activation of PIP 5-kinases, and PITP for the delivery of PI to the lipid kinases. More important will be spatial organization of the synthetic machinery to enable these lipids to be formed.

ACKNOWLEDGMENTS

Work in the author's laboratory is supported by the Wellcome Trust and I thank them for their generous support.

REFERENCES

Allen, V., Swigart, P., Cheung, R., Cockcroft, S., & Katan, M. (1997). Regulation of inositol lipid-specific phospholipase Cδ by changes in Ca^{2+} ion concentrations. Biochem. J. 327, 545–552.

Apgar, J.R. (1995). Activation of protein kinase C in rat basophilic leukemia cells stimulates increased production of phosphatidylinositol 4-phosphate and phosphatidylinositol 4,5-bisphosphate: correlation with actin polymerization. Mol. Biol. Cell. 6, 97–108.

Atkinson, T.P., & Yang, Q. (1996). Translocation of phospholipase C-γ2 induced by in vitro activation of protein tyrosine kinase activity in mast cells lysates. Cell. Signal. 8, 461–465.

Balla, T., Downing, G.J., Jaffe, H., Kim, S., Zolyomi, A., & Catt, K.J. (1997). Isolation and molecular cloning of wortmannin-sensitive bovine type III phosphatidylinositol 4-kinases. J. Biol. Chem. 272, 18358–18366.

Banno, Y., Nakashima, T., Kumada, T., Ebisawa, K., Nonomura, Y., & Nozawa, Y. (1992). Effects of gelsolin on human platelet cytosolic phosphoinositide-phospholipase C isozymes. J. Biol. Chem. 267, 6488–6494.

Berridge, M.J. (1993). Inositol trisphosphate and calcium signalling. Nature 361, 315–325.

Bianchini, L., Todderud, G., & Grinstein, S. (1993). Cytosolic [Ca^{2+}] homeostasis and tyrosine phosphorylation of phospholipase Cγ2 in HL60 granulocytes. J. Biol. Chem. 268, 3357–3363.

Bubb, M.R., Baines, I.C., & Korn, E.D. (1998). Localization of actobindin, profilin I, profilin II, and phosphatidylinositol-4,5-bisphosphate (PIP_2) in *Acanthamoeba castellanii.* Cell Motil. Cytoskel. 39, 134–146.

Carpenter, C.L., & Cantley, L.C. (1996). Phosphoinositide kinases. Curr. Opin. Cell Biol. 8, 153–158.

Cerione, R.A., & Zheng, Y. (1996). The Dbl family of oncogenes. Curr. Opin. Cell Biol. 8, 216–222.

Chardin, P., Paris, S., Antonny, B., Robineau, S., Beraud-Dufour, S., Jackson, C.L., & Chabre, M. (1996). A human exchange factor for ARF contains Sec7- and pleckstrin-homology domains. Nature 384, 481–484.

Chong, L.D., Traynor Kaplan, A., Bokoch, G.M., & Schwartz, M.A. (1994). The small GTP-binding protein rho regulates a phosphatidylinositol 4-phosphate 5-kinase in mammalian cells. Cell 79, 507–513.

Cockcroft, S. (1987). Polyphosphoinositide phosphodiesterase: regulation by a novel guanine nucleotide binding protein, G_p. Trends Biochem. Sci. 12, 75–78.

Cockcroft, S. (1997). Phospholipase D: regulation by GTPases and protein kinase C and physiological relevance. Prog. Lipid Res. 35, 345–370.

Cockcroft, S. (1998). Phosphatidylinositol transfer proteins: a requirement in signal transduction and vesicle traffic. BioEssays. 20, 423–432.

Cockcroft, S., & Thomas, G.M.H. (1992). Inositol lipid specific phospholipase C isozymes and their differential regulation by receptors. Biochem. J. 288, 1–14.

Cockcroft, S., Taylor, J.A., & Judah, J.D. (1985). Subcellular localisation of inositol lipid kinases in rat liver. Biochim. Biophys. Acta 845, 163–170.

Cockcroft, S., Thomas, G.M.H., Fensome, A., Geny, B., Cunningham, E., Gout, I., Hiles, I., Totty, N.F., Troung, O., & Hsuan, J.J. (1994). Phospholipase D: a downstream effector of ARF in granulocytes. Science 263, 523–526.

Colley, W.C., Sung, T., Roll, R., Jenco, J., Hammond, S.M., Altshuller, Y., Bar-Sagi, D., Morris, A.J., & Frohman, M.A. (1997). Phospholipase D2, a distinct phospholipase D isoform with novel regulatory properties that provokes cytoskeletal reorganization. Curr. Biol. 7, 191–201.

Cox, D., Chang, P., Zhang, Q., Reddy, P.G., Bokoch, G.M., & Greenberg, S. (1997). Requirements for both Rac1 and Cdc42 in membrane ruffling and phagocytosis in leukocytes. J. Exp. Med. 186, 1487–1494.

Cross, M.J., Roberts, S., Ridley, A.J., Hodgkin, M.N., Stewart, A., Claesson-Welsh, L., & Wakelam, M.J.O. (1996). Stimulation of actin stress fibre formation mediated by activation of phospholipase D. Curr. Biol. 6, 588–597.

Cunningham, E., Thomas, G.M.H., Ball, A., Hiles, I., & Cockcroft, S. (1995). Phosphatidylinositol transfer protein dictates the rate of inositol trisphosphate production by promoting the synthesis of PIP_2. Curr. Biol. 5, 775–783.

Cutler, N.S., Heitman, J., & Cardenas, M.E. (1997). STT4 is an essential phosphatidylinositol 4-kinase that is a target of wortmannin in *Saccharomyces cerevisiae*. J. Biol. Chem. 272, 27671–27677.

de Camilli, P., Emr, S.D., McPherson, P.S., & Novick, P. (1996). Phosphoinositides as regulators in membrane traffic. Science 271, 1533–1539.

Del Vecchio, R.L., & Pilch, P.F. (1991). Phosphatidylinositol 4-kinase is a component of glucose transporter (GLUT 4)-containing vesicles. J. Biol. Chem. 266, 13278–13283.

della Bianca, V., Grzeskowiak, M., Lissandrini, D., & Rossi, F. (1991). Source and role of diacylglycerol formed during phagocytosis of opsonized yeast particles and associated respiratory burst in human neutrophils. Biochem. Biophys. Res. Commun. 177, 948–955.

DiNubile, M.J., & Huang, S. (1997). High concentrations of phosphatidylinositol-4,5-bisphosphate may promote actin filament growth by three potential mechanisms: inhibiting capping by neutrophil lysates, severing actin filaments and removing capping protein-β_2 from barbed ends. Biochim. Biophys. Acta 1358, 261–278.

Divecha, N., & Irvine, R.F. (1995). Phospholipid signaling. Cell 80, 269–278.

Downing, G.J., Kim, S., Nakanishi, S., Catt, K.J., & Balla, T. (1996). Characterization of a soluble adrenal phosphatidylinositol 4-kinase reveals wortmannin sensitivity of type III phosphatidylinositol kinases. Biochemistry 35, 3587–3594.

Drayer, A.L., & van Haastert, P.J.M. (1992). Molecular cloning and expression of a phosphoinositide-specific phospholipase C of *Dictyostelium discoideum*. J. Biol. Chem. 267, 18387–18392.

Dusi, S., Donini, M., della Bianca, V., & Rossi, F. (1994). Tyrosine phosphorylation of phospholipase C-γ2 is involved in the activation of phosphoinositide hydrolysis by Fc receptors in human neutrophils. Biochem. Biophys. Res. Commun. 201, 1100–1108.

Endemann, G., Dunn, S.N., & Cantley, L.C. (1987). Bovine brain contains two types of phosphatidylinositol kinase. Biochemistry 26, 6845–6852.

Endemann, G.C., Graziani, A., & Cantley, L.C. (1991). A monoclonal antibody distinguishes two types of phosphatidylinositol 4-kinase. Biochem. J. 273, 63–66.

Essen, L., Perisic, O., Cheung, R., Katan, M., & Williams, R.L. (1996). Crystal structure of a mammalian phosphoinositide-specific phospholipase Cδ. Nature 380, 595–602.

Exton, J.H. (1997a). Phospholipase D: enzymology, mechanisms of regulation, and function. Physiol. Revs. 77, 303–320.

Exton, J.H. (1997b). New developments in phospholipase D. J. Biol. Chem. 272, 15579–15582.

Falasca, M., Logan, S.K., Lehto, V.P., Baccante, G., Lemmon, M.A., & Schlessinger, J. (1998). Activation of phospholipase Cγ by PI 3-kinase-induced PH domain-mediated membrane targeting. EMBO J. 17, 414–422.

Fallman, M., Gullberg, M., Hellberg, C., & Andersson, T. (1992). Complement receptor-mediated phagocytosis is associated with accumulation of phosphatidylcholine-derived diglyceride in human neutrophils. Involvement of phospholipase D and direct evidence for a positive feedback signal of protein kinase C. J. Biol. Chem. 267, 2656–2663.

Feng, J., Rhee, S.G., & Im, M. (1996). Evidence that phospholipase d1 is the effector in the G_H (transglutaminase II)-mediated signaling. J. Biol. Chem. 271, 16451–16454.

Fensome, A., Cunningham, E., Prosser, S., Tan, S.K., Swigart, P., Thomas, G., Hsuan, J., & Cockcroft, S. (1996). ARF and PITP restore GTPγS-stimulated protein secretion from cytosol-depleted HL60 cells by promoting PIP_2 synthesis. Curr. Biol. 6, 730–738.

Fensome, A., Whatmore, J., Morgan, C.P., Jones, D., & Cockcroft, S. (1998). ARF and Rho proteins mediate FMLP-dependent activation of PLD in human neutrophils. J. Biol. Chem. 273, 13157–13164.

Ferguson, K.M., Lemmon, M.A., Schlessinger, J., & Sigler, P.B. (1995). Structure of the high affinity complex of inositol trisphosphate with a phospholipase C pleckstrin homology domain. Cell 83, 1037–1046.

Frank, S., Upender, S., Hansen, S.H., & Casanova, J.E. (1998). ARNO is a guanine nucleotide exchange factor for ADP-ribosylation factor 6. J. Biol. Chem. 272, 23–27.

Fukumi, K., Endo, T., Imamura, M., & Takenawa, T. (1994). a-actinin and vinculin are PIP_2-binding proteins involved in signaling by tyrosine kinases. J. Biol. Chem. 269, 1518–1522.

Garcia-Bustos, J.F., Marini, F., Stevenson, I., Frei, C., & Hall, M.N. (1994). PIK1, an essential phosphatidylinositol 4-kinase associated with the yeast nucleus. EMBO J. 13, 2352–2361.

Gilmore, A.P., & Burridge, K. (1996). Regulation of vinculin binding to talin and actin by phosphatidylinositol-4,5-bisphosphate. Nature 381, 531–535.

Greenberg, S., Chang, P., & Silverstein, S.C. (1993). Tyrosine phosphorylation is required for Fc receptor-mediated phagocytosis in mouse macrophages. J. Exp. Med. 177, 529–534.

Greenberg, S., Chang, P., Wang, D.C., Xavier, R., & Seed, B. (1996). Clustered syk tyrosine kinase domains trigger phagocytosis. Proc. Natl. Acad. Sci. USA 93, 1103–1107.

Guillemain, I., & Exton, J.H. (1997). Effects of brefeldin A on phosphatidylcholine phospholipase D and inositol phospholipid metabolism in HL-60 cells. Eur. J. Biochem. 249, 812–819.

Ha, K., & Exton, J.H. (1993). Activation of actin polymerization by phosphatidic acid derived from phosphatidylcholine in IIC9 fibroblasts. J. Cell Biol. 123, 1789–1796.

Hackam, D.J., Rotstein, O.D., Schreiber, A., Zhang, W., & Grinstein, S. (1997). Rho is required for the initiation of calcium signaling and phagocytosis by Fcγ receptors in macrophages. J. Exp. Med. 186, 955–966.

Hall, A. (1998). Rho GTPases and the actin cytoskeleton. Science 279, 509–514.

Hammond, S.M., Altshuller, Y.M., Sung, T., Rudge, S.A., Rose, K., Engebrecht, J., Morris, A.J., & Frohman, M.A. (1995). Human ADP-ribosylation factor-activated phosphatidylcholine-specific phospholipase D defines a new and highly conserved gene family. J. Biol. Chem. 270, 29640–29643.

Hammond, S.M., Jenco, J.M., Nakashima, S., Cadwallader, K., Gu, Q., Cook, S., Nozawa, Y., Prestwich, G.D., Frohman, M.A., & Morris, A.J. (1997). Characterization of two alternatively spliced forms of phospholipase D1. Activation of the purified enzymes by phosphatidylinositol 4,5-bisphosphate, ADP-ribosylation factor, and Rho family monomeric GTP-binding proteins and protein kinase C-a. J. Biol. Chem. 272, 3860–3868.

Hammonds-Odie, L.P., Jackson, T.R., Profit, A.A., Blader, I.J., Turck, C.W., Prestwich, G.D., & Theibert, A.B. (1996). Identification and cloning of centaurin-a. J. Biol. Chem. 271, 18859–18868.

Han, J., Luby-Phelps, K., Das, B., Shu, X., Xia, Y., Mosteller, R.D., Krishna, U.M., Falck, J.R., White, M.A., & Broek, D. (1998). Role of substrates and products of PI 3-kinase in regulating activation of Rac-related guanosine triphosphatases by Vav. Science 279, 558–563.

Hara, S., Swigart, P., Jones, D., & Cockcroft, S. (1997). The first 5 amino acids of the carboxy terminus of phosphatidylinositol transfer protein α (PITPα) play a critical role in inositol lipid signaling: transfer activity of PITP is essential but not sufficient for restoration of phospholipase C signaling. J. Biol. Chem. 272, 14909–14913.

Harlan, J.E., Hajduk, P.J., Yoon, H.S., & Fesik, S.W. (1994). Pleckstrin homology domains bind to phosphatidylinositol-4,5-bisphosphate. Nature 371, 168–170.

Hartwig, J.H., Bokoch, G.M., Carpenter, C.L., Janmey, P.A., Taylor, L.A., Toker, A., & Stossel, T.P. (1995). Thrombin receptor ligation and activated rac uncap actin filament barbed ends through phosphoinositide synthesis in permeabilised human platelets. Cell 82, 643–653.

Hay, J.C., & Martin, T.F.J. (1993). Phosphatidylinositol transfer protein required for ATP-dependent priming of Ca^{2+}-activated secretion. Nature 366, 572–575.

Hay, J.C., Fisette, P.L., Jenkins, G.H., Fukami, K., Takenawa, T., Anderson, R.E., & Martin, T.F.J. (1995). ATP-dependent inositide phosphorylation required for Ca^{2+}-activated secretion. Nature 374, 173–177.

Hemmings, B.A. (1997). Akt signaling: linking membrane events to life and death decisions. Science 275, 628–630.

Hinchcliffe, K.A., Irvine, R.F., & Divecha, N. (1996). Aggregation-dependent, integrin-mediated increases in cytoskeletal associated $PtdInsP_2$ (4,5) levels in human platelets are controlled by translocation of PtdIns 4-P 5-kinase C to the cytoskeleton. EMBO J. 15, 6516–6524.

Hirayama, T., Ohto, C., Mizoguchi, T., & Shinozaki, K. (1995). A gene encoding a phosphatidylinositol-specific phospholipase C in induced by dehydration and salt stress in *Arabidopsis thaliana*. Proc. Natl. Acad. Sci. USA 92, 3903–3907.

Homma, Y., & Emori, Y. (1995). A dual functional signal mediator showing RhoGAP and phospholipase C-δ stimulating activities. EMBO J. 14, 286–291.

Ishihara, H., Shibasaki, Y., Kizuki, N., Katagiri, H., Yazaki, Y., Asano, T., & Oka, Y. (1996). Cloning of cDNAs encoding two isoforms of 68-kDa Type I phosphatidylinositol 4-phosphate 5-kinase. J. Biol. Chem. 271, 23611–23614.

Janmey, P.A. (1994). Phosphoinositides and calcium as regulators of cellular actin assembly and disassembly. Annu. Rev. Physiol. 56, 169–191.

Janmey, P.A., & Stossel, T.P. (1987). Modulation of gelsolin function by phosphatidylinositol 4,5-bisphosphate. Nature 325, 362–364.

Janmey, P.A., & Stossel, T.P. (1989). Gelsolin-polyphosphoinositide interaction. Full expression of gelsolin-inhibiting function by polyphosphoinositides in vesicular form and inactivation by dilution, aggregation, or masking of the inositol head group. J. Biol. Chem. 264, 4825–4831.

Janmey, P.A., Iida, K., Yin, H.L., & Stossel, T.P. (1987). Polyphosphoinositide micelles and polyphosphoinositide-containing vesicles dissociate endogenous gelsolin-actin complexes and promote actin assembly from the fast-growing end of actin filaments blocked by gelsolin. J. Biol. Chem. 262, 12228–12236.

Janmey, P.A., Lamb, J., Allen, P.G., & Matsudaira, P.T. (1992). Phosphoinositide-binding peptides derived from the sequences of gelsolin and villin. J. Biol. Chem. 267, 11818–11823.

Jenkins, G.H., Fisette, P.L., & Anderson, R.A. (1994). Type I phosphatidylinositol 4-phosphate 5-kinase isoforms are specifically stimulated by phosphatidic acid. J. Biol. Chem. 269, 11547–11554.

Jones, D.H., Morgan, C., and Cockcroft, S. Phospholipase D and membrane traffic: potential roles in regulated exocytosis, membrane delivery and vesicle budding. *Biochim.*

Jones, D.H, Morgan, C., and Cockcroft, S. (1999). Phospholipase D and membrane traffic: potential roles in regulated exocytosis, membrane delivery and vesicle budding. Biochim. Biophys. Acta. (In press).

Kapeller, R., & Cantley, L.C. (1994). Phosphatidylinositol 3-kinase. BioEssays 16, 565–575.

Katan, M., & Williams, R.L. (1997). Phosphoinositide-specific phospholipase C: structural basis for catalysis and regulatory mechanisms. Sem. Cell Dev. Biol. 8, 287–296.

Kauffmann-Zeh, A., Klinger, R., Endemann, G., Waterfield, M.D., Wetzker, R., & Hsuan, J.J. (1994). Regulation of human type II phosphatidylinositol kinase activity by epidermal growth-factor dependent phosphorylation and receptor association. J. Biol. Chem. 269, 31243–31251.

Kauffmann-Zeh, A., Thomas, G.M.H., Ball, A., Prosser, S., Cunningham, E., Cockcroft, S., & Hsuan, J.J. (1995). Requirement for phosphatidylinositol transfer protein in epidermal growth factor signalling. Science 268, 1188–1190.

Kawakami, N., Shimohama, S., Hayakawa, T., Sumida, Y., & Fujimoto, S. (1996). Tyrosine phosphorylation and translocation of phospholipase Cγ2 in polymorphonuclear leukocytes treated with pervanadate. Biochim. Biophys. Acta 1314, 167–174.

Kim, H.K., Kim, J.W., Zilberstein, A., Margolis, B., Kim, J.G., Schlessinger, J., & Rhee, S.G. (1991). PDGF stimulation of inositol phospholipid hydrolysis requires PLC-γ1 phosphorylation on tyrosine residues 783 and 1254. Cell 65, 435–441.

Klarlund, J.K., Guilherme, A., Holik, J.J., Virbasius, J.V., Chawla, A., & Czech, M.P. (1997). Signaling by phosphoinositide-3,4,5-trisphosphate through proteins containing pleckstrin and Sec7 homology domains. Science 275, 1927–1930.

Klarlund, J.K., Rameh, L.E., Cantley, L.C., Buxton, J.M., Holik, J.J., Sakelis, C., Patki, V., Corvera, S., & Czech, M.P. (1998). Regulation of GRP1-catlyzed ADP ribosylation factor guanine nucleotide exchange by phosphatidylinositol 3,4,5-trisphosphate. J. Biol. Chem. 273, 1859–1862.

Kobayashi, T., Stang, E., Fang, K.S., de Moerloose, P., Parton, R.G., & Gruenberg, J. (1998). A lipid associated with the antiphospholipid syndrome regulates endosome structure and function. Nature 392, 193–197.

Kodaki, T., & Yamashita, S. (1997). Cloning, expression and characterization of a novel phospholipase D complementary DNA from rat brain. J. Biol. Chem. 272, 11408–11413.

Kolanus, W., Nagel, W., Schiller, B., Zeitmann, L., Godar, S., Stockinger, H., & Seed, B. (1996). $\alpha L\beta 2$ integrin/LFA-1 binding to ICAM-1 induced by cytohesin-1, a cytoplasmic regulatory molecule. Cell 86, 233–242.

Ktistakis, N.T., Brown, H.A., Waters, M.G., Sternweis, P.C., & Roth, M.G. (1996). Evidence that phospholipase D mediates ADP ribosylation factor-dependent formation of Golgi coated vesicles. J. Cell Biol. 134, 295–306.

Kun, B., Roth, M.G., & Ktistakis, N.T. (1997). Phosphatidic acid formation by phospholipase D is required for transport from the endoplasmic reticulum to the golgi complex. Curr. Biol.7, 301–307.

Kusner, D.J., Hall, C.F., & Schlesinger, L.S. (1996). Activation of phospholipase D is tightly coupled to the phagocytosis of Mycobacterium tuberculosis or opsonized zymosan by human macrophages. J. Exp. Med. 184, 585–595.

Lassing, I., & Lindberg, U. (1985). Specific interaction between phosphatidylinositol 4,5-bisphosphate and profilactin. Nature 314, 472–474.

Laudanna, C., Campbell, J.J., & Butcher, E.C. (1996). Role of Rho in chemoattractant-activated leukocyte adhesion through integrins. Science 271, 981–983.

Lee, S.B., Shin, S.H., Hepler, J.R., Gilman, A.G., & Rhee, S.G. (1993). Activation of phospholipase C-b2 mutants by G-protein αq and $\beta\gamma$ subunits. J. Biol. Chem. 268, 25952–25957.

Lemmon, M.A., Falasca, M., Ferguson, K.M., & Schlessinger, J. (1997). Regulatory recruitment of signalling molecules to the cell membrane by pleckstrin homology domains. Trends Cell Biol. 7, 237–242.

Liao, F., Shin, H.S., & Rhee, S.G. (1993). In vitro tyrosine phosphorylation of PLC-$\gamma 1$ and PLC-$\gamma 2$ by src-family protein tyrosine kinases. Biochem. Biophys. Res. Commun. 191, 1028–1033.

Loijens, J.C., & Anderson, R.A. (1996). Type I phosphatidylinositol-4-phosphate 5-kinases are distinct members of this novel lipid kinase family. J. Biol. Chem. 271, 32937–32943.

Ma, L., Cantley, L.C., Janmey, P.A., & Kirschner, M.W. (1998). Corequirement of specific phosphoinositides and small GTP-binding protein Cdc42 in inducing actin assembly in *Xenopus* egg extracts. J. Cell Biol. 140, 1125–1136.

Marcil, J., Harbour, D., Naccache, P.H., & Bourgoin, S. (1998). Human phospholipase D1 can be tyrosine-phosphorylated in HL60 granulocytes. J. Biol. Chem. 272, 20660–20664.

Martelli, A.M., Gilmour, R.S., Bertagnolo, V., Neri, L.M., Manzoli, L., & Cocco, L. (1992). Nuclear localization and signalling activity of phosphoinositidase C(β) in Swiss 3T3 cells. Nature 358, 242–245.

Massenburg, D., Han, J., Liyanage, M., Patton, W.A., Rhee, S.G., Moss, J., & Vaughan, M. (1994). Activation of rat brain phospholipase D by ADP-ribosylation factors 1, 5, and 6: separation of ADP-ribosylation factor-dependent and oleate-dependent enzymes. Proc. Natl. Acad. Sci. USA 91, 11718–11722.

Meyers, R., & Cantley, L.C. (1997). Cloning and characterisation of a wortmannin-sensitive human phosphatidylinositol 4-kinase. J. Biol. Chem. 272, 4384–4390.

Mosier, M., Six, D.A., & Dennis, E.A. (1998). Group IV cytosolic phospholipase A_2 binds with high affinity and specificity to phosphatidylinositol 4,5-bisphosphate resulting in dramatic increases in activity. J. Biol. Chem. 273, 2184–2191.

Musacchio, A., Gibson, T., Rice, P., Thompson, J., & Saraste, M. (1993). The PH domain: a common piece in structural patchwork of signalling proteins. Trends Biochem. Sci. 18, 343–348.

Nakagawa, T., Goto, K., & Kondo, H. (1996). Cloning, expression, and localization of 230-kDa phosphatidylinositol 4-kinase. J. Biol. Chem. 271, 12088–12094.

Nakanishi, S., Catt, K.J., & Balla, T. (1995). A wortmannin-sensitive phosphatidylinositol 4-kinase that regulates hormone-sensitive pools of inositol phospholipids. Proc. Natl. Acad. Sci. USA 92, 5317–5321.

Nimnual, A.S., Yatsula, B.A., & Bar-Sagi, D. (1998). Coupling of ras and rac guanosine triphosphatases through the Ras exchanger Sos. Science 279, 560–562.

Ohashi, M., Jan de Vries, K., Frank, R., Snoek, G., Bankaitis, V., Wirtz, K., & Huttner, W.B. (1995). A role for phosphatidylinositol transfer protein in secretory vesicle formation. Nature 377, 544–547.

Panaretou, C., Domin, J., Cockcroft, S., & Waterfield, M.D. (1997). Characterization of p150, an adapter protein for the human phosphatidylinositol (PtdIns) 3-kinase. Substrate presentation by phosphatidylinositol transfer protein to the p150-PtdIns 3-kinase complex. J. Biol. Chem. 272, 2477–2485.

Park, D., Jhon, D., Lee, C., Ryu, S.H., & Rhee, S.G. (1993). Removal of the carboxy-terminal region of phospholipase C-β1 by calpain abolishes activation by Gαq. J. Biol. Chem. 268, 3710–3714.

Paterson, H.F., Savopoulos, J.W., Perisic, O., Cheung, R., Ellis, M.V., Williams, R.L., & Katan, M. (1995). Phospholipase C δ1 requires a PH domain for interaction with the plasma membrane. Biochem. J. 312, 661–666.

Pertile, P., Liscovitch, M., Chalifa, V., & Cantley, L.C. (1995). Phosphatidylinositol 4,5-bisphosphate synthesis is required for activation of phospholipase D in U937 cells. J. Biol. Chem. 270, 5130–5135.

Pitcher, J.A., Touhara, K., Payne, E.S., & Lefkowitz, R.J. (1995). Pleckstrin homology domain-mediated membrane association and activation of the β-adrenergic receptor kinase requires co-ordinate interaction with Gβγ-subunits and lipid. J. Biol. Chem. 270, 11707–11710.

Prosser, S., Sarra, R., Swigart, P., Ball, A., & Cockcroft, S. (1997). Deletion of 24 amino acids from the C-terminus of phosphatidylinositol transfer protein causes loss of phospholipase C-mediated inositol lipid signalling. Biochem. J. 324, 19–23.

Rameh, L.E., Chen, C., & Cantley, L.C. (1995). Phosphatidylinositol (3,4,5)P_3 interacts with SH2 domains and modulates PI 3-kinase association with tyrosine-phosphorylated proteins. Cell 83, 821–830.

Rameh, L.E., Tolias, K.F., Duckworth, B.C., & Cantley, L.C. (1997). A new pathway for synthesis of phosphatidylinositol-4,5-bisphosphate. Nature 390, 192–196.

Ren, X., Bokoch, G.M., Traynor-Kaplan, A., Jenkins, G.H., Anderson, R.A., & Schwartz, M.A. (1996). Physical association of the small GTPase Rho with a 68-kDa phosphatidylinositol 4-phosphate 5-kinase in Swiss 3T3 cells. Mol. Biol. Cell 7, 435–442.

Rhee, S.G., & Bae, Y.S. (1997). Regulation of phosphoinositide-specific of phospholipase C isozymes. J. Biol. Chem. 272, 15045–15048.

Rothman, J.E. (1994). Mechanisms of intracellular protein transport. Nature 372, 55–63.

Schiavo, G., Gu, Q.M., Prestwich, G.D., Sollner, T.H., & Rothman, J.E. (1996). Calcium-dependent switching of the specificity of phosphoinositide binding to synaptotagmin. Proc. Natl. Acad. Sci. USA 93, 13327–13332.

Serrander, L., Fallman, M., & Stendahl, O. (1996). Activation of phospholipase D is an early event in integrin-mediated signalling leading to phagocytosis in human neutrophils. Inflammation 20, 439–450.

Shaw, G. (1996). The pleckstrin homology domain: an intriguing multifunctional protein module. BioEssays 18, 35–46.

Singer, W.D., Brown, H.A., & Sternweis, P.C. (1997). Regulation of eukaryotic phosphtaidylinositol-specific phospholipase C and phospholipase D. Annu. Rev. Biochem. 66, 475–509.

Srinivasan, S., Seaman, M., Nemoto, Y., Daniell, L., Suchy, S.F., Emr, S., de Camilli, P., & Nussbaum, R. (1997). Disruption of three phosphatidylinositol polyphosphate 5-phosphatase genes from Saccharomyces cerevisiae results in pleiotropic abnormalities of vacuole morphology, cell shape and osmohomeostasis. Eur. J. Cell Biol. 74, 350–360.

Stauffer, T.P., Ahn, S., & Meyer, T. (1998). Receptor-induced transient reduction in plasma membrane PtdIns(4,5)P_2 concentration monitored in living cells. Curr. Biol. 8, 343–346.

Steed, P.M., Nagar, S., & Wennogle, L.P. (1996). Phospholipase D regulation by a physical interaction with the actin binding protein gelsolin. Biochemistry 35, 5229–5237.

Stephens, L., Jackson, T., & Hawkins, P.T. (1993). Agonist-stimulated synthesis of phosphatidylinositol(3,4,5)-trisphosphate: a new intracellular signaling system? Biochim. Biophys. Acta 1179, 27–75.

Stossel, T.P. (1993). On the crawling of animal cells. Science 260, 1086–1094.

Stoyanov, B., Volinia, S., Hanck, T., Rubio, I., Loubtchenkov, M., Malek, D., Stoyanova, S., Vanhaesebroeck, B., Dhand, R., Nurnberg, B., Gierschik, P., Seedorf, S., Hsuan, J.J., Waterfield, M.D., & Wetzker, R. (1995). Cloning and characterization of a G protein-activated human phosphoinositide-3 kinase. Science 269, 690–693.

Stutchfield, J., & Cockcroft, S. (1993). Correlation between secretion and phospholipase D activation in differentiated HL60 cells. Biochem. J. 293, 649–655.

Sun, H., Kwiatkowska, K., Wooten, D.C., & Yin, H.L. (1995a). Effects of CapG overexpression on agonist-induced motility and second messenger generation. J. Cell Biol. 129, 147–156

Sun, H., Lin, K., & Yin, H.L. (1997). Gelsolin modulates phospholipase C activity in vivo through phospholipid binding. J. Cell Biol. 138, 811–820.

Sun, H.Q., Kwiatkowska, K., & Yin, H.L. (1995b). Actin monomer binding proteins. Curr. Opin. Cell Biol. 7, 102–110.

Suzuki, K., Yamaguchi, T., Tanaka, T., Kawanishi, T., Nishimaki Mogami, T., Yamamoto, K., Tsuji, T., Irimura, T., Hayakawa, T., & Takahashi, A. (1995). Activation induces dephosphorylation of cofilin and its translocation to plasma membranes in neutrophil-like differentiated HL-60 cells. J. Biol. Chem. 270, 19551–19556.

Thomas, G.M.H., Cunningham, E., Fensome, A., Ball, A., Totty, N.F., Troung, O., Hsuan, J.J., & Cockcroft, S. (1993). An essential role for phosphatidylinositol transfer protein in phospholipase C-mediated inositol lipid signalling. Cell 74, 919–928.

Toker, A., & Cantley, L.C. (1997). Signalling through the lipid products of phosphoinositide-3-OH kinase. Nature 387, 673–676.

Tolias, K.F., Cantley, L.C., & Carpenter, C.L. (1995). Rho family GTPases bind to phosphoinositide kinases. J. Biol. Chem. 270, 17656–17659.

Tolias, K.F., Couvillon, A.D., Cantley, L.C., & Carpenter, C.L. (1998). Characterization of a Rac1- and RhoGDI-associated lipid kinase signaling complex. Mol. Cell Biol. 18, 762–770.

Touhara, K., Inglese, J., Pitcher, J.A., Shaw, G., & Lefkowitz, R.J. (1994). Binding of G protein $\beta\gamma$-subunits to pleckstrin homology domains. J. Biol. Chem. 269, 10217–10220.

Tuscher, O., Lorra, C., Bouma, B., Wirtz, K.W.A., & Huttner, W.B. (1997). Cooperativity of phosphatidylinositol transfer protein and phospholipase D in secretory vesicle formation from the TGN-phosphoinositides as a common denominator. FEBS Lett. 419, 271–275.

van Blitterswijk, W.J., & Hilkmann, H. (1993). Rapid attenuation of receptor-induced diacylglycerol and phosphatidic acid by phospholipase D-mediated transphosphatidylation: formation of bisphosphatidic acid. EMBO. J. 12, 2655–2662.

Vanhaesebroek, B., Leevers, S.J., Panayotou, G., & Waterfield, M.D. (1997). Phosphoinositide 3-kinases: a conserved family of signal transducers. Trends Biochem. Sci. 22, 267–272.

Wang, D., Miller, R., Shaw, R., & Shaw, G. (1996). The pleckstrin homology domain of human β1 II spectrin is targeted to the plasma membrane in vivo. Biochem. Biophys. Res. Commun. 225, 420–426.

Way, G., & Cockcroft, S. (1999). Activation of exocytosis by cross-linking of the IgE receptor is dependent on ARF-regulated phospholipase D in RBL-2H3 cells: Evidence that the mechanism of activation is dependent on PIP_2 synthesis. (Submitted).

Weekes, J., Barry, S.T., & Critchley, D.R. (1996). Acidic phospholipids inhibit the intramolecular association between the N- and C-terminal regions of vinculin, exposing actin-binding and protein kinase C phosphorylation sites. Biochem. J. 314, 827–832.

West, M.A., Bright, N.A., & Robinson, M.S. (1997). The role of ADP-ribosylation factor and phospholipase D in adapter recruitment. J. Cell Biol. 138, 1239–1254.

Whatmore, J., Morgan, C.P., Cunningham, E., Collison, K.S., Willison, K.R., & Cockcroft, S. (1996). ADP-ribosylation factor1-regulated phospholipase D is localized at the plasma membrane and intracellular organelles in in HL60 cells. Biochem. J. 320, 785–794.

Wiedemann, C., Schafer, T., & Burger, M.M. (1996). Chromaffin granule-associated phosphatidylinositol 4-kinase activity is required for stimulated secretion. EMBO J. 15, 2094–2101.

Willars, G.B., Nahorski, S.R., & Challiss, R.A.J. (1998). Differential regulation of muscarinic acteylcholine receptor-sensitive polyphosphoinositide pools and consequences for signaling in human neuroblastoma cells. J. Biol. Chem. 273, 5037–5046.

Williams, R.L., & Katan, M. (1996). Structural views of phosphoinositide-specific phospholipase C: signalling the way ahead. Structure 4, 1387–1394.

Wong, K., & Cantley, L.C. (1994). Cloning and characterization of a human phosphatidylinositol 4-kinase. J. Biol. Chem. 269, 28878–28884.

Wong, K., Meyers, R., & Cantley, L.C. (1997). Subcellular locations of phosphatidylinositol 4-kinase isoforms. J. Biol. Chem. 272, 13236–13241.

Wu, D., Jiang, H., Katz, A., & Simon, M.I. (1993). Identification of critical regions on phospholipase C-β1 required for activation by G-proteins. J. Biol. Chem. 268, 3704–3709.

Yoko-O, T., Matsui, Y., Yagisawa, H., Nojima, H., Uno, I., & Toh-E, A. (1993). The putative phosphoinositide-specific phospholipase C gene. PLC1, of the yeast Saccaromyces cerevisae is important for cell growth. Proc. Natl. Acad. Sci. USA 90, 1804–1808.

Yonezawa, N., Nishida, E., Iida, K., Yahara, I., & Sakari, H. (1990). Inhibitions of the interactions of cofilin, destrin anddeoxyribonuclease I with actin by phosphoinositides. J. Biol. Chem. 265, 8382–8386.

Zheng, J., Cahill, S.M., Lemmon, M.A., Fushman, D., Schlessinger, J., & Cowburn, D. (1996). Identification of the binding site for acidic phospholipids on the PH domain of dynamin: implications for stimulation of GTPases activity. J. Mol. Biol. 255, 14–21.

Zhou, M., Ravichandran, K.S., Olejniczak, E.T., Petros, A.M., Meadows, R.P., Sattler, M., Harlan, J.E., Wade, W.S., Burakoff, S.J., & Fesik, S.W. (1995). Structure and ligand recognition of the phosphotyrosine binding domain of Shc. Nature 378, 584–592.

Zigmond, S.H., Joyce, M., Borleis, J., Bokoch, G.M., & Devreotes, P.N. (1997). Regulation of actin polymerization in cell-free systems by GTPγS and Cdc42. J. Cell Biol. 138, 363–374.

Zvelebil, M.J., Macdougall, L., Leevers, S., Volinia, S., Vanhaesebroeck, B., Gout, I., Panayotou, G., Domin, J., Srein, R., Pages, F., Koga, H., Salim, K., Linacre, J., Das, P., Panaretou, C., Wetzker, R., & Waterfield, M. (1996). Structural and functional diversity of phosphoinositide 3-kinases. Phil. Trans. R. Soc. Lond. B. 351, 217–223.

SECTION IV

THE PATHWAY

PATHWAYS THROUGH THE MACROPHAGE VACUOLAR COMPARTMENT

Joel A. Swanson

I. INTRODUCTION

The macrophage vacuolar compartment consists of several kinds of vesicular organelles, distinguishable by their membranes, their contents, and the other organelles with which they communicate. The compartment receives membrane and solute content both by endocytosis at the plasma membrane and by vesicular

Advances in Cell and Molecular Biology of Membranes and Organelles
Volume 5, pages 267-284.

ISBN: 1-55938-999-0

transport from the Golgi apparatus. Within the vacuolar compartment, molecules mix and redistribute by organelle fusion and cytoskeleton-associated organelle motility. The distinct identities of the component organelles are maintained in the face of this impressive rate of mixing by a similarly rapid lateral segregation of membrane molecules, together with vesicle fission reactions that restore or maintain compartment identities.

The essential route through the vacuolar compartment consists of three component activities: entry, transit and recycling. Entry, which includes the processes of phagocytosis, pinocytosis and receptor-mediated endocytosis, occurs when plasma membrane invaginates to enclose extracellular molecules or particles into intracellular vesicles. Subsequent membrane fusion and fission reactions typically deliver the contents of these vesicles to endosomes and then to lysosomes, where most macromolecules are degraded by acid hydrolases. The exchange of membrane and content that occurs en route to lysosomes, and the mechanisms that reestablish compartment identities, allow some internalized membrane and solutes to recycle to the cell surface. The net effect is that extracellular solutes and particles are delivered efficiently to lysosomes through a series of compartments whose distinct identities are maintained even as the exogenous passengers move through them.

This review examines the various pathways through the vacuolar compartment. After describing its elements and their essential interactions, a model is presented to explain its dynamics. The aim is to describe the general flow of membrane, solutes, water and macromolecules into and through the macrophage.

II. THE VACUOLAR COMPARTMENT

The discovery and characterization of the vacuolar compartment began with Metchnikoff's description of phagocytosis in the late nineteenth century and has continued to the present day (Rabinovitch, 1995). From time-lapse cinematography of rat peritoneal macrophages, Lewis observed that pinosomes formed by invagination of cell membrane, then migrated toward the nucleus (Lewis, 1931). Later, Zanvil Cohn and colleagues defined other essential properties of the macrophage vacuolar compartment. Using phase-contrast and electron microscopy, together with biochemistry, they showed that phase-dense granules that fused with newly formed phagosomes and pinosomes were lysosomes (Cohn and Benson, 1965; Cohn et al., 1966). Working primarily with mouse peritoneal macrophages, they identified components of serum, as well as other compounds, that increased pinocytosis and that accelerated formation of the granular lysosomes (Cohn and Benson, 1965; Cohn, 1966). Subsequent quantitative morphological studies by Steinman et al. (1976) refined the ultrastructural descriptions of pinosomes and lysosomes and measured their dynamics. Using horseradish peroxidase (HRP) as an electron microscopic histochemical label for endocytic vesicles, they showed

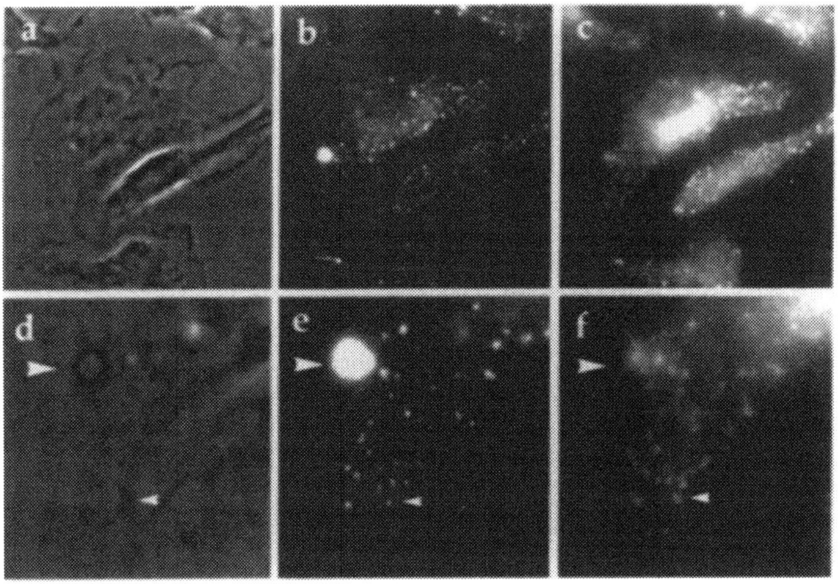

Figure 1. Macrophages labeled with the solute probe fluorescein dextran (average molecular weight 3,000, 1 mg/ml) and the ligand Texas Red-labeled transferrin (50 μg/ml) for 5 minutes to label pinosomes and endosomes. Panels **d–f** show a magnified region of panels **a–c**. (**a** and **d**) Phase-contrast images. (**b** and **e**) Fluorescein dextran-labeled pinosomes and endosomes. (**c** and **f**) Transferrin-labeled endosomes. The large arrowheads indicate a macropinosome, and the small arrowheads indicate colocalization of fluorescein dextran and transferrin in an endosome.

by stereological methods that pinocytosis delivers large quantities of surface membrane and extracellular fluid into a relatively small lysosomal compartment. Thus, by 1976 the vacuolar compartment had been shown to consist of phagosomes, pinosomes and lysosomes that exchanged contents at remarkably high rates.

The discovery and characterization of endosomes helped explain the dynamics of the vacuolar compartment. Endosomes receive endocytosed ligands and solutes from the plasma membrane, and sort them efficiently back to the plasma membrane or to other vesicular compartments. They were initially described in macrophages as acidic organelles that were labeled transiently by brief incubation with mannose-rich glycoproteins or monovalent antibodies against Fc receptor (Stahl et al., 1980; Tietze et al., 1982; Mellman et al., 1984). Similarly, early endosomes can be labeled by brief incubations with fluorescent probes of pinocytosis or receptor-mediated endocytosis (Figure 1).

Endocytic compartments in macrophages conform only approximately to those of other cells. They form a line of communication running from the plasma membrane to early endosomes, then to late endosomes and lysosomes; these organelles have a characteristic pH of 6.5, 6.0 and 5.0, respectively (Mellman, 1996). Currently, molecular markers for early endosomes in macrophages include the transferrin receptor, EEA1, cathepsin H and Rab5 (Buys et al., 1984; Desjardins et al., 1994; Mu et al., 1995; Claus et al., 1998). Late endosomes contain lysosone-associated membrane proteins-1 and -2 (LAMP-1, LAMP-2; a.k.a. lgp-A, lgp-B), cathepsin S, macrosialin and Rab7 (Rabinowitz et al., 1992; Claus et al., 1998). Lysosomes contain LAMPs, macrosialin, the bulk of the cell's acid hydrolases, and content probes loaded by endocytosis and a long chase period (Rabinowitz et al., 1992; Claus et al., 1998). The cation-independent, mannose 6-phosphate receptor (CI-MPR), a marker of late endosomes in other cells, has not been a reliable compartment marker in macrophages, as its localization differs considerably between in different studies (Pitt et al., 1992; Racoosin and Swanson, 1993; Desjardins et al., 1994).

The distinction between late endosomes and lysosomes is not sharp, however, as the most abundant endocytic organelle in many macrophages is a lysosome that has some features of late endosomes. Elongated tubuloreticular lysosomes were identified in bone marrow-derived macrophages, J774.1 macrophages and peritoneal macrophages treated with phorbol myristate acetate (PMA) (Swanson et al., 1987; Knapp and Swanson, 1990; Luo and Robinson, 1992). They can be labeled by incubating macrophages with fluorescent probes of endocytosis or by immunolocalization of cathepsins D or L, or LAMP-1 (Swanson et al., 1987; Araki et al., 1996) (Figure 2). Tubular lysosomes are associated with microtubules (Swanson et al., 1987), and their extension along microtubules is mediated by the mechanoenzyme kinesin (Hollenbeck and Swanson, 1990). They are designated lysosomes because they contain acid hydrolases, they are pH 5.0 to 5.5, and they retain fluorescent probes of fluid-phase pinocytosis for at least 24 hours following endocytosis (Swanson et al., 1987). They are rich in LAMP-1, LAMP-2 and macrosialin, markers of both late endosomes and lysosomes (Rabinowitz et al., 1992; Racoosin and Swanson, 1993). However, they also interact with pinosomes and phagosomes within minutes of their formation, a trait characteristic of late endosomes (Knapp and Swanson, 1990; Racoosin and Swanson, 1993). Rabinowitz et al. (1992) described tubuloreticular compartments by electron microscopy, and identified regions containing CI-MPR and Rab7, again consistent with these compartments being late endosomes. Contrary to this, Racoosin et al. (1993) observed CI-MPR and Rab7 on macropinosomes and a population of small vesicles, but not on tubular lysosomes. Because of these differences, there is disagreement about whether the compartment labeled by brief incubation in fluid-phase probes, followed by a chase period of 30–90 minutes, should be called a late endosome or a lysosome. The tubular lysosomes defined by morphological criteria (Figure 2) correspond best with what Claus et al. (1998) refer to as late endosomes. The

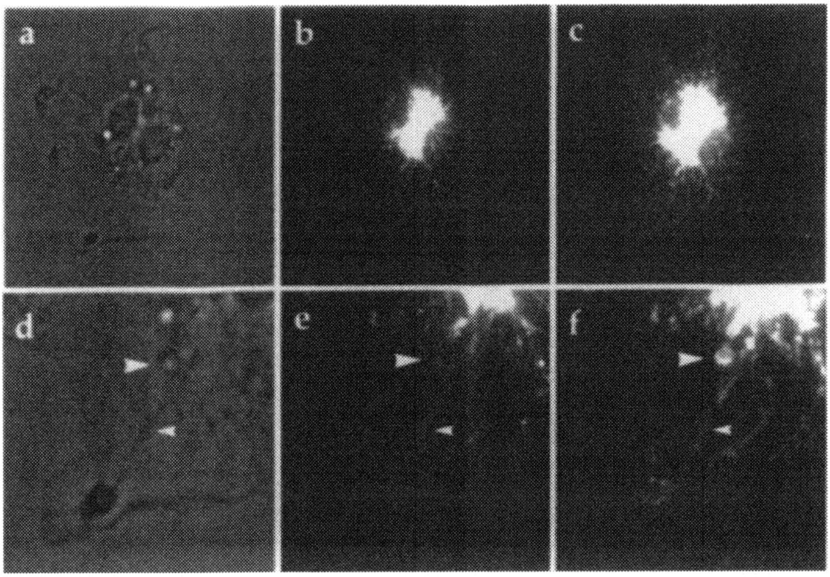

Figure 2. Tubular lysosomes and their interactions with macropinosomes. The macrophage was pulse-labeled with fluorescein dextran (average molecular weight 10,000, lysine-fixable; 1 mg/ml) for 5 minutes, then incubated 60 minutes in unlabeled medium before preparation for immunofluorescent localization of LAMP-1 (as in Racoosin and Swanson, 1993). Panels **d–f** show a magnified region of panels **a–c**. (**a** and **d**) Phase-contrast images. (**b** and **e**) Fluorescein dextran-labeled tubular lysosomes. (**c** and **f**) Texas Red fluorescence showing distribution of LAMP-1. A tubular lysosome containing fluorescein dextran and LAMP-1 is indicated by the small arrowheads. The macropinosome indicated by the large arrowheads has acquired LAMP-1, but not fluorescein dextran. The macropinosome at the bottom of panel **d** is partially wrapped by a tubular lysosome.

nomenclature difficulties indicate that the macrophage vacuolar compartment does not conform neatly to the definitions of late endosomes and lysosomes obtained from other kinds of cells.

Accumulating evidence indicates that there are different kinds of lysosomes inside macrophages. Tassin et al. (1990), identified two kinds of lysosomes in bone marrow-derived macrophages: a perinuclear population containing readily degradable molecules internalized by endocytosis, and a peripheral population containing undegradable molecules. Hoock et al. (1997), localized ankyrin 3 to phase-dense, granular lysosomes that, unlike tubular lysosomes, could only be labeled through endocytosis using long incubations with fluid-phase probes. Moreover, lysosomes isolated by fractionation of J774 macrophages display considerable heterogeneity in their profile of acid hydrolases and their propensity for

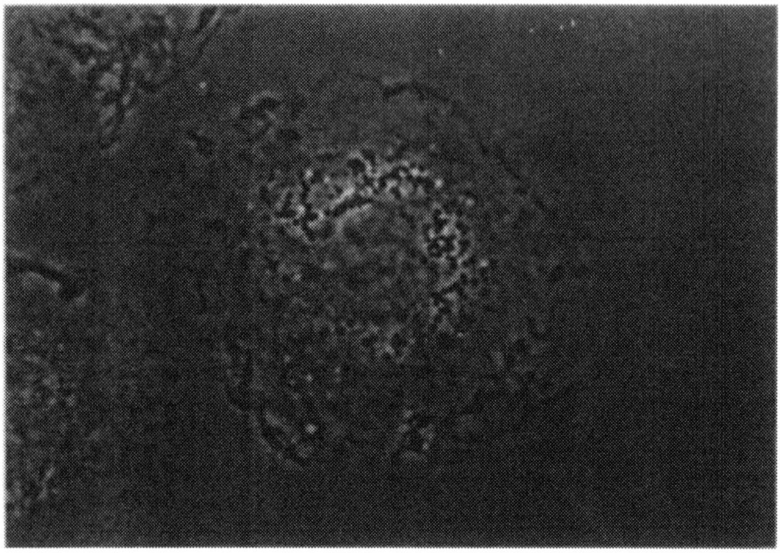

Figure 3. Phase contrast of a bone marrow-derived mouse macrophage activated with interferon-γ (100 U/ml) and LPS (100 ng/ml). The perinuclear, phase-dense, granular lysosomes are similar to those described by Cohn and Benson (1965). They are uncommon in unactivated bone marrow-derived mouse macrophages.

inducible secretion (Claus et al., 1998). Tubular or vesicular, phase-neutral lysosomes are always abundant in bone marrow-derived macrophages, whereas phase-dense granules, such as described by Cohn and Benson (1965), accumulate only when macrophages are activated by interferon-γ and lipopolysaccharide (LPS) (Figure 3). Those granules may be a distinct form of lysosome peculiar to some stages in macrophage development, or they may represent a late stage in normal lysosome morphogenesis, the formation of which can be accelerated by culture conditions.

In summary, the macrophage vacuolar compartment consists of phagosomes; macro- and micropinosomes; early and late endosomes; tubulovesicular, phase-neutral lysosomes; and granular, phase-dense lysosomes. These subdivisions are based largely on morphological criteria. The boundaries of these subcompartments are not sharp, but grade one into another. Their dynamics must be such that they maintain distinguishing traits even as they exchange or transfer some of their contents.

III. ENTRY

Phagocytosis, a necessary component of many macrophage responses to infection, is the most prominent route of entry into macrophages. Most, if not all, phagocytosis by macrophages is initiated and directed by cell surface receptors. The best characterized are the Fc receptors, which recognize the Fc portion of IgG (Ravetch, 1994). Particles opsonized with IgG are engulfed by a strong endocytic response that entails active extension of pseudopodia (Kaplan, 1977). Many other classes of receptor mediate phagocytosis as well, including complement receptors, mannose receptors and scavenger receptors (Sung et al., 1983b; Wright and Griffin, 1985; Dunne et al., 1994).

Some pathogenic microorganisms that are capable of evading delivery to lysosomes can be directed into lysosomes by opsonization with IgG. *Toxoplasma gondii* enters macrophages in vacuoles that do not fuse with lysosomes. Opsonization of *T. gondii* with IgG can direct phagosomes containing them toward lysosomes (Jones and Hirsch, 1972). Similarly, phagosomes containing *Mycobacterium tuberculosis* do not fuse with lysosomes, except when the bacteria are opsonized with IgG (Armstrong and Hart, 1973a,b). Thus, Fc receptor ligation can direct phagosomes to lysosomes.

Solute entry by pinocytosis occurs through formation of variously sized vesicles. Macropinosomes can be as large as 5 μm in diameter (Figure 1). Micropinocytosis occurs through small vesicles consisting mostly of clathrin-coated vesicles (90–120 nm diameter), caveolae (70 nm diameter), and perhaps other kinds of uncoated micropinosomes as well (Kiss and Rohlich, 1984; Kiss and Geuze, 1997).

Rates of fluid influx by pinocytosis are remarkably high. The quantitative studies of pinocytosis by Steinman et al. (1976) indicated that peritoneal macrophages internalize the equivalent of the entire cell surface area every 33 minutes and the cell volume every 2 hours. These rates were probably too high, because they were measured using HRP, which can also enter via mannose receptor-mediated endocytosis (Sung et al., 1983a; Swanson et al., 1985). Nonetheless, similarly high rates were obtained with other probes of pinocytosis, including ^{14}C-sucrose, fluorescein dextran and lucifer yellow, which show little or no adsorptive endocytosis (Besterman et al., 1981; Swanson et al., 1985; Berthiaume et al., 1995).

Differences in rates of pinocytosis reflect differences primarily in the rate of macropinocytosis. Transformed macrophage cell lines, such as J774 macrophages, exhibit sustained, consititutive macropinocytosis. However, in primary cultures of macrophages, pinocytic influx can be increased by stimulation with phorbol esters or macrophage colony-stimulating factor (M-CSF), and most of this increase appears to be due to macropinocytosis (Racoosin and Swanson, 1989; Swanson, 1989). In contrast, stimulation of macropinocytosis does not alter rates of transferrin or acetylated low-density lipoprotein receptor-mediated endocytosis (Racoosin and Swanson, 1992), suggesting that rates of clathrin-coated vesicle or caveolae formation do not vary in these circumstances.

Like all eukaryotic cells examined to date, macrophages internalize a variety of soluble ligands by receptor-mediated endocytosis; including transferrin, acetylated low-density lipoprotein, mannose-rich molecules, interferon-γ, M-CSF and soluble immune complexes (Goldstein et al., 1979; Stahl et al., 1980; Mellman and Plutner, 1984; Guilbert and Stanley, 1986; Celada and Schreiber, 1987; Racoosin and Swanson, 1992). Rates of ligand entry by receptor-mediated endocytosis vary. Some receptors remain at the surface, others enter no further than endosomes, and others go directly to lysosomes (Goldstein et al., 1979; Tietze et al., 1982; Buys et al., 1984). This reflects the sorting functions of endosomes, and the less well-characterized sorting properties of the plasma membrane. In general, however, receptor-mediated endocytosis is a much more efficient mechanism of internalization than pinocytosis, because binding to receptors increases concentrations of molecules in endocytic vesicles far higher than can be achieved by solute trapping in the enclosed fluid volume of pinosomes.

IV. TRANSIT

Most, but not all, of the molecules and particles that enter macrophages by endocytosis are eventually delivered into lysosomes. Early studies identified conditions that altered rates of phagosome-lysosome fusion. Later, as other intracellular routes for soluble ligands were described, such as recycling from endosomes, it became evident that before fusing with lysosomes, phagosomes and macropinosomes interact with nonlysosomal endocytic compartments. The rates of the interactions with endosomes and lysosomes, and the degree of the mixing of organelle contents varies with the contents of endocytic vesicles.

A number of experimental and pathological conditions affect the fusion of phagosomes with lysosomes. Hart and colleagues measured phagosome–lysosome fusion using lysosomes prelabeled with acridine orange or with thorotrast, an electron-dense endocytic tracer, and found that phagosomes containing *M. tuberculosis* did not fuse with lysosomes (Armstrong and Hart, 1973b). They identified polyanionic suramin and sulfatides as molecules that mimicked this inhibition of fusion, and suggested that their production by mycobacteria in phagosomes inhibited fusion with lysosomes (Hart and Young, 1975; Goren et al., 1976). However, Goren et al. (1987) later challenged the interpretation of these studies and provided evidence that the rate of probe transfer from lysosomes into phagosomes depended in part on the molecules used to measure fusion. Presently, the most sensitive assays of phagosome-lysosome fusion measure phagosome acquisition of LAMP-1 or -2, acid hydrolases, or fluorescent macromolecules or colloidal gold preloaded by endocytosis (Harding and Geuze, 1992; Pitt et al., 1992; Rabinowitz et al., 1992; Oh and Swanson, 1996; Desjardins et al., 1997; Claus et al., 1998).

Such methods have confirmed that *M. tuberculosis*, as well as other pathogenic microorganisms including *Legionella pneumophila* and *T. gondii*, enter macrophages in phagosomes that do not fuse with lysosomes (Xu et al., 1994; Clemens and Horwitz, 1995; Sibley, 1995). Compounds such as bafilomycin A_1, ammonium chloride and chloroquine, that raise the pH of acidic compartments, do inhibit phagosome-lysosome fusion (Claus et al., 1998). The efficacy of other putative inhibitors of phagosome-lysosome fusion, including sulfatides and dextran sulfate, remain uncertain.

Although earlier studies indicated that endocytosis delivers a representative sample of plasma membrane to lysosomes, later studies indicated that the profile of membrane proteins in endocytic vesicles changes *en route* to lysosomes. Muller, Mellman and coworkers (Mellman et al., 1980; Muller et al., 1980a) iodinated proteins of phagosomes, pinosomes and plasma membranes, and showed that the principle constituents of membranes in all of these compartments were similar to each other. Pitt et al. (1992) showed that although the bulk constituents of phagosomes do not change much with time, the profile of specific marker molecules does change. Phagosomes containing *Staphylococcus aureus* gradually acquired several markers of endocytic compartments, including LAMP-1 and lysosomal enzymes. Other proteins were depleted from the phagosomes. Several other studies showed that phagosomes and macropinosomes resemble early endosomes shortly after formation, then late endosomes and lysosomes at later times (Rabinowitz et al., 1992; Racoosin and Swanson, 1993; Desjardins et al., 1994).

It is puzzling that although the route appears to be constant, the rates of phagosome delivery to lysosomes vary dramatically from one experimental system to another. Macropinosomes and most phagosomes reach lysosomes within 30 minutes of formation (Pitt et al., 1992; Racoosin and Swanson, 1993; deChastellier et al., 1995; Oh and Swanson, 1996), yet some phagosomes containing latex particles progress on a time scale of hours (Desjardins et al., 1994). Phagosome progression may be related to the interaction between the particle surface and the phagosomal membrane. Membranes of phagosomes containing Mycobacteria adhere closely to the bacterium, and those phagosomes do not fuse with lysosomes. deChastellier and Thilo (1997) have proposed that particle surfaces influence the rate of progression through the vacuolar compartment, and that a hydrophobic surface such as a Mycobacterial cell wall inhibits recycling or maturation of phagosomal membranes which is prerequisite to fusion with lysosomes.

Fusion between endocytic organelles sometimes results in only partial transfer of lumenal contents. Wang and Goren (1987) found that different methods for measuring phagosome-lysosome fusion yielded quantitative differences in the observed rates of fusion. After preloading lysosomes with different sizes of fluorescent probes, they observed that smaller fluorescent probes reported phagosome-lysosome fusion earlier than did larger molecules like fluorescein dextran. Later studies quantified this size-selective transfer between endocytic compartments (Berthiaume et al., 1995; Desjardins et al., 1997). The size-dependent transfer of solutes indicates that fusion of endocytic organelles creates narrow, aqueous bridges that allow limited content exchange. The pyranhalysis described by Willingham and Pastan (1978), in which lysosomes engaged macropinosomes transiently and reversibly, like pyranha fish devouring prey, may indicate the dynamics that allow size-selective transfer of contents between organelles.

Transient and incomplete fusions may provide a mechanism for maintaining distinct compartment identities even as contents pass between them, a phenomenon Desjardins has called kiss-and-run (Desjardins, 1995). The extent to which different compartments retain their identities while exchanging contents remains to be determined.

Once a phagolysosome is formed, it may continue to fuse with lysosomes. Oh and Swanson (1996) found that a variety of particles opsonized with IgG were all delivered quickly to lysosomes, but that continued interactions of those phagolysosomes with other endocytic compartments depended on the kind of particle contained in the phagosome. Phagolysosomes containing polystyrene microspheres were removed from circulation such that they were relatively inaccessible to exogenously added probes. This is another indication that particle surface properties can influence interactions between phagolysosomes and lysosomes.

The high constitutive rates of pinocytosis measured by Steinman et al. (1976) indicated that internalized water and membrane-permeant solutes must move across membranes. This general flow was delineated by Cohn and Ehrenreich (1969), who showed that high concentrations of membrane-impermeant solutes could impede that flow. They allowed macrophages to internalize high concentrations of sucrose or other nondegradable disaccharides by pinocytosis. Because sucrose could not diffuse across vacuolar membranes, it accumulated in lysosomes and caused them to expand by osmotic pressure. This expansion could be reversed by allowing endocytosis of the sucrose-degrading enzyme invertase, which hydrolyzed sucrose to glucose and fructose in the vacuoles and caused them to collapse from loss of the monosaccharides and the osmotically obligated water. These experiments indicated that in the absence of impermeant solutes like sucrose, water internalized by pinosomes crosses membrane into cytosol, then crosses the plasma membrane out of the cell. Under normal circumstances, most solutes and macromolecules would be degraded by acid hydrolases to subunits that diffuse across membranes or through specific membrane transporters.

Larger molecules can sometimes enter macrophage cytoplasm from phagosomes or macropinosomes. Rock and colleagues showed that proteins associated with phagocytic particles can be delivered across phagosomal membranes into cytosol, where they may be processed and returned to the cell surface for presentation in association with major histocompatibility complex (MHC) class I molecules (Kovacsovics-Bankowski and Rock, 1995; Rock, 1996). The efficiency of protein delivery across phagosomal membranes depends on the kind of particle ingested; solid particles deliver protein more efficiently than deformable particles do (Sousa and Germain, 1995; Oh et al., 1997). Norbury et al. (1995) have provided evidence that macropinosomes can also facilitate delivery of extracellular proteins into cytoplasm for processing and presentation via MHC class I-restricted antigen presentation. The mechanism underlying this route remains unknown.

V. RECYCLING

Direct morphological evidence for membrane recycling from endocytic compartments was obtained by Muller et al. (1980b). Beads coated with lactoperoxidase were used to iodinate membrane proteins within phagosomes, then radioactive molecules were localized by autoradiography. Label was initially localized to phagosomes, but chase periods of only a few minutes showed a major redistribution of label to the plasma membrane.

Although HRP did not indicate recycling from macrophages, other probes of fluid-phase pinocytosis did. Besterman et al. (1981) detected two distinguishable rates of solute recycling, which indicated a rapidly recycling compartment that emptied with a half-time of six minutes, and a slowly recycling compartment, emptying with a half-time of several hours. Studies using lucifer yellow supported this two compartment model, and indicated that some probe was irreversibly retained in another compartment (Swanson et al., 1985). The morphology of these compartments has not been defined precisely, but they most likely correspond to endosomes (rapidly recycling compartment), late endosomes or lysosomes (slowly recycling compartment), and storage or granular lysosomes (nonrecycling compartment).

Quantitative studies of pinocytosis led to a model to explain how macrophages accumulate solutes continuously without substantial loss of cell surface membrane. Berthiaume et al. (1995) measured pinocytosis and recycling using different sizes of solute probes and found that small probes recycled from cells more efficiently than did larger probes. This supported a model in which recycling vesicles are smaller than endocytic vesicles (Swanson, 1989). Accordingly, many small vesicles recycling to the surface could return membrane without returning the entire volume of endocytosed water and solutes. The water would instead cross endocytic membranes into cytosol, then leave the cell across the plasma membrane; the solutes would either cross membranes, be retained within the vacuolar compartment, or recycle with the small volume enclosed by the recycling vesicles.

To what extent do lysosomes recycle their contents to the cell surface? Unanue and colleagues showed that antigen processed in lysosomes could be returned to the macrophage cell surface for antigen presentation (Unanue, 1993). Moreover, lysosomal enzymes are released from macrophages during phagocytosis of zymosan, which has been shown to be linked to ligation of mannose receptors (Ohsumi and Lee, 1987), and by conditions that increase lysosomal pH (Griffiths, 1996; Tapper, 1996; Claus et al., 1998).

VI. DYNAMICS

Although the component organelles have been identified, their relationships to the general flow of membrane and solutes are not fully established. The best that can be offered presently is an interpretive summary of flow through the vacuolar compartment.

A successful model for flow must be able to explain how cells maintain the different intracellular compartments (early endosomes, late endosomes and lysosomes). Secondly, it should explain how the compartment can accommodate large endocytic loads, whose default pathway entails maturation through early and late endosome-like stages to a lysosome. Finally, it must explain the recycling of fluids and membranes that accompanies endocytosis.

Sequestration is a central feature of the macrophage's role in disposal of foreign or undegradable materials. The cell is capable of clearing and permanently sequestering large quantities of material, and of returning fragments of it to the cell surface for antigen presentation to lymphocytes. The high rates of constitutive flux of water and membrane through the cell may reflect an idling activity necessary for these high capacity responses.

The various endocytic compartments maintain their distinct identities despite continuous interactions with other compartments, so they must possess efficient mechanisms for reestablishing their identities after fusion. Each compartment may be defined by the combination of sorting signals for its constituent molecules, as well as the sorting signals for the molecules of compartments with which they interact. The net effect would be a steady-state distribution of marker proteins at different levels along the route. For example, the early endosome may be defined in part by the sorting signals that control the distribution of transferrin receptors, as well as those that retain LAMP in late endosomes. This model for movement through the vacuolar compartment is similar to some suggested for the Golgi apparatus, which hold that the heirarchical organization of Golgi cisternae is maintained by sorting mechanisms that retain the molecules comprising each level, even as other, secreted molecules pass readily through those levels (Mironov et al., 1997).

Phagocytosis delivers particles and pinocytosis delivers solutes to lysosomes through a series of early and late endosome-like stages. The passage can be considered a maturation process because there are transitions when individual phagosomes or pinosomes lose features of one compartment and acquire features of the next. Nonetheless, for such maturation to occur there must be some stable or constant compartments with which the maturing organelles interact.

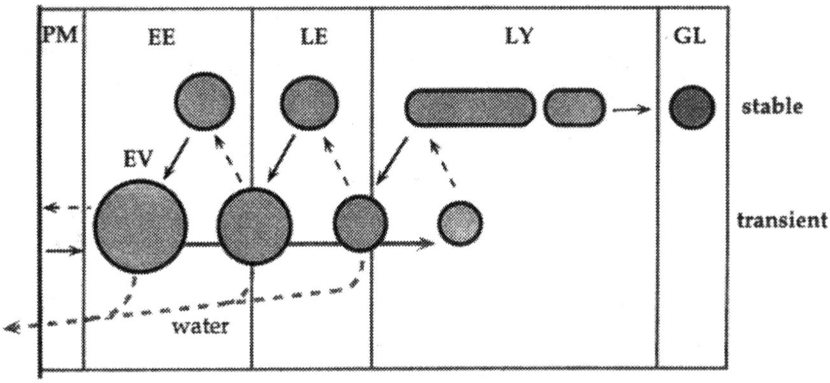

Figure 4. Schematic diagram of the dynamics of flow through the macrophage vacuolar compartment. PM, plasma membrane; EE, early endosome; LE, late endosome; LY, lysosomes, including tubular lysosomes; GL, granular lysosome; EV, endocytic vesicle, including phagosomes, macropinosomes and micropinosomes. Solid arrows indicate vesicle fusion events that combine markers, dashed arrows indicate vesicle fission events that segregate markers. Water flux out of pinosomes is indicated by the wide, dashed line. The vacuolar compartment consists of stable elements that retain markers over time (e.g., resident proteins), and transient elements, whose profile of markers changes with time inside the macrophage. Early endosomes and late endosomes would be considered stable in that they continually resegregate their markers. Pinosomes and phagosomes would acquire features of early endosomes by vesicle fusion, then lose those features by a resegregation process that coincides with or precedes the acquisition of markers for late endosomes. Lysosomes are stable compartments that can recycle to a limited extent, through the retrograde pathways that restore compartment identities. Granular lysosomes may form only when the contents of lysosomes cannot be disposed of by degradative hydrolysis or recycling.

Thus, there appear to be both stable and transient elements to trafficking through the vacuolar compartment (Figure 4). Macropinosomes and phagosomes, and perhaps also smaller endocytic vesicles, are transient. Early endosomes, late endosomes and lysosomes are stable, in that they retain their defining markers over time. Their stability may simply reflect the steady-state distribution of their distinguishing markers, which gives the appearance of a separate population of organelles. Sorting mechanisms continually reestablish their identities. Vesicles may merge together completely then resegregate into distinct organelles, or they may retain their distinguishing features throughout fusion and fission by a kiss-and-run type of mechanism. Either way, the processes that reestablish compartment identities could serve to carry membrane and some content retrograde toward the cell surface.

The functional lysosome is a dynamic, highly plastic organelle that interacts with incoming vesicles soon after their formation. This lysosome may mature into a granular lysosome that interacts less with incoming material. Several studies have indicated that lysosomes can change as they age inside cells, declining in their ability to interact with incoming endocytic vesicles (Montgomery et al., 1991; Desjardins et al., 1997). Molecules internalized by endocytosis may reach a granular lysosome only if they are not degraded first in the tubulovesicular lysosome. Thus, under most circumstances, macromolecules would be degraded fully, and the smaller molecules derived within would cross lysosomal membranes into cytoplasm. Phagocytosed particles could be disposed of by degradative hydrolysis, with some recycling of degradation products to antigen-loading compartments and the plasma membrane (Harding and Geuze, 1992). Molecules that cannot be degraded in lysosomes may accumulate and persist, perhaps forming the granular lysosome. In this way, it would be similar to the telolysosomes or residual bodies described in early morphological studies as lysosome-derived organelles that are removed from circulation (Daems et al., 1972).

The processes that allow phagosomes to mature through a series of endocytic stages could be altered to various extents by the physical properties of ingested particles. If the organelles of the vacuolar compartment are defined by sorting processes that remove or retain component molecules, then one can imagine how particle surfaces could interfere with this sorting to various extents. For example, interactions between a latex bead and the phagosomal membrane could influence the ability of that membrane to fuse with other compartments, to sort specific membrane proteins into recycling vesicles, or to exchange contents after fusion with other organelles. Pathogenic microorganisms could alter these dynamics in a similar way. A differential effect on the retention of marker proteins could lead to a phagosome with an antigenic profile unlike any compartment in the uninfected cell. Thus, the surface of *M. tuberculosis* may not inhibit phagosome-lysosome fusion strictly, but instead force an aberrant maturation of its phagosome, causing it to accumulate transferrin receptor and LAMPs, and preventing it from acquiring the proton ATPase (Sturgill-Koszycki et al., 1996).

The nature of the vesicle fusion machinery is being defined in macrophages as well. Its components almost certainly regulate progression through the vacuolar compartment, and may be manipulated by intracellular pathogens to change vesicular trafficking patterns. Defining the mechanisms that regulate vesicle fusion and fission should help determine what controls the rate of movement through these compartments and how movement is directed.

Much remains to be learned about how the macrophage vacuolar compartment works. The movements of pinosomes and phagosomes, viewed by time-lapse video microscopy, appear far more graceful than the models presently used to explain their dynamics.

ACKNOWLEDGMENTS

Work in this laboratory is supported by a grant from the NIH. I thank Albert Tsang for assistance with the preparation of the micrographs, and Dr. Michele Swanson for scientific and editorial advice.

REFERENCES

Araki, N., Johnson, M.T., & Swanson, J.A. (1996). A role for phosphoinositide 3-kinase in the completion of macropinocytosis and phagocytosis in macrophages. J. Cell Biol. 135, 1249–1260.

Armstrong, J.A., & Hart, P.D. (1973a). Phagosome-lysosome interactions in cultured macrophages infected with virulent tubercle bacilli. Reversal of the usual nonfusion pattern and observations on bacterial survival. J. Exp. Med. 142, 1–16.

Armstrong, J.A., & Hart, P.D. (1973b). Response of cultured macrophages to *Mycobacterium tuberculosis*, with observations on fusion of lysosomes with phagosomes. J. Exp. Med. 134, 713–740.

Berthiaume, E.P., Medina, C., & Swanson, J.A. (1995). Molecular size-fractionation during endocytosis in macrophages. J. Cell Biol. 129, 989–998.

Besterman, J.M., Airhart, J.A., Woodworth, R.C., & Low, R.B. (1981). Exocytosis of pinocytosed fluid in cultured cells: kinetic evidence for rapid turnover and compartmentation. J. Cell Biol. 91, 716–727.

Buys, S., Keogh, E.A., & Kaplan, J. (1984). Fusion of intracellular membrane pools with cell surfaces of macrophages stimulated by phorbol esters and calcium ionophores. Cell 38, 569–576.

Celada, A., & Schreiber, R.D. (1987). Internalization and degradation of receptor-bound interferon-gamma by murine macrophages. Demonstration of receptor recycling. J. Immunol. 139, 147–153.

Claus, V., Jahraus, A., Tjelle, T., Berg, T., Kirschke, H., Faulstich, H., & Griffiths, G. (1998). Lysosomal enzyme trafficking between phagosomes, endosomes, and lysosomes in J774 macrophages. Enrichment of cathepsin H in early endosomes. J. Biol. Chem. 273, 9842–9851.

Clemens, D.L., & Horwitz, M.A. (1995). Characterization of the *Mycobacterium tuberculosis* phagosome and evidence that phagosomal maturation is inhibited. J. Exp. Med. 181, 257–270.

Cohn, Z.A. (1966). The regulation of pinocytosis in mouse macrophages. I. Metabolic requirements as defined by the use of inhibitors. J. Exp. Med. 124, 557–571.

Cohn, Z.A., & Benson, B. (1965). The *in vitro* differentiation of mononuclear phagocytes. II. The influence of serum on granule formation, hydrolase production and pinocytosis. J. Exp. Med. 121, 835–848.

Cohn, Z.A., & Ehrenreich, B.A. (1969). The uptake, storage and intracellular hydrolysis of carbohydrates by macrophages. J. Exp. Med. 129, 201–225.

Cohn, Z.A., Fedorko, M.E., & Hirsch, J.G. (1966). The in vitro differentiation of mononuclear phagocytes. V. The formation of macrophage lysosomes. J. Exp. Med. 123, 757–766.

Daems, W.T., Wisse, E., & Brederoo, P. (1972). Electron microscopy of the vacuolar apparatus. In: Lysosomes: A Laboratory Handbook, Vol. I (Dingle, J.T., Ed.), pp. 64–112. North-Holland Publishing, Amsterdam.

deChastellier, C., & Thilo, L. (1997). Phagosome maturation and fusion with lysosomes in relation to surface property and size of the phagocytic particle. Eur. J. Cell Biol. 74, 49–62.

deChastellier, C., Lang, T., & Thilo, L. (1995). Phagocytic processing of the macrophage endoparasite, Mycobacterium avium, in comparison to phagosomes which contain Bacillus subtilis or latex beads. Eur. J. Cell Biol. 68, 167–182.

Desjardins, M. (1995). Biogenesis of phagolysosomes: the 'kiss and run' hypothesis. Trends Cell Biol. 5, 183–186.

Desjardins, M., Huber, L.A., Parton, R.G., & Griffiths, G. (1994). Biogenesis of phagolysosomes proceeds through a sequential series of interactions with the endocytic apparatus. J. Cell Biol. 124, 677–688.

Desjardins, M., Nzala, N.N., Corsini, R., & Rondeau, C. (1997). Maturation of phagosomes is accompanied by changes in their fusion properties and size-selective acquisition of solute materials from endosomes. J. Cell Sci. 110, 2303–2314.

Dunne, D.W., Resnick, D., Greenberg, J., Krieger, M., & Joiner, K.A. (1994). The type I macrophage scavenger receptor binds to gram-positive bacteria and recognizes lipoteichoic acid. Proc. Natl. Acad. Sci. USA 91, 1863–1867.

Goldstein, J.L., Ho, Y.K., Basu, S.K., & Brown, M.S. (1979). Binding site on macrophages that mediates uptake and degradation of acetylated low density lipoprotein, producing massive cholesterol deposition. Proc. Natl. Acad. Sci. USA 76, 333–337.

Goren, M.B., D'Arcy Hart, P., Young, M.R., & Armstrong, J.A. (1976). Prevention of phagosome-lysosome fusion in cultured macrophages by sulfatides *of Mycobacterium tuberculosis*. Proc. Natl. Acad. Sci. USA 73, 2510–2514.

Goren, M.B., Vatter, A.E., & Fiscus, J. (1987). Polyanionic agents as inhibitors of phagosome-lysosome fusion in cultured macrophages: evolution of an alternative interpretation. J. Leukoc. Biol. 41, 111–121.

Griffiths, G.M. (1996). Secretory lysosomes-A special mechanism of regulated secretion in haemopoietic cells. Trends Cell Biol. 6, 329–332.

Guilbert, L.J., & Stanley, E.R. (1986). The interaction of ^{125}I-colony-stimulating factor-1 with bone marrow-derived macrophages. J. Biol. Chem. 261, 4024–4032.

Harding, C.V., & Geuze, H.J. (1992). Class II MHC molecules are present in macrophage lysosomes and phagosomes that function in the phagocytic processing of Listeria monocytogenes for presentation to T cells. J. Cell Biol. 119, 531–542.

Hart, P.D., & Young, M.R. (1975). Interference with normal phagosome-lysosome fusion in macrophages, using ingested yeast cells and suramin. Nature 256, 47–49.

Hollenbeck, P.J., & Swanson, J.A. (1990). Radial extension of macrophage tubular lysosomes supported by kinesin. Nature 346, 864–866.

Hoock, T.C., Peters, L.L., & Lux, S.E. (1997). Isoforms of ankyrin-3 that lack the NH_2-terminal repeats associate with mouse macrophage lysosomes. J. Cell Biol. 136, 1059–1070.

Jones, T.C., & Hirsch, J.G. (1972). The interaction between *Toxoplasma Gondii* and mammalian cells. II. The absence of lysosomal fusion with phagocytic vacuoles containing living parasites. J. Exp. Med. 136, 1173–1194.

Kaplan, G. (1977). Differences in the mode of phagocytosis with Fc and C3 receptors in macrophages. Scand. J. Immunol. 6, 797–807.

Kiss, A.L., & Geuze, H.J. (1997). Caveolae can be alternative endocytotic structures in elicited macrophages. Eur. J. Cell Biol. 73, 19–27.

Kiss, A.L., & Rohlich, P. (1984). Receptor-mediated pinocytosis of IgG and immune complex in rat peritoneal macrophages: an electron microscopic study. Eur. J. Cell Biol. 34, 88–95.

Knapp, P.E., & Swanson, J.A. (1990). Plasticity of the tubular lysosomal compartment in macrophages. J. Cell Sci. 95, 433–439.

Kovacsovics-Bankowski, M., & Rock, K.L. (1995). A phagosome-to-cytosol pathway for exogenous antigens presented on MHC class I molecules. Science 267, 243–246.

Lewis, W.H. (1931). Pinocytosis. Johns Hopkins Hosp. Bull. 49, 17–27.

Luo, Z., & Robinson, J.M. (1992). Co-localization of an endocytic marker and acid phosphatase in a tubular/reticular compartment in macrophages. J. Histochem. Cytochem. 40, 93–103.

Mellman, I. (1996). Endocytosis and molecular sorting. Annu. Rev. Cell Dev. Biol. 12, 575–625.

Mellman, I., & Plutner, H. (1984). Internalization and degradation of macrophage Fc receptors bound to polyvalent immune complexes. J. Cell Biol. 98, 1170–1177.

Mellman, I., Plutner, H., & Ukkonen, P. (1984). Internalization and rapid recycling of macrophage Fc receptors tagged with monovalent antireceptor antibody: possible role of a prelysosomal compartment. J. Cell Biol. 98, 1163–1169.

Mellman, I.S., Steinman, R.M., Unkeless, J.C., & Cohn, Z.A. (1980). Selective iodination and polypeptide composition of pinocytic vesicles. J. Cell Biol. 86, 712–722.

Mironov, A.A., Weidman, P., & Luini, A. (1997). Variations on the intracellular transport theme: maturing cisternae and trafficking tubules. J. Cell Biol. 138, 481–484.

Montgomery, R.R., Webster, P., & Mellman, I. (1991). Accumulation of indigestible substances reduces fusion competence of macrophage lysosomes. J. Immunol. 147, 3087–3095.

Mu, F.-T., Callaghan, J.M., Steele-Mortimer, O., Stenmark, H., Parton, R.G., Campbell, P.L., McCluskey, J., Yeo, J.-P., Tock, E.P.C., & Toh, B.H. (1995). EEA1, an early endosome-associated protein. J. Biol. Chem. 270, 13503–13511.

Muller, W.A., Steinman, R.M., & Cohn, Z.A. (1980a). The membrane proteins of the vacuolar system. I. Analysis by a novel method of intralysosomal iodination. J. Cell Biol. 86, 292–303.

Muller, W.A., Steinman, R.M., & Cohn, Z.A. (1980b). The membrane proteins of the vacuolar system. II. Bidirectional flow between secondary lysosomes and plasma membrane. J. Cell Biol. 86, 304–314.

Norbury, C.C., Hewlett, L.J., Prescott, A.R., Shastri, N., & Watts, C. (1995). Class I MHC presentation of exogenous soluble antigen via macropinocytosis in bone marrow macrophages. Immunity 3, 783–791.

Oh, Y.-K., & Swanson, J.A. (1996). Different fates of phagocytosed particles after delivery into macrophage lysosomes. J. Cell Biol. 132, 585–593.

Oh, Y.-K., Harding, C.V., & Swanson, J.A. (1997). The efficiency of antigen delivery from macrophage phagosomes into cytoplasm for MHC class I-restricted antigen presentation. Vaccine 15, 511–518.

Ohsumi, Y., & Lee, Y.C. (1987). Mannose-receptor ligands stimulate secretion of lysosomal enzymes from rabbit alveolar macrophages. J. Biol. Chem. 262, 7955–7962.

Pitt, A., Mayorga, L.S., Stahl, P.D., & Schwarz, A.L. (1992). Alterations in the protein composition of maturing phagosomes. J. Clin. Invest. 90, 1978–1983.

Rabinovitch, M. (1995). Professional and non-professional phagocytes: an introduction. Trends Cell Biol. 5, 85–87.

Rabinowitz, S., Horstmann, H., Gordon, S., & Griffiths, G. (1992). Immunocytochemical characterization of the endocytic and phagosomal compartments in peritoneal macrophages. J. Cell Biol. 116, 95–112.

Racoosin, E.L., & Swanson, J.A. (1989). Macrophage colony stimulating factor (rM-CSF) stimulates pinocytosis in bone marrow-derived macrophages. J. Exp. Med. 170, 1635–1648.

Racoosin, E.L., & Swanson, J.A. (1992). M-CSF-induced macropinocytosis increases solute endocytosis but not receptor-mediated endocytosis in mouse macrophages. J. Cell Sci. 102, 867–880.

Racoosin, E.L., & Swanson, J.A. (1993). Macropinosome maturation and fusion with tubular lysosomes in macrophages. J. Cell Biol. 121, 1011–1020.

Ravetch, J.V. (1994). Fc receptors: rubor redux. Cell 78, 553–560.

Rock, K.L. (1996). A new foreign policy: MHC class I molecules monitor the outside world. Immunol. Today 17, 131–137.

Sibley, L.D. (1995). Invasion of vertebrate cells by *Toxoplasma gondii*. Trends Cell Biol. 5, 129–132.

Sousa, C.R., & Germain, R.N. (1995). Major histocompatibility complex class I presentation of peptides derived from soluble exogenous antigen by a subset of cells engaged in phagocytosis. J. Exp. Med. 182, 841–851.

Stahl, P., Schlesinger, P.H., Sigardson, E., Rodman, J.S., & Lee, Y.C. (1980). Receptor-mediated pinocytosis of mannose glycoconjugates by macrophages: characterization and evidence for receptor recycling. Cell 19, 207–215.

Steinman, R.M., Brodie, S.E., & Cohn, Z.A. (1976). Membrane flow during pinocytosis. A stereological analysis. J. Cell Biol. 68, 665–687.

Sturgill-Koszycki, S., Schaible, U.E., & Russell, D.G. (1996). Mycobacterium-containing phagosomes are accessible to early endosomes and reflect a transitional state in normal phagosome biogenesis. EMBO J. 15, 6960–6968.

Sung, S.-S.J., Nelson, R.S., & Silverstein, S.C. (1983a). The role of the mannose/N-acetylglucosamine receptor in the pinocytosis of horseradish peroxidase by mouse peritoneal macrophages. J. Cell Physiol. 116, 21–25.

Sung, S.-S.J., Nelson, R.S., & Silverstein, S.C. (1983b). Yeast mannans inhibit binding and phagocytosis of zymosan by mouse peritoneal macrophages. J. Cell Biol. 96, 160–166.

Swanson, J., Bushnell, A., & Silverstein, S.C. (1987). Tubular lysosome morphology and distribution within macrophages depend on the integrity of cytoplasmic microtubules. Proc. Natl. Acad. Sci. USA 84, 1921–1925.

Swanson, J.A. (1989). Phorbol esters stimulate macropinocytosis and solute flow through macrophages. J. Cell Sci. 94, 135–142.

Swanson, J.A., Yirinec, B.D., & Silverstein, S.C. (1985). Phorbol esters and horseradish peroxidase stimulate pinocytosis and redirect flow of pinocytosed fluid through macrophages. J. Cell Biol. 100, 851–859.

Tapper, H. (1996). The secretion of preformed granules by macrophages and neutrophils. J. Leukoc. Biol. 59, 613–622.

Tassin, M.-T., Lang, T., Antoine, J.C., Hellio, R., & Ryter, A. (1990). Modified lysosomal compartment as a carrier of slowly and non-degradable tracers in macrophages. Eur. J. Cell Biol. 52, 219–226.

Tietze, C., Schlesinger, P., & Stahl, P. (1982). Mannose-specific endocytosis receptor of alveolar macrophages: demonstration of two functionally distinct intracellular pools of receptor and their roles in receptor recycling. J. Cell Biol. 92, 417–424.

Unanue, E.R. (1993). Macrophages, antigen-presenting cells, and the phenomena of antigen handling and presentation. In: Fundamental Immunology, Third Ed. (Paul, W.E., Ed.), pp. 111–144. Raven Press, New York.

Wang, Y.-L., & Goren, M.B. (1987). Differential and sequential delivery of fluorescent lysosomal probes into phagosomes in mouse peritoneal macrophages. J. Cell Biol. 104, 1749–1754.

Willingham, M.C., & Yamada, S.S. (1978). A mechanism for the destruction of pinosomes in cultured fibroblasts: piranhalysis. J. Cell Biol. 78, 480–487.

Wright, S.D., & Griffin, F.M. (1985). Activation of phagocytic cells' C3 receptors for phagocytosis. J. Leukoc. Biol. 38, 327–339.

Xu, S., Cooper, A., Sturgill-Koszycki, S., van Heyningen, T., Chatterjee, D., Orme, I., Allen, P., & Russell, D.G. (1994). Intracellular trafficking in *Mycobacterium tuberculosis* and *Mycobacterium avium*-infected macrophages. J. Immunol. 153, 2568–2578.

SEQUENTIAL MATURATION OF PHAGOSOMES PROVIDES UNIQUE TARGETS FOR PATHOGENS

Carmen Alvarez-Dominguez, Luis Mayorga, and Philip D. Stahl

I. INTRODUCTION

It is now widely appreciated that the rules that govern membrane trafficking along the secretory and endocytic pathways are also applicable to the internalization of particles by phagocytes and the maturation of phagosomes. Just as in other forms

Advances in Cell and Molecular Biology of Membranes and Organelles
Volume 5, pages 285-297.

ISBN: 1-55938-999-0

of intracellular transport, rapid exchange of phagosomal proteins is mediated by vesicle fusion and vesicle budding to and from the limiting membrane of the newly formed and the mature phagosome. Understanding phagosomal membrane protein exchange and the factors that regulate it is central to understanding how pathogens survive within host cells. Moreover, the process is also key to understanding endosymbiosis commonly found among lower eukaryotes. In higher organisms, phagocytosis is exclusively performed by specialized cells such as neutrophils and mononuclear phagocytes. Internalization, where engulfment of large particles proceeds after engagement of specific receptors, appears to take several forms depending on the cell surface proteins engaged, the size of the particle and the cellular machinery employed. Following closure, the newly formed phagosome undergoes a series of carefully orchestrated membrane fusion and budding events resulting in both the influx and efflux of material from the phagosome. Phagosome maturation transforms the plasma-membrane derived phagosome into a phagolysosome (Silverstein et al., 1989). Not long ago, the process of phagolysosome formation was considered unidirectional and simple; i.e., a phagosome fuses with a degradative lysosome becoming a new vesicle, the phagolysosome. It is now clear that the process is complex and certainly multidirectional and dynamic (Beron et al., 1995a). A substantial body of evidence indicates that phagosomes fuse with endosomes, both early and late (Mayorga et al., 1991; Pitt et al., 1992a,b; Rabinowitz et al., 1992; Desjardins et al., 1994a,b; de Chastellier et al., 1995).

Studies on the composition of phagosomes, isolated at different times following particle internalization, indicate that the protein composition of phagosomes is markedly altered during transformation into phagolysosomes. Rapid fluctuations in phagosome composition, membrane proteins and content, point to selective membrane budding and vesicle fusion events as the responsible agents (Pitt et al., 1992b; Desjardins et al., 1994a,b). Indeed, the development of an assay to measure phagosome-endosome fusion *in vitro* opened up the possibility of identifying factors responsible to phagosomal membrane trafficking. Initial studies showed that phagosome-endosome fusion *in vitro* has requirements similar to endosome-endosome fusion: viz., energy, cytosolic and membrane-bound proteins. The similarity of phagosome-endosome fusion with endosome-endosome fusion, apart from directly confirming that phagosomes fuse with early endosomes, suggests a key role for the endocytic apparatus in phagosome maturation. Early results with *in vitro* assays suggested the possibility that several GTP-binding proteins were responsible for the regulation of early phagosome-endosome fusion events (Mayorga et al., 1991). More recently, the presence of several small GTPases (i.e., Rab5 and Rab7) and heterotrimeric G-proteins on phagosomes was documented and their role in fusion was confirmed (Desjardins et al., 1994a,b; Beron et al., 1995b). Moreover, these studies were also informative when viewed in the context of changes in phagosomal membrane protein composition with time. Thus, small GTPases are recruited to and cycle off the phagosomal membrane, possibly marking discreet periods during which specific trafficking events occur during phagosome maturation (Dejardins et al., 1994b; Mordue and Sibley, 1997).

Recently, cell-free assays were developed to measure phagosome-endosome fusion with phagosomes containing live microorganisms (Alvarez-Dominguez et al., 1996). This analysis revealed that, unlike dead microorganisms, live pathogens are able to influence phagosome-endosome fusion by targeting the regulatory proteins of the fusion machinery. The strategy to delay phagosome maturation by interdicting the fusion or fission process may be common among intracellular pathogens (Alvarez-Dominguez et al., 1997).

In this review, we attempt to draw new insights on phagosome maturation. The hypothesis, suggested by the work of several groups, is that phagosome maturation is a highly ordered, sequential process. Specifically, the interaction of phagosomes with early endocytic compartments sets into motion a process whose completion is required for interaction of phagosomes with later compartments (e.g., late endosomes and lysosomes). The sequential nature of phagosome maturation and the requirement for obligatory intermediates might well create special opportunities for intracellular pathogens. Indeed, in the immediate future, phagosome maturation might be defined in terms of the mechanisms utilized by specific pathogens to create biological niches for intracellular survival. Interference might be achieved by elaborating lipids or factors that modulate the function of key regulatory proteins in the host. Just as virus assembly was used to dissect the secretory pathway in mammalian cells, intracellular pathogens will guide the way to identify the steps required for phagosome maturation.

II. SEQUENTIAL STEPS INVOLVED IN PHAGOSOME MATURATION

As noted above, the cell surface proteins/receptors mediating entry may alter the intracellular itinerary of a pathogen or particle. Apart from events that occur in the plane of the membrane (e.g., that might be due to the nature of the particle or microorganism) and the modulation of the actin cytoskeleton (Mordue and Sibley, 1997), the first step of phagosome maturation appears to be the fusion and intermingling of the phagosome with the endocytic compartment. The changes observed in the composition of maturing phagosomes with time suggest that phagosomes do not simply accumulate protein by promiscuously fusing with multiple fusion partners. Rather, the "phago-endosome" is a dynamic compartment where proteins rapidly appear and disappear selectively, indicating that both fusion and fission processes are likely to play a central role. In fact, it has been reported that membrane proteins and content markers may selectively recycle, possibly via the formation of tubules or carrier vesicles (Muller et al., 1980; Pitt et al., 1992a; Claus et al., 1998; Tjelle et al., 1998). Examination of the compositional changes in phagosomes during maturation reveals that early phagosomes, due to extensive interactions with early endosomes, share many characteristics with endosomes. Early endosomal markers, such as the mannose receptor and α-adaptin, appear early on and selectively disappear with time. Markers typical of late endosomes (i.e., lysosomal membrane-asociated protein-1, vacuolar proton pump ATPase, M6P receptor or hydrolytic enzymes such as cathepsin-D and small GTPases such as Rab7) seemingly replace early endosomal markers with time (Pitt et al., 1992b; Desjardins et al., 1994a,b). Thus, the early phagosome gradually transforms into a vesicle that presents features of late endosomes, with a higher content of hydrolytic enzymes.

Since a substantial portion of the proton pump ATPase resides in late endosomes, fusion of phagosomes with elements of the late compartment results in a gradual drop in the intraphagosomal pH. In fact, the observation that phagosomes gradually lower their pH has been know for many years (Hirsch, 1965). Acidification is clearly critical for the killing of certain microbial agents. However, additional work pointed out that low pH alone is not toxic for many bacteria but induces their killing by facilitating the generation of hydrogen peroxide and superoxide (Garcia del Portillo and Finlay, 1995). It may also facilitate fusion with lysosomes. In endosomes, accumulation of H^+-ATPase is accompanied by removal of the Na^+/K^+-ATPase, which facilitates the reduction in pH. It is not clear whether similar mechanisms operate in early phaogosomes. Early phagosomes display a pH around 6.0 whereas late phagosomes are more acidic, perhaps as low as pH 4.0–5.5 (Bovier et al., 1994).

Does the extensive remodeling of protein and lipid components of the maturing phagosome activate factors required for fusion with the late endosomal compartment? Alternatively, does the process remove, sequester or otherwise inactivate inhibitory proteins? For example, the retention of one or more Rab5 isoforms and other factors may lead to continued interaction of phagosomes with early endosomes. With time, a second small GTPase, Rab7, appears and is subsequently also released (Desjardins et al., 1994a,b). Fusion with a lysosomal compartment appears to be specific and closely regulated. Lysosomal fusion requires one or more GTPases but no specific Rab has been localized to lysosomes (Tjelle et al., 1996; Ward et al., 1997; Mullock et al., 1998). Phagosomes containing undigestible particles mature slowly, remain interactive with the endosomal compartment and appear to have an intracellular itinerary different from phagosomes containing digestible particles (de Castellier and Thilo, 1997). These reports raise an interesting caveat, i.e., data from phagosomes isolated using non-digestible particles must be viewed in context. However, phagosomes harboring live pathogens such as *Mycobacterium* ssp. or *Listeria monocytogenes* also interrupt the maturation process and comparisons between such phagosomes containing non-degradable material may be highly instructive (Sturgill-Koszycki et al., 1994; Alvarez-Dominguez et al., 1997). In summary, phagosome maturation may be divided into a series of steps (or possible targets for pathogens) including (i) fusion with the early endocytic compartment and/or subcomponents of the early compartment, (ii) termination of productive docking interactions with early endosomes, (iii) acquisition of molecules that allow for docking and fusion with elements of the late endocytic compartments, (iv) acquisition of H^+-ATPase and molecules that drive acidification and superoxide formation and, (v) fusion with lysosomes. There are also several GTPases including rac and members of the rap-family that control the association and function of NADPH-oxidase, a key enzyme for the generation of oxygen radicals involved in pathogen killing (DeLeo and Quinn, 1996).

Fusion with lysosomes transforms the phagosome into a low pH, highly degradative compartment. Progression through the endocytic route from late endosomes to lysosomes appears to require Rab7 (Press et al., 1998). Multiple reports using colloidal-gold markers to load lysosomes have demonstrated that phagosomes and late endosomes can fuse with lysosomes (Rabinowitz et al., 1992; de Chastellier and Thilo, 1997; Funato et al., 1997). Reconstitution of phagosome-lysosome fusion in Streptolysin O- (SLO-) permeabilized cells has revealed that these events depend on energy, cytosolic proteins, microtubules and several GTPases (including Rab GTPases) as well as fusogenic factors such as NEM-sensitive fusion protein (NSF) (Funato et al., 1997). Moreover, late endosome-lysosome and lysosome-lysosome fusion require Rab GTPases (Ward et al., 1997; Mullock et al., 1998). However, Rab5 and Rab7 do not localize to lysosomes (Tjelle et al., 1996; Mullock et al., 1998). Fusion with a terminal compartment devoid of Rab GTPases would suggest an asymmetric fusion event perhaps with Rabs associated with one fusion partner and not the other. However, it remains unclear whether phagolysosome biogenesis requires fusion of the phagosome directly with preexisting secondary lysosomes or whether a phagosome becomes lysosome-fusion-competent by first becoming lysosome-like via multiple interactions with the late endosome and elements of the trans-Golgi network (Rabinowitz et al., 1992; Beron et al., 1995). Lysosome-lysosome fusion is well documented (Funato et al., 1997; Ward et al., 1997) and it is possible that a phagosome can only fuse with a mature secondary lysosome after it, itself, becomes lysosome-like.

III. PROTEINS REGULATING PHAGOSOME MATURATION

Initial reports with the cell-free fusion assay suggested that several GTPases are involved in phagosome-endosome fusion (Mayorga et al., 1991). Recently, using a modification of the *in vitro* fusion assay, the small GTPase Rab5, was shown to be a key regulator of phagosome-endosome fusion (Alvarez-Dominguez et al., 1996). NSF and SNAP, generic fusion factors, were also shown to be involved (Alvarez-Dominguez et al., 1996). Since Rab5 is rate limiting for endocytosis (Bucci et al., 1992), the regulation of phagosome-endosome fusion by Rab5 creates interesting questions about the boundaries between the early phagocytic and early endocytic pathways. Phagosomal membranes must have some capacity to assist translocation of Rab5 from the cytosol to the membrane and to facilitate GDP-GTP exchange, since Rab5 is only active in the GTP form. GTP hydrolysis, on the other hand, is not necessary for the fusion of phagosomes with endosomes

(Alvarez-Dominguez et al., 1996). Since endosome-endosome fusion is symmetric with respect to Rab5 (Rab5:GTP is required on both fusion partners) (Barbieri et al., 1998) one would expect that appropriately primed endosome and phagosome fusion partners will also be required for phagosome-endosome docking and fusion. As discussed later, phagosomes and endosomes may be considered equivalent structures. Alternatively, the newly formed phagosome may be viewed as a unique structure whose membrane exchange relies on the endocytic apparatus. An implicit requirement of the latter is that only certain elements of the endocytic apparatus might be expected to interact with the newly formed phagosome. Rab5 is expressed as 3 isoforms, Rab5a, 5b and 5c (Bucci et al., 1995), the products of separate genes (Barbosa et al., 1996). It is possible that one or combinations of more than one of the isoforms plays a selective role in targeting endocytic vesicles to phagosomes while other isoforms, or combinations thereof, regulate other parts of the early endocytic pathway. This arrangement would keep the endocytic and phagocytic pathways separate and regulatable.

Are sequentially acting RabGTPases (e.g., Rab5 and Rab7) coupled during phagosome maturation? Regulation of phagosome-late endosome fusion events have not been well characterized, but presumably Rab7 is a key regulatory molecule (Feng et al., 1995; Press et al., 1998). Although still at an early stage of analysis, the accumulation of Rab GTPases by phagosomes harboring *Mycobacterium* ssp. and *L. monocytogenes*, two intracellular pathogens known to block the maturation process at the endosome, may be instructive. *Mycobacterium* ssp. phagosomes accumulate Rab5, the key early acting GTPase, but fail to accumulate Rab7, an immediate downstream GTPase (Via et al., 1997). *L. monocytogenes* appears to block the maturation process a step further along the maturation route, since *L. monocytogenes* phagosomes accumulate both Rab5 and Rab7 (Alvarez-Dominguez et al., 1996). The acquisition of fusion competence may be a characteristic of each stage of the phagosome maturation process. Moreover, the loss of fusion competence may be equally important. Acquisition of fusion competence may be controlled by the dissociation or inactivation of factors such as Rab5 and Rab7 (Desjardins et al., 1994a,b; Alvarez-Dominguez et al., 1997; Via et al., 1997). The findings that phagosomes that retain Rab5 (i.e., *Mycobacterium* ssp.) or Rab5-Rab7 (i.e. *L. monocytogenes*) fail to mature, is consistent with the argument that the dissociation of these proteins (and possibly associated regulatory factors) from the phagosomal membrane corresponds to the removal of a barrier. Removal of a barrier may correspond with the formation of vesicles or tubules to allow for selective egress of membrane proteins from the phagosome. Phagosomes appear to have all the necessary machinery to form coated vesicles (including coatamer proteins) (Beron and Stahl, unpublished observations), and tubules and membrane egress from the phagosome has been reported by several groups (Pitt et al., 1992a; Tjelle et al., 1998). Thus, the removal of a barrier for maturation may take several forms. A model summarizing these suggestions is shown in Figure 1.

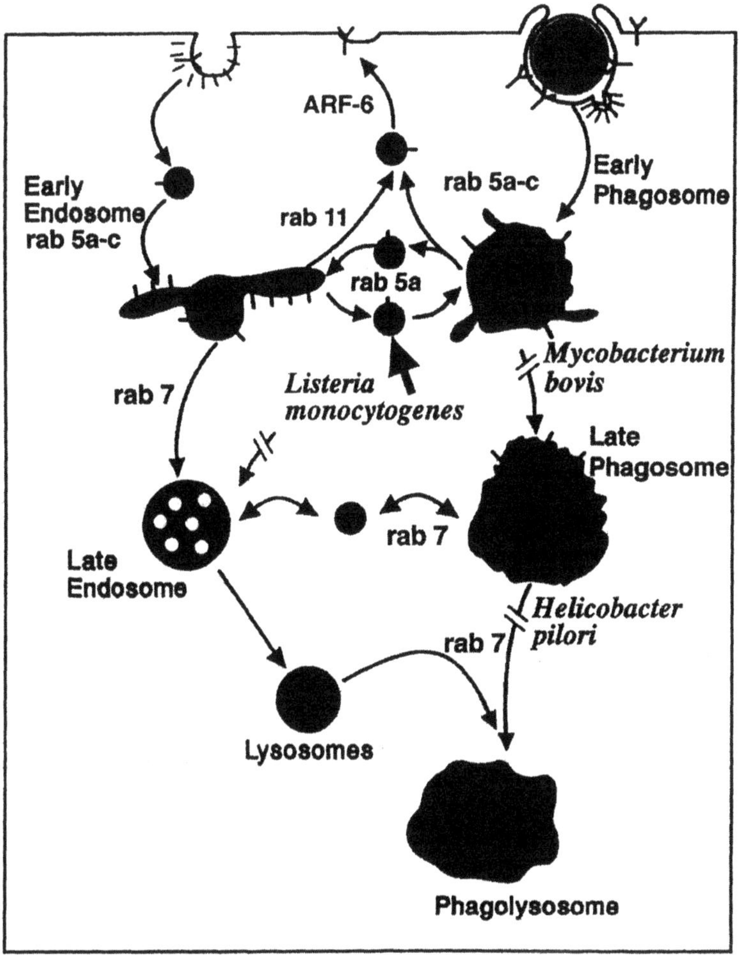

Figure 1. After phagosome formation, fusion of early phagosomes with early endosomes depends on the function of Rab5a. The other isoforms of Rab5 (b and c) may also be involved. Movement beyond this step is apparently blocked by *Mycobacterium bovis*. *M. bovis* phagosomes contain Rab5 but are devoid of Rab7 (Via et al., 1997). Progression of early phagosomes to late phagosomes depends on fusion with elements of the late endosomal compartment, putatively a Rab7 dependent process. Movement beyond this step is blocked by *Listeria monocytogenes*. Phagosomes containing live *L. monocytogenes* accumulate Rab5a and Rab7 and have an acidic pH (Alvarez-Dominguez et al., 1996, 1997). The latter, coupled with the fact that *L. monocytogenes* phagosomes are devoid of mature cathepsin D and mannose-6-phosphate receptors, suggests that fusion with late endosomes is selective, i.e., access to late compartments bearing the H^+-ATPase is permitted whereas access to late endosomes containing lysosomal enzymes cathepsin D or mannose-6-phosphate receptors is prevented. Further progression of the late phagosomes and transformation to phagolysosomes also appears to depend on the function of Rab7, as reflected in the block generated by *Helicobacter pilori*, whose phagosomes are devoid of Rab5, but contain Rab7.

IV. PATHOGENS THAT EXPLOIT SPECIFIC TARGETS DURING PHAGOSOME MATURATION

The mechanisms that certain microorganisms have evolved to interfere with phagosome maturation may offer unique insight for studying membrane trafficking (Silverstein et al., 1989; Garcia del Portillo and Finlay, 1995). Although much literature on the morphology of intracellular infection has existed for some time, only recently have the intracellular targets begun to be characterized. Thus, inhibition of fusion with early endosomes is characteristic of *Toxoplasma gondii* phagosomes, since endosomal markers fail to localize in their membranes. *T. gondii* phagosomes also fail to fuse with the biosynthetic compartment (Mordue and Sibley, 1997). The absence of endosomal or lysosomal markers (class I and class II-major histocompatibility complex, alkaline phosphatase, TfR, lgp or CD63) on *Legionella pneumophila* phagosomes suggests that they fail to fuse with endosomes or lysosomes and thrive in a specialized compartment separated from the endocytic route (Garcia del Portillo and Finlay, 1995a,b). Analysis of enriched-phagosomes harboring *Mycobacterium* ssp. revealed the lack of H^+-ATPase in their membranes suggesting a failure to fuse with late endosomes. Since *Mycobacterium* ssp. phagosomes are competent to exchange material with the early endocytic compartment, they appear to reside in a compartment arrested in an early transitional stage with limited proteolytic activity (Clemens and Hortwitz, 1995; Sturgill-Koszycki et al., 1996). *Mycobacterium bovis* phagosomes were found to acquire and retain Rab5, but failed to associate with Rab7. Thus, the block in phagosomal maturation appears to correspond with transport to or fusion with elements of the late endosome (Via et al., 1997). Phagosomes harboring *L. monocytogenes* accumulate Rab5 and Rab7, yet they fail to interact with late endosomes containing M6P receptor or lamp-1 molecules (Alvarez-Dominguez et al., 1997). *L. monocytogenes* phagosomes do access H^+-ATPase and contain limited proteolytic activity since the immature form of cathepsin D is transported to these phagosomes. However, conversion to the mature form is prevented. Thus, *Listeria* appears to interrupt the maturation process at a point between Rab5 and Rab7 which creates a unique "early-late" compartment. *Salmonella typhimurium* phagosomes contain lgp-proteins putatively due to fusion with a subclass of late endosomes but the process bypasses late endosomes containing M6P receptor. *S. typhimurium* vesicles also display limited fusion with secondary lysosomes and with early endosomes. No defined target in the docking/fusion machinery has been identified (Garcia del Portillo and Finlay, 1995b).

Transport beyond the late endosomal compartment is blocked by *Helicobacter pylori* due to the activity of the bacterial toxin, VacA, that causes an extensive vacuolation of the cells. *Helicobacter* phagosomes lack the early endosomal markers Rab5 and TfR and lysosomal markers such as cathepsin D. However, they selectively accumulate Rab7 and contain H^+-ATPase. Although Rab5 was absent from these large vesicles, overexpression of Rab5 dominant negative mutants (S34N and N133I) partially inhibited the formation of VacA vacuoles. These findings suggest that efficient membrane flow from plasma membrane through early endosomes to late endosomes is critical for vacuole formation. Rab7 dominant negative mutants (T22N and N125I) also inhibited vacuole formation (Papini et al., 1994, 1997). Putative intracellular targets of these bacteria are summarized in Figure 1.

Some pathogens avoid phagosome acidification by failing to fuse with certain late endosomal compartments *Mycobacterium* ssp. phagosomes have a reported pH of >6.3–6.5 whereas *S. typhimurium* phagosomes have a pH between 4.0 and 5.0 (Rathman et al., 1996). The pH of phagosomes containing *T. gondii* is neutral (pH 6.8–7.0). *L. pneumophila* also inhibits phagosome acidification (Garcia del Portillo and Finlay, 1995). All these observations suggest that the lack of phagosome acidification correlates with inhibition of phagosome maturation. However, the ability of a compartment to acidify is complex and may involve delivery of proton pump molecules via fusion with selected subcompartments of the late endocytic pathway, by assembly of functional pump molecules in the membrane and by the possible presence of rectifying anion channels that modulate electrogenic proton pumps.

V. PARALLELS BETWEEN ENDOSOME MATURATION AND PHAGOSOME MATURATION

Do all the steps described for phagosome maturation find their counterpart in the endosome-maturation process? Is there a common theme that connects the two processes? Two views of endosome transport are the so-called maturation model and the vesicular transport model. The maturation model depicts an endosome as a dynamic compartment that recycles receptors (and other molecules) by vesicle budding and that acquires new receptors by fusion with newly formed endosomes. The ability to fuse with newly formed endosomes decreases as the endosome sorts membrane proteins and receptors. From these two premises, the model predicts that endosomes mature as they become more and more enriched in ligands (and de-enriched in recycling receptors) whereupon they become (a) inaccessible to newly formed endosomes and (b) fusogenic with late endocytic compartments. The vesicular transport model (i.e., stable early endosome model) (Aniento et al., 1996) proposes the existence of a stable early endocytic compartment (e.g., similar to a Golgi cisterna) that receives material via fusion with newly formed endocytic vesicles. The compartment also sorts proteins and forms vesicles (including multivesicular bodies) for transport to the cell surface and to late endosome compartments. It is likely that elements of both models are operative. The existence of a semipermanent early endosomal sorting compartment, access to which requires Rab5, seems consistent with all the available data. The notion that portions of the sorting compartment undergo maturation and move to the cell center where access to late endosomes occurs, is also supported by the data. It is possible that endosomal membrane fragments remain behind during the maturation process and provide the basis for endosome renewal via interaction with incoming endocytic vesicles. The nature of this nucleating activity is unknown but a prediction would be that it contains a Rab GTPase receptor and SNARE molecules at minimum. During phagocytosis the newly formed phagosome behaves like an early endosome and accesses some portion of the endocytic compartment via Rab5. From the perspective of the maturation model, the newly formed early phagosome would be virtually equivalent to an early endocytic sorting compartment. All of the elements of the early endosome might be found in the early phagosome because homotypic fusion via Rab5 permits complete access. As the phagosome matures, just like the maturing endosome, additional membrane is added via vesicle fusion and membrane proteins are removed by budding and fission events. Viewed in this way, phagosome maturation and endosome maturation would be very similar if not identical processes. On the other hand, if the newly formed phagosome interacts with a stable endocytic compartment one might expect a partial intermingling of the components. It is possible that the endosomal apparatus has stable subcompartments which have specific functions. One or more of the endocytic subcompartments could specialize in phagosome maturation whereas others might subserve housekeeping endocytic transport and membrane recycling or delivery of proteins to compartments enriched in class II molecules. Rab GTPases often exist as isoforms which are very similar to each other (Bucci et al., 1992, 1995). Identifying endocytic subcompartments and determining their role in phagosome maturation and other functions provides a challenge for the future.

VI. FURTHER DIRECTIONS

Interferon-γ (IFN-γ) is one of the most potent activators of phagocytosis and killing. Several reports have demonstrated that IFN-γ accelerates the fusion of phagosomes harboring various pathogens with lysosomes. Following IFN-γ activation, *S. typhimurium* phagosome-lysosome fusion is enhanced and *Salmonella* killing activity is increased (Ishibashi & Arai, 1990). Similar findings have been reported with *Mycobacterium avium*. These observations correspond well with the bacteriostatic and bactericidal activities of activated macrophages. They also may reflect changes in membrane trafficking pathways that mediate phagosome maturation (Schaible et al., 1998). Recently, the transcription factor NF-IL6, which controls the expression of G-CSF as well as other cytokines, has been demonstrated to play a critical role in activating phagosome-endosome fusion and in controlling bacterial growth. Mice lacking NF-IL6 [NF-IL6 (-/-)] demonstrate high susceptibility to *Brucella* infection due to inhibition of *Brucella* phagosome fusion with endosomes (Pizarro-Cerda et al., 1998).

Although many recent studies correlate cytokine signaling with phagosome-endosome or phagosome-lysosome interactions, no reports have appeared directly linking proteins of the docking/fusion machinery with cytokine signal transduction. T cell receptor (TCR) signaling of T lymphocytes has been linked to Rab5 controlled TCR endocytosis (Andre et al., 1997).

Natural resistance to certain infections has recently been associated with the presence of Nramp1 on phagosomes. Nramp1 appears to control the replication of intracellular parasites by altering the intravacuolar environment of the phagosome. Acquisition of Nramp1 by phagosomes is similar to the acquisition of lamp-1. However, Nramp1 is specific in that infection by *Mycobacterium* or *Salmonella* are affected by the protein whereas infections by *Listeria* or *Legionella* are not (Gruneheid et al., 1997).

These studies on phagosome maturation predict that additional regulatory proteins will be discovered, perhaps as targets for pathogens whose survival strategies include the selective disruption of membrane trafficking.

REFERENCES

Alvarez-Dominguez, C., Barbieri, A., Beron W., Wandinger-Ness, A., & Stahl, P.D. (1996). Phagocytosed Listeria monocytogenes influences rab5-regulated in vitro phagosome-endosome fusion. J. Biol. Chem. 271, 13834–13843.

Alvarez-Dominguez, C., Roberts, R., & Stahl, P.D. (1997). Internalized Listeria monocytogenes modulates intracellular trafficking and delays maturation of the phagosome. J. Cell Sci. 110, 731–743.

Andre, P., Boretto, J., Hueber, A.O., Reguier-Vigoroux, A., Gorvel J.P., Ferrier, P., & Chavrier, P. (1997). A dominant-negative mutant of rab5 GTPase enhances T cell signalling by interfering with TCR downmodulation in transgenic mice. J. Immunol. 159, 5253–5263.

Aniento, F., Gu, F., Parton, R., & Gruenberg, J. (1996). An endosomal βcop is involved in the pH-dependent formation of transport vesicles destined for late endosomes. J. Cell Biol. 133, 29–41.

Barbieri, M.A., Hoffenberg, S., Roberts, R., Mukhopadhyay, A., Pomrehn, A., Dickey, B.F., & Stahl, P.D. (1998). Evidence for a symmetrical requirement for Rab5-GTP in *in vitro* endosome-endosome fusion. J. Biol. Chem. 273, 25850–25855.

Barbosa, M.D.F.S., Wakeland, E.K., Zerial, M., & Kingsmore, S.F. (1996). Genetic maping of the Rab5a and Rab5b genes on mouse chromosomes 17 and 2, respectively. Mamm. Genome 7, 166–167.

Beron, W., Alvarez-Dominguez, C., Mayorga, L., & Stahl, P.D. (1995a). Membrane trafficking along the phagocytic pathway. Trends Cell Biol. 5, 100–104.

Beron, W., Colombo, M.I., Mayorga, L., & Stahl, P.D. (1995b). In vitro reconstitution of phagosome©endosome fusion: evidence for regulation by heterotrimeric GTPases. Arch. Biochem. Biophys. 317, 337–342.

Bovier, G., Benoliel, A.M., Foa, C., & Bougrand, P. (1994). Relationship between phagosome acidification, phagosome-lysosome fusion and mechanism of particle ingestion. J. Leukoc. Biol. 55, 729–734.

Bucci, C., Lutcke, A., Steele-Mortimer, O., Olkkonen, V.M., Dupree, P., Chiariello, M., Bruni, C.B., Simons, K., & Zerial, M. (1995). Co-operative regulation of endocytosis by three rab5 isoforms. FEBS Lett. 366, 65–71.

Bucci, C., Parton, R.G., Mather, I.H., Stannenberg, H., Simons, K., Hoflack, B., & Zerial, M. (1992). The small GTPase rab5 functions as a regulatory factor in the early endocytic pathway. Cell 70, 715–728.

Claus, V., Jahraus, A., Tjelle, T., Berg, T., Kirschke, H., Faulstich, H., & Griffiths, G. (1998). Lysosomal enzyme trafficking between phagosomes, endosomes and lysosomes in J774 macrophages. J. Biol. Chem. 273, 9842–9851.

Clemens, D.L., & Hortwitz, M.A. (1995). Characterization of the Mycobacterium tuberculosis phagosome and evidence that phagosomal maturation is inhibited. J. Exp. Med. 181, 257–270.

de Chastellier, C., & Thilo, L. (1997). Phagosome maturation and fusion with lysosomes in relation to surface property and size of the phagocytic particle. Eur. J. Cell Biol. 74, 49–62.

de Chastellier, C., Lang, T., & Thilo, L. (1995). Phagocytic processing of the macrophage endoparasite Mycobacterium avium, in comparision to phagosomes which contain Bacillus subtilis or latex beads. Eur. J. Cell Biol. 68, 167–182.

DeLeo, F.R., & Quinn, M.T. (1996). Assembly of the phagocyte NADPH-oxidase: molecular interaction of oxidase proteins. J. Leukoc. Biol. 60, 677–691.

Desjardins, M., Celis, J.E., van Meer, G., Dieplinger, H., Jahraus, A., Griffiths, G., & Huber, LA. (1994a). Molecular characterization of phagosomes. J. Biol. Chem. 269, 32194–32200.

Desjardins, M., Huber, L., Parton, R.G. & Griffiths, G. (1994b). Biogenesis of phagolysosomes proceeds through a sequential series of interactions with the endocytic apparatus. J. Cell Biol. 124, 677–688.

Feng, Y., Press, B., & Wandinger-Ness, A. (1995). Rab7: an important regulator of late endocytic membrane traffic. J. Cell Biol. 131, 1435–1452.

Funato, K., Beron, W., Yang, C.Z., Mukhopadhyay, A., & Stahl, P.D. (1997). Reconstitution of phagosome-lysosome fusion in streptolysin O-permabilized cells. J. Biol. Chem. 272, 16147–16151.

Garcia del Portillo, F., & Finlay, B. (1995a). Targetting of Salmonella typhimurium to vesicles containing lysosomal membrane glycoproteins bypasses compartments with mannose 6-phosphate receptors. J. Cell Biol. 129, 81–97.

Garcia del Portillo, F., & Finlay, B. (1995b). The varied lifestyles of intracellular pathogens within eukaryotic vacuolar compartments. Trends Microbiol. 3, 373–380.

Gruneheid, S., Pinner, E., Desjardins, M., & Gros, P. (1997). Natural resistance to infection with intracellular pathogens: the Nramp1 protein is recruited to the membrane of the phagosome. J. Exp. Med. 185, 717–730.

Hirsch, J.G. (1965). Phagocytosis. Ann. Rev. Microbiol. 19, 339–350.

Ishibashi, Y., & Arai, T. (1990). Effect of IFN-γ on phagosome-lysosome fusion in Salmonella typhimurium infected murine macrophages. FEMS Microbiol. Immunol. 64, 75–82.

Mayorga, L., Bertini, F., & Stahl, PD. (1991). Fusion of newly formed phagosomes with endosomes in intact cells and in a cell free system. J. Biol. Chem. 266, 6511–6517.

Mordue, D.G., & Sibley, L.D. (1997). Intracellular fate of vacuoles containing Toxoplasma gondii is determined at the time of formation and depends on the mechanism of entry. J. Immunol. 159, 4452–4459.

Muller, W.A., Steinman, R.M., & Cohn, Z.A. (1980). The membrane proteins of the vacuolar system. II. Bidirectional flow between secondary lysosomes and plasma membrane. J. Cell Biol. 86, 304–314.

Mullock, B.M., Bright, N.A., Fearon, C.W., Gray, S.R., & Luzio, J.P. (1998). Fusion of lysosoms with late endosomes producesa hybrid organelle of intermediate density and is NSF dependent. J. Cell Biol. 140, 591–601.

Papini, E., de Bernard, M., Milia, E., Bugnoli, M., Zerial, M., Rappuoli, R., & Montecucco, C. (1994). Cellular vacuoles induced by Helicobacter pylori originate from late endosomal compartments. Proc. Natl. Acad. Sci. USA 91, 9720–9724.

Papini, E., Satin, B., Bucci, C., de Bernard, M., Telford, J.L., Manetti, R., Rappuoli, R., Zerial, M., & Montecucco, C. (1997). The small GTPase binding protein rab7 is essential for cellular vacuolation induced by Helicobacter pylori cytotoxin. EMBO J. 16, 15–24.

Pitt, A., Mayorga, L., Schwartz, A.L., & Stahl, P.D. (1992a). Transport of phagosomal components to an endosomal compartment. J. Biol. Chem. 267, 126–132.

Pitt, A., Mayorga, L., Stahl, P.D., & Schwartz, A.L. (1992b). Alterations in the protein composition of maturing phagosomes. J. Clin. Invest. 90, 1978–1983.
Pizarro-Cerda, J., Desjardins, M., Moreno, A., Akira, S., & Gorvel, J.P. (1999). Modulation of endocytosis in nuclear factor IL-6(-/-) macrophages is responsible for a high susceptibility to intracellular bacterial infection. J. Immunol. 162, 3519–3526.
Press, B., Feng, Y., Hoflack, B., & Wandiger-Ness, A. (1998). Mutant Rab7 causes the accumulation of cathepsin D and cation-independent mannose 6-phosphate receptor in an early endocytic compartment. J. Cell Biol. 140, 1075–1089.
Rabinowitz, S., Horstmann, H., Gordon, S., & Griffiths, G. (1992). Immunocytochemical characterization of the endocytic and phagolysosomal compartments in peritoneal macrophages. J. Cell Biol. 116, 95–112.
Rathman, M., Sjaastad, M.D., & Falkow, S. (1996). Acidification of phagosomes containing Salmonella typhimurium in murine macrophages. Infect. Immun. 64, 2765–2773.
Schaible, U.E., Sturgill-Koszycki, S., Schlesinger, P.H., & Russell, D.G. (1998). Cytokine activation leads to acidification and increases maturation of Mycobacterium avium-containing phagosome in murine macrophages. J. Immunol. 160, 1290–1296.
Silverstein, S.C., Greenberg, S., DiVirgilio, F., & Steinberg, T.H. (1989). Phagocytosis. In: Fundamental Immunology (Paul, W.E., Ed.), pp. 703–719. Raven Press, New York.
Sturgill-Koszycki, S., Schaible, U.E., & Russell, D.G. (1996). Mycobacterium-containing phagosomes are accesible to early endosomes and reflect a transitional state in normal phagosome biogenesis. EMBO J. 15, 6960–6968.
Sturgill-Koszycki, S., Schlesinger, P.H., Chakraborty, P., Haddix, P.L., Collins, H.L., Fok, A.K., Allen, R.D., Gluck, S.L., Heuser, J., & Russell, D.G. (1994). Lack of acidification in Mycobacterium phagosomes produced by exclusion of the vesicular proton-ATPase. Science 263, 678–681.
Tjelle, T.E., Brech, A., Juvet, L.K., Griffiths, G., & Berg, T. (1996). Isolation and characterization of early endosomes, late endosomes and terminal lysosomes: their role in protein degradation. J. Cell Sci. 109, 2905–2914.
Tjelle, T.E., Saigal, B., Froystad, M., & Berg, T. (1998). Degradation of phagosomal components in late endocytic organelles. J. Cell Sci. 111, 141–148.
Via, L.E., Deretic, D., Ulmer, R.J., Hibler, N.S., Huber, L.A., & Deretic, V. (1997). Arrest of mycobacterial phagosome maturation is caused by a block in vesicle fusion between stages controlled by rab5 and rab7. J. Biol. Chem. 272, 13326–13331.
Ward, D.M., Leslie, J.D., & Kaplan, J. (1997). Homotypic lysosome fusion in macrophages: analysis using an in vitro assay. J. Cell Biol. 139, 665–673.
Yamashiro, D.J., Tycko, B., Fluss, S.R., & Maxfield, F.R. (1984). Segregation of tranferrin to a mildly acidic (pH 6.4) para-Golgi compartment in the recycling pathway. Cell 37, 789–800.

PHAGOSOMAL ACIDIFICATION
MECHANISMS AND FUNCTIONAL SIGNIFICANCE

David J. Hackam, Ori D. Rotstein, and Sergio Grinstein

Advances in Cell and Molecular Biology of Membranes and Organelles
Volume 5, pages 299-319.

ISBN: 1-55938-999-0

I. INTRODUCTION

In order to function as an effective microbicidal organelle, the early phagosome must mature by fusing with endosomes and lysosomes. In concert with phagosomal maturation there is a progressive acidification of the interior of the phagosome, which confers and/or enhances its microbicidal capabilities by several mechanisms. Firstly, the acidification can have a direct bactericidal or bacteriostatic effect, inasmuch as various microorganisms are unable to grow at low pH. In addition, the acidic milieu can activate microbicidal enzymes, such as pH-sensitive proteases or lipases that are delivered to the phagosome during fusion with lysosomes. Lastly, the increased acidification will promote the dismutation of superoxide to hydrogen peroxide, which in turn functions as the substrate for myeloperoxidase, another potent microbicidal agent delivered by phago-lysosomal fusion. The critical role of phagosomal acidification in bacterial killing is perhaps best illustrated by the failure of macrophages to eliminate many species of mycobacteria. These bacteria survive and replicate within phagosomes, at least in part by impairing the process of phagosomal acidification. An understanding of the mechanisms responsible for phagosomal acidification is therefore essential in the analysis of the antimicrobial features of phagocytes.

Because the nascent phagosome is in essence an invagination of the plasma membrane, it is expected to contain the proton transport systems found on the surface of phagocytes. These include the sodium-hydrogen exchanger (NHE) and H^+-conductive channels, as well as a variety of exchange systems and channels that translocate bicarbonate, a base equivalent. During the course of their maturation, phagosomes acquire the vacuolar type proton ATPase (V-ATPase). The following section will present an overview of these pH regulatory systems, and then discuss their contribution to the regulation of phagosomal acidification. Because phagosomal acidification proceeds normally in nominally bicarbonate-free media, systems transporting this anion will not be reviewed in great detail.

II. WHY REGULATE CYTOSOLIC PH DURING PHAGOCYTOSIS?

The resting intracellular pH of leukocytes (pH_i) is approximately 7.2 (Roos et al., 1981). Maintenance of the cytosolic pH near neutrality is essential for effective microbial killing: impairment of cytosolic pH regulation in phagocytes results in decreases in respiratory burst, cellular migration, and lysosomal activity (see Hackam et al., 1996a, for a review). Intracellular acid equivalents are generated as products of metabolism, particularly during activation. In phagocytes, protons are released by the NADPH oxidase during its redox cycle. In addition, the attendant activation of the NADPH-regenerating hexose monophosphate shunt liberates H^+ and CO_2 (Grinstein et al., 1986). The latter hydrates to form carbonic acid. It has been calculated that, if H^+ extrusion failed to occur, the pH_i of fully activated neutrophils could drop by as much as five units within minutes of activation (Lukacs et al., 1993). In addition to the protons generated within the cells, extracellular sources of acid could further accentuate the acid stress experienced by phagocytes, particularly at inflammatory sites. In abscess cavities, for instance, the pH has been reported to be as low as 5.7 (Bryant et al,. 1980). This effect is compounded in settings containing metabolic products such as short-chain fatty acids which serve to shuttle protons into the cell (Rotstein et al., 1987, 1988). For example, *Bacteroides* species, the most prevalent anaerobic bacterial species in intra-abdominal infection, release these weak acids into the local milieu and appear to inhibit the respiratory burst of neutrophils by inducing cytosolic acidification (Rotstein et al., 1987). Phagocytes tend to compensate for this acidic burden by modulating the activity of a family of proton (equivalent) transporters described below.

III. THE SODIUM–PROTON EXCHANGER

The extrusion of intracellular protons in exchange for extracellular sodium ions is one of the principal mechanisms whereby eukaryotic cells, including phagocytes, regulate cytoplasmic pH. This task is performed by a family of antiporters, NHEs, which are characterized by being electroneutral and sensitive to inhibition by the diuretic amiloride and its analogues. A convergence of studies have revealed the existence of a family of NHE isoforms and provided insights into their structural features, subcellular distribution and physiological activity.

A. NHE Isoforms

Genetic and Structural Considerations

In 1989, Sardet and colleagues described the isolation of a cDNA encoding a mammalian sodium-proton exchanger (Sardet et al., 1989). The nucleotide sequence encoded a protein of approximately 90 kDa, which is now referred to as NHE1. Using low-stringency screening of cDNA libraries with probes based on the sequence of NHE1, five other isoforms have been identified (NHE2–NHE6) (Orlowski et al., 1993; Tse et al., 1993; Wang et al., 1993; Klanke et al., 1995). These are each derived from distinct genes, and exhibit 35–55% amino acid identity to each other (Orlowski et al., 1997; Wakabayashi et al., 1997). The predicted membrane topology of all isoforms is similar in that each has 10–12 membrane spanning domains encompassing the amino terminal half of the molecule, and a large cytoplasmic region at the carboxy-terminal half.

Isoform Characteristics

The various isoforms differ with respect to both tissue and subcellular distribution, and in their sensitivity to inhibitors. For instance, NHE1 is expressed in virtually all cells, while NHE2, NHE3 and NHE4 expression is largely restricted to the gastrointestinal tract and kidney (Orlowski et al., 1993; Tse et al., 1993; Wang et al., 1993). NHE5 is expressed in various non-epithelial tissues, including brain, spleen, testis and skeletal muscle (Klanke et al., 1995; Orlowski et al., 1997). To the extent that it is abundant in the spleen, NHE5 may be present in phagocytes, but this has not yet been confirmed experimentally. Differences exist in the subcellular distribution of the isoforms. NHE1 is localized in the basolateral membrane of polarized renal and intestinal epithelial cells, whereas NHE3 is concentrated on the apical surface (Bookstein et al., 1994; Hoogerwerf et al., 1996). The NHE isoforms may also be readily distinguished by their pharmacological properties. NHE1 is approximately 50 to 500 times more sensitive to inhibition by amiloride and its analogues than is NHE3 (Wakabayashi et al., 1997). Potency of inhibition for the other isoforms follows the order: NHE1 > NHE2 >> NHE3 > NHE4 (Orlowski et al., 1997).

As mentioned, NHE1–5 are thought to function predominantly at the plasma membrane. However, Van Dyke (1995) and Hensley et al. (1990) have reported Na^+/H^+ activity in endomembranes as well, and suggested a role for these transporters along the endolysosomal pathway. Accordingly, NHE3 has been demonstrated by immunohistochemical means to be present on subapical vesicles, as well as on perinuclear organelles that colocalize with markers of recycling endosomes (Biemesderfer et al., 1997). Finally, NHE6 was recently reported to be exclusively intracellular, localizing in mitochondria. Therefore, a role for NHE in phagosomal pH regulation must be contemplated (see below).

B. Functional Properties

Direction of Transport

In its physiological or forward mode, the NHE exchanges extracellular sodium for intracellular protons (Figure 1B). Of note, the stoichiometry of exchange is one Na^+ for one H^+, so that the activity of NHE transport is electroneutral. Transport through the NHE is entirely reversible, the direction of net transport being dictated by the combined gradients of Na^+ and H^+. The large inward Na^+ gradient that prevails across the plasmalemma is generated primarily by the Na^+/K^+-ATPase, which exports intracellular Na^+ and imports K^+ in a 3:2 ratio. Importantly, whereas forward transport is stimulated by elevating the extracellular Na^+ concentration, it is inhibited by increasing the extracellular H^+ concentration. This is thought to result from competition between Na^+ and H^+ on the

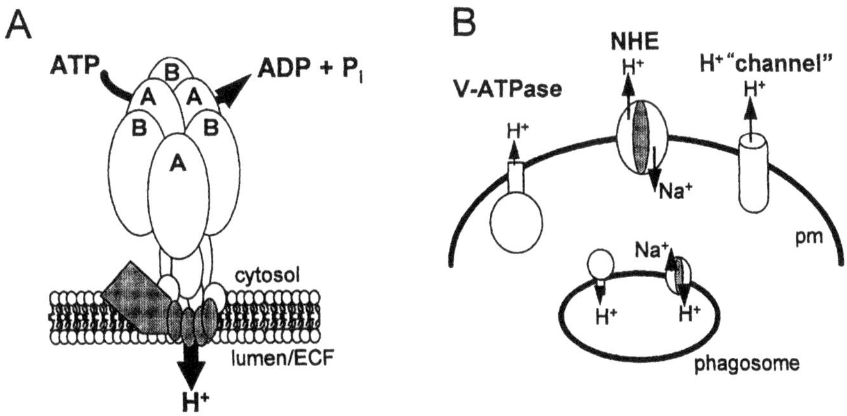

Figure 1. Mechanisms of proton transport. **A.** Schematic representation of vacuolar type (V) H^+-ATPase. **B.** Schematic distribution of V-ATPase, Na^+/H^+-exchanger and proton conductive (channel) pathway on the plasma membrane and the phagosome. ECF, extracellular fluid; pm, plasma membrane. See text for details.

external binding site and can be predicted to inhibit forward NHE activity as the phagosome acidifies.

Determination of the Rate of Transport

Despite the large inward Na^+ gradient established by the Na^+/K^+-ATPase, the rate of Na^+/H^+ exchange is dictated primarily by the intracellular pH (Demaurex et al., 1994). Once the cytosolic pH drops below a certain threshold level or set point, proton extrusion is markedly increased (Aronson et al., 1982). This feature is consistent with a role of the exchanger in protecting the cells against cytosolic acidification. The pronounced sensitivity of the antiport to pH_i has been attributed to the existence of an allosteric modifier site that controls the rate of countertransport (Aronson et al., 1982; Aronson, 1985). When protonated, the modifier site activates cation exchange. As pH_i approaches neutrality, deprotonation of the allosteric site curtails the activity of the antiporter, preventing alkalization of the cytosol beyond the physiological set point.

A variety of other factors have been shown to modulate the rate of NHE activity. In phagocytes, these include, the chemotactic peptide formylmethionylleucylphenylalanine (fMLP), phorbol myristoyl acetate (PMA), and the chemokine leukotriene B_4 (Molski et al., 1980; Grinstein et al., 1984; Sumomoto et al., 1988). Other determinants of NHE activity include cytosolic ATP content, calcium and calcium-binding proteins, tyrosine phosphorylation, GTP-binding proteins and mechanical stress. While several of these could impact on the activation of NHE during the phagocytic process, a full discussion of each of these factors is beyond the scope of this section. For interested readers, further information on this topic can be found in recent reviews (Orlowski et al., 1997; Wakabayashi et al., 1997).

IV. VACUOLAR-TYPE PROTON ATPASE (V-ATPASE)

The vacuolar class or V-class of proton pumps are specialized transporters that utilize cytoplasmic ATP to drive acid equivalents across biological membranes (Forgac, 1989). These pumps are responsible for the acidification of a variety of intracellular compartments, including endosomes, lysosomes, and Golgi-derived vesicles. In addition to their endomembrane distribution, studies have suggested the presence of V-ATPases in the plasma membrane of renal epithelia, osteoclasts and phagocytes. In renal epithelia and osteoclasts, acid extrusion fulfills a tissue-specific biological role, such as acidification of the urine or bone resorption, respectively. In contrast, the plasmalemmal proton pumps of phagocytes are involved in cytoplasmic pH homeostasis, as will be described below. Most importantly, V-ATPases are thought to be the principal source of phagosomal acidification.

A. Structural Considerations

Electron microscopic studies have demonstrated the V-ATPase to have a cytoplasmic stalk, which attaches a ball complex to a transmembrane domain (Figure 1). The cytoplasmic stalk and ball complex is termed V_1, while the membrane sector is designated V_o (Finbow et al., 1997). The V-ATPase forms a heteromeric complex containing at least 13 subunits of aggregate molecular weight 560–740 kDa (Arai et al., 1988). By convention, the subunits in V_1 are designated A-G in decreasing order of mass, while three different subunits comprise the V_o sector. The catalytic (ATPase) activity of the enzyme complex has been shown to reside in the V_1 complex. ATP binds to subunits A (67–73 kDa) and B (55–60 kDa), although ATPase activity is restricted to subunit A (Arai et al., 1987). Proton conductance across the membrane is served by the V_o domain, which has been shown to be composed of a symmetrical hexameric arrangement of 16 kDa (C) subunits (Holzenburg et al., 1993), in association with two other larger proteins. Image analysis of the 16 kDa array has revealed the presence of a central pore, which presumably serves as the site of proton translocation through the membrane (Holzenburg et al., 1993).

B. Functional Considerations

V-ATPase Activity in Phagocytes

Proton pumping by the V-ATPase is an ATP-consuming, electrogenic process, and can operate against the passive electrochemical proton gradient. Pumping is further characterized by its sensitivity to the macrolide antibiotics bafilomycin A_1 and concanamycin, which are very potent and seemingly specific inhibitors. Studies have demonstrated that the V-ATPases are present and functional on the plasma membrane of phagocytes (Finbow et al., 1997; Swallow et al., 1988, 1989, 1990). The V-ATPase may therefore play an important role in cytosolic pH regulation under conditions of acidic extracellular pH, as occurs in abscesses and tumors, where other proton transporters such as NHE and HCO_{3^-}-dependent exchange would be inhibited. Indeed, when the relative activities of these mechanisms were assessed in phagocytes at low extracellular pH (Swallow et al., 1993), the V-ATPases were shown to be the predominant pathway, contributing importantly to pH_i maintenance and preserving the ability of the cells to produce superoxide (Swallow et al., 1993).

ATPases present in the plasmalemma can be expected to be incorporated into the forming phagosome from the outset. Additional pumps are delivered by fusion with V-ATPase rich endomembrane vesicles. The relative contribution of these processes is discussed below.

Activation of V-ATPase Activity

Under resting conditions, a bafilomycin-sensitive efflux of H^+ can be measured in elicited peritoneal macrophages, consistent with constitutive V-ATPase activity in the plasma membrane of these cells. In contrast, unstimulated neutrophils demonstrate little pump activity, even after an acid load is imposed. V-ATPase-induced extrusion of cytosolic H^+ is only observed when neutrophils are activated by agonists of protein kinase C (PKC), such as PMA, or by fMLP (Nanda et al., 1992). The precise location of the functional V-ATPase may also differ between cell types. In neutrophils, cell activation was associated with a translocation of pumps to the plasma membrane (Nanda et al., 1996), while measures to prevent exocytosis in macrophages had no appreciable effect on V-ATPase mediated pH recovery in acid-loaded macrophages (Swallow et al., 1990).

Modulation of V-ATPases at Inflammatory Sites

V-ATPase activity in phagocytes has been shown to be modulated by specific factors in the cellular environment, particularly those present at inflammatory sites. Nitric oxide (NO), the highly reactive nitrogen intermediate formed by the oxidation of L-arginine which affects a variety of phagocyte functions, was found to reduce V-ATPase activity in macrophages. Likewise, chronic incubation of cells in the oxidant metabolite hydrogen peroxide (H_2O_2), which is a normal product of activated phagocytes, caused a dose-dependent decrease in proton pump activity (Brisseau et al., 1994). Augmentation of proton pump activity has also been demonstrated. The inflammatory mediator interleukin-1 caused a time- and dose-dependent increase in V-ATPase activity of macrophages, which involved de novo mRNA and protein synthesis (Brisseau et al., 1996).

V. H^+ CONDUCTIVE PATHWAYS

Stimulation of phagocytes is associated with the activation of a plasma membrane conductive pathway specific for protons (Swallow et al., 1989, 1990; Swallow et al., 1993). Electrophysiological analysis, coupled with perfusion of the cytosolic milieu revealed subsequently that this conductive pathway can be elicited by depolarization of the plasma membrane, even when the cells are otherwise quiescent (Figure 1B). The conductive pathway was shown to be highly directional: it can conduct protons only from inside the cell into the extracellular space, i.e., it is exquisitely outward-rectifying (Swallow et al., 1989). Interestingly, the conductive pathway is acutely activated by acidification of the cytosol yet markedly inhibited by extracellular acidification. These features may provide a mechanism for the regulated extrusion of acid generated during metabolic activation of phagocytes, while preventing deleterious acid influx during the sojourn of phagocytes in acidic environments. It is also noteworthy that both the voltage and the pH-sensitivity of the conductance are shifted when PKC is stimulated (Nanda et al., 1991), or when arachidonic acid is added (Kapus et al., 1994). The direction of the shift is such that acid extrusion would be favored during the course of metabolic activation.

The presence or function of the putative H^+ channels on the phagosomal membrane has not been studied. A conductive H^+ flux could contribute to the acidification of the phagosome or lead to its dissipation, depending on the prevailing transmembrane potential and H^+ gradient. To our knowledge, no determinations of the electrical properties of the phagosome have been reported.

VI. OTHER PH REGULATORY MECHANISMS IN LEUKOCYTES

Three additional pH regulatory mechanisms acting at the plasma membrane of phagocytes deserve mention. An electroneutral anion exchanger transports Cl^- for HCO_3^-, in an analogous fashion to the sodium proton exchanger (Madshus, 1988). Transport is driven by the concentration gradient for the anions, which under physiological conditions would induce Cl-influx with HCO_3^- efflux. The efflux of a base is equivalent to net uptake of protons, so that the anion exchanger can be expected to participate in the recovery from cytosolic alkalosis. In the case of the phagosome, Cl-/HCO_3^- exchange would be expected to shunt the acidification generated by the pump. It is unclear whether Cl-/HCO_3^- exchangers are present and/or active in the phagosomal membrane.

A second type of Cl-/HCO_{3-} exchanger exists in the plasma membrane of some phagocytes. It differs from the conventional anion exchanger described above in its dependence on extracellular Na^+, which is cotransported to the cell interior. Coupling to Na^+ alters the direction of net transport of acid equivalents; under physiological conditions, this system mediates the electroneutral exchange of extracellular Na^+ and HCO_3^- for intracellular Cl- (and possibly H^+). Thus, this system is a net acid extruder and may participate in phagosomal acidification.

VII. DETECTION OF PROTON TRANSPORT ACROSS PHAGOSOMAL MEMBRANES

In order to assess the factors contributing to the regulation of phagosomal acidification, strategies are required to detect the transport of protons across the phagosomal membrane. Initial studies in this area were performed at the light or electron microscopic level, and relied on the pH-dependent distribution of weak acids or bases across the phagosomal membrane (D'Arcy Hart et al., 1972; Gordon et al., 1980). Such experiments were usually end-point determinations and could not be used for dynamic measurements. These limitations were solved by the development of pH-sensitive fluorescent compounds, generally fluorescein derivatives. For optimal determination of phagosomal pH, the pH-sensitive compound must be attached to the surface of the phagocytic particle, so that it senses the pH of the environment to which the particle is exposed. To ensure the accuracy of the pH recordings, however, certain variables must be borne in mind. First, it is essential to distinguish between particles that have entered phagosomes and those that are merely adhered to the cell surface. This is particularly problematic when populations of cells are analyzed and the associated particles are not individually identified. Secondly, it is essential that the probes provide accurate readings at the very acidic pH levels often reached by mature phagolysosome. This requires that the pK_a of the probe be in the range of the values measured. For simple fluorescein derivatives ($pK_a \approx 6.5$) the organellar pH can be over 1.5 units lower than the range where the probe is optimally sensitive. Finally, the probe must withstand the hydrolytic and oxidative environment of the phagosome, remaining fluorescent, pH-sensitive and coupled to the phagocytic particle throughout the course of the experiment.

With these caveats in mind, we have opted to study proton transport across the phagosomal membrane using fluorescence ratio imaging (Figure 2). This technique offers several advantages. It is extremely versatile with respect to the types of cells (macrophages, neutrophils, heterologously transfected fibroblasts) and the phagocytic particles (latex beads, yeast, bacteria) that can be monitored (Hackam et al., 1997a). It allows for the simultaneous determination of phagosomal acidification in multiple cells that are analyzed individually, providing statistically meaningful information. In addition, the morphology of the cells and particles can be monitored continuously, providing important indicators of cellular status and of the location of the particles. Finally, microscopic ratio imaging can be combined

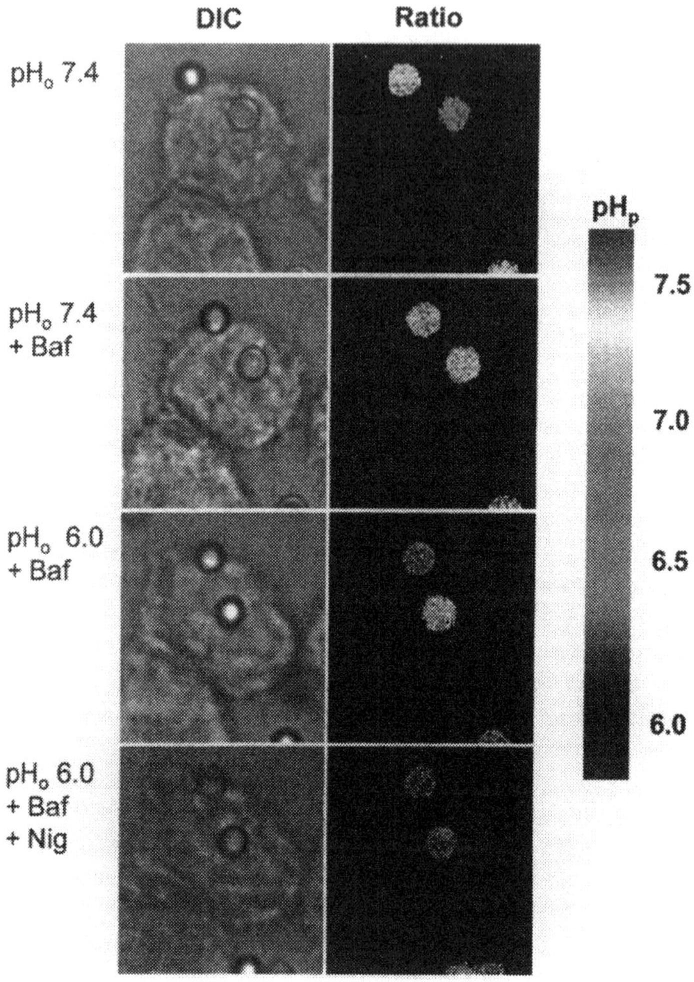

Figure 2. Measurement of phagosomal pH (pHp). J774 macrophages were allowed to interact with fluoresceinated, opsonized 3 mm diameter latex beads, some of which became internalized. The cells were then visualized using differential interference contrast optics (left panels), while the fluorescence was measured with alternating excitation at 440 and 490 nm. The fluorescence ratio (right panels) was used for the measurement of the pH in the vicinity of the bead surface. A pseudocolor pH scale is shown to the right. One extracellular and one internalized bead are shown, to facilitate understanding of the experimental protocol, which proceeds sequentially from top to bottom. Top line: cells incubated under physiological conditions, extracellular pH (pH_o) 7.4. Second line from top: image acquired 12 minutes after addition of 200 nM bafilomycin. Third line from top: bafilomycin-treated cells were exposed to medium pH_o 6.0. Bottom line: cells exposed to K^{+}-rich medium, pH_o 6.0. Adapted from Hackam et al., (1997b) with permission.

with other single cell methodologies, such as microinjection, extending the range of applications. The principal disadvantage of this methodology is the dependence on highly specialized and costly imaging equipment, which requires ongoing maintenance and technical support.

A representative experiment designed to determine the pH of phagocytosed particles is illustrated in Figure 2, which proceeds sequentially from top to bottom. The pH scale to the right of the figure provides calibration of the pH of the fluoresceinated particles, as displayed in the pseudocolor ratio image. Because differential interference contrast microscopy could not reliably discriminate between extra- and intracellular particles, three functional criteria were routinely used. First, the pH of particles within phagosomes was elevated by addition of bafilomycin, the inhibitor of the V-ATPase or NH_4Cl (not shown), which traverses the plasma and phagosomal membranes as NH_3 and becomes protonated in the phagosomal lumen. Neither bafilomycin nor NH_4Cl are expected to affect extracellular beads. Secondly, abrupt changes in extracellular pH affected extracellular beads, but had no acute effect on intraphagosomal particles (Figure 2). Finally, addition of nigericin altered phagosomal pH (pH_p), but had no effect on extracellular beads. Therefore, sensitivity to added inhibitors, to NH_4Cl or to extracellular acidification can be used to identify intraphagosomal particles.

As with any membrane-bound organelle, the pH of the phagosome represents the net balance between proton accumulation and proton leakage. When either one of these parameters is altered, the magnitude of the resulting changes in pH will be dictated by the buffering capacity of the phagosomal compartment. A complete assessment of the factors responsible for regulating phagosomal pH requires knowledge of these parameters.

VIII. FACTORS REGULATING PHAGOSOMAL PH

A. The V-ATPase

The factors responsible for the acidification observed as phagosomes mature were determined by several independent means. Using a pharmacological approach, it has been established that the acidification is completely blocked using bafilomycin A_1 or concanamycin, known inhibitors of the V-ATPase (e.g. Figure 3). Second, using cells treated to selectively permeabilize the plasma membrane (without affecting the phagosomal membrane), it was determined that phagosomal acidification was dependent upon ATP and Mg, known substrates for the V-ATPase (Hackam et al., 1998a). In view of these findings, the principal mechanism responsible for acidification of the phagosomal lumen appears to be active, ATP-dependent proton transport.

Two possible sources could provide the proton pumps which become incorporated into the phagosome. Since at least some of these transporters are thought to

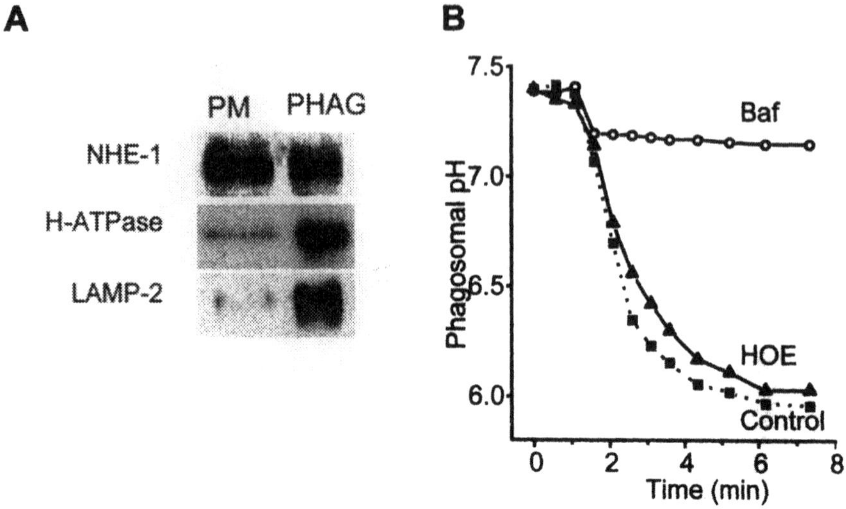

Figure 3. Mechanisms of phagosomal acidification. **A**. Demonstration that NHE1 is present in the phagosomal membrane. Purified plasmalemmal (PM) and phagosomal (PHAG) membranes (20 μg protein each) were subjected to electrophoresis and blotted onto nitrocellulose. The blots were then probed with polyclonal antibodies to NHE1 or to the 39 kDa subunit of the V-ATPase, or with monoclonal anti-LAMP-2 antibody. **B**. Phagosomal pH was monitored as in Figure 2 to determine the effect of pump and antiporter inhibitors on pH_p. Time course of acidification of latex beads in untreated (control) J774 cells (squares), in cells treated with 1 μM HOE 694 (triangles), or in cells exposed to bafilomycin and concanamycin (100 nM each). Adapted from Hackam et al., (1997b) with permission.

be present on the surface of macrophages, they become engulfed as part of the plasma membrane that will constitute the nascent phagosome. Alternatively, the pumps may be delivered to the phagosomal membrane by fusion with endomembrane organelles known to be rich in pumps, namely endosomes and lysosomes. Several sets of observations suggest that endomembrane fusion is the predominant source of V-ATPases. First, when purified phagosomes where isolated from macrophages that had internalized phagocytic particles, the density of pumps as determined by immunoblotting was shown to increase over time (Pitt et al., 1992; Desjardins et al., 1994). Additional information is provided from studies on purified phagosomes containing mycobacterial species. A collection of studies have shown that these phagosomes do not fully mature after they are formed (Sturgill-Koszycki et al., 1994; Xu et al., 1994; Hackam et al., 1997b). Instead, they remain suspended at a stage where they interact with early and recycling endosomes, yet are unable to interact with distal components of the internalization network, such as late endosomes and lysosomes (Sturgill-Koszycki et al., 1996). Of note, the density of pumps on these phagosomes is substantially reduced. Their pH never reaches the very acidic range attained by phagosomes containing inert particles, which mature more completely (Sturgill-Koszycki et al., 1994). This implies that the extent to which the phagosomal membrane interacts with the endosomal network is the chief determinant of pump delivery and of luminal acidification.

What regulates the delivery of V-ATPases to the phagosome? The answer to this question will require a more complete understanding of vesicular traffic and is essential to comprehend how phagosomal acidification is thwarted by intracellular pathogens. Several recent observations in other systems provide some insight into the possible mechanisms involved. It is most likely that fusion will involve molecules of the soluble NSF attachment protein (SNAP) receptor (SNARE) family and that N-ethylmaleimide-sensitive factor (NSF) and SNAP will catalyze or prime the reaction (Hackam et al., 1996b; Hackam et al., 1998b). Small molecular weight GTPases of the rab family and very likely to regulate the process. Divalent cations including calcium (Jaconi et al., 1990; Mayorga et al., 1994; Zimmerli et al., 1996), and phosphoinositides (Ninomiya et al., 1994) will surely prove to be important triggering agents, and proximity of the phagosome with its target compartments is likely to be governed by translocation along microtubules (Blocker et al., 1996, 1997). In addition, biophysical parameters may be critical determinants of phagosomal maturation through fusion. By analogy, the transition between early and late endosomes was found to be dependent upon a gradient of decreasing pH along the endosomal pathway: dissipation of acidification precludes normal traffic between these compartments (Gruenberg et al., 1995). By analogy, it is conceivable that phagosomal maturation, and therefore the accumulation of proton pumps, requires acidification of either endosomes or of the phagosome itself. This seemingly leads to a circular argument: is phagosomal acidification the cause or a consequence of phagosomal maturation? Resolution of this dilemma requires definition of whether: (a) pumps pre-existing in the plasmalemma contribute significantly to the early phase of acidification, and (b) acidification is essential for fusion with early endosomes, or only with late endosomes and/or lysosomes. More detailed kinetic studies will be required to define these issues. Clearly, our understanding of the mechanisms governing delivery of V-ATPases to phagosomes remains in its infancy.

B. Na^+/H^+ Antiporter

Because the plasma membrane is internalized during phagocytosis, we have considered whether the NHE could participate in the regulation of phagosomal pH. In an initial series of studies, we assessed the activity of the antiporter in phagosomes by inducing neutrophils to engulf bacteria labeled with a pH-sensitive fluorescent dye (Grinstein et al., 1984). Because these studies were performed before single cell imaging techniques were available to us, population studies were performed. To remove extracellularly adhering bacteria, cells were washed extensively and treated with lysozyme, an enzyme capable of digesting the bacterial wall without affecting the neutrophil membrane. Though imperfect, these studies indicated that the rate and extent of phagosomal acidification were not affected by: (a) addition of methyl-isobutyl-amiloride, a permeable and very potent amiloride analogue, at concentrations several-fold higher than required to inhibit the antiporter of the surface membrane, and (b) omission of Na^+ from the medium in which phagocytosis occurred. These findings indicated that intrinsic Na^+/H^+ exchange was unlikely to contribute to phagosomal acidification. However, these studies did not address whether such antiporters were present on phagosomes, whether they could function, and if so, why they were found to be incapable of contributing to phagosomal acidification.

To address these questions, additional studies were performed using microscopic imaging and biochemical techniques (Hackam et al., 1997b). First, we examined which of the known isoforms of NHE were present in macrophages. Using a combined biochemical, molecular and pharmacological approach, we determined that J774 cells, which have a monocytic phenotype, express only NHE1, the ubiquitous isoform of the exchanger. To address whether NHE1 was incorporated into phagosomal membranes, J774 cells were allowed to internalize latex particles, and purified phagosomes were isolated by density gradient centrifugation. The purity of the resulting population was assessed by transmission electron microscopy (see Hackam et al., 1996b for representative micrograph). These mature phagosomes were rich in lysosomal associated membrane protein-2 (LAMP-2) and V-ATPases, indicating extensive fusion with endolysosomes (Figure 3A). Importantly, NHE1 was clearly detectable in the phagosomal membrane, at a density that was comparable to that found in the plasma membrane (Figure 3A).

On the phagosome, the NHE could conceivably affect phagosomal pH. Along with the opsonized particle, the nascent phagosome traps a variable volume of the extracellular milieu, which is usually Na^+ rich. The phagosomal lumen would therefore be expected to contain Na^+ at a concentration significantly higher than that of the cytoplasm. The extent to which NHE can theoretically contribute to phagosomal acidification can be estimated on thermodynamic grounds, based on the knowledge that the exchange reaction is electroneutral, with a 1:1 stoichiometry. Because at equilibrium

$$\frac{[Na^+]phag}{[Na^+]cytosol} = \frac{[H^+]phag}{[H^+]cytosol} \tag{1}$$

the ratio of cytosolic to phagosomal (luminal) Na^+ will dictate to what extent the phagosomal pH (pH_p) can deviate from the cytosolic pH. Assuming that the phagosomal $[Na^+]$ is maintained at extracellular levels (approximately 140 mM) and that cytoplasmic $[Na^+]$ approximates 10–15 mM (Roos et al., 1981), pH_p could in principle become up to two pH units more acidic than the cytosol, approaching the values reported in many studies (Alpuche-Aranda et al., 1992; Bouvier et al., 1994; Steinberg et al., 1994). Thus, NHE could conceivably contribute importantly to phagosomal acidification. Yet, when the rate and extent of acidification of the phagosome were compared under conditions where NHE was functional or inactivated, little difference was found (Figure 3B). These findings imply that antiport activity does not contribute measurably to acidification of the phagosome.

Although the antiporters do not seem to participate in phagosomal acidification, they are nevertheless present and active in the phagosomal membrane. This was demonstrated by forcing NHE to operate in the reverse mode (i.e., exchanging luminal H^+ for cytoplasmic Na^+). In these experiments the cytoplasmic Na^+ concentration was raised by incubating cells in medium containing 140 mM NaCl and ouabain, an inhibitor of the Na^+ pump. The phagosome was allowed to acidify spontaneously, then the V-type pumps were acutely blocked by addition of bafilomycin and/or concanamycin. Under these conditions, reverse NHE activity would be expected to increase pH_p. Importantly, the rate of alkalinization was significantly greater in Na^{+-} loaded cells compared to low $[Na^+]$ controls, and this difference was sensitive to the amiloride analogue HOE 694 (Figure 4A).

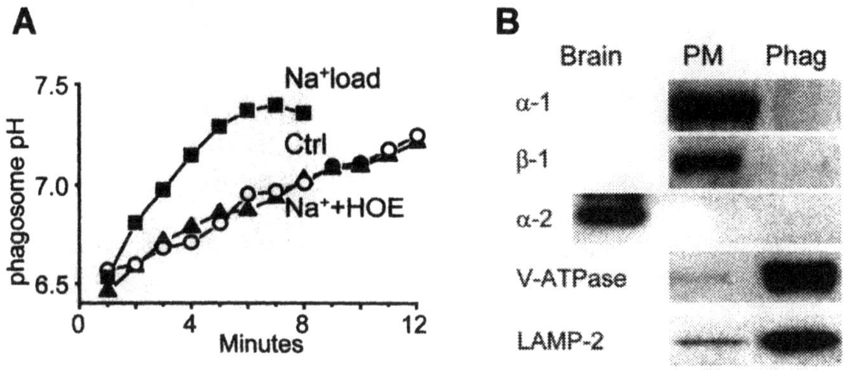

Figure 4. Assessment of NHE activity in the phagosomal membrane. **A.** Rate of dissipation of phagosomal acidification upon addition of bafilomycin and concanamycin (100 nM each). Cells were otherwise untreated (open circles) or had been Na^+-loaded for 60 minutes. Where specified, 1 μM HOE 694 was present during the last phase of Na^+-loading, as well as during phagocytosis and dissipation. **B.** Assessment of the presence of the Na^+/K^+-ATPase in surface and phagosomal membranes. Purified plasmalemmal (PM) and phagosomal (Phag) membranes obtained 30 or 60 min after phagocytosis (20 μg protein each), or brain microsomes (Brain) were subjected to electrophoresis and blotted onto nitrocellulose. The blots were then probed with polyclonal antibodies to the α-1, β-1 and α-2 subunits of the Na^+/K^+-ATPase, the 39 kDa subunit of the V-ATPase and LAMP-2. Adapted from Hackam et al. (1997b) with permission.

From these data, we concluded that the NHE was present and capable of functioning on phagosomal membranes.

Why, then, is the NHE unable to contribute to phagosomal acidification? A significant sustained contribution of NHE to phagosomal acidification requires the continued existence of a Na^+ gradient across the phagosomal membrane. The latter would have to be maintained by sustained active Na^+ pumping. Remarkably, direct immunochemical assessment of the abundance of $\alpha1$, $\alpha2$ and $\beta1$ subunits of Na^+/K^+-ATPase revealed that mature phagosomes are virtually devoid of these pumps (see Figure 4B). This finding seems to explain why, despite the presence of active NHE1 in the phagosomal membrane, the antiporter fails to contribute measurably to steady-state phagosomal acidification. In the absence of Na^+/K^+-ATPases, the combined H^+ and Na^+ gradients may not only fail to support luminal acidification, but may actually favor dissipation of the H^+ gradient generated by the V-ATPase.

C. Other Potential Regulators of Phagosomal pH

Two other processes may contribute significantly to the establishment and maintenance of the phagosomal pH. First, the H^+ conductance described earlier could in principle promote luminal acidification, if a sufficiently large potential develops across the phagosomal membrane (lumen negative). It has been demonstrated that, at least in neutrophils, a marked depolarization occurs across the plasma membrane of activated cells, ostensibly as a result of activation of the electrogenic NADPH oxidase. To the extent that the oxidase is particularly abundant in the phagosome, an even larger depolarization could in fact develop across the phagosomal membrane, promoting the activation of the conductance and driving net H^+ flow into the lumen. To date, however, the potential contribution of the H^+ channels to phagosomal pH remains untested.

The products of the oxidase have been claimed to promote the dissipation of the acidification initiated by the H^+ pumps. Segal et al. (1981) found that shortly after sealing of the phagosomal membrane, pH_p in fact became *alkaline* before a secondary acidification was noticeable. The initial alkalosis was attributed to the products of the NADPH oxidase, since it was not observed in patients suffering from chronic granulomatous disease, who lack a functional oxidase complex. Specifically, it was suggested that dismutation of superoxide anions to peroxide consumes H^+, accounting for the elevation in pH. The precise time course of these events and their contribution vis-à-vis the H^+ pump and H^+ conductance merit further analysis.

IX. IMPORTANCE OF V-ATPASE–MEDIATED PHAGOSOMAL ACIDIFICATION TO INTRACELLULAR INFECTION: THE ROLE OF NRAMP

Recent studies have shed light upon the importance of V-ATPase-mediated phagosomal acidification for the antimicrobial properties of phagocytes. As described earlier, mycobacteria thrive within phagosomes by minimizing the incorporation of V-ATPases, thus preventing complete phagosomal acidification. Interestingly, macrophages from different strains of mice are differentially susceptible to this effect, in a manner that correlates with their resistance to mycobacterial infection. The genetic locus regulating susceptibility to mycobacterial infection has been identified at the *bcg* allele, and the gene product is called Nramp, which stands for natural resistance associated macrophage protein. While Nramp does not affect the ability of macrophages to internalize mycobacteria, it has a dramatic effect on their ability to perform intracellular killing. Given the importance of phagosomal acidification to mycobacterial killing (see above), we investigated whether there were differences in the pH of mycobacterial phagosomes in Nramp-expressing (wild-type) mice compared to mice where the *bcg* locus had been knocked out by recombination (generated by Dr. P. Gros, McGill University, Montreal, Canada). Interestingly, the phagosomal pH was significantly more acidic in wild-type mice compared to their knockout counterparts (Hackam, et al., 1998a). This effect was specific for live, virulent mycobacteria, as inert particles and heat-killed mycobacteria formed phagosomes that became very acidic. Importantly, the increased acidification was associated with enhanced delivery of V-ATPases to the phagosomal membrane. These data illustrate the importance of phagosomal acidification to mycobacterial killing, and suggest that regulation of phagosomal pH may participate in the determination of resistance to infections by mycobacteria and possibly other organisms.

X. CONCLUDING REMARKS AND FUTURE DIRECTIONS

Taken together, the available evidence strongly indicates that phagosomal acidification is a key determinant of the microbicidal activity of macrophages and neutrophils. This is highlighted by the failure of such cells to kill internalized mycobacterial species that survive within cells, at least in part, by preventing phagosomal acidification. The acidic milieu in the phagosomal lumen is generated and maintained largely, and perhaps entirely, through the activity of ATP-dependent proton pumps of the vacuolar or V type. While a modest number of pumps may be present on the plasmalemma as it invaginates to form a phagosome, the vast majority of V-ATPases are delivered to the phagosome subsequently, as it matures within the cell along a pathway analogous to endosomal maturation. Ongoing and future work will undoubtedly disclose the molecular mechanisms governing the traffic of pump-containing vesicles to the phagosome, and the strategies that some microbes have developed to subvert this process, thereby acquiring the ability to become intracellular parasites.

ACKNOWLEDGMENTS

Research in the laboratories of S. G. and O. D. R. is supported by the Medical Research Council of Canada. D. J. H. is the recipient of a Medical Research Council of Canada Fellowship and an Ethicon-Society of University Surgeons Surgical Research Award. S. G. is an International Scholar of the Howard Hughes Medical Institute and is the current holder of the Pitblado Chair in Cell Biology.

REFERENCES

Alpuche Aranda, C.M., Swanson, J.A., Loomis, W.P., & Miller, S.I. (1992). Salmonella typhimurium activates virulence gene transcription within acidified macrophage phagosomes. Proc. Natl. Acad. Sci. USA 89, 10079-10083.

Arai, H., Burn, M., & Forgac, M. (1987). Inhibition of the coated vesicle proton pump and labelling of a 17,000 dalton polypeptide by N,N′-dicyclohexylcarbodiimide. J. Biol. Chem. 262, 11006-11011.

Arai, H., Terres, G., Pink, S., & Forgac, M. (1988). Topography and subunit stoichiometry of the coated vesicle proton pump. J. Biol. Chem. 263, 8796-8802.

Aronson, P.S. (1985). Kinetic properties of the plasma membrane Na+/H+ exchanger. Annu. Rev. Physiol. 47, 545-560.

Aronson, P.S., Nee, J., & Suhm, M.A. (1982). Modifier role if internal H+ in activating the Na+-H+ exchanger in renal microvillus membrane vesicles. Nature 299, 161-163.

Biemesderfer, D., Rutherford, P.A., Nagy, T., Pizzonia, J.H., Abu-Alfa, A.K., & Aronson, P.S. (1997). Monoclonal antibodies for high-resolution localization of NHE3 in adult and neonatal rat kidney. Am. J. Physiol. 273, 289-299.

Blocker, A., Severin, F.F., Burkhardt, J.K., Bingham, J.B., Yu, H., Olivo, J., Schroer, T.A., Hyman, A.A., & Griffiths, G. (1997). Molecular requirements for bidirectional movements of phagosomes along microtubules. J. Cell Biol. 137, 113-129.

Blocker, A., Severin, F.F., Haberman, A., Hyman, A.A., Griffiths, G., & Burkhardt, J.K. (1996). Microtubule-associated protein-dependent binding of phagosomes to microtubules. J. Biol. Chem. 271, 3803-3811.

Bookstein, C., Depaoli, A.M., Xie, Y., Niu, P., Musch, M.W., Rao, M.C., & Cheng, E.B. (1994). Na/H exchangers, NHE1 and NHE3 or rat intestine. Expression and localization. J. Clin. Invest. 93, 106-113.

Bouvier, G., Benoliel, A.M., Foa, C., & Bongrand, P. (1994). Relationship between phagosome acidification, phagosome-lysosome fusion, and mechanism of particle ingestion. J. Leukoc. Biol. 55, 729-734.

Brisseau, G.F., Grinstein, S., Hackam, D.J., Nordstrom, T., Manolson, M.F., Khine, A.A., & Rotstein, O.D. (1996). Interleukin-1 increases vacuolar type H+-ATPase activity in murine peritoneal macrophages. J. Biol. Chem 271, 2005-2011.

Brisseau, G.F., Tsai, O., Nordstrom, T., Marshall, J.C., Grinstein, S., & Rotstein, O.D. (1994). Oxidant stress inhibits pH regulatory mechanisms in murine peritoneal macrophages. Surgery 116, 268-274; discussion 274-275.

Bryant, R.E., Rashad, A.L., Mazza, J.A., & Hammond, D. (1980). Beta-lactamase activity in human pus. J. Infect. Dis. 142, 594-601.

D'Arcy Hart, P., Armstron, J.A., Brown, C.A., & Draper, P. (1972). Ultrastructural study of the behaviour of macrophages toward parasitic mycobacteria. Infect. Immun. 5, 803-807.

Demaurex, N., & Grinstein, S. (1994). Na/H antiport: modulation by ATP and role in cell volume regulation. J. Exp. Biol. 196, 389-404.

Desjardins, M., Huber, L.A., Parton, R.G., & Griffiths, G. (1994). Biogenesis of phagolysosomes proceeds through a sequential series of interactions with the endocytic apparatus. J. Cell. Biol. 124, 677-688.

Finbow, M.E., & Harrison, M.A. (1997). The vacuolar H+-ATPase: a universal proton pump of eukaryotes. Biochem. J. 324, 697-712.

Forgac, M. (1989). Structure and function of a vacuolar class of ATP-driven proton pumps. Physiol. Rev. 69, 765-769.

Gordon, A.H., D'Arcy Hart, P., & Young, M.R. (1980). Ammonia inhibits phagosome-lysosome fusion in macrophages. Nature 286, 79-80.

Grinstein, S., & Furuya, W. (1984). Amiloride sensitive Na/H exchange in human neutrophils: mechanism of activation by chemotactic factors. Biochem. Biophys. Res. Commun. 122, 755-762.

Grinstein, S., & Furuya, W. (1986). Cytoplasmic pH regulation in phorbol ester-activated human neutrophils. Am. J. Physiol. 251, C55-C65.

Gruenberg, J., & Maxfield, F.R. (1995). Membrane transport in the endocytic pathway. Curr. Opin. Cell Biol. 7, 552-563.

Hackam, D.J., Grinstein, S., & Rotstein, O.D. (1996a). Intracellular pH regulation in leukocytes: mechanisms and functional significance. Shock 5, 17-21.

Hackam, D.J., Rotstein, O.D., Bennett, M.K., Klip, A., Grinstein, S., & Manolson, M.F. (1996b). Characterization and subcellular localization of target membrane soluble NSF attachment protein receptors (t-SNAREs) in macrophages. Syntaxins 2,3 and 4 are present on phagosomal membranes. J. Immunol. 156, 4377-4383.

Hackam, D.J., Rotstein, O.D., Schreiber, A.D., Zhang, W.J., & Grinstein, S. (1997a). Rho is required for the initiation of calcium signaling and phagocytosis by Fcg receptors in macrophages. J. Exp. Med. 186, 955-966.

Hackam, D.J., Rotstein, O.D., Schreber, A., Trumble, W.S., & Grinstein, S. (1998b). V-SNARE dependent secretion is required for phagocytosis. Proc. Nat. Acad. Sci. U.S.A. 95, 11691–11696.

Hackam, D.J., Rotstein, O.D., Zhang, W.J., Demaurex, N., Woodside, M., Tsai, O., & Grinstein, S. (1997b). Regulation of phagosomal acidification: Differential targeting of Na^+/H^+ exchangers, Na+/K+-ATPases, and vacoular-type H+-ATPases. J Biol Chem 272, 29810-29820.

Hackam, D.J., Rotstein, O.D., Zhang, W.J., Gruenheid, S., Gros, P., & Gruenstein, S. (1998a). Resistance to intracellular injection: Mutation of *Nramp1* impairs phagosomal acidification. J. Exp. Med. 188, 351–364.

Hensley, C., Bradley, M., & Mircheff, A.K. (1990). Subcellular distribution of Na+/H+ antiport activity in rat renal cortex. Kidney Int. 37, 707-716.

Holzenburg, A., Jones, P.C., Franklin, T., Palli, T., Heinberg, T., Marsh, D., Findlay, J.B.C., & Finbow, M.E. (1993). Evidence for a common structure for a class of membrane channels. Eur. J. Biochem. 213, 21-30.

Hoogerwerf, W.A., Tsao, S.C., Devuyst, O., Levine, S.A., Yun, C.H.C., Yip, J.W., Cohen, M.E., Wilson, P.D., Lazenby, A.J., Tse, C.M., et al., (1996). NHE2 and NHE3 are human and rabbit intestinal brush-border proteins. Am. J. Physiol. 270, G29-G41.

Jaconi, M.E., Lew, D.P., Carpentier, J.L., Magnusson, M., Sjogren, M., & Stendahl, O. (1990). Cytosolic free calcium elevation mediates the phagosome-lysosome fusion during phagocytosis in human neutrophils. J. Cell Biol. 110, 1555-1564.

Kapus, A., Romanek, R., & Grinstein, S. (1994). Arachidonic acid stimulates the plasma membrane H+ conductance of macrophages. J. Biol. Chem. 269, 4736-4745.

Klanke, C.A., Su, Y.R., Callen, D.F., Wang, Z., Meneton, P., Baird, N., Kandasamy, R.A., Orlowski, J., Otterud, B.E., & Leppert, M. (1995). Molecular cloning and physical and genetic mapping of a novel human Na+/H+ exchanger (NHE5/SLC9A5) to chromosome 16q22.1. Genomics 25, 615-622.

Lukacs, G., Rotstein, O.D., & Grinstein, S. (1995). An overview of intracellular pH regulation: Role of H^+-ATPases. In: Organellar Proton-ATPases (Nelson, N., Ed.), pp. 29-47. R.G. Landes, Austin, TX.

Lukacs, G.L., Kapus, A., Nanda, A., Romanek, R., & Grinstein, S. (1993). Proton conductance of the plasma membrane: properties, regulation, and functional role. [Review]. Am.J. Physiol. 265, C3-C14.

Madshus, I.H. (1988). Regulation of intracellular pH in eukaryotic cells. Biochem. J. 250, 1-8.

Mayorga, L.S., Beron, W., Sarrouf, M.N., Colombo, M.I., Creutz, C., & Stahl, P.D. (1994). Calcium dependent fusion among endosomes. J. Biol. Chem. 269, 30927-30934.

Molski, T.F.P., Naccache, P.H., Volpe, M., Wolpert, L.M., & Sha'afi, R.I. (1980). Specific modulation of the intracellular pH of rabbit neutrophils by chemotactic factors. Biochem. Biophys. Res. Commun. 94, 508-514.

Nanda, A., & Grinstein, S. (1991). Protein kinase C activates an H+ (equivalent) conductance in the plasma membrane of human neutrophils. Proc. Natl. Acad. Sci. USA 88, 10816-10820.

Nanda, A., Brumell, J.H., Nordstrom, T., Kjeldsen, L., Sengelov, H., Borregaard, N., Rostein, O.D., & Grinstein, S. (1996). Activation of proton pumping in human neutrophils occurs by exocytosis of vesicles bearing vacuolar-type H^+-ATP-ases. J. Biol. Chem. 271, 15963-15970.

Nanda, A., Gukovskaya, A., Tseng, J., & Grinstein, S. (1992). Activation of vacuolar-type proton pumps by protein kinase C. Role in neutrophil pH regulation. J. Biol. Chem 267, 22740-22746.

Ninomiya, N., Hazeki, K., Fukui, Y., Seya, T., Okada, T., Hazeki, O., & Ui, M. (1994). Involvement of phosphatidylinositol 3-kinase in Fc gamma receptor singalling. J. Biol. Chem. 269, 22732-22737.

Orlowski, J., & Grinstein, S. (1997). Na/H exchangers of mammalian cells. J. Biol. Chem. 272, 22373-22376.

Orlowski, J., Kandasamy, R.A., & Shull, G.E. (1993). Molecular cloning of putative members of the Na/H exchanger gene family. cDNA cloning, deduced amino acid sequence, and mRNA tissue expressin of the rat Na/H exchanger NHE-1 and two structurally related proteins. J. Biol. Chem. 267, 9331-9339.

Pitt, A., Mayorga, L.S., Stahl, P.D., & Schwartz, A.L. (1992). Alterations in the protein composition of maturing phagosomes. J. Clin. Invest. 90, 1978-1983.

Roos, A., & Boron, W.F. (1981). Intracellular pH. Physiol. Rev. 61, 296-434.

Rotstein, O.D., Nasmith, P.E., & Grinstein, S. (1987a). The Bacteroides by-product succinic acid inhibits neutrophil respiratory burst by reducing intracelular pH. Infect. Immun. 55, 864-870.

Rotstein, O.D., Nasmith, P.E., & Grinstein, S. (1988). pH-dependent impairment of the neutrophil respiratory burst by the Bacteroides byproduct succinate. Clin. Invest. Med. 11, 259-265.

Rotstein, O.D., Wells, C.L., Pruett, T.L., Sorenson, J.J., & Simmons, R.L. (1987b). Succinic acid production by *Bacteroides fragilis*: a potential virulence factor. Arch. Surg. 122, 93-98.

Sardet, C., Franchi, A., & Pouyssegur, J. (1989). Molecular cloning, primary structure, and expression of the human growth factor activatable Na^+/H^+ antiporter. Cell 56, 271-280.

Segal, A.W., Geisow, M., Garcia, R., Harper, A., & Miller, R. (1981). The respiratory burst of phagoyctic cells is associated with a rise in vacuolar pH. Nature 290, 406-409.

Steinberg, T.H., & Swanson, J.A. (1994). Measurement of phagosome-lysosome fusion and phagosome pH. Meth. Enzymol. 236, 147-160.

Sturgill-Koszycki, S., Schlesinger, P.H., Chakraborty, P., Haddix, P.L., Collins, H.L., Fok, A.K., Allen, R.D., Gluck, S.L., Heuser, J., & Russell, D.G. (1994). Lack of acidification in *Mycobacterium* phagosomes produced by exclusion of the vesicular proton-ATPase. Science 263, 678-681.

Sturgill-Koszycki, S., Shiable, U.E., & Russell, D.G. (1996). Mycobacterium-containing phagosomes are accessible to early endosomes and reflect a transitional state in normal phagosome biogenesis. EMBO J. 15, 6960-6968.

Sumomoto, H., Sato, M., Takeshig, K., Cragoe, E.J., & Minakami, S. (1988). Cytoplasmic pH change induced by leukotriene B4 in human neutrophils. Biochem. Biophys. Acta 970, 31-38.

Swallow, C.J., Grinstein, S., & Rotstein, O.D. (1988). Cytoplasmic pH regulation in macrophages by an ATP-dependent and N,N′-dicyclohexylcarbodiimide-sensitive mechanism. Possible involvement of a plasma membrane proton pump. J. Biol. Chem. 263, 19558-19563.

Swallow, C.J., Grinstein, S., & Rotstein, O.D. (1990). A vacuolar type H^+-ATPase regulates cytoplasmic pH in murine macrophages. J. Biol. Chem 265, 7645-7654.

Swallow, C.J., Grinstein, S., Sudsbury, R.A., & Rotstein, O.D. (1993). Relative roles of Na+/H+ exchange and vacuolar-type H^+ ATPases in regulating cytoplasmic pH and function in murine peritoneal macrophages. J. Cell. Physiol. 157, 453-460.

Swallow, C.J., Rotstein, O., & Grinstein, S. (1989). Mechanisms of cytoplasmic pH recovery in acid-loaded macrophages. J. Surg. Res.46, 588-592.

Tse, C.M., Levine, S., Yun, C., Brant, S., Counillon, L.T., Pouyssegur, J., & Donowitz, M. (1993). Structure/function studies of the epithelial isoforms of the mammalian Na+/H+ exchanger gene family. J. Membr. Biol. 135, 93-1-8.

Van Dyke, R.W. (1995). Na+/H+ exchange modulates acidification of early rat liver endocytic vesicles. Am. J. Physiol. 269, C943-C954.

Wakabayashi, S., Shigekawa, M., & Pouyssegur, J. (1997). Molecular physiology of vertebrate Na+/H+ exchangers. Physiol. Rev. 77, 51-74.

Wang, Z., Orlowski, J., & Shull, G.E. (1993). Primary structure and functional expression of a novel gastrointestinal isoform of the rat Na/H exchanger. J. Biol. Chem. 268, 11925-11928.

Xu, S., Cooper, A., Sturgill-Koszycki, S., van Heyningen, T., Chatterjee, D., Orme, I., Allen, P., & Russell, D.G. (1994). Intracellular trafficking in *Mycobacterium tuberculosis* and *Mycobacterium avium*-infected macrophages. J. Immunol. 153, 2568-2578.

Zimmerli, S., Majeed, M., Gustavsson, M., Stendahl, O., Sanan, D.A., & Ernst, J.D. (1996). Phagosome-lysosome fusion is a calcium-independent event in macrophages. J. Cell Biol. 132, 49-61.

THE PHAGOCYTE ACTIN CYTOSKELETON

Hui-Qiao Sun, Keng-Mean Lin, Masaya Yamamoto, and Helen L. Yin

Advances in Cell and Molecular Biology of Membranes and Organelles
Volume 5, pages 321-342.

ISBN: 1-55938-999-0

I. THE CORTICAL ACTIN NETWORK

Ultrastructural studies show that actin filaments in the macrophage cortical cytoplasm are organized into a three-dimensional network (Hartwig and Shevlin, 1986; Hartwig and Yin, 1988) which gives the cytoplasm the gel-like consistency noted by early light microscopists. Actin filaments are attached to the plasma membrane to form a submembranous meshwork that stabilizes the plasma membrane. Many agonists that generate movements induce a rapid and transient increase in polymerized actin. For example, when neutrophils are stimulated with the chemotactic peptide N-formyl-methionyl-leucyl-phenylalanine (fMLP), phagocytizable particles, or phorbol esters, there is a transient increase in actin polymerization (Sheterline et al., 1984, Omannet al., 1989). Actin polymerization initiates the extension of the pseudopod, and generates protrusive force. Thus, the assembly and disassembly of the actin filaments are critical for the motile functions of leukocytes. Actin polymerization and network remodeling are modulated by actin regulatory proteins which respond to second messengers generated during agonist stimulation.

II. ACTIN POLYMERIZATION DYNAMICS

A variety of studies show that pseudopod extension is initiated by nucleated actin filament assembly at the leading edge of the cell (Welch et al., 1997). In resting neutrophils, about 50% of the actin is polymerized into filaments, and this value increases to 70–80% after stimulation. Filament growth is associated with an increase in actin nucleating activity, and growth can be blocked by cytochalasins, which bind to the fast polymerizing end of an actin filament ([+] end, also called "barbed" end with respect to the orientation of heavy meromyosin arrowheads bound to the filaments). A variety of approaches show that the (+) end of actin filaments are concentrated at the cortical region immediately under the plasma membrane, and actin polymerizes in this region following agonist stimulation (Symons and Mitchison, 1991; Eddy et al., 1997).

Another prominent feature of the cortical actin cytoskeleton is that after polymerization at the plasma membrane, the actin molecules travel down the filaments in a process called treadmilling. Treadmilling occurs due to the preferential addition of ATP.actin monomers to the (+) end, and the preferential depolymerization of ADP.actin monomers from the (–) end. The energy for treadmilling is derived from the hydrolysis of ATP.actin to ADP.actin soon after actin is incorporated into the filaments. The rate of treadmilling of pure actin is limited by the off rate of actin from the (–) end. However, the rate of turnover of actin filaments *in vivo* and in cell extracts is two to three orders of magnitude faster than predicted from *in vitro* values (Theriot and Mitchison, 1991; Theriot et al., 1992), suggesting that the process is greatly accelerated by accessory proteins.

These observations led to the following questions: How does the resting cell prevent complete polymerization of its actin? How is nucleated actin assembly initiated during agonist stimulation? How is the site of actin polymerization restricted to the subplasmalemmal region to generate membrane protrusions? How is the turnover of actin filaments in the dynamic cytoskeleton regulated?

III. A LARGE PROPORTION OF ACTIN IS PREVENTED FROM POLYMERIZING IN RESTING CELLS

Actin monomers in resting neutrophils do not spontaneously polymerize because they are sequestered by actin monomer binding proteins. Many actin monomer binding proteins have been described, and it is becoming clear that they have common as well as unique properties (Sun et al., 1995a). Until recently, β-thymosins, profilins and actin depolymerizing factor (ADF)/cofilin are three families of low molecular weight proteins which are thought to contribute to the monomer pool. Now it is clear that profilin and ADF/cofilin have additional functions important for generating a dynamic actin cytoskeleton. These aspects will be discussed in detail later.

IV. STIMULUS-POLYMERIZATION COUPLING

Leukocyte chemotaxis and phagocytosis require the directed extension of pseudopodia towards the chemotactic source or phagocytic particle. There is a tight coupling between membrane protrusion during pseudopod extension and explosive actin polymerization at the plasma membrane (Welch et al. 1997; Carlier, 1998). Actin polymerization is induced by agonist-mediated increase in the number of actin nuclei with exposed (+) ends, which compete effectively with monomer sequestering proteins for actin monomers. Because actin monomers do not spontaneously form de novo nuclei rapidly, nucleation is facilitated by nucleating proteins (assisted de novo nucleation). Nuclei can also be generated by uncapping of preexisting nuclei which are previously not available for polymerization because their (+) ends are blocked by filament end capping proteins (uncapping/nucleated assembly). The number of nuclei can also be increased by severing preexisting filaments (severing/nucleated assembly). These three mechanisms allow the cell to specify the temporal and spatial program for actin assembly. Cells may use different combinations of these mechanisms in response to different messengers, to generate an expanded repertoire of cytoskeletal responses. In platelets, the severing/uncapping mechanism predominates following thrombin stimulation (Hartwig et al., 1995; Barkalow et al., 1996), while *Dictyostelium* responds to chemotactic stimuli by severing or de novo nucleation, and capping is used to terminate rather than initiate actin polymerization (Eddy et al., 1997).

Actin regulatory proteins which create actin nuclei in response to signaling are likely to be important in stimulus-polymerization coupling. Relatively little is known about proteins which accelerate de novo actin nucleation. Much more is known about barbed end capping proteins and filament severing proteins. These proteins can be classified into two major groups: the Ca^{2+}-dependent gelsolin family of capping and/or severing proteins (Liu et al., 1988; Yu et al., 1990), which includes gelsolin and CapG (Liu et al., 1998; Yu et al., 1990), and the Ca^{2+}-independent capping protein which caps but does not sever (Schafer et al., 1996). Both families of proteins can be dissociated from filament barbed ends (uncapping) by phosphatidylinositol 4,5-bisphosphate (PIP_2) (Janmey and Stossel, 1987; Yu et al., 1990; Schafer et al., 1996), suggesting a common mechanism for initiating actin filament assembly. They coexist in many cells; in platelets, barbed end filament assembly during platelet activation is due to severing by gelsolin and subsequent capping of the severed filaments by capping protein (Barkalow et al., 1996).

Overexpression of either class of capping proteins in mammalian cells increases cell motility (Cunningham et al., 1991; Hug et al., 1995; Sun et al., 1995). This phenotype is consistent with the model that cell motility is enhanced by dynamic capping and uncapping of actin nuclei (Carlier, 1998). (+) end capping increases the steady state actin monomer concentration as a result of actin depolymerizing from the (–) ends. During agonist stimulation, actin monomers are funneled to the small number of locally uncapped filaments at the plasma membrane. Since the rate of polymerization is dependent on the concentration of available actin monomers, the large number of monomers drives rapid elongation from the selectively uncapped filaments to produce the motile phenotype. The *in vivo* results confirm the importance of capping proteins in actin dynamics and highlight the complexity of the actin cycle in cells.

V. RHO GTPASES SWITCH ON THE ACTIN CYTOSKELETON

In the past decade, small GTPases (G-proteins) of the Ras superfamily have emerged as key regulators of the actin cytoskeleton and their interactions with the plasma membrane (Ridley and Hall, 1992). Ridley and Hall reported that Rho is activated by lysophosphatidic acid to cause stress fiber assembly and focal adhesion formation in fibroblasts. At the same time, Rac is identified as the switch which responds to growth factor stimulation to promote assembly of actin filaments in the lamellipodia and membrane ruffles (Ridley et al., 1992). Subsequently, Cdc42 is found to induce thin actin-rich protrusions from the plasma membrane called filopodia (Kozma et al., 1995; Nobes and Hall, 1995). There is significant crosstalk between these signals, allowing them to coordinate multiple cytoskeletal responses to execute complex functions.

Machesky and Hall (Machesky and Hall, 1997) monitored actin polymerization in fibroblasts after small G-protein activation. They found that Rac-induced membrane ruffles are triggered by localized actin polymerization at the plasma membrane, which is independent of integrin complex assembly. Rho-induced stress fiber assembly is not associated with increased actin assembly. Instead, it is due to integrin-independent bundling of actomyosin filaments, and the subsequent integrin-dependent contraction of the bundles to form stress fibers.

The phagocyte cytoskeleton also responds to small G-proteins. Rac or Cdc42 activation leads to lamellipodial protrusion or filopodia formation, respectively, and both induce the formation of focal adhesions (Allen et al., 1997). On the other hand, Rho promotes the formation of contractile filaments without increasing macrophage adhesion (Aepfelbacher et al., 1996; Allen et al., 1997). Because contraction is not balanced by strong adhesion to the substrate as with fibroblasts, macrophages round up.

Small G-proteins are implicated in phagocytosis and receptor-mediated endocytosis. Expression of the dominant negative forms of Rac or Cdc42 inhibits phagocytosis and the accumulation of actin filaments in phagocytic cups (Cox et al., 1997). Rho is required for Fcγ receptor-mediated calcium signaling and phagocytosis in macrophages. Inhibition of Rho by C3 exotoxin prevents receptor clustering, and decreases the amount of polymerized actin and inhibited phagocytosis (Hackam et al., 1997). Rho and Rac inhibit transferrin receptor-mediated endocytosis by regulating clathrin-coated vesicle formation (Lamaza et al., 1996). Although cytochalasin does not affect pinocytosis in this study, subsequent studies show that another actin depolymerizing drug, latrunculin, has biphasic effects on endocytosis, establishing the importance of the actin cytoskeleton in endocytosis (Lamaze et al., 1997).

Exocytosis, a process accompanied by complex changes in the actin cytoskeleton (Norman et al., 1994) is likewise regulated by small and heterotrimeric G-proteins (Norman et al., 1994). Controlled actin depolymerization at the cell cortex facilitates exocytosis but excessive actin depolymerization prevents the process (Muellem et al., 1995).

VI. SMALL G-PROTEIN EFFECTS IN SEMI-INTACT CELLS AND CELL-FREE LYSATES

It is known for some time that GTPγS stimulates actin polymerization in permeabilized neutrophils, presumably by activating heterotrimeric G-proteins and/or small G-proteins (Tardif et al., 1995). The relative contributions of these G-proteins to the actin polymerization response have now been examined using cell-free lysates. Zigmond et al. (1997) found that although *D. Dictyostelium* mutants lacking heterotrimeric G-protein subunits have impaired chemotactic response to cAMP, GTPγS still stimulates actin polymerization in their lysates. Therefore, the direct targets of GTPγS in the mutant lysates that stimulate actin polymerization are likely to be small G-proteins. CdC42 induces actin polymerization in *Dictyostelium* as well as neutrophil lysates (Zigmond et al., 1997), suggesting that it may be the switch for polymerization. Paradoxically, Rac, which causes actin polymerization in fibroblasts, is reported to have no effect in this study. In contrast, another group finds that Rac induces the dissociation of gelsolin from actin (which is diagnostic of gelsolin uncapping from filament (+) ends) in similar neutrophil extracts (Arcaro, 1998).

The importance of small G-proteins in stimulus-polymerization coupling can also be demonstrated using permeabilized cell models. Hartwig et al. (1995) showed that GTPγS induces actin polymerization in semi-intact platelets by promoting uncapping, and GTP-activated Rac has an identical effect. Mackay et al. (1997) found that GTPγS induces cortical actin polymerization, stress fibers and focal adhesion formation by activating endogenous Rac and Rho.

VII. THE SEARCH FOR DOWNSTREAM CYTOSKELETAL EFFECTORS OF SMALL G-PROTEINS

The profound effects of the small G-proteins on the actin cytoskeleton prompted intense searches for their direct downstream effectors. Results from yeast two hybrid screens, affinity chromatography and genetic screens reveal many potential targets which bind preferentially to the GTP-activated, but not GDP-bound, small G-proteins (Van Aelst and D'Souza-Schorey, 1997). In general, the Rho target proteins are distinct from that of Rac and Cdc42, as would be consistent with Rho's unique effects on the cytoskeleton. In contrast, Rac and Cdc42 have common, as well as distinct, targets. With a few exceptions, these direct targets are not bona fide actin regulatory proteins; many are kinases that can potentially modulate further downstream actin regulatory proteins. For the purpose of this review, we will focus on two such kinases.

VIII. RHO KINASES, DOWNSTREAM EFFECTORS OF RHO

ROKα/Rho kinase and its close relative p160ROCK are serine/threonine kinases which bind GTP-Rho and are thus activated (Leung et al., 1996). Rho kinases stimulate stress fiber formation by increasing myosin light-chain phosphorylation (Kimura et al., 1996). This is accomplished in two ways: first, by phosphorylating the myosin binding subunit of myosin light-chain phosphatase to decrease phosphatase activity. Second, by phosphorylating the myosin light-chain to activate myosin directly. Overexpression of constitutively active Rho kinase promotes stress fiber and focal adhesion formation (Leung et al., 1996). These results strongly implicate Rho kinase as an important Rho effector. Rho kinase also phosphorylates the ezrin/radixin/moesin (ERM) family of crosslinking proteins which link the actin cytoskeleton to integrin membrane receptors (Tsukita and Yonemura, 1997). ERM phosphorylation precedes focal adhesion formation and stress fiber assembly in Rho-activated cells (Matsui et al., 1998). Reconstitution studies show that ERM is required for Rho- and Rac-dependent cytoskeletal responses in permeabilized cells (Mackay et al., 1997), establishing ERM as a significant downstream effector of small G-proteins.

IX. PHOSPHATIDYLINOSITOL 4-PHOSPHATE 5-KINASES, DOWNSTREAM EFFECTORS OF RAC AND RHO

Phosphatidylinositol 4-phosphate 5-kinases (PIP5K) are important in phosphoinositide signaling because it converts phosphatidylinositol 4-phosphate (PI(4)P) to phosphatidylinositol 4,5-bisphosphate (PIP_2). Furthermore, type I PIP5K can also utilize phosphoinositides which are phosphorylated at the D-3 position of the inositol ring to generate $PI(3,4)P_2$ and $PI(3,4,5)P_3$ (Zhang et al., 1997). These phosphoinositides have been implicated in a wide array of cellular responses, including cytoskeletal changes and regulation of actin binding proteins. For example, PIP_2 is central to several signaling pathways:

PIP_2 is hydrolyzed by phospholipase C (PLC) to generate two important second messengers, Ins $(1,4,5)P_3$ which mobilizes intracellular Ca^{2+} stores, and diacylglycerol which activates protein kinase C. In addition, PIP_2 is converted by the agonist-activated phosphoinositide 3-kinase (PI3K) to generate novel D-3 phosphoinositides, such as $PI(3,4,5)P_3$ (PIP_3) and $PI(3,4)P_2$ (De Camilli et al., 1996; Toker and Cantley, 1997) (Figure 1).

Type I PIP5K binds Rac (Tolias et al., 1995) and Rho (Ren et al., 1996) in vitro in a GTP-independent manner. Although the lack of dependence on GTP is not consistent with the usual behavior of many other direct effectors of the small G-proteins, this association is likely to be functionally significant. Rho or Rac activation increases PIP_2 synthesis in cells or cell lysates. PIP_2 synthesis is increased during integrin-mediated cell attachment and Rho stimulates PIP5K

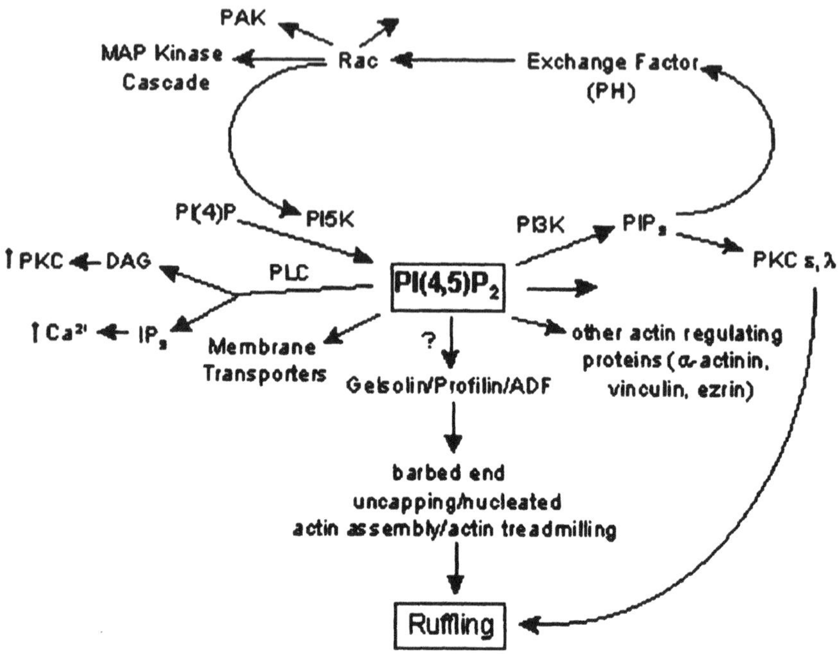

Figure 1. Phosphoinostide regulation pathways

activity in cell lysates (Ren et al., 1996). Rac increases PIP_2 synthesis in permeabilized platelets and promotes barbed end actin assembly as well (Hartwig et al., 1995). Rac and Cdc42 stimulate PIP_2 synthesis in neutrophil lysates (Zigmond et al., 1997). Overexpression of PIP5K induces the formation of short actin filament bundles (Shibasaki et al., 1997). Although the PIP5K effects do not completely recaptitulate that of Rho or Rac activation (the PIP5K-induced bundles do not resemble stress fibers, and there is a decrease rather than increase in focal adhesions), these results strongly implicate PIP5K as a downstream effector of the small G-proteins.

X. PI(4,5)P_2 AS A REGULATOR OF THE ACTIN CYTOSKELETON

The relation between actin rearrangement and PIP5K activation has received much attention because PIP_2 modulates the activity of many actin regulatory proteins in vitro. The role of PIP_2 as a potential regulator of the cytoskeleton was first described in 1985, when Lassing and Lindberg (1985) showed that PIP_2 inhibits profilin:actin interactions. Subsequently, gelsolin was also identified as a

PIP_2-regulated protein (Janmey and Stossel, 1987). The list has grown to include capping protein (Schafer et al., 1996); cofilin (Yonezawa et al., 1990); α-actinin, a stress fiber and focal adhesion associated protein (Fukami et al., 1992); vinculin, a focal adhesion protein (Gilmore and Burridge, 1996); ERM (Mirao et la., 1996) and N-WASP (Miki et al., 1996, 1998), an actin depolymerizing protein analogous to the Wiskott-Aldrich syndrome protein which binds Cdc42. The latter four proteins are activated by PIP_2, while the others are inactivated by PIP_2.

There is still considerable disagreement as to whether phosphoinositides are the physiological triggers for actin polymerization. In some cells, the time course of actin polymerization parallels that of the increase in PIP_2. In platelets, thrombin induced actin polymerization occurs at a time when PIP_2 level increases (Hartwig et al., 1995). Furthermore, PIP_2 added to semi-intact platelets elicits actin polymerization, and this effect is blocked by a gelsolin-derived PIP_2 binding peptide. These studies show conclusively that PIP_2 induces filament polymerization, presumably by uncapping actin filament nuclei. However, in neutrophils, agonist-induced actin polymerization occurs at a time when PIP_2 level is reduced rather than increased (Eberle et al., 1990), as would be expected if uncapping and monomer desequestration are initiated by PIP_2. To explain this discrepancy, it is often hypothesized that local PIP_2 availability can be enhanced by compartmentalization or differential turnover, even as the bulk PIP_2 mass is reduced. Alternatively, it is also possible that PIP_2 binding by capping proteins is further regulated by additional signals generated during agonist stimulation. There is now experimental evidence to support both possibilities (Glaser et al., 1996; Pike and Casey, 1996; Lin et al., 1997).

PIP_2 involvement in cytoskeletal regulation is also supported by some, but not all, experiments to manipulate PIP_2 content in cells and cell-free models. Microinjection of anti-PIP_2 prevents stress fiber and focal adhesion formation (Fukami et al., 1988; Gilmore and Burridge, 1996). Overexpression of PIP5K induces the formation of short actin filament bundles (48), while overexpression of the PIP_2 phosphatase (synaptojanin), which dephosphorylates PIP_2, reduces the number of actin stress fibers (Sakisaka et al., 1997). PIP_2 induces actin polymerization in semi-intact platelets, and a PIP_2 binding peptide from gelsolin blocks both Rac (which increases PIP_2 synthesis, see above) and PIP_2-induced actin assembly (Hartwig et al., 1995). On the other hand, although Rac and Cdc42 both induce PIP_2 synthesis in neutrophil lysates, only Cdc42 stimulates actin polymerization, and this effect is not blocked by anti-PIP_2 (Zigmond et al., 1997). However, in another study, Rac induces the dissociation of gelsolin:actin complexes, although the gelsolin-derived PIP_2 binding peptide does not block this process (Arcaro, 1998). In either case, PIP_2 involvement cannot be ruled out because an effect on PIP_2 content has not been documented.

XI. ACTIN REGULATORY PROTEINS

We will highlight recent findings about three actin regulatory proteins which are likely to be involved in agonist-stimulated actin dynamics. These proteins are potential small G-protein effectors, although they do not bind directly to the G-proteins. They can potentially be regulated by phosphoinositides.

XII. PROFILIN PROMOTES ACTIN POLYMERIZATION IN MULTIPLE WAYS

Profilin is the first actin monomer binding protein to be identified. In spite of a wealth of biochemical and structural data, the role of profilin in cells has been an enigma until very recently, because its behavior is not always consistent with that of a simple actin monomer binding protein. Although profilin acts like a monomer binding protein when filament (+) ends are capped (i.e., in resting conditions) to prevent actin polymerization, profilin can also stimulate actin polymerization. This dichotomy can now be explained by the breakthrough finding that actin complexed with profilin has a dramatically higher affinity for filament (+) ends than actin monomers alone (Pantaloni and Carlier, 1993). Therefore, profilin:actin complexes are the preferred polymerizing species in agonist-stimulated cells. The impact of facilitated profilin:actin addition to the (+) end is amplified by the availability of a large store of actin monomers which can be shuttled from the β-thymosin sequestered actin monomer pool to profilin (Carlier and Pantaloni, 1997). Thus, two different actin monomer binding proteins cooperate to promote rapid filament assembly following filament uncapping.

Many in vivo observations are consistent with positive and negative regulation of actin assembly by profilin. Mutations in the profilin gene disrupt multiple actin-dependent processes in Drosophila and yeast. Microinjection of profilin into mammalian cells depolymerizes actin, whereas profilin-actin complexes have the opposite effect (Cao et al., 1992). Profilin overexpression in stably transfected cells depolymerizes actin stress fibers but increases polymerized actin at the cell periphery by prolonging the half-life of cortical filaments (Finkel et al., 1994). Based on *in vitro* observations, it is reasonable to postulate that the switch between the positive and negative roles of profilin *in vivo* may depend on the stoichiometry of the interacting components and whether actin filaments are capped.

Besides binding actin, profilin also binds many other partners. These include a complex containing unconventional actins (also called actin related proteins, Arp2/3) (Machesky et al., 1997) and many proline-rich proteins, such as the vasodilator-stimulated phosphoprotein (VASP), which is found in focal adhesions (Reinhard et al., 1995), its relative Mena (Gertler et al., 1996) and formin-related proteins involved in development and morphogenesis (Watanabe et al., 1997). Interesting, the latter group of proteins also binds to the GTP-activated small

G-proteins Rho (Watanabe et al., 1997) and Cdc42 (Evangalista et al., 1997), suggesting that profilin is an indirect downstream effector of the small G-proteins. The profilin polyproline binding domain does not resemble that of other well-known proline targeting motifs, such as SH3 and WW domains. Molecular modeling of the recently solved profilin structure in complex with an L-proline decamer suggests that the polyproline peptide can accommodate two profilin-actin complexes (Mahoney et al., 1997), providing a potential mechanism for amplifying signals for the targeting and/or assembly of the cytoskeletal signaling complex.

XIII. GELSOLIN AS A DOWNSTREAM EFFECTOR OF RAC

Gelsolin is an actin filament severing and capping protein (Liu et al., 1998) which can effectively remodel the cytoskeleton. The combined effect of gelsolin is to promote the formation of a large number of short actin filaments that are capped at their (+) end. Uncapping can release a large number of nuclei for explosive polymerization. Gelsolin is activated by Ca^{2+} to sever and cap actin filaments (Yin and Stossel, 1980). Paradoxically, EGTA does not dissociate gelsolin from filament ends, because a Ca^{2+} molecule is trapped in the gelsolin:actin complex and is inaccessible to EGTA (Bryan and Kurth, 1984). Phosphoinositides, such as PIP_2, are the only known agents which can dissociate gelsolin from actin *in vitro* (Janmey and Stossel, 1987). Therefore, PIP_2 may be the physiological stimulus for gelsolin uncapping from filament barbed ends.

In contrast to the behavior of purified gelsolin:actin complexes *in vitro*, gelsolin:actin complexes in leukocytes dissociate following agonist stimulations (Chapommier et al., 1987; Deaton et al., 1992), and in response to changes in intracellular Ca^{2+} concentration (Chaponnier et al., 1987; Arcaro, 1998). Rac induces gelsolin:actin complex dissociation in leukocyte extracts, establishing gelsolin as a downstream effector of Rac (Arcaro, 1998). The relation between gelsolin and Rac is indirect, because gelsolin does not bind Rac, and Rac has no effect on purified gelsolin:actin complexes.

The relation between gelsolin and Rac is further supported by the finding that dermal fibroblasts isolated from gelsolin null mice do not ruffle their plasma membrane in response growth factors which normally activate Rac (Witke et al., 1995), and Rac expression is increased fivefold in the null cells (Azuma et al., 1998). The ruffling defect occurs downstream of Rac activation, because Rac is able to translocate normally to the plasma membrane following growth factor stimulation. The nonruffling phenotype of the gelsolin null cells can be rescued by introduction of gelsolin (Azuma et al., 1998). These studies clearly establish that gelsolin is an essential effector of Rac-mediated actin dynamics.

Gelsolin null cells isolated from other tissues also display other motility defects (Lu et al., 1997; Lascola et al., 1998). However, not all motile functions are disrupted, reaffirming that the cells have a wide repertoire of cytoskeletal proteins to execute specific functions, and there may be considerable overlap to compensate for the loss of a single actin regulatory protein.

Gelsolin binds PIP_2 with μM affinity (Lin et al., 1997). Gelsolin also binds PI(4)P, PI(3,4,5)P_3, and PI(3,4)P_2 (Hartwig et al., 1996; Arcaro, 1998), with a preference for PIP_2 (Lu et al., 1998). The PIP_2 binding sites on gelsolin has been mapped by deletion mutagenesis and peptide analogue studies. Two regions are identified, and they are rich in positively charged amino acids which can bind the phosphates headgroups (Yu et al., 1992). The recently solved high resolution crystal structure of gelsolin shows that these regions map to a common flat, solvent-exposed surface in an actin binding domain of gelsolin (Burtnick et al., 1997). The gelsolin PIP_2 binding domain is different from the pleckstrin homology domain (Harlan et al., 1994), even though they have similar affinity for phosphoinositides.

XIV. PHOSPHOINOSITIDES REGULATE GELSOLIN AND PROFILIN

As discussed above, the current hypothesis about how PIP_2 induces explosive cortical assembly during agonist stimulation is as follows: PIP_2 uncaps by dissociating capping proteins such as gelsolin from actin filament ends. PIP_2 also dissociates actin monomers bound to profilin, allowing monomers to add onto the newly exposed filament ends (Carlier and Pantaloni, 1997). There are still several unresolved issues concerning this hypothesis. The first is that in most cells, actin polymerizes at a time when PIP_2 level is reduced (Eberle et al., 1990; Dadabay et al., 1991) rather than increased, as would be expected if uncapping and monomer desequestration are initiated by PIP_2. Another complication is that since profilin:actin complexes are the preferred actin polymerizing species (Pantaloni and Carlier, 1993), profilin has to be dissociated from PIP_2 so it can bind actin, at the same time that gelsolin has to bind PIP_2 to become uncapped. The apparently contradictory demands for PIP_2 at the site of coordinated actin assembly cannot be readily explained by a simple model involving a change in PIP_2 concentration only.

We have recently obtained evidence which can potentially help resolve some of these issues. First, PIP_2 binding to gelsolin is enhanced significantly by Ca^{2+} (Lin et al., 1997), explaining how gelsolin can uncap filaments even as membrane PIP_2 is hydrolyzed by PLC and phosphorylated by PI3K. Second, a direct comparison of gelsolin and profilin (Lu et al., 1996; Lin et al., 1997) shows that they have similar affinity for PIP_2 in EGTA, but gelsolin binds with much higher affinity in Ca^{2+}. Gelsolin can therefore displace profilin from PIP_2 to simultaneously uncap actin filaments and activate profilin. Third, profilin binds D-3 lipids with higher affinity than PIP_2 (Lu et al., 1996), while gelsolin prefers PIP_2 (Lu et al., 1998). Since receptor-mediated signaling increases D-3 phosphoinositide concentration, profilin may switch to these lipids, especially when faced with increasing competition for PIP_2 from gelsolin.

Profilin regulation by D-3 lipids is consistent with growing evidence for a role of PI3K in membrane ruffling (Figure 1). PIP_3 induces membrane ruffling by several mechanisms. It activates Rac by increasing Rac guanine nucleotide exchange (Rodriguez-Viciana et al., 1997). PIP_3 may also stimulate ruffling by activating the Ca^{2+}-insensitive protein kinase C (Derman et al., 1997). The effects of PIP_3 are very cell type and stimulus specific. Wortmannin, which inhibits PI3K, blocks transforming growth factorβ- but not fMLP- induced actin assembly (Thelen et al., 1995). Likewise, phorbol ester-induced actin assembly in plateletes was also wortmannin insensitive (Hartwig et al., 1996).

Based on the findings discussed above, and the currently available information about gelsolin and profilin behavior in cells, we propose a new model for how the two proteins are involved in uncapping and nucleated actin assembly during cell activation.

XV. A MODEL FOR THE COORDINATED REGULATION OF GELSOLIN AND PROFILIN

In resting cells, some gelsolin caps actin filaments (10–30%) (Lind et al., 1987; Hartwig et al., 1989; Barkalow et al., 1996), some is attached to the plasma membrane (<5%) (Hartwig et al., 1989; Chen et al., 1996), while the bulk is cytosolic (13,90). Gelsolin is able to cap filaments in spite of low ambient Ca^{2+}, because once gelsolin caps in a previous cycle, it is not dissociated simply by lowering Ca^{2+} (Chaponnier et al., 1987). Gelsolin does not bind PIP_2 because most of the PIP_2 is sequestered by profilin, which is more than times as abundant (Sun et al., 1996). The predominant association of profilin with the plasma membrane has been demonstrated experimentally (Ostrander et al., 1995). The combination of gelsolin's capping of filament barbed ends and profilin's inactivation by PIP_2 prevents actin polymerization at the barbed end (Figure 2).

In stimulated cells, a rise in cytosolic Ca^{2+} activates gelsolin to sever actin filaments, generating many barbed end capped nuclei. Ca^{2+} also increases gelsolin's affinity for PIP_2, promoting gelsolin to dissociate from filament ends near the plasma membrane. There is thus an increase in membrane-associated gelsolin (Hartwig et al., 1989) and a decrease in gelsolin:actin complexes.

Ca^{2+}/gelsolin displaces profilin from PIP_2. A portion of the profilin binds to the newly synthesized D-3 lipids, but the bulk falls off the membrane because D-3 lipids are much less abundant than PIP_2. This allows profilin to capture actin monomers stored in the ß-thymosin reservoir, and to drive actin assembly from the newly uncapped barbed ends (Pantaloni and Carlier, 1993). Multiple rounds of severing, uncapping and facilitated actin addition fuel the explosive amplification of filament growth.

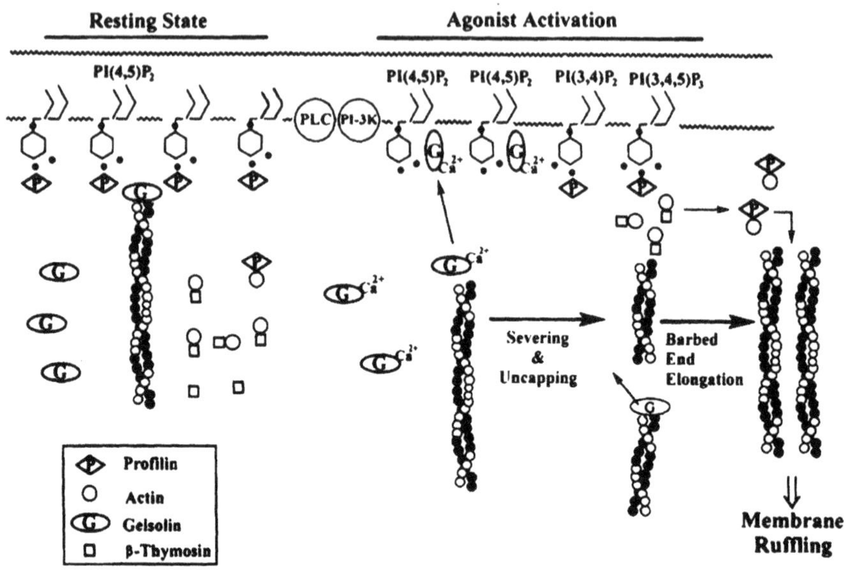

Figure 2. Coordinated regulation of gelsolin and profilin during agonist stimulated actin assembly

This model is oversimplified, but can provide a testable conceptual framework to approach the question of how the cytoskeleton is regulated during agonist stimulation.

XVI. THE INVOLVEMENT OF GELSOLIN IN OTHER CELLULAR FUNCTIONS

Recently, gelsolin was identified as an immediate and prominant substrate for caspase-3, a critical protease in apoptosis (Kothakota et al., 1997). Gelsolin cleaved by caspase-3 no longer requires Ca^{2+} to sever actin filaments. The constitutively active gelsolin can therefore indiscrimately dismantle the membrane cytoskeleton to cause membrane blebbing, a hallmark of apoptosis. In support of the role of gelsolin in apoptosis, the gelsolin null neutrophils have a delayed onset of apoptosis (Kothakota et al., 1997). This finding highlights the importance of inhibiting gelsolin severing to preserve the integrity of the cell, and its selective activation to remodel the cytoskeleton under normal conditions.

There is emerging evidence to suggest that PIP_2 binding to actin modulating proteins may have implications beyond a direct effect on the cytoskeleton. This was first demonstrated for profilin (Goldschmidt-Clermont et al., 1991). Profilin inhibits PLCγ, but not PLCβ. Phosphorylation of PLCγ by growth factor receptor kinases reduces inhibition, switching off the negative regulation by profilin. Gelsolin (Sun et al., 1997) and its relative CapG (Sun et al., 1995b) can stimulate as well as inhibit a wider spectrum of PLCs than profilin, and can modulate the activity of PLCγ (Baldassare et al., 1997). Moderate gelsolin or CapG overexpression has profound effects on platelet derived growth factor and bradykinin-activated PLC activity in cells (Sun et al., 1995, 1997). Thus, gelsolin and CapG may have more impact on membrane signaling than profilin. Gelsolin has also been reported to regulate PI3K (Sinh et al., 1996; Lu et al., 1998) *in vitro.*

The dual activities of gelsolin may allow it to integrate signal transduction and cytoskeletal responses. The possibility of crosstalk between divergent PIP_2-binding proteins is particularly relevant as more of these proteins are identified.

XVII. ADF/COFILIN PROMOTES ACTIN FILAMENT TURNOVER

The question of how *in vivo* actin filament treadmilling is accelerated by two to three orders of magnitude compared with purified actin has been resolved (Carlier et al., 1997). Rapid turnover is due to facilitation by ADF/cofilin. ADF and cofilin are ubiquitous 15–20 kDa proteins which are originally identified as actin monomer binding proteins. Subsequently, they are also shown to bind actin filaments and induce depolymerization (Moon and Drubin, 1995). ADF and cofilin are products of different genes, and they have distinct patterns of expression and regulation (Minamide et al., 1997; Obinata et al., 1997). However, for the sake of simplicity, we will refer to these proteins collectively as cofilin.

Genetic studies establish that cofilin is essential for the viability of many lower eukaryotes. Cofilin is localized in the yeast cortical actin cytokeleton, which exists as patches in the cell membrane. The actin patches move around rapidly, and the actin filaments within the patches turnover rapidly *in vivo* (Lappalainen and Drubin, 1997). Turnover is dependent on cofilin, because temperature sensitive cofilin mutants do not depolymerize their actin filaments at the restrictive temperature (Lappalainen and Drubin, 1997). The involvement of cofilin in dynamic actin filament disassembly is also supported by the finding that cofilin increases the rate of Listeria propulsion in cell extracts, by enhancing the rate of actin depolymerization (Carlier et al., 1997; Rosenblatt et al., 1997).

These studies establish that the primary effect of cofilin is to enhance the treadmilling rate of actin filaments, by promoting depolymerization of actin from the (–) end. As a result of increased actin depolymerization, the steady-state concentration of actin monomers increases, and this in turn increases the rate of barbed end growth at steady-state. The effect of cofilin can therefore synergize with that of capping proteins to increase the size of the actin monomer pool under resting conditions, and to increase the rate of barbed end filament growth after agonist stimulation. The increase in polymerized actin may account for the increase in actin bundles observed after cofilin overexpression (Aizawa et al., 1997).

There is still considerable controversy as to how cofilin increases actin depolymerization at the (–) end (Carlier, 1998). Theoretically, increased net actin depolymerization can occur in three ways. First, cofilin can bind actin monomers to drain actin from filament (–) ends. However, this mechanism is not sufficient to account for the rapid rate of actin filament treadmilling observed *in vivo* and in cell extracts. Second, cofilin binds to actin filaments and accelerates the rate of actin dissociation from the (–) end. Carlier et al. (1997) found that cofilin increases the (–) end off rate 25-fold, suggesting that this may be the primary mechanism of action. Cryoelectron microscopy showed that cofilin induces large scale changes in the twist of the actin filament helix (McGough et al., 1997), suggesting that the conformational change may facilitate actin dissociation from the (–) end. Third, a previously popular hypothesis is that cofilin severs actin filaments to create more ends for depolymerization (Moon and Drubin, 1995). Severing by cofilin has been demonstrated by multiple techniques and is observed in real time by video microscopy. However, cofilin is much less effective than gelsolin in severing, and severing appears to be restricted to bends within the filament, rather than throughout the entire filament as in the case of gelsolin. These differences suggest that cofilin's primary function is not severing, although this mechanism cannot be discounted at present.

In vitro, cofilin binding to actin is altered by pH, phosphorylation and phospholipids (101). At pH <7.2, cofilin binds along actin filaments cooperatively to change the pitch of the filaments (McGough et al., 1997). At higher pH, filaments rapidly disassemble (Moon and Drubin, 1995). Since cell activation is often associated with a change in cytosolic pH, the sensitivity of cofilin to pH may be an important regulatory mechanism. Cofilin-actin interaction is inhibited by phospholipids; however, unlike profilin and gelsolin, phosphatidylinositol, in addition to phosphoinositides, is also inhibitory (Yonezawa, 1990). Since these lipids are located at the plasma membrane where actin polymerizes during agonist stimulation, they may inhibit cofilin to bias depolymerization away from the leading edge.

Phosphorylation appears to be a major regulatory mechanism for all vertebrate cofilins, even though yeast cofilin is not phosphorylated (Moon and Drubin, 1995). Phosphorylation of vertebrate cofilin at the Ser3 position inhibits binding to actin monomers and filaments (Agnew et al., 1995), and dephosphorylation is observed following agonist stimulation of a variety of cells. In the neutrophil-like differentiated HL60 cells, opsonized zymosin, fMLP and arachidonic acid induces rapid dephosphorylation, and subcellular fractionation show that the dephosphorylated cofilin is recovered in a higher proportion in plasma membrane and phagolysosome fractions (Suzuki et al., 1995). Cofilin is also markedly dephosphorylated in neutrophils after fMLP or phorbol ester stimulation, and translocates to the actin-rich cell cortex (110). Inhibitor studies suggest that in resting neutrophils, cofilin is phosphorylated by a protein kinase C independent pathway, and dephosphorylated by okadaic acid sensitive phosphatases (Djafarzadeh and Niggli, 1997). cAMP, Ca^{2+}, and GTPγS also dephosphorylate cofilin (Moon and Drubin, 1995; Meberg et al., 1998), implicating a G-protein and a Ca^{2+}-dependent mechanism for regulating cofilin. The kinases and phosphatases that contriubte to cofilin regulation have not been identified and multiple players are likely to be involved (Meberg et al., 1998).

ACKNOWLEDGMENTS

H.L.Y. is grateful to Thomas P. Stossel for his profound influence on her scientific development and thanks him for his comments about this review. This work is supported by the National Institutes of Health grants #GM51112 and #NS31430, and a Welch Foundation grant #I-1200.

REFERENCES

Aepfelbacher, M., Essler, M., Huber, E., Czech, A., & Weber, P.C. (1996). Rho is a negative regulator of human monocyte spreading. J. Immunol. 157, 5070-5075.

Agnew, B.J., Minamide, L.S., & Bamburg, J.R. (1997). Reactivation of phosphorylated actin depolymerizing factor and identification of the regulatory site. J. Biol. Chem. 270, 17582-17587.

Aizawa, H., Fukui, Y., & Yahara, I. (1997). Live dynamics of Dictyostelium cofilin suggests a role in remodeling actin latticework into bundles. J. Cell Sci. 110, 2333-2344.

Allen, W.E., Jones, G.E., Pollard, J.W., & Ridley, A.J. (1997). Rho, Rac and Cdc42 regulate actin organization and cell adhesion in macrophages. J. Cell Sci. 110, 707-720.

Arcaro, A. (1998). The small GTP-binding protein Rac promotes the dissociation of gelsolin from actin filaments in neutrophils. J. Biol. Chem. 273, 805-813.

Azuma, T., Witke, W., Stossel, T.P., Hartwig, J.H., & Kwiatkowski, D.J. (1998). Gelsolin is a downstream effector of rac for fibroblast motility. EMBO J. 17, 1362-1370.

Baldassare, J.J., Henderson, P.A., Tarver, A., & Fisher, G.J. (1997). Thrombin activation of human platelets dissociates a complex containing gelsolin and actin from phosphatidylyinositide-specific phospholipase $C\gamma_1$. Biochem. J. 324, 283-287.

Barkalow, K., Witke, W., Kwiatkowski, D.J., & Hartwig, J.H. (1996). Coordinated regulation of platelet actin filament barbed ends by gelsolin and capping protein. J. Cell Biol. 134, 389-399.

Bryan, J., & Kurth, M.C. (1997). Actin-gelsolin interactions. J. Biol. Chem. 259, 7480-7487.

Burtnick, L.D., Koepf, E.K., Grimes, J., Jones, E.Y., Stuart, D.I., McLaughlin, P.J., & Robinson, R.C. (1997). The crystal structure of plasma gelsolin: implications for actin severing, capping and nucleation. Cell 90, 661-670.

Cao, L.-G., Babcock, G.G., Rubenstein, P.A., & Wang, Y.-L. (1992). Effects of profilin and profilactin on actin structure and function in living cells. J. Cell Biol. 117, 1023-1029.

Carlier, M.-F. (1998). Control of actin dynamics. Curr. Opin. Cell Biol. 10, 45-51.

Carlier, M.F., & Pantaloni, D. (1997). Control of actin dynamics in cell motility. J. Mol. Biol. 269, 459-467.

Carlier, M.F., Laurent, V., Santolini, J., Melki, R., Didry, D., Xia, G., Hong, Y., Chua, N., & Pantaloni, D. (1997). Actin depolymerizing factor (ADF/cofilin) enhances the rate of filament turnover: implication in actin-based motility. J. Cell Biol. 136, 1307-1322.

Chaponnier, C., Yin, H.L., & Stossel, T.P. (1987). Reversibility of gelsolin/actin interaction in macrophages. Evidence of Ca^{2+}-dependent and Ca^{2+}-independent pathways. J. Exp. Med. 165, 97-106.

Chen, P., Murphy-Ullrich, J.E., & Wells, A. (1996). A role for gelsolin in actuating epidermal growth factor receptor-mediated cell motility. J. Cell Biol. 134, 689-698.

Cox, D., Chang, P., Zhang, Q., Reddy, P.G., Bokoch, G.M., & Greenberg, S. (1997). Requirements for both Rac1 and Cdc42 in membrane ruffling and phagocytosis in leukocytes. J. Exp. Med. 186, 1487-1494.

Cunningham, C.C., Stossel, T.P., & Kwiatkowski, D.J. (1991). Enhanced motility in NIH 3T3 fibroblasts that overexpress gelsolin. Science 251, 1233-1236.

Dadabay, C.Y., Patton, E., Cooper, J.A., & Pike, L.J. (1991). Lack of correlation between changes in polyphosphoinositide levels and actin/gelsolin complexes in A431 cells treated with epidermal growth factor. J. Cell Biol. 112, 1151-1156.

De Camilli, P., Emr, S.D., McPherson, P.S., & Novick, P. (1996). Phosphoinositides as regulators in membrane traffic [Review]. Science 271, 1533-1539.

Deaton, J.D., Guerrero, T., & Howard, T.H. (1992). Role of gelsolin interaction with actin in regulation and creation of actin nuclei in chemotactic peptide activated polymorphonuclear neutrophils. Mol. Biol. Cell 3, 1427-1435.

Derman, M.P., Toker, A., Hartwig, J.H., Spokes, K., Falck, J.R., Chen, C., Cantley, L.C., & Cantley, L.G. (1997). The lipid products of phosphoinositide 3-kinase increase cell motility through protein kinase C. J. Biol. Chem. 272, 6465-6470.

Djafarzadeh, S., & Niggli, V. (1997). Signaling pathways involved in dephosphorylation and localization of the actin-binding protein cofilin in stimulated human neutrophils. Exp. Cell Res. 236, 427-435.

Eberle, M., Traynor-Kaplan, A.E., Sklar, L.A., & Norgauer, J. (1990). Is there a relationship between phosphatidylinositol trisphosphate and F-actin polymerization in human neutrophils? J. Biol. Chem. 265, 16725-16728.

Eddy, R.J., Han, J., & Condeelis, J.S. (1997). Capping protein terminates but does not initiate chemoattractant-induced actin assembly in *Dictyostelium*. J. Cell Biol. 139, 1243-1253.

Evangalista, M., Blundell, K., Longtine, M.S., Chow, C.J., Adames, N., Pringle, J.R., Peter, M., & Boone, C. (1997). Bni1p, a yeast Formin linking Cdc42p and the actin cytoskeleton during polarized morphogenesis. Science 276, 118-122.

Finkel, T., Theriot, J.A., Dise, K.R., Tomaselli, G.F., & Goldschmidt-Clermont, P.J. (1994). Dynamic actin structures stabilized by profilin. Proc. Natl. Acad. Sci. USA 91, 1510-1514.

Fukami, K., Furuhashi, K., Inagaki, M., Endo, T., Hatano, S., & Takenawa, T. (1992). Requirement of phosphatidylinositol 4,5-bisphosphate for α-actinin function. Nature 359, 150-152.

Fukami, K., Matsuoka, K., Nakanishi, O., Yamakawa, A., Kawai, S., & Takenawa, T. (1988). Antibody to phosphatidylinositol 4,5-bisphosphate inhibits oncogene-induced mitogenesis. Proc. Natl. Acad. Sci. USA 85, 9057-9061.

Gertler, F.B., Niebuhr, K., Reinhard, M., Wehland, J., & Soriano, P. (1996). Mena, a relative of VASP and drosophila enabled, is implicated in the control of microfilament dynamics. Cell 87, 227-239.

Gilmore, A.P., & Burridge, K. (1996). Regulation of vinculin binding to talin and actin by phosphatidylinositol-4-5-bisphosphate. Nature 381, 531-535.

Glaser, M., Wanaski, S., Buser, C.A., Boguslavsky, V., Rashidzada, W., Morris, A., Rebecchi, M., Scarlata, S.F., Runnels, L.W., Prestwich, G.D., Chen, J., Aderem, A., Ahn, J., & McLaughlin, S. (1996). Myristoylated alanine-rich C kinase substrate (MARCKS) produces reversible inhibition of phospholipase C by sequestering phosphatidylinositol 4,5-bisphosphate in lateral domains. J. Biol. Chem. 271, 26187-26193.

Goldschmidt-Clermont, P.J., Kim, J.W., Machesky, L.M., Rhee, S.G., & Pollard, T.D. (1991). Regulation of phospholipase C-γ1 by profilin and tyrosine phosphorylation. Science 251, 1231-1233.

Hackam, D.J., Rotstein, O.D., Schreiber, A., Zhang, W.J., & Grinstein, S. (1997). Rho is required for the initiation of calcium signaling and phagocytosis by Fcgamma receptors in macrophages. J. Exp. Med. 186, 955-966.

Harlan, J.E., Hajduk, P.J., Yoon, H.S., & Fesik, S.W. (1994). Pleckstrin homology domains bind to phosphatidylinositol-4,5-bisphosphate. Nature 371, 168-170.

Hartwig, J.H., & Shevlin, P.A. (1986). The architecture of actin filaments and the ultrastructural location of actin-binding protein in the periphery of lung macrophages. J. Cell Biol. 103, 1007-1020.

Hartwig, J.H., & Yin, H.L. (1988). The organization and regulation of the macrophage actin skeleton. Cell Motil. Cytoskeleton 10, 117-125.

Hartwig, J.H., Bokoch, G.M., Carpenter, C.L., Janmey, P.A., Taylor, L.A., Toker, A., & Stossel, T.P. (1995). Thrombin receptor ligation and activated Rac uncap actin filament barbed ends through phosphoinositide synthesis in permeabilized human platelets. Cell 82, 643-653.

Hartwig, J.H., Chambers, K.A., & Stossel, T.P. (1989). Association of gelsolin with actin filaments and cell membranes of macrophages and platelets. J. Cell Biol. 108, 467-479.

Hartwig, J.H., Kung, S., Kovacsovics, T., Janmey, P.A., Cantley, L.C., Stossel, T.P., & Toker, A. (1996). D3 phosphoinositides and outside-in integrin signaling by glycoprotein IIb-IIIa mediate platelet actin assembly and filopodial extension induced by phorbol 12-myristate 13-acetate. J. Biol. Chem. 271, 32986-32993.

Hirao, M., Sato, N., Kondo, T., Yonemura, S., Monden, M., Sasaki, T., Takai, Y., & Tsukita, S. (1996). Regulation mechanism of ERM (Ezrin/Radixin/Moesin) protein/plasma membrane association: possible involvement of phosphatidylinositol turnover and rho-dependent signaling pathway. J. Cell Biol. 135, 37-51.

Hug, C., Jay, P.Y., Reddy, I., McNally, J.G., Bridgman, P.C., Elson, E.L., & Cooper, J.A. (1995). Capping protein levels influence actin assembly and cell motility in dictyostelium. Cell 81, 591-600.

Janmey, P.A., & Stossel, T.P. (1987). Modulation of gelsolin function by phosphatidylinositol 4,5-bisphosphate. Nature 325, 362-364.

Kimura, K., Ito, M., Amano, M., Chihara, K., Fukata, Y., Nakafuku, M., Yamamori, B., Feng, J., Nakano, T., Okawa, K.I., A., & Kaibuchi, K. (1996). Regulation of myosin phosphatase by Rho and Rho-associated kinase (Rho-kinase). Science 273, 245-248.

Kothakota, S., Azuma, T., Reinhard, C., Klippel, A., Tang, J., Chu, K., McGarry, T.J., Kirschner, M.W., Koths, K., Kwiatkowski, D.J., & Williams, L.T. (1997). Caspase-3-generated fragments of gelsolin: effector of morphological change in apoptosis. Science 278, 294-298.

Kozma, R., Ahmed, S., Best, A., & Lim, L. (1995). The ras-related protein Cdc42Hs and bradykinin promote formation of peripheral actin microspikes and filopodia in swiss 3T3 fibroblasts. Mol. Cell. Biol. 15, 1942-1952.

Lamaza, C., Chuang, T.-H., Terlecky, L.J., Bokoch, G.M., & Schmid, S.L. (1996). Regulation of receptor-mediated endocytosis by Rho and Rac. Nature 382, 177-179.

Lamaze, C., Fujimoto, L.M., Yin, H.L., & Schmid, S.L. (1997). The actin cytoskeleton is required for receptor-mediated endocytosis in mammalian cells. J. Biol. Chem. 272, 20332-20335.

Lappalainen, P., & Drubin, D.G. (1997). Cofilin promotes rapid actin filament turnover in vivo. Nature 388, 78-82.

Lascola, C.D., Nelson, D.J., & Kraig, R.P. (1998). Cytoskeletal actin gates a Cl-channel in neocortical astrocytes. J. Neurosci. 18, 1679-1692.

Lassing, I., & Lindberg, U. (1985). Specific interaction between phosphatidylinositol 4,5 bisphosphate and profilactin. Nature 314, 604-606.

Leung, T., Chen, X., Manser, E., & Lim, L. (1996). The p160 RhoA-binding kinase ROKα is a member of a kinase family and is involved in the reorganization of the cytoskeleton. Mol. Cell. Biol. 16, 5313-5327.

Lin, K., Wenegieme, E.F., Lu, P., Chen, C., & Yin, H.L. (1997). Gelsolin binding to phosphatidylinositol 4,5 bisphosphate is modulated by calcium and pH. J. Biol. Chem. 272, 20443-20450.

Lind, S.E., Janmey, P.A., Chaponnier, C., Herbert, T.-J., & Stossel, T.P. (1987). Reversible binding of actin to gelsolin and profilin in human platelet extracts. J. Cell Biol. 105, 833-842.

Liu, Y.T., Rozelle, A.L., & Yin, H.L. (1998). The gelsolin family of actin-filament severing and capping proteins. In: G Proteins, Cytoskeleton and Cancer (Aktories, K., Kohama, K., & Maruta, H., Eds.). Lander BioScience, Austin, TX.

Lu, M., Witke, W., Kwiatkowski, D.J., & Kosik, K.S. (1997). Delayed retraction of filopodia in gelsolin null mice. J. Cell Biol. 138, 1279-1287.

Lu, P.-J., Shieh, W.-R., Rhee, S.G., Yin, H.L., & Chen, C.-S. (1996). Lipid products of phosphoinositide 3-kinase bind human profilin with high affinity. Biochemistry 35, 14027-14034.

Lu, P.J., Hsu, A., Wang, D., Yan, H., Yin, H.L., & Chen, C.S. (1998). Phosphoinositide 3-kinase signaling in rat liver nuclei. Biochemistry 37, 5738–5745.

Machesky, L.M., & Hall, A. (1997). Role of actin polymerization and adhesion to extracellular matrix in Rac- and Rho-induced cytoskeletal reorganization. J. Cell Biol. 138, 913-926.

Machesky, L.M., Reeves, E., Wientjes, F., Mattheyse, F.J., Grogan, A., Totty, N.F., Burlingame, A.L., Hsuan, J.J., & Segal, A.W. (1997). Mammalian actin-related protein 2/3 complex localizes to regions of lamellipodial protrusion and is composed of evolutionarily conserved proteins. Biochem. J. 328, 105-112.

Mackay, D.J., Esch, F., Furthmayr, H., & Hall, A. (1997). Rho- and rac-dependent assembly of focal adhesion complexes and actin filaments in permeabilized fibroblasts: an essential role for ezrin/radixin/moesin proteins. J. Cell Biol. 138, 927-938.

Mahoney, N.M., Janmey, P.A., & Almo, S.C. (1997). Structure of the profilin-poly-L-proline complex involved in mrophogenesis and cytoskeletal regulation. Nat. Struct. Biol. 43, 953-960.

Matsui, T., Maeda, M., Doi, Y., Yonemura, S., Amano, M., Kaibuchi, K., & Tsukita, S. (1998). Rho-kinase phosphorylates COOH-terminal threonines of ezrin/radixin/moesin (ERM) proteins and regulates their head-to-tail association. J. Cell Biol. 140, 647-657.

McGough, A., Pope, B., Chiu, W., & Weeds, A.G. (1997). Cofilin changes the twist of F-actin: Implications for actin filament dynamics and cellular function. J. Cell Biol. 138, 771-781.

Meberg, P.J., Ono, S., Minamide, L.S., Takahashi, M., & Bamburg, J.R. (1998). Actin depolymerizing factor and cofilin phosphorylation dynamics: response to signals that regulate neurite extension. Cell Motil. Cytoskeleton 39, 172-190.

Miki, H., Miura, K., & Takenawa, T. (1996). N-WASP, a novel actin-depolymerizing protein, regulates the cortical cytoskeletal rearrangement in a PIP_2-dependent manner downstream of tyrosine kinases. EMBO J. 15, 5326-5335.

Miki, H., Sasaki, T., Takai, Y., & Takenawa, T. (1998). Induction of filopodium formation by a WASP-related actin-depolymerizing protein N-WASP. Nature 391, 93-96.

Minamide, L.S., Painter, W.B., Schevzov, G., Gunning, P., & Bamburg, J.R. (1997). Differential regulation of actin depolymerizing factor and cofilin in response to alterations in the actin monomer pool. J. Biol. Chem. 272, 8303-8309.

Moon, A., & Drubin, D.G. (1997). The ADF/cofilin proteins: Stimulus responsive modulators of actin dynamics. Mol. Biol. Cell 6, 1423-1431.

Muallem, S., Kwiatkowska, K., Xu, X., & Yin, H.L. (1995). Actin filament disassembly is a sufficient final trigger for exocytosis in non-excitable cells. J. Cell Biol. 128, 589-598.

Nobes, C.D., & Hall, A. (1995). Rho, Rac, and Cdc42 GTPases regulate the assembly of multimolecular focal complexes associated with actin stress fibers, lamellipodia, and filopodia. Cell 81, 53-62.

Norman, J.C., Price, L.S., Ridley, A.J., Hall, A., & Koffer, A. 1994). Actin filament organization in activated mast cells is regulated by heterotrimeric and small GTP-binding proteins. J. Cell Biol. 126, 1005-1015.

Obinata, T., Nagaoka-Yasuda, R., Ono, S., Kusano, K., Mohri, K., Ohtaka, Y., Yamashiro, S., Okada, K., & Abe, H. (1997). Low molecular-weight G-actin binding proteins involved in the regulation of actin assembly during myofibrillogenesis. Cell Struct. Funct. 22, 181-189.

Omann, G.M., Porasik, M.M., & Sklar, L.A. (1989). Oscillating actin polymerization/depolymerization responses in human polymorphonuclear leukocytes. J. Biol. Chem. 264,16355-16358.

Ostrander, D.B., Gorman, J.A., & Carman, G.M. (1995). Regulation of profilin localization in *Saccharomyces cerevisiae* by phosphoinositide metabolism. J. Biol. Chem. 270, 27045-27050.

Pantaloni, D., & Carlier, M.-F. (1993). How profilin promotes actin filament assembly in the presence of thymosin ß4. Cell 75, 1007-1014.

Pike, L.J., & Casey, L. (1996). Localization and turnover of phosphatidylinositol 4,5-bisphosphate in cavenolin-enriched membrane domains. J. Biol. Chem. 271, 26453-26456.

Reinhard, M., Geihl, K., Abel, K., Haffner, C., Jarchau, T., Hoppe, V., Jockusch, B.M., & Walter, U. (1995). The proline-rich focal adhesion and microfilament protein VASP is a ligand for profilin. EMBO J. 14, 1583-1589.

Ren, X.-D., Bokoch, G.M., Traynor-Kaplan, A., Jenkins, G.H., Anderson, R.A., & Schwartz, M.A. (1996). Physical association of the small GTPase Rho with a 68-kDa phosphatidylinositol 4-phosphate 5-kinase in Swiss 3T3 cells. Mol. Biol. Cell 7, 435-442.

Ridley, A.J., & Hall, A. (1992). The small GTP-binding protein rho regulates the assembly of focal adhesions and actin stress fibers in response to growth factors. Cell 70, 389-399.

Ridley, A.J., Paterson, H.F., Johnson, C.L., Diekmann, D., & Hall, A. (1992). The small GTP-binding protein rac regulates growth factor-induced membrane ruffling. Cell 70, 401-410.

Rodriguez-Viciana, P., Warne, P.H., Khwaga, A., Marte, B.M., Pappin, D., Das, P., Waterfield, M.D., Ridley, A., & Downward, J. (1997). Role of phosphoinositide 3-OH kinase in cell transformation and control of the actin cytoskeleton by Ras. Cell 69, 457-467.

Rosenblatt, J., Agnew, B.J., Abe, H., Bamburg, J.R., & Mitchison, T. (1997). Xenopus actin depolymerizing factor/cofilin (XAC) is responsible for the turnover of actin filaments in listeria monocytogenes tails. J. Cell Biol. 136, 1323-1332.

Sakisaka, T., Itoh, T., Miura, K., & Takenawa, T. (1997). Phosphatidylinosiol 4,5-bisphosphate phosphatase regulates the rearrangement of actin filaments. Mol. Cell. Biol. 17, 3841-3849.

Schafer, D.A., Jennings, P.B., & Cooper, J.A. (1996). Dynamics of capping protein and actin assembly in vitro: uncapping barbed ends by polyphosphoinositides. J. Cell Biol. 135,169-179.

Sheterline, P., Rickard, J.E., & Richards, R.C. (1984). Fc receptor-directed phagocytic stimuli induce transient actin assembly at an early stage of phagocytosis in neutrophil leukocytes. Eur. J. Cell Biol. 34, 80-87.

Shibasaki, Y., Ishihara, H., Kizuki, N., Asano, T., Oka, Y., & Yazaki, Y. (1997). Massive actin polymerization induced by phosphatidylinositol-4-phosphate 5-kinase *in vivo*. J. Biol. Chem. 272, 7578-7581.

Singh, S.S., Chauhan, A., Murakami, N., & Chauhan, V.P. (1996). Profilin and gelsolin stimulate phosphatidylinositol 3-kinase activity. Biochemistry 35, 16544-16549.

Sun, H., Kwiatkowska, K., & Yin, H.L. (1996). ß-thymosins are not simple actin monomer buffering proteins. J. Biol. Chem. 271, 9223-9230.

Sun, H., Lin, K., & Yin, H.L. (1997). Gelsolin modulates phospholipase C activity *in vivo* through phospholipid binding. J. Cell Biol. 138, 811-820.

Sun, H.-Q., Kwiatkowska, K., & Yin, H.L. (1995a). Actin monomer binding proteins. Curr. Opin. Cell Biol. 7, 102-110.

Sun, H.-Q., Kwiatkowska, K., Wooten, D.C., & Yin, H.L. (1995b). Effects of CapG overexpression on agonist-induced motility and second messenger generation. J. Cell Biol. 129, 147-156.

Suzuki, K., Yamaguchi, T., Tanaka, T., Kawanishi, T., Nishimaki-Mogami, T., Yamamoto, K., Tsuji, T., Irimura, T., Hayakawa, T., & Takahashi, A. (1995). Activation induces dephosphorylation of cofilin and its translocation to plasma membranes in neutrophil-like differentiated HL-60 cells. J. Biol. Chem. 270, 19551-19556.

Symons, M.H., & Mitchison, T.J. (1991). Control of actin polymerization in live and permeabilized fibroblasts. J. Cell Biol. 114, 503-513.

Tardif, M., Huang, S., Redmond, T., Safer, D., Pring, M., & Zigmond, S.H. (1995). Actin polymerization induced by GTPyS in permeabilized neutrophils is induced and maintained by free barbed ends. J. Biol. Chem. 270, 28075-28083.

Thelen, M., Uguccioni, M., & Bosiger, J. (1995). PI 3-kinase-dependent and independent chemotaxis of human neutrophil leukocytes. Biochem. Biophys. Res. Comm. 217, 1255-1262.

Theriot, J.A., & Mitchison, T.J. (1991). Actin microfilament dynamics in locomoting cells. Nature 352, 126-131.

Theriot, J.A., Mitchison, T.J., Tilney, L.G., & Portnoy, D.A. (1992). The rateof actin-based motility of intracellular *Listeria monocytogenes* equals the rate of actin polymerization. Nature 357, 257-260.

Toker, A., & Cantley, L.C. (1997). Signalling through the lipid products of phosphoinositide-3-OH kinase. Nature 387, 673-676.

Tolias, K.F., Cantley, L.C., & Carpenter, C.L. (1995). Rho family GTPases bind to phosphoinositide kinases. J. Biol. Chem. 270, 17656-17659.

Tsukita, S., & Yonemura, S. (1997). ERM (ezrin/radixin/moesin) family: from cytoskeleton to signal transduction. Curr. Opin. Cell Biol. 9, 70-75.

Van Aelst, L., & D'Souza-Schorey, C. (1997). Rho GTPases and signaling networks. Genes Dev. 11, 2295-2322.

Watanabe, N., Madaule, P., Watanabe, G., Kakizuka, A., Saito, Y., Nakao, K., Jockusch, B.M., & Narumiya, S. (1997). p140mDia, a mammalian homolog of Drosophila diaphanous, is a traget protein for Rho small GTPase and is a ligand for profilin. EMBO J. 16, 3044-3056.

Welch, M.D., Mallavarapu, A., Rosenblatt, J., & Mitchison, T.J. (1997). Actin dynamics *in vivo*. Curr. Opin. Cell Biol. 9, 54-61.

Witke, W., Sharpe, A.H., Hartwig, J.H., Azuma, T., Stossel, T.P., & Kwiatkowski, D.J. (1995). Hemostatic, inflammatory, and fibroblast responses are blunted in mice lacking gelsolin. Cell 81, 41-51.

Yin, H.L., & Stossel, T.P. (1980). Control of cytoplasmic actin gel-sol transformation by gelsolin, a calcium-dependent protein. Nature 218, 583-586.

Yonezawa, N., Nishida, E., Iida, K., Yahara, I., & Sakai, H. (1990). Inhibition of the interactions of cofilin, destrin, and deoxyribonuclease I with actin by phosphoinositides. J. Biol. Chem. 265, 8382-8386.

Yu, F.-X., Johnston, P.A., Sudhof, T.C., & Yin, H.L. (1990). gCap39, a calcium ion- and polyphosphoinositide-regulated actin capping protein. Science 250, 1413-1415.

Yu, F.-X., Sun, H.-Q., Janmey, P.A., & Yin, H.L. (1992). Identification of a polyphosphoinositide-binding sequence in an actin monomer-binding domain of gelsolin. J. Biol. Chem. 267, 14616-14621.

Zhang, X., Loijens, J.C., Boronenkov, I.V., Parker, G.J., Norris, F.A., Chen, J., Thum, O., Prestwich, G.D., Majerus, P.W., & Anderson, R.A. (1997). Phosphatidylinositol-4-phosphate 5-kinase isozymes catalyze the synthesis of 3-phosphate-containing phosphatidylinositol signaling molecules. J. Biol. Chem. 272, 17756-17761.

Zigmond, S.H., Joyce, M., Borleis, J., Bokoch, G.M., & Devreotes, P.N. (1997). Regulation of actin polymerization in cell-free systems by GTPgammaS and Cdc42. J. Cell Biol. 138, 363-374.

SECTION V

RESPONSES

NRAMP1
A NOVEL MACROPHAGE PROTEIN WITH A KEY FUNCTION IN RESISTANCE TO INTRACELLULAR PATHOGENS

Samantha Gruenheid, Emil Skamene, and Philippe Gros

Advances in Cell and Molecular Biology of Membranes and Organelles
Volume 5, pages 345-362.

ISBN: 1-55938-999-0

I. HOST DETERMINANTS OF SUSCEPTIBILITY TO INFECTION: A GENETIC APPROACH

The widespread emergence of antibiotic-resistant variants of a large number of human pathogens and the resulting failure of current antimicrobial therapies in an increasing number of common infections are important limitations to the treatment of many infectious diseases. A greater understanding of host defense mechanisms against such infectious agents may open new windows of opportunity for intervention in these infections. One useful strategy for the identification of such mechanisms is the study of "host resistance" factors. These may manifest themselves as genetic differences either in human populations during epidemics and in endemic areas of disease, or in inbred strains of laboratory animals affecting susceptibility to experimental infections. Identifying and characterizing the genetic component of such innate differences, including the cloning of the relevant genes and corresponding proteins may reveal novel aspects of the host antimicrobial defenses and may suggest how the determinants of microbial pathogenicity can overcome such defenses.

In humans, susceptibility to infectious diseases usually behaves as a complex trait, with genetic heterogeneity, variable penetrance and expressivity. In addition, virulence factors of the microbe, host immune status, prior exposure, vaccination and other factors all can affect the outcome of host-microbe interaction. Thus, the search for individual major gene effects in humans that influence the susceptibility to infection, disease progression, and ultimate outcome has been inherently difficult. An attractive alternative to this complex situation is to identify potential major gene effects in animal models of the disease, and then use these "genetic hits" to look for a parallel effect of the human homologues of these genes in association or formal linkage studies in the field. Genetic analysis of the susceptibility to infectious diseases in well-characterized inbred mouse strains presents many advantages compared to gene discovery in humans. Inbred strains of mice are a unique resource, providing an unlimited number of genetically identical individuals homozygous at each locus, with each strain representing a specific assortment of the gene and allele pools present in the wild. Once differences in resistance or susceptibility to infection between different mouse strains are identified, informative crosses can be set up in a prospective, directed fashion to identify and eventually isolate the loci involved. Another advantage of the laboratory mouse as an animal model is the ease with which experimental conditions for infection can be controlled, minimizing environmental effects and allowing functional dissection of the phenotypic expression of the gene under study. Finally, the possibility of introducing germ line mutations in the mouse provides the opportunity of directly assessing the importance of candidate genes on the outcome of infection.

The power of mouse genetics for the identification of host resistance loci, including the parallel analysis of these genes in humans, is illustrated by the identification and characterization of *Nramp1*, a novel phagocyte specific protein. This protein plays a key role in mouse and human defenses against intracellular infectious diseases, and identifies a novel biochemical mechanism of macrophage antimicrobial function.

II. THE BCG/ITY/LSH LOCUS AND THE NRAMP1 GENE

In the mouse, innate resistance or susceptibility to infection with several mycobacterial species—including *Mycobacterium lepraemurium, M. bovis* (BCG), *M. intracellulare*, and *M. avium* is under genetic control of the chromosome 1 locus *Bcg* (Skamene et al., 1980; Malo and Skamene, 1994; McLeod et al., 1995). The *Bcg* gene is present in two allelic forms in inbred mouse strains, Bcg^r (resistant, dominant) or Bcg^s (susceptible, recessive), and is expressed phenotypically as a differential growth rate of these bacteria in the spleen and liver during the early phase of infection (Forget et al., 1981; Gros et al., 1981). Initial mapping studies (Skamene et al., 1982) suggested that *Bcg* may be identical to two other previously mapped loci, *Ity* and *Lsh*, which control resistance to infection with other antigenically unrelated intracellular parasites, *Salmonella typhimurium* (Lissner et al., 1983) and *Leishmania donovani* (Croker et al., 1984), respectively. Experiments *in vivo* (Gros et al., 1983) and *in vitro* with explanted cell populations have indicated that the macrophage is the site of phenotypic expression of the genetic difference at *Bcg/Ity/Lsh* (Gros et al., 1983; Lissner et al., 1983; Croker et al., 1984; Stach et al., 1984; Stokes et al., 1986; Denis et al., 1990; DeChastellier et al., 1993). These experiments have indicated that *Bcg* plays a key role in early host defense mechanisms by influencing the bactericidal or bacteriostatic activity of the macrophage.

A positional cloning approach was adopted to clone the *Bcg/Ity/Lsh* gene and identify the encoded protein. This approach involved first a segregation analysis in informative populations of backcross or F2 mice, where the inheritance of the phenotype under study (in this case resistance to infection) was compared with the inheritance of informative markers dispersed throughout the genome, generating a high resolution linkage map in the vicinity of *Bcg* (Malo et al., 1993b). This genetic map was then converted to a physical map after isolation and characterization of overlapping DNA fragments which cover the entire genetic interval (Malo et al., 1993a). Once the physical interval was cloned, transcription units were isolated from the interval by exon amplification and these were further analyzed as candidates for *Bcg*. Among six candidates identified in the minimal physical interval for *Bcg*, one corresponded to a novel gene that was expressed exclusively in spleen and liver whose expression was enriched in macrophage populations derived from them. This gene was given the name *Nramp1* (natural resistance associated macrophage protein 1). The *Nramp1* gene was found to encode a highly hydrophobic protein of predicted molecular mass 60 kDa with characteristics of an integral membrane protein, including 12 puta-

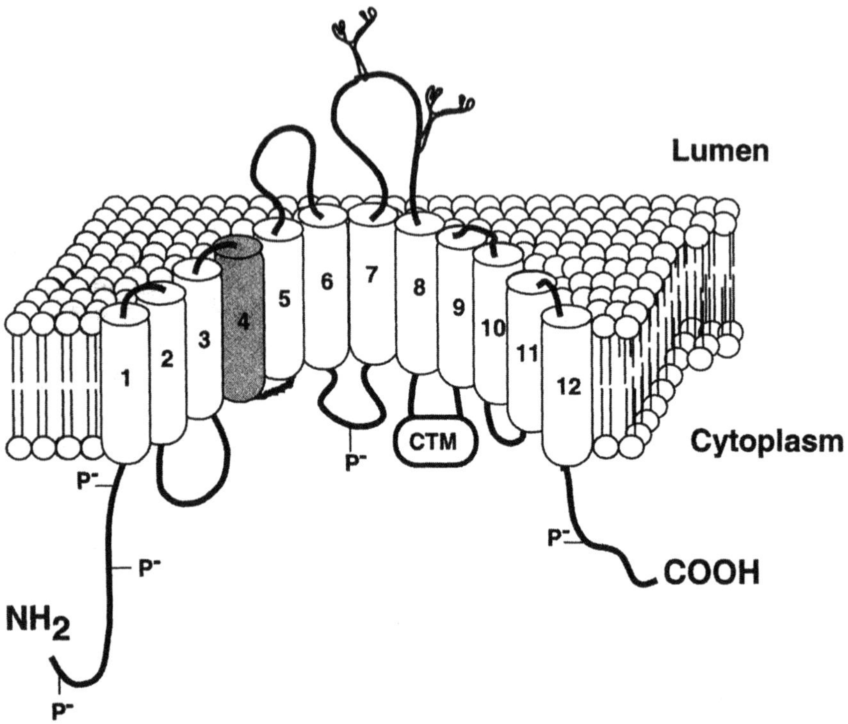

Figure 1. Schematic representation of the membrane-associated arrangement of the Nramp proteins. The amino-(NH_2) and carboxy-termini (COOH) of the protein are identified on the cytoplasmic side of the membrane, and the individual predicted transmembrane domains (numbered 1 to 12) are shown. The predicted transmembrane domain 4, which is mutated in mice bearing the *Bcg*^{s} allele of *Nramp1* and the microcytic anemia (*mk*) mutant, and in the Belgrade rat is shaded (see text). Possible phosphorylation (P^-) and glycosylation (in luminal loop 4) sites are identified. A consensus transport motif (CTM) found in several bacterial and eukaryotic membrane transport proteins is depicted in the proposed cytoplasmic loop 4. This figure is modified from Vidal et al., 1993 and Govoni and Gros (1998).

tive transmembrane (TM) domains, a glycosylated extracytoplasmic loop, several predicted phosphorylation sites and a consensus transport signature found in a large number of prokaryotic and eukaryotic membrane transporters (Fig 1) (Vidal et al., 1993). A unique aspect of the Nramp1 sequence is the presence of several charged amino acid residues within the highly hydrophobic predicted TM domains. Such charged residues are thermodynamically unfavored in TM domains, and are usually found in ion transporters and channels where they may line a hydrophilic transmembrane pore forming a transport path. Together, these

Table 1. Effect of Elimination of *Nramp1* Function in 129sv ($Nramp1^{-/-}$) Mice on Resistance and Susceptibility to Different Types of Infections

Under Nramp1 *Control*	*Not Under* Nramp1 *Control*
Mycobacterium bovis[1]	*Pseudomonas aeruginosa*[3]
Mycobacterium avium[1]	*Listeria monocytogenes*[3,4]
Salmonella typhimurium	*Legionella pneumohila*[3]
Leishmania donovani	*Staphylococcus aureus*[5]
Toxoplasma gondii[2]	*Bacillus subtilis*[1]

Note: Animals were assayed for resistance to experimental infection *in vivo* or explanted macrophages were infected *in vitro* and the extent of intracellular replication was monitored. Data are from (Vidal et al., 1995) and from personal communications from Drs. C. DeChastellier (1), R. McLeod (2), F. Gervais (3), D. Malo (4), and S. Tarkowskie (5).

results suggest that Nramp1 is a membrane transport protein. Sequencing of the *Nramp1* mRNA from 27 Bcg^r and Bcg^s inbred mouse strains revealed that susceptibility to infection was associated with a non-conservative glycine to aspartic acid substitution at position 169 (G169D), within the fourth predicted TM domain of the protein (Malo et al., 1994). Although the addition of a charged residue into a transmembrane domain would be predicted to disrupt the tertiary structure of the protein, additional proof was required to confirm that this polymorphism at *Nramp1* was associated with a loss-of-function and with susceptibility to infection in Bcg^s mice. To formally demonstrate that *Nramp1* and *Bcg* are indeed the same gene, a mouse mutant (knockout) lacking a functional *Nramp1* gene was created by gene targeting experiments in embryonic stem cells (Vidal et al., 1995). Although the mutant $Nramp1^{-/-}$ mice created on the 129sv background (Bcg^r) had normal appearance and longevity, they no longer showed natural resistance to experimental infection with *Mycobacterium*, *Salmonella* and *Leishmania* (Vidal et al., 1995). In the converse experiment, a transgene coding for the resistant allele of *Nramp1* (Bcg^r, G169) was introduced onto the genetic background of a susceptible mouse strain (Bcg^s, D169). These transgenic mice became resistant to infection with *Mycobacterium* and *Salmonella*, demonstrating that *Nramp1* and *Bcg* are indeed allelic (Govoni et al., 1996). Additional studies have shown that, when compared to their $Nramp1^{wt}$ littermates, $Nramp1^{-/-}$ mice also become susceptible to *Toxoplasma gondii* and *M. avium* indicating that Nramp1 plays a role in the defense against these pathogens as well. However, mutations at *Nramp1* have no effect on natural resistance or susceptibility to *Legionella pneumophila, Listeria monocytogenes, Pseudomonas aeruginosa, Staphylococcus aureus,* and *Bacillus subtilis.*

III. *NRAMP1* AND ITS ROLE IN HUMAN DISEASE

The association of human *NRAMP1* polymorphisms with disease susceptibility has been most extensively studied in leprosy. Although *NRAMP1* alleles do not seem to contribute to disease susceptibility in certain populations analyzed (Shaw et al., 1993; Levee et al., 1994; Roger et al., 1997), positive results were obtained in a recent large study of susceptibility to leprosy in South Vietnam (Abel et al., 1998). In this study, segregation analyses were conducted on 285 Vietnamese and 117 Chinese families with leprosy. There was evidence of a codominant major gene in the Vietnamese families but no evidence for a major gene in the Chinese families. In the Vietnamese families alone, linkage was observed for both intragenic *NRAMP1* ($P < 0.05$) and extended *NRAMP1* haplotypes ($P < 0.02$). These results support the hypothesis that *NRAMP1* or a gene closely linked to it plays a role in susceptibility to leprosy in the Vietnamese families.

Very recently, the association of *NRAMP1* polymorphic variants with susceptibility to another mycobacterial infection was demonstrated in a case-control study of tuberculosis in The Gambia, West Africa (Bellamy et al., 1998). In this study, the distribution of 4 polymorphic alleles of *NRAMP1* was determined in 410 adults with smear-positive pulmonary tuberculosis, as well as in 417 ethnically-matched healthy controls. Patients with polymorphisms in intron 4 and in the 3' untranslated region of the gene were particularly overrepresented in the tuberculous population, as opposed to those with most common *NRAMP1* genotypes (odds ratio, 4.09; 95% confidence interval, 1.86–9.12; chi-square = 14.58, $P < 0.001$). The authors concluded that genetic variations at *NRAMP1* affect susceptibility to tuberculosis in West Africans. Together, these data show that human *NRAMP1* may be an important determinant of innate susceptibility or progression of at least two major infectious diseases caused by *Mycobacteria*. Thus, the study of genetic basis of susceptibility to mycobacterial infections in mice has enabled the identification of a genetic component of host response to these infections in humans as well.

IV. *NRAMP1* EXPRESSION IN PROFESSIONAL PHAGOCYTES

Detailed studies in mice and humans have revealed a striking pattern of tissue and cell specific expression of *Nramp1/NRAMP1* mRNAs. In mouse tissues, *Nramp1* mRNA is detected exclusively in the spleen and liver (Vidal et al., 1993). Using a panel of bone marrow-derived hemopoietic precursor cell colonies corresponding either to undifferentiated progenitors or to mature lymphoid, erythroid and myeloid lineages, *Nramp1* mRNA was shown to be exclusively expressed in mature myeloid monocyte/macrophage and granulocytic lineages. Using the murine macrophage cell line RAW264.7 as a model, we observed that both bacterial lipopolysaccharide (LPS) and interferon-γ (IFN-γ) treatments can upregulate *Nramp1* mRNA expression, and that maximum *Nramp1* expression is induced by pre-treatment with IFN-γ followed by LPS exposure (Govoni et al., 1997). This provides strong evidence that *Nramp1* expression is upregulated as phagocytic cells are primed and activated during infection and/or during inflammatory responses.

In humans, *NRAMP1* gene expression is detected in lungs, liver and spleen, but is most abundant in peripheral blood leukocytes (Cellier et al., 1997). Studies in purified blood cell populations showed that polymorphonuclear leukocytes are by far the major site of *NRAMP1* expression, followed to a lesser degree by monocytes. In addition, Northern blot analysis of mRNA from cultured cells demonstrated that increased *NRAMP1* expression was associated with differentiation towards mature macrophages and granulocytes. This pattern of lineage-specific expression was seen in primary cells but could also be replicated using the HL-60 model cell line differentiated *in vitro* towards either the monocytic or granulocytic pathway. Thus, *Nramp1* expression is restricted to professional phagocytes.

V. CHARACTERIZATION OF THE NRAMP1 PROTEIN

To further investigate *Nramp1* function in phagocytic cells, we have raised polyclonal antisera directed against the N- and C-terminal portions of the Nramp1 protein (Vidal et al., 1996). In addition, experiments have been performed in mammalian cells expressing a recombinant Nramp1 protein in which a short epitope tag was inserted at the carboxy-terminus; that can be easily detected in transfected cells using a commercially available monoclonal antibody directed against this tag (Vidal et al., 1996). Western blot and immunoprecipitation analyses of both endogenous and transfected protein reveal Nramp1 as an integral membrane protein resistant to extraction with urea, that migrates as a broad band with an apparent molecular mass of approximately 80–100 kDa. Glycosidase sensitivity experiments demonstrate that almost 50% of its apparent molecular mass is due to extensive glycosylation with complex oligosaccharides (Vidal et al., 1993). Additionally, treatment of macrophages with ^{32}P-orthophosphate results in a strong labeling of the protein, indicating that Nramp1 is also phosphorylated *in vivo*. The potential role of this phosphorylation in Nramp1 function and the kinase involved are still unknown, although the Nramp1 cDNA sequence contains consensus phosphorylation sites for protein kinase C (Vidal et al., 1993).

Surprisingly, when immunoprecipitation is performed on macrophages from resistant (*Bcg^r*, G169) and susceptible (*Bcg^s*, D169) inbred mouse strains, the mature Nramp1 protein is not detected in macrophages from susceptible mice, suggesting that the D169 mutation affects normal processing and results in targeting of the D169 protein for degradation (Vidal et al., 1996). Immunofluorescence studies in primary macrophages from normal and *Nramp1$^{-/-}$* mutants together with studies in RAW264.7 transfectants expressing an epitope-tagged Nramp1 have

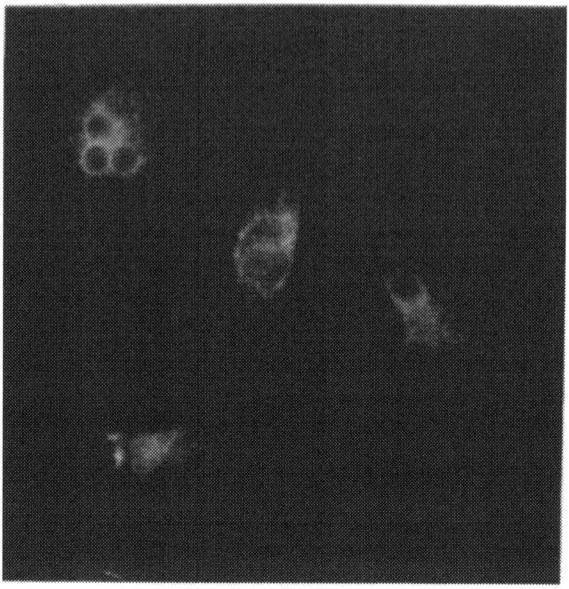

Figure 2. Nramp1 association with latex bead-containing phagosomes. Peritoneal macrophages were isolated from *Nramp1* wild-type mice and cultured on glass coverslips. The cells were fed with medium containing 3 μm latex beads for one hour and then further incubated for one hour to allow phagosome maturation. The cells were then fixed and subjected to indirect immunofluorescence with anti-Nramp1 antiserum.

shown that Nramp1 is not expressed at the plasma membrane, but rather is restricted to a subcellular membranous compartment of these cells. Double-label immunofluorescence studies demonstrated that Nramp1 colocalizes with Lamp-1 (lysosomal-associated membrane protein) in a late endocytic compartment (late endosome/lysosome) (Gruenheid et al., 1997). The localization of Nramp1 to the endosomal/lysosomal network raised the possibility that Nramp1 might be recruited to the phagosomal membrane during infection. Phagosomes are known to sequentially interact and acquire proteins from both early and late endosomes before their final fusion with terminal lysosomes (Pitt et al., 1992; Desjardins et al., 1994). Using the model system of latex bead-containing phagosomes (Korn, 1974), Nramp1 association with the phagosomal membrane was investigated. Latex beads are inert, nondegradable particles that are readily phagocytosed by macrophages and macrophage cell lines. They are available in sizes that can be easily visualized by microscopy for immunofluorescence applications. Additionally, latex-bead phagosomes can easily be purified from cell homogenates by virtue of their low density by floatation on sucrose gradients (Desjardins et al., 1994). Nramp1 was shown to be associated with the phagosome by its presence in high levels in purified phagosomal fractions, as well as by a dramatic ring-like Nrampl Immunofluorescence staining around latex bead phagosomes (Fig 2) (Gruenheid et al., 1997). In time-course studies, Nramp1 was shown to be recruited to the phagosomal membrane and remain associated with this structure during its maturation to phagolysosome with kinetics similar to those seen for Lamp-1 (Gruenheid et al., 1997). Recently, Nramp1 recruitment to the membrane of parasite-containing phagosomes during infection has been confirmed with phagosomes containing *Salmonella* and *Yersinia* (Govoni and Gros, 1999) species. The targeting of Nramp1 from endocytic vesicles to the phagosomal membrane supports the hypothesis that Nramp1 controls the replication of intracellular parasites by directly (or indirectly) altering the intravacuolar environment of the microbe-containing phagosome.

A good correlation has been noted between the subcellular localization of Nramp1, its kinetics of association with the phagosome, and the known intracellular survival strategies adopted by microbes differentially affected by *Nramp1* mutations *in vivo* (Table 1). One common characteristic of the phagosome containing *Mycobacteria*, *Salmonella*, and *Leishmania*, for example, is that they all seem to acquire various amounts of the late endosomal marker Lamp-1 at some stage (Pitt et al., 1992; Garcia-del Portillo et al., 1993, 1995; Desjardins et al., 1994; Sturgill-Koszycki et al., 1994; Clemens et al., 1995), which colocalizes with Nramp1. On the other hand, the lack of an effect of *Nramp1* mutations on infections with *Listeria* and *Legionella*, both *in vivo* and *in vitro* in explanted macrophages (Gervais and Gros, unpublished), is consistent with the unique intracellular behavior of these bacteria, as *Listeria* escapes from the phagosome within 20 minutes (Campbell, 1994), and the *Legionella* phagosome does not acquire endosomal or lysosomal markers, including Lamp-1 (Clemens and Horwotz, 1995). Finally, loss of Nramp1 function does not seem to affect the replication and/or pathogenesis *in vivo* of nonintracellular pathogens such as *S. aureus*, *B. subtilis*, and *P. aeruginosa* (Table 1).

VI. NRAMP1 AND PHAGOSOMAL ACIDIFICATION

In order to study the effect of *Nramp1* on the intraphagosomal environment, we used microscopic imaging technology to examine the properties of phagosomes in wild-type and *Nramp1*$^{-/-}$ mice (Hackam et al., 1998). In these studies, peritoneal macrophages were isolated and allowed to phagocytose either inert latex beads, heat-killed *M. bovis* (strain BCG), or live BCG. The pH inside phagosomes was measured by conjugation of pH-sensitive dyes (fluorescein isothiocyanate and Oregon green) to the particles before phagocytosis followed by microfluorescence ratio imaging of the particles within the phagosomes after phagocytosis. While the phagosomes containing heat-killed BCG or latex beads were found to acidify fully and to the same extent in both wild-type and *Nramp1*$^{-/-}$ macrophages (pH 5.5), phagosomes containing live BCG displayed attenuated acidification in the *Nramp1*$^{-/-}$ macrophages (pH 6.5), yet acidified normally in the wild-type macrophages (pH 5.5). Furthermore, BCG-containing phagosomes in macrophages from a naturally susceptible mouse strain (*Bcg*^{s}, Nramp1 D169) displayed attenuated acidification comparable to that seen in *Nramp1*$^{-/-}$ macrophages. Further investigation revealed that the decreased acidification of the BCG-phagosomes in the *Nramp1*$^{-/-}$ macrophages was correlated with decreased fusion of these phagosomes to the late-endocytic compartments, as determined by quantification of Lamp-2 (a late endosomal/lysosomal marker) around the BCG-phagosome. The decreased acidification of the BCG-phagosomes in the *Nramp1*$^{-/-}$ macrophages was also associated with decreased vacuolar type proton ATPase (V-ATPase) activity. It is not yet clear if the observed decreased acidification is a cause or an effect of the decreased fusion of the phagosome with late endocytic compartments, since the (V-ATPase) is found mainly in the late endocytic compartments. In addition, some studies have suggested that V-ATPase-mediated acidification may be required for phagosomal maturation (van Weert et al., 1995; Hammond et al., 1998).

The attenuated acidification and decreased fusion of BCG-containing phagosomes observed in the *Nramp1*$^{-/-}$ macrophages is consistent with previous reports that live Mycobacteria impair or inhibit phagosomal maturation and acidification. Although this phenomenon was described over 20 years ago (Armstrong and Hart, 1975), the mechanism by which Mycobacteria are able to inhibit phagosomal maturation is not yet known. It is important to note that these previous studies were without exception performed in cell lines or mice that are now known to be homozygous for the susceptible allele of *Nramp1* (*Bcg*^{s}, D169). The new and most interesting result that has emerged from our study with *Nramp1* wild-type and *Nramp1*$^{-/-}$ mice is that in the presence of functional Nramp1, the mycobacterial block to phagosomal maturation does not appear to take place. The mechanism by which Nramp1 counters the mycobacterial block to acidification remains to be determined. Further studies using microscopic imaging should allow a closer examination of the interplay between the phagosome, mycobacteria, and Nramp1. It will also be possible to address whether this parasite-mediated block of phagosomal maturation, and the Nramp1-mediated neutralization of this block, is common to all parasites affected by Nramp1, or is a phenomenon unique to *M. bovis* (BCG).

VII. THE NRAMP GENE FAMILY AND DIVALENT CATION TRANSPORT

The biochemical mechanism of Nramp1 action including the possible substrate transported by this protein remain to be determined. However, important clues of its function have come from the recent discovery and study of additional *Nramp* genes as well as homologues from different species. In mammals, a second *Nramp* gene, *Nramp2*, has been cloned (Gruenheid et al., 1995; Vidal et al., 1995). *Nramp2* codes for a highly homologous protein (78% identical) which is ubiquitously expressed, as opposed to the phagocyte-specific *Nramp1*. The function of Nramp2 was also unknown until recently, when two independent reports showed by very different methods that Nramp2 functions as a transporter of divalent cations, including iron. In the first study, *Nramp2* was isolated by expression cloning from a rat cDNA library in *Xenopus* oocytes in a screen for iron transport proteins. This study showed that Nramp2 (designated DCT1 in this report) mediates electrogenic transport of Fe^{2+}, Mn^{2+}, Zn^{2+} and other divalent cations (Gunshin et al., 1997). The transport was shown to be pH-dependent and coupled to proton symport, which suggests that divalent cation transport by Nramp2 could be driven by the cotransport of a proton down its concentration gradient. mRNA expression studies in this report showed ubiquitous expression of Nramp2, with a striking upregulation of the mRNA levels in the proximal intestine following chronic iron depletion (Gunshin et al., 1997).

In a concurrent report, a mutation in *Nramp2* was identified as the cause of iron deficiency in a mouse model of microcytic anemia, the *mk* mouse (Fleming et al., 1997b). Soon after, it was also determined that *Nramp2* is also mutated in a rat mutant with an *mk*-like phenotype, the *Belgrade* rat (Fleming et al., 1997a). The *mk* mouse and the *Belgrade* rat both show deficient iron uptake in the intestine and in red blood cell precursors known as reticulocytes (Edwards and Hoke, 1972, 1975; Thomas and Valberg, 1972; Garrick et al., 1978). In addition, *Belgrade* rats are reported to have abnormalities in manganese metabolism (Thomas and Valberg, 1972) and *mk* mice show signs of zinc deficiency (Russell et al., 1970). Surprisingly, both the *mk* mouse and the *Belgrade* rat carry the same mutation in *Nramp2*, a glycine to arginine substitution in the fourth predicted transmembrane domain of the protein (G185R). Transfection studies demonstrated that transient overexpression of wild-type (G185) Nramp2 in 293T cells causes a dramatic increase in Fe^{2+} uptake in these cells, while expression of the mutated form of the protein (D185) did not affect iron uptake (Fleming et al., 1997a). These results provide compelling evidence that *Nramp2* is the long-sought-after transferrin-independent intestinal iron uptake system in mammals. Additionally, it is likely that Nramp2 is involved in the uptake of other divalent cations in the intestine, as well as in divalent cation uptake in tissues other than the intestine.

Nramp homologues encoding remarkably conserved proteins have also been identified in many phylogenetically distant organisms through various genome sequencing projects, as well as by conventional cloning methods. *Nramp* homologues have been described in *Drosophila melanogaster* (70% identity), *Caenorhabditis elegans* (68% identity), plants (*OsNramp* family, 50–60% identity), *Saccharomyces cerevisiae* (40–45% identity) and several microbes, including several species of *Mycobacteria* (37% identical) as well as several other gram-negative and gram-positive bacteria (Cellier, personal communication). The remarkable degree of sequence conservation amongst *Nramp* family members is suggestive of a parallel conservation of function (Cellier et al., 1995). Recent studies have provided strong support to this proposal. Yeast have two *Nramp* homologues, *SMF1* and *2*, and combined loss-of-function at both loci leads to hypersensitivity to ethyleneglycotetraacetic acid (EGTA) as well as hypersensitivy to alkaline pH (Supek et al., 1996). SMF1 was recently shown to function as a Mn^{2+} transporter since deletion of *SMF1* causes decreased cellular uptake of Mn^{2+} and overexpression of the protein results in increased Mn^{2+} uptake. When we tested whether structural similarity between the mammalian Nramp and the yeast Smf proteins would result in functional complementation in yeast, Nramp2, but not Nramp1, was found to complement both the hypersensitivity to EGTA and hypersensitivity to alkaline pH characteristic of the *SMF1/SMF2* mutant yeast (Pinner et al., 1997). Complementation by Nramp2 was specific and required a functional protein, as independent mutations at residues highly conserved in the Nramp family abrogated Nramp2 complementation. Since Mn^{2+} was the only divalent cation capable of completely suppressing both the EGTA and pH phenotypes, these results suggest that Nramp2 can transport Mn^{2+} in yeast (Pinner et al., 1997).

Other studies support a conservation of divalent cation transport function among Nramp proteins. The fly *Nramp* homologue, known as *malvolio*, has been shown to be important for taste discrimination behavior in the fly (Rodrigues et al., 1995). Whereas wild-type flies will avoid media containing 100 mM sodium chloride and strongly favor media containing sugars, flies with a mutation in *malvolio* show an increased acceptance of salt, and a greatly reduced preference for sugars. However, when the flies are fed with media supplemented with excess Fe^{2+} or Mn^{2+} for a minimum of two hours before testing, normal taste behavior is restored (Orgad and Nelson, 1998). This suggests that malvolio is important for Fe^{2+}/Mn^{2+} metabolism in the fly, and that these ions are somehow involved in taste perception in *Drosophila.* Finally, the Nramp protein of *Escherichia coli,* which is the most distant relative on the evolutionary scale from mammalian Nramp2, has recently been shown to promote ion transport (Cellier, personal communication).

VIII. IMPLICATIONS FOR *NRAMP1*

How do these findings help us understand the mechanism of action of Nramp1 in the phagosomal membrane? Since Nramp2 shares divalent cation transport func-

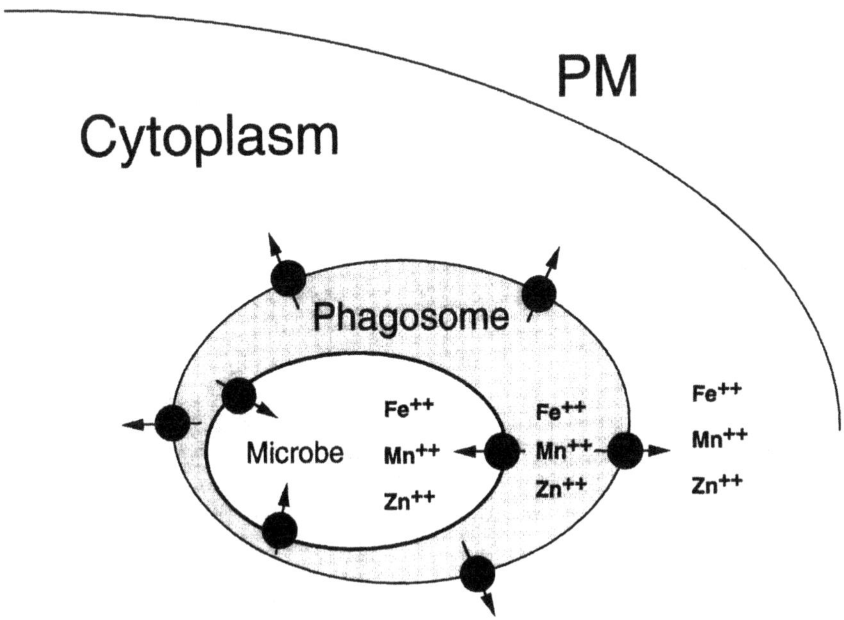

Figure 3. Proposed model of action of the Nramp1 transporter and possible bacterial homologue. A hypothetical model for competition of the same divalent cation substrates by the mammalian and bacterial *Nramp* proteins within the phagosome compartment of the macrophage is shown. This figure is adapted from Govoni and Gros (1998).

tion with its distant relatives *Drosophila* malvolio, yeast SMF, and even bacterial Nramp, it is tempting to speculate that the more closely related Nramp1 will also be involved in divalent cation transport. We (Cellier et al., 1996) (Figure 3) and others (Supek et al., 1996) have hypothesized that Nramp1 may act to deplete the phagosomal environment of either Fe^{2+} or of other divalent cations such as Mn^{2+}, Mg^{2+}, Zn^{2+}, or Ca^{2+} that are metabolically essential for microbial survival and replication (Crosa, 1997). This hypothesis is attractive, since both the microbes and the host have developed sophisticated and competing systems to scavenge such cations and metals from within the intraphagosomal space (Kontoghiorghes and Weinberg, 1995; Crosa, 1997). Such microbial import or scavenging systems are known to behave as major virulence determinants and their elimination from the microbial genome abrogates the ability to replicate intracellularly (Litwin and Calderwood, 1993). Yet, despite considerable efforts, transport activity of Nramp1 has not been demonstrated in any of the available assays. We and others have not yet been able to demonstrate Nramp1-mediated transport of divalent cations in yeast, *Xenopus* oocytes, or transfected mammalian cells. This suggests that

Nramp1 may need additional regulatory signals or macrophage-specific proteins to be constitutively active or to be triggered for activity.

The question that remains is how the putative Nramp1-mediated divalent cation depletion could affect intraphagosomal pathogen survival and, more specifically, how could this activity lead to the disruption of the BCG-mediated block of phagosomal maturation? Several possibilities must be considered. It is possible that divalent cations could be directly involved in the fusion and maturation process of phagosomes, and that Nramp1-mediated changes in divalent cation concentrations could only be critical for phagosome maturation during the course of infection, when parasites are competing for available ions. Indeed, Zn^{2+} has been shown to be necessary in an *in vitro* endosome-endosome fusion assay (Aballay wt al., 1995). A second possibility is that divalent cations may be necessary for the expression of pathogenicity determinants in the BCG bacterium, including the expression of molecules that are required to modulate fusogenic properties of the bacterial phagosome. Competition for such divalent cations by Nramp1 and a bacterial homologue may ultimately decide the fusogenic fate of the phagosome. Yet another possibility involves a more general and indirect effect of divalent cations on bacterial survival. Divalent cations are essential cofactors for many enzymes including bacterial superoxide dismutase (SOD) and catalase, two enzymes that neutralize the cytotoxic effects of the oxidative burst in macrophages, including the elimination of oxygen radicals and superoxide ions from the phagosomal space. Additionally, iron-sensitive operons have been described in many microbes including *Salmonella*, suggesting that divalent cation availability can directly regulate microbial gene expression. Alternatively, the regulation could be more indirect. In the absence of functional SOD and catalase to neutralize the harsh environment of the phagosome, the pathogen may not be able to attenuate phagosomal maturation.

IX. CONCLUSION

In conclusion, the study of a mouse model of genetic control of susceptibility to infections has led to the cloning of a gene that encodes a novel macrophage-specific membrane transporter. The parallel analysis of the human counterpart of this gene in association studies has demonstrated the parallel importance of this locus as a major predisposition factor for disease susceptibility in endemic areas of tuberculosis and leprosy. The identified Nramp family encodes divalent cation transporters, suggesting that modulation of divalent cation concentration in the microenvironment of the phagosome is critical for macrophages to control microbial replication. In addition, these studies suggest a new potential biochemical site for intervention in infectious diseases.

REFERENCES

Aballay, A., Sarrouf, M.N., Colombo, M.I., Stahl, P.D., & Mayorga, L.S. (1995). Zn2+ depletion blocks endosome fusion. Biochem. J. 312(pt 3), 919-923.

Abel, L., Sanchez, F.O., Oberti, J., Thuc, N.V., Hoa, L.V., Lap, V.D. Skamene, E., Lagrange, P.H., & Schurr, E. (1998). Susceptibility to leprosy is linked to the human NRAMP1 gene. J. Infect. Dis. 77, 133-145.

Armstrong, J.A., & Hart, P.D. (1975). Phagosome-lysosome interactions in cultured macrophages infected with virulent tubercle bacilli. Reversal of the usual nonfusion pattern and observations on bacterial survival. J. Exp. Med. 142, 1-16.

Bellamy, R., Ruwende, C., Corrah, T., McAdam, K.P., Whittle, H.C., & Hill, A.V. (1998). Variations in the *NRAMP1* gene and susceptibility to tuberculosis in West Africans. N. Engl. J. Med. 338, 640-644.

Campbell, P.A. (1994). Macrophage-Listeria interactions. Immunol. Series 60, 313-328.

Cellier, M., Belouchi, A., & Gros, P. (1996). Resistance to intracellular infections: comparative genome analysis of NRAMP. Trends Genet. 12, 201-204.

Cellier, M., Prive, G., Belouchi, A., Kwan, T.., Rodrigues, V., Chia, W., & Gros, P. (1995). The natural resistance associated macrophage protein (Nramp) defines a new family of membrane proteins conserved throughout evolution. Proc. Natl. Acad. Sci. USA 92, 10089-10094.

Cellier, M., Shustik, C., Dalton, W., Rich, E., Hu, J., Malo, D., Schurr, E., & Gros, P. (1997). The human *NRAMP1* gene, as a marker of professional primary phagocytes: studies in blood cells, and in HL-60 promyelocytic leukemia. J. Leukoc. Biol. 61, 96-105.

Clemens, D.L., & Horwitz, M,A. (1995). Characterization of the Mycobacterium tuberculosis phagosome and evidence that phagosomal maturation is inhibited. J. Exp. Med. 181, 257-270.

Crocker, P.R., Blackwell, J.M., & Bradley, D.J. (1984). Expression of the natural resistance gene *Lsh* in resident liver macrophages. Infect. Immun. 43, 1033-1040.

Crosa, J.H. (1997). Signal transduction and transcriptional and post-transcriptional control of iron regulated genes in bacteria. Microbiol. Molec. Biol. Rev. 61, 319-336.

De Chastellier, C., Frehel, C., Offreso, C., & Skamene, E. (1993). Implication of phagosome-lysosome fusion in restriction of Mycobacterium avium growth in bone marrow macrophages from genetically resistant mice. Infect. Immun. 61, 3775-3784.

Denis, M., Forget, A., Pelletier, M., Gervais, F., & Skamene, E. (1990). Killing of *Mycobacterium smegmatis* by macrophages from genetically susceptible and resistant mice. J. Leukoc. Biol. 47, 25-30.

Desjardins, M., Huber, A., Parton, R.G., & Griffiths, G. (1994). Biogenesis of phagolysosomes proceeds through a series of sequential interactions with the endocytic apparatus. J. Cell Biol. 124, 677-688.

Edwards, J.A., & Hoke, J.E. (1972). Defect of intestinal mucosal iron uptake in mice with hereditary microcytic anemia. Proc. Soc. Exp. Biol. Med. 141, 81-84.

Edwards, J.A., & Hoke, J.E.. (1975). Red cell iron uptake in hereditary microcytic anemia. Blood 46, 235-245.

Fleming, M.D., Romano, M.A., Su, M.A, Garrick, L.M., Garrick, M.D., & Andrews, N.C. (1997a). Nramp2 is mutated in the anemic Belgrade (b) rat—evidence of a role for Nramp2 in endosomal iron transport. Proc. Natl. Acad. Sci. USA 95, 1148-1153.

Fleming, M.D., Trenor, C.C., Su, M.A., Foernzler, D., Beier, D.R., Dietrich, W.F., & Andrews, N.C. (1997b). Microcytic anemia mice have a mutation in Nramp2, a candidate iron transporter gene. Nature Genet. 6, 383-386.

Forget, A., Skamene, E., Gros, P., Miailhe, A.C., & Turcotte, R. (1981). Differences in response among inbred strains of mice to infection with small doses of *Mycobacterium bovis* (BCG). Infect. Immun. 32, 42-50.

Garcia-del Portillo, F., & Finlay, B.B. (1995). Targeting of *Salmonella typhimurium* to vesicles containing lysosomal membrane glycoproteins bypasses compartments with mannose 6-phosphate receptors. J. Cell Biol. 129, 81-97.

Garcia-del Portillo, F., Zwick, M.B., Leung, K.Y., & Finlay, B.B. (1993). *Salmonella* induces the formation of filamentous structures containing lysosomal membrane glycoproteins in epithelial cells. Proc. Natl. Acad. Sci. USA 90, 10544-10548.

Garrick, L.M., Edwards, J.A., & Hoke, J.E. (1978). The effect of hemin on globin synthesis and iron uptake by reticulocytes of the Belgrade rat. FEBS Lett. 93, 109-114.

Govoni, G., & Gros, P. (1998). Macrophage Nramp1 and its role in resistance to microbial infections. Inflamm. Res. 47, 277-284.

Govoni, G., Gauthier, S., Billia, F., Iscove, N.N., & Gros, P. (1997). Cell-specific and inducible Nramp1 gene expression in mouse macrophages in vitro and in vivo. J. Leukoc. Biol. 62, 277-286.

Govoni, G., Vidal, S., Gauthier, S., Skamene, E., Malo, D., & Gros, P. (1996). The *Bcg/Ity/Lsh* locus: genetic transfer of resistance to infections in C57BL/6J mice transgenic for the Nramp1^{Gly169} allele. Infect. Immun. 64, 2923-2929.

Gros, P., Skamene, E., & Forget, A. (1981). Genetic control of natural resistance to *Mycobacterium bovis* (BCG) in mice. J. Immunol. 127, 2417-2423.

Gros, P., Skamene, E., & Forget, A. (1983). Cellular mechanism of genetically controlled host resistance to *Mycobacterium bovis* (BCG). J. Immunol. 131, 1966-1972.

Gruenheid, S., Cellier, M., Vidal, S., & Gros, P. (1995). Identification and characterization of a second mouse *Nramp* gene. Genomics 25, 514-525.

Gruenheid, S., Pinner, E., Desjardins, M., & Gros, P. (1997). Natural resistance to infection with intracellular parasites: the Nramp1 protein is recruited to the membrane of the phagosome. J. Exp. Med. 185, 717-730.

Gunshin, H., Mackenzie, B., Berger, U.B., Gunshin, Y., Romero, M.F., Boron, W.F., Nussberger, S., Gollan, J.L., & Hediger, M.A. (1997). Cloning and characterization of a mammalian proton-coupled metal-ion transporter. Nature 388, 482-488.

Hackam, D.J., Rotstein, O., Zhang, W.J., Gruenheid, S., Gros, P., & Grinstein, S. (1998). Host resistance to intracellular infection: mutation of natural resistance-associated macrophage protein1 (Nramp1) impairs phagosomal acidification. J. Exp. Med. 188.

Hammond, T.G., Goda, F.O., Navar, G.L., Campbell, W.C., Majewski, R.R., Galvan, D.L., Pontillon, F., Kaysen, J.H., Goodwin, TJ., Paddock, S.W., & Verroust, P.J. (1998). Membrane potential mediates H(+)-ATPase dependence of "degradative pathway" endosomal fusion. J. Membr. Biol. 162, 157-167.

Kontoghiorghes, G.J., & Weinberg, E.D. (1995). Iron: mammalian defense systems, mechanisms of disease and chelation therapy approaches. Blood Rev. 9, 33-45.

Korn, E.D. (1974). The isolation of the amoeba plasma membrane and the use of latex beads for the isolation of phagocytic vacuole (phagosome) membranes from amoebae including the culture techniques for amoebae. Meth. Enzymol. 31(pt A), 686-698.

Levee, G., Liu, J., Gicquel, B., Chanteau, S., & Schurr, E. (1994). Genetic control of susceptibility to leprosy in French Polynesia: no evidence for linkage with markers on telomeric human chromosome 2q. Int. J. Lep. 62, 499-511.

Lissner, C.R., Swanson, R.N., & O'Brien, A.D. (1983). Genetic control of the innate resistance of mice to *Salmonella typhimurium*: expression of the *Ity* gene in peritoneal and splenic macrophages isolated *in vitro*. J. Immunol. 131, 3006-3013.

Litwin, C.M., & Calderwood, S.B. (1993). Role of iron in regulation of virulence genes. Clin. Microbiol. Rev. 6, 137-149.

Malo, D., & Skamene, E. (1994). Genetic control of resistance to infection. Trends Genet. 10, 365-371.

Malo, D., Vidal, S., Lieman, J.H., Ward, D.C., & Gros, P. (1993a). Physical delineation of the minimal chromosomal segment encompassing the murine host resistance locus *Bcg*. Genomics 17, 667-675.

Malo, D., Vidal, S., Skamene, E., & Gros, P. (1993b). High resolution linkage map in the vicinity of the host resistance locus *Bcg* on mouse chromosome 1. Genomics 16, 655-663.

Malo, D., Vogan, K., Vidal, S., Hu, J., Cellier, M., Schurr, E., Fuks, A., Bumstead, N., Morgan, K., & Gros, P. (1994). Haplotype mapping and sequence analysis of the mouse *Nramp* gene predict susceptibility to infection with intracellular parasites. Genomics 23, 51-61.

McLeod, R., Buschman, E., Arbuckle, L.D., & Skamene, E. (1995). Immunogenetics in the analysis of resistance to intracellular parasites. Curr. Opin. Immunol. 7, 539-552.

Orgad, S., & Nelson, H. (1998). Metal ions suppress the abnormal taste behavior of the Drosophila mutant malvolio. J. Exp. Biol. 201, 115-120.

Pinner, E., Gruenheid, S., Raymond, M., & Gros, P. (1997). Functional complementation of the yeast divalent cation transporter family SMF by NRAMP2, a member of the mammalian natural resistance-associated macrophage protein family. J. Biol. Chem. 272, 28933-28938.

Pitt, A., Mayorga, L.S., Stahl, P.D., & Schwartz, A.L. (1992). Alterations in the protein composition of maturing phagosomes. J. Clin. Invest. 90,1978-1983.

Rodrigues, V., Cheah, P.Y., Ray, K., & Chia, W. (1995). Malvolio, the Drosophila homologue of mouse NRAMP-1 (Bcg), is expressed in macrophages and in the nervous system and is required for normal taste behaviour. EMBO J. 4, 3007-3020.

Roger, M., Levee, G., Chanteau, S., Gicquel, B., & Schurr, E. (1997). Leprosy susceptibility is not linked to the human natural resistance-associated macrophage protein gene (*NRAMP1*) in French Polynesia. Int. J. Lepr. Other Mycobact. Dis. 65, 197-202.

Russell, E.S., McFarland, E.C., & Kent, E.L. (1970). Low viability, skin lesions, and reduced fertility associated with microcytic anemia in the mouse. Transplant Proc. 2, 144-151.

Shaw, M.A., Atkinson, S., Dockrell, H., Hussain, R., Lins-Lainson, Z., Shaw, J., Ramos, F., Silveira, F., Mehdi, S.Q., Kaukab, F., et al. (1993). An RFLP map for 2q33-q37 from multicase mycobacterial and leishmanial disease families: no evidence for an *Lsh/Ity/Bcg* gene homologue influencing susceptibility to leprosy. Ann. Hum. Genet. 57, 251-271.

Skamene, E., Gros, P., Forget, A., Kongshavn, P.A., St. Charles, C., & Taylor, B.A. (1982). Genetic regulation of resistance to intracellular pathogens. Nature 297, 506-509.

Skamene, E., Kongshavn, P.A.L., & Landy, M. (1980). In: Genetic Control of Natural Resistance to Infection and Malignancy. Academic Press, New York.

Stach, J.L., Gros, P., Forget, A., & Skamene, E. (1984). Phenotypic expression of genetically controlled natural resistance to *Mycobacterium bovis* (BCG). J. Immunol. 132, 888-892.

Stokes, R.W., Orme, I., & Collins, F.M. (1986). Role of mononuclear phagocytes in expression of resistance and susceptibility to Mycobacterium avium infections in mice. Infect. Immun. 54, 811-819.

Sturgill-Koszycki, S., Schlessinger, P.H., Chakraborty, P., Haddix, P.L., Collins, H.L., Fok, A.K. Allen, R.D., Gluck, S.L., Heuser, J., & Russell, D.G. (1994). Lack of acidification in Mycobacterium phagosomes produced by exclusion of the vesicular proton ATPase. Science 263, 678-681.

Supek, F., Supekova, L., Nelson, H., & Nelson, N. (1996). A yeast manganese transporter related to the macrophage protein involved in conferring resistance to Mycobacterium. Proc. Natl. Acad. Sci. USA 93, 5105-5110.

Thomson, A.B., & Valberg, L.S. (1972). Intestinal uptake of iron, cobalt, and manganese in the iron-deficient rat. Am. J. Physiol. 223, 1327-1329.

van Weert, A.W.M., Dunn, K.W., Geuze, H.J., Maxfield, F.R., & Stoorvogel, W. (1995). Transport from late endosomes to lysosomes, but not sorting of integral membrane proteins in endosomes depends on the vacuolar proton pump. J. Cell Biol. 130, 821-834.

Vidal, S., Belouchi, A.M., Cellier, M., et al. (1995a). Cloning and characterization of a second human *NRAMP* gene on chromosome 12q13. Mamm. Genome 6, 224-230.

Vidal, S., Tremblay, M., Govoni, G., Gauthier, S., Sebastiani, G., Malo, D., Skamene, E., Olivier, M., Jothy, S., & Gros, P. (1995b). The *Ity/Lsh/Bcg* locus: natural resistance to infection with intracellular parasites is abrogated by disruption of the *Nramp 1* gene. J. Exp. Med. 182, 655-666.

Vidal, S.M., Malo, D., Vogan, K., Skamene, E., & Gros, P. (1993). Natural resistance to infection with intracellular parasites: isolation of a candidate for *Bcg*. Cell 73, 469-486.

Vidal, S.M., Pinner, E., Lepage, P., Gauthier, S., & Gros, P. (1996). Natural resistance to intracellular infections: *Nramp1* encodes a membrane phosphoglycoprotein absent in macrophages from susceptible ($Nramp1^{D169}$) mouse strains. J. Immunol. 157, 3559-3568.

UPTAKE AND PRESENTATION OF PHAGOCYTOSED ANTIGENS BY DENDRITIC CELLS

Matthew L. Albert, Shannon Turley, Wendy Garrett, Ira Mellman, Kayo Inaba, Nina Bhardwaj, and Ralph M. Steinman

Advances in Cell and Molecular Biology of Membranes and Organelles
Volume 5, pages 363-378.

ISBN: 1-55938-999-0

I. HISTORICAL OVERVIEW: THREE DIFFERENT SYSTEMS THAT ARE INVOLVED IN PHAGOCYTOSIS *IN VIVO*

A. The Reticuloendothelial System

In 1913, Landau and Aschoff formulated the term "reticuloendothelial system" (RES) to define a "special type of cell of wide distribution in mammals" (Aschaff, 1924). The cells of the RES were held to have a common origin and exhibited a branching morphology and phagocytic activity. The main cell types in the RES were "histiocytes," cells lining the blood and lymphatic sinuses, and the reticular cells of the spleen and lymph nodes, but endothelial cells and fibrocytes were also included. The key functional criterion was the uptake of vital dyes usually following intravenous injection. The vital dyes included carmine, trypan blue, and pyrrhol blue, with each giving a distinct granular pattern in the endocytic cell (Aschoff, 1924; Maximow, 1924, 1927). Today, the term "histiocyte" is best replaced by "macrophage," while the uptake of dyes and small colloids would be classified as pinocytosis rather than phagocytosis. Nonetheless, the function of the RES was to clear the blood stream and tissues of a variety of substances, including colloidal carbon, the main tracer used for quantitative work.

B. The Mononuclear Phagocyte System

The role of the RES in defense reactions and in the clearance of foreign bodies and effete red blood cells was of great interest when the field of modern cell biology emerged in the 1950s and 1960s. In fact, this area was so active that the current Society and *Journal of Leukocyte Biology* were named the RES society and *Journal of the Reticuloendothelial Society.* However, the new cell biology allowed investigators to study phagocytosis and the endocytic pathway with greater precision and to visualize, by electron microscopy, clear-cut distinctions in the endocytic activity and ultrastructure of endothelial cells, fibroblasts, and macrophages. All these cells had previously been considered to have a common origin and function to form the RES. A particularly useful tracer was horseradish peroxidase (HRP), since it could be visualized with cytochemistry at the light (Straus,

1969) and electron microscopic levels (Graham et al., 1966) and could be quantitated by its enzymatic activity (Steinman and Cohn, 1972a,b). With cell biological approaches, it became evident that macrophages were specialized to exhibit a high level of endocytic activity. Furthermore, only the macrophage had a bone marrow origin (van Furth and Cohn, 1968). Therefore the lumping of endothelial cells, fibroblasts, sinusoidal lining cells, ill-defined reticular cells, and macrophages into a single system became unsatisfactory.

During the 1969 Leiden Conference on Mononuclear Phagocytes, the first of many such meetings organized by Ralph van Furth, a reclassification of cells was proposed, culminating in a new system called the mononuclear phagocyte system (MPS) (van Furth et al., 1972). The cells of the MPS, the macrophages of Metchnikoff, were derived from the bone marrow via the blood monocyte (van Furth et al., 1972). Macrophages exhibited a well-developed lysosomal apparatus, robust phagocytic and pinocytic activities particularly for microorganisms and immune complexes, firm trypsin-resistant adherence to glass and plastic, and slow cellular turnover. Eventually it became possible to generate large numbers of macrophages from proliferating progenitors, using macrophage colony-stimulating factor (M-CSF).

The mouse peritoneal macrophage and the human blood monocyte were the main objects for the early study of macrophage cell biology (Steinman and Moberg, 1994). Many of the principles of endocytosis were uncovered with these cells, and later their robust secretory activities became evident (Nathan, 1987). Interestingly, major depots of dendritic cells (DCs) were omitted from the MPS, e.g., the DCs in the epidermis, airway lining, and the T cell areas of lymphoid organs (van Furth et al., 1972). These were not considered as part of the MPS, since their existence had yet to be uncovered and since phagocytic activity was not prominent in regions that were later found to be rich in DCs.

C. The Dendritic Cell System

A totally independent line of investigation, the generation of immunity in culture, led to the identification of a distinctive system, the dendritic cell system (DCs). Mishell and Dutton had learned to induce primary antibody responses from mouse spleen suspensions. The requisite cells were divided into nonadherent, radiosensitive lymphocytes and adherent, radioresistant "accessory cells" presumed to be macrophages. Steinman and Cohn (1973, 1974) pursuing these adherent cells, observed that the accessory populations contained an additional cell type that had not been previously described. These DCs, so named because of their many processes that actively formed and retracted in culture, were typically nonphagocytic. The DCs were replete with granular mitochondria but had few lysosomes. In addition, the cells lacked such characteristic surface features of macrophages as firm glass adherence, a plasma membrane ATPase, and Fc receptors. When accessory cells were isolated from separate pieces of splenic red pulp and white pulp, the latter were the main source of DCs (Steinman and Cohn, 1974). In electron microscopic studies, the counterparts of DCs were found in the white pulp and were noted to be poorly phagocytic for electron dense, colloidal thorium dioxide (Steinman et al., 1975).

Separate populations of splenic DCs and macrophages could be prepared on the basis of their different surface properties. DCs were nonadherent, lacked Fc receptors, but could be lysed with a monoclonal antibody 33D1 and complement (Steinman et al., 1979, 1983). Once DCs and macrophages were separated from one another, it became evident that the DCs were uniquely rich in MHC class II products, termed Ia antigens at the time (Steinman et al., 1979; Nussenzweig et al., 1981). Although macrophages were initially held to be a major MHC II-rich cell, most rodent macrophages in the steady-state expressed little MHC II (Steinman et al., 1979). These were the days when user friendly fluorescence-activated cell sorter instruments were not available, but the DCs clearly had abundant MHC II by immunofluorescence, complement dependent cytotoxicity, and direct binding studies with radioiodinated, monoclonal anti-MHC II antibodies (Steinman et al., 1980; Nussenzweig et al., 1981).

Of great interest was the fact that DCs were potent stimulators of allogeneic or MHC-mismatched T cells in the mixed leukocyte reaction (MLR) (Steinman and Witmer, 1978; Steinman et al., 1983). The MLR was originally called the mixed lymphocyte reaction because it was thought that any cell that expressed Ia antigens or MHC II products, especially B cells, could stimulate allogeneic T cells. However, DCs were active MLR stimulators while MHC II positive B cells and macrophages were not (Steinman and Witmer, 1978; Steinman et al., 1983). The experiments showed that after a T cell had been activated by the DC, either in the MLR or in the primary antibody response, then the T cell responded vigorously to antigens presented on the same B cells and macrophages that were weak initiators of immunity (Inaba and Steinman, 1984, 1985). These new features, potency and capacity to initiate immunity in culture, provided functional evidence that DCs represented a new pathway for white cell differentiation.

In one of the last volumes of the *Journal of the Reticuloendothelial Society*, a DC system was proposed that included the Langerhans cells (LCs) of the skin, the interdigitating cells of the T cell areas in lymphoid tissues and thymic medulla, the veiled cells in afferent lymph, and the isolated DCs (Tew et al., 1982). Later the interstitial DCs in organs like heart and liver portal triads were included (Hart and Fabre, 1981). All these cells exhibited features that were very unlike the properties of the RES and MPS. The DC system showed weak endocytic activity, high expression of MHC II products, low Fc receptors, poor adherence to glass and plastic and, when tested, potent MLR stimulating activity. The proposal was also made that this system is a dynamic one, with cells like the LCs and interstitial DCs being able to move from nonlymphoid tissues via the lymph to the T cell areas of lymphoid organs.

Subsequently DCs were successfully generated in large numbers from proliferating progenitors in blood (Inaba et al., 1992b) and bone marrow (Inaba et al., 1992a). A key requisite cytokine was granulocyte M-CSF (GM-CSF). The DCs developed in distinct cell aggregates that were full of immature DCs, as described below. In other parts of the culture, distinct granulocyte and macrophage colonies were developing as well. The macrophage colonies were flatter and much more firmly adherent than the aggregates of immature DCs. When the immature DCs were cultured in M-CSF or G-CSF, they would not differentiate into macrophages or granulocytes, respectively (Inaba et al., 1992a,b). In other words, the GM-CSF stimulated cultures were exhibiting several different pathways of white cell differentiation, one of these being the DC pathway.

A large number of studies used endocytic activity *in vivo* to distinguish the DC system from the MPS and RES. Steinman and Cohn (1974) injected colloidal carbon, HRP, and HRP anti-HRP, but found little and sometimes no tracer in the DCs isolated from these animals a day later. Fossum injected colloidal carbon and ferritin tagged with rhodamine (Fossum, 1980; Fossum et al., 1984); macrophages and sinus lining endothelial cells endocytosed the carbon and ferritin, but DCs took up very little. When the fate of antigens was followed in the presence of antibody, uptake into DCs again did not take place. Instead native antigen was retained on distinctive follicular dendritic cells (FDCs) in the germinal centers. By electron microscopy, the antigen clearly was extracellular on the FDC (Chen et al., 1978; Fossum, 1980) as indicated first by Nossal and co-workers (1968). The LCs of the epidermis were also only weakly endocytic (Nordquist et al., 1966; Wolff and Schreiner, 1970; Sagbeil, 1972). Eikelenboom (1978, 1980) reported likewise for the DCs in splenic T cell areas.

One exception was the observation that DCs in the T cell area could contain phagocytic inclusions (Fossum et al., 1984) that we would now term apoptotic bodies. Later, when Fossum and colleagues followed the rapid clearance of allogeneic leukocytes in rats, they noted avid uptake by DCs which contained phagocytic inclusions (Fossum and Rolstad, 1986). An earlier study had shown phagocytic inclusions in thymic DCs in rats undergoing thymic involution (Duijvestijn et al., 1982). When Pugh et al. (1983) described the DCs in mesenteric lymph, they also noted inclusions that stained for DNA. These findings have become much clearer as a result of recent studies showing that immature DCs exhibit phagocytic activity for apoptotic cells (Albert et al., 1998a; Inaba et al., 1998).

In summary, the study of phagocytosis *in vivo* has continued actively from the time of Metchnikoff. Distinct RES, MPS and DC systems have been delineated. Below we describe how the DC system, distinguished initially in part by its lack of phagocytic activity, can use phagocytosis to present antigens. A critical distinction is to consider the immature stage of the DC life history.

II. IMMATURE AND MATURE STAGES OF DENDRITIC CELL DEVELOPMENT

A. The Langerhans Cell as an Immature Dendritic Cell

For many, it was mysterious how DCs could present antigens so efficiently on MHC II and yet be relatively weak at endocytic activity. Actually the same is true of B cells which poorly internalize most tracers including particulates, the exception being soluble antigens that bind to the B cell receptor or Ig. There are several possible explanations for this mystery (Steinman and Swanson, 1995). The key point is to understand that DCs develop from immature cells with endocytic activity and a vacuolar system that is specialized to form MHC II–peptide complexes.

The clue to this developmental regulation of endocytosis and antigen processing came with epidermal LCs (Schuler and Steinman, 1985; Romani et al., 1989). Much like splenic DCs, the LCs expressed high levels of MHC class II, originated from bone marrow progenitors, and potently stimulated T cells in the MLR. However, strong MLR stimulatory activity was apparent only after the LCs had been cultured. During culture the LCs matured: the cells enlarged, expressed more MHC II, and lost features found on LCs within the epidermis such as Fc receptors and the F4/80 macrophage antigen (Schuler and Steinman, 1985). The reverse situation occurred when LCs were used to present soluble protein antigens to T-T hybridomas. The latter are chronically activated cells that are less dependent upon costimulation by the antigen presenting cell. Freshly isolated LCs actively presented proteins to T-T hybrids, but cultured LCs did not (Romani et al., 1989). Therefore, it was postulated that immature DCs, such as LCs, were specialized to pick up antigens in the periphery; then endocytic activity was lost as the cells matured and acquired T cell stimulatory function.

B. Other Immature DC Populations for Cell Biological Studies

The distinction between immature and mature DCs became evident in several other contexts: mouse spleen, rat lung, human blood monocytes, and bone marrow-derived DCs in several species. In each case the immature DC was endocytic while the mature DC was weak or inactive. The immature to mature transition proved to be a key control point for the development of immunity, since the mature cells also expressed many other features that would lead to strong T cell-mediated immunity. These features included: expression of several molecules for T cell adhesion and costimulation (CD40, 54, 58, 80, 86), resistance to the immunosuppressive effects of interleukin (IL)-10, and induction of certain chemokine receptors like CCR7 that guide the mature DC to chemokines (SLC and MIP-3β) expressed constitutively in the T cell areas.

Of relevance to this chapter is the fact that immature DCs were found to be capable of phagocytic activity, including microrganisms. Reis e Sousa showed phagocytosis by immature epidermal LCs (Reis e Sousa et al., 1993), as Inaba et al. (1993) found with immature DCs from GM-CSF stimulated, mouse bone marrow cultures. Matsuno demonstrated uptake of latex particles by immature DCs in rat liver (Matsuno et al., 1996). In each case, phagocytic activity ceased upon maturation (Inaba et al., 1993; Reise Sousa et al., 1993; Matsumo et al., 1996). Bone marrow cultures were particularly important, since they provided the field with large numbers of developing DCs for cell biological and immunological studies (Pierre et al., 1997; Inaba et al., 1998). A similarly important system was the human blood monocyte stimulated with GM-CSF and IL-4. This cell was also active in endocytosis (Sallusto et al., 1995; Bender et al., 1996; Romani et al., 1996) and could rapidly mature into a classical, T cell stimulatory DC (Sallusto et al., 1995; Bender et al., 1996; Romani et al., 1996).

C. Some Cell Biological Features of Immature DCs

The endocytic and processing capacities of immature DCs are starting to be understood. All the known types of uptake pathways are evident: (a) fluid phase pinocytosis of tracers like lucifer yellow, (b) adsorptive uptake of mannosylated BSA via the macrophage mannose receptor (MMR) and of immune complexes via Fc receptors (FcR), (c) macropinocytosis, and d) uptake of an array of microbes. The latter include protozoa like *Leishmania* (Moll et al., 1995) but primarily the amastigote stage (von Stebut et al., 1998), yeasts (Reis e Sousa et al., 1993), chlamydia (Su et al., 1998), mycobacteria (Inaba et al., 1993), bacteria (Winzler et al., 1997), and several viruses that are known to infect cells through an endocytic route such as influenza (Bhardwaj et al., 1994) and adenovirus (Zhong et al., 1998). Immature DCs also begin to express a multilectin receptor called DEC-205 that is homologous to the MMR (Jiang et al., 1995). The ligands for DEC-205 have not been identified yet, but antibodies to DEC-205 are rapidly internalized in coated pits, as has been described for the MMR.

Once internalized, endocytic substrates (including phagocytosed particles) are delivered to MHC class II-rich, lysosomal compartments or MIICs (Inaba et al., 1998). The MIIC is the likely place where processed peptides gain access to newly synthesized MHC II products chaperoned through the endocytic system by the invariant chain. The immature DC briskly synthesizes MHC II products and invariant chain (Pure et al., 1990; Kampgen et al., 1991). Sorting signals in the cytosolic tail of the invariant chain target the class II invariant chain complex to the endocytic system, which contains proteases and the peptide editing molecule, H-2M or HLA-DM (Cresswell, 1994). Our current studies show that MHC II-peptide complexes are first seen in MIICs when DCs are successfully processing antigens.

Beyond the regulation of endocytic activity during DC maturation, as described above, there also is regulation at other levels. A good example is from Pierre and Mellman (1998) who found that immature DCs are rich in cystatin C, a cysteine protease inhibitor. This is significant because a cysteine protease, cathepsin S, is the main protease for degrading the invariant chain. Degradation in turn is important for at least two reasons: (a) proteolysis liberates the MHC II molecule from its tethering to the endocytic system by the cytoplasmic domain of the invariant chain, and (b) it produces the CLIP peptide that is readily displaced by antigenic peptides through the action of H-2M and H-2O. Immature DCs place the cystatin C, normally a secretory product, into MIICs, but when a maturation stimulus is received, the cystatin C levels drop. This presumably enhances cathepsin S activity and the formation of MHC class II peptide complexes that can transit to the cell surface.

III. UPTAKE AND PROCESSING OF APOPTOTIC AND NECROTIC CELLS

A. Contrasting Immunologic Consequences Following Phagocytosis of Apoptotic Cells by Macrophages and DCs

Cells undergo apoptotic death under several circumstances *in vivo*: the normal turnover of many tissues, morphogenesis, and several stages of lymphocyte development and function. Macrophages efficiently engulf dying cells. The two best examples in the immune system are the clearance of thymocytes in the thymic cortex and B lymphoblasts in the germinal center. In these examples, the macrophages are scavenging lymphocytes that do not receive survival signals during positive selection. Developing T cells and somatically mutating B cells are not rescued from programmed cell death, because the lymphocytes lack reactivity with self major histocompatibility complex (MHC) or the original immunizing antigen, respectively. The macrophages handle so many dying cells that they can exhibit numerous phagocytic inclusions called tingible bodies.

The recognition of dying cells is highly regulated from both sides of this engagement. Early markers of apoptotic death signal the phagocyte to engulf the dying cell. These include the externalization of phosphatidylserine, which is normally exposed on the inner leaflet of the plasma membrane (Fadok et al., 1992), as well as other undefined carbohydrates unique to the surface of apoptotic material (Savill, 1998). These signals trigger phagocytosis via numerous receptors (Platt et al., 1998): CD14, vitronectin receptors, and various scavenger receptors like SR-A, CD36 and C1qR. In this way, the macrophage sequesters potentially inflammatory and noxious factors within the apoptotic cell, and tissue injury is prevented (reviewed in Savill, 1997). In addition to the sequestration of the dying cell, phagocytosis primes macrophages to make cytokines such as IL-10 and transforming growth factor (TGF)-β upon subsequent exposure to lipopolysaccharides (LPS) (Voll et al., 1997; Fadok et al., 1998). These cytokines have immunosuppressive function.

It is now evident that immature DCs phagocytose apoptotic cells, bodies and/or blebs (Albert et al., 1998b). In contrast to macrophages, however, uptake leads to antigen presentation rather than total destruction of the ingested cell. This activity likely explains the *in vivo* phenomenon termed cross priming, first identified by Bevan (1976) but later found in other instances including the presentation of tumor and self antigens (Huang et al., 1994). In Bevan's experiments, minor histocompatibility antigens expressed by injected leukocytes were presented to the immune system by the MHC products of the recipient animal, i.e., the antigens crossed the MHC barrier and were presented on the host's antigen presenting cells. Cross priming originally was described for MHC class I products, but can also be applied to MHC class II. Priming to peptides from transplants (standard organ grafts and also xenografts) on recipient MHC class II products has been documented repeatedly (Moses et al., 1990; Benichou et al., 1992; Fangmann et al., 1992; Lee et al., 1994; Liu et al., 1996). This cross priming is also called the "indirect pathway of allograft rejection" (Lechler and Batchelor, 1982), and it may be very important in the chronic rejection that takes place with so many allografts.

When both macrophages and DCs are present in a culture and an antigenic apoptotic meal is administered, the macrophage "wins" (scavenging, not antigen presentation, occurs; Albert et al., 1998b). Therefore if DCs are to present apoptotic cells, the latter probably have to be delivered to DCs at sites where there are relatively few macrophages, e.g., in the T cell areas as we shall discuss below.

B. Mechanisms Underlying the Uptake and Processing of Apoptotic Cells on MHC Class I

The uptake of apoptotic cells is restricted to the immature stage of DC development (Albert et al., 1998a). It is blocked by cytochalasin D, chelation of divalent ions with EDTA, and low temperature (Albert et al., 1998a). One of the components of the receptor machinery is the integrin $\alpha_v\beta_5$ (Albert et al., 1998a). This receptor is expressed primarily on immature DCs and not on monocytes, macrophages and mature DCs. In contrast, macrophages utilize a highly related $\alpha_v\beta_3$ integrin. The remarkable thing is that macrophages phagocytose apoptotic cells better than immature DCs, but they do not form MHC class I peptide complexes to a sufficient extent to be recognized by effector CTLs (Albert et al., 1998a).

Because different but very related vitronectin receptors are used by macrophages and DCs to engulf apoptotic cells, the receptor may determine the subsequent fate of the ingested cell. In other words, $\alpha_v\beta_3$ directs apoptotic cells for degradation while $\alpha_v\beta_5$ results in processing onto MHC class I. These different consequences may relate to the use of different intracellular vacuoles, one for degradation and one for presentation. This possibility also emerges from recent data of Amigorena and colleagues (Regnault et al., 1998). They showed that Fc receptor-mediated uptake of ovalbumin–anti-ovalbumin immune complexes into DCs, but not B cells, is accompanied by presentation on MHC class I. Therefore,

DCs are specialized to present peptides from dying cells and immune complexes. Possibly, DCs contain a transporter that allows peptides to access the cytoplasm and the traditional endogenous MHC I pathway, or the endocytic vacuole itself is the site for loading peptides onto MHC class I derived from the biosynthetic and/or recycling pathways.

Following acquistion of antigen, a maturation signal allows the DC to become strongly stimulatory for T cells, to resist immunosuppression, and to localize in the T cell areas. Ongoing studies indicate that maturation is not induced by uptake of apoptotic cells alone. An additional signal is required, such as inflammatory cytokines in a monocyte conditioned medium or microbial products like LPS. This feature will be discussed more in the next two sections.

C. Presentation of Dying Cells on Both MHC Class I and Class II Products

The presentation of dying cells on the MHC products of DCs has now been studied for MHC class II using a more cell biological approach. We took advantage of a monoclonal antibody, termed Y-Ae, that recognizes I-A^b molecules presenting a peptide derived from I-E (Rudensky et al., 1991). When I-A^b DCs were fed I-E^+ B lymphoblasts, the DCs began to express the Y-Ae epitope, direct evidence for the transfer of antigens from dying cells to DCs (Inaba et al., 1998). The transfer only occurred with immature DCs and was preceded by phagocytosis. During phagocytosis, most of the B cell fragments were found in MIICs. Both necrotic and apoptotic cells were successfully presented on the MHC II molecules of maturing DCs, which in turn acted as potent stimulators of antigen-specific T cells.

Because the Y-Ae antibody permitted quantification by flow cytometry, one could determine the dose of preprocessed I-E peptide that gave the same loading of MHC class II products as the dose of I-E protein in the dying cell. It was determined that 1 μM preprocessed peptide was comparable to 0.3 nM I-E protein (Inaba et al., 1998). This striking difference in efficiency, at least 3,000-fold, is attributable in part to the fact that phagocytosis of dying cells enhances the efficiency of uptake relative to soluble peptide. Comparisons of fluid phase and adsorptive uptake some years ago showed the latter to be a few thousand-fold more efficient in peritoneal macrophages (Steinman and Cohn, 1972a). Yet phagocytosis only delivers the I-E protein into the immature DC. Once in the DC, likely within the MIICs, there must be efficient conversion of the protein to MHC-peptide complexes that are stably expressed on the cell surface.

CD4+ helper cells are required for the presentation of apoptotic cell-derived antigens on MHC class I to CD8+ T cells (Bennett et al., 1997). A likely explanation is the fact that activated CD4 cells express CD40L (Bennett et al., 1998; Ridge et al., 1998; Schoenberger et al., 1998), and CD40L is a major stimulus for DCs (Caux et al., 1994). Ligation of CD40 on DCs induces the production of IL-12, turnover neurosis factor (TNF) and MIP-1, upregulates the expression of MHC II and costimulatory molecules, and enhances DC viability. Another TNF family member on activated T cells is TRANCE. This also sustains the viability of DCs by stimulating bcl-x_L (Wong et al., 1997). Therefore the capacity of DCs to present apoptotic cells on both MHC class I and II may allow the DC to elicit both CD4+ helper and CD8+ killer T cells, with the former maintaining DC viability and cytokine synthesis.

D. Presentation of Dying Cells for Purposes of Tolerance Rather than Immunity

Uptake of apoptotic cells may also be important for tolerance or "cross tolerance" as described by Kurts and colleagues (1998). In their experiments, sequences for the OVA protein were expressed under the tissue-specific insulin promoter in transgenic mice. When OVA-specific, MHC class I restricted, TCR transgenic T cells were adoptively transferred into mice that were expressing OVA peptide in their islet cells, the T cells accumulated and expanded in the draining lymph nodes. These T cells were responding to antigens presented on bone marrow-derived cells, not the islet cells, and the response was followed by tolerance or deletion via a Fas-dependent mechanism (Kurts et al., 1996, 1997, 1998). The authors proposed that a leukocyte captures exogenous antigens for both MHC class I and II presentation, transports these to the draining lymph nodes, and presents the peptides to naive $CD4^+$ and $CD8^+$ T cells. It is likely that apoptotic islet cells were responsible for transferring antigens, since presentation of OVA peptides was enhanced if the OVA-expressing cells were targeted by activated CTLs. We speculate that the key presenting cell is the DC because of its special capacities of capturing antigens in the periphery and migrating to the T cell areas of the lymphoid organs for presentation on both MHC class I and II.

One idea would be that DCs in tissues both phagocytose apoptotic cells and then directly tolerize T cells because, in the steady-state, the DCs lack costimulatory functions that are needed to produce immunity. The difficulty with this idea is that migratory DCs are short lived once they reach the lymphoid tissues. Even though thousands of DCs traffic each hour in any afferent lymphatic, DCs are not found in efferent lymph. Therefore the life span of migratory DCs is likely to be very short, leaving them no time to induce tolerance. Recently we used the Y-Ae monoclonal, described above, to show that DCs, injected via the subcutaneous route, are efficiently processed by recipient DCs in the draining lymph node. In other words, when one injects I-E^+ DCs into I-A^b mice, most of the DCs in the draining lymph node express the Y-Ae epitope two days later (Inaba et al., 1998). There is evidence, admittedly far from complete, that the resident DCs in the T cell areas are functionally and perhaps developmentally distinct (Saunders et al., 1996; Wu et al., 1996). These lymphoid DCs or DC2 may regulate immunity, for example, by directly tolerizing the peripheral T cell or by inducing the production of cytokines like IL-10. Since peripheral tolerance is an active process that seems to take days to initiate (Kurts et al., 1997), it may be important that DCs pick up apoptotic bodies in peripheral tissues and then undergo processing by long-lived, functionally distinct, regulatory or tolerogenic DCs in the T cell areas.

IV. DISCUSSION

It has been known for some time that there are two separate pathways of white cell differentiation that yield macrophages and DCs. Macrophages are specialized for high capacity, continued endocytic activity followed by thorough digestion, while DCs only endocytose for a brief stage of their life history efficiently targeting the internalized substrates for presentation of MHC-peptide complexes.

Much of the early literature failed to detect significant accumulation of endocytic substrates by the DCs of the T cell areas *in vivo*. However, there are three features that need to be taken into consideration. First, DCs restrict their endocytic activity for many substrates to the immature stage of development. Second, certain substrates are internalized much better than others, e.g., apoptotic cells. Third, the best way to detect endocytosis is not to look for the antigen per se, but to look for the MHC-peptide complexes that arise from that antigen. These principles hold true for another presenting cell, the B cell, although the latter is not yet known to have the capacity to phagocytose many different microorganisms and dying cells, as is the case with immature DCs.

The targeting of phagocytic stimuli to DCs should prove valuable for controlling T cell responses to cellular and microbial antigens. In the steady-state, phagocytosis of apoptotic cells by DCs may lead to tolerance, while in inflammatory states, uptake of microbes and necrotic cells may be pivotal to strong T cell-mediated immunity.

REFERENCES

Albert, M.L., Pearce, S.F.A., Francisco, L.M., et al. (1998a). Immature dendritic cells phagocytose apoptotic cells via $\alpha_v\beta_5$ and CD36, and cross-present antigens to cytotoxic T lymphocytes. J. Exp. Med. 188, 1359-1368.

Albert, M.L., Sauter, B., & Bhardwaj, N. (1998b). Dendritic cells acquire antigen from apoptotic cells and induce class I-restricted CTLs. Nature 392, 86-89.

Aschoff, L. (1924). In: Lectures in Pathology (Hoeber, P.B., Ed.). New York.

Bender, A., Sapp, M., Schuler, G., Steinman, R.M., & Bhardwaj, N. (1996). Improved methods for the generation of dendritic cells from nonproliferating progenitors in human blood. J. Immunol. Methods 196, 121-135.

Benichou, G., Takizawa, P.A., Olson, C.A., McMillan, M., & Sercarz, E.E. (1992). Donor major histocompatibility complex [MHC] peptides are presented by recipient MHC molecules during graft rejection. J. Exp. Med. 175, 305-308.

Bennett, S.R.M., Carbone, F.R., Karamalis, F., Flavell, R.A., Miller, J.F.A.P., & Heath, W.R. (1998). Help for cytotoxic-T-cell responses is mediated by CD40 signalling. Nature 393, 478-480.

Bennett, S.R.M., Carbone, F.R., Karamalis, F., Miller, J.J.A.P., & Heath, W.R. (1997). Induction of a CD8+ cytotoxic T lymphocyte response by cross-priming requires cognate CD4+ T cell help. J. Exp. Med. 186, 65-70.

Bevan, M.J. (1976). Cross-priming for a secondary cytotoxic response to minor H antigens with H-2 congenic cells which do not cross-react in the cytotoxic assay. J. Exp. Med. 143, 1283-1288.

Bhardwaj, N., Bender, A., Gonzalez, N., Bui, L.K., Garrett, M.C., & Steinman, R.M. (1994). Influenza virus-infected dendritic cells stimulate strong proliferative and cytolytic responses from human CD8+ T cells. J. Clin. Invest. 94, 797-807.

Caux, C., Massacrier, C., Vanbervliet, B., et al. (1994). Activation of human dendritic cells through CD40 cross-linking. J. Exp. Med. 180, 1263-1272.

Chen, L.L., Adams, J.C., & Steinman, R.M. (1978). Anatomy of germinal centers in mouse spleen with special reference to "follicular dendritic cells." J. Cell Biol. 77, 148-164.

Cresswell, P. (1994). Assembly, transport, and function of MHC class II molecules. Annu. Rev. Immunol. 12, 259-293.

Duijvestijn, A.M., Kohler, Y.G., & Hoefsmit, E.C.M. (1982). Interdigitating cells and macrophages in acute involuting rat thymus. Cell. Tissue Res. 224, 291-301.

Eikelenbloom, P. (1978). Dendritic cells in the rat spleen follicles. Cell Tissue Res. 190, 79-87.

Eikelenbloom, P. (1980). Characterization of non-lymphoid cells in the white pulp of the mouse spleen: an in vivo and in vitro study. Cell Tissue Res. 195, 445-460.

Fadok, V.A., Bratton, D.L., Konowal, A., Freed, P.W., Westcott, J.Y., & Henson, P.M. (1998). Macrophages that have ingested apoptosis cells in vitro inhibit proinflammatory cytokine production through autocrine/paracrine mechanisms involving TGF-ß, PGE2, and PAF. J. Clin. Invest. 101, 890-898.

Fadok, V.A., Savill, J.S., Haslett, C., et al. (1992). Different populations of macrophages use either the vitronectin receptor or the phosphatidylserine receptor to recognize and remove apoptotic cells. J. Immunol. 149, 4029-4035.

Fangmann, J., Dalchau, R., & Fabre, J.W. (1992). Rejection of skin allografts by indirect allorecognition of donor class I major histocompatibility complex peptides. J. Exp. Med. 175, 1521-1529.

Fossum, S. (1980). The architecture of rat lymph nodes. IV. Distribution of ferritin and colloidal carbon in the draining lymph nodes after foot-pad injection. Scand. J. Immunol. 12, 433-441.

Fossum, S., & Rolstad, B. (1986). The roles of interdigitating cells and natural killer cells in the rapid rejection of allogeneic lymphocytes. Eur. J. Immunol. 16, 440-450.

Fossum, S., Rolstad, B., & Ford, W.L. (1984). Thymus independence, kinetics and phagocytic ability of interdigitating cells. Immunobiol. 168, 403-413.

Graham, R.C., Jr., & Karnovsky, M.J. (1966). The early stages of absorption of injected horseradish peroxidase in the proximal tubules of mouse kidney: ultrastructural cytochemistry by a new technique. J. Histochem. Cytochem. 14, 291.

Hart, D.N.J., & Fabre, J.W. (1981). Demonstration and characterization of Ia-positive dendritic cells in the interstitial connective tissues of rat heart and other tissues, but not brain. J. Exp. Med. 154, 347-361.

Heath, W.R., Kurts, C., Miller, J.F.A.P., & Carbone, F. (1998). Cross-tolerance: a pathway for inducing tolerance to peripheral tissue antigens. J. Exp. Med. 187, 1549-1553.

Huang, A.-L, Golumbek, P., Ahmadzadeh, M., Jafee, E., Pardoll, D., Levitsky, H. (1994). Role of bone marrow-derived cells in presenting MHC class I–restricted tumor antigens. Science. 264, 961-965.

Inaba, K., & Steinman, R.M. (1984). Resting and sensitized T lymphocytes exhibit distinct stimulatory [antigen-presenting cell] requirements for growth and lymphokine release. J. Exp. Med. 160, 1717-1735.

Inaba, K., & Steinman, R.M. (1985). Protein-specific helper T lymphocyte formation initiated by dendritic cells. Science 229, 475-479.

Inaba, K., Inaba, M., Naito, M., & Steinman, R.M. (1993). Dendritic cell progenitors phagocytose particulates, including Bacillus Calmette-Guerin organisms, and sensitize mice to mycobacterial antigens in vivo. J. Exp. Med. 178, 479-488.

Inaba, K., Inaba, M., Romani, N., et al. (1992a). Generation of large numbers of dendritic cells from mouse bone marrow cultures supplemented with granulocyte/macrophage colony-stimulating factor. J. Exp. Med. 176, 1693-1702.

Inaba, K., Steinman, R.M., Witmer-Pack, M., et al. (1992b). Identification of proliferating dendritic cell precursors in mouse blood. J. Exp. Med. 175, 1157-1167.

Inaba, K., Turley, S., Yamaide, F., et al. (1998). Efficient presentation of phagocytosed cellular fragments on the MHC class II products of dendritic cells. J. Exp. Med. In Press.

Jiang, W., Swiggard, W.J., Heufler, C., et al. (1995). The receptor DEC-205 expressed by dendritic cells and thymic epithelial cells is involved in antigen processing. Nature 375, 151-155.

Kampgen, E., Koch, N., Koch, F., et al. (1991). Class II major histocompatibility complex molecules of murine dendritic cells: synthesis, sialylation of invariant chain, and antigen processing capacity are down-regulated upon culture. Proc. Natl. Acad. Sci. USA 88, 3014-3018.

Kurts, C., Heath, W.R., Carbone, F.R., Allison, J., Miller, J.F.A.P., & Kosaka, H. (1996). Constitutive class I-restricted exogenous presentation of self antigens in vivo. J. Exp. Med. 184, 923-930.

Kurts, C., Heath, W.R., Kosaka, H., Miller, J.F.A.P., & Carbone, F.R. (1998). The peripheral deletion of autoreactive CD8+ T cells induced by cross-presentation of self-antigens involves signaling through CD95 (Fas, Apo-1). J. Exp. Med. 188, 415-420.

Kurts, C., Kosaka, H., Carbone, F.R., Miller, J.F.A.P., & Heath, W.R. (1997). Class I-restricted cross-presentation of exogenous self antigens leads to deletion of autoreactive CD8+ T cells. J. Exp. Med. 186, 239-245.

Lechler, R.I., & Batchelor, J.R. (1982). Restoration of immunogenicity to passenger cell-depleted kidney allografts by the addition of donor strain dendritic cells. J. Exp. Med. 155, 31-41.

Lee, R.S., Grusby, M.J., Glimcher, L.H., Winn, H.J., & Auchincloss, H., Jr. (1994). Indirect recognition by helper cells can induce donor-specific cytotoxic T lymphocytes in vivo. J. Exp. Med. 179, 865-872.

Liu, Z., Colovai, A.I., Tugulea, S., et al. (1996). Indirect recognition of donor HLA-DR peptides in organ allograft rejection. J. Clin. Invest. 98, 1150-1157.

Matsuno, K., Ezaki, T., Kudo, S., & Uehara, Y. (1996). A life stage of particle-laden rat dendritic cells in vivo: their terminal division, active phagocytosis and translocation from the liver to hepatic lymph. J. Exp. Med. 183, 1865-1878.

Maximow, A.A. (1924). Relation of blood cells to connective tissues and endothelium. Physiol. Rev. 4, 533-563.

Maximow, A.A. (1927). Morphology of mesenchymal reactions. Arch. Pathol. Lab. Med. 4, 557-560.

Moll, H., Flohe, S., & Rollinghoff, M. (1995). Dendritic cells in Leishmania major-immune mice harbor persistent parasites and mediate an antigen-specific T cell immune response. Eur. J. Immunol. 25, 693-699.

Moses, R.D., Pierson, R.N., III. Winn, H.J., & Auchincloss, H., Jr. (1990). Xenogeneic proliferation and lymphokine production are dependent on CD4+ helper T cells and self antigen-presenting cells in the mouse. J. Exp. Med. 172, 567-575.

Nathan, C.F. (1987). Secretory products of macrophages. J. Clin. Invest. 79, 319-326.

Nordquist, R.E., Olson, R.L., & Everett, M.A. (1966). The transport, uptake, and storage of ferritin in human epidermis. Arch.Dermatol. 94, 482-490.

Nossal, G.J.V., Abbot, A., Mitchell, J., & Lummus, Z. (1968). Antigen in immunity. XV. Ultrastructural features of antigen capture in primary and secondary lymphoid follicles. J. Exp. Med. 127, 277-296.

Nussenzweig, M.C., Steinman, R.M., Unkeless, J.C., Witmer, M.D., Gutchinov, B., & Cohn, Z.A. (1981). Studies of the cell surface of mouse dendritic cells and other leukocytes. J. Exp. Med. 154, 168-187.

Pierre, P., & Mellman, I. (1998). Developmental regulation of invariant chain proteolysis controls MHC class II trafficking in mouse dendritic cells. Cell 93, 1135-1145.

Pierre, P., Turley, S.J., Gatti, E., et al. (1997). Developmental regulation of MHC class II transport in mouse dendritic cells. Nature 388, 787-792.

Platt, N., da Silva, R.P., & Gordon, S. (1998). Recognising death: the phagocytosis of apoptotic cells. Trends Cell Biol. 8, 365-372.

Pugh, C.W., MacPherson, G.G., & Steer, H.W. (1983). Characterization of nonlymphoid cells derived from rat peripheral lymph. J. Exp. Med. 157, 1758-1779.

Pure, E., Inaba, K., Crowley, M.T., et al. (1990). Antigen processing by epidermal Langerhans cells correlates with the level of biosynthesis of major histocompatibility complex class II molecules and expression of invariant chain. J. Exp. Med. 172, 1459-1469.

Regnault, A., Lankar, D., Lacabanne, V., et al. (1998). Fc gamma R-mediated induction of dendritic cell maturation and MHC class I-restricted antigen presentation after immune complex internalization. Submitted.

Reis e Sousa, C., Stahl, P.D., & Austyn, J.M. (1993). Phagocytosis of antigens by Langerhans cells in vitro. J. Exp. Med. 178, 509-519.

Ridge, J.P., Di Rosa, F., & Matzinger, P. (1998). A conditioned dendritic cell can be a temporal bridge between a CD4+ T helper and a T-killer cell. Nature 393, 474-478.

Romani, N., Koide, S., Crowley, M., et al. (1989). Presentation of exogenous protein antigens by dendritic cells to T cell clones: intact protein is presented best by immature, epidermal Langerhans cells. J. Exp. Med. 169, 1169-1178.

Romani, N., Reider, D., Heuer, M., et al. (1996). Generation of mature dendritic cells from human blood: an improved method with special regard to clinical applicability. J. Immunol. Methods 196, 137-151.

Rudensky, A.Y., Rath, S., Preston-Hurlburt, P., Murphy, D.B., & Janeway, C.A., Jr. (1991). On the complexity of self. Nature 353, 660-662.

Sagbeil, R.W. (1972). In vivo and in vitro uptake of ferritin by Langerhans cells in the epidermis. J. Invest. Dermatol. 58, 47-54.

Sallusto, F., Cella, M., Danieli, C., & Lanzavecchia, A. (1995). Dendritic cells use macropinocytosis and the mannose receptor to concentrate antigen in the major histocompatibility class II compartment. Downregulation by cytokines and bacterial products. J. Exp. Med. 182, 389-400.

Saunders, D., Lucas, K., Ismaili, J., et al. (1996). Dendritic cell development in culture from thymic precursor cells in the absence of granulocyte/macrophage colony-stimulating factor. J. Exp. Med. 184, 2185-2196.

Savill, J. (1997). Recognition and phagocytosis of cells undergoing apoptosis. Br. Med. Bull. 53, 491-508.

Savill, J. (1998). Phagocytic docking without shocking. Nature 392, 442-443.

Schoenberger, S.P., Toes, R.E.M., van der Voort, E.I.H., Offringa, R., & Melief, C.J.M. (1998). T-cell help for cytotoxic T lymphocytes is mediated by CD40-CD40L interactions. Nature 393, 480-483.

Schuler, G., & Steinman, R.M. (1985). Murine epidermal Langerhans cells mature into potent immunostimulatory dendritic cells in vitro. J. Exp. Med. 161, 526-546.

Steinman, R.M., & Cohn, Z.A. (1972a). The interaction of particulate horseradish peroxidase (HRP)-anti HRP immune complexes with mouse peritoneal macrophages in vitro. J. Cell Biol. 55, 616-634.

Steinman, R.M., & Cohn, Z.A. (1972b). The interaction of soluble horseradish peroxidase with mouse peritoneal macrophages in vitro. J. Cell Biol. 55, 186-204.

Steinman, R.M., & Cohn, Z.A. (1973). Identification of a novel cell type in peripheral lymphoid organs of mice. I. Morphology, quantitation, tissue distribution. J. Exp. Med. 137, 1142-1162.

Steinman, R.M., & Cohn, Z.A. (1974). Identification of a novel cell type in peripheral lymphoid organs of mice. II. Functional properties in vitro. J. Exp. Med. 139, 380-397.

Steinman, R.M., & Moberg, C.L. (1994). Editorial. A tribute to Zanvil A. Cohn. An appreciation of the Physician-Scientist. J. Exp. Med. 179, 1-30.

Steinman, R.M., & Swanson, J. (1995). Commentary. The endocytic activity of dendritic cells. J. Exp. Med. 181, 283-288.

Steinman, R.M., & Witmer, M.D. (1978). Lymphoid dendritic cells are potent stimulators of the primary mixed leukocyte reaction in mice. Proc. Natl. Acad. Sci. USA 75, 5132-5136.

Steinman, R.M., Adams, J.C., & Cohn, Z.A. (1975). Identification of a novel cell type in peripheral lymphoid organs of mice. IV. Identification and distribution in mouse spleen. J. Exp. Med. 141, 804-820 (1975).

Steinman, R.M., Gutchinov, B., Witmer, M.D., & Nussenzweig, M.C. (1983). Dendritic cells are the principal stimulators of the primary mixed leukocyte reaction in mice. J. Exp. Med. 157, 613-627.

Steinman, R.M., Kaplan, G., Witmer, M.D., & Cohn, Z.A. (1979). Identification of a novel cell type in peripheral lymphoid organs of mice. V. Purification of spleen dendritic cells, new surface markers, and maintenance in vitro. J. Exp. Med. 149, 1-16.

Steinman, R.M., Nogueira, N., Witmer, M.D., Tydings, J.D., & Mellman, I.S. (1980). Lymphokine enhances the expression and synthesis of Ia antigen on cultured mouse peritoneal macrophages. J. Exp. Med. 152, 1248-1261.

Straus, W. (1969). Use of horseradish peroxidase as a marker protein for studies of phagolysosomes, permeability and immunology. Meth. Achievm. Exp. Path. 4, 54-91.

Su, H., Messer, R., Whitmire, W., Fischer, E., Portis, J.C., & Caldwell, H.D. (1998). Vaccination against chlamydial genital tract infection following immunization with dendritic cells pulsed ex vivo with non-viable chlamydiae. J. Exp. Med. 188, 809-818.

Tew, J.G., Thorbecke, J., & Steinman, R.M. Dendritic cells in the immune response: characteristics and recommended nomenclature. J. Reticuloendothel. Soc. 31, 371-380.

van Furth, R., & Cohn, Z.A. (1968). The origin and kinetics of mononuclear phagocytes. J. Exp. Med. 128, 415-435.

van Furth, R., Cohn, Z.A., Hirsch, J.G., Humphrey, J.H., Spector, W.G., & Langevoort, H.L. (1972). The mononuclear phagocyte system: a new classification of macrophages, monocytes, and their precursor cells. Bull. World Health Organ. 46, 845-852.

Voll, R.E., Herrmann, M., Roth, E.A., Stach, C., & Kalden, J.R. (1997). Immunosuppressive effects of apoptotic cells. Nature 390, 350-351.

von Stebut, E., Belkaid, Y., Jakob, T., Sacks, D.L., & Udey, M.C. (1998). Uptake of Leishmania major amastigotes results in activation and interleukin 12 release from murine skin-derived dendritic cells: implications for the initiation of anti-Leishmania immunity. J. Exp. Med. 188, 1547-1552.

Winzler, C., Rovere, P., Rescigno, M., et al. (1997). Maturation stages of mouse dendritic cells in growth factor-dependent long-term cultures. J. Exp. Med. 185, 317-328.

Wolff, K., & Schreiner, E. (1970). Uptake, intracellular transport, and degradation of exogenous protein by Langerhans cells. J. Invest. Dermatol. 54, 37-52.

Wong, B.R., Josien, R., Lee, S.Y., et al. (1997). TRANCE (tumor necrosis factor [TNF]-related activation-induced cytokine), a new TNF family member predominantly expressed in T cells, is a dendritic cell specific survival factor. J. Exp. Med. 186, 2075-2080.

Wu, L., Li, C.-L., & Shortman, K. (1996). Thymic dendritic cell precursors: relationship to the T lymphocyte lineage and phenotype of the dendritic cell progeny. J. Exp. Med. 184, 903-911.

Zhong, L., Granelli-Piperno, A., Choi, Y., & Steinman, R.M. (1998). Recombinant adenovirus is an efficient and non-perturbing genetic vector for human dendritic cells. Submitted.

PROCESSING AND PRESENTATION OF PHAGOCYTOSED ANTIGENS TO THE IMMUNE SYSTEM

Jean Pieters

Advances in Cell and Molecular Biology of Membranes and Organelles
Volume 5, pages 379-406.

ISBN: 1-55938-999-0

I. INTRODUCTION

Despite the continuous exposure to a wide variety of potentially harmful microorganisms, vertebrate organisms usually do not succumb to infections. This ability to survive is largely due to the existence of various host defense mechanisms. Two levels of defense mechanisms against infectious organisms are known to operate: innate (or inborn) immunity and adaptive immunity. The innate immune system, also present in nonvertebrates, relies on the functioning of germline-encoded molecules to identify and eliminate potential harmful substances. In contrast, the adaptive immune system, which appeared much later in evolution, consists of a system of B and T lymphocytes that posses antigen specific receptor molecules. These receptors result from somatic gene rearrangements that occur during the lifetime of the individual, which stands in contrast to the components of the innate immune systems, which have been selected over long periods of evolution (Janeway, 1989).

Innate immunity is predominantly mediated through the activity of macrophages, dendritic cells, neutrophils and natural killer cells (Brown et al., 1994). Whereas natural killer cells lyse mainly infected cells (Ljunggren and Karre, 1990; Colonna, 1996), macrophages, dendritic cells and neutrophils posses several mechanisms to internalize and destroy a wide variety of extracellular microorganisms (Stahl et al., 1980; Ezekowitz et al., 1990; Gordon, 1995; Silverstein, 1995). Besides these different cell types, the innate immune system consists of acute phase proteins present in serum such as the constituents of the complement cascade and C reactive protein, that either directly or via accessory molecules bind to invading microorganisms to remove them from the circulation (Kuby, 1992). Acting in concert, these mechanisms ensure an immediate defense against invading microorganisms. One drawback of the innate immune system, however, is that it has co-evolved with various microbes that it is supposed to clear. Therefore microorganisms have had the possibility to evolve to circumvent these host defense mechanisms, to become pathogenic (Falkow, 1991).

More specific ways to deal with invading microbes are ensured through the action of the adaptive immune system, which comprises of B and T lymphocytes besides the above-mentioned macrophages and dendritic cells. Rather than immediately killing the invaders, the adaptive immune system relies on the ability of a cell harboring an infectious agent to signal to its surroundings that it contains material foreign to the host, often derived from infectious organisms. This signaling is mediated via different classes of cell surface molecules that are encoded in the so-called major histocompatibility complex (MHC), a set of genes involved in immune recognition (Zinkernagel and Doherty, 1974a,b). The function of these cell surface molecules is to display small peptide fragments (or epitopes) to T lymphocytes (Hansen et al., 1993).

Several classes of MHC molecules exist (see Figure 1). MHC class I molecules usually present peptides derived from antigens located in the cytosol (e.g., derived from viruses) and are expressed on virtually all cell types (Heemels and Ploegh, 1995). MHC class II molecules are expressed only on specialized antigen presenting cells (APCs) and present peptides derived from exogenous antigens (i.e., most bacteria) (Cresswell, 1994; Pieters, 1997b). In addition, a series of nonclassical MHC molecules have been characterized, some of which may have a role in the adaptive immune response (Groh et al., 1998).

Complexes of MHC molecules loaded with peptides are recognized by T cells, resulting in the induction of a specific (i.e., raised against the particular ligand presented) immune response that can be mounted over a period of several days. If successful, these adapted responses will lead to the elimination of the infectious organism. Elimination of a viral infection can occur through lysis of the infected cell by cytotoxic T cells recognizing the MHC class I-(viral) peptide complex, whereas a bacterial infection can be eradicated through the action of helper T cells that are activated upon recognition of a MHC class II-peptide complex (see also Figure 1). Helper T cells can further activate the microbicidal activities of macrophages (T helper 1 cells), or assist in the induction of an antibody response by stimulating B lymphocytes to differentiate into antibody producing cells (T helper 2 cells) (Mosmann and Coffman, 1989).

The two types of T cells—cytotoxic and helper T—have probably evolved to cope with distinct types of microorganisms that invade different compartments of vertebrate host cells. On one hand, some microorganisms are obligate intracellular pathogens, invading the cytosol and making use of host cellular machinery for their replication. It is therefore important that the cells harboring these organisms are rapidly destroyed, removing the infected cells which harbor the invading microorganisms, and this function is carried out by the cytotoxic T cells. On the other hand, many infectious agents such as most bacteria reside and replicate extracellularly. In that case, killing of only those cells that have taken up the infectious agents would leave the vast majority of bacteria untouched. Therefore, generation of an antibody response through the action of T helper cells that results in high levels of serum immunoglobulins will more effectively combat these infections.

The cells of the adaptive immune system, the B and T lymphocytes, bearing appropriate receptors for a given antigen or peptide are normally present at extremely low frequencies in the pools of circulating lymphocytes. The power of

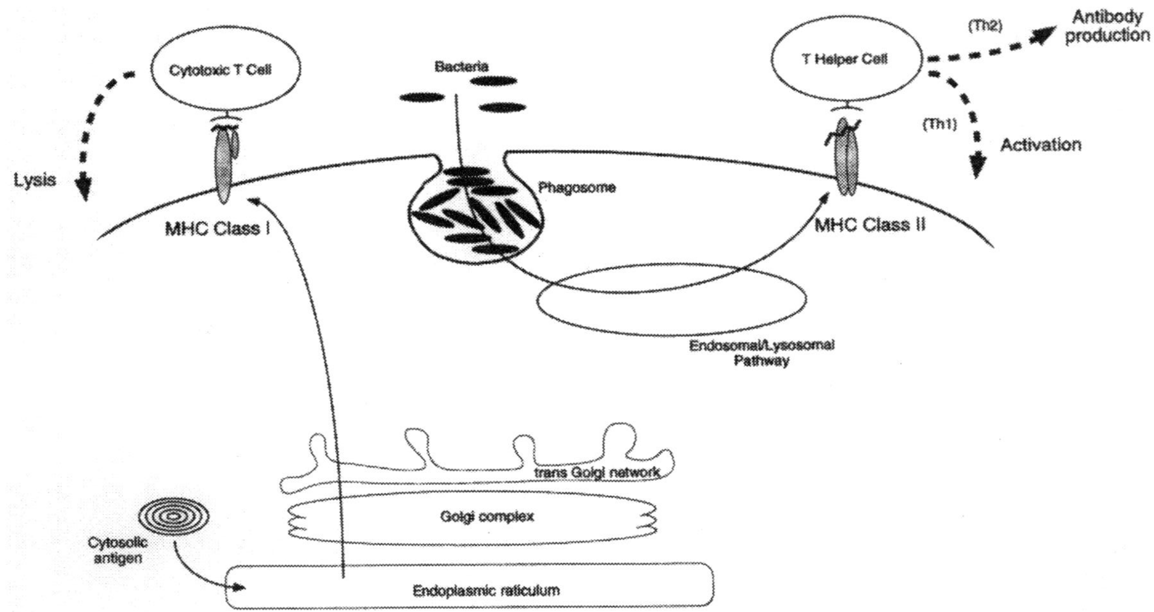

Figure 1. Activation of the adaptive immune system by MHC class I and class II molecules. MHC class I molecules present peptides derived from cytosolically located antigens at the cell surface to cytotoxic T cells. Activation of cytotoxic T cells leads to rapid lysis of the presenting cell **(left side)**. Antigens that reside extracellularly and are taken up through phagocytosis are degraded in the endosomal/lysosomal system and presented by MHC class II molecules **(right side)**. MHC class II-peptide complexes are recognized by T helper cells that can either activate the presenting cell (T helper 1 cells, Th1) or stimulate the production of antibodies by B cells (T helper 2 cells, Th2).

the adaptive immune system is that it has the ability during the initial phase of an infection to dramatically expand the number of those B and T cells effective against a particular pathogen.

It is important to remember that innate and adaptive immune systems have had the chance to adapt to each other over the course of vertebrate evolution. As a result, components such as the complement system function both in innate as well as the adaptive immune response (Farries and Atkinson, 1991). Also, cytokines produced by macrophages as a result of an encounter with pathogenic bacteria can have a marked effect on the outcome of a T cell response. Conversely, T and B cells produce factors that influence the functioning of macrophages and neutrophils (Brown et al., 1994). Furthermore, exciting recent research suggests that the development of the adaptive immune system relied on components of the innate immune system, which have been conserved over hundreds of millions of years (Lemaitre et al., 1996; Medzhitov et al., 1997; Merdzhitov and Janeway, 1998). Thus, the traditional distinction between innate and adaptive immunity is rapidly fading which allows us to study host defense mechanisms in a much broader sense then has been the case in the past decades (Bachmann and Zinkernagel, 1997; Merdzhitov and Janeway, 1997).

In this chapter, I will review how antigens, in particular those derived from invading microbes, are internalized and presented to the cells of the adaptive immune system. Most microbes that enter cells through phagocytosis are extracellular and therefore the focus of this chapter will be on how antigens are presented by MHC class II molecules to T helper cells. The long-standing interaction of various bacteria with vertebrates has also resulted in the emergence of strategies used by certain bacteria to circumvent efficient recognition and destruction by the immune system, and I will briefly describe our current knowledge of some of these mechanisms.

II. UPTAKE AND INTERNALIZATION OF ANTIGENS IN ANTIGEN PRESENTING CELLS

For the cells involved in both the innate and adaptive immune defense systems to be effective, microbes or microbial derived material first must be internalized. Thus, a limited number of cell types have to recognize a wide variety of microorganisms. Indeed, it appears that these different cell types use a limited set of mechanisms and/or cell surface receptors for the uptake of microbes or material derived from infectious material. It is in fact likely that the cells involved in host defense recognize patterns, rather than specific molecular structures, allowing them to recognize and internalize a wide variety of microorganisms using a restricted set of receptor molecules (Janeway, 1989; Bachmann and Zinkernagel, 1997; Merdzhitov and Janeway, 1997).

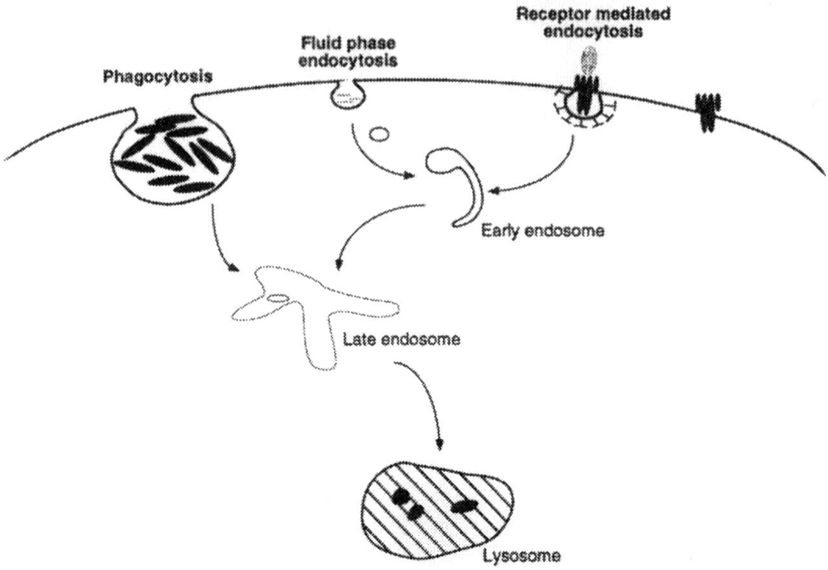

Figure 2. Possible pathways for antigen internalization. Particule material such as viable and killed bacteria are taken up by phagocytosis. In addition, most cells can take up fluids as well as soluble molecules via fluid phase endocytosis, whereas a more specific way to internalize molecules occurs through receptor mediated endocytosis. After internalization, material is transported along the endosomal/lysosomal pathways where it can be degraded. Note that some pathogens, such as mycobacteria, have evolved mechanisms to avoid lysosomal delivery and can survive within the phagosome (see also section VII).

The different modes of uptake known today are represented in Figure 2. First, entire microorganisms can be internalized by phagocytes such as neutrophils and macrophages through phagocytosis. This results in the formation of phagosomes, large (~ 250–500 nm) vesicles formed upon invagination of the plasma membrane during uptake. Second, microorganisms secrete metabolic products in the circulation, which can be subsequently internalized via fluid phase uptake or receptor mediated endocytosis, depending on the type of substance. Third, microbes partially degraded by phagocytes might be released into the circulation, and these products may re-enter dendritic cells, macrophages as well as B lymphocytes.

Uptake of material by these various mechanisms leads to its transfer from the extracellular space to the endosomal/lysosomal network. This is a collection of internal membranes, connected to each other as well as to the outside of the cell that is used by cells to acquire nutrients from the extracellular fluids by degrading incoming material (Mellman, 1996). In antigen presenting cells, the endosomal pathway is used to generate epitopes for presentation to T cells (Cresswell and Marks, 1987; Pieters, 1997a). Regardless of the mode of uptake, material that is internalized is bound to encounter an increasingly acidic and degradative environment (Kornfeld and Mellman, 1989), although exceptions exists (see later).

As the adaptive immune system recognizes peptide fragments and not intact proteins, the digestion of incoming material has to be carefully balanced, such that proteinous material is not degraded completely, but can be presented at the surface of antigen presenting cells in a recognizable form. To achieve both the proper destruction of microbes or microbial derived material during the initial phase of the infection as well as the presentation of T cell epitopes for initiating a cell-mediated immune response, the different types of antigen presenting cells (B cells, macrophages and dendritic cells) use distinct mechanisms for the internalization and digestion of microbes and antigens.

A. B Lymphocytes

B cells are not phagocytic, and accordingly they internalize microbial derived products, rather than particulate material. B cells mainly capture antigens via cell surface immunoglobulins or B cell receptors. B cell receptors are clonally distributed on B lymphocytes, and therefore each B lymphocyte expresses an immunoglobulin molecule with a defined specificity. Such antigens are subsequently transferred into endocytic organelles where they can be degraded, and from where peptides are transported to MHC class II compartments for presentation to T helper cells (Lanzavecchia, 1985; Pieters, 1997b; Watts, 1997). Apart from antigen internalization, ligand binding to the B cell receptor can lead to the activation of an intracellular signaling cascade which in turn leads to proliferation of B cells into plasma cells (which are the antibody producing cells) a process assisted by T lymphocytes when triggered with MHC class II-peptide complexes (see Figure 1) (Liu et al., 1997). Thus, B cells are very effective in the immune response during later stages of an infection, but are of limited value during the initial phase.

B. Macrophages

Macrophages are the primary cells involved in uptake and clearance of circulating microorganisms (Unanue, 1984; Gordon, 1995). Macrophages are extremely efficient both in phagocytosis as well as in the degradation of phagocytosed organisms and use a variety of mechanisms to carry out these functions (see Figure 2). The distinct mechanisms used to internalize intact microbes as well as microbial products are discussed in separate chapters of this volume, and therefore I will only briefly summarize them here. Apart from the nonspecific engulfment of fluids or macropinocytosis, particle internalization is mediated via a number of specific receptors (Steinman and Swanson, 1995; Allen and Aderem, 1996). These include several classes of Fc receptors (see Daeron, 1997 and the chapter by S. Greenberg), complement receptors and mannose receptors (see Stahl et al., 1980; Ezekowitz et al., 1990; Stahl, 1992 and chapter by R. Ezekowitz and I. Fraser). Thus, in contrast to B cells, macrophages are able to internalize large amounts of antigens immediately after infection. This makes these cells not only efficient in the initial clearance of microbes, but also in generating an adaptive immune response through the presentation of antigens on their cell surface (see later) (Ezekowitz et al., 1991; Tomlinson, 1993; Gaynor et al., 1995).

C. Dendritic Cells

Dendritic cells are bone marrow derived and are distributed as precursor cells through the blood to peripheral organs. During migration, these precursor dendritic cells acquire the capacity to internalize antigens while at the same time the expression of MHC class II complexes is induced (Engering et al., 1998). After arrival in peripheral tissues, these cells differentiate into "immature" dendritic cells and are able to capture large amounts of antigen, both through receptor mediated and nonreceptor mediated processes (Sallusto and Lanzavecchia, 1994; Steinman and Swanson, 1995). The mechanisms for antigen uptake resemble those used by macrophages and include phagocytosis (Inaba et al., 1993; Svensson et al., 1997), macropinocytosis (Sallusto et al., 1995; Steinman and Swanson, 1995), Fc receptor mediated uptake (Maurer and Stingl, 1995), and mannose receptor mediated uptake (Engering et al., 1997). During an infection, these so-called "immature" dendritic cells are stimulated into "mature" dendritic cells, which results in a series of phenotypic alterations resulting in a downregulation of endocytic capacity as well as an upregulation of T cell costimulatory molecules (Sallusto et al., 1995; Banchereau and Steinman, 1998). The capacity to internalize a wide variety of antigens in a regulated fashion most probably contributes to the enormous efficiency of dendritic cells to induce effective immune responses (Steinman, 1991; Banchereau and Steinman, 1998).

III. ANTIGEN PROCESSING

The internalization of microbes or microbial-derived material into the endosomal/lysosomal network serves two purposes. First, it exposes native antigens to proteolysis that is needed for the generation of antigenic peptides, and secondly, it brings the degraded antigens into contact with newly synthesized MHC class II complexes transported to these sites (see later). By itself, the uptake of bacteria by phagocytes into phagosomes can already produce a strong inflammatory response that leads to a rapid killing of the microorganisms in the phagosome. Fusion of the phagosome with other endosomal/lysosomal organelles then leads to a further degradation of bacterial proteins. The exact proteases that are involved in the generation of antigenic fragments are only beginning to be identified (Chapman, 1998; Manoury et al., 1998).

Proteases known to reside within endosomes and lysosomes are the members of the cathepsin family (Chapman, 1998). The major endosomal proteolytic activities are thought to be represented by the aspartate proteases cathepsin D and the cysteine proteases cathepsin L and S. Surprisingly, genetic deletion of these endosomal protease activities by gene targeting does not seem to have a profound effect on the types of epitopes generated (Deussing et al., 1998; Riese et al., 1998); the main effects of protease deficiency detected so far are on the intracellular transport pathways of MHC class II complexes to organelles where antigenic peptides are generated (see later). Either the proteases that are specifically involved in antigen degradation have so far escaped identification or the activity of the distinct cathepsins is quite redundant (Chen et al., 1997; Riese et al., 1998). Recently, an antigen processing activity was found in B lymphocytes that is distinct from the cathepsins (Manoury et al., 1998). This protease, an asparaginyl endopeptidase, might be specific for microbial antigens, suggesting that certain proteases co-evolved with the immune system to deal with invading pathogens (Deussing et al., 1998; Nakagawa et al., 1998).

The mechanisms that dictate the generation of a given peptide (epitope) that can be presented by class II molecules have not been fully elucidated. Not only might the proteases in different cell types differ (Barbey et al., 1995), but the destination of a given antigen may differ according to its mode of uptake (Santoro et al., 1994; Ferrari et al., 1997). In addition, the activation state of an antigen presenting cell under inflammatory conditions may alter its endocytic capacities as well as the expression of certain proteases. Therefore the generation of productive class II peptide complexes may be highly regulated by distinct mechanisms in different cell types.

IV. ANTIGEN PRESENTATION BY MHC MOLECULES

The endocytic system thus functions in degrading antigens into a form that can be presented to T cell receptors on T lymphocytes at the cell surface of antigen presenting cells, i.e., as small peptide fragments of ~9–25 amino acid residues (Demotz et al., 1989; Rotzschke et al., 1990; Rammensee et al., 1993). How these fragments are loaded onto and presented by the MHC molecules to the T cell receptors has been clarified over the past 20 years (Heemels and Ploegh, 1995; Pieters, 1997b).

A. MHC Class I Pathway

MHC class I molecules exist as a complex, which is formed upon insertion of a heavy chain associated with β2 microglobulin into the membrane of the endoplasmic reticulum (ER) (Figure 3). Antigens present in the cytoplasm and derived from foreign organisms are recognized by the cellular proteolytic machinery, and subsequently serve as substrate for the proteasome, an important cytosolic proteolytic machinery (Monaco, 1995; Baumeister et al., 1998). Once degraded by the proteasome, small peptide fragments are imported into the ER (Lehner and Cresswell, 1996) where peptides are bound by MHC class I molecules, a process that may be assisted by chaperone molecules that reside in the ER (Nicchitta, 1998). Binding of peptides to the class I complex triggers their export from the ER and subsequent transport though the secretory pathway for deposition at the plasma membrane (Townsend et al., 1989) (see also Figure 1). MHC class I molecules thus function as signaling molecules at the cell surface of virtually all cells. On the cell surface, peptides are exposed for recognition by $CD8^+$ cytotoxic T cells. If T cells recognize the peptide as derived from a foreign molecule, it will activate its cytotoxic capacity via signal transduction processes initiated via the T cell receptor, leading to lysis of the infected cell.

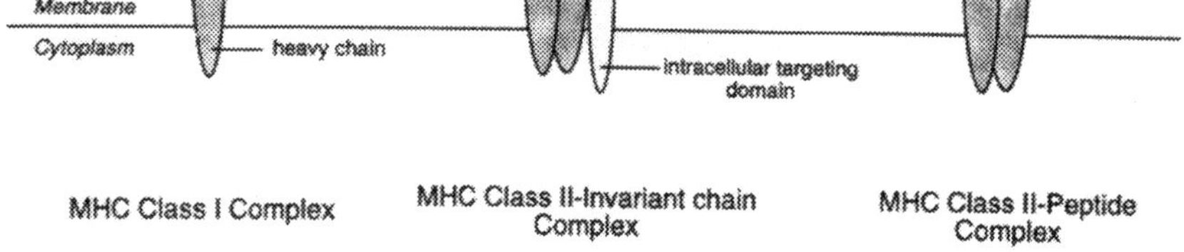

Figure 3. Schematic representation of the MHC class I and MHC class II complexes. The left panel shows the MHC class I molecule, which associaties with β2 microglobulin and can bind antigenic peptides that are imported into the endoplasmic reticulum. The middle panel depicts the MHC class II complex, consisting of an α-chain, a β-chain and the invariant chain. It should be noted that whereas MHC class I molecules bind peptides of ~9 amino acid residues, MHC class II molecules are able to bind relatively large polypeptide fragments after removal of the invariant chain. A model of the MHC class II-peptide complex is represented in the right panel. Adapted from Pieters, 1997.

B. MHC Class II Pathway

The MHC class II complex consists of an α chain and a β chain that assemble in the ER with a third molecule, called the invariant chain (Ii) (Cresswell, 1994; Pieters, 1997b) (Figure 3). Although Ii is not encoded within the MHC, it performs a number of chaperone functions to ensure a proper peptide presentation by MHC class II molecules. First, Ii aids in the folding and egress of class II molecules out of the ER (Anderson and Miller, 1992). Second, the lumenal domain of Ii prevents peptide loading of class II molecules by competing for the same binding site (Roche and Cresswell, 1990; Teyton et al., 1990; Ghosh et al., 1995), and third, Ii provides a targeting signal for transport of the class II complex into the endocytic pathway (Bakke and Dobberstein, 1990; Lotteau et al., 1990; Pieters et al., 1993). The intracellular pathway followed by MHC class II molecules is summarized in Figure 4. From their site of synthesis, the ER, the complex is transported through the Golgi complex to the trans-Golgi network. From there, the class II complex is routed to the so-called MHC class II compartments that are specialized organelles belonging to the endocytic pathway in antigen presenting cells (Pieters, 1997a). Targeting of the class II/Ii complex to MHC class II compartments is mediated by sorting signals in the Ii cytoplasmic tail that cause the class II complex to deviate from the secretory pathway and be transported into the endosomal/lysosomal pathway.

V. SUBCELLULAR ORGANELLES INVOLVED IN MHC CLASS II RESTRICTED ANTIGEN PRESENTATION

After transport of the class II/Ii complex to post-Golgi organelles, this complex has to be converted into peptide bearing class II molecules, a process requiring first the degradation of the associated invariant chain and, second, the actual loading of antigenic peptide followed by transport of the peptide-class II complex to the cell surface. It appears that these events occur in a highly coordinated fashion, involving a sequential transport of newly synthesized class II complexes through distinct organelles. Some of these organelles are part of the normal endocytic network, while others have a dedicated function in the biogenesis of MHC class II-peptide complexes.

A. MHC Class II Compartments

MHC class II compartments were originally identified in B cells and melanoma cells as multivesicular organelles that were enriched in MHC class II molecules (Peters et al., 1991; Pieters et al., 1991). Whereas some of these class II positive organelles contained endocytic marker molecules, the bulk of the class

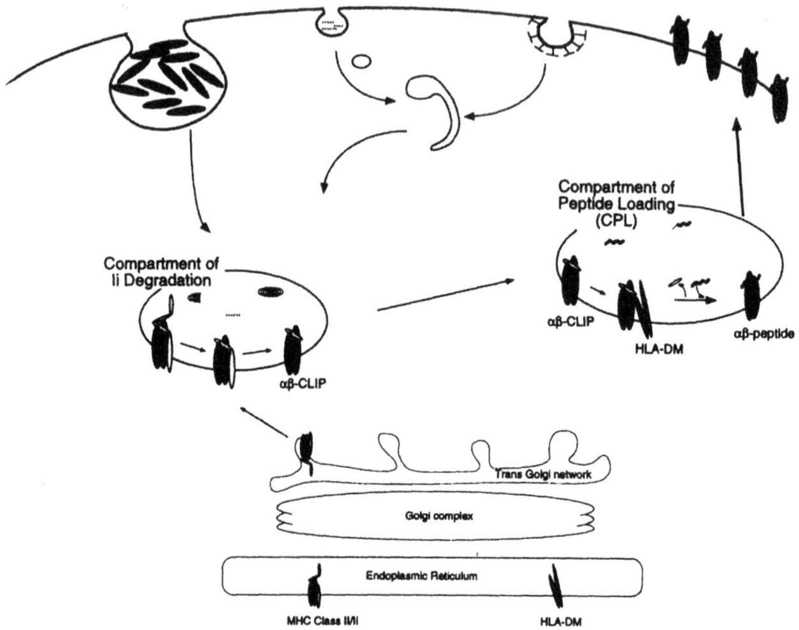

Figure 4. The MHC class II loading pathway. Antigenic peptide loading of MHC class II molecules requires the sequential transit of newly synthesized complexes through two distinct MHC class II compartments. After assembly in the endoplasmic reticulum, the class II/Ii complex is targeted at the trans-Golgi network to MHC class II compartments by virtue of targeting signals that reside in the Ii cytoplasmic tail. Upon arrival in a post-Golgi, endocytic organelle, Ii is progressively degraded from its lumenal, C-terminal region through the action of cathepsisn (Compartment of Ii Degradation), leading to the generation of MHC class II molecules complexed with CLIP (for class II associated invariant chain peptide). Subsequently, the class II-CLIP complex is transported to a distinct organelle that contains HLA-DM. Here, CLIP is exachanged for antigenic peptides in a reaction catalyzed by HLA-DM. From this compartment of peptide loading, termed CPL, peptide loaded class II molecules are transported to the cell surface. The CPL, in contrast to the compartment of invariant chain degradation, is not directly accessible to antigens internalized via fluid phase; antigents taken up through receptor mediated endocytosis can, however be rapidly transported to the CPL, either in an intact from or after degradation into peptides.

II positive structures were distinct from endosomes and lysosomes. In the absence of any other specific marker, these class II positive structures were called "multivesicular bodies" (Pieters et al., 1991) or "MIIC" (Peters et al., 1991), based on their morphological appearance and class II content, respectively. These morphological observations were later confirmed by the biochemical isolation and characterization of MHC class II compartments from a variety of cell types (Amigorena et al., 1994; Tulp et al., 1994; West et al., 1994), clearly establishing the class II compartments as separate entities, distinct from conventional endosomes and lysosomes.

After the arrival of the class II complex in the MHC class II compartments, the composition of the complex drastically changes. Since class II molecules are still loaded with the invariant chain which occupies the peptide binding site, Ii has to be exchanged for antigenic peptides (Blum and Cresswell, 1988; Cresswell, 1996). It was recognized a number of years ago that proteolytic digestion by endosomal proteases partially digest Ii, yielding a complex of MHC class II molecules with an Ii breakdown product, subsequently termed CLIP (for class II associated invariant chain peptide) (Blum and Cresswell, 1988; Pieters et al., 1991). More recently, the mechanism by which CLIP is exchanged for antigenic peptides has been unraveled, and was found to involve another set of chaperones, the class II–like molecules human leukocyte antigen-DM (HLA-DM) (Mellins et al., 1990; Kelly et al., 1991). In cells lacking HLA-DM, most class II molecules remain associated with the CLIP peptide, indicating a function of HLA-DM in peptide loading of MHC class II molecules (Mellins et al., 1990; Fung-Leung et al., 1996; Martin et al., 1996; Miyazaki et al., 1996). HLA-DM is found within MHC class II containing organelles, and a series of *in vitro* reconstitution experiments showed that HLA-DM catalyzes the removal of CLIP from class II and edits the peptide repertoire that appears at the cell surface (Denzin and Cresswell, 1995; Sherman et al., 1995; Sloan et al., 1995; Kropshofer et al., 1996).

A puzzling feature of the DM-assisted peptide loading of class II molecules was the fact that HLA-DM appeared to be targeted to MHC class II compartments using a tyrosine-based sorting signal, whereas the invariant chain targeted the class II complex to these organelles via a di-leucine containing sorting signal (Pieters et al., 1993; Lindstedt et al., 1995; Marks et al., 1995). This discrepancy was recently solved when it was shown that the MHC class II compartments, as originally defined to function in Ii degradation and antigenic peptide loading, actually consist of two physically and functionally distinct subcellular compartments each of which performs one of these functions; one characterized by the presence of Ii, the other containing HLA-DM. In the former, invariant chain positive organelles, Ii is partially degraded to yield class II-CLIP complexes. These complexes are then transported to the HLA-DM positive class II compartments, where antigenic peptides are loaded onto MHC class II molecules and are therefore termed CPL, for compartment of peptide loading (Figure 4) (Ferrari et al., 1997).

Thus, the generation of newly synthesized class II/Ii molecules involves a sequential transport of these complexes through a series of post-Golgi organelles. From the trans-Golgi network, the class II complex is targeted to endocytic, Ii containing organelles by virtue of the presence of the sorting signals within the Ii polypeptide. After digestion of Ii by endosomal proteases, the class II complex is delivered to the HLA-DM positive organelles. In these HLA-DM positive organelles, that are distinct from the endocytic pathway, the class II complexes become readily loaded with antigenic peptide prior to their deposition at the cell surface (Figure 4).

Endocytosed material does not have direct access to the HLA-DM positive CPL (Ferrari et al., 1997) and this may represent an important mechanism to ensure efficient antigen presentation. Indeed, MHC class II molecules have been shown to bind not only peptides, but also large polypeptides, once the invariant chain is removed (Stern et al., 1994; Busch et al., 1996). Therefore, if HLA-DM dependent peptide loading would occur in the same organelles where antigen is internalized and the invariant chain is removed from the class II complex, it might be impossible to load enough class II molecules with appropriate peptides to ensure adequate expression at the cell surface (Germain and Rinker, 1993; Lanzavecchia, 1996). The separation of the peptide loading event from the bulk endocytic pathway may prevent unproductive class II-peptide complex formation and thereby provide a mechanism to ensure the selective recruitment of peptide-receptive MHC class II molecules (i.e., class II-CLIP) and HLA-DM within the same subcellular location.

Interestingly, while the HLA-DM positive compartment of peptide loading is not accessible to proteins internalized through fluid phase endocytosis, antigen internalized via surface immunoglobulin is readily transferred to these organelles (Ferrari et al., 1997). This suggests that mechanisms exist to separate incoming antigens to distinct locations. One possibility is that fluid phase endocytosed and phagocytosed antigens are degraded as soon as they encounter proteases along the endocytic pathway. The resulting peptides are then transferred to the compartment of peptide loading by as yet unknown mechanisms, possibly involving chaperone molecules (Vanbuskirk et al., 1989). In the case of receptor mediated uptake, which usually results in an enhanced efficiency of class II loading (Lanzavecchia, 1985; Bonnerot et al., 1995; Engering et al., 1997), distinct mechanisms may be involved in the internalization, processing and transport to peptide loading organelles (Mitchell et al., 1995; Aluvihare et al., 1997).

The peptide repertoire presented by class II molecules can be further shaped by the activity of another class II-like molecule HLA-DO which is expressed on thymic epithelium and in B lymphocytes (Karlsson et al., 1991). HLA-DO colocalizes with MHC class II and associates with HLA-DM molecules intracellularly (Liljedahl et al., 1996; Denzin et al., 1997). In B lymphocytes, HLA-DO may inhibit HLA-DM function by favoring the presentation of antigens internalized by surface immunoglobulin molecules, but the precise effect of HLA-DO in other cells remains unclear (Liljedahl et al., 1998). The exact location of HLA-DO is unknown, but since its association with HLA-DM is required for exit from the ER, DO is expected to localize in the HLA-DM positive CPL, where it may assist DM in editing the peptide repertoire presented by MHC class II molecules.

B. Endocytic Organelles

In most antigen presenting cells the majority of MHC class II molecules are located in MHC class II compartments which are the main sites for the generation of MHC class II-peptide complexes. However, other organelles have been reported to have a role in class II restricted antigen presentation. Most notably, early endosomal organelles that receive recycling class II molecules from the cell surface have been suggested to play a role in the presentation of a subset of epitopes (Pinet et al., 1995). However, it cannot be excluded that in these endosomal organelles the class II complexes are further targeted to the MHC class II compartments where the actual peptide loading takes place. Interestingly, by following the fate of peptide loaded molecules from their site of assembly (the CPL) to the plasma membrane, these complexes were found to recycle extensively between endocytic organelles and the HLA-DM positive compartments (Ferrari et al., 1997). Such recycling might ensure a more complete loading of class II molecules with antigenic peptides prior to their deposition at the cell surface resulting in the expression of a sufficient number of peptide-MHC class II complexes at the cell surface for a proper T cell stimulation (Valitutti et al., 1996).

C. Phagosomes

Microorganisms internalized by phagocytes into phagosomes are usually delivered to compartments of the endosomal/lysosomal pathway. Within these compartments, degradation may result in the generation of antigenic peptides that can be transported to the CPL and loaded onto MHC class II molecules. However, as will be discussed below, some pathogens remain within a phagosome that resists fusion with endosomal/lysosomal organelles. The extent of class II localization in these phagosomes is unclear, and may vary according to the organisms inhabiting it (Clemens and Horwitz, 1993, 1995). In addition, extensive processing is not likely to occur within these phagosomes, and the presence of chaperones such as HLA-DM molecules remain to be established. Thus, while certain microorganisms that are internalized by phagocytosis will clearly be delivered to the class II loading pathway, those pathogens retained in organelles resisting fusion are likely to escape recognition by MHC class II restricted T cells.

VI. REGULATION OF MHC CLASS II RESTRICTED ANTIGEN PRESENTATION

Presentation of epitopes to T cells by MHC class II molecules can be regulated at several stages. Although class II MHC molecules are constitutively expressed on B lymphocytes there is an augmented expression of class II following B cell stimulation (Reith et al., 1995). Moreover, activation of B cells results in the expression of several surface molecules that enhance their interaction with T cells, the so-called costimulatory molecules (Lanzavecchia, 1990; Linsley et al., 1991). In macrophages, both the class II expression as well as costimulatory activity is induced after infection, most probably as a result of cytokines secreted by T lymphocytes recruited to the site of inflammation (Celada and Maki, 1991; Liu and Janeway, 1992; Stout, 1993).

That dendritic cells can regulate antigen capture capacity has already briefly been mentioned. Furthermore, when "immature" dendritic cells are challenged with inflammatory products, they undergo a series of phenotypic alterations (i.e., they "mature") that results in an increased capacity to present antigenic peptides to T lymphocytes (Cella et al., 1997; Pierre et al., 1997). A key factor in generating such a highly potent antigen presenting cell from immature dendritic cells, is the coordinated regulation of the endocytic capacity and trafficking routes of MHC class II molecules. In immature dendritic cells, MHC class II complexes are predominantly located intracellularly and the presence of inflammatory agents induce a rapid and transient increase in the synthesis of class II molecules. In addition, in mature dendritic cells the MHC class II complexes become long lived, with their half-life increasing from ~10 hours to over 100 hours (Cella et al., 1997). The mechanisms responsible for the modulation of MHC class II half-life are unknown. However, given the function of the MHC class II associated invariant chain as a targeting and retention signal for class II complexes, it is likely that the regulated endosomal proteolysis of the invariant chain contributes to changes in class II stability and location (Pieters et al., 1991).

Together, the changes that occur during the maturation of dendritic cells result in a large increase of class II complexes expressed on the cell surface that have been rapidly loaded with peptides acquired during the initial phase of an infection. The dramatic increase in half-life of these loaded class II complexes allows the presentation of epitopes over a relatively wide time window. In this way, antigens captured at the site of inflammation in the peripheral tissues can be presented after migration of the dendritic cells to the lymphoid areas, a process that is quite important for the induction of a T helper response during an *in vivo* infection (Cella et al., 1997; see also the chapter by Albert et al.).

VII. EVASION OF IMMUNE RECOGNITION BY PHAGOCYTOSED MICROBES

As discussed above, vertebrates have developed sophisticated mechanisms to deal with invading microorganisms. However, many microbes are known to be able to circumvent the host defense mechanisms at various levels, and therefore have the potential to become pathogenic.

The strategies that pathogens use to subvert the various host-defense mechanisms are still poorly characterized. Known mechanisms include targeting to a specific site in the host to escape rapid destruction (Jones et al., 1992; Hultgren et al., 1993), attachment to the surface of cells that are unable to efficiently destroy them (Falkow, 1991; Bliska et al., 1993), and finding a niche within the cells that will allow them to survive as well as manipulate and remodel these sites to serve their own needs (Beverley, 1996; Galan and Bliska, 1996; Finlay and Cossart, 1997). To escape MHC class II restricted immune recognition, pathogens can manipulate the vacuoles in which they reside to avoid generation class II-peptide complexes; some even escape from the vacuolar/endocytic system into the cytoplasm.

A. Escape into the Cytoplasm

Several microorganisms gain access to host cells via internalization from the plasma membrane, and rapidly escape from the endomembrane system into the cytoplasm. This is the case for *Trypanosoma cruzi* (see chapter by N. Andrews), *Shigella flexneri* (see chapter by P. Sansonetti) and *Listeria monocytogenes* (see chapter by P. Cossart). Among these pathogens, *L. monocytogenes* represents an organism for which both the cell biology as well as the induction of immune responses is well-studied (Theriot, 1995; Finlay and Cossart, 1997). Inside the phagosome, it has the capacity to retard the maturation of the phagosomes into a phagolysosome, by an as yet unknown mechanism (Alvarez-Dominguez et al., 1997). The success of *Listeria* as a pathogen is most probably due to its capacity to lyse the phagosomal vacuole at a stage that precedes fusion with lysosomal vesicles, a process mediated by the pore forming protein listeriolysin O. This results in its escape into the cytoplasm, where the bacteria can nucleate actin to propel themselves to neighboring cells (Kocks et al., 1992; Theriot, 1995). Escape from the vacuole at an early stage would thus prevent the degradation of *Listeria* as well as the formation of peptide MHC class II complexes. While this mechanisms allows *Listeria* to survive under certain conditions, both class I and class II restricted responses are induced that are quite effective (Zinkernagel, 1974; Brunt et al., 1990; Kaufmann, 1993). Interestingly, the pathway followed by *Listeria* has recently been exploited as a system for delivering antigens to the class I loading pathway to elicit specific immune responses (Darji et al., 1997a, b).

Presentation of phagocytosed and processed antigens by class I molecules has also been reported to occur for a variety of other particulate substrates (Rock et al., 1990; Pfeifer et al., 1993; Kovacsovics-Bankowski and Rock, 1995), as well as pathogens, including *Mycobacterium tuberculosis* (Mazzaccaro et al., 1996). Whether or not this loading occurs after translocation of material into the cytoplasm or perhaps via a direct loading in the phagosome, which can also contains class I molecules that may be internalized from the plasma membrane (Clemens and Horwitz, 1995), is not fully clear.

B. Inhibition of Lysosomal Degradation

One group of pathogens that enter their host cells through phagocytosis thrive within the acidic and hostile environment of the phagolysosome (e.g., *Coxiella burnetii, Leishmania spp*., and *Francisella tularensis*). The mechanisms used by these microbes to avoid destruction in the lysosome are still largely unknown (Antoine et al., 1998). Although the immune system must have developed means to deal with this class of pathogens, it remains to be established how antigens are degraded and the immune system is activated (Fruth et al., 1993; Baca et al., 1994).

C. Prevention of Phagosome–Lysosome Fusion

A number of microorganisms that enter their host cells through phagocytosis are able to survive within the phagosome, by inhibiting their delivery from phagosomes into lysosomes. One of the most studied microbes that uses this strategy are the mycobacteria (see chapter by C. de Castellier). In the early seventies, D'Arcy Hart and collegues observed that live mycobacteria administered to macrophages remain inside a phagosome that does not mature into a lysosome, whereas dead bacilli are readily transported to lysosomal organelles (Armstrong and D'Arcy Hart, 1971). More recently, these observations were extended by the demonstration that phagosomes harboring viable mycobacteria fail to acidify (Gordon et al., 1980; Crowle et al., 1991; Sturgill-Koszycki et al., 1994) and lack late endosomal and lysosomal markers (Clemens and Horwitz, 1995; Barker et al., 1997; Hasan et al., 1997). Endosomal molecules that are in fact found in the mycobacterial phagosomes (e.g., transferrin, a necessary component for the growth of mycobacteria; (Snow, 1970; Fiss et al., 1994) might arrive there either through co-internalization with the bacteria from the plasma membrane or through co-internalization with the bacteria from the plasma membrane or through selective fusion of a subset of endocytic vesicles with phagosomes (Sturgill-Koszycki et al., 1994; Alvarez-Dominguez et al., 1997). In contrast to those containing live mycobacteria, phagosomes containing dead bacilli or inert particles do not seem to differ in their composition from endosomes and lysosomes (Desjardins et al., 1994; Hasan et al., 1997; Funato et al., 1997; Hasan and Pieters, 1998). Thus, phagosomes containing live or dead bacilli are structurally different. The capacity of mycobacteria to resist lysosomal delivery not only causes the bacilli to survive, but also allows them to escape from immune recognition. Indeed, presentation of mycobacterial-derived peptides by MHC class II molecules is impaired after infection of monocytes with live mycobacteria (Pancholi et al., 1993; Gercken et al., 1994).

What are the molecular mechanisms involved in inhibition of phagosome-lysosome fusion as exploited by living mycobacteria? By analyzing differences between phagosomes harboring viable mycobacteria and those containing killed bacilli, a molecule was recently identified in macrophages that is responsible for the intraphagosomal survival of mycobacteria. This protein is a member of the WD repeat protein family (Neer et al., 1994), termed TACO (for tryptophane aspartate containing coat protein) and is a component of the macrophage cortical cytoskeleton that is recruited around the nascent phagosome. Importantly, TACO remains associated only with those phagosomes harboring viable mycobacteria, but rapidly dissociates from phagosomes containing killed bacilli. Retention of TACO is actively mediated by living bacteria and, as a result of being coated with TACO, these phagosomes are unable to fuse with lysosomes, ensuring the intraphagosomal survival of mycobacteria (Ferrari et al., 1999).

If macrophages express a protein that allows mycobacteria to survive intracellularly, how can a challenge with these pathogens be countered? One reason is that one of the most important cells involved in clearing of bacterial infections, the Kupffer cells in the liver, do not express TACO. As expected, mycobacteria that are internalized by Kupffer cells immediately are transferred to vacuoles containing lysosomal membrane proteins and are completely degraded within 12 to 16 hours (Wardle, 1987; Ferrari et al., 1999). Although Kupffer cells provide an efficient clearance site for mycobacteria, one of the results of mycobacterial infections can be the formation of granulomas in different organs. Granulomas are thought to be part of the host response against some bacterial infections by containment of the infectious agents to a restricted site (Gordon et al., 1994). Granulomas are often found in the liver and largely consist of macrophages harboring viable bacilli. Interestingly, the macrophages present in granulomas are distinct from Kupffer cells, largely cosisting of infiltrated, TACO-expressing macrophages (Nibbering et al., 1989; Ferrari et al., 1999).

Together, these findings strengthen the idea that after a mycobacterial infection, as is true for many other infectious diseases (Falkow et al., 1992), a balance is established between the host and the pathogen: on the one hand, degradation of mycobacteria by various cell types (e.g., Kupffer cells, neutrophiles,, and activated macrophages) will result in the generation of antigenic peptides that can be loaded into the MHC class II pathway whereas, on the other hand, by residing inside TACO-positive macrophages, mycobacteria manage to survive within the host.

Other pathogenic organisms such as *Legionella pneumophila, Chlamidia trachomatis and Toxoplasma gondii*, also occupy a phagosome that fails to acidify, and may therefore use mechanisms similar to mycobacteria to evade immune recognition. It is however unclear whether the vacuoles that are occupied by these pathogens are similar in nature to those in which mycobacteria survive (Sibley, 1995; Dobrowolski and Sibley, 1996). Interestingly, in the case of *T. gondii* the mode of uptake may determine the intracellular fate of these parasites: when the parasites are coated with antibodies ("opsonized") prior to uptake, they are internalized directly into the endosomal/lysosomal pathway rather than residing in a distinct vacuole (Joiner et al., 1990) (see also the chapters by D. Sibley and K. Joiner). As expected, this leads to a difference in the formation of peptide-class II complexes, possibly allowing nonopsonized parasites to evade recognition by class II restricted T cells (Mordue and Sibley, 1997).

VIII. CONCLUDING REMARKS

Much has been learned over the past decades about the mechanisms governing the presentation of antigenic peptides to T lymphocytes. Model systems employing protein antigens such as ovalbumin have been useful to analyze the cell surface receptors, the intracellular processes as well as the organelles involved in the formation of peptide-MHC complexes.

MHC restricted processing and presentation, however, is usually initiated through the invasion of a wide variety of microorganisms into different types of cells. At the same time, many of these microbes need the host cell environment for successful replication. That a delicate balance is maintained by the immune system, which limits the growth of such microorganisms, is illustrated by the rapid progression of infectious disease under immune compromised conditions.

Furthermore, we are beginning to learn more about non-MHC restricted host defense mechanisms, such as the activation of T cell subsets through lipid moieties derived from bacteria. Recently the components of innate immunity are receiving more attention. As these represent the first cells to encounter infectious agents they obviously play a crucial role in host defense. However, these can also serve as targets for pathogens to evade immune recognition, making use of normal cell biological processes that are ongoing within the host. Most of the mechanisms that microbes have evolved in order to survive within host cells and evade immune recognition remain unknown. Analyzing the strategies employed by pathogens is likely not only to increase the knowledge of vertebrate immunity, but may further deepen our insight into host cell biology in the coming years.

ACKNOWLEDGMENTS

I thank Anneke Engering, Manfred Kopf and Ed Palmer for critical reading and comments on the manuscript. The Basel Institute for Immunology was founded and is supported by Hoffmann-La Roche Ltd., Basel, Switzerland.

REFERENCES

Allen, L.A., & Aderem, A. (1996). Mechanisms of phagocytosis. Curr. Opin. Immunol. 8, 36-40.

Aluvihare, V.R., Khamlichi, A.A., Williams, G.T., Adorini, L., & Neuberger, M. S. (1997). Acceleration of intracellular targeting of antigen by the B-cell antigen receptor: importance depends on the nature of the antigen- antibody interaction. EMBO J. 16, 3553-3562.

Alvarez-Dominguez, C., Roberts, R., & Stahl, P.D. (1997). Internalized Listeria monocytogenes modulates intracellular trafficking and delays maturation of the phagosome. J. Cell Sci. 110, 731-743.

Amigorena, S., Drake, J.R., Webster, P., & Mellman, I. (1994). Transient accumulation of new MHC molecules in a novel endocytic compartment in B lymphocytes. Nature 369, 113-120.

Anderson, M.S., & Miller, J. (1992). Invariant chain can function as a chaperone protein for class II major histocompatibility complex molecules. Proc. Natl. Acad. Sci. USA 89, 2282-2286.

Antoine, J.C., Prina, E., Lang, T., & Courret, N. (1998). The biogenesis and properties of the parasitophorous vacuoles that harbour Leishmania in murine macrophages. Trends Microbiol. 6, 392-401.

Armstrong, J., & D'Arcy Hart, P. (1971). Response of cultured macrophages to *Mycobacterium tuberculosis*, with observations on fusion of lysosomes with phagosomes. J. Exp. Med. 134, 713-740.

Babbitt, B.P., Allen, P.M., Matsueda, G., Haber, E., & Unanue, E.R. (1985). Binding of immunogenic peptides to Ia histocompatibility molecules. Nature 317, 359-361.

Baca, O.G., Li, Y.P., & Kumar, H. (1994). Survival of the Q fever agent Coxiella burnetii in the phagolysosome. Trends Microbiol. 2, 476-480.

Bachmann, M.F., & Zinkernagel, R.M. (1997). Neutralizing antiviral B cell responses. Ann. Rev. Immunol. 15, 235-270.

Bakke, O., & Dobberstein, B. (1990). MHC class II-associated invariant chain contains a sorting signal for endosomal compartments. Cell 63, 707-716.

Banchereau, J., & Steinman, R. (1998). Dendritic cells and the control of immunity. Nature 392, 245-252.

Barbey, C., Watts, C., & Corradin, G. (1995). Antigen-processing organelles from DRB1*1101 and DRB1*1104 B cell lines display a differential degradation activity. Eur. J. Immunol. 25, 30-36.

Barker, L.P., George, K.M., Falkow, S., & Small, P.L. (1997). Differential trafficking of live and dead Mycobacterium marinum organisms in macrophages. Infect. Immun. 65, 1497-1504.

Baumeister, W., Walz, J., Zuhl, F., & Seemuller, E. (1998). The proteasome: paradigm of a self-compartmentalizing protease. Cell 92, 367-380.

Beverley, S.M. (1996). Hijacking the cell: parasites in the driver's seat. Cell 87, 787-789.

Bliska, J.B., Galan, J.E., & Falkow, S. (1993). Signal transduction in the mammalian cell during bacterial attachment and entry. Cell 73, 903-920.

Blum, J.S., & Cresswell, P. (1988). Role for intracellular proteases in the processing and transport of class II HLA antigens. Proc. Natl. Acad. Sci. USA 85, 3975-3979.

Bonnerot, C., Lankar, D., Hanau, D., Spehner, D., Davoust, J., Salamero, J., & Fridman, W.H. (1995). Role of B cell receptor Ig alpha and Ig beta subunits in MHC class II-restricted antigen presentation. Immunity 3, 335-347.

Brown, E., Atkinson, J.P., & Fearon, D.T. (1994). Innate immunity: 50 ways to kill a microbe. Curr. Opin. Immunol. 6, 73-74.

Brunt, L.M., Portnoy, D.A., & Unanue, E.R. (1990). Presentation of Listeria monocytogenes to CD8+ T cells requires secretion of hemolysin and intracellular bacterial growth. J. Immunol. 145, 3540-3546.

Busch, R., Cloutier, I., Sekaly, R.-P., & Hammerling, G.J. (1996). Invariant chain protects class II histocompatibility antigens from binding intact polypeptides in the endoplasmic reticulum. EMBO J. 15, 418-428.

Celada, A., & Maki, R.A. (1991). IFN-gamma induces the expression of the genes for MHC class II I-A beta and tumor necrosis factor through a protein kinase C-independent pathway. J. Immunol. 146, 114-120.

Cella, M., Engering, A., Pinet, V., Pieters, J., & Lanzavecchia, A. (1997). Inflammatory stimuli induce accumulation of MHC class II complexes on dendritic cells. Nature 388, 782-787.

Chapman, H. (1998). Endosomal proteolysis and MHC class II function. Curr. Opin. Immunol. 10, 93-102.

Chen, J.M., Dando, P.M., Rawlings, N.D., Brown, M.A., Young, N.E., Stevens, R.A., Hewitt, E., Watts, C., & Barrett, A.J. (1997). Cloning, isolation, and characterization of mammalian legumain, an asparaginyl endopeptidase. J. Biol. Chem. 272, 8090-8098.

Clemens, D.L., & Horwitz, M.A. (1993). Hypoexpression of major histocompatibility complex molecules on Legionella pneumophila phagosomes and phagolysosomes. Infect. Immun. 61, 2803-2812.

Clemens, D.L., & Horwitz, M.A. (1995). Characterization of the Mycobacterium tuberculosis phagosome and evidence that phagosomal maturation is inhibited. J. Exp. Med. 181, 257-270.

Colonna, M. (1996). Natural killer cell receptors specific for MHC class I molecules. Curr. Opin. Immunol. 8, 101-109.

Cresswell, P. (1994). Assembly, transport, and function of MHC class II molecules. Annu. Rev. Immunol. 12, 259-293.

Cresswell, P. (1996). Invariant chain structure and MHC class II function. Cell 84, 505-507.

Cresswell, P., Blum, J.S., Kelner, D.N., & Marks, M.S. (1987). Biosynthesis and processing of class II histocompatibility antigens. Crit. Rev. Immunol. 7, 31-53.

Crowle, A.J., Dahl, R., Ross, E., & May, M.H. (1991). Evidence that vesicles containing living, virulent Mycobacterium tuberculosis or Mycobacterium avium in cultured human macrophages are not acidic. Infect. Immun. 59, 1823-1831.

Daeron, M. (1997). Fc receptor biology. Annu. Rev. Immunol. 15, 203-234.

Darji, A., Chakraborty, T., Wehland, J., & Weiss, S. (1997a). TAP-dependent major histocompatibility complex class I presentation of soluble proteins using listeriolysin. Eur. J. Immunol. 27, 1353-1359.

Darji, A., Guzman, C.A., Gerstel, B., Wachholz, P., Timmis, K.N., Wehland, J., Chakraborty, T., & Weiss, S. (1997b). Oral somatic transgene vaccination using attenuated S. typhimurium. Cell 91, 765-775.

Demotz, S., Grey, H.M., Appella, E., & Sette, A. (1989). Characterization of a naturally processed MHC class II-restricted T- cell determinant of hen egg lysozyme. Nature 342, 682-684.

Denzin, L.K., & Cresswell, P. (1995). HLA-DM induces CLIP dissociation from MHC class II alpha beta dimers and facilitates peptide loading. Cell 82, 155-165.

Denzin, L.K., Sant'Angelo, D.B., Hammond, C., Surman, M.J., & Cresswell, P. (1997). Negative regulation by HLA-DO of MHC class II- restricted antigen processing. Science 278, 106-109.

Desjardins, M., Huber, L.A., Parton, R.G., & Griffiths, G. (1994). Biogenesis of phagolysosomes proceeds through a sequential series of interactions with the endocytic apparatus. J. Cell Biol. 124, 677-688.

Deussing, J., Roth, W., Saftig, P., Peters, C., Ploegh, H., & Villadangos, J. (1998). Cathepsins B and D are dispensible for major histocompatibility complex class II-mediated antigen presentation. Proc. Natl. Acad. Sci. USA 95, 4516-4521.

Dobrowolski, J.M., & Sibley, L.D. (1996). Toxoplasma invasion of mammalian cells is powered by the actin cytoskeleton of the parasite. Cell 84, 933-939.

Engering, A.J., Cella, M., Fluitsma, D.M., Hoefsmit, E.C.M., Lanzavecchia, A., & Pieters, J. (1997). The mannose receptor functions as a high capacity and broad specificity antigen receptor in human dendritic cells. Eur. J. Immunol. 27, 2417-2425.

Engering, A.J., Richters, C.D., Fluitsma, D.M., van Pelt, A.M., Kamperdijk, E.W., Hoefsmit, E.C., & Pieters, J. (1998). MHC class II and invariant chain biosynthesis and transport during maturation of human precursor dendritic cells. Int. Immunol. 10, 1713-1723. (In press).

Ezekowitz, R.A., Sastry, K., Bailly, P., & Warner, A. (1990). Molecular characterization of the human macrophage mannose receptor: demonstration of multiple carbohydrate recognition-like domains and phagocytosis of yeast in Cos-1 cells. J. Exp. Med. 172, 1785-1794.

Ezekowitz, R.A., Williams, D.J., Koziel, H., Armstrong, M.Y., Warner, A., Richards, F.F., & Rose, R.M. (1991). Uptake of Pneumocystis carinii mediated by the macrophage mannose receptor. Nature 351, 155-158.

Falkow, S. (1991). Bacterial entry into eukaryotic cells. Cell 65, 1099-1102.

Falkow, S., Isberg, R.R., & Portnoy, D.A. (1992). The interaction of bacteria with mammalian cells. Annu. Rev. Cell. Biol. 8, 333-363.

Farries, T.C., & Atkinson, J.P. (1991). Evolution of the complement system. Immunol. Today 12, 295-300.

Ferrari, G., Knight, A.M., Watts, C., & Pieters, J. (1997). Distinct intracellular compartments involved in invariant chain degradation and antigenic peptide loading of major histocompatibility complex (MHC) class II molecules. J. Cell Biol. 139, 1433-1446.

Ferrari, G., Naito, M., Langen, H., & Pieters, J. (1999). A coat protein on phagosomes involved in the intracellular survival of mycobacteria. Cell 97, 435-447.

Finlay, B.B., & Cossart, P. (1997). Exploitation of mammalian host cell functions by bacterial pathogens. Science 276, 718-725.

Fiss, E.H., Yu, S., & Jacobs, W.R., Jr. (1994). Identification of genes involved in the sequestration of iron in mycobacteria: the ferric exochelin biosynthetic and uptake pathways. Mol. Microbiol. 14, 557-569.

Fruth, U., Solioz, N., & Louis, J.A. (1993). Leishmania major interferes with antigen presentation by infected macrophages. J. Immunol. 150, 1857-1864.

Funato, K., Beron, W., Yang, C.Z., Mukhopadhyay, A., & Stahl, P.D. (1997). Reconstitution of phagosome-lysosome fusion in streptolysin O-permeabilized cells. J. Biol. Chem. 272, 16147-16151.

Fung-Leung, W.-P., Surh, C.D., Liljedahl, M., Pang, J., Leturcq, D., Peterson, P.A., Webb, S.R., & Karlsson, L. (1996). Antigen presentation and T cell development in H2-M-deficient mice. Science 271, 1278-1281.

Galan, J.E., & Bliska, J.B. (1996). Cross-talk between bacterial pathogens and their host cells. Ann. Rev. Cell Dev. Biol. 12, 221-255.

Gaynor, C.D., McCormack, F.X., Voelker, D.R., McGowan, S.E., & Schlesinger, L.S. (1995). Pulmonary surfactant protein A mediates enhanced phagocytosis of Mycobacterium tuberculosis by a direct interaction with human macrophages. J. Immunol. 155, 5343-5351.

Gercken, J., Pryjma, J., Ernst, M., & Flad, H.-D. (1994). Defective antigen presentation by Mycobacterium tuberculosis-infected monocytes. Infect. Immun. 62, 3472-3278.

Germain, R.N., & Rinker, A.G., Jr. (1993). Peptide binding inhibits protein aggregation of invariant-chain free class II dimers and promotes surface expression of occupied molecules. Nature 363, 725-728.

Ghosh, P., Amaya, M., Mellins, E., & Wiley, D.C. (1995). The structure of an intermediate in class II MHC maturation: CLIP bound to HLA-DR3. Nature 378, 457-462.

Gordon, A.H., Hart, P.D., & Young, M.R. (1980). Ammonia inhibits phagosome-lysosome fusion in macrophages. Nature 286, 79-80.

Gordon, S., Keshav, S., & Stein, M. (1994). BCG-induced granuloma formation in murine tissues. Immunobiology 191, 369-377.

Gordon, S. (1995). The macrophage. Bioessays 17, 977-986.

Groh, V., Steinle, A., Bauer, S., & Spies, T. (1998). Recognition of stress-induced MHC molecules by intestinal epithelial gammadelta T cells. Science 279, 1737-1740.

Hansen, T.H., Carreno, B.M., & Sachs, D.H. (1993). The major histocompatibility complex. In: Fundamental Immunology (Paul, W.E., Ed.), pp. 577-628. Raven Press, Ltd., New York.

Hasan, Z., & Pieters, J. (1998). Subcellular fractionation by organelle electrophoresis: separation of phagosomes containing heat-killed yeast particles. Electrophoresis 19, 1179-1184.

Hasan, Z., Schlax, C., Kuhn, L., Lefkovits, I., Young, D., Thole, J., & Pieters, J. (1997). Isolation and characterization of the mycobacterial phagosome: segregation from the endosomal/lysosomal pathway. Molec. Microbiol. 24, 545-553.

Heemels, M.T., & Ploegh, H. (1995). Generation, translocation, and presentation of MHC class I-restricted peptides. Annu. Rev. Biochem. 64, 463-491.

Hultgren, S.J., Abraham, S., Caparon, M., Falk, P., St. Geme, J.W.d., & Normark, S. (1993). Pilus and nonpilus bacterial adhesins: assembly and function in cell recognition. Cell 73, 887-901.

Inaba, K., Inaba, M., Naito, M., & Steinman, R. (1993). Dendritic cell progenitors phagocytose particulates, including Bacillus Calmette-Guerin organisms, and sensitize mice to mycobacterial antigens in vivo. J. Exp. Med. 178, 479-488.

Janeway, C.A., Jr. (1989). Approaching the asymptote? Evolution and revolution in immunology. Cold Spring Harb. Symp. Quant. Biol. 54(pt 1), 1-13.

Jensen, P.E. (1990). Regulation of antigen presentation by acidic pH. J. Exp. Med. 171, 1779-1784.

Joiner, K.A., Fuhrman, S.A., Miettinen, H.M., Kasper, L.H., & Mellman, I. (1990). Toxoplasma gondii: fusion competence of parasitophorous vacuoles in Fc receptortransfected fibroblasts. Science 249, 641-646.

Jones, B.D., Lee, C.A., & Falkow, S. (1992). Invasion by *Salmonella typhimurium* is affected by the direction of flagellar rotation. Infect. Immun. 60, 2475-2480.

Karlsson, L., Surh, C.D., Sprent, J., & Peterson, P.A. (1991). A novel class II MHC molecule with unusual tissue distribution. Nature 351, 485-488.

Kaufmann, S.H. (1993). Immunity to intracellular bacteria. Annu. Rev. Immunol. 11, 129-163.

Kelly, A.P., Monaco, J.J., Cho, S., & Trowsdale, J. (1991). A new human HLA class II-related locus, DM. Nature 353, 517-573.

Kocks, C., Gouin, E., Tabouret, M., Berche, P., Ohayon, H., & Cossart, P. (1992). L. monocytogenes-induced actin assembly requires the actA gene product, a surface protein. Cell 68, 521-531.

Kornfeld, S., & Mellman, I. (1989). The biogenesis of lysosomes. Annu. Rev. Cell Biol. 5, 483-525.

Kovacsovics-Bankowski, M., & Rock, K.L. (1995). A phagosome-to-cytosol pathway for exogenous antigens presented on MHC class I molecules. Science 267, 243-246.

Kropshofer, H., Vogt, A.B., Moldenhauer, G., Hammer, J., Blum, J.S., & Hammerling, G.J. (1996). Editing of the HLA-DR-peptide repertoire by HLA-DM. EMBO J. 15, 6144-6154.

Kuby, J. (1992). Immunology. W.H. Freeman and Company, New York.

Lanzavecchia, A. (1985). Antigen-specific interaction between T and B cells. Nature 314, 537-539.

Lanzavecchia, A. (1990). Receptor-mediated antigen uptake and its effect on antigen presentation to class II-restricted T lymphocytes. Annu. Rev. Immunol. 8, 773-793.

Lanzavecchia, A. (1996). Mechanisms of antigen uptake for presentation. Curr. Opin. Immunol. 8, 348-354.

Lehner, P.J., & Cresswell, P. (1996). Processing and delivery of peptides presented by MHC class I molecules. Curr. Opin. Immunol. 8, 59-67.

Lemaitre, B., Nicolas, E., Michaut, L., Reichhart, J.M., & Hoffmann, J.A. (1996). The dorsoventral regulatory gene cassette spatzle/Toll/cactus controls the potent antifungal response in Drosophila adults. Cell 86, 973-983.

Liljedahl, M., Kuwana, T., Fungleung, W.P., Jackson, M., Peterson, P.A., & Karlsson, L. 1996). HLA-DO is a lysosomal resident which requires association with HLA-DM for efficient intracellular transport. EMBO J. 15, 4817-4824.

Liljedahl, M., Winqvist, O., Surh, C.D., Wong, P., Ngo, K., Teyton, L., Peterson, P.A., Brunmark, A., Rudensky, A.Y., Fung-Leung, W.P., & Karlsson, L. (1998). Altered antigen presentation in mice lacking H2-O. Immunity 8, 233-243.

Lindstedt, R., Liljedahl, M., Peleraux, A., Peterson, P.A., & Karlsson, L. (1995). The MHC class II molecule H2-M is targeted to an endosomal compartment by a tyrosine-based targeting motif. Immunity 3, 561-572.

Linsley, P.S., Brady, W., Grosmaire, L., Aruffo, A., Damle, N.K., & Ledbetter, J.A. (1991). Binding of the B cell activation antigen B7 to CD28 costimulates T cell proliferation and interleukin 2 mRNA accumulation. J. Exp. Med. 173, 721-730.

Liu, Y., & Janeway, C.A., Jr. (1992). Cells that present both specific ligand and costimulatory activity are the most efficient inducers of clonal expansion of normal CD4 T cells. Proc. Natl. Acad. Sci. USA 89, 3845-3849.

Liu, Y.J., de Bouteiller, O., & Fugier-Vivier, I. (1997). Mechanisms of selection and differentiation in germinal centers. Curr. Opin. Immunol. 9, 256-262.

Ljunggren, H.G., & Karre, K. (1990). In search of the "missing self": MHC molecules and NK cell recognition [see comments]. Immunol. Today 11, 237-244.

Lotteau, V., Teyton, L., Peleraux, A., Nilsson, T., Karlsson, L., Schmid, S.L., Quaranta, V., & Peterson, P.A. (1990). Intracellular transport of class II MHC molecules directed by invariant chain. Nature 348, 600-605.

Manoury, B., Hewitt, E.W., Morrice, N., Dando, P.M., Barrett, A.J., & Watts, C. (1998). An asparaginyl endopeptidase processes a microbial antigen for class II MHC presentation. Nature 396, 695-699.

Marks, M.S., Roche, P.A., van Donselaar, E., Woodruff, L., Peters, P.J., & Bonifacino, J.S. (1995). A lysosomal targeting signal in the cytoplasmic tail of the beta chain directs HLA-DM to MHC class II compartments. J. Cell. Biol. 131, 351-369.

Martin, W.D., Hicks, G.G., Mendiratta, S.K., Leva, H.I., Ruley, H.E., & Vankaer, L. 1996). H2-M mutant mice are defective in the peptide loading of class II molecules, antigen presentation, and T cell repertoire selection. Cell 84, 543-550.

Maurer, D., & Stingl, G. (1995). Immunoglobulin E-binding structures on antigen-presenting cells present in skin and blood. J. Invest. Dermatol. 104, 707-710.

Mazzaccaro, R.J., Gedde, M., Jensen, E.R., Vansanten, H.M., Ploegh, H.L., Rock, K.L., & Bloom, B.R. (1996). Major histocompatibility class I presentation of soluble antigen facilitated by Mycobacterium tuberculosis infection. Proc. Natl. Acad. Sci. USA 93, 11786-11791.

Medzhitov, R., Preston-Hurlburt, P., & Janeway, C.A., Jr. (1997). A human homologue of the *Drosophila* Toll protein signals activation of adaptive immunity. Nature 388, 394-397.

Mellins, E., Smith, L., Arp, B., Cotner, T., Celis, E., & Pious, D. (1990). Defective processing and presentation of exogenous antigens in mutants with normal HLA class II genes. Nature 343, 71-74.

Mellman, I. (1996). Endocytosis and molecular sorting. Ann. Rev. Cell Dev. Biol. 12, 575-625.

Merdzhitov, R., & Janeway, C. (1997). Innate immunity: impact on the adaptive immune response. Curr. Opin. Immunol. 9, 4-9.

Merdzhitov, R., & Janeway, C. (1998). An ancient system of host defense. Curr. Opin. Immunol. 10, 12-15.

Mitchell, R.N., Barnes, K.A., Grupp, S.A., Sanchez, M., Misulovin, Z., Nussenzweig, M.C., & Abbas, A.K. (1995). Intracellular targeting of antigens internalized by membrane immunoglobulin in B lymphocytes. J. Exp. Med. 181, 1705-1714.

Miyazaki, T., Wolf, P., Tourne, S., Waltzinger, C., Dierich, A., Barois, N., Ploegh, H., Benoist, C., & Mathis, D. (1996). Mice lacking H2-M complexes, enigmatic elements of the MHC class II peptide-loading pathway. Cell 84, 531-541.

Monaco, J.J. (1995). Pathways for the processing and presentation of antigens to T cells. J. Leukoc. Biol. 57, 543-547.

Mordue, D.G., & Sibley, L.D. (1997). Intracellular fate of vacuoles containing Toxoplasma gondii is determined at the time of formation and depends on the mechanism of entry. J. Immunol. 159, 4452-4459.

Mosmann, T.R., & Coffman, R.L. (1989). TH1 and TH2 cells: different patterns of lymphokine secretion lead to different functional properties. Annu. Rev. Immunol. 7, 145-173.

Nakagawa, T., Roth, W., Wong, P., Nelson, A., Farr, A., Deussing, J., Villadangos, J.A., Ploegh, H., Peters, C., & Rudensky, A.Y. (1998). Cathepsin L: critical role in Ii degradation and CD4 T cell selection in the thymus. Science 280, 450-453.

Neer, E.J., Schmidt, C.J., Nambudripad, R., & Smith, T.F. (1994). The ancient regulatory-protein family of WD-repeat proteins. Nature 371, 297-300.

Nibbering, P.H., van der Heide, G.A., & van Furth, R. (1989). Immunocytochemical analysis of cellular responses to BCG. Clin. Exp. Immunol. 75, 147-154.

Nicchitta, C.V. (1998). Biochemical, cell biological and immunological issues surrounding the endoplasmic reticulum chaperone GRP94/gp96 [In Process Citation]. Curr. Opin. Immunol. 10, 103-109.

Pancholi, P., Mirza, A., Bhardwaj, N., & Steinman, R.M. (1993). Sequestration from immune CD4+ T cells of mycobacteria growing in human macrophages. Science 260, 984-986.

Peters, P.J., Neefjes, J.J., Oorschot, V., Ploegh, H.L., & Geuze, H.J. (1991). Segregation of MHC class II molecules from MHC class I molecules in the Golgi complex for transport to lysosomal compartments. Nature 349, 669-676.

Pfeifer, J.D., Wick, M.J., Roberts, R.L., Findlay, K., Normark, S.J., & Harding, C.V. (1993). Phagocytic processing of bacterial antigens for class I MHC presentation to T cells. Nature 361, 359-362.

Pierre, P., Turley, S.J., Gatti, E., Hull, M., Meltzer, J., Mirza, A., Inaba, K., Steinman, R.M., & Mellman, I. (1997). Developmental regulation of MHC class II transport in mouse dendritic cells. Nature 388, 787-792.

Pieters, J. (1997a). MHC class II compartments: specialized organelles of the endocytic pathway in antigen presenting cells. Biol. Chem. 378, 751-758.

Pieters, J. (1997b). MHC class II restricted antigen presentation. Curr. Opin. Immunol. 9, 89-96.

Pieters, J., Bakke, O., & Dobberstein, B. (1993). The MHC class II-associated invariant chain contains two endosomal sorting signals within its cytoplasmic tail. J. Cell Sci. 106, 831-846.

Pieters, J., Horstmann, H., Bakke, O., Griffiths, G., & Lipp, J. (1991). Intracellular transport and localization of major histocompatibility complex class II molecules and associated invariant chain. J. Cell Biol. 115, 1213-1223.

Pinet, V., Vergelli, M., Martin, R., Bakke, O., & Long, E.O. (1995). Antigen presentation mediated by recycling of surface HLA-DR molecules. Nature 375, 603-606.

Rammensee, H.G., Falk, K., & Rotzschke, O. (1993). MHC molecules as peptide receptors. Curr. Opin. Immunol. 5, 35-44.

Reith, W., Steimle, V., Durand, B., Kobr, M., & Mach, B. (1995). Regulation of MHC class II gene expression. Immunobiology 193, 248-253.

Riese, R., Mitchell, R., Viladangos, J., Karp, E., De Sanctis, G., Ploegh, H., & Chapman, H. (1998). Cathepsin S activity regulates antigen presentation and immunity. J. Clin. Invest. 101, 2351-2363.

Roche, P.A., & Cresswell, P. (1990). Invariant chain association with HLA-DR molecules inhibits immunogenic peptide binding. Nature 345, 615-618.

Rock, K.L., Gamble, S., & Rothstein, L. (1990). Presentation of exogenous antigen with class I major histocompatibility complex molecules. Science 249, 918-921.

Rotzschke, O., Falk, K., Deres, K., Schild, H., Norda, M., Metzger, J., Jung, G., & Rammensee, H.G. (1990). Isolation and analysis of naturally processed viral peptides as recognized by cytotoxic T cells. Nature 348, 252-254.

Sallusto, F., & Lanzavecchia, A. (1994). Efficient presentation of soluble antigen by cultured human dendritic cells is maintained by granulocyte/macrophage colony-stimulating factor plus interleukin 4 and downregulated by tumor necrosis factor alpha. J. Exp. Med. 179, 1109-1118.

Sallusto, F., Cella, M., Danieli, C., & Lanzavecchia, A. (1995). Dendritic cells use macropinocytosis and the mannose receptor to concentrate macromolecules in the major histocompatibility complex class II compartment: downregulation by cytokines and bacterial products. J. Exp. Med. 182, 389-400.

Santoro, L., Drouet, C., Reboul, A., Mach, J.P., & Colomb, M.G. (1994). Covalent binding of C3b to monoclonal antibodies selectively up-regulates heavy chain epitope recognition by T cells. Eur. J. Immunol. 24, 1620-1626.

Sherman, M.A., Weber, D.A., & Jensen, P.E. (1995). DM enhances peptide binding to class II MHC by release of invariant chain-derived peptide. Immunity 3, 197-205.

Sibley, L. (1995). Invasion of vertebrate cells by Toxoplasma gondii. Trends Cell Biol. 6, 129-132.

Silverstein, S.C. (1995). Phagocytosis of microbes: insights and prospects. Trends Cell Biol. 5, 141-142.

Sloan, V.S., Cameron, P., Porter, G., Gammon, M., Amaya, M., Mellins, E., & Zaller, D.M. (1995). Mediation by HLA-DM of dissociation of peptides from HLA-DR. Nature 375, 802-806.

Snow, G.A. (1970). Mycobactins: iron-chelating growth factors from mycobacteria. Bacteriol. Rev. 34, 99-125.

Stahl, P., Schlesinger, P.H., Sigardson, E., Rodman, J.S., & Lee, Y.C. (1980). Receptor-mediated pinocytosis of mannose glycoconjugates by macrophages: characterization and evidence for receptor recycling. Cell 19, 207-215.

Stahl, P.D. (1992). The mannose receptor and other macrophage lectins. Curr. Opin. Immunol. 4, 49-52.

Steinman, R.M. (1991). The dendritic cell system and its role in immunogenicity. Annu. Rev. Immunol. 9, 271-296.

Steinman, R.M., & Swanson, J. (1995). The endocytic activity of dendritic cells. J. Exp. Med. 182, 283-288.

Stern, L.J., Brown, J.H., Jardetzky, T.S., Gorga, J.C., Urban, R.G., Strominger, J.L., & Wiley, D.C. (1994). Crystal structure of the human class II MHC protein HLA-DR1 complexed with an influenza virus peptide. Nature 368, 215-221.

Stout, R.D. (1993). Macrophage activation by T cells: cognate and non-cognate signals. Curr. Opin. Immunol. 5, 398-403.

Sturgill-Koszycki, S., Schaible, U.E., & Russell, D.G. (1996). Mycobacterium-containing phagosomes are accessible to early endosomes and reflect a transitional state in normal phagosome biogenesis. EMBO J. 15, 6960-6968.

Sturgill-Koszycki, S., Schlesinger, P.H., Chakraborty, P., Haddix, P.L., Collins, H.L., Fok, A.K., Allen, R.D., Gluck, S.L., Heuser, J., & Russell, D.G. (1994). Lack of acidification in Mycobacterium phagosomes produced by exclusion of the vesicular proton-ATPase. Science 263, 678-681.

Svensson, M., Stockinger, B., & Wick, M.J. (1997). Bone marrow-derived dendritic cells can process bacteria for MHC-I and MHC-II presentation to T cells. J. Immunol. 158, 4229-4236.

Teyton, L., O'Sullivan, D., Dickson, P.W., Lotteau, V., Sette, A., Fink, P., & Peterson, P.A. (1990). Invariant chain distinguishes between the exogeneous and endogeneous antigen presentation pathways. Nature 348, 39-44.

Theriot, J.A. (1995). The cell biology of infection by intracellular bacterial pathogens. Ann. Rev. Cell Dev. Biol. 11, 213-239.

Tomlinson, S. (1993). Complement defense mechanism. Curr. Opin. Immunol. 5, 83-89.

Townsend, A., Ohlen, C., Bastin, J., Ljunggren, H.-G., Foster, L., & Karre, K. (1989). Association of class I major histocompatibility heavy and light chains induced by viral peptides. Nature 340, 443-448.

Tulp, A., Verwoerd, D., Dobberstein, B., Ploegh, H.L., & Pieters, J. (1994). Isolation and characterization of the intracellular MHC class II compartment. Nature 369, 120-126.

Unanue, E.R. (1984). Antigen-presenting function of the macrophage. Ann. Rev. Immunol. 2, 395-428.

Valitutti, S., Muller, S., Dessing, M., & Lanzavecchia, A. (1996). Different responses are elicited in cytotoxic T lymphocytes by different levels of T cell receptor occupancy. J. Exp. Med. 183, 1917-1921.

Vanbuskirk, A., Crump, B.L., Margoliash, E., & Pierce, S.K. (1989). A peptide binding protein having a role in antigen presentation is a member of the HSP70 heat shock family. J. Exp. Med. 170, 1799-1809.

Wardle, E.N. (1987). Kupffer cells and their function. Liver 7, 63-75.

Watts, C. (1997). Capture and processing of exogenous antigens for presentation on MHC molecules. Ann. Rev. Immunol. 15, 821-850.

West, M.A., Lucocq, J.M., & Watts, C. (1994). Antigen processing and class II MHC peptideloading compartments in human B-lymphoblastoid cells. Nature 369, 147-151.

Zinkernagel, R.M. (1974). Restriction by H-2 gene complex of transfer of cell-mediated immunity to Listeria monocytogenes. Nature 251, 230-233.

Zinkernagel, R.M., & Doherty, P.C. (1974a). Immunological surveillance against altered self components by sensitised T lymphocytes in lymphocytic choriomeningitis. Nature 251, 547-548.

Zinkernagel, R.M., & Doherty, P.C. (1974b). Restriction of in vitro T cell-mediated cytotoxicity in lymphocytic choriomeningitis within a syngeneic or semiallogeneic system. Nature 248, 701-702.

ANTIMICROBIAL MECHANISMS OF MACROPHAGES

Michael U. Shiloh and Carl F. Nathan

Advances in Cell and Molecular Biology of Membranes and Organelles
Volume 5, pages 407-439.

ISBN: 1-55938-999-0

I. INTRODUCTION/CRITERIA FOR ESTABLISHING HOST DEFENSE MECHANISMS

Metchnikoff won a Nobel Prize in 1908 for his discovery of phagocytosis by macrophages and his inference that this constituted a major mechanism of host defense. In the studies that followed, there was reason to anticipate that the antimicrobial mechanisms of macrophages would prove to be multiple: the diversity of genomes competing to replicate at each other's expense could be expected to select for species with several biochemically distinct mechanisms for eradicating pathogens. Moreover, the mechanisms used by macrophages have been found to be distributed among multiple types of cells, both phagocytes and nonphagocytes. Thus the macrophage has no monopoly on any antimicrobial pathway. Nonetheless, for infections by nonviral, obligate or facultative intracellular pathogens, the macrophage retains the centrality proposed by Metchnikoff. No other cell has the macrophage's combination of firepower, migratory capacity and longevity, nor occupies so pivotal a position at the interface between innate and adaptive immunity.

Ninety years after Metchnikoff's award, it is surprising how little of the macrophage's antimicrobial machinery has been defined in molecular detail. The discrepancy between the cell's importance for host survival on the one hand, and the brevity of our list of its molecular weapons on the other, may itself be a consequence of microbial diversity. Among multiple pathogens, some may evolve resistance to any given antimicrobial mechanism. The host counters with multiple antimicrobial systems. The inhibition of any one of them may not reveal a

Table 1. Criteria for Establishing the Role of a Gene in Host Defense

	ROI	*RNI*	*Iron*	*Acid*	*PCD*	*Peptide*	PLA_2
Correlation of gene expression with host resistance	Yes	Yes	Yes	Yes	Yes	ND	Yes
Exacerbation of infection with pharmacologic inhibitors	Yes	Yes	Yes	Not clear	ND	ND	ND
Antimicrobial actions of exogenous products	Yes	Yes	Yes	Yes	ND	Yes	Yes
Decreased host resistance upon host gene disruption	Yes Hu, Mu	Yes Mu only	Not clear	ND	ND	ND	ND
Microbial resistance genes	Yes	Yes	Yes	Yes	Yes	Yes	ND

Note: Abbreviations: Acid, phagosomal acidification; Iron, iron limitation; Hu, human; Mu, murine; ND, not determined; PCD, programmed cell death; Peptide, antimicrobial peptides; PLA_2, phospholipase A_2; ROI, reactive oxygen intermediates; RNI reactive nitrogen intermediates.

phenotype with the relatively small panel of microbes to which experimenters manage to resort.

If the redundancy of host defenses is a consequence of microbial diversity, it seems fitting that microbial genetics may provide an end-run around the experimental limitations imposed by that redundancy. Genetic selections applied to pathogens can reveal what genes they require (Hensel et al., 1995), or express (Mahan et al., 1995; Valdivia and Falkow, 1996), when infecting and resisting the host. Improved gene expression technologies are complementing such approaches. A particularly productive experiment is to simultaneously manipulate the genes of the macrophage that affect the expression of a putative antimicrobial pathway, and the genes of the pathogen that encode putative pathways of resistance (De Groote et al., 1997; Ehrt et al., 1997; Shiloh et al., 1999). Unfortunately, tools for facile genetic manipulation are not yet available for certain pathogens, much less for many hosts.

This chapter provides an analytical overview of what is now known about macrophage antimicrobial pathways, updating an article written 15 years ago with a similar goal (Nathan, 1983). Some of the individual pathways are covered from a different perspective in other chapters, while reviews elsewhere deal with the related topics of macrophage activation and deactivation (Nathan, 1991; Reiner, 1994; Vodovotz and Bogdan, 1994).

Five experimental criteria have helped establish that a macrophage antimicrobial mechanism is operative against a particular pathogen (Table 1):

1. Correlation of host gene expression or protein activity with resistance to infection. Expression of a gene during infection of the host or the macrophage, histochemical evidence for expression of the protein at the site of infection, or elaboration of enzyme products subsequent to interaction of the organism with the host meet the simplest criterion for participation of a given pathway in host defense. The criterion is not necessarily invalidated when disease progresses despite the presence of the putative protective mechanism; all infectious disease reflects the evasion or breakdown of mechanisms that otherwise serve to prevent it.
2. Exacerbation of infection with inhibitors. Pharmacologic inhibition of a putative pathway of host defense supports the role of the targeted mechanism when its inhibition worsens the course of infection. The power of this intervention is proportional to the specificity of the inhibitor and its lack of toxicity and mechanism-independent side effects. The inhibitors of inducible nitric oxide synthase (iNOS) most commonly used in experiments with infected mice or macrophages, aminoguanidine and N-methyl-L-arginine (NMA), can inhibit all three NOS isoforms; aminoguanidine inhibits other enzymes as well, and NMA may conceivably interfere with aspects of L-arginine metabolism unrelated to NO production. On the other hand, in the case of pleiotropic pathways like NO production, administration of an inhibitor at a chosen time during infection may more directly address the role of the pathway in the effector function of macrophages than does its lifelong genetic inactivation. Heritable inactivation of iNOS can disrupt such regulatory processes as natural killer (NK) cell function (Diefenbach et al., 1998) and T cell clonal expansion (Liew, 1995).
3. Antimicrobial actions of gene products or enzyme products furnished exogenously. Experiments of this type can address whether an antimicrobial product is sufficient to kill the organism in question and can explore whether there is synergy between distinct pathways. A shortcoming is the difficulty of knowing how to furnish the candidate product in physiologically meaningful concentrations, combinations and conditions.
4. Decreased host resistance upon disruption of the candidate gene. Increased host susceptibility to a pathogen after selective disruption of a gene encoding a postulated element of the host defense pathway provides strong evidence for the physiologic importance of that path. Although generally considered the most important kind of evidence, gene knockout has limitations above and beyond the time and cost involved. First, experiments are confined to a very limited range of non-human species, usually mice. Spontaneous deficiencies in genes directly encoding candidate antimicrobial pathways have only rarely been recognized in man. Major species differences in host range and host response therefore limit the list of human pathogens that can be studied in a phys-

iologically relevant setting. Second, the phenotype imparted by deficiency of a gene product at the level of the whole organism may or may not be explained by its deficiency at the level of the macrophage, even if the macrophage has a phenotype. Third, as noted above, lifelong deficiency of a given gene product may have secondary consequences, or lead to compensatory mechanisms, independent of a direct impact on macrophage effector mechanisms. Fourth, inbred mice harbor unidentified genes that can profoundly affect the course of infection, the effects of which are inadequately controlled in many studies of knockout mice in which wild-type and genetically altered individuals are compared. Matings based on tests for the altered allele may unknowingly fix allelic differences to which the target gene is linked.

A frequently cited route around some of these roadblocks could be provided by techniques for inducible inactivation of genes in targeted cell populations in adult mice, such as via the Cre-loxP recombination system (Rajewsky et al., 1996). The Cre recombinase would be placed under the control of a promoter whose activity is relatively restricted to macrophages. The paucity of promoters known to be so restricted, and the heterogeneity of promoter usage in different populations of macrophages, may help explain why no one has yet reported the application of the Cre-loxP method to the study of macrophage antimicrobial mechanisms.

5. Correlation between microbial resistance to killing by macrophages, and the expression of microbial resistance genes to defined host products. The notion that the host and its pathogens have co-evolved suggests that the identification of microbial resistance genes imparting resistance to specific host pathways, whose deletion leads to attenuated virulence, can be suggestive of the selective pressure placed upon the microbe by the host.

No one of these criteria alone provides sufficient evidence; one would prefer evidence of all five types. Below we summarize the extent to which these criteria have been met for leading candidates among putative antimicrobial mechanisms of the macrophage.

II. FORMATION OF REACTIVE OXYGEN INTERMEDIATES (ROI)

A. Coordinate Expression of ROI Production and Macrophage Bactericidal Activity

The high-output path of ROI production in phagocytes is the respiratory burst; this is the best-characterized of the cell-based antimicrobial weapons in the host arsenal. Over six decades, the "extra respiration of phagocytosis" (Baldridge and Gerard, 1933) has come to be attributed to the regulated assembly of a pentameric protein complex at the cell membrane which catalyzes the transfer of electrons from NADPH to molecular oxygen (De Leo and Quinn, 1996), generating superoxide ($\bullet O_2^-$). Superoxide can participate with appropriately chelated transition metals in a Haber-Weiss reaction to produce hydroxyl radical ($OH\bullet$), or dismutate to hydrogen peroxide (H_2O_2), from which additional ROI arise.

Study of the coordinate expression of ROI production with microbicidal activity of macrophages began with the observation that immunologic activation of macrophages in ways known to enhance their antimicrobial capacity was accompanied by an enhanced capacity for the respiratory burst (Nathan and Root, 1977). Head-to-head comparisons of macrophage antimicrobial activity and respiratory burst capacity began with macrophages obtained from mice immunized and boosted with heat killed *Trypanosoma cruzi*. These cells became trypanocidal at the same time that they acquired the ability to secrete large quantities of ROI upon stimulation with the parasite itself, or as a substitute, with the phorbol ester PMA (Nathan et al., 1979). Similar experiments with macrophages infected with *Toxoplasma gondii* extended the correlation of microbicidal activity and capacity to produce ROI (Murray et al., 1979). Many other examples have subsequently been compiled, and many facultative or obligate intracellular pathogens capable of eliciting immunologic activation of macrophages have been shown to trigger the respiratory burst of activated human or mouse macrophages (Nathan et al., 1980).

B. Antimicrobial Activity of Exogenous ROI

Corroborating the results with macrophage-pathogen coculture experiments, ROI treatment of toxoplasma with superoxide, peroxide or ROI generating systems in cell free conditions was toxoplasmacidal (Murray and Cohn, 1979). Exogenous ROI have been demonstrated to be cidal for many other organisms, including viruses, bacteria, fungi, and helminths (Klebanoff, 1998).

C. Inhibition of ROI Formation

Inhibition studies of the respiratory burst were initiated concurrently with the correlation studies described above. Early inhibition experiments were performed with superoxide dismutase (SOD) and catalase to reduce the intracellular concentrations of superoxide and hydrogen peroxide, respectively. ROI scavengers such as mannitol, benzoate, histidine or diazabicyclooctane were also used as inhibitors of hydroxyl radical or singlet oxygen. Since these inhibitors are not enzymes, their specificities are difficult to predict. In a study using *T. gondii*, Murray et al. (1979) showed that catalase, SOD, mannitol, and diazabicyclooctane were individually capable of converting microbistatic macrophages into permissive hosts and of diminishing the toxoplasmacidal activity of microbicidal macrophages. Studies with enzymatic inhibitors were limited by dependence on pinocytosis for their uptake. The requirement for pretreatment with high concentrations of enzymes was interpreted in this light, but may also have reflected a role for non-enzymatic scavenging, for example, reaction of NO with the heme in catalase (Chan et al., 1992). Conversely, efforts to inhibit endogenous catalase with azide may have led to the generation of reactive nitrogen intermediate (RNI) from the peroxidatic action of catalase upon azide in the presence of H_2O_2 (Klebanoff and Nathan, 1993).

More recently, the flavoprotein inhibitor diphenylene iodonium (DPI) has gained favor as an inhibitor of the respiratory burst in phagocytes. In human neutrophils, for example, DPI inhibited both the respiratory burst and microbicidal activity against *Staphylococcus aureus* and *Escherichia coli* (Hampton and Winterbourn, 1995). Unfortunately, DPI inhibits other flavoproteins as well. Compared to the concentrations necessary to inhibit the respiratory burst, much lower concentrations block NOS (Stuehr et al., 1991), while higher concentrations inhibit mitochondrial NADH:ubiquinone reductase (Holland et al., 1973).

D. Effect of Genetic Deficiency in Components of the Respiratory Burst

The critical importance of the generation of ROI by the respiratory burst is underscored by the phenotype of patients with chronic granulomatous disease (CGD), a genetic deficiency of any one of four subunits of the respiratory burst NADPH oxidase. The hallmarks of CGD are recurrent suppurative bacterial infections in the face of chronic inflammation with granuloma formation. The most common mutations in CGD, afflicting two-thirds of the patients, display X-linked recessive inheritance and affect the $gp91^{phox}$ subunit. Mouse knockouts in either $gp91^{phox}$ (Pollock et al., 1995) or $gp47^{phox}$ (Jackson et al., 1995) recapitulate the phenotype observed in humans. Consistent with this phenotype, mutant mouse macrophage cell lines which produced little to no ROI were unable to kill the intracellular protozoan parasite *Leishmania donovani* (Murray, 1981) or *Listeria monocytogenes* (Inoue et al., 1995). Although CGD patients suffer recurrent infections, the variety of pathogens is limited (Mouy et al., 1989). With antibiotic treatment, patients often survive to adulthood, and some patients are not identified until they are adults. Cells from patients in some kindreds retain a portion of respiratory burst activity. Study of such families suggests that as little as 5–10% of a normal respiratory burst capacity is sufficient to generate a normal phenotype (Ezekowitz et al., 1988). Nonetheless, even in patients whose respiratory burst activity is undetectable, monocytes are capable of killing some intracellular pathogens, albeit at reduced rates (reviewed in Nathan, 1983). These observations suggest that although absence of the respiratory burst is usually life-threatening in the absence of medical intervention, additional host defense mechanisms can partially compensate.

E. Microbial Resistance to ROI

Microbial resistance mechanisms to ROI can be classified as those that allow pathogens to avoid triggering the respiratory burst, those that allow pathogens to inhibit or diminish the burst, those that allow pathogens to catabolize the products of the burst, and those that repair molecular damage the burst causes.

T. gondii enters nonactivated macrophages without triggering the burst (Wilson et al., 1980). This evasive mechanism is countered and the trophozoites killed either by coating the protozoa with antibody or by activating the macrophages (Borges and Johnson, 1975; Jones et al., 1975).

Bordetella pertussis secretes an adenylyl cyclase that enters the cytosol of the phagocyte by an unknown route to produce a massive elevation in cAMP. The latter forestalls the respiratory burst by unknown means (Confer and Eaton, 1982), although recent work has demonstrated that the *B. pertussis* adenylate cyclase is necessary for induction of apoptosis in murine alveolar macrophages (Khelef et al., 1993). *Francisella tularensis* releases an acid phosphatase that blocks the respiratory burst (Reilly et al., 1996). Although the mechanism is unknown, one hypothesis is that this enzyme may reverse the phosphorylation of NADPH oxidase components necessary for their mutual assembly (Heyworth et al., 1991). Pyocyanin, a pigment produced by *Pseudomonas aeruginosa*, inhibits the respiratory burst by depleting intracellular NADPH, the reducing cosubstrate for O_2 (Muller et al., 1989). *Yersinia spp.* export a tyrosine phosphatase (YopH) via the type III secretion system which inhibits the respiratory burst (Bliska and Black, 1995). This inhibition is mediated by selective dephosphorylation of $p130^{Cas}$, and to a lesser extent, FAK (Black and Bliska, 1997), two components of focal adhesions. In neutrophils, tyrosine phosphorylation of focal adhesion proteins coincides with activation of the respiratory burst, and pharmacological inhibition of tyrosine kinases simultaneously inhibits phosphorylation of focal adhesion proteins and the respiratory burst (Fuortes et al., 1993).

Microbial genes involved in catabolism of ROI (Farr and Kogoma, 1991) include those controlled by the oxyR and soxRS regulons in *E. coli* and *Salmonella typhimurium*. Among these are catalase, glutathione reductase and SOD. Using mice deficient in $gp91^{phox}$, De Groote et al. (1997) confirmed the complementarity of host-pathogen defense mechanisms for ROI by demonstrating that *S. typhimurium* deficient in periplasmic CuZn-SOD, whose virulence was attenuated in wild-type mice, displayed a virulent phenotype in gp91^{phox}-deficient mice. However, the distribution of oxyR, soxRS and related bacterial ROI-resistance genes is by no means universal. For example, glutathione and oxyR are lacking in *Mycobacterium tuberculosis* (Deretic et al., 1995; Newton et al., 1996); SOD is widespread, but soxRS is not. Recently, the first viral antioxidant was identified as a human poxvirus selenoprotein encoding a putative glutathione peroxidase; this gene protected transfected cells from death induced by UV irradiation or exposure to hydrogen peroxide (Shisler et al., 1988). DNA repair systems, such as recBC, help to correct some of the most important molecular damage inflicted by ROI (Buchmeier et al., 1995).

III. FORMATION OF REACTIVE NITROGEN SPECIES BY NITRIC OXIDE SYNTHASE

A. Correlation of Reactive Nitrogen Intermediate Production with Macrophage Bactericidal Activity

The role of RNIs in host defense has been studied extensively since the observation that macrophages produce nitrite and nitrate in response to lilppolysaccharide (LPS) (Stuehr and Marletta, 1985) and that a molecule with the properties of NO is an intermediate (Miwa et al., 1987). The enzymatic source of NO in activated macrophages was purified, cloned, and identified as a homodimeric protein containing FAD, FMN, tetrahydrobiopterin, and tightly bound calmodulin (Stuehr et al., 1991; Hevel et al., 1991; Cho et al., 1992; Xie et al., 1992). Later, heme was identified as another critical cofactor (Pufahl and Marletta, 1993). This complex enzyme uses electrons from NADPH to catalyze the deimination of arginine, producing nitric oxide and citrulline (Stuehr, 1997). The enzyme was named iNOS, for (NO) synthase whose activity is independent of elevations in intracellular Ca^{2+} and whose expression is immunologically inducible (Xie et al., 1992). An alternate designation for iNOS is NOS2, reflecting its place in the sequence of NOSs cloned (neuronal NOS: NOS1; endothelial NOS: NOS3).

Mice

The initial evidence that RNI generated by iNOS contribute to macrophage antimicrobial activity came from two sources: the temporal correlation of NOS activity and NO production with macrophage microbicidal/static activity; and the enhancement of microbial growth in macrophage co-cultures attendant on withdrawing arginine or adding reversible NOS inhibitors in the form of arginine analogues (reviewed in Nathan and Hibbs, 1991). Evidence both *in vitro* and *in vivo* has continued to come primarily from murine systems and now extends to numerous pathogens, including viruses, bacteria, fungi, protozoa and helminths

(Bogdan, 1997). For some organisms such as *L. monocytogenes*, evidence both for and against a role for RNI has emerged from *in vitro* and *in vivo* systems. These discrepancies may reflect differences in the timing of experimental endpoints, which affect the relative importance of macrophages versus T cells. Alternatively, they may be mediated by the incomplete action or imperfect specificity of NOS inhibitors. Such concerns were a major motivation for the derivation of mice deficient in iNOS (see below).

Humans

A major barrier to the acceptance of RNI as microbicidal molecules has been controversy regarding the ability of human macrophages to express iNOS, or to do so at functional levels. There have been two concerns. First, many investigators have been unable to induce iNOS reproducibly in human monocytes from healthy donors with stimuli which readily induce iNOS in mouse macrophages. However, the mouse cells that constitute the point of reference have been mature macrophages collected from tissue sites, usually the peritoneal cavity, often after injection of inflammatory stimuli, or transformed macrophages resembling tissue macrophages. There have been no reports of induction of iNOS in monocytes from the blood of mice or in macrophages derived from mouse blood monocytes by *in vitro* culture. Thus, what some have interpreted as a species difference between mice and people may instead reflect a difference in the state of differentiation of the cell populations being compared.

In fact, there are numerous reports of iNOS expression in human mononuclear phagocytes. The most robust expression has been found either in tissue macrophages or in monocyte-derived macrophages collected from patients with inflammatory or infectious diseases. These conditions include alcoholic and viral hepatitis, cirrhosis, acute respiratory distress syndrome, bronchiectasis, bronchopneumonia, tuberculosis, obliterative bronchiolitis in transplanted lungs, atherosclerosis, giant cell arteritis, myocardial infarction, rheumatoid and osteoarthritis, joint prosthesis-associated inflammation, diverse cancers, malaria, IgA nephropathy, proliferative glomerulonephritis, Crohn's disease, ulcerative colitis, diverticulitis, *Helicobacter pylori* gastritis, celiac disease, and multiple sclerosis (MacMicking et al., 1997a; Weinberg, 1998). To date, such evidence has been presented in over 50 reports in peer-reviewed journals; these reports include multiple independent confirmations. Whether or not the mononuclear phagocytes are iNOS positive at the time of collection, their expression of iNOS can be induced or enhanced by cytokines, microbes or microbial products more readily than is the case for cells from the blood of healthy subjects.

One of the most striking examples involved alveolar macrophages from patients with idiopathic pulmonary fibrosis. These cells were iNOS-negative, but became strongly iNOS-positive when infected with mycobacteria *in vitro*. Mycobacterial infection did not induce iNOS in alveolar macrophages from presumably normal lobes of subjects who were bronchoscoped because of localized pulmonary disorders (Nozaki et al., 1997). Thus, two kinds of stimuli were required: unknown factors associated with idiopathic pulmonary fibrosis, and unknown factors associated with ingestion of mycobacteria. Simply adding interferon and LPS *in vitro* to monocytes from normal donors does not seem to mimic such a combination of signals.

As one would expect, in studies where iNOS expression has been nil or minimal, NOS inhibitors have had no impact on the antimicrobial activity of the human macrophages examined (usually, this activity has been modest). In contrast, in some circumstances where substantial iNOS expression has been documented, pharmacological inhibition of iNOS has led to markedly diminished killing of the pathogens by the human macrophages *in vitro*. Such results have been reported for leishmania (Vouldoukis et al., 1995) and mycobacteria (Nozaki et al., 1997), two of the pathogens whose control in mice appears to be most strongly dependent upon iNOS (MacMicking et al., 1997b; Diefenbach et al., 1998).

A conservative interpretation of these observations is that we have little ability to culture human blood monocytes from uninfected, uninflamed donors in a way that induces them to differentiate and respond as they do *in vivo* after migrating into infected or inflamed tissues. We may be using inappropriate extracellular matrices, cytokines, autacoids, or culture conditions. The deficiency lies in our understanding, not in the cells' capacity for gene expression. There is no question that with appropriate stimuli, human macrophages can express iNOS. When they do, iNOS is in some cases a necessary mechanism for the inhibition of certain pathogens *in vitro*. As in the mouse (Nathan, 1997), it should be anticipated that there will be other pathogens for whose control iNOS will prove to be unimportant or even detrimental in the same sorts of experiments.

Another basis for skepticism about the potential antimicrobial role of iNOS in humans has been its detection in pulmonary alveolar macrophages from patients with active tuberculosis (Nicholson et al., 1996). Some have argued that this proves the irrelevance of iNOS to the control of human tuberculosis, notwithstanding overwhelming evidence to the contrary in mice (MacMicking et al., 1997b; McKinney, personal communication). It is surprising to encounter this argument, because it is based on a misunderstanding of the concept of disease. The same argument would reject the utility of every element of the immune system that is expressed in any infectious disease. By definition, infectious disease occurs when the immune system *fails*. In contrast, in about 90% of immunocompetent people infected with *M. tuberculosis*, the immune system works, and tuberculosis does not occur. Assuming for the sake of argument that iNOS may participate in control of tuberculosis in most infected individuals, why might it fail to exert the usual level of control in a proportion of those infected? Possibilities include relative insufficiency of iNOS, resistance of the mycobacteria to RNI, or loss of another host factor that is necessary but not sufficient for control of mycobacterial replication.

Whether iNOS actually does play a role in the control of human tuberculosis or any other infectious disease is not established by finding the enzyme at the site of infection. The role of iNOS in the control of infectious disease within human macrophages *in vivo* is likely to remain an open question until patients are identified with primary deficiencies in iNOS.

B. Antimicrobial Activity of Exogenous RNI

Addition of exogenous RNI in the form of NO or NO-generating compounds has exerted static or cidal effects against a diverse array of pathogens, including viruses, bacteria, protozoa, helminths and fungi (Bogdan, 1997; MacMicking et al., 1997a and references therein). This evidence is critical because active iNOS not only produces NO, but also consumes arginine and NADPH, oxidizes tetrahydrobiopterin, and generates citrulline; in theory any of these effects could contribute to iNOS-dependent antimicrobial activity. A shortcoming of these experiments is that lack of efficacy of RNI tested alone could lead to underestimation of their contribution, if other factors, such as ROI, interact with RNI synergistically. A striking example is the colicidal synergy between NO and H_2O_2 (Pacelli et al., 1995). The burden of evidence with exogenous RNI suggests that the antimicrobial actions of RNI can plausibly account for the observed correlation between iNOS expression and macrophage microbicidal activity. However, this does not exclude important indirect roles of RNI *in vivo*, such as in the activation of NK cells (Diefenbach et al., 1998).

C. Genetic Deficiency of iNOS

Primary genetic deficiency of iNOS in human subjects has not been described. However, in the mouse, the significance of iNOS has been established using mice homozygous for a disrupted iNOS gene. The impact of iNOS deficiency on the course of infection varies from detrimental to advantageous, depending on the organism studied (reviewed in Nathan, 1997). iNOS deficient mice display markedly increased susceptibility to *M. tuberculosis*, *Leishmania major*, *L. monocytogenes*, *T. gondii*, and ectromelia virus. iNOS-deficient mice infected with *M. tuberculosis* demonstrated susceptibility that was greater than or equal to the most susceptible knockout mice studied to date (MacMicking et al., 1997b). *L. major* infected iNOS-deficient mice demonstrated enhanced susceptibility as early as 24 hours post-infection, when only a few of the leishmania-containing cells were iNOS-positive. At this early time in the infection, RNI appeared to be serving as signals regulating the function of NK cells and macrophages, rather than as toxins for the protozoa. Specifically, iNOS was essential for early production of interleukin (IL)-12 and interferonγ (IFNγ), maturation of NK cells, and suppression of transforming growth factor (TGF)-β production (Diefenbach et al., 1998).

D. Microbial Resistance to RNI

To counter the toxicity of RNI, various microbial pathogens have evolved RNI-resistance mechanisms. Although these mechanisms are not as well-characterized as those operative against ROI, they fall into the same four categories described above for ROI. Indeed, as noted below, many of the same genes implicated in ROI resistance mediate microbial resistance to RNI as well. However, just as some ROI resistance genes, like catalase, do not confer resistance to RNI (Chan et al., 1992), it can be anticipated that some genes will confer resistance specifically to RNI and not ROI. A precedent is the flavohemoglobin of *S. typhimurium* (Crawford and Goldberg, 1998). How the latter protects bacteria from RNI is unknown; one hypothesis is that it serves as an NO sensor, signaling the induction of other genes whose products confer protection (Crawford and Goldberg, 1998).

Similarly, in *E. coli* and *S. typhimurium*, the transcription factor oxyR senses hydrogen peroxide by undergoing intramolecular disulfide bonding. This imparts an ability to act as a transcription factor that induces the expression of several ROI resistance genes, such as those encoding catalase, glutathione reductase and alkyl hydroperoxide reductase. When glutathione is deficient, oxyR can be activated by S-nitrosylation, leading to enhanced resistance to the bactericidal action of S-nitrosothiols in *E. coli* (Hausladen et al., 1996; Zheng and Storz, 1998). Likewise, soxR, a sensor of superoxide accumulation, is activated by NO (Hidalgo et al., 1997) to induce soxS, which in turn induces genes that confer resistance to NO. Deletion of soxR in *E. coli* enhanced the killing of *E. coli* by NO gas and by macrophages in a NOS-dependent manner (Nunoshiba et al., 1995). However, the virulence of *S. typhimurium* deficient in either oxyR or soxRS was not attenuated; *S. typhimurium* may use alternative means to regulate its RNI resistance (Miller et al., 1989; Fang et al., 1997).

One such mechanism is the expression of low molecular weight thiols, such as glutathione and homocysteine. Salmonella deficient in *metL*, encoding an enzyme in the homocysteine synthesis pathway, were hypersusceptible to RNI and attenuated for virulence in mice (De Groote et al., 1996). Not all important intracellular pathogens contain glutathione, however, and there is little information on microbial levels of homocysteine or cysteine.

The most widely distributed RNI-resistance gene identified to date, conserved in organisms as diverse as plants, actinomycetes, purple bacteria, yeast, protozoa, nematodes, and mammals, was discovered in enteric bacteria as the gene encoding the small subunit of alkyl hydroperoxide reductase, termed AhpC. The holoenzyme, a heterotetramer comprised of two molecules of AhpC and two molecules of a distinct gene product, AhpF, catalyzes the reduction of organic hydroperoxides to alcohols. On its own, AhpC, a 22 kDa protein, was thought to have no function. Rhee and colleagues discovered that an AhpC homologue in yeast could protect a test enzyme (glutamine synthetase) against metal-catalyzed oxidation in the presence of an exogenous thiol, dithiothreitol (Chae et al., 1994b). Since *ahpC* homologues are ubiquitously distributed (Montemartini et al., 1998) while *ahpF* homologues appear to be much more restricted, we speculated that *ahpC* homologues may serve a distinct function, the need for which is also ubiquitous, such as resistance to RNI. Indeed, salmonella deficient in *ahpC* proved to be profoundly hypersusceptible to RNI, a defect from which they could be rescued by expression of *ahpC* cloned either from salmonella or from *M. tuberculosis*. When *M. tuberculosis ahpC* was stably expressed in human cells, it protected them from the apoptotic death caused by transient transfection with iNOS (Chen et al., 1998). Thus, *ahpC* may be the component of the oxyR regulon responsible for oxyR-regulated resistance to RNI, but since the expression of *ahpC* homologues is much more widespread than that of oxyR, oxyR-independent means must exist for regulating their expression.

In recognition that the distribution of *ahpC* homologues exceeds that of known alkyl hydroperoxidase activity, it has recently been proposed that *ahpC* homologues be designated peroxiredoxins (Chae et al., 1994b). Our working hypothesis is that peroxiredoxins may be able to partner with a variety of other proteins that serve to reduce them after their oxidation by either ROI or RNI. For example, in the experiments cited above with a peroxiredoxin from yeast, dithiothreitol presumably served to take the place of the natural reducing partner, later proposed to be thioredoxin (Chae et al., 1994a). We are currently searching for the hypothesized reducing partner for AhpC when it acts to protect *S. typhimurium* from RNI. Identification of such a partner will permit elucidation of the mechanism by which AhpC protects organisms against RNI.

M. tuberculosis may express several additional RNI-resistance mechanisms besides *ahpC*. This has been suggested by experiments in which candidate RNI resistance genes were cloned by gain-of-function selection strategies in genetically tractable, heterologous bacterial hosts exposed to levels of RNI sufficient to kill all the nontransformed hosts and almost all the transformed hosts. The first example of such a candidate identified in *E. coli* transformed with an *M. tuberculosis* genomic library was *noxR1*, which encodes a basic, 15.5 kDa protein of unknown function. Expression of *noxR1* conferred upon *E. coli*, *S. typhimurium* and *Mycobacterium smegmatis* an enhanced ability to resist RNI and ROI, whether the bacteria were exposed to exogenous compounds in medium or to endogenous products in macrophages (Ehrt et al., 1997). Additional candidate NOXR genes have been cloned from *M. tuberculosis* and are being characterized (Ruan et al., 1999).

Pyocyanin, an inhibitor of the respiratory burst, is nitrosylated in the presence of NO, thereby decreasing the amount of RNI available to interact with other molecules important for microbial survival (Warren et al., 1990). Another microbial RNI resistance gene encodes Cu,Zn-SOD. Located in the periplasm of *S. typhimurium* this isoform of SOD prevents the formation of peroxynitrite by decomposing one of its precursors, superoxide (De Groote et al., 1997). Prevention of formation of more potent RNI from less potent RNI illustrates one mechanism of action for an RNI resistance gene product. However, SOD differs from many of the above candidates in that it is not thought to act directly upon RNI.

Microbial targets of RNI (including RNI that arise upon interaction with products of the respiratory burst) include DNA, which undergoes deamination and strand breaks; lipids, which are peroxidized; heme iron and iron-sulfur clusters, whose nitrosylation (in)activates hemoproteins and iron-sulfur cluster enzymes; and cysteine, methionine and tyrosine residues, whose nitrosylation or nitration can occasionally enhance but usually diminishes the activity of the protein in which these amino acids are modified (reviewed in Fang, 1997). Some RNI resistance mechanisms consist in repairing such injuries. A striking example is the recBC system of DNA repair in *S. typhimurium* (Buchmeier et al., 1995; Shiloh et al., 1999).

The multiplicity of RNI resistance mechanisms raises interesting questions about specificity of RNI in terms of subcellular location and preferential attack on certain molecular targets. Using bacterial genes as a mirror, we seem to be learning that host-derived RNI may constitute a multiple array of individually distinct threats to microbial survival.

E. Effect of Combined gp91^{phox}, iNOS Deficiency in Mice

Although production of ROI and RNI are powerful effector systems, recent experiments have raised the possibility that other mechanisms may exist that are at least as important. Thus, mice rendered genetically deficient in certain cytokines (tumor necrosis factor (TNF), IL-6) or transcription factors (NF-IL-6, ICSBP, IRF2) are highly susceptible to infection with listeria, yet their macrophages *in vitro* are either normal in RNI and ROI production or only slightly deficient (Kopf et al., 1994; Tanaka et al., 1995; Fehr et al., 1997). In addition, NF-IL-6 deficient macrophages killed bacteria poorly *in vitro* (Tanaka et al., 1995), suggesting that the observed defects *in vivo* were not due solely to factors extrinsic to the macrophage's microbicidal machinery, such as decreased ability of macrophages to congregate at infectious foci, impairment in execution of the complement cascade, or incomplete differentiation of neutrophils.

To assess directly the contribution of additional mechanism(s), we generated mice doubly deficient in gp91^{phox} and iNOS (Shiloh et al., 1999). Double knockout mice are markedly susceptible to spontaneous infection with several organisms, including *E. coli*, and succumb readily upon *in vivo* challenge with both wild-type and attenuated strains of *S. typhimurium*, a facultative intracellular pathogen. In addition, some double knockout mice infected with a virulent strain of *L. monocytogenes* had spleen colony-forming units (CFU) values that were greater than or equal to 3 log higher than control mice. However, others reduced the number of listeria as effectively as wild-type mice. Moreover, double knockout mice infected with an attenuated listeria strain completely eradicated inocula even of 10^8 per mouse. Explanted sodium periodate elicited peritoneal macrophages from double knockout mice were inefficient at killing salmonella, virulent listeria and *E. coli*, but their killing of attenuated listeria and *E. coli* was substantial. With respect to the organisms studied, these studies suggest that ROI and RNI are usually the predominant antimicrobial products of mice *in vivo* and their macrophages *in vitro*. However, other mechanisms exist that are moderately effective against less virulent organisms and occasionally quite effective even against a virulent strain of listeria.

IV. IRON LIMITATION

A. Correlation of Decreased Iron with Improved Host Defense and Reversal with Inhibitors

Most microbes, like eukaryotes, require iron for essential functions, including deoxyribonucleotide synthesis via ribonucleotide reductase, reduction of oxygen for ATP synthesis, and incorporation into heme. Microbes synthesize siderophores to chelate and reduce extracellular iron, which is primarily found in its oxidized, ferric form (Fe III) (Neilands, 1995). The host contests microbes for access to iron. Host iron content is tightly regulated, especially during bacterial infection (Weinberg, 1984). For example, activated human monocyte-derived macrophages inhibit intracellular multiplication of *Legionella pneumophila* by limiting the availability of iron; bactericidal activity is reversed by iron donors (Byrd and Horwitz, 1989). Treatment of *S. typhimurium*- or *L. monocytogenes*-infected mice with desferoxamine, an iron chelator, enhances mouse survival (Jones et al., 1977) and increases the LD_{50} (Sword, 1966).

However, control of intracellular bacterial growth requires a delicate balance between limiting the iron available for bacterial replication and maintaining sufficient intracellular iron for host metabolic and antimicrobial pathways. Both NADPH oxidase and iNOS depend on heme for electron transport. Exposure of mammalian cells to RNI can cause them to lose iron (Hibbs et al., 1988), and mouse macrophages expressing iNOS are restricted in iNOS activity when heme becomes limiting (Albakri and Stuehr, 1996). These considerations may underlie the observation that macrophages containing either too little or too much iron are not listericidal (Alford et al., 1990).

B. Mechanisms of Iron Restriction

One mechanism of intracellular iron restriction is the downregulation of transferrin receptors (Byrd and Horwitz, 1989) and destruction of ferritin (Byrd and Horwitz, 1993) by IFNγ-activated macrophages. However, recent evidence suggests an alternative mechanism that may serve to limit iron in one critical intracellular compartment—the phagosome—while permitting the level of iron in the cytosol to remain adequate to support the macrophage's iron-dependent antimicrobial enzymes.

The *Ity/Lsh/Bcg* mouse locus has long been known to control the extent of resistance to *S. typhimurium* (*Ity*), *L. donovani* (*Lsh*) and *Mycobacterium bovis* (*Bcg*) both for the whole mouse and for isolated macrophages (Lissner et al., 1985). The *Bcg* gene occurs in two allelic forms in inbred mouse strains, Bcg^r (resistant, dominant) and Bcg^s (susceptible, recessive), and displays autosomal dominant inheritance. In 1993 the *Bcg* gene was cloned and its product termed natural resistance-associated macrophage protein 1 (Nramp1) (Vidal et al., 1993). A single nucleotide variation leading to a point mutation from glycine to aspartic acid at amino acid 169 segregates with the Bcg^s phenotype (Vidal et al., 1993; Malo et al., 1994). Deletion of the Nramp1 allele (Vidal et al., 1995b) from Bcg^r mice confirmed that the Nramp1$^{Gly \rightarrow Asp169}$ allele is a null allele and that the *Bcg*, *Lsh* and *Sty* genes are the same locus. At the protein level, Nramp1 is absent from Nramp1$^{Gly \rightarrow Asp169}$ strains (Vidal et al., 1996). In wild-type strains, Nramp1 is a membrane protein predominantly expressed in phagocytes, where it is recruited to the phagosomal membrane (Gruenheid et al., 1997). Nramp1 expression is increased following exposure to LPS and IFNγ with kinetics similar to those of iNOS (Govoni et al., 1997). Following the identification of Nramp1, a second ubiquitously expressed Nramp1 homologue was identified in mice (Gruenheid et al., 1995), rats (Gunshin et al., 1997), and humans (Vidal et al., 1995a) and named Nramp2.

In 1996, an Nramp1 homologue from *Saccharomyces cerevisiae* was identified in a screen for yeast genes that complement a mutant with a defect in metal homeostasis (Supek et al., 1996). Deletion of the gene, SMF1, rendered the yeast sensitive to low manganese concentrations, while its overexpression dramatically increased manganese uptake. The homology to Nramp led the authors to suggest that Nramp may function to limit phagosomal manganese during infection (Supek et al., 1996, 1997). Recent work on a rat Nramp2 homologue with 92% homology to human Nramp2 supports the notion that Nramp1 may function to transport cations, and directs attention to iron as a possible cation of central importance. When expressed in frog oocytes, rat Nramp2 served as a proton-coupled, membrane potential-dependent, active transporter of several divalent cations, including iron and manganese (Gunshin et al., 1997). Fe^{2+} uptake was also stimulated in oocytes expressing mouse Nramp1, albeit with lower apparent affinity for Fe^{2+} than Nramp2 (Gunshin et al., 1997). Shortly thereafter, identification of a mutation leading to iron-deficiency anemia in mice and rats indicated that Nramp2 controls the uptake of iron by the intestine (Fleming et al., 1997, 1998). Nramp2 also appears essential for normal transport of iron from endosomes into the cytosol, and thus is presumed to be localized in the endosomal membrane, although direct evidence for this has not yet been provided and the cells to which it may pertain have not been identified (Fleming et al., 1998). A G185R mutation in mouse or rat Nramp2 disrupts both of the putative iron transport functions described above

(Fleming et al., 1997, 1998). The foregoing findings raise the possibility that Nramp1 may be a phagocyte-specific homologue of Nramp2 having some of the same functions. If so, then Nramp1 may simultaneously provide intracellular iron for host defense (e.g., synthesis of heme for gp91^{phox} and iNOS) while depriving microbes in the phagosome of iron and manganese.

In the mouse, while Nramp1 clearly affects the outcome of infection with bacillus Calmette Guerin (BCG) (Vidal et al., 1995b), it seems to have little bearing on infection with *M. tuberculosis* (Medina et al., 1996). The situation in humans is not yet clear. Studies in Brazilian populations found no association between susceptibility to tuberculosis and Nramp1 (Blackwell et al., 1997). On the other hand, four human Nramp1 polymorphisms in West African populations were associated with an increased risk for tuberculosis (Bellamy et al., 1998).

V. ACIDIFICATION OF THE PHAGOSOME

A. Correlation of Phagosome Acidification with Host Defense

Metchnikoff's suggestion that acidification of the phagosome is an antimicrobial mechanism (Metchnikoff, 1893; Tauber and Chernyak, 1989) gained support from two observations: the confirmation by Rous (1925) that phagocytes can acidify their phagosomal compartment and the recognition that the stomachs of normal hosts are sterile while stomachs of achlorhydric hosts are permissive for bacterial survival (reviewed in Giannella et al., 1973). Recent work suggests that acidity alone does not fully explain the sterilizing power of gastric juice; rather, it appears to be the combination of acid with RNI (Dykhuizen et al., 1996; McKnight et al., 1997). Rather than diminish the possible antimicrobial role of acid in the phagosome, this latter insight instead suggests that acid may serve also as a cofactor for RNI in the phagosome of the activated macrophage.

Further endorsement for the role of acidification in host defense awaited microscopic techniques that permitted the accurate determination of pH (Ohkuma and Poole, 1978) and were applied at the same time as the visualization of ingested pathogens. Phagocytosis of several microbes including *E. coli* (Horwitz and Maxfield, 1984), *T. gondii* (Sibley et al., 1985), and *Candida albicans* (Watanabe et al., 1991) led to phagosomal acidification that coincided with killing of the intracellular organism.

B. Microbial Responses to Acidification

Acidic conditions impose a stress on bacterial viability, and bacteria have evolved complex homeostatic mechanisms to regulate their cytoplasmic pH. Here we will not discuss the microbe's adaptation to acid, which in *E. coli* involves the induction of amino acid decarboxylases and undefined rpoS-dependent processes.

Instead, we will focus on the pathogens' efforts to prevent acidification of the phagosome by the macrophage.

The initial observation that *Legionella pneumophila* inhibits acidification of its phagosome (Horwitz and Maxfield, 1984) was followed by similar observations for a diverse group of organisms, including *T. gondii* (Sibley et al., 1985), *Nocardia asteroides* (Black et al., 1986), and *Mycobacterium avium-intracellulare* (MAI) (Sturgill-Koszycki et al., 1994). The ability of *T. gondii* to inhibit acification is lost upon its interaction with antibody; opsonized *T. gondii* is found in acidified phagosomes and loses viability, while unopsonized *T. gondii* resists macrophage killing and resides in nonacifidied phagosomes (Sibley et al., 1985). For MAI, inhibition of acidification is countered by macrophage activation. Thus, in nonactivated macrophages infected with MAI, which fail to restrict MAI growth (Schaible et al., 1998), vesicular proton-ATPase is relatively excluded from the maturing phagosome during its fusion with endosomes/lysosomes (Sturgill-Koszycki et al., 1994). In contrast, in IFNγ/LPS activated macrophages, the MAI-containing phagosome acquires vesicular proton ATPases and acidifies, and the viability of MAI falls (Schaible et al., 1998). In an analogous manner, *S. typhimurium* delays the acidification of its phagosome until it can express specific acid tolerance genes, at which point it disregards the successful phagosomal acidification to pH of less than 5.0 (Alpuche Aranda et al., 1992). *L. monocytogenes* avoids residing in an acidified compartment by secreting the pore-forming protein listeriolysin, which is catalytically inactive at neutral pH and maximally active at pH 5.5 (Geoffroy et al., 1987) and thus specifically lyses the acidified phagolysosome (Beauregard et al., 1997).

C. Pharmacologic Inhibition of Acidification

Despite the implications of the correlative experiments described above, attempts to block phagosomal acidification with pharmacological agents have not supported the hypothesis that phagosomal acidification is directly microbicidal. Treatment of macrophages with the lysosomotropic agents chloroquine or ammonium chloride inhibited rather than promoted the intracellular replication of *L. pneumophila* (Byrd and Horwitz, 1991) and *F. tularensis* (Fortier et al., 1995), apparently by limiting the availability of iron; acidic pH is critical for the release of ferric iron from endocytosed transferrin (Dautry-Varsat et al., 1983). Similarly, interference with acidification by treatment with the vesicular ATPase-specific inhibitor bafilomycin A led to diminished intracellular growth of *S. typhimurium*, most likely by forestalling its acid-induced acid tolerance response (Rathman et al., 1996). The implication is that when induced by acid, the acid tolerance response confers protection against stresses potentially more microbicidal than acid.

Thus, it seems likely that the degree of acidification of the phagosome attained by activated macrophages is not sufficient in and of itself to constitute a major mechanism of killing for most pathogens, but instead is likely to serve important ancillary functions. In addition to those noted above, acidification of the phagosome may serve to enhance the reactivity of nitrite (Nathan, 1995) and peroxynitrite (Beckman and Koppenol, 1996). Acidification of the phagosome may also reflect a mechanism to dispose of cytoplasmic acid, since excess cytoplasmic acidity can reduce superoxide production by macrophages (Swallow et al., 1990).

VI. PROGRAMMED CELL DEATH (APOPTOSIS)

A. Correlation of Macrophage Apoptosis with Microbial Killing

Recent evidence has demonstrated apoptosis of macrophages following their infection by certain bacterial pathogens. The initial observation of apoptosis of macrophages infected by *Shigella flexneri* (Zychlinsky et al., 1992) has been extended to include strains of salmonella, yersinia, bordetella, legionella, and mycobacteria (reviewed in Zychlinsky and Sansonetti, 1997). Although most instances of infection-induced apoptosis have been limited to *in vitro* experiments, apoptosis has been observed in macrophages from the broncheoalveolar spaces (Klingler et al., 1997) and caseating granulomas (Keane et al., 1997) of patients with active pulmonary tuberculosis and in rectal biopsies from patients with shigellosis (Islam et al., 1997). It is unclear whether apoptosis of infected macrophages is a bacterial virulence mechanism because it kills cells that might otherwise kill the pathogen or, in contrast, is a host defense mechanism because it deprives the pathogen of an intracellular environment in which to replicate while shielded from antibody and complement (Zychlinsky, 1993). Both may be correct, depending on the state of activation of the macrophages.

Treatment of BCG-infected macrophages with ATP induced apoptosis and led to diminished viability of intracellular BCG (Molloy et al., 1994). This phenomenon was subsequently attributed to the interaction of ATP with the purinergic P2Z receptor (Lammas et al., 1997). Lammas et al. (1997) suggested that ATP-mediated apoptosis of IFNγ-primed, BCG-infected macrophages may represent a novel antimycobacterial mechanism that acts independently of other apoptotic pathways. Although it has been argued that lysed cells may release the ATP needed to mediate such a reaction, the dilution of intracellular ATP upon cell lysis is potentially enormous; the action of extracellular phosphatases may further limit the physiologic relevance of the phenomenon.

Also unclear is the degree to which the lysis of bacterially-infected macrophages by T cells may represent a form of host defense in and of itself, and to what extent the process may be incidental to the recognition of antigen and production of cytokines by the T cells. Such cytokines may recruit and activate additional macrophages to deal with the pathogens that are released. The issue has been studied chiefly with mycobacteria. Some anti-mycobacterial T cells recognize peptide epitopes and are restricted by major histocompatibility complex molecules; others recognize glycolipids and are restricted by CD1. Distinct subpopulations of T cells restricted by CD1 can lyse macrophages infected with *M. tuberculosis*. In one study, when the CD1-dependent T cells were $CD8^+$, the viability of the mycobacteria inside the macrophages was reduced, but when the CD1-dependent T cells were $CD8^-$, lysis of infected macrophages was not associated with a reduction in mycobacterial viability (Stenger et al., 1997). CD1-restricted T cells reactive with mycobacterial glycolipids are predominantly Th1 in phenotype; their lysis of infected macrophages is accompanied by secretion of IFNγ. Although β2-microglobulin deficient mice, which lack a CD8 T cell response, experienced more rapidly progressive infection with *M. tuberculosis* than wild-type mice (Flynn et al., 1992), infection with BCG or *M. tuberculosis* was no more extensive in mice that were deficient in perforin or Fas-receptor than in wild-type mice (Laochumroonvorapong et al., 1997). Since the major mechanism available for T cells to induce apoptosis is thought to require Fas, and the major mechanism for T cells to induce ATP release from target cells is thought to be cytolysis dependent on perforin, these observations suggest that $CD8^+$ cells are important for recovery from tuberculosis in mice for reasons other than inducing apoptosis of infected macrophages.

In vivo, apoptotic cells are rapidly phagocytized and destroyed by macrophages (Savill et al., 1989a, b). Apoptotic, *M. avium*-infected macrophages are bound by noninfected macrophages; by an unknown mechanism, this leads to diminished CFU (Fratazzi et al., 1997). Treatment of the MAI-infected macrophages with anti-TNF antibody and IL-10 prior to addition of fresh macrophages prevented both the apoptosis and the mycobactericidal effect (Fratazzi et al., 1997). Compared to wild-type mice, TNF p55 receptor knockout mice experienced more extensive infection by *M. tuberculosis* (Flynn et al., 1995) or *L. monocytogenes* (Pfeffer et al., 1993; Rothe et al., 1993). Thus, apoptosis of macrophages, possibly mediated by their own TNF, may indirectly constitute an antimicrobial mechanism by retaining microbes within a membrane-bound compartment and signalling noninfected macrophages to ingest the packaged pathogens.

Additional studies of microbial pathogenesis would be of interest in mice and macrophages deficient in mediators and regulators of apoptosis, such as FasL, FasR, Bcl-2, p53, Il-1β converting enzyme (ICE, caspase 1), and other caspases. The bactericidal capacity of monocytes obtained from humans with FasR defects (Canale-Smith syndrome) would also be informative (Orlinick et al., 1997).

B. Microbial Inhibitors of Apoptosis

Herpesvirus, poxvirus and baculovirus express a variety of anti-apoptotic proteins (Spriggs, 1996). *L. donovani* (Moore and Matlashewski, 1994) and chlamydia (Fan et al., 1998) block apoptosis of infected host cells, but the microbial genes involved have not been identified. When pathogens express genes that inhibit host cell apoptosis, the interpretation is encouraged that apoptosis constitutes a mechanism of host defense.

VII. ANTIMICROBIAL PEPTIDES

A. Suggestive Evidence for Macrophage Antimicrobial Peptides

Neutrophils of various species contain many antimicrobial peptides and proteins, such as α-defensins and bactericidal/permeability-increasing protein (BPI) (see chapter by Elsbach). However, there are major species and cell type variations with respect to expression of these molecules. Thus, α-defensins and BPI have not been identified in murine PMN. Furthermore, with the exception of rabbit alveolar macrophages (Patterson-Delafield et al., 1980), α-defensins have not been isolated from mononuclear phagocytes, nor has the β-defensin that was recently discovered in epithelial cells (Diamond et al., 1991; Schonwetter et al., 1995; Goldman et al., 1997).

Neutrophil serprocidins (for *ser*ine *pro*tease homologues with microbi*cid*al activity), homologous to T cell granzymes and mast cell protease, are packaged in azurophil granules. Human monocytes leave the bone marrow with azurophil granules remaining from the differentiation stage preceding the split between monocytes and neutrophils. These granules are lost during the differentiation of monocytes to macrophages.

Lysozyme, an enzyme found in neutrophil granules and body fluids, is also secreted by macrophages. Lysozyme acts on the peptidoglycan bacterial cell wall and fungal chitin. Lysozyme's microbicidal activity appears to be limited to nonpathogenic gram-positive bacteria and some fungi (Elsbach et al., in press), perhaps testifying to its efficacy. The decisive test of lysozyme's efficacy would be the response of lysozyme-deficient mice to infection with various organisms, both pathogenic and nonpathogenic.

Serum leukocyte protease inhibitor (SLPI) was shown to have antimicrobial activity (Hiemstra et al., 1996) before it was recognized to be a product of macrophages as well as epithelial cells (Jin et al., 1997). However, intracellular accumulation of SLPI is undetectable in macrophages (Zhu and Ding, unpublished observations), and it is not clear that SLPI participates in macrophage killing of intracellular pathogens.

Recently, antimicrobial peptides have been purified from murine macrophage cell lines and peritoneal macrophages (Hiemstra et al., 1993; Hiemstra and van Furth, 1994). These MUMPs (for *mu*rine *mi*crobicidal *p*roteins) resemble H1 and H2 histones and are bactericidal for a diverse group of organisms. Whether MUMPs make contact with microbes during an encounter of the macrophage with susceptible pathogens and how they kill microbes remain unknown.

B. Microbial Resistance to Host Antimicrobial Peptides

Though the importance of antimicrobial peptides for macrophage bacterial killing is still uncertain, the general importance of antimicrobial peptides for host defense is suggested by the existence of bacterial resistance systems to host peptides. Several mechanisms by which bacteria have been suggested to resist antimicrobial peptides include destruction or inhibition of peptide biological activity, modification of peptide binding or insertion into bacterial membranes, and restoration of pertubations induced by peptide activity (reviewed in Groisman, 1994).

VIII. PHOSPHOLIPASE

Many biological activities have been attributed to phospholipase A_2 (PLA_2) (reviewed in Chilton, 1996), including potent antimicrobial activity. For example, a PLA_2 isolated from sterile rabbit inflammatory fluid has potent microbicidal activity against gram-positive organisms (Weinrauch et al., 1996). A PLA_2 recently isolated from murine Paneth cells but also present in the spleen is potently bactericidal against *E. coli*, *S. typhimurium* and *L. monocytogenes* (Harwig et al., 1995). PLA_2's mechanism of cidal activity is unclear, although it has been suggested that group II PLA_2, which has a high affinity for bacterial phospholipids, can bind directly to bacteria because of its cationic nature. Alternatively, the free fatty acid products of phospholipases are directly cidal for *L. monocytogenes in vitro* (Akaki et al., 1997).

Several different types of PLA_2 have been identified in macrophages, including secretory (Group II) PLA_2s, a cytosolic, Ca^{2+}-dependent (Group IV) PLA_2 and a cytosolic, Ca^{2+}-independent PLA_2 (Dennis, 1994). Although several of these PLA_2 probably serve signaling functions, the possibility remains that one or more may contribute to macrophage bactericidal activity. Phospholipase inhibitors are available, but their potential nonspecificity may hamper characterization of the role of specific phospholipases in host defense. Mice deficient in individual phospholipases are available either through natural mutation, as is the case for secretory, group II PLA_2 (Kennedy et al., 1995) or through homologous recombination, as is the case for cytosolic, Ca^{2+}-dependent (Group IV) PLA_2 (Bonventre et al., 1997). Investigation of bacterial survival in knockout mice and macrophages should help define the role of phospholipases in host defense.

IX. SUMMARY

The innate immune system has responded to microbial diversity by evolving multiple microbicidal mechanisms. Each mechanism may be essential for survival from a particular set of pathogens, while toward most microbes, several mechanisms may be effectively redundant. At the same time, against yet other pathogens, host defense may depend on synergistic interactions between distinct mechanisms. For example, RNI and ROI can react with each other to form even more bactericidal species, such as peroxynitrite (Brunelli et al., 1995). Some such interactions may be subtle. For example, among bactericidal RNIs are S-nitrosothiols (Upchurch et al., 1995). To kill bacteria, S-nitrosothiols must be taken up by the microbe and decomposed (De Groote et al., 1995). The decomposition of nitrosothiols can be catalyzed by free metals such as iron (McAninly et al., 1993) whose release from transferrin is dependent on acidification (Dautry-Varsat et al., 1983). In some settings, antimicrobial activity may even require the macrophage to kill itself. For all these reasons, the rigorous identification of individual macrophage microbicidal mechanisms by multiple criteria is experimentally challenging. Indeed, in a complex, integrated system, a strictly reductionist analysis can be misleading.

When we last surveyed the evidence for identification of macrophage microbicial pathways (Nathan, 1983), production of ROI had just become experimentally established as a major mechanism; other mechanisms were hypothesized, but their roles were not defined. Fifteen years later, production of RNI has become experimentally established as the second major mechanism. Other mechanisms are not merely hypothesized but proven to exist, through the use of macrophages from mice doubly deficient in production of both ROI and RNI. However, the specific identification of these additional mechanism(s) remains incomplete. This challenge will probably best be met through the application of genetic approaches both at the level of the host and the pathogen.

ACKNOWLEDGMENTS

We thank Denise Marciano and Aihao Ding for critically reviewing this manuscript. This work was funded by National Institutes of Health grant HL-51967 (C. F. N.) and by NIH Medical Scientist Training program grant GM-07739 (M. U. S.). Additional support to M. U. S. was provided by the Iris L. and Leverett S. Woodward Medical Scientist Fellowship.

REFERENCES

Akaki, T., Sato, K., Shimizu, T., Sano, C., Kajitani, H., Dekio, S., & Tomioka, H. (1997). Effector molecules in expression of the antimicrobial activity of macrophages against Mycobacterium avium complex: roles of reactive nitrogen intermediates, reactive oxygen intermediates, and free fatty acids. J. Leukoc. Biol. 62, 795-804.

Albakri, Q.A., & Stuehr, D.J. (1996). Intracellular assembly of inducible NO synthase is limited by nitric oxide-mediated changes in heme insertion and availability. J. Biol. Chem. 271, 5414-5421.

Alford, C.E., Amaral, E., & Campbell, P.A. (1990). Listericidal activity of human neutrophil cathepsin G. J. Gen. Microbiol. 136, 997-100.

Alpuche Aranda, C.M., Swanson, J.A., Loomis, W.P., & Miller, S.I. (1992). Salmonella typhimurium activates virulence gene transcription within acidified macrophage phagosomes. Proc. Natl. Acad. Sci. USA 89, 10079-10083.

Baldridge, C.W., & Gerard, R.W. (1933). The extra respiration of phagocytosis. Am. J. Physiol. 103, 235-236.

Beauregard, K.E., Lee, K.D., Collier, R.J., & Swanson, J.A. (1997). pH-dependent perforation of macrophage phagosomes by listeriolysin O from Listeria monocytogenes. J. Exp. Med. 186, 1159-1163.

Beckman, J.S., & Koppenol, W.H. (1996). Nitric oxide, superoxide, and peroxynitrite: the good, the bad, and ugly. Am. J. Physiol. 271, C1424-1437.

Bellamy, R., Ruwende, C., Corrah, R., McAdam, K.P.W.J., Whittle, H.C., & Hill, A.V.S. (1998). Variations in the NRAMP1 gene and susceptibility to tuberculosis in west africans. N. Engl. J. Med. 338, 640-644.

Black, C.M., Paliescheskey, M., Beaman, B.L., Donovan, R.M., & Goldstein, E. (1986). Acidification of phagosomes in murine macrophages: blockage by Nocardia asteroides. J. Infect. Dis. 154, 952-958.

Black, D.S., & Bliska, J.B. (1997). Identification of p130Cas as a substrate of Yersinia YopH (Yop51), a bacterial protein tyrosine phosphatase that translocates into mammalian cells and targets focal adhesions. EMBO J. 16, 2730-2744.

Blackwell, J.M., Black, G.F., Peacock, C.S., Miller, E.N., Sibthorpe, D., Gnananandha, D., Shaw, J.J., Silveira, F., Lins-Lainson, Z., Ramos, F., Collins, A., & Shaw, M.A. (1997). Immunogenetics of leishmanial and mycobacterial infections: the Belem Family Study. Philos. Trans. R. Soc. Lond. B. Biol. Sci. 352, 1331-1345.

Bliska, J.B., & Black, D.S. (1995). Inhibition of the Fc receptor-mediated oxidative burst in macrophages by the Yersinia pseudotuberculosis tyrosine phosphatase. Infect. Immun. 63, 681-685.

Bogdan, C. (1997). Of microbes, macrophages and nitric oxide. Behring Inst. Mitt. 99, 58-72.

Bonventre, J.V., Huang, Z., Taheri, M.R., O'Leary, E., Li, E., Moskowitz, M.A., & Sapirstein, A. (1997). Reduced fertility and postischaemic brain injury in mice deficient in cytosolic phospholipase A2. Nature 390, 622-625.

Borges, J.S., & Johnson, W.D., Jr. (1975). Inhibition of multiplication of Toxoplasma gondii by human monocytes exposed to T-lymphocyte products. J. Exp. Med. 141, 483-496.

Brunelli, L., Crow, J.P., & Beckman, J.P. (1995). The comparative toxicity of nitric oxide and peroxynitrite to *Escherichia coli*. Arch. Bioch. Biophys. 316, 327-334.

Buchmeier, N.A., Libby, S. J., Xu, Y., Loewen, P.C., Switala, J., Guiney, D.G., & Fang, F.C. (1995). DNA repair is more important than catalase for Salmonella virulence in mice. J. Clin. Invest. 95, 1047-1053.

Byrd, T.F., & Horwitz, M.A. (1989). Interferon gamma-activated human monoctyes downregulate transferrin receptors and inhibit the intracellular multiplication of *Legionella pneumophila* by limiting the availability of iron. J. Clin. Invest. 83, 1457-1465.

Byrd, T.F., & Horwitz, M.A. (1991). Chloroquine inhibits the intracellular multiplication of *Legionella pneumophila* by limiting the availability of iron. J. Clin. Invest. 88, 351-357.

Byrd, T.F., & Horwitz, M.A. (1993). Regulation of transferrin receptor expression and ferritin content in human mononuclear phagocytes. Coordinate upregulation by iron transferrin and downregulation by interferon gamma. J. Clin. Invest. 91, 969-976.

Chae, H.Z., Chung, S.J., & Rhee, S.G. (1994a). Thioredoxin-dependent peroxide reductase from yeast. J. Biol. Chem. 269, 27670-27678.

Chae, H.Z., Robison, K., Poole, L.B., Church, G., Storz, G., & Rhee, S.G. (1994b). Cloning and sequencing of thiol-specific antioxidant from mammalian brain: alkyl hydroperoxide reductase and thiol-specific antioxidant define a large family of antioxidant enzymes. Proc. Natl. Acad. Sci. USA 91, 7017-7021.

Chan, J., Xing, Y., Magliozzo, R.S., & Bloom, B.R. (1992). Killing of virulent Mycobacterium tuberculosis by reactive nitrogen intermediates produced by activated murine macrophages. J. Exp. Med. 175, 1111-1122.

Chen, L., Xie, Q.-w., & Nathan, C. (1998). Alkyl hydroperoxide reductase subunit C (AhpC) protects bacterial and human cells against reactive nitrogen intermediates. Mol. Cell 1, 795-805.

Chilton, F. (1996). Would the real role(s) for secretory PLA2s please stand up [editorial; comment]. J. Clin. Invest. 97, 2161-2162.

Cho, H.J., Xie, Q.W., Calaycay, J., Mumford, R.A., Swiderek, K.M., Lee, T.D., & Nathan, C. (1992). Calmodulin is a subunit of nitric oxide synthase from macrophages. J. Exp. Med. 176, 599-604.

Confer, D.L., & Eaton, J.W. (1982). Phagocyte impotence caused by an invasive bacterial adenylate cyclase. Science 217, 948-950.

Crawford, M.J., & Goldberg, D.E. (1998). Role for the *Salmonella* flavohemoglobin in protection from nitric oxide. J. Biol. Chem. 273, 12543-12547.

Dautry-Varsat, A., Ciechanover, A., & Lodish, H.F. (1983). pH and the recycling of transferrin during receptor-mediated endocytosis. Proc. Natl. Acad. Sci. USA 80, 2258-2262.

De Groote, M.A., Granger, D., Xu, Y., Campbell, G., Prince, R., & Fang, F.C. (1995). Genetic and redox determinants of nitric oxide cytotoxicity in a *Salmonella typhimurium* model. Proc. Natl. Acad. Sci. USA 92, 6399-6403.

De Groote, M.A., Ochsner, U.A., Shiloh, M.U., Nathan, C., McCord, J.M., Dinauer, M.C., Libby, S.J., Vazquez-Torres, A., Xu, Y., & Fang, F.C. (1997). Periplasmic superoxide dismutase protects Salmonella from products of phagocyte NADPH-oxidase and nitric oxide synthase. Proc. Natl. Acad. Sci. USA 94, 13997-14001.

De Groote, M.A., Testerman, T., Xu, Y., Stauffer, G., & Fang, F.C. (1996). Homocysteine antagonism of nitric oxide-related cytostasis in *Salmonella typhimurium*. Science 272, 414-417.

De Leo, F.R., & Quinn, M.T. (1996). Assembly of the phagocyte NADPH oxidase: molecular interaction of oxidase proteins. J. Leukoc. Biol. 60, 677-691.

Dennis, E.A. (1994). Diversity of group types, regulation, and function of phospholipase A2. J. Biol. Chem. 269, 13057-13060.

Deretic, V., Philipp, W., Dhandayuthapani, S., Mudd, M.H., Curcic, R., Garbe, T., Heym, B., Via, L.E., & Cole, S.T. (1995). *Mycobacterium tuberculosis* is a natural mutant with an inactivated oxidative-stress regulatory gene: implications for sensitivity to isoniazide. Mol. Microbiol. 17, 889-900.

Diamond, G., Zasloff, M., Eck, H., Brasseur, M., Maloy, W.L., & Bevins, C.L. (1991). Tracheal antimicrobial peptide, a cysteine-rich peptide from mammalian tracheal mucosa: peptide isolation and cloning of a cDNA. Proc. Natl. Acad. Sci. USA 88, 3952-3956.

Diefenbach, A., Schindler, H., Donhauser, N., Lorenz, E., Laskay, T., MacMicking, J., Röllinghoff, M., Gresser, I., & Bogdan, C. (1998). Type1 interferon (IFNa/b) and type 2 nitric oxide synthase regulate the innate immune response to a protozoan parasite. Immunity 8, 77-87.

Dykhuizen, R.S., Frazer, R., Duncan, C., Smith, C.C., Golden, M., Benjamin, N., & Leifert, C. (1996). Antimicrobial effect of acidified nitrite on gut pathogens: importance of dietary nitrate in host defense. Antimicrob. Agents Chemother. 40, 1422-1425.

Ehrt, E., Shiloh, M.U., Ruan, J., Choi, M., Gunzberg, S., Nathan, C., Xie, Q.-w., & Riley, L.W. (1997). A novel antioxidant gene from *Mycobacterium tuberculosis*. J. Exp. Med. 186, 1885-1896.

Ezekowitz, R.A., Dinauer, M.C., Jaffe, H.S., Orkin, S.H., & Newburger, P.E. (1988). Partial correction of the phagocyte defect in patients with X-linked chronic granulomatous disease by subcutaneous interferon gamma. N. Engl. J. Med. 319, 146-151.

Fan, T., Lu, H., Hu, H., Shi, L., McClarty, G.A., Nance, D.M., Greenberg, A.H., & Zhong, G. (1998). Inhibition of apoptosis in chlamydia-infected cells: blockade of mitochondrial cytochrome c release and caspase activation. J. Exp. Med. 187, 487-496.

Fang, F.C. (1997). Mechanisms of nitric oxide-related antimicrobial activity. J. Clin. Invest. 99, 2818-2825.

Fang, F.C., Vasquez-Torres, A., & Xu, Y. (1997). The transcriptional regulator SoxS is required for resistance of Salmonella typhimurium to paraquat but not for virulence in mice. Infect. Immun. 65, 5371-5375.

Farr, S.B., & Kogoma, T. (1991). Oxidative stress responses in Escherichia coli and Salmonella typhimurium. Microbiol. Rev. 55, 561-585.

Fehr, T., Schoedon, G., Odermatt, B., Holtschke, T., Schneemann, M., Bachmann, M.F., Mak, T.W., Horak, I., & Zinkernagel, R.M. (1997). Crucial role of interferon consensus sequence binding protein, but neither of interferon regulatory factor 1 nor of nitric oxide synthesis for protection against murine listeriosis. J. Exp. Med. 185, 921-931.

Fleming, M.D., Romano, M.A., Su, M.A., Garrick, L.M., Garrick, M.D., & Andrews, N.C. (1998). Nramp2 is mutated in the anemic Belgrade (b) rat: evidence of a role for Nramp2 in endosomal iron transport. Proc. Natl. Acad. Sci. USA 95, 1148-1153.

Fleming, M.D., Trenor, C.C.R., Su, M.A., Foernzler, D., Beier, D.R., Dietrich, W.F., & Andrews, N.C. (1997). Microcytic anaemia mice have a mutation in Nramp2, a candidate iron transporter gene. Nat. Genet. 16, 383-386.

Flynn, J.L., Goldstein, M.M., Chan, J., Triebold, K.J., Pfeffer, K., Lowenstein, C.J., Schreiber, R., Mak, T.W., & Bloom, B.R. (1995). Tumor necrosis factor-alpha is required in the protective immune response against Mycobacterium tuberculosis in mice. Immunity 2, 561-572.

Flynn, J.L., Goldstein, M.M., Triebold, K.J., Koller, B., & Bloom, B.R. (1992). Major histocompatibility complex class I-restricted T cells are required fore resistance to *Mycobacterium tuberculosis* infection. Proc. Natl. Acad. Sci. USA 89, 12013-12017.

Fortier, A.H., Leiby, D.A., Narayanan, R.B., Asofoadjei, E., Crawford, R.M., Nacy, C.A., & Meltzer, M.S. (1995). Growth of Francisella tularensis LVS in macrophages: the acidic intracellular compartment provides essential iron required for growth. Infect. Immun. 63, 1478-1483.

Fratazzi, C., Arbeit, R.D., Carini, C., & Remold, H.G. (1997). Programmed cell death of Mycobacterium avium serovar 4-infected human macrophages prevents the mycobacteria from spreading and induces mycobacterial growth inhibition by freshly added, uninfected macrophages. J. Immunol. 158, 4320-4327.

Fuortes, M., Jin, W.W., & Nathan, C. (1993). Adhesion-dependent protein tyrosine phosphorylation in neutrophils treated with tumor necrosis factor. J. Cell Biol. 120, 777-784.

Geoffroy, C., Gaillard, J.L., Alouf, J.E., & Berche, P. (1987). Purification, characterization, and toxicity of the sulfhydryl-activated hemolysin listeriolysin O from Listeria monocytogenes. Infect. Immun. 55, 1641-1646.

Giannella, R.A., Broitman, S.A., & Zamcheck, N. (1973). Influence of gastric acidity on bacterial and parasitic enteric infections. Ann. Int. Med. 78, 271-276.

Goldman, M.J., Anderson, G.M., Stolzenberg, E.D., Kari, U.P., Zasloff, M., & Wilson, J.M. (1997). Human beta-defensin-1 is a salt-sensitive antibiotic in lung that is inactivated in cystic fibrosis. Cell 88, 553-560.

Govoni, G., Gauthier, S., Billia, F., Iscove, N.N., & Gros, P. (1997). Cell-specific and inducible Nramp1 gene expression in mouse macrophages in vitro and in vivo. J. Leukoc. Biol. 62, 277-286.

Groisman, E.A. (1994). How bacteria resist killing by host-defense peptides. Trends Microbiol. 2, 444-449.

Gruenheid, S., Cellier, M., Vidal, S., & Gros, P. (1995). Identification and characterization of a second mouse Nramp gene. Genomics 25, 514-525.

Gruenheid, S., Pinner, E., Desjardins, M., & Gros, P. (1997). Natural resistance to infection with intracellular pathogens: the Nramp1 protein is recruited to the membrane of the phagosome. J. Exp. Med. 185, 717-730.

Gunshin, H., Mackenzie, B., Berger, U.V., Gunshin, Y., Romero, M.F., Boron, W.F., Nussberger, S., Gollan, J.L., & Hediger, M.A. (1997). Cloning and characterization of a mammalian proton-coupled metal-ion transporter. Nature 388, 482-488.

Hampton, M., & Winterbourn, C. (1995). Modification of neutrophil oxidant production with diphenyleneiodonium and its effect on bacterial killing. Free Radic. Biol. Med. 18, 633-639.

Harwig, S.S., Tan, L., Qu, X.D., Cho, Y., Eisenhauer, P.B., & Lehrer, R.I. (1995). Bactericidal properties of murine intestinal phospholipase A2. J. Clin. Invest. 95, 603-610.

Hausladen, A., Privalle, C.T., Keng, T., De Angelo, J., & Stamler, J.S. (1996). Nitrosative stress: activation of the transcription factor OxyR. Cell 86, 719-729.

Hensel, M., Shea, J.E., Gleeson, C., Jones, M.D., Dalton, E., & Holden, D.W. (1995). Simultaneous identification of bacterial virulence genes by negative selection. Science 269, 400-403.

Hevel, J.M., White, K.A., & Marletta, M.A. (1991). Purification of the inducible murine macrophage nitric oxide synthase. Identification as a flavoprotein. J. Biol. Chem. 266, 22789-22791.

Heyworth, P.G., Curnutte, J.T., Nauseef, W.M., Volpp, B.D., Pearson, D.W., Rosen, H., & Clark, R.A. (1991). Neutrophil nicotinamide adenine dinucleotide phosphate oxidase assembly. Translocation of p47-phox and p67-phox requires interaction between p47-phox and cytochrome b558. J. Clin. Invest. 87, 352-356.

Hibbs, J.B., Jr., Taintor, R.R., Vavrin, Z., & Rachlin, E.M. (1988). Nitric oxide: a cytotoxic activated macrophage effector molecule [published erratum appears in Biochem. Biophys. Res. Commun. 1989 158, 624]. Biochem. Biophys. Res. Commun. 157, 87-94.

Hidalgo, E., Ding, H., & Demple, B. (1997). Redox signal transduction: mutations shifting [2Fe-2S] centers of the SoxR sensor-regulator to the oxidized form. Cell 88, 121-129.

Hiemstra, P.S., & van Furth, R. (1994). Antimicrobial mechanisms: antimicrobial polypeptides of mononuclear phagocytes. Immunol. Ser. 60, 197-202.

Hiemstra, P.S., Eisenhauer, P.B., Harwig, S.S., van den Barselaar, M.T., van Furth, R., & Lehrer, R.I. (1993). Antimicrobial proteins of murine macrophages. Infect. Immun. 61, 3038-3046.

Hiemstra, P.S., Maassen, R.J., Stolk, J., Heinzel-Wieland, R., Steffens, G.J., & Dijkman, J.H. (1996). Antibacterial activity of antileukoprotease. Infect. Immun. 64, 4520-4524.

Holland, P.C., Clark, M.G., Bloxham, D.P., & Lardy, H.A. (1973). Mechanism of action of the hypoglycemic agent diphenyleneiodonium. J. Biol. Chem. 248, 6050-6056.

Horwitz, M.A., & Maxfield, F.R. (1984). Legionella pneumophila inhibits acidification of its phagosome in human monocytes. J. Cell Biol. 99, 1936-1943.

Inoue, S., Itagaki, S., & Amano, F. (1995). Intracellular killing of Listeria monocytogenes in the J774.1 macrophage-like cell line and the lipopolysaccharide (LPS)-resistant mutant LPS1916 cell line defective in the generation of reactive oxygen intermediates after LPS treatment. Infect. Immun. 63, 1876-1886.

Islam, D., Veress, B., Bardhan, P.K., Lindberg, A.A., & Christensson, B. (1997). In situ characterization of inflammatory responses in the rectal mucosae of patients with shigellosis. Infect. Immun. 65, 739-749.

Jackson, S.H., Gallin, J.I., & Holland, S.M. (1995). The p47phox mouse knock-out model of chronic granulomatous disease. J. Exp. Med. 182, 751-758.

Jin, F.Y., Nathan, C., Radzioch, D., & Ding, A. (1997). Secretory leukocyte protease inhibitor: a macrophage product induced by and antagonistic to bacterial lipopolysaccharide. Cell 88, 417-426.

Jones, R.L., Peterson, C.M., Grady, R.W., Kumbaraci, T., Cerami, T., & Graziano, J.H. (1977). Effects of iron chelators and iron overload on *Salmonella* infection. Nature 267, 63-65.

Jones, T.C., Len, L., & Hirsch, J.G. (1975). Assessment in vitro of immunity against Toxoplasma gondii. J. Exp. Med. 141, 466-482.

Keane, J., Balcewicz-Sablinska, M.K., Remold, H.G., Chupp, G.L., Meek, B.B., Fenton, M.J., & Kornfeld, H. (1997). Infection by Mycobacterium tuberculosis promotes human alveolar macrophage apoptosis. Infect. Immun. 65, 298-304.

Kennedy, B.P., Payette, P., Mudgett, J., Vadas, P., Pruzanski, W., Kwan, M., Tang, C., Rancourt, D.E., & Cromlish, W.A. (1995). A natural disruption of the secretory group II phospholipase A2 gene in inbred mouse strains. J. Biol. Chem. 270, 22378-22385.

Khelef, N., Zychlinsky, A., & Guiso, N. (1993). Bordetella pertussis induces apoptosis in macrophages: role of adenylate cyclase-hemolysin. Infect. Immun. 61, 4064-4071.

Klebanoff, S.J., & Nathan, C.F. (1993). Nitrite production by stimulated human polymorphonuclear leukocytes supplemented with azide and catalase. Biochem. Biophys. Res. Commun. 197, 192-196.

Klingler, K., Tchou-wong, K.-M., Brändli, O., Aston, C., Kim, R., Chi, C., & Rom, W.N. (1997). Effects of mycobacteria on regulation of apoptosis in mononuclear phagocytes. Infec. Immun. 65, 5272-5278.

Kopf, M., Baumann, H., Freer, G., Freudenberg, M., Lamers, M., Kishimoto, T., Zinkernagel, R., Bluethmann, H., & Kohler, G. (1994). Impaired immune and acute-phase responses in interleukin-6-deficient mice. Nature 368, 339-342.

Lammas, D.A., Stober, C., Harvey, C.J., Kendrick, N., Panchalingam, S., & Kumararatne, D.S. (1997). ATP-induced killing of mycobacteria by human macrophages is mediated by purinergic P2Z(P2X7) receptors. Immunity 7, 433-444.

Laochumroonvorapong, P., Wang, J., Liu, C.C., Ye, W., Moreira, A.L., Elkon, K.B., Freedman, V.H., & Kaplan, G. (1997). Perforin, a cytotoxic molecule which mediates cell necrosis, is not required for the early control of mycobacterial infection in mice. Infect. Immun. 65, 127-132.

Liew, F.Y. (1995). Regulation of lymphocyte functions by nitric oxide. Curr. Opin. Immunol. 7, 396-399.

Lissner, C.R., Weinstein, D.L., & O'Brien, A.D. (1985). Mouse chromosome 1 Ity locus regulates microbicidal activity of isolated peritoneal macrophages against a diverse group of intracellular and extracellular bacteria. J. Immunol. 135, 544-547.

MacMicking, J., Xie, Q.W., & Nathan, C. (1997a). Nitric oxide and macrophage function. Annu. Rev. Immunol. 15, 323-350.

MacMicking, J.D., North, R.J., La Course, R., Mudgett, J.S., Shah, S.K., & Nathan, C.F. (1997b). Identification of nitric oxide synthase as a protective locus against tuberculosis. Proc. Natl. Acad. Sci. USA 94, 5243-5248.

Mahan, M.J., Tobias, J.W., Slauch, J.M., Hanna, P.C., Collier, R.J., & Mekalanos, J.J. (1995). Antibiotic-based selection for bacterial genes that are specifically induced during infection of a host. Proc. Natl. Acad. Sci. USA 92, 669-673.

Malo, D., Vogan, K., Vidal, S., Hu, J., Cellier, M., Schurr, E., Fuks, A., Bumstead, N., Morgan, K., & Gros, P. (1994). Haplotype mapping and sequence analysis of the mouse Nramp gene predict susceptibility to infection with intracellular parasites. Genomics 23, 51-61.

McAninly, J., Williams, D.L.H., Askew, S.C., Butler, A.R., & Russell, C. (1993). Metal ion catalysis in nitrosothiol (RSNO) decomposition. J. Chem. Soc. Chem. Comm. 1758-1759.

McKnight, G.M., Smith, L.M., Drummond, R.S., Duncan, C.W., Golden, M., & Benjamin, N. (1997). Chemical synthesis of nitric oxide in the stomach from dietary nitrate in humans. Gut 40, 211-214.

Medina, E., Rogerson, B.J., & North, R.J. (1996). The *Nramp1* antimicrobial resistance gene segregates independently of resistance to virulant *Mycobacterium tuberculosis*. Immunology 88, 479-481.

Miller, S.I., Kukral, A.M., & Mekalanos, J.J. (1989). A two-component regulatory system (phoP phoQ) controls Salmonella typhimurium virulence. Proc. Natl. Acad. Sci. USA 86, 5054-5058.

Miwa, M., Stuehr, D.J., Marletta, M.A., Wishnok, J.S., & Tannenbaum, S.R. (1987). Nitrosation of amines by stimulated macrophages. Carcinogenesis 8, 955-958.

Molloy, A., Laochumroonvorapong, P., & Kaplan, G. (1994). Apoptosis, but not necrosis, of infected monocytes is coupled with killing of intracellular bacillus Calmette-Guerin. J. Exp. Med. 180, 1499-1509.

Montemartini, M., Nogoceke, E., Singh, M., Steinert, P., Flohe, L., & Kalisz, H.M. (1998). Sequence analysis of the tryparedoxin peroxidase gene from Crithidia fasciculata and its functional expression in Escherichia coli. J. Biol. Chem. 273, 4864-4871.

Moore, K.J., & Matlashewski, G. (1994). Intracellular infection by Leishmania donovani inhibits macrophage apoptosis. J. Immunol. 152, 2930-2937.

Mouy, R., Fischer, A., Vilmer, E., Seger, R., & Griscelli, C. (1989). Incidence, severity and prevention of infections in chronic granulomatous disease. J. Pediatr. 114, 555-560.

Muller, P.K., Krohn, K., & Muhlradt, P.F. (1989). Effects of pyocyanine, a phenazine dye from Pseudomonas aeruginosa, on oxidative burst and bacterial killing in human neutrophils. Infect. Immun. 57, 2591-2596.

Murray, H.W. (1981). Interaction of Leishmania with a macrophage cell line. Correlation between intracellular killing and the generation of oxygen intermediates. J. Exp. Med. 153, 1690-1695.

Murray, H.W., & Cohn, Z.A. (1979). Macrophage oxygen-dependent antimicrobial activity. I. Susceptibility of Toxoplasma gondii to oxygen intermediates. J. Exp. Med. 150, 938-949.

Murray, H.W., Juangbhanich, C.W., Nathan, C.F., & Cohn, Z.A. (1979). Macrophage oxygen-dependent antimicrobial activity. II. The role of oxygen intermediates. J. Exp. Med. 150, 950-964.

Nathan, C. (1991). Mechanisms and modulation of macrophage activation. Behring Inst. Mitt. 200-207.

Nathan, C. (1995). Natural resistance and nitric oxide. Cell 82, 873-876.

Nathan, C. (1997). Perspective series: nitric oxide and nitric oxide synthases. Inducible nitric oxide synthase: what difference does it make? J. Clin. Invest. 100, 2417-2423.

Nathan, C., Nogueira, N., Juangbhanich, C., Ellis, J., & Cohn, Z. (1979). Activation of macrophages in vivo and in vitro. Correlation between hydrogen peroxide release and killing of Trypanosoma cruzi. J. Exp. Med. 149, 1056-1068.

Nathan, C.F. (1983). Mechanisms of macrophage antimicrobial activity. Trans. R. Soc. Trop. Med. Hyg. 77, 620-630.

Nathan, C.F., & Hibbs, J.B., Jr. (1991). Role of nitric oxide synthesis in macrophage antimicrobial activity. Curr. Opin. Immunol. 3, 65-70.

Nathan, C.F., & Root, R.K. (1977). Hydrogen peroxide release from mouse peritoneal macrophages: dependence on sequential activation and triggering. J. Exp. Med. 146, 1648-1662.

Nathan, C.F., Murray, H.W., & Cohn, Z.A. (1980). The macrophage as an effector cell. N. Engl. J. Med. 303, 622-626.

Neilands, J.B. (1995). Siderophores: structure and function of microbial iron transport compounds. J. Biol. Chem. 270, 26723-26726.

Newton, G.L., Arnold, K., Price, M.S., Sherrill, C., Delcardayre, S.B., Aharonowitz, T., Cohen, G., Davies, J., Fahey, R.C., & Davis, C. (1996). Disruption of thiols in microorganisms: nycothiol is a major thiol in most actinomycetes. J. Bacteriol. 178, 1990-1995.

Nicholson, S., Bonecini-Almeida, M.d.G., Lapa e Silva, J.R., Nathan, C., Xie, Q.W., Mumford, R., Weidner, J.R., Calaycay, J., Geng, J., Boechat, N., & et al. (1996). Inducible nitric oxide synthase in pulmonary alveolar macrophages from patients with tuberculosis. J. Exp. Med. 183, 2293-2302.

Nozaki, Y., Hasegawa, Y., Ichiyama, S., Nakashima, I., & Shimokata, K. (1997). Mechanism of nitric oxide-dependent killing of Mycobacterium bovis BCG in human alveolar macrophages. Infect. Immun. 65, 3644-3647.

Nunoshiba, T., De Rojas-Walker, T., Tannenbaum, S R., & Demple, B. (1995). Roles of nitric oxide in inducible resistance of Escherichia coli to activated murine macrophages. Infect. Immun. 63, 794-798.

Ohkuma, S., & Poole, B. (1978). Fluorescence probe measurement of the intralysosomal pH in living cells and the perturbation of pH by various agents. Proc. Natl. Acad. Sci. USA 75, 3327-3331.

Orlinick, J.R., Vaishnaw, A., Elkon, K.B., & Chao, M.V. (1997). Requirement of cysteine-rich repeats of the Fas receptor for binding by the Fas ligand. J. Biol. Chem. 272, 28889-28894.

Pacelli, R., Wink, D.A., Cook, J.A., Krishna, M.C., De Graff, W., Friedman, N., Tsokos, M., Samuni, A., & Mitchell, J.B. (1995). Nitric oxide potentiates hydrogen peroxide-induced killing of Escherichia coli. J. Exp. Med. 182, 1469-1479.

Patterson-Delafield, J., Martinez, R., & Lehrer, R. (1980). Microbicidal cationic proteins in rabbit alveolar macrophages: a potential host defense mechanism. Infect. Immun. 30, 180-192.

Pfeffer, K., Matsuyama, T., Kundig, T.M., Wakeham, A., Kishihara, K., Shahinian, A., Wiegmann, K., Ohashi, P.S., Kronke, M., & Mak, T.W. (1993). Mice deficient for the 55 kd tumor necrosis factor receptor are resistant to endotoxic shock, yet succumb to L. monocytogenes infection. Cell 73, 457-467.

Pollock, J.D., Williams, D.A., Gifford, M.A., Li, L.L., Du, X., Fisherman, J., Orkin, S.H., Doerschuk, C.M., & Dinauer, M.C. (1995). Mouse model of X-linked chronic granulomatous disease, an inherited defect in phagocyte superoxide production. Nat. Genet. 9, 202-209.

Pufahl, R.A., & Marletta, M.A. (1993). Oxidation of NG-hydroxy-L-arginine by nitric oxide synthase: evidence for the involvement of the heme in catalysis. Biochem. Biophys. Res. Commun. 193, 963-970.

Rajewsky, K., Gu, H., Kühn, R., Betz, U.A.K., Müller, W., Roes, J., & Schwenk, F. (1996). Conditional gene targeting. J. Clin. Invest. 98, 600-603.

Rathman, M., Sjaastad, M.D., & Falkow, S. (1996). Acidification of phagosomes containing Salmonella typhimurium in murine macrophages. Infect. Immun. 64, 2765-2773.

Reilly, T.J., Baron, G.S., Nano, F.E., & Kuhlenschmidt, M.S. (1996). Characterization and sequencing of a respiratory burst-inhibiting acid phosphatase from Francisella tularensis. J. Biol. Chem. 271, 10973-10983.

Reiner, N.E. (1994). Altered cell signaling and mononuclear phagocyte deactivation during intracellular infection. Immunol. Today 15, 374-381.

Rothe, J., Lesslauer, W., Lotscher, H., Lang, Y., Koebel, P., Kontgen, F., Althage, A., Zinkernagel, R., Steinmetz, M., & Bluethmann, H. (1993). Mice lacking the tumour necrosis factor receptor 1 are resistant to TNF-mediated toxicity but highly susceptible to infection by Listeria monocytogenes. Nature 364, 798-802.

Rous, P. (1925). The relative reaction within living mammalian tissues. II. On the mobilization of acid material within cells, and the reaction as influenced by the cell state. J. Exp. Med. 41, 399-411.

Ruan, J., St. John, G., Ehrt, S., Riley, L., & Nathan, C. (1999). *noxR3*, a novel gene from *Mycobacterium tuberculosis,* protects *Salmonella typhimurium* from nitrosative and oxidative stress. Infect. Immun. 67, 3276-3283.

Savill, J.S., Henson, P.M., & Haslett, C. (1989a). Phagocytosis of aged human neutrophils by macrophages is mediated by a novel "charge-sensitive" recognition mechanism. J. Clin. Invest. 84, 1518-1527.

Savill, J.S., Wyllie, A.H., Henson, J.E., Walport, M.J., Henson, P.M., & Haslett, C. (1989b). Macrophage phagocytosis of aging neutrophils in inflammation. Programmed cell death in the neutrophil leads to its recognition by macrophages. J. Clin. Invest. 83, 865-875.

Schaible, U.E., Sturgill-Koszycki, S., Schlesinger, P.H., & Russell, D.G. (1998). Cytokine activation leads to acidification and increases maturation of Mycobacterium avium-containing phagosomes in murine macrophages. J. Immunol. 160, 1290-1296.

Schonwetter, B.S., Stolzenberg, E.D., & Zasloff, M.A. (1995). Epithelial antibiotics induced at sites of inflammation. Science 267, 1645-1648.

Shiloh, M.U., MacMicking, J.D., Nicholson, S., Brause, J.E., Potter, S., Marino, M., Fang, F., Dinauer, M., & Nathan, C. (1999). Phenotype of mice and macrophages deficient in both phagocyte oxidase and inducible nitric oxide synthase. Immunity 10, 29-38.

Shisler, J.L., Senkevich, T.G., Berry, M.J., & Moss, B. (1988). Ultraviolet-induced cell death blocked by a selenoprotein from a human dermatotropic poxvirus. Science 279, 102-105.

Sibley, L.D., Weidner, E., & Krahenbuhl, J.L. (1985). Phagosome acidification blocked by intracellular Toxoplasma gondii. Nature 315, 416-419.

Spriggs, M.K. (1996). One step ahead of the game: viral immunomodulatory molecules. Annu. Rev. Immunol. 14, 101-130.

Stenger, S., Mazzaccaro, R.J., Uyemura, K., Cho, S., Barnes, P.F., Rosat, J.P., Sette, A., Brenner, M.B., Porcelli, S.A., Bloom, B.R., & Modlin, R.L. (1997). Differential effects of cytolytic T cell subsets on intracellular infection. Science 276, 1684-1687.

Stuehr, D.J. (1997). Structure-function aspects in the nitric oxide synthases. Annu. Rev. Pharmacol. Toxicol. 37, 339-359.

Stuehr, D.J., & Marletta, M.A. (1985). Mammalian nitrate biosynthesis: mouse macrophages produce nitrite and nitrate in response to *Escherichia coli* lipopolysaccharide. Proc. Natl. Acad. Sci. USA 82, 7738-7742.

Stuehr, D.J., Fasehun, O.A., Kwon, N.S., Gross, S.S., Gonzalez, J.A., Levi, R., & Nathan, C.F. (1991). Inhibition of macrophage and endothelial cell nitric oxide synthase by diphenyleneiodonium and its analogs. FASEB J. 5, 98-103.

Sturgill-Koszycki, S., Schlesinger, P.H., Chakraborty, P., Haddix, P.L., Collins, H.L., Fok, A.K., Allen, R.D., Gluck, S.L., Heuser, J., & Russell, D.G. (1994). Lack of acidification in Mycobacterium phagosomes produced by exclusion of the vesicular proton-ATPase. Science 263, 678-681.

Supek, F., Supekova, L., Nelson, H., & Nelson, N. (1996). A yeast manganese transporter related to the macrophage protein involved in conferring resistance to mycobacteria. Proc. Natl. Acad. Sci. USA 93, 5105-5110.

Supek, F., Supekova, L., Nelson, H., & Nelson, N. (1997). Function of metal-ion homeostasis in the cell division cycle, mitochondrial protein processing, sensitivity to mycobacterial infection and brain function. J. Exp. Biol. 200, 321-330.

Swallow, C.J., Grinstein, S., Sudsbury, R.A., & Rotstein, O.D. (1990). Modulation of the macrophage respiratory burst by an acidic environment: the critical role of cytoplasmic regulation by proton extrusion pumps. Surgery 108, 363-368; discussion 368-369.

Sword, C.P. (1966). Mechanisms of pathogenesis of *Listeria monocytogenes* infection. I. Influence of iron. J. Bacteriol. 92, 536-542.

Tanaka, T., Akira, S., Yoshida, K., Umemoto, M., Yoneda, Y., Shirafuji, N., Fujiwara, H., Suematsu, S., Yoshida, N., & Kishimoto, T. (1995). Targeted disruption of the NF-IL6 gene discloses its essential role in bacteria killing and tumor cytotoxicity by macrophages. Cell 80, 353-361.

Tauber, A.I., & Chernyak, L. (1989). The birth of immunology. II. Metchnikoff and his critics. Cell. Immunol. 121, 447-473.

Upchurch, G.R., Jr., Welch, G.N., & Loscalzo, J. (1995). S-nitrosothiols: chemistry, biochemistry, and biological actions. Adv Pharmacol 34, 343-349.

Valdivia, R.H., & Falkow, S. (1996). Bacterial genetics by flow cytometry: rapid isolation of Salmonella typhimurium acid-inducible promoters by differential fluorescence induction. Mol. Microbiol. 22, 367-378.

Vidal, S., Belouchi, A.M., Cellier, M., Beatty, B., & Gros, P. (1995a). Cloning and characterization of a second human NRAMP gene on chromosome 12q13. Mamm. Genome 6, 224-230.

Vidal, S., Tremblay, M.L., Govoni, G., Gauthier, S., Sebastiani, G., Malo, D., Skamene, E., Olivier, M., Jothy, S., & Gros, P. (1995b). The Ity/Lsh/Bcg locus: natural resistance to infection with intracellular parasites is abrogated by disruption of the Nramp1 gene. J. Exp. Med. 182, 655-666.

Vidal, S.M., Malo, D., Vogan, K., Skamene, E., & Gros, P. (1993). Natural resistance to infection with intracellular parasites: Isolation of a candidate for Bcg. Cell 73, 469-485.

Vidal, S.M., Pinner, E., Lepage, P., Gauthier, S., & Gros, P. (1996). Natural resistance to intracellular infections: Nramp1 encodes a membrane phosphoglycoprotein absent in macrophages from susceptible (Nramp1 D169) mouse strains. J. Immunol. 157, 3559-3568.

Vodovotz, Y., & Bogdan, C. (1994). Control of nitric oxide synthase expression by transforming growth factor-beta: implications for homeostasis. Prog. Growth Factor Res. 5, 341-351.

Vouldoukis, I., Riveros-Moreno, V., Dugas, B., Ouaaz, F., Becherel, P., Debre, P., Moncada, S., & Mossalayi, M.D. (1995). The killing of Leishmania major by human macrophages is mediated by nitric oxide induced after ligation of the Fc epsilon RII/CD23 surface antigen. Proc. Natl. Acad. Sci. USA 92, 7804-7808.

Warren, J.B., Loi, R., Rendell, N.B., & Taylor, G.W. (1990). Nitric oxide is inactivated by the bacterial pigment pyocyanin. Biochem. J. 266, 921-923.

Watanabe, K., Kagaya, K., Yamada, T., & Fukazawa, Y. (1991). Mechanism for candidacidal activity in macrophages activated by recombinant gamma interferon. Infect. Immun. 59, 521-528.

Weinberg, E.D. (1984). Iron withholding: A defense against infection and neoplasia. Physiol. Rev. 64, 65-102.

Weinrauch, Y., Elsbach, P., Madsen, L.M., Foreman, A., & Weiss, J. (1996). The potent anti-*Staphylococcus aureus* activity of a sterile rabbit inflammatory fluid is due to a 14-kD phospholipase A2. J. Clin. Invest. 97, 250-257.

Wilson, C.B., Tsai, V., & Remington, J.S. (1980). Failure to trigger the oxidative metabolic burst by normal macrophages: possible mechanism for survival of intracellular pathogens. J. Exp. Med. 151, 328-346.

Xie, Q.W., Cho, H.J., Calaycay, J., Mumford, R.A., Swiderek, K.M., Lee, T.D., Ding, A., Troso, T., & Nathan, C. (1992). Cloning and characterization of inducible nitric oxide synthase from mouse macrophages. Science 256, 225-228.

Zheng, M., & Storz, G. (1998). Activation of the OxyR transcription factor by reversible disulfide bond formation. Science 279, 1718-1721.

Zychlinsky, A. (1993). Programmed cell death in infectious diseases. Trends Microbiol. 1, 114-117.

Zychlinsky, A., & Sansonetti, P. (1997). Perspectives series: host/pathogen interactions. Apoptosis in bacterial pathogenesis. J. Clin. Invest. 100, 493-495.

Zychlinsky, A., Prevost, M.C., & Sansonetti, P.J. (1992). *Shigella flexneri* induces apoptosis in infected macrophages. Nature 358, 167-169.

COMPONENTS AND ORGANIZATION OF THE NADPH OXIDASE OF PHAGOCYTIC CELLS

ITS ROLE IN MICROBIAL KILLING AND IN THE MOLECULAR PATHOLOGY OF CHRONIC GRANULOMATOUS DISEASE

Anthony W. Segal, Frans Wientjes, R.W. Stockley, and Lodewijk V. Dekker

Advances in Cell and Molecular Biology of Membranes and Organelles
Volume 5, pages 441-483.

ISBN: 1-55938-999-0

I. INTRODUCTION

The NADPH oxidase generates superoxide in the professional phagocytic cells, neutrophils, monocytes and macrophages, and a variety of other cells derived from the myeloid lineage as well as B lymphocytes. The superoxide is generated by a short electron transport chain incorporated within a flavocytochrome that spans the membrane of the wall of the phagocytic vacuole. This flavocytochrome is the best characterized of a newly described family that transport electrons across the plasma membrane of plants, yeast, and higher animals. The flavocytochrome is selectively activated in this region of the plasma membrane by a group of specialized cytoplasmic phox proteins under the regulation of various signal transduction pathways. Deficiency of superoxide production because of genetic lesions resulting in deficiencies of the flavocytochrome b, or cytosolic phox proteins, results in chronic granulomatous disease (CGD). This syndrome is characterized by an abnormal predisposition to infection with bacteria and fungi, and to an abnormal inflammatory response.

In this chapter we will review current knowledge of the structure and function of the NADPH oxidase and show how much of this information has been revealed through an understanding of the molecular pathology of CGD.

II. THE RESPIRATORY BURST THAT GENERATES SUPEROXIDE

The "extra respiration of phagocytosis" that reflects activity of the NADPH oxidase was first noticed by Baldridge and Gerard in 1933 when examining the phagocytosis of bacteria by canine neutrophils (Baldridge and Gerard, 1933). It was not until 1959 that it was realized that this additional oxygen consumption was not simply due to the added energy requirements of phagocytosis, because it was completely resistant to inhibitors of mitochondrial function (Sbarra and Karnovsky, 1959). The oxygen is reduced to the superoxide free radical by the addition of a single electron (Babior et al., 1973) (See Figure 1). This respiratory

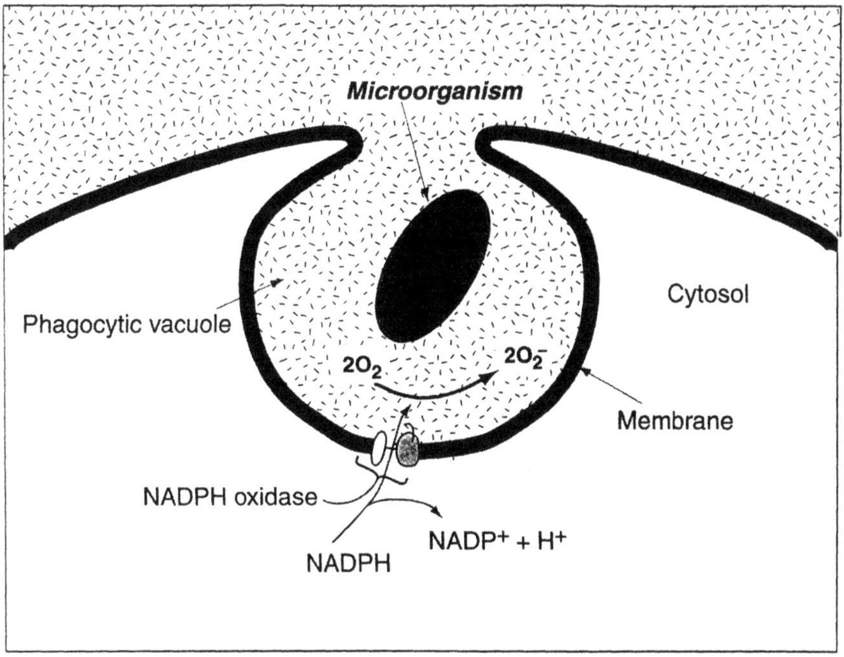

Figure 1. The NADPH oxidase becomes located in the wall of the phagocytic vacuole. It pumps electrons from NADPH into the vacuole to form superoxide.

burst was then shown to be required for neutrophils to kill bacteria normally. In the absence of oxygen bacteria are killed inefficiently, even though phagocytosis and degranulation of cytoplasmic granules are normal under these conditions (Selvaraj and Sbarra, 1966; Mandell, 1974). The most definitive evidence for the physiological and clinical importance of the respiratory burst was provided by the syndrome of CGD. CGD is a syndrome (a heterogeneous group of disorders with a common phenotype) the unifying features of which are a severe predisposition to infection with bacteria and fungi and the inability of their cells to kill bacteria *in vitro* (Clem and Klebanoff, 1966) and it is coupled to the absence of the respiratory burst (Klebanoff and Rosen, 1978; reviewed in Thrasher et al., 1994).

The oxygen is consumed by the NADPH oxidase, an electron transport chain in the wall of the phagocytic vacuole that is activated by a number of cytosolic factors. At a molecular level the common cellular lesion in CGD is caused by defects in the various components of the oxidase. These different molecular phenotypes have provided valuable models with which to dissect out and understand the different molecular components of the oxidase system, their interaction and regulation.

It has been claimed that similar proteins to those in the microbicidal phagocytes are also present in other cells including fibroblasts and in plant cells. The relevance of these observations will be discussed under the relevant sections.

III. COMPONENTS OF THE NADPH OXIDASE

A. The Flavocytochrome b Electron Transport Chain

The electron transporting component is a flavocytochrome that is activated in the wall of the phagocytic vacuole. It is an $\alpha_1\beta_1$ heterodimer (Wallach and Segal, 1996).

The β Subunit

Also called gp91^{phox} (gp for glycoprotein, 91 from a rough molecular weight, and phox from phagocyte oxidase) it is heavily glycosylated (Harper et al., 1985) and contains 570 amino acids with a predicted molecular mass of 65 kDa (Royer-Pokora et al., 1986; Teahan et al., 1987). It can be grossly divided into 2 main functional domains: i) a relatively hydrophilic C-terminal half, flavin and NADPH binding globular domain which is exposed to the cytosol. ii) a hydrophobic, membrane spanning, glycosylated, heme binding, N-terminus. Figure 2 gives a schematic representation of gp91^{phox}

The C terminal half has strong homology with the ferredoxin NADP reductase (FNR) family of reductases and has been modeled on the coordinates used to work out the crystal structure of FNR (Taylor et al., 1993). In this model it is predicted to bind FAD (Segal et al., 1992) and also NADPH, where the NADPH binding site must be exposed to the cytosol. It's closest homology is with FRE1, a ferric reductase present in *Saccharomyces cerevisiae* (Roman et al., 1993). As compared with other members of the family, both proteins have a 20 amino acid insert which in the model forms a short helix predicted to occupy the NADPH binding site in the unstimulated state (Cohen et al., 1981), and thereby inhibit activity.

The N-terminal half is much more hydrophobic, with a number of predicted transmembrane helices which probably bind the two hemes (see below) and act as a channel for electrons through the membrane. The location of the glycosylation sites by site directed mutagenesis in an *in vitro* translation system increases the confidence we can apply to the model, with particular reference to the organization of the transmembrane helices (Wallach and Segal, 1997).

The gene coding for this protein called CYBB is located on the short arm of the X chromosome at Xp21.1. It encompasses 30 kb and contains 13 exons. Defects in this gene account for about two-thirds of the cases of CGD (Roos et al., 1996). The mutations have been identified in approximately 300 families with X-linked CGD. In the vast majority the mutation involves structurally important regions of

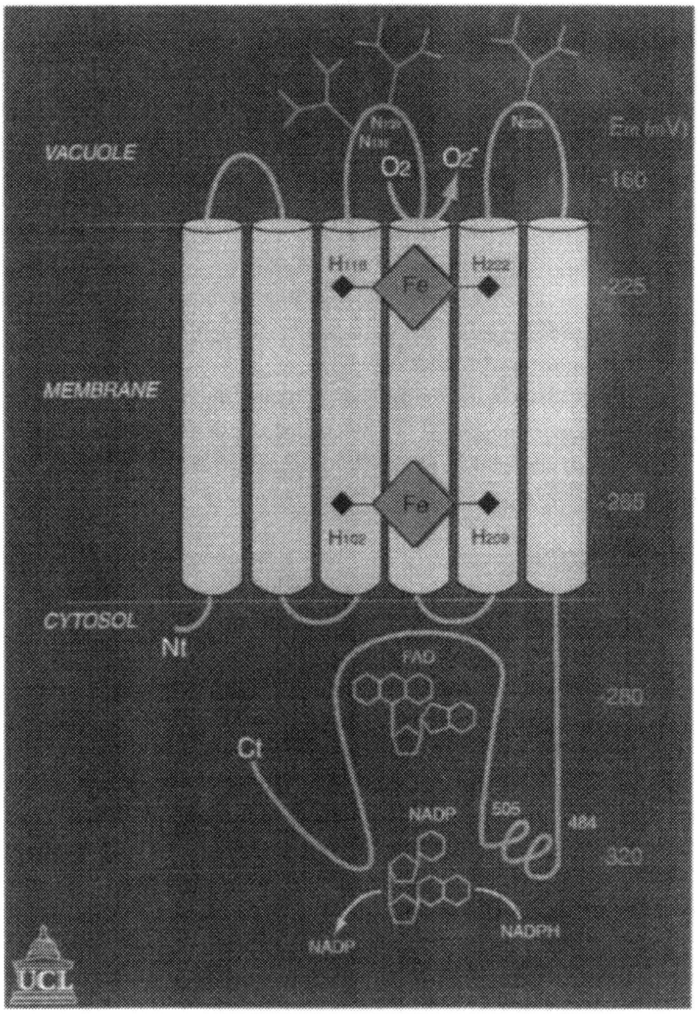

Figure 2. Schematic representation of the structure of the electron transporting component of the NADPH oxidase. This is a flavocytochrome b (called b-558 because the peak of absorbance of its α absorption band is at 558 nm). This flavocytochrome has two main domains, an N-terminal hydrophobic region that houses the two hemes, and the more hydrophilic C-terminal half that binds the FAD and NADPH. The asparagine residues to which carbohydrate is attached, and which help to orientate the transmembrane helices, are shown. The midpoint potentials of the redox molecules, from NADPH at -320 mV to superoxide at -160 mV, are shown on the right. Activation of electron transport might be activated by the withdrawl of the region between amino acids 484 to 515 from the NADPH binding site, thereby allowing the substrate to react with the FAD.

the molecule, resulting in complete loss of both subunits, which are clearly required for mutual stability (Parkos et al., 1987; Segal, 1987). In approximately 5% of mutations is the protein conserved and in these cases analysis of the biological and biochemical consequences can be very informative, as these mutations are likely to involve regulatory and catalytic rather than structural domains.

The α Subunit (Also Called p22^{phox})

This subunit has a molecular mass of 22 kDa, and is very hydrophobic, most of the molecule being composed of two hydrophobic, possibly transmembranous, helices (Parkos et al., 1987). The C-terminus has a proline rich tail that appears to be important for the binding of activating cytosolic factors which fail to translocate to the membrane if this region is mutated (Dinauer et al., 1991; Leusen et al., 1994). The gene for this polypeptide, CYBA is located on 16q24. Defects in this protein are responsible for a small proportion (approximately 5%) of autosomal recessively inherited CGD.

The Substrate Is NADPH

The substrate for the oxidase is NADPH, as its name would suggest. The NADPH is generated by the hexose monophosphate shunt (Zatti and Rossi, 1965). K_m values for the reaction are typically in the region of 30–40 μM (Bender et al., 1983; Yamashita et al., 1985). The NADPH analogues 2-azido-NADPH (Segal et al., 1992) and pyridoxal 5'-diphospho-5'-adenosine (Ravel and Lederer, 1993) have been used to affinity label the β subunit of the flavocytochrome, thus confirming the location of the NADPH binding site as that predicted from the homology data for the FNR family.

The Flavin

Flavins are the redox cofactors usually linked to NAD(P)H and are capable of coupling a two-electron donor to a one-electron transporter, in this case heme. The participation of a flavoprotein in this system was postulated as long ago as 1964 by Cagan and Karnovsky but for a long time the identity of the flavoprotein component remained elusive. This was in large part due to the weak binding of the cofactor and its ready dissociation from the apoprotein upon detergent solubilization.

The observation that superoxide generation by the detergent solubilized oxidase was stimulated by the addition of exogenous FAD (Babior and Kipnes, 1977) and inhibited by the analogue 5-carba-5-deaza FAD (Light et al., 1981) provided early evidence that a flavoprotein was an essential component of the oxidase. Diphenyl iodonium, a potent inhibitor of electron transport which is believed to act at the level of the flavoprotein, was also found to inhibit NADPH-dependent superoxide production (Cross, 1987). Perhaps the most compelling evidence, however, was the detection of a FAD semiquinone free-radical by electron paramagnetic resonance (EPR) after the addition of NADPH to membranes from activated but not unstimulated neutrophils (Kakinuma et al., 1986). Analysis of the flavin content of these membranes revealed almost exclusively noncovalently bound FAD (Kakinuma et al., 1986). Binding of FAD to the β subunit was demonstrated directly using the photoaffinity ligand $[^3H]NAP_4$-FAD (Doussiere et al., 1995).

Importantly the FAD content of neutrophil membranes was shown to be depleted for the cells of patients with X-linked, but not autosomal recessive, CGD (Segal et al., 1992), indicating a linkage between the absence of a heme spectrum, the lack of the two subunits of what we now know to be the flavocytochrome b_{558}, and gross diminution in the amount of FAD.

Binding of the flavin cofactor to the β subunit was indicated initially by the homology between the C-terminal half of gp91^{phox} and members of the FNR flavoenzyme family (Taylor et al., 1993). A comparison between the alignment of the distal 250 amino acids in the hydrophilic region of the flavocytochrome with the FNR family of reductases indicated general homology. Direct homology in the region of FAD-isoalloxazine binding was weak but the amino acids (HPFTLTS) corresponded closely in terms of charge, size and function; these structural differences might account for the relative instability of FAD binding in flavocytochrome b_{558}.

The most detailed analysis including phylogenetic tree construction (Taylor et al., 1993) identified the most related sequence as that of *S. cerevisae* ferric reductase (Dancis et al., 1992). Recently a homologous ferric reductase from *S. pombe* has been cloned and sequenced (Roman et al., 1993). Similar molecules have been found in plants (Groom et al., 1996) and *Dictyostelium*.

There is only one crystal structure of a member of the NAD(P)H reductase family; FNR from spinach leaves (Karplus et al., 1991). This has been used to construct a three-dimensional model of the homologous region of the β subunit of the oxidase (Taylor et al., 1993), which stretches from residue Leu 279 to the C-terminus. The major difference between the FNR structure and the model of flavocytochrome b-$_{558}$ (residues 484–504) and FRE1 was the presence of a 20 amino acid insert in the latter two molecules that was absent from FNR. In the model this sequence was predicted to form a short helix overlying the nucleotide binding cleft, which might have to be displaced to allow NADPH binding. Activation could involve the removal of this helix by the cytosolic factors to enable substrate binding. The regulatory role of this portion of the molecule is supported by the finding of a few rare patients with X-linked CGD in whom the flavocytochrome proteins are conserved, but nonfunctional. In some of these cases the translocation of cytosolic NADPH oxidase factors to the membranes are disturbed (Roos et al., 1996).

The Hemes

The NADPH oxidase is a b type cytochrome as evidenced by spectroscopy in alkaline pyridine which reveals the spectrum of a protoheme pyridine hemochrome (Segal and Jones, 1979). The heme is present in a molar ratio of 2:1 to FAD (Segal et al., 1992; Shatwell et al., 1996). Evidence from EPR and Raman spectroscopy indicated a low spin, six-coordinate heme with axial imidazole or imidazolate ligands (Hurst et al., 1991). The midpoint potential of the heme(s) is very low. It was originally determined at -245 mV (Cross et al., 1981) but subsequent analysis indicated that this was in fact a composite of two hemes with $E_{m\ 7.0}$ of -225 and -265 mV (Shatwell et al., 1996). In a patient with an $Arg_{54} \rightarrow$ Ser mutation in the β subunit and retention of both subunits, the value of the lower potential heme was depressed to -300 mV (Shatwell et al., 1996), below that of the flavin semiquinone at -280 mV. It seems probable that the two hemes have different functions, with the lower potential heme acting as a reductase and the other as an oxidase, binding oxygen and reducing it to superoxide. This would explain the observation that a maximum of only 40% of the heme is involved in the binding of CO (Cross et al., 1981, 1982).

The precise location of the hemes is uncertain, but recent evidence suggests that they are both located on the β subunit. After the strong homology had been demonstrated between the β subunit and FRE1, the yeast ferric reductase, overexpression and gene deletion models showed the latter to be a low potential flavocytochrome b (Shatwell et al., 1996). Subsequently it was found that there were two pairs of histidines conserved in other cytochromes that were predicted to be located in transmembrane helices about 12 amino acids apart, an ideal organization to locate the two hemes at either side of the membrane. Site directed mutagenesis of any one of the four histidines in *S. cerevisiae* (Finegold et al., 1996) led to the loss of all the cytochrome b heme spectrum. Because all rather than one half of the heme was lost, it is not possible to distinguish a general disorganization of the molecule due to a loss of one of the hemes from the specific targeted disruption of a heme binding ligand.

Expression of the β subunit in cultured COS cells in the absence of the α subunit has shown this molecule to house the hemes (Yu et al., 1998).

This β subunit is the prototype for a family of flavocytochromes that transport electrons across the plasma membrane. In addition to FRE1, homologues have been described in plants (Groom et al., 1996) in which a defence response appears to involve the production of hydrogen peroxide but not superoxide, by a flavocytochrome with a much higher potential than that of the phagocyte oxidase. The superoxide generating system of fibroblasts (Emmendorffer et al., 1993) seems to involve a different cytochrome system (Meier et al., 1993).

B. Cytosolic Factors

Three Phox Protein Components in the Cytosol

The discovery that cytosolic factors are necessary for activation of the NADPH oxidase came from the development in the mid-1980s of a cell-free assay system for measuring the oxidase activity. It was found by several investigators that in broken cell preparations no oxidase activity was detectable but that this activity could be induced by addition of amphiphilic compounds as arachidonic acid and sodium dodecyl sulfate (SDS) (Bromberg and Pick, 1984, 1985; Curnutte, 1985; McPhail et al., 1985). This was an important finding because it directly showed that the particulate fraction of cell homogenates which contain the membrane-bound flavocytochrome could not be activated to generate superoxide unless the cytosol fraction was also present (Curnutte et al., 1987; Babior et al., 1988). Neutrophil cytosol from many autosomal recessive (AR)-CGD patients lacked this cytosolic contribution (Fujita et al., 1987). The cell-free assay allowed further characterization of the cytosolic factors by fractionation of cytosol and by complementation analysis of material derived from different AR-CGD patients.

The first biochemical abnormality which was recognized in AR-CGD was a failure to phosphorylate a 47 kDa protein upon activation with phorbol myristate acetate (PMA), a protein kinase C (PKC) stimulant (Segal et al., 1985). Fractionation studies of neutrophil cytosol coupled with complementation analysis of cytosols from AR-CGD cytosols showed that this protein was one of several cytosolic components which act synergistically in the activation of the NADPH oxidase in the cell-free assay (Curnutte et al., 1988; Nunoi et al., 1988; Volpp et al., 1988; Bolscher et al., 1989, 1990; Pick et al., 1989; Sha'ag and Pick, 1990). Identification of two of these components was greatly facilitated by the fortuitous development of antibodies against them in a rabbit antiserum prepared against the eluate from a GTP affinity column (Volpp et al., 1988). Although the GTP column was used to search for a GTP-binding protein involved in the activation of the oxidase (see below), the main immunogenic proteins found in the eluate were proteins with apparent molecular weights of 47 and 67 kDa, respectively. The 47 kDa protein was absent in most patients with AR-CGD and corresponded with the missing phosphoprotein, whereas the 67 kDa protein was missing in a small subgroup of AR-CGD patients. Both proteins have been purified from several sources (Tanaka et al., 1993; Teahan et al., 1990; Pilloud-Dagher and Vignais, 1991) and have been named $p47^{phox}$ and $p67^{phox}$.

The above-mentioned antiserum allowed the cloning and sequencing of the cDNAs coding for these proteins (Lomax et al., 1989; Volpp et al., 1989a,b; Leto et al., 1990; Rodway et al., 1990). The $p47^{phox}$ gene codes for a protein of 390 amino acids with a predicted molecular weight of ca 44 kDa and the $p67^{phox}$ gene for a protein of 526 amino acids and a predicted molecular weight of ca 61 kDa. In the neutrophil cytosol, these proteins occur in a complex with a molecular weight of about 240–300 kDa (Park et al., 1992; Wientjes et al., 1993) and it was likely that they might be complexed to other protein molecules. By co-immunoprecipitation, it was found that they were complexed with a 40 kDa protein, named $p40^{phox}$ (Wientjes et al., 1993). Other groups also found this protein, either by copurification with $p47^{phox}$ and $p67^{phox}$ (Wientjes et al., 1993) or as above, by co-immunoprecipitation (Tsunawaki et al., 1994). The gene for this protein has been cloned and sequenced as well (Wientjes et al., 1993) and codes for a protein of 339 amino acids and a predicted molecular weight of 37 kDa. $p40^{phox}$ and $p47^{phox}$ show a 22% identity over the N-terminal two-thirds of their sequence with homology approaching 67% when conservative substitutions are taken into account. All three proteins have a distinct domain structure that is described in the next section.

The three phox proteins described here all translocate to the membrane upon activation. This occurs both in intact neutrophils and in cell-free systems (Clark et al., 1990; Doussière et al., 1990). The flavocytochrome is their docking site because they do not translocate in the absence of flavocytochrome (X-CGD) (Heyworth et al., 1991; Wientjes et al., 1993). It is not clear whether the proteins move to the membrane as a complex or individually. Although they exist as a complex in the cytosol, the main interaction seems to be between $p67^{phox}$ and $p40^{phox}$ because i) part of $p47^{phox}$ is uncomplexed (Park et al., 1992; Wientjes et al., 1993); ii) $p67^{phox}$ and $p40^{phox}$ remain stably associated during purification procedures (Wientjes et al., 1993); iii) levels of $p40^{phox}$ are diminished in $p67^{phox}$-deficient neutrophils (Wientjes et al., 1993). There are indications that $p47^{phox}$ is the primary mover because in $p47^{phox}$-deficient cells $p67^{phox}$ does not translocate whereas in the reverse situation $p47^{phox}$ does (Kleinberg et al., 1990; Heyworth et al., 1991) In the cell-free assay, $p47^{phox}$ is not absolutely required for activation: high concentrations of $p67^{phox}$ and *rac* (see below) can substitute for it (Freeman & Lambeth, 1996). This suggests that the role of $p47^{phox}$ is to bring $p67^{phox}$ in the right place for activation of electron transport. See the section "Interactions Between the Cytosolic Factors" for a more detailed discussion of the interactions of the cytosolic factors with the flavocytochrome.

Concomitant with activation and translocation, phosphorylation takes place. This has been most thoroughly investigated for $p47^{phox}$ which becomes phosphorylated in a series of steps (Okamura et al., 1988a, b; Rotrosen and Leto, 1990) the first of which occur in the cytosol, but later phosphorylations take place at the membrane and are dependent on the presence of the flavocytochrome (Heyworth et al., 1989, 1991; Rotrosen and Leto, 1990). It is not clear what the functions of the phosphorylations are in activation. The seven serine residues in the C-terminal part of $p47^{phox}$ which become phosphorylated have been identified (El Benna et al., 1994a, 1996a) and by site-directed mutagenesis it has been determined that serine-379 is the crucial one: its replacement by alanine inhibited translocation and activation after transfection into p47^{phox} B lymphocytes (Faust et al., 1995).

In the cell-free assay, activation of the oxidase can take place without phosphorylation events, probably because the negatively charged amphiphiles substitute for phosphate groups and allow interaction of components at the membrane by charge neutralization. Recently, cell-free assays have been described in which amphiphiles were replaced by phosphorylation of p47^{phox} (Park and Ahn, 1995; Park et al., 1997). For a more thorough discussion on the role of p47^{phox} phosphorylation in the activation process, see the section on phosphorylation of oxidase components.

p67^{phox} (Dusi and Rossi, 1993; El Benna et al., 1997), p40^{phox} (Fuchs et al., 1997), and the flavocytochrome (Garcia and Segal, 1988) become phosphorylated as well but these phosphorylation events are less well investigated and their role in NADPH oxidase activation is even less well-understood. p47^{phox} and p67^{phox} are thought to activate the electron transport through the flavocytochrome. However, p40^{phox} is not needed in the cell-free assay. Its role in the NADPH oxidase is not clear. Recently, some conflicting observations have been published regarding its function. An antibody against the C-terminus of p40^{phox} caused dissociation of p40^{phox} and p67^{phox} and caused inhibition of the oxidase indicating a role for p40^{phox} in the activation mechanism (Tsunawaki et al., 1996). On the other hand, cotransfection of p40^{phox} in K562 cells together with the other phox proteins caused inhibition of superoxide production from which it was concluded that p40^{phox} downregulates NADPH oxidase activity (Sathyamoorthy et al., 1997). More work is clearly needed in this area to elucidate the precise roles of the different phox proteins.

Involvement of a Small GTP-Binding Protein

Although p47^{phox} and p67^{phox} were able to reconstitute some NADPH oxidase activity in the cell-free assay, the fractionation studies mentioned above also indicated the presence of a third factor which was GTP dependent: the addition of GTP-γ-S greatly enhanced the cell-free oxidase activity. This factor was subsequently identified by two different groups as the small GTP-binding protein *rac*1 (Abo and Pick, 1991; Abo et al., 1991) and *rac*2 (Knaus et al., 1991). These proteins are 92% homologous and, whereas *rac*1 has a wider tissue distribution, *rac*2 is relatively specific for myeloid cells (Didsbury et al., 1989) and might be the one involved in NADPH oxidase activation in neutrophils. Alternatively, the isolation of the two different forms of *rac* might reflect a species difference because *rac*1 was isolated from guinea pig macrophages (Abo and Pick, 1991; Abo et al., 1991) and *rac*2 from human neutrophils (Knaus et al., 1991).

Although some researchers could not find a correlation between *rac*2 translocation to the membrane and activation of the NADPH oxidase (Philips et al., 1995), a number of *in vivo* experiments clearly implicated *rac* in NADPH oxidase activation: introduction of *rac* antisense oligonucleotides in B lymphocytes inhibited the oxidase (Dorseuil et al., 1992; Voncken et al., 1995); neutrophils of *bcr* knock-out mice showed an increase of *rac*2 membrane translocation and of superoxide production upon activation (98) (*bcr* is a GTPase activating protein for *rac* and lower levels of *bcr* can therefore be expected to give rise to higher levels of active *rac*); in differentiated HL60 cells transfected with dominant-negative mutant forms of *rac* superoxide production was inhibited (Gabig et al., 1995).

The *rac* proteins belong to the Ras superfamily with which they share approximately 30% sequence homology and more specifically to the Rho subfamily. They posses the C-terminal CAAX motif which localizes them to the plasma membrane. *rac* was purified as a heterodimer, complexed to *rho*-GDI (GDP dissociation inhibitor). Dissociation of *rac* from this inhibitory protein and its translocation to the membrane are necessary for activation of the NADPH oxidase (Abo et al., 1994). Detailed studies, using site-directed mutagenesis (Diekmann et al., 1994; Freeman et al., 1994, 1996; Nisimoto et al., 1997) or *rac-cdc*42*hs* chimeric proteins (Kwong and Leto, 1995), showed the presence of two regions in *rac* involved in interaction with the NADPH oxidase: the region homologous to the *ras* effector region (amino acids 26–40) and a 12-amino acid region not present in *ras*, the "insert" region (amino acids 123–135). For a more detailed discussion on the interactions of *rac* with the NADPH oxidase components, see the section "Interactions Between the Cytosolic Factors."

Domain Structure of the Cytosolic Phox Proteins

The cloning and characterization of $p47^{phox}$, $p67^{phox}$ and $p40^{phox}$ revealed that they contain specific domains important for protein-protein interactions and NADPH oxidase activation. Four classes of domain have been characterized, including the SH3 domain, PC motif, tetratrico-peptide repeats and the phox domain. The spatial organization of these domains is summarized in Figure 3.

SH3 Domains

This domain was first discovered in the Src family of kinases and termed Src-homology 3 domain (SH3). Biochemical and structural studies have revealed the SH3 domain as a 60 residue motif which forms a β-barrel comprising of 5 antiparallel β-sheets. This domain was found to bind the proline-rich motifs R/KXXPXXP or PXXPXR/K which form a type II left-handed poly-proline helix (Guillory et al., 1997). Within the phox proteins, $p47^{\text{phox}}$ and $p67^{phox}$ contain two SH3 domains whereas $p40^{phox}$ contains a single SH3 domain. Potential target proline-rich regions have been found in $p47^{phox}$, $p67^{phox}$ and in the flavocytochrome, and are responsible for some of the protein-protein interactions found between components of the NADPH oxidase. It is worth noting that there are many pro-

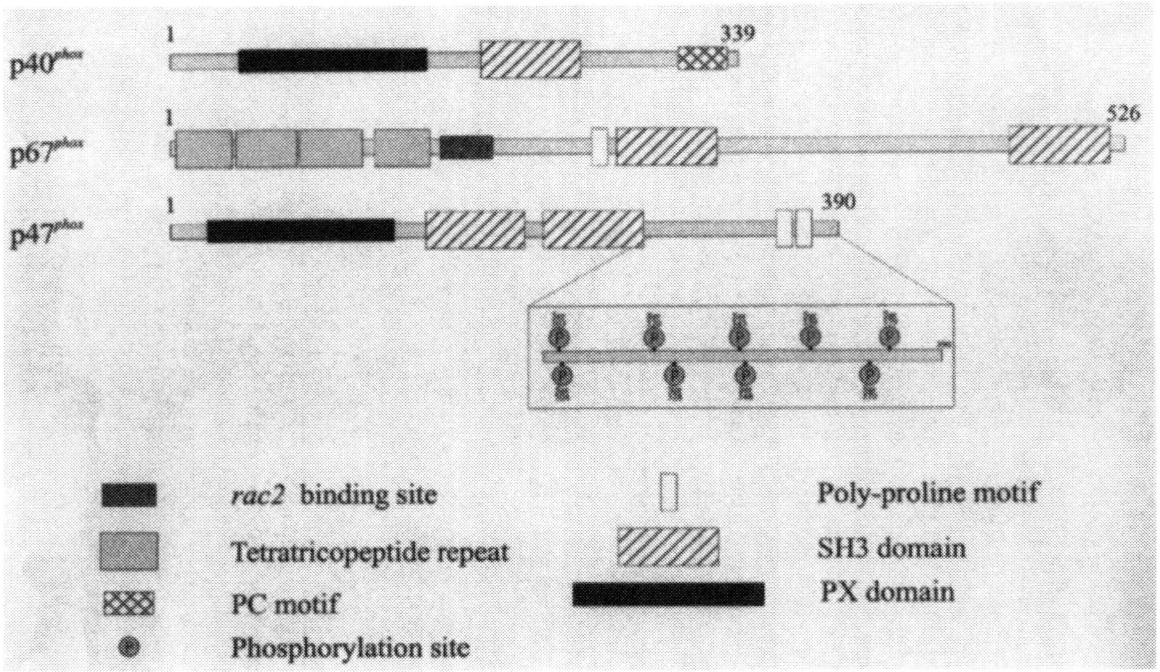

Figure 3. The domain structure of the cytosolic phox proteins. Schematic showing the known domains as elucidated by sequence homology and expression and functional studies of truncated proteins.

line-rich regions within the phox proteins and accurate mapping of all relevant sites is incomplete.

Tetratricopeptide Repeat

The tetratricopeptide repeat (TPR) is a degenerate 34 residue repeat which has been identified in a wide variety of proteins and has been implicated in protein-protein interactions and polyunsaturated fatty acid binding such as arachidonic acid. It has been postulated by multiple sequence alignments that $p67^{phox}$ contains four TPR domains (Ponting, 1996) which are found at the N-terminus in a region distinct from the *rac* binding site. Recent structural analysis of three TPR domains from protein phosphatase 5 (Barford et al., 1998) show each repeat to consist of a pair of antiparallel α-helices which may form a super-helical structure with neighboring TPR domains. A single site mutation found within TPR3 (Gly 78 to Glu substitution) (de Boer et al., 1994), which leads to nonfunctional $p67^{phox}$, is predicted to disrupt the TPR folding and potential TPR mediated interactions. This would account for the loss of function because, although the role of this region is unclear, it is necessary for NADPH oxidase activation.

PC Motif/Octicosapeptide Repeat

Multiple sequence alignment revealed a 28 residue repeat at the C-terminus of $p40^{phox}$ (Ponting, 1996) which has been found in various other proteins including PKC isoforms λ and ζ. This motif is thought to be important for protein-protein interactions and subsequent biochemical studies and site-directed mutational analysis have highlighted the importance of this region in the binding of $p67^{phox}$ (Nakamura et al., 1998).

Phox Domain

There exists a region of homology between the N-terminus of $p40^{phox}$ and $p47^{phox}$ which on further analysis reveals an approximately 100 residue domain which is also found in other proteins including Cpk-like PtdIns 3-kinases (Ponting, 1996). This region has been termed the PX domain and may contain an SH3 binding poly-proline motif but its role is unknown at this time.

Interactions between the Cytosolic Factors

With so many SH3 domains and proline-rich sequences (PRSs) around there are many possible interactions between the phox proteins and many have indeed been reported. Some, but by no means all, of these interactions act through SH3 domains binding to PRSs. A number of investigators using various techniques such as blot overlays, binding to proteins on beads, and the yeast two-hybrid sys-

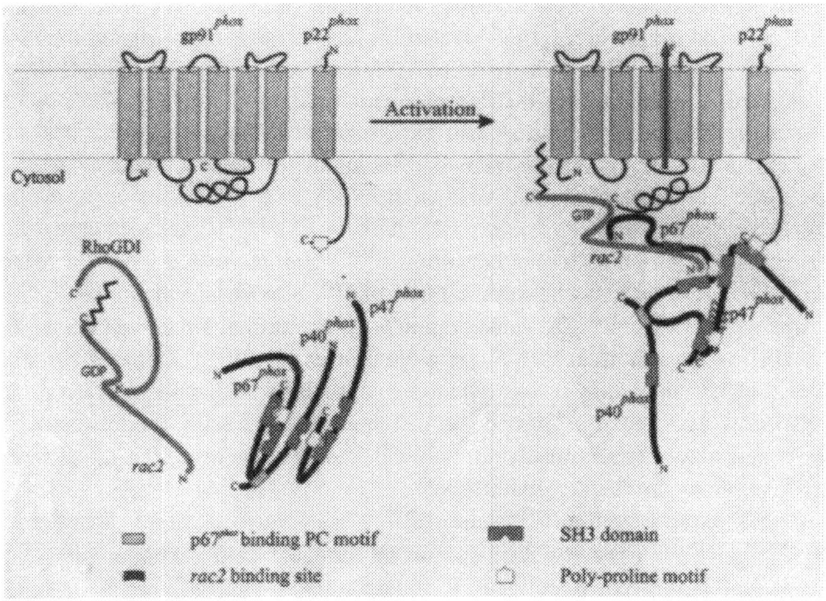

Figure 4. Model depicting organization of the NADPH oxidase components in an active and inactive state. This summarizes most of the information elucidated by biochemical studies on the spatial organization of the cytosolic *phox* proteins with respect to the membrane bound flavocytochrome. The model presented here shows the potential path of activation by which p67^{phox} and *rac*2 activate the flavocytochrome directly by binding to gp91^{phox} and allowing its substrate NADPH to bind and initiate the electron transport chain. The point of interaction is postulated to the residues 484 to 505 which may form an α-helical region within gp91^{phox}. For full details of the all the interactions see text.

tem, have found that the C-terminal SH3 domain of p67^{phox} binds to the C-terminus of p47^{phox} (amino acids 358–390) (Finan et al., 1994; Leto et al., 1994; Leusen et al., 1995; Fuchs et al., 1996). The binding sequence in the p47^{phox} terminus is the proline-rich motif in this region and the minimal sequence required has been determined as QPAVPPRPS where the proline-interlocked arginine is crucial for binding (Finan et al., 1994).

The tandem SH3 domains of p47^{phox} seem to play a very important role in the activation process. This domain is capable of binding p47^{phox}'s own C-terminus as well as p22^{phox}, the flavocytochrome small subunit, as well as p67^{phox} (Sumimoto et al., 1994, 1996; de Mendez et al., 1996, 1997). In resting state p47^{phox} seems to be folded back onto itself with its C-terminal PRS binding to its own SH3 domain. Activators such as arachidonic acid and SDS are capable of unmasking the SH3 domains allowing it to bind to the PRS in p22^{phox} (amino acids 149–162) (Leto et al., 1994; Sumimoto et al., 1994). The mutation Pro156—Glu, found in a p22^{phox}-related CGD patient (118) disrupted the binding of the p47^{phox} SH3 domains to p22^{phox} (Leto et al., 1994; Sumimoto et al., 1994). An arachidonic acid/SDS dependent binding of p47^{phox} SH3s to the N-terminal half of p67^{phox} has also been reported (Sumimoto et al., 1994; de Mendez et al., 1996). It is not known at the moment what the binding region in the p67^{phox} N-terminus is and whether the PRS in p67^{phox} is involved or not. Recently, the specificity of the two SH3 domains of p47^{phox} has been unraveled: the first (most N-terminal) one binds to p22^{phox} (Sumimoto et al., 1996; de Mendez et al., 1997) whereas the second binds p67^{phox} (de Mendez et al., 1997).

The p67^{phox}–p47^{phox} PRS interaction is not required for activation of the NADPH oxidase in the cell-free assay (Leusen et al., 1995) and the C-terminal half of p67^{phox} including its two SH3 domains can be deleted without effect on the cell-free oxidase activity. However, both domains are necessary for oxidase activity in an *in vivo* model (transfection of p67^{phox} constructs into p67^{phox} -deficient EBV cell lines (de Mendez et al., 1994).

p40^{phox} binds to both p67^{phox} and p47^{phox} (Wientjes et al., 1996) although the p40^{phox}–p67^{phox} interaction is of significantly greater affinity with a dissociation constant of 43 nM. The p40^{phox}–p47^{phox} interaction is again a SH3-PRS one. The SH3 domain of p40^{phox} was shown to bind to the C-terminus of p47^{phox} (Fuchs et al., 1995, 1996; Ito et al., 1996; Wientjes et al., 1996) and the same PRS as in the p67^{phox} SH3 binding was involved (Fuchs et al., 1996; Ito et al., 1996). Competition between the two SH3 domains for this PRS was observed (Ito et al., 1996). The interacting regions between p40^{phox} and p67^{phox} are as follows. Some interaction of p40^{phox} with the N-terminal half of p67^{phox} was observed (Wientjes et al., 1996; Sathyamoorthy et al., 1997) but the main binding area on p67^{phox} for p40^{phox} was shown to be the p67^{phox} inter-SH3 region (amino acids 300–460) (Fuchs et al., 1996; Nakamura et al., 1998; Sathyamoorthy et al., 1997). The p67^{phox}–binding region on p40^{phox} was shown to be the C-terminal half of the protein (amino acids 260–339) (Fuchs et al., 1995, 1996; Wientjes et al., 1996; Nakamura et al., 1998). Although the complete C-terminal region of p40^{phox} was necessary for binding to the inter-SH3 region of p67^{phox}, the PC motif (amino acids 285–306) is especially important (Nakamura et al., 1998).

Binding to the Flavocytochrome

As mentioned above, p47^{phox} associates with p22^{phox}, the small subunit of the flavocytochrome, through a SH3-PRS interaction. There are also indications for interactions of p47^{phox} with gp91^{phox}, the large subunit. These indications were not obtained by *in vitro* binding studies or by the yeast two-hybrid system but mainly by work with peptides. Early work showed that peptides corresponding to the C-terminus of the gp91^{phox} and p22^{phox} inhibited activation of the NADPH oxidase and translocation of p47^{phox} and p67^{phox} to the membrane in the cell-free assay (Kleinberg et al., 1990; Nakanishi et al., 1992). Furthermore, a point mutation in gp91^{phox} has been described, substituting Asp500 for Glu, that resulted in defective translocation of p47^{phox} and p67^{phox} although normal amounts of the flavocytochrome were present in the membrane (Leusen et al., 1994b). A peptide with this mutation was subsequently shown to inhibit translocation of the cytosolic factors and activation of the oxidase in the cell-free system.

Other indications for a multiple-site interaction between p47^{phox} and the flavocytochrome came from experiments with random-sequence peptide display libraries (DeLeo et al., 1995a, b, 1996). In these studies gp91^{phox} sequences Ser86-Lys93 and Phe450-Lys457 were identified as binding regions for p47^{phox} and the corresponding synthetic peptides inhibited translocation of the cytosolic factors and activation of the oxidase.

The peptide sequence A323YRRNSVRL332 in the C-terminal part of p47^{phox}, which was found earlier to be inhibitory as a synthetic peptide in the cell-free assay (Nauseef et al., 1993), appeared to be involved in binding to gp91^{phox} and was found in the same type of experiment to bind to p67^{phox} as well (DeLeo et al., 1996). *In vitro* binding experiments, described in the same paper, showed competition of gp91^{phox} and p67^{phox} for binding to p47^{phox} but only after activation by amphiphiles. The authors concluded that "p67^{phox} and the flavocytochrome b utilize a common binding site in p47^{phox} presumably at distinct stages during the activation process" (DeLeo et al., 1996). It might be of relevance that the serine found in this sequence is one of the phosphorylated serines (Ser328). Phosphorylation might change the specificity of binding and shift it from gp91^{phox} to p67^{phox} or vice versa.

It is thus becoming clear that in the NADPH oxidase system there are no simple one-domain-one-domain protein interactions but that activation/deactivation of the oxidase is a dynamic multiple-site interaction process. Figure 4 is a model for these shifting multiple interactions during the activation process. We should keep in mind, however, that the details of this process are still poorly understood and that there are many unanswered questions, for example what the functions are of the numerous other domains in the phox proteins (e.g., the N-terminal p67^{phox} SH3 domain, the p67^{phox} PRS, the phox domain) and what the physiological roles are of the different interactions which have been shown to take place.

Interactions with the Cytoskeleton and with rac

There is circumstantial evidence that the cytosolic oxidase factors interact with the cytoskeleton: p47^{phox} and p67^{phox} are partially located in the residue which remained as insoluble material after detergent extraction of neutrophils (Nauseef et al., 1991; Woodman et al., 1991). Further, the small GTP-binding protein *rac* is involved in membrane ruffling and cytoskeletal rearrangements in fibroblasts (Ridley et al., 1992) and is also implicated in oxidase activation (see above). In a recent study we showed a more specific interaction: coronin, an actin-binding protein was found to be associated with the cytosolic oxidase factors. *In vitro* binding studies indicated a p40^{phox}-coronin interaction in which the C-terminal half of p40^{phox} was involved (Grogan et al., 1997). Double labeling in immuno-electron microscopy showed the association p67^{phox}-coronin on cytoskeletal fibers of the submembranous cytoskeleton. Interestingly, oxidase activation had a strong cytoskeletal rearrangement effect (Grogan et al., 1997): cytosolic oxidase components as well as actin and coronin underwent massive relocalizations. These cytoskeletal rearrangements did not occur in cells deficient in p47^{phox} or p67^{phox} suggesting a regulatory role for these proteins in cytoskeletal rearrangements which accompany neutrophil activation.

Another binding partner for a cytosolic phox protein is *rac*. This protein also translocates to the membrane upon activation but contradictory data have been published on the role of the oxidase components in the translocation (Quinn et al., 1993; El Benna et al., 1994b; Heyworth et al., 1994; Dusi et al., 1995; Philips et al., 1995). *In vitro* binding studies showed it to bind to the N-terminal 199 amino acids of p67^{phox} (Diekmann et al., 1994; Prigmore et al., 1995). In human neutrophils it is probably *rac*2: it is more abundant than *rac*1 in myeloid cells (Abo et al., 1994) and in the yeast two-hybrid system *rac*2 bound to p67^{phox} six times more strongly than *rac*1 (Dorseuil et al., 1996). The region in *rac* necessary for binding to p67^{phox} was the effector region: mutations herein abolished binding to p67^{phox} and activation of the oxidase (Diekmann et al., 1994; Freeman et al., 1994, 1996; Kwong and Leto, 1995; Nisimoto et al., 1997). *Rac*'s "insert" region was also implicated in oxidase regulation (Nisimoto et al., 1997) where mutations in this region do not affect p67^{phox} binding but do inhibit NADPH oxidase activation. This implies that residues in *rac* may play a direct role in activation and possibly binding to the flavocytochrome.

A mutation in p67^{phox} deleting Lys58 (Leusen et al., 1996) has been reported to abolish *rac* binding but, in the light of the domain structure of this region of p67^{phox}, it seems more likely that this is the result of p67^{phox} mis-folding.

Recently, the *rac*-binding region of p67^{phox} was mapped to the sequence Lys170-Leu199 (Ahmed et al., 1998). Interestingly, deletion of the p67^{phox} PRS or of the C-terminal SH3 domain resulted in a ca 10-fold increase in *rac* binding as well as *in vitro* phosphorylation of p67^{phox} by p21-activated kinase (PAK). It was suggested that p67^{phox} folds back on itself, just like p47^{phox}, via a SH3-PRS interaction thereby making the *rac* binding site cryptic. This would explain some other observations as well: i) C-terminally truncated forms of p47^{phox} and p67^{phox} were able to activate the cell-free system without amphiphiles (Hata et al., 1998), and ii) deletion of the PRS of p47^{phox} resulted in a three-fold decrease of oxidase activity in the cell-free assay, but this negative effect was abolished by using a p67^{phox} mutant lacking the C-terminal SH3 domain (Hata et al., 1997). This domain seemed to have a negative regulatory role which is suppressed by interaction with the p47^{phox} PRS. Both observations could be explained by the hypothesis that the p67^{phox} C-terminal SH3 domain loops back to the p67^{phox} PRS.

Cytosolic Factors in Other Cells Including Plants

p47^{phox} and p67^{phox} are much more abundent in B lymphocytes than is the flavocytochrome (Chetty et al., 1995). This could relate to the possible role that these proteins play in the regulation of the cytoskeleton (Grogan et al., 1997) rather than exclusively to activation of the NADPH oxidase. A cytoskeletal function might also account for the presence of these cytosolic proteins in fibroblasts (Jones et al., 1994) from which the flavocytochrome is absent (Meier et al., 1993).

The description of p47^{phox} and p67^{phox} in plants that react with antibodies to the human proteins (Dwyer et al., 1996) is another matter. Unfortunately these proteins have not been characterized beyond their immunoreactivity, which could be nonspecific. At a recent meeting in Antwerp on Plasma Membrane Redox systems, a number of groups announced that they had been unable to repeat these results (Segal, personal communication).

IV. ACTIVATION MECHANISMS

A. Activation in the Cell Free Assay

Activity in the "cell free" system can only been reconstituted in the presence of certain amphiphiles such as SDS or arachidonate, and requires the presence of *rac* in its GTP-bound state. The dual dependence on GTP and amphiphiles highlights the two pathways thought to regulate the oxidase, one involving the small GTP-binding protein *rac* and the other resulting from conformational changes in the constituents of the complex provoked by the amphiphile.

GTP-Dependence of Oxidase Activation

GTP-loaded *rac* is required for effective reconstitution using purified components whereas GDP-bound *rac* is ineffective (Abo et al., 1991, 1992; Abo and Segal, 1995). Such a strict GTP dependence is less clear from the reconstitution of crude subcellular fractions, however a GTP binding protein may still be critically involved in the activation process. Indeed addition of GTPγS in this crude cell free system leads to further activation of the oxidase and endogenous GTP present in reconstituted subcellular fractions may account for activity in the absence of added GTP (Philips et al., 1995). The *rac* requirement in the purified system suggests a direct action on one or more of the components, be it p47^{phox}, p67^{phox} or the cytochrome. Direct binding of *rac* to p67^{phox} has been observed concomitant with oxidase activation (Diekmann et al., 1994). Perhaps *rac* binding results in a conformational change in p67^{phox} allowing productive protein interactions with the other components. In addition membrane binding of p67^{phox} may be mediated by *rac* since membrane interaction of *rac* occurs upon activation (Kreck et al., 1996) (see previous sections for more detailed discussion of *rac* interactions with oxidase components).

Amphiphile Dependence of Oxidase Activation

While the GTP dependence of NADPH oxidase activation finds a logical explanation in the involvement of GTP loaded *rac*, the stimulation of activity by amphiphiles is less straightforward. Several recent publications have provided plausible explanations for the amphiphile effect which center around the fact that they induce conformational changes in the oxidase components much in the same way as normal cellular activation mechanisms would.

An important observation has been that it is possible to activate the oxidase in a reconstituted system in the absence of amphiphiles by presenting prephosphorylated $p47^{phox}$ (Park et al., 1997). This suggests that the biological action of the amphiphile is very similar to that of the phosphorylation event, i.e., it induces a productive conformational change in $p47^{phox}$. The fact that the amphiphile provokes oxidase activation in the absence of ATP suggests that its action does not require kinase activity and confirms that it essentially mimics the action of kinases acting on $p47^{phox}$ (Park et al., 1997). Direct evidence for this has been provided by *in vitro* studies using purified $p47^{phox}$ in which amphiphiles induce conformational changes which are very similar to those occurring on phosphorylation (Park and Babior, 1997; Swain et al., 1997). It was observed recently that, like prephosphorylated $p47^{phox}$, a $p47^{phox}$ mutant lacking its C-terminus (the main area of phosphorylation; see below) supports oxidase activation in the cell free system in the absence of amphiphiles (Hata et al., 1998). Thus at the molecular level amphiphiles or phosphorylation open up the $p47^{phox}$ structure by displacing the C-terminus thereby allowing interaction with other oxidase components. This modification appears in principle sufficient to drive oxidase activation. It should be noticed that oxidase activation can occur even in the absence of $p47^{phox}$ at suprastoichiometric levels of $p67^{phox}$ and *rac* (Freeman and Lambeth, 1996; Koshkin et al., 1996). This has led to the suggestion that $p47^{phox}$ is not so much an essential part of the core enzyme function but rather facilitates the assembly of the complex and could as such be regarded as an adaptor protein. It is clearly essential in the *in vivo* context since many CGD cases are attributed to the absence of this oxidase component.

Oxidase activity can be induced by prephosphorylated $p47^{phox}$ in the absence of ATP, however the level of activation is not the same as that induced by amphiphiles (Park et al., 1997). Thus the amphiphiles may target components other than $p47^{phox}$ and perhaps mimic conformational changes in these as well. In this respect the observation that $p67^{phox}$ as well as $p40^{phox}$ are phosphoproteins is of interest although the contribution of the phosphorylation to structural changes in these proteins is less defined. Alternatively, rather than mimicking a phosphorylation event resulting in a conformational change in the oxidase components the amphiphiles may themselves be necessary for the activation process by binding to and activating the oxidase in a specific manner. The observation that phospholipase A_2 activity, resulting in the generation of arachidonic acid in the cell, is important for activation of the NADPH oxidase is in this respect relevant (Dana et al., 1998).

B. Phosphorylation of Oxidase Components

Phosphorylation of $p47^{phox}$ occurs at its C-terminal end in which 9 or 10 serine phosphorylation sites have been identified by phosphopeptide sequencing and site-directed mutagenesis (El Benna et al., 1994a, 1996a). Several kinases are known to phosphorylate $p47^{phox}$ *in vitro*, among them PKC, cAMP-dependent protein kinase A (PKA), PAK, a number of proline-directed kinases, an unknown phosphatidic acid (PA)-activated kinase, and four renaturable kinases that are activated by phosphatidylinositol-3-kinase (PI-3-kinase) (El Benna et al., 1994a, 1996a, b; Ding et al., 1995, 1996; Knaus et al., 1995; Prigmore et al., 1995; Waite et al., 1997). Phosphopeptide maps of $p47^{phox}$ suggest that phosphorylation of $p47^{phox}$ in activated neutrophils occurs at serine 303, 304, 320, 328, 345, 348, 359 and/or 370 and 379 (El Benna et al., 1996a). Of these sites, serine 345 and 348 are consensus sites for proline directed kinases and these residues are phosphorylated *in vitro* by two proline-directed kinases—extracellular signal-regulated kinase (ERK) and p38—but not by a third member of this family, the Jun N-terminal kinase (JNK) (El Benna et al., 1996a, b). Phosphorylation by PKC *in vitro* occurs at all residues except serine 345 and 348 while phosphorylation by PKA was reported to occur at serines 320, 359 and/or 370 and perhaps at serine 328 (El Benna et al., 1996a). PAK-1 and PAK-2 phosphorylated a $p47^{phox}$ peptide based on the sequence around serine 328 but not other $p47^{phox}$ peptides (Knaus et al., 1995). Serine 328 was phosphorylated in phorbol ester-stimulated neutrophils suggesting that this site may be targeted by both PKC and PAK or that PAK is activated by PKC activation. On the whole only limited information is available as to the phosphorylation of these individual serines upon stimulation *in vivo*. The mapping was performed on phorbol ester activated neutrophils (El Benna et al., 1996a) which to some extent biases the profile towards PKC driven events, however it was reported that f met Leu Phe (fMLP) stimulation leads to a very similar phosphorylation profile (El Benna et al., 1996a).

The contribution of phosphorylation of the individual sites to the activation of $p47^{phox}$ is not clear. A $p47^{phox}$ mutant in which all ser residues at the C-terminus were replaced with alanines does not support PMA-activated oxidase suggesting this mutant is in a locked conformation which cannot be opened by phosphorylation (Faust et al., 1995). Replacement of ser379 alone with alanine resulted in a similar nonproductive mutant. Interestingly, SDS treatment renders this mutant effective in a reconstituted system, arguing that amphiphile treatment mimics a phosphorylation event and can overcome the effect of the mutation (Park et al.,

1997). Replacement of other individual serines with alanines resulted in mutants which could still support oxidase activation (Faust et al., 1995). Thus phosphorylation of ser379 is either the critical modification of the p47^{phox} C-terminus leading to the intramolecular change or it is a permissive event for the action of the other phosphorylation sites in the protein. The development of activating glutamic acid mutants which mimic the phosphorylation event may be more revealing in this respect.

It was observed recently that in addition to p47^{phox}, p40^{phox}, p67^{phox} (El Benna et al., 1997; Fuchs et al., 1997) and both subunits of the flavocytochrome (Garcia and Segal, 1988) are also phosphoproteins. p67^{phox} phosphorylation occurs upon stimulation of neutrophils with fMLP and PMA and it was suggested that PKC is (one of) the relevant p67^{phox} kinase(s). p40^{phox} phosphorylation is also induced upon PMA and fMLP treatment of neutrophils, however *in vitro* kinase assays suggest that in this case the kinase involved may be casein kinase II. At present the contribution of phosphorylation of these components to NADPH oxidase activation is not known.

C. Cellular Activation Pathways

Since in the cell free system NADPH oxidase activation is regulated by a combination of small GTP binding proteins and protein kinases, pathways resulting in activation of these have been under intense study. They are targeted by a wide range of transmembrane receptor systems among them receptors for chemotactic peptide, cytokines, complement, leukotrine B4, adenosine and IgG (Thrasher et al., 1994; Wientjes and Segal, 1995). Pharmacological tools have been invaluable in elucidating signal transduction pathways leading to oxidase activation (Figure 5). It has long been recognized that phorbol esters, cell permeant irreversible activators of PKC, stimulate the oxidase (Segal and Allison, 1978; Badwey et al., 1980). Moreover, PKC inhibitors affect oxidase activation as do inhibitors of several proline-directed kinases, tyrosine kinases and phosphatase inhibitors. Oxidase activation is also sensitive to pertussis toxin, implicating the heterotrimeric G-protein Gi, and to wortmannin, implicating PI-3 kinase.

In the following section some aspects of neutrophil signaling relevant to activation of the NADPH oxidase will be discussed. Other chapters within this book address individual pathways in more detail, therefore this is not intended as an extensive review of all possible signal transduction pathways.

D. Early Consequences of Receptor Activation—Tyrosine Kinases, PI-3 Kinase, Phospholipase C, and PKC

Receptor-ligand interaction leads to the activation of a large number of signaling pathways. The proximal events resulting from this interaction are dependent on the type of receptor. Serpentine receptors such as the fMLP receptor induce

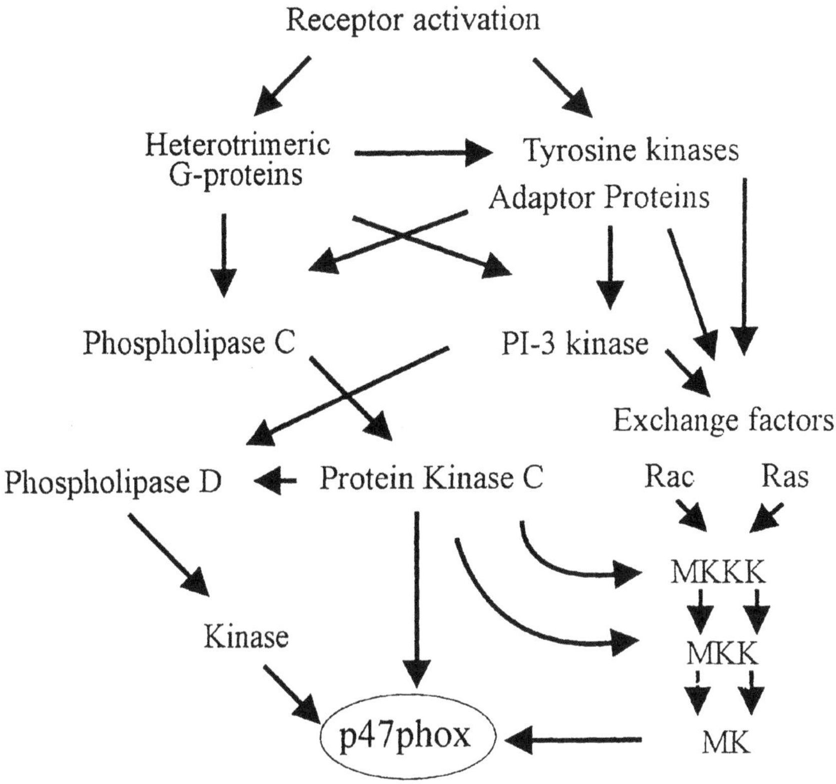

Figure 5. A schematic showing proposed signal transduction pathways leading from receptor activation to phosphorylation of p47^{phox}. No individual MAP kinase pathways are shown but the characteristic three layers of kinases are indicated (MK, MAP kinase; MKK, MAP kinase kinase; MKKK, MAP kinase kinase Kinase). For details see text.

activation of the heterotrimeric G-proteins Gi and Gq, while receptors like the β-adrenergic receptor activate the heterotrimeric G-protein Gs (Simon et al., 1991; Katz et al., 1992; Lee et al., 1992; Jiang et al., 1996; Offermanns and Simon, 1996). Fcγ receptors employ tyrosine kinases and tyrosine phosphorylation as part of their signaling response (de Haas et al., 1995; Daeron, 1997; Ravetch, 1997). The signaling events induced by these different classes of receptors overlap considerably. For instance serpentine receptors as well as Fcγ receptors activate phospholipase C, PI-3 kinase and tyrosine kinases (Naccache et al., 1990; Grinstein and Furuya, 1991; Liang and Huang, 1995; Jiang et al., 1996; Daeron, 1997). Receptors coupled to Gs activate adenylate cyclase, however these receptors can also affect other signaling systems through their Gβγ subunits (Tang and Gilman, 1991, 1992; Tang et al., 1992; Harry et al., 1997).

One of the earlier events elicited by a number of receptor systems is the activation of PI-3 kinase (Vanhaesebroeck et al., 1997a). Activation of this enzyme leads to the formation of phosphatidylinositol-3,4,5-trisphosphate (PI-3,4,5-P3). The importance of PI-3 kinase for the respiratory burst is highlighted by the fact that a highly specific PI-3 kinase inhibitor—wortmannin—is one of the most potent inhibitors of the respiratory burst (Baggiolini et al., 1987; Baggiolini and Wymann, 1990) although it should be kept in mind that this inhibititory effect is restricted to the burst induced by fMLP or opsonised zymosan. Wortmannin does not affect the respiratory burst induced by phorbol ester. Neutrophils contain several PI-3 kinase isoforms which can be activated directly by Gβγ subunit of heterotrimeric G-proteins or by recruitment to receptor tyrosine kinases (Stephens et al., 1994; Daeron, 1997; Vanhaesebroeck et al., 1997a, b). PI-3 kinase itself regulates a number of signaling processes including the activation of kinases such as protein kinase B and phosphoinositide-dependent kinase, small GTP-binding proteins such as *ras* and *rac* which themselves regulate cascades of kinases and furthermore PI-3 kinase regulates phospholipase D. Thus the effects of wortmannin on the oxidase may be related to PI-3 kinase regulating multiple signaling.

Like the isoforms of PI-3 kinase, phospholipase C isoforms are regulated by serpentine receptors as well as Fcγ receptors (Lias et al., 1992; Rhee and Bae, 1997). Phospholipase C-β binds to Gβγ subunits of the heterotrimeric G-proteins Gi and to Gαq whereas phospholipase C-γ is activated directly by Fcγ receptors through an SH2 domain interaction (Katz et al., 1992; Lee et al., 1992; Liao et al., 1992; Jiang et al., 1996; Offermanns and Simon, 1996). Phospholipase C-γ activation by serpentine receptors may also occur (Rhee and Bae, 1997). Phospholipase C activation leads to formation of membrane diacylglycerol which activates PKC, itself an important activator of the NADPH oxidase. Furthermore inositol-1,4,5-trisphosphate is generated by the action of phospholipase C resulting in mobilisation of Ca^{2+} from intracellular stores. Ca^{2+} is important for the respiratory burst induced by receptor activation since depletion of intracellular Ca^{2+} results in failure of the burst.

The contribution of PKC to oxidase activation has long been recognized but remains ill-defined. Clearly, PKC is able to phosphorylate components of the oxidase and PKC stimulation by phorbol ester leads to a respiratory burst (Segal and Allison, 1978; El Benna et al., 1996a). However, the use of PKC inhibitors, in particular staurosporine has proved very confusing and inconclusive. Phorbol ester stimulated oxidase activation is invariably inhibited by staurosporine compatible with an action of PKC in the activation pathway (Twomey et al., 1990). By contrast, receptor-stimulated respiratory burst is not sensitive to staurosporine and in fact priming and stimulation of the fMLP stimulated respiratory burst by staurosporine have been reported (Twomey et al., 1990). It has been argued that these actions are in part related to effects of the inhibitor other than on PKC

(Combadiere et al., 1990; Krause et al., 1992). The concentration and treatment length was critical to the effect that was observed, a phenomenon perhaps related to the specificity profile of the inhibitor. The use of more selective compounds developed from staurosporine—including indocarbazoles and bisindolylmaleimides—yielded more conclusive results and suggested that PKC may mediate fMLP-stimulated oxidase activation (Krause et al., 1992; Kessels et al., 1993; Wenzel-Seifert et al., 1994; Cabamis et al., 1996). PKC may impinge on the oxidase in many ways including by direct phosphorylation of oxidase components (El Benna et al., 1996a). PKC may also stimulate the oxidase indirectly. Targets for PKC in the neutrophil that have been implicated include phospholipase D and several of the microtubule-associated kinases such as ERK and p38 (see below). PKC also modifies the actin cytoskeleton and influences receptor-effector coupling through phospholipase C and adenylate cyclase.

Very little is known regarding the PKC isotype involved in the oxidase response. Neutrophils contain a number of PKC isotypes which respond to various stimuli (Kent et al., 1996; Sergeant and McPhail, 1999). Preliminary data from our own laboratory suggest that PKC isotypes respond differently to various agonists suggesting that they are involved in different neutrophil functions. Alternatively they may be a means by which different receptors drive a similar response.

E. Distal Effects of Receptor Activation—Rac, Proline-Directed Kinases, and Phospholipase D

Although the involvement of *rac* in oxidase activation is now generally accepted (see above), several issues remain unresolved. Perhaps the most imminent question is the mechanism by which *rac* becomes activated, which in neutrophils is largely uncharacterized. It is likely that in common with other small GTP-binding proteins, exchange factors are involved in *rac* activation. A *rac* exchange activity activated by the subunits of the heterotrimeric G-protein Gi has been described in neutrophil extracts, activation of which leads to *rac*-mediated dissociation of actin/gelsolin (Arcaro and Wymann, 1993). *Rac* can be stimulated in neutrophils by fMLP when activation of the rac target PAK is used as a read-out (Knaus et al., 1995). fMLP activation of PAK is inhibited by pertussis toxin, suggesting the involvement of the heterotrimeric G-protein Gi, and by wortmannin, suggesting involvement of PI-3-kinase (Knaus et al., 1995). Such regulation is compatible with recent literature suggesting activation of several *rac* exchange factors by the product of PI-3 kinase, PI-3,4,5-P3 (Parker, 1995). The mechanism by which PI-3,4,5-P3 affects *rac* exchange factors may involve interaction of this lipid with pleckstrin homology (PH) domains (Parker, 1995; Klarlund et al., 1997). This mechanism is involved in regulation of the *rac* exchange factors vav and sos. The action of the lipids on the vav PH domain reduces autoinhibition and allows phosphorylation by the tyrosine kinase Lck which is necessary for exchange activity (Han et al., 1998). The *rac* exchange activity of sos is also regulated by its PH domain and it appeared that ras activation (itself regulated by the ras exchange domain in sos) is necessary to unmask *rac* exchange activity of sos. Ras is thought to do this by activating PI-3 kinase (Rodriguez-Viciana et al., 1994). Although these mechanisms have been established in cells that do not show oxidase activity, they may well have relevance to the oxidase system. For instance in neutrophils a molecular complex is formed after fMLP activation which includes PI-3 kinase, the tyrosine kinase lyn, the adaptor protein shc and perhaps the grb-2/sos complex which binds to shc (Ptasznik et al., 1995, 1996). Such a complex may well result in *rac* activation in the cell and subsequently in *rac*-mediated events. Other cascades may also be initiated through this complex, most notably the ERK cascade through sos and ras.

As mentioned, *rac* has a direct function in activating the oxidase, however it may in addition activate the oxidase indirectly. For instance *rac* may impinge on the oxidase through activation of protein kinases such as PAK-1 and -2, which themselves phosphorylate $p47^{phox}$ (see above). It may also activate proline directed kinases in the JNK cascade (although JNK was negative in phosphorylating $p47^{phox}$ *in vitro*) and, furthermore, *rac* may support oxidase activation through modification of the cytoskeleton.

In appreciating the necessity for active *rac* for oxidase activity, it is not clear how *rac* would become active when phorbol esters are used to induce a respiratory burst. These agents stimulate PKC directly. Their effect on the oxidase suggests that *rac* is activated by PKC for which little evidence is available at present other than perhaps through a pathway involving phospholipase D (see below). Alternatively circumstances may exist in the cell in which activation of *rac* is not necessary for oxidase activation or endogenous *rac* present in the cell may be in a state that allows its participation in the oxidase activation process. It seems unlikely that phorbol esters would activate *rac* independently of PKC since the activation of the oxidase by phorbol esters is inhibited by a wide range of PKC inhibitors.

As indicated above, $p47^{phox}$ contains a number of consensus phosphorylation sites for proline-directed kinases implying regulation by one or more members of the mitogen-activated protein kinase (MAP) kinase family (ERK, p38, JNK stress-activated protein kinase [SAPK]). Many members of this family are present in neutrophils and their activation in various studies correlates with oxidase activation (Torres et al., 1993; Werthen et al., 1994; Nick et al., 1997; Rane et al., 1997; Sue et al., 1997). The pathway leading to ERK phosphorylation in neutrophils is very well-characterized. It is initiated by the small GTP-binding protein ras and involves the sequential phosphorylation of three layers of kinases. Ras activates two kinases, raf-1 and MAP/ERK kinase kinase-1 (MEKK-1), which in turn activate the MAP/ERK kinases (MEK-1 and MEK-2) which activate ERK-1 and ERK-2 (Kyriakis and Avruch, 1996). In neutrophils raf-1 and MEKK-1 activation by fMLP are both inhibited by low concentrations of wortmannin but not by a selective inhibitor of PKC suggesting that PI-3 kinase is a proximal stimulus of the cascade while PKC is not (Worthen et al., 1994; Ardi et al., 1996). MEK-1 and

MEK-2 were inhibited by wortmannin with MEK-1 requiring higher concentrations of the inhibitor than MEK-2 (Grinstein et al., 1994; Downey et al., 1996; Sue et al., 1997). PKC inhibition also reduces MEK-1 and MEK-2 activation by fMLP suggesting that PKC may act on the cascade at the level of MEK (Grinstein et al., 1994; Downey et al., 1996). In a recent study it was observed that one of the PKC isotypes, PKC-ζ, activates MEK but not raf while other isotypes stimulate raf (Schonwasser et al., 1998). Thus different PKC isotypes may stimulates the MAP-kinase cascade at different levels. Whether or not PKC-ζ is selectively activated by PI-3,4,5-P3, as has been widely reported, is still a matter of debate (Nakanishi et al., 1993; Toker et al., 1994; Palmer et al., 1995).

It was noticed that the respiratory burst is inhibited at wortmannin concentrations that would not affect MEK-1 (Sue et al., 1997) but would inhibit MEK-2 (Downey et al., 1996), arguing that perhaps MEK-2 is an intermediate in the oxidase activation. Regulation of MEK-2 by PKC and PI-3 kinase suggests that multiple stimuli converge at this level in the regulation of ERK. Some insight into the relevance of the activation of this cascade for oxidase activation has been gained by the use of a recently developed inhibitor of MEK. This inhibitor was ineffective in one study (Coffer et al., 1998), however two other reports showed an inhibition of the fMLP induced respiratory burst (Avdi et al., 1996; Rane et al., 1997) suggesting that under certain circumstances a pathway involving MEK activation may mediate the oxidase response. Whether MEK-1 or MEK-2 are relevant is not clear from this approach since the inhibitor, although apparently selective for MEK, does not discriminate between MEK-1 and -2.

A second MAP kinase cascade in neutrophils leads to phosphorylation of p38, the mammalian homologue of the yeast protein HOG (Kyrakis et al., 1996). This pathway is also stimulated by fMLP and is sensitive to wortmannin and selective inhibitors of PKC (Krump et al., 1997; Nick et al., 1997). In contrast to ERK, p38 activation by fMLP was reported to be insensitive to pertussis toxin (Nick et al., 1997). p38 is phosphorylated by map kinase-3 and -4 in transfected cells of which MKK3 is possibly relevant in neutrophils (Nick et al., 1997). The relevance of p38 for oxidase activation was assessed using a specific inhibitor and it was shown that the fMLP induced respiratory burst but not the phorbol ester induced burst was inhibited (Nick et al., 1997). This suggests that oxidase activation may require a p38 mediated event, however circumstances exist in which a p38 independent response may exist.

PI-3 kinase, PKC and small GTP-binding proteins have been implicated in the activation of phospholipase D and may mediate its activation by fMLP receptors (Reinhold et al., 1990; Lopez et al., 1995; Exton, 1997). PKC contributes to phospholipase D activation through a curious mechanism that does not involve a catalytic action but a direct interaction with the regulatory domain (Singer et al., 1996). A noncatalytic action of PKC is interesting in relation to the lack of effect of PKC inhibitors on fMLP induced oxidase under certain circumstances (see above). Small GTP binding proteins like RhoA and ADP ribosylation factor

(ARF), and perhaps also *rac*, bind and activate phospholipase D (218). PI-3 kinase activates phospholipase D through its product PI-3,4,5-P3 which activates cloned phospholipase C isoforms directly (Exton, 1997). PI-3 kinase may also affect phospholipase D indirectly through ARF. The final result of phospholipase D activation is the formation of PA from phosphatidyl choline. PA has been implicated in oxidase activation since its formation in neutrophils is highly correlated with the respiratory burst and inhibition of PA formation by ethanol (through the transphosphatidylation reaction) inhibited the burst (Qualliotine Mann et al., 1993; L'Meureux et al., 1995). PA may act through the activation of one or more recently described p47^{phox} kinase(s) or through a direct action on the oxidase (McPhail et al., 1995; Waite et al., 1997). Alternatively, PA may cause the dissociation of *rac* and GDI (McPhail et al., 1995) by competing with the lipid modified tail on *rac* for its binding to the hydrophobic pocket in GDI (Nauseef et al., 1993).

F. Summary of Activation Systems

Although the activation pathways leading to the respiratory burst are not fully clarified, some patterns are becoming apparent (Figure 5). Important early events include the activation of tyrosine kinases and the phospholipid modifying enzymes PI-3 kinase and phospholipase C. These in turn lead to activation of several serine kinases including PKC. PKC in particular is very important for oxidase activation since direct activation of PKC leads to a powerful respiratory burst. A consequence of these early events appears to be the activation of *rac*, ras, proline-directed kinases and phospholipase D. This results in activation of kinases responsible for p47^{phox} phosphorylation and the conditions are in place for the activation of the enzyme.

V. FUNCTION OF THE NADPH OXIDASE

The importance of the system to host immunity is exemplified by the clinical phenotype arising from defective function (Curnutte, 1993). *In vitro*, neutrophils from patients with CGD have markedly impaired microbicidal activity against specific microorganisms, yet retain the ability to efficiently kill many others. Similarly, normal neutrophils incubated under anaerobic conditions demonstrate impaired microbicidal activity against a restricted range of bacteria (Taylor et al., 1993).

The dogma is that this oxygen-dependent killing is mediated directly through oxygen free radicals and the reaction products of hydrogen peroxide with myeloperoxidase (Weiss, 1989).

The superoxide is generated within the phagocytic vacuole after this has closed around the microbe (Segal and Meshulam, 1979). Spontaneous dismutation between two molecules of superoxide occurs optimally at pH 4.8, and is rapid at physiological pH. The dismutation results in the formation of one molecule of peroxide O_2^{2-} and regenerates a molecule of oxygen. At physiological pH the O_2^{2-} is protonated to form H_2O_2.

Although both H_2O_2 and O_{2-} are produced in large amounts by stimulated phagocytes, and concentrations might reach the high millimolar range in the phagocytic vacuole, the ability of these molecules to react directly with biological materials is probably limited (Weiss, 1989).

The role of myeloperoxidase, present in abundance in the azurophil granules, is of interest. The accepted theory is that it utilizes H_2O_2 as a substrate to catalyze the oxidation of halide ions to hypohalous acids, which may then interact with nitrogen containing compounds to form reactive and potentially microbicidal halamine species. It is not entirely clear, however, that the halogenation reaction is necessarily the major or natural function of myeloperoxidase. When an experiment was conducted with radioactive iodide to demonstrate that bacteria were iodinated when phagocytosed by neutrophils, no evidence was found that they were. Instead it was found that the proteins that were iodinated came from the neutrophils rather than the bacteria (Segal et al., 1983). Myeloperoxidase has the ability to react directly with superoxide, and its reaction with hydrogen peroxide can be predominantly as a catalase rather than as a peroxidase, depending upon the concentration of H_2O_2 (Winterbourn et al., 1985). It might therefore act to remove superoxide and H_2O_2 rather than to produce toxic products from them. Where these reduced oxygen species are present at lower concentration, for example in the extracellular matrix, secreted myeloperoxidase might act in a different way to generate hypohalous acids. The generation by neutrophils of other more reactive oxygen free radicals, such as hydroxyl radicals and singlet oxygen, and more recently of radicals derived from nitric oxide, remains speculative (Someya et al., 1993).

An alternative possibility is that the oxidase acts to promote the more effective killing of microbes by proteins maintained within the cytoplasmic granules and released into the vacuole. There are clearly killing systems other than those of the microbicidal products of the oxidase. Patients with CGD are largely free of infection, unlike individuals without neutrophils, who rapidly succumb. In addition, proteins extracted from cytoplasmic granules kill bacteria in the test tube, but only at pHs above 7.0 (Odeberg and Olsson, 1975). Why are these proteins not effective in the phagocytic vacuoles in CGD cells? The possibility existed that it was because conditions within their vacuoles were suboptimal for the activity of these granule contents. We therefore measured the pH within the phagocytic vacuoles. Contrary to the accepted belief, we found that in normal cells the vacuolar pH rose to 7.8–8.0 with kinetics that followed those of the respiratory burst, before slowly falling into the acid range. In contrast, in CGD cells, or in normal cells under anaerobic conditions, the pH fell rapidly to 6–6.5 (Segal et al., 1981; Cech and Lehrer, 1984).

The mechanism of this rise in pH in normal vacuole would appear to be through the pumping by the NADPH oxidase of electrons, unassociated with protons, across the wall of the phagocytic vacuole to form superoxide which then dismutates to peroxide. The superoxide and peroxide are then protonated, consuming protons within the vacuole. The protons released into the cytoplasm cause a drop in cytoplasmic pH and are then exchanged for sodium in the extracellular fluid (Grinstein and Furuya, 1986).

The contents of the cytoplasmic granules are maintained at a pH of about 5 by a proton pump. When the neutral proteinases like elastase, proteinase 3 and cathepsin G are released into the vacuole where the pH has been optimized by the NADPH oxidase, they are activated and kill and digest the engulfed microbe. In CGD, or anaerobic cells the pH is suboptimal and killing and digestion (Segal et al., 1981) are compromised. This accounts for some of the features of CGD that were previously unexplained. These include the fact that the phagocytic vacuoles of normal cells become very large whereas those in CGD cells are relatively much smaller. This swelling in normal cells can be explained by the fact that the products of digestion are osmotically active, thereby sucking water into the vacuole and causing it to swell. Because enzymatic digestion is less efficient in the CGD cells their vacuoles swell less. The other feature is the granulomatous tissue reaction that has given its name to the syndrome of CGD. Granulomas occur in response to the presence of undigested particles in the tissues. The fact that microbial tissue can produce this has now been shown in gp91^{phox} knockout mice in which the injection of heat killed *Aspergillus* led to a granulomatous reaction not seen in control mice (Morgenstern et al., 1997). This mechanism of pH control may represent the most important microbicidal function of the NADPH-oxidase. The relative importance of the reactive oxygen species and granule enzymes remains to be established; the development of knockout mice lacking the enzymes will assist in clarifying these issues.

ACKNOWLEDGMENTS

We thank the Wellcome Trust, The Medical Research Council, the Arthritis and Rheumatism Council and the CGD Research Trust for support.

REFERENCES

Abo, A., & Pick, E. (1991). Purification and characterization of a third cytosolic component of the superoxide-generating NADPH oxidase of macrophages. J. Biol. Chem. 266, 23577-23585.

Abo, A., & Segal, A.W. (1995). Reconstitution of cell-free NADPH oxidase activity by purified components. Methods Enzymol. 256, 268-278.

Abo, A., Boyhan, A., West, I., Thrasher, A.J., & Segal, A.W. (1992). Reconstitution of neutrophil NADPH oxidase activity in the cell-free system by four components: p67-phox, p47-phox, p21rac1, and cytochrome b-245. J. Biol. Chem. 267, 16767-16770

Abo, A., Pick, E., Hall, A., Totty, N., Teahan, C.G., & Segal, A.W. (1991). The small GTP-binding protein, p21rac 1, is involved in activation of the NADPH oxidase. Nature 353, 668-670.

Abo, A., Webb, M.R., Grogan, A., & Segal, A.W. (1994). Activation of NADPH oxidase involves the dissociation of p21*rac* from its inhibitory GDP/GTP exchange protein (rhoGDI) followed by its translocation to the plasma membrane. Biochem. J. 298, 585-591.

Ahmed, A., Prigmore, E., Govind, S., Veryard, C., Kozma, R., Wientjes, F.B., Segal, A.W., & Lim, L. (1998). Cryptic Rac binding and PAK phosphorylation sites of NADPH oxidase component p67phox. J. Biol. Chem. 273, 15693-15701.

Arcaro, A., & Wymann, P. (1993). Wortmannin is a potent phosphatidylinositol 3-kinase inhibitor: the role of phosphatidylinositol 3,4,5-trisphosphate in neutrophil responses. Biochem. J. 296, 297-301.

Avdi, N.J., Winston, B.W., Russel, M., Young, S.K., Johnson, G.L., & Worthen, G.S. (1996). Activation of MEKK by formyl-methionyl-leucyl-phenylalanine in human neutrophils. Mapping pathways for mitogen-activated protein kinase activation. J. Biol. Chem. 271, 33598-33606.

Babior, B.M., & Kipnes, R.S. (1977). Superoxide forming enzyme from human neutrophils: evidence for a flavin requirement. Blood 50, 517-524.

Babior, B.M., Kipnes, R.S., & Curnutte, J.T. (1973). Biological defence mechanisms: the production by leukocytes of superoxide, a potential bactericidal agent. J. Clin. Invest. 52, 741-744.

Babior, B.M., Kuver, R., & Curnutte, J.T. (1988). Kinetics of activation of the respiratory burst oxidase in a fully soluble system from human neutrophils. J. Biol. Chem. 263, 1713-1718.

Badwey, J.A., Curnutte, J.T., Robinson, J.M., Lazdins, J.K., Briggs, R.T., Karnovsky, M.J., & Karnovsky, M.L. (1980). Comparative aspects of oxidative metabolism of neutrophils from human blood and guinea pig peritonea: magnitude of the respiratory burst, dependence upon stimulating agents, and localization of the oxidases. J. Cell Physiol. 105, 541-545.

Baggiolini, M., & Wymann, M.P. (1990). Turning on the respiratory burst. Trends Biochem. Sci. 15, 69-72.

Baggiolini, M., Dewald, B., Schnyder, J., Ruch, W., Cooper, P.H., & Payye, T.G. (1987). Inhibition of the phagocytosis-induced respiratory burst by the fungal metabolite wortmannin and some analogues. Exp. Cell Res. 169, 408-418.

Baldridge, C.W., & Gerard, R.W. (1933). The extra respiration of phagocytosis. Am. J Physiol. 103, 235-236.

Barford, D., Cohen, P.T., & Das, A.K. (1998). The structure of the tetracosipeptide repeats of protein phosphatase 5: implications for TPR-mediated protein-protein interactions. EMBO J. 17, 1192-1199.

Bender, J.G., McPhail, L.C., & Van Epps, D.E. (1983). Exposure of human neutrophils to chemotactic factors potentiates activation of the respiratory burst enzyme. J. Immunol. 130, 2316-2323.

Bolscher, B.G.J.M., Denis, S.W., Verhoeven, A.j., & Roos, D. (1990). The activity of one soluble component of the cell-free NADPH:O2 oxidoreductase of human neutrophils depends on guanosine 5'-O-(3-thio)triphosphate. J. Biol. Chem. 265, 15782-15787.

Bolscher, B.G.J.M., van Zwieten, R., Kramer, I.M., Weening, R.S., Verhoeven, A.J., & Roos, D. (1989). A phosphoprotein of Mr 47 000, defective in autosomal chronic granulomatous disease, copurifies with one of two soluble components required for NADPH:O2 oxidoreductase activity in human neutrophils. J. Clin. Invest. 83, 757-763.

Bromberg, Y., & Pick, E. (1984). Unsaturated fatty acids stimulate NADPH-dependent superoxide production by cell-free system derived from macrophages. Cell Immunol. 88, 213-221.

Bromberg, Y., & Pick, E. (1985). Activation of NADPH-dependent superoxide production in a cell-free system by sodium dodecyl sulfate. J. Biol. Chem. 260, 13539-13545.

Cabanis, A., Gressier, B., Brunet, C., Dine, T., Luyckx, M., Cazin, M., & Cazin, J.C. (1996). Effect of the protein kinase C inhibitor GF 109 203X on elastase release and respiratory burst of human neutrophils. Gen. Pharmacol. 27, 1409-1414.

Cagan, R.H., & Karnovsky, M.L. (1964). Enzymic basis of the respiratory stimulation during phagocytosis. Nature 204, 255-257.

Cech, P., & Lehrer, R.I. (1984). Phagolysosomal pH of human neutrophils. Blood 63, 88-95.

Chetty, M., Thrasher, A.J., Abo, A., & Casimir, C.M. (1995). Low NADPH oxidase activity in Epstein-Barr-virus-immortalized B-lymphocytes is due to a post-transcriptional block in expression of cytochrome b558. Biochem. J. 306, 141-145.

Chuang, T.H., Bohl, B.P., & Bokoch, G.M. (1993). Biologically active lipids are regulators of Rac.GDI complexation. J. Biol. Chem. 268, 26206-26211.

Clark, R.A., Volpp, B.D., Leidal, K.G., & Nauseef, W.M. (1990). Two cytosolic components of the human neutrophil respiratory burst oxidase translocate to the plasma membrane during cell activation. J. Clin. Invest. 85, 714-721.

Clem, W.H., & Klebanoff, S.J. (1966). Inhibitory effect of saliva on glutamic acid accumulation by Lactobacillus acidophilus and the role of the lactoperoxidase-thiocyanate system. J. Bacteriol. 91, 1848-1853.

Coffer, P.J., Geijsen, N., M'rabet, L., Schweizer, R.C., Maikoe, T., Raaijmakers, J.A., Lammers, J.W., & Koenderman, L. (1998). Comparison of the roles of mitogen-activated protein kinase kinase and phosphatidylinositol 3-kinase signal transduction in neutrophil effector function. Biochem. J. 329, 121-130.

Cohen, M.B., Isturiz, R.E., Malech, H.L., Root, R.K., Wilfert, C.M., Gutman, L., & Buckley, R.H. (1981). Fungal infection in chronic granulomatous disease. Am. J. Med. 71, 59-62.

Combadiere, C., El Benna, J., Pedruzzi, E., Hakim, J., & Perianin, A. (1993). Stimulation of the human neutrophil respiratory burst by formyl peptides is primed by a protein kinase inhibitor, staurosporine. Blood 82, 2890-2898.

Combadiere, C., Hakim, J., Giroud, J.P., & Perianin, A. (1990). Staurosporine, a protein kinase inhibitor, up-regulates the stimulation of human neutrophil respiratory burst by N-formyl peptides and platelet activating factor. Biochem. Biophys. Res. Commun. 168, 65-70.

Cross, A.R. (1987). The inhibitory effects of some iodonium compounds on the superoxide generating system of neutrophils and their failure to inhibit diaphorase activity. Biochem. Pharmacol. 36, 489-493.

Cross, A.R., Higson, F.K., Jones, O.T., Harper, A.M., & Segal, A.W. (1982). The enzymic reduction and kinetics of oxidation of cytochrome b-245 of neutrophils. Biochem. J. 204, 479-485.

Cross, A.R., Jones, O.T., Harper, A.M., & Segal, A.W. (1981). Oxidation-reduction properties of the cytochrome b found in the plasma-membrane fraction of human neutrophils. A possible oxidase in the respiratory burst. Biochem. J. 194, 599-606.

Curnutte, J.T. (1985). Activation of human neutrophil nicotinamide adenine dinucleotide phosphate, reduced (triphosphopyridine nucleotide, reduced) oxidase by arachidonic acid in a cell-free system. J. Clin. Invest. 75, 1740-1743.

Curnutte, J.T. (1993). Chronic granulomatous disease: the solving of a clinical riddle at the molecular level. Clin. Immunol. Immunopathol. 67, 2-15.

Curnutte, J.T., Berkow, R.L., Roberts, R.L., Shurin, S.B., & Scott, P.J. (1988). Chronic granulomatous disease due to a defect in the cytosolic factor required for nicotinamide adenine dinucleotide phosphate oxidase activation. J Clin. Invest. 81, 606-610.

Curnutte, J.T., Kuver, R., & Babior, B.M. (1987). Activation of the respiratory burst oxidase in a fully soluble system from human neutrophils. J. Biol. Chem. 262, 6450-6452.

Daeron, M. (1997). Fc receptor biology. Annu. Rev. Immunol. 15, 203-234.

Dana, R., Leto, T.L., Malech, H.L., & Levy, R. (1998). Essential requirement of cytosolic phospholipase A_2 for activation of the phagocyte NADPH oxidase. J. Biol. Chem. 273, 441-445.

Dancis, A., Roman, D.G., Anderson, G.J., Hinnebusch, A.G., & Klausner, R.D. (1992). Ferric reductase of Saccharomyces cerevisiae: molecular characterisation, role in iron uptake and transcriptional control by iron. Proc. Natl. Acad. Sci. USA 89, 3869-3873.

de Boer, M., Hilarius-Stokman, P.M., Hossle, J.-P., Verhoeven, A.J., Graf, N., Kenney, R.T., Seger, R., & Roos, D. (1994). Autosomal recessive chronic granulomatous disease with absence of the 67-kD cytosolic NADPH oxidase component: identification of mutation and detection of carriers. Blood 83, 531-536.

de Haas, M., Vossebeld, P.J., von dem Borne, A.E., & Roos, D. (1995). Fc gamma receptors of phagocytes. J. Lab. Clin. Med. 126, 330-341.

de Mendez, I., Adams, A.G., Sokolic, R.A., Malech, H.L., & Leto, T.L. (1996). Multiple SH3 domain interactions regulate NADPH oxidase assembly in whole cells. EMBO J. 15, 1211-1220.

de Mendez, I., Garrett, M.C., Adams, A.G., & Leto, T.L. (1994). Role of p67-phox SH3 domains in assembly of the NADPH oxidase system. J. Biol. Chem. 269, 16326-16332.

de Mendez, I., Homayounpour, N., & Leto, T.L. (1997). Specificity of p47phox SH3 domain interactions in NADPH oxidase assembly and activation. Mol. Cell Biol. 17, 2177-2185.

DeLeo, F.R., Nauseef, W.M., Jesaitis, A.J., Burrit, J.B., Clark, R.A., & Quinn, M.T. (1995a). A domain of p47phox that interacts with human neutrophil flavocytochrome b558. J. Biol. Chem. 270, 26246-26251.

DeLeo, F.R., Ulman, K.V., Davis, A.R., Jutila, K.L., & Quinn, M.T. (1996). Assembly of the human neutrophil NADPH oxidase involves binding of p67phox and flavocytochrome b to a common funtional domain in p47phox. J. Biol. Chem. 271, 17013-17020.

DeLeo, F.R., Yu, L., Burritt, J.B., Loetterle, L.R., Bond, C.W., Jesaitis, A.J., & Quinn, M.T. (1995b). Mapping sites of interaction of p47-phox and flavocytochrome b with random-sequence peptide phage display libraries. Proc. Natl. Acad. Sci. USA 92, 7110-7114.

Didsbury, J., Weber, R.F., Bokoch, G.W., Evans, T., & Snyderman, R.J. (1989). rac, a novel ras-related family of proteins that are botulinum toxin substrates. J. Biol. Chem. 264, 16378-16382.

Diekmann, D., Abo, A., Johnston, C., Segal, A.W., & Hall, A. (1994). Interaction of Rac with p67*phox* and regulation of phagocytic NADPH oxidase activity. Science 265, 531-533.

Dinauer, M.C., Pierce, E.A., Erickson, R.W., Muhlebach, T.J., Messner, H., Orkin, S.H., Seger, R.A., & Curnutte, J.T. (1991). Point mutation in the cytoplasmic domain of the neutrophil p22-phox cytochrome b subunit is associated with a nonfunctional NADPH oxidase and chronic granulomatous disease. Proc. Natl. Acad. Sci. USA 88, 11231-11235.

Ding, J., Knaus, U.G., Lian, J.P., Bokoch, G.M., & Badwey, J.A. (1996). The reraturable 69 and 63 kDa protein kinases that undergo rapid activation in chemoattractant-stimulated guinea pig neutrophils are p21-activated kinases. J. Biol. Chem. 271, 24869-24873.

Ding, J., Vlahos, C.J., Liu, R., Brown, R.F., & Badwey, J.A. (1995). Antagonists of phosphatidylinositol 3-kinase block activation of several novel protein kinases in neutrophils. J. Biol. Chem. 270, 11684-11691.

Dorseuil, O., Reibel, L., Bokoch, G.M., Camonis, J., & Gacon, G. (1996). The rac target NADPH oxidase p67phox interacts preferentially with rac2 rather than rac1. J. Biol. Chem. 271, 83-88.

Dorseuil, O., Vazquez, A., Lang, P., Bertoglio, J., Gacon, G., & Gerald, L. (1992). Inhibition of superoxide production in B lymphocytes by Rac antisense oligonucleotides. J. Biol. Chem. 267, 20540-20542.

Doussière, J., Buzenet, G., & Vignais, P.V. (1995). Photoaffinity labeling and photoinactivation of the O2(-)-generating oxidase of neutrophils by an azido derivative of FAD. Biochemistry 34, 1760-1770.

Doussière, J., Pilloud, M.-C., & Vignais, P.V. (1990). Cytosolic factors in bovine neutrophil oxidase activation. Partial purification and demonstration of translocation to a membrane fraction. Biochemistry 29, 2225-2232.

Downey, G.P., Butler, J.R., Brumell, J., Borregaard, N., Kjeldsen, L., Sue, A.K., & Grinstein, S. (1996). Chemotactic peptide-induced activation of MEK-2, the predominant isoform in human neutrophils. Inhibition by wortmannin. J. Biol. Chem. 271, 21005-21011.

Dusi, S., & Rossi, F. (1993). Activation of NADPH oxidase of human neutrophils involves the phosphorylation and the translocation of cytosolic p67phox. Biochem. J. 296, 367-371.

Dusi, S., Donini, M., & Rossi, F. (1995). Mechanisms of NADPH oxidase activation in human neutrophils: p67phox is required for the translocation of rac 1 but not of rac 2 from cytosol to the membranes. Biochem. J. 308, 991-994.

Dwyer, S.C., Legendre, L., Low, P.S., & Leto, T.L. (1996). Plant and human neutrophil oxidative burst complexes contain immunologically related proteins. Biochim. Biophys. Acta 1289, 231-237.

El Benna, J., Dang, P.M.C., Gaudry, M., Fay, M., Morel, F., Hakim, J., & Gougerot-Pocidalo, M. (1997). Phosphorylation of the respiratory burst oxidase subunit p67phox during human neutrophil activation. J. Biol. Chem 272, 17204-17208.

El Benna, J., Faust, L.P., & Babior, B.M. (1994a). The phosphorylation of the respiratory burst oxidase component p47phox during neutrophil activation. Phosphorylation of sites recognized by protein kinase C and by proline-directed kinases. J. Biol. Chem. 269, 23431-23436.

El Benna, J., Faust, L.P., Johnson, J.L., & Babior, B.M. (1996a). Phosphorylation of the respiratory burst oxidase subunit p47phox as determined by two-dimensional phosphopeptide mapping. Phosphorylation by protein kinase C, protein kinase A, and a mitogen-activated protein kinase. J. Biol. Chem. 271, 6374-6378.

El Benna, J., Han, J., Park, J.W., Schmid, E., Ulevitch, R.J., & Babior, B.M. (1996b). Activation of p38 in stimulated human neutrophils: phosphorylation of the oxidase component p47phox by p38 and ERK but not by JNK. Arch. Biochem. Biophys. 334, 395-400.

El Benna, J., Ruedi, J.M., & Babior, B.M. (1994b). Cytosolic guanine nucleotide-binding protein Rac2 operates in vivo as a component of the neutrophil respiratory burst oxidase. Transfer of Rac2 and the cytosolic oxidase components p47phox and p67phox to the submembranous actin cytoskeleton during oxidase activation. J. Biol. Chem. 269, 6729-6734.

Emmendorffer, A., Roesler, J., Elsner, J., Raeder, E., Lohmann Matthes, M.L., & Meier, B. (1993). Production of oxygen radicals by fibroblasts and neutrophils from a patient with x-linked chronic granulomatous disease. Eur. J. Haematol. 51, 223-227.

Exton, J.H. (1997). New developments in phospholipase D. J. Biol. Chem. 272, 15579-15582.

Faust, L.R., El Benna, J., Babior, B.M., & Chanock, S.J. (1995). The phosphorylation targets of p47phox, a subunit of the respiratory burst oxidase. Functions of the individual target serines as evaluated by site-directed mutagenesis. J. Clin. Invest. 96, 1499-1505.

Finan, P., Shimizu, Y., Gout, I., Hsuan, J., Truong, O., Butcher, C., Bennett, P., Waterfield, M.D., & Kellie, S. (1994). An SH3 domain and proline-rich sequence mediate an interaction between two components of the phagocyte NADPH oxidase complex. J. Biol. Chem. 269, 13752-13755.

Finegold, A.A., Shatwell, K.P., Segal, A.W., Klausner, R.D., & Dancis, A. (1996). Intramembrane bisheme motif for transmembrane electron transport conserved in a yeast iron reductase and the human NADPH oxidase. J. Biol. Chem. 271, 31021-31024.

Freeman, J.L., & Lambeth, J.D. (1996). NADPH oxidase activity is independent of p47phox in vitro. J. Biol. Chem. 271, 22578-22582.

Freeman, J.L., Abo, A., & Lambeth, J.L. (1996). Rac "insert region" is a novel effector region that is implicated in the activation of NADPH oxidase, but not pak65. J. Biol. Chem. 271, 19794-19801.

Freeman, J.L., Kreck, M.L., Uhlinger, D.J., & Lambeth, J.D. (1994). Ras effector-homologue region on Rac regulates protein associations in the neutrophil respiratory burst oxidase complex. Biochemistry 33, 13431-13435.

Fuchs, A., Bouin, A., Rabilloud, T., & Vignais, P.V. (1997). The 40-kDa component of the phagocyte NADPH oxidase (p40phox) is phosphorylated during activation in differentiated HL60 cells. Eur. J. Biochem. 249, 531-539.

Fuchs, A., Dagher, M.C., & Vignais, P.V. (1995). Mapping the domains of interaction of p40*phox* with both p47*phox* and p67*phox* of the neutrophil oxidase complex using the two-hybrid system. J. Biol. Chem. 270, 5695-5697.

Fuchs, A., Dagher, M.C., Faure, J., & Vignais, P.V. (1996). Topological organization of the cytosolic activating complex of the superoxide-generating NADPH oxidase. Pinpointing the sites of interaction between p47phox, p67phox and p40phox using the two-hybrid system. Biochim. Biophys. Acta 1312, 39-47.

Fujita, I., Takeshige, K., & Minakami, S. (1987). Characterization of the NADPH-dependent superoxide production activated by sodium dodecyl sulfate in a cell-free system of pig neutrophils. Biochim. Biophys. Acta 931, 41-48.

Gabig, T.G., Crean, C.D., Mantel, P.L., & Rosli, R. (1995). Function of wild-type or mutant Rac2 and Rap1a GTPases in differentiated HL60 cell NADPH oxidase activation. Blood 85, 804-811.

Garcia, R.C., & Segal, A.W. (1988). Phosphorylation of the subunits of cytochrome b_{-245} upon triggering of the respiratory burst of human neutrophils and macrophages. Biochem. J 252, 901-904.

Grinstein, S., & Furuya, W. (1986). Cytoplasmic pH regulation in phorbol ester-activated human neutrophils. Am. J Physiol. 251, C55-C65.

Grinstein, S., & Furuya, W. (1991). Tyrosine phosphorylation and oxygen consumption induced by G proteins in neutrophils. Am. J. Physiol. 260, C1019-C1027.

Grinstein, S., Butler, J.R., Furuya, W., L'Allemain, G., & Downey, G.P. (1994). Chemotactic peptides induce phosphorylation and activation of MEK-1 in human neutrophils. J. Biol. Chem. 269, 19313-19320.

Grogan, A., Reeves, E., Keep, N.H., Wientjes, F., Totty, N., Burlingame, A.L., Hsuan, J., & Segal, A.W. (1997). Cytosolic phox proteins interact with and regulate the assembly of coronin in neutrophils. J. Cell. Sci. 110, 3071-3081.

Groom, Q.J., Torres, M.A., Fordham Skelton, A.P., Hammond Kosack, K.E., Robinson, N.J., & Jones, J.D. (1996). rbohA, a rice homologue of the mammalian gp91phox respiratory burst oxidase gene. Plant J. 10, 515-522.

Guillory, R.J., Bax, B., & Najmudin, S. (1997). Structural analysis of the SH2 and SH3 domain: modules that regulate protein interactions. J. Biochem. Mol. Biol. Biophys. 1, 73-88.

Han, J., Luby-Phelps, K., Das, B., Shu, X., Xia, Y., Mosteller, R.D., Krishna, U.M., Falck, J.R., White, M.A., & Broek, D. (1998). Role of substrates and products of PI 3-kinase in regulating activation of Rac-related guanosine triphosphatases by Vav. Science 279, 558-560.

Harper, A.M., Chaplin, M.F., & Segal, A.W. (1985). Cytochrome b-245 from human neutrophils is a glycoprotein. Biochem. J. 227, 783-788.

Harry, A., Chen, Y., Magnusson, R., Iyengar, R., & Weng, G. (1997). Differential regulation of adenylyl cyclases by Galphas. J. Biol. Chem. 272, 19017-19021.

Hata, K., Ito, T., Takeshige, K., & Sumimoto, H. (1998). Anionic amphiphile-independent activation of the phagocyte NADPH oxidase in a cell-free system by p47phox and p67phox, both in C terminally truncated forms. J. Biol. Chem. 273, 4232-4236.

Hata, K., Takeshige, K., & Sumimoto, H. (1997). Roles for proline-rich regions of p47phox and p67phox in the phagocyte NADPH oxidase activation in vitro. Biochem. Biophys. Res. Commun. 241, 226-231.

Heyworth, P.G., Bohl, B.P., Bokoch, G.M., & Curnutte, J.T. (1994). Rac translocates independently of the neutrophil NADPH oxidase components p47*phox* and p67*phox*. J. Biol. Chem. 269, 30749-30752.

Heyworth, P.G., Curnutte, J.T., Nauseef, W.M., Volpp, B.D., Pearson, D.W., Rosen, H., & Clark, R.A. (1991). Neutrophil nicotinamide adenine dinucleotide phosphate oxidase assembly. Translocation of p47-phox and p67-phox requires interaction between p47-phox and cytochrome b558. J. Clin. Invest. 87, 352-356.

Heyworth, P.G., Shrimpton, C.F., & Segal, A.W. (1989). Localization of the 47 kDa phosphoprotein involved in the respiratory-burst NADPH oxidase of phagocytic cells. Biochem. J. 260, 243-248.

Hurst, J.K., Loehr, T.M., Curnutte, J.T., & Rosen, H. (1991). Resonance raman and electron paramagnetic resonance structural investigations of neutrophil cytochrome b558. J. Biol. Chem. 266, 1627-1634.

Ito, T., Nakamura, R., Sumimoto, H., Takeshige, K., & Sakaki, Y. (1996). An SH3 domain-mediated interaction between the phagocyte NADPH oxidase factors p40phox and p47phox. FEBS Lett. 385, 229-232.

Jiang, H., Kuang, Y., Wu, Y., Smrcka, A., Simon, M.I., & Wu, D. (1996). Pertussis toxin-sensitive activation of phospholipase C by the C5a and fMet-Leu-Phe receptors. J. Biol. Chem. 271, 13430-13434.

Jones, S.A., Wood, J.D., Coffey, J.M., & Jones, O.T. (1994). The functional expression of p47-phox and p67-phox may contribute to the generation of superoxide by an NADPH oxidase-like system in human fibroblasts. FEBS Lett. 355, 178-182.

Kakinuma, K., Kaneda, M., Chiba, T., & Ohnishi, T. (1986). Electron spin resonance studies on a flavoprotein in neutrophil plasma membranes. Redox potentials of the flavin and its participation in NADPH oxidase. J. Biol. Chem. 261, 9426-9432.

Karplus, P.A., Daniels, M.J., & Herriott, J.R. (1991). Atomic structure of ferredoxin-NADP+ reductase: prototype for a structurally novel flavoenzyme family. Science 251, 60-66.

Katz, A., Wu, D., & Simon, M.I. (1992). Subunits beta gamma of heterotrimeric G protein activate beta 2 isoform of phospholipase C. Nature 360, 686-689.

Kent, J.D., Sergeant, S., Burns, D.J., & McPhail, L.C. (1996). Identification and regulation of protein kinase C-delta in human neutrophils. J. Immunol. 157, 4641-4647.

Kessels, G.C., Krause, K.H., & Verhoeven, A.J. (1993). Protein kinase C activity is not involved in N-formylmethionyl-leucyl-phenylalanine-induced phospholipase D activation in human neutrophils, but is essential for concomitant NADPH oxidase activation: studies with a staurosporine analogue with improved selectivity for protein kinase C. Biochem. J. 292, 781-785.

Klarlund, J.K., Guilherme, A., Holik, J.J., Virbasius, J.V., Chawla, A., & Czech, M.P. (1997). Signaling by phosphoinositide-3,4,5-trisphosphate through proteins containing pleckstrin and Sec7 homology domains [see comments]. Science 275, 1927-1930.

Klebanoff, S.J., & Rosen, H. (1978). The role of myeloperoxidase in the microbicidal activity of polymorphonuclear leukocytes. Ciba. Found. Symp. 65, 263-284.

Kleinberg, M.E., Malech, H.L., & Rotrosen, D. (1990). The phagocytic 47-kilodalton cytosolic oxidase protein is an early reactant in activation of the respiratory burst. J. Biol. Chem. 265, 15577-15583.

Knaus, U.G., Heyworth, P.G., Evans, T., Curnutte, J.T., & Bokoch, G.M. (1991). Regulation of phagocyte oxygen radical production by the GTP-binding protein Rac 2. Science 254, 1512-1515.

Knaus, U.G., Morris, S., Dong, H.J., Chernoff, J., & Bokoch, G.M. (1995). Regulation of human leukocyte p21-activated kinases through G protein-coupled receptors. Science 269, 221-223.

Koshkin, V., Lotan, O., & Pick, E. (1996). The cytosolic component p47(phox) is not a sine qua non participant in the activation of NADPH oxidase but is required for optimal superoxide production. J. Biol. Chem. 271, 30326-30329.

Krause, K.H., van Delden, C., Huggler, E., Monod, A., & Lew, D.P. (1992). Differential effects on neutrophil activation of staurosporin and its protein kinase C-selective derivative cgp 41231. Eur. J. Pharmacol. 227, 221-224.

Kreck, M.L., Freeman, J.L., Abo, A., & Lambeth, J.D. (1996). Membrane association of Rac is required for high activity of the respiratory burst oxidase. Biochemistry 35, 15683-15692.

Krump, E., Sanghera, J.S., Pelech, S.L., Furuya, W., & Grinstein, S. (1997). Chemotactic peptide N-formyl-met-leu-phe activation of p38 mitogen-activated protein kinase (MAPK) and MAPK-activated protein kinase-2 in human neutrophils. J. Biol. Chem. 272, 937-944.

Kwong, C.H., & Leto, T.L. (1995). Characterization of the effector-specifying domain of rac involved in NADPH oxidase activation. J. Biol. Chem. 270, 19868-19872.

Kyriakis, J.M., & Avruch, J. (1996). Sounding the alarm: protein kinase cascades activated by stress and inflammation. J. Biol. Chem. 271, 24313-24316.

Lee, C.H., Park, D., Wu, D., Rhee, S.G., & Simon, M.I. (1992). Members of the Gq alpha subunit gene family activate phospholipase C beta isozymes. J. Biol. Chem. 267, 16044-16047.

Leto, T.L., Adams, A.G., & de Mendez, I. (1994). Assembly of the phagocyte NADPH oxidase: binding of Src homology 3 domains to proline-rich targets. Proc. Natl. Acad. Sci. USA 91, 10650-10654.

Leto, T.L., Lomax, K.J., Volpp, B.D., Nunoi, H., Sechler, J.M., Nauseef, W.M., Clark, R.A., Gallin, J.I., & Malech, H.L. (1990). Cloning of a 67-kD neutrophil oxidase factor with similarity to a noncatalytic region of p60c-src. Science 248, 727-730.

Leusen, J.H., Bolscher, B.G., Hilarius, P.M., Weening, R.S., Kaulfersch, W., Seger, R.A., Roos, D., & Verhoeven, A.J. (1994a). 156Pro→Gln substitution in the light chain of cytochrome b558 of the human NADPH oxidase (p22-phox) leads to defective translocation of the cytosolic proteins p47-phox and p67-phox. J. Exp. Med. 180, 2329-2334.

Leusen, J.H., de Boer, M., Bolscher, B.G., Hilarius, P.M., Weening, R.S., Ochs, H.D., Roos, D., & Verhoeven, A.J. (1994b). A point mutation in gp91-phox of cytochrome b558 of the human NADPH oxidase leading to defective translocation of the cytosolic proteins p47-phox and p67-phox. J. Clin Invest. 93, 2120-2126.

Leusen, J.H., De Klein, A., Hilarius, P.M., Ahlin, A., Palmblad, J., Smith, C.I., Diekmann, D., Hall, A., Verhoeven, A.J., & Roos, D. (1996). Disturbed interaction of p21-rac with mutated p67-phox causes chronic granulomatous disease. J. Exp. Med. 184, 1243-1249.

Leusen, J.H., Fluiter, K., Hilarius, P.M., Roos, D., Verhoeven, A.J., & Bolscher, B.G. (1995). Interactions between the cytosolic components p47phox and p67phox of the human neutrophil NADPH oxidase that are not required for activation in the cell-free system. J. Biol. Chem. 270, 11216-11221.

L'Heureux, G.P., Bourgoin, S., Jean, N., McColl, S.R., & Naccache, P.H. (1995). Diverging signal transduction pathways activated by interleukin-8 and related chemokines in human neutrophils: interleukin-8, but not NAP-2 or GRO alpha, stimulates phospholipase D activity. Blood 85, 522-531.

Liang, L., & Huang, C.K. (1995). Tyrosine phosphorylation induced by cross-linking of Fc gamma-receptor type II in human neutrophils. Biochem. J. 306, 489-495.

Liao, F., Shin, H.S., & Rhee, S.G. (1992). Tyrosine phosphorylation of phospholipase C-gamma 1 induced by cross-linking of the high-affinity or low-affinity Fc receptor for IgG in U937 cells. Proc. Natl. Acad. Sci. USA 89, 3659-3663.

Light, D.R., Walsh, C., O Callaghan, A.M., Goetzl, E.J., & Tauber, A.I. (1981). Characteristics of the cofactor requirements for the superoxide-generating NADPH oxidase of human polymorphonuclear leukocytes. Biochemistry 20, 1468-1476.

Lomax, K.J., Leto, T.L., Nunoi, H., Gallin, J.I., & Malech, H.L. (1989). Recombinant 47-kilodalton cytosol factor restores NADPH oxidase in chronic granulomatous disease. Science 245, 409-412.

Lopez, I., Burns, D.J., & Lambeth, J.D. (1995). Regulation of phospholipase D by protein kinase C in human neutrophils. Conventional isoforms of protein kinase C phosphorylate a phospholipase D-related component in the plasma membrane. J. Biol. Chem. 270, 19465-19472.

Mandell, G.L. (1974). Bactericidal activity of aerobic and anaerobic polymorphonuclear neutrophils. Infect. Immun. 9, 337-341.

McPhail, L.C., Qualliotine Mann, D., & Waite, K.A. (1995). Cell-free activation of neutrophil NADPH oxidase by a phosphatidic acid-regulated protein kinase. Proc. Natl. Acad. Sci. USA 92, 7931-7935.

McPhail, L.C., Shirley, P.S., Clayton, C.C., & Snyderman, R. (1985). Activation of the respiratory burst enzyme from human neutrophils in a cell-free system. J. Clin. Invest. 75, 1735-1739.

Meier, B., Jesaitis, A.J., Emmendorffer, A., Roesler, J., & Quinn, M.T. (1993). The cytochrome b-558 molecules involved in the fibroblast and polymorphonuclear leucocyte superoxide-generating NADPH oxidase systems are structurally and genetically distinct. Biochem. J. 289, 481-486.

Morgenstern, D.E., Gifford, M.A., Li, L.L., Doerschuk, C.M., & Dinauer, M.C. (1997). Absence of respiratory burst in X-linked chronic granulomatous disease mice leads to abnormalities in both host defense and inflammatory response to Aspergillus fumigatus. J. Exp. Med. 185, 207-218.

Naccache, P.H., Gilbert, C., Caon, A.C., Gaudry, M., Huang, C.-K., Bonak, V.A., Umezawa, K., & McColl, S.R. (1990). Selective inhibition of human neutrophil functional responsiveness by erbastin, an inhibitor of tyrosine protein kinase. Blood 76, 2098-2104.

Nakamura, R., Sumimoto, M., Mizuki, K., Hata, K., Ago, T., Kitajima, S., Takeshige, K., Sakaki, Y., & Ito, T. (1998). The PC motif: A novel and evolutionarily conserved sequence involved in interaction between evolutionarily conserved sequence involved in interaction between $P40^{phox}$ and $P67^{phox}$, SH3 domain-containing cytosolic factors of the phagocytic NADPH oxidase. Eur. J. Biochem. 251, 583-589.

Nakanishi, A., Imajoh-Ohmi, S., Fujinawa, T., Kikuchi, H., & Kanegasaki, S. (1992). Direct evidence for interaction between COOH-terminal regions of cytochrome b558 subunits and cytosolic 47-kDa protein during activation of an O2-generating system in neutrophils. J. Biol. Chem. 267, 17072-17074.

Nauseef, W.M., Volpp, B.D., McCormick, S., Leidal, K.G., & Clark, R.A. (1991). Assembly of the neutrophil respiratory burst oxidase. J. Biol. Chem. 266, 5911-5917.

Nakanishi, H., Brewer, K.A., & Exton, J.H. (1993). Activation of the zeta isozyme of protein kinase C by phosphatidylinositol 3,4,5-trisphosphate. J. Biol. Chem. 268, 13-16.

Nauseef, W.M., McCormick, S., Renee, J., Leidal, K.G., & Clark, R.A. (1993). Functional domain in an arginine-rich carboxyl-terminal region of p47phox. J. Biol. Chem. 268, 23646-23651.

Nick, J.A., Avdi, N.J., Young, S.K., Knall, C., Gerwins, P., Johnson, G.L., & Worthen, G.S. (1997). Common and distinct intracellular signaling pathways in human neutrophils utilized by platelet activating factor and FMLP. J. Clin. Invest. 99, 975-986.

Nisimoto, Y., Freeman, J.L.R., Motalebi, S.A., Hirshberg, M., & Lambeth, J.D. (1997). Rac binding to p67phox. Structural basis for interaction of the rac1 effector region and insert region with components of the respiratory burst oxidase. J. Biol. Chem. 272, 18834-18841.

Nunoi, H., Rotrosen, D., Gallin, J.I., & Malech, H.L. (1988). Two forms of autosomal chronic granulomatous disease lack destinct neutrophil cytosol factors. Science 242, 1298-1301.

Odeberg, H., & Olsson, I. 1975. Antibacterial activity of cationic proteins from human granulocytes. J. Clin. Invest. 56, 1118-1124.

Offermanns, S., & Simon, M.I. (1996). Organization of transmembrane signalling by heterotrimeric G proteins. Cancer Surv. 27, 177-198.

Okamura, N., Curnutte, J.T., Roberts, R.L., & Babior, B.M. (1988a). Relationship of protein phosphorylation to the activation of the respiratory burst in human neutrophils. Defects in the phosphorylation of a group of closely related 48-kDa proteins in two forms of chronic granulomatous disease. J. Biol. Chem. 263, 6777-6782.

Okamura, N., Malawista, S.E., Roberts, R.L., Rosen, H., Ochs, H.D., Babior, B.M., & Curnutte, J.T. (1988b). Phosphorylation of the oxidase-related 48K phosphoprotein family in the unusual autosomal cytochrome-negative and X-linked cytochrome-positive types of chronic granulomatous disease. Blood 72, 811-816.

Palmer, R.H., Dekker, L.V., Woscholski, R., Le Good, J.A., Gigg, R., & Parker, P.J. (1995). Activation of PRK1 by phosphatidylinositol 4,5-bisphosphate and phosphatidylinositol 3,4,5-trisphosphate. A comparison with protein kinase C isotypes. J. Biol. Chem. 270, 22412-22416.

Park, J.-W., Ma, M., Ruedi, J.M., Smith, R.M., & Babior, B.M. (1992). The cytosolic components of the respiratory burst oxidase exist as a Mr-240,000 complex that acquires a membrane-binding site during activation of the oxidase in a cell-free system. J. Biol. Chem. 267, 17327-17332.

Park, J.W., & Ahn, S.M. (1995). Translocation of recombinant p47phox cytosolic component of the phagocyte oxidase by in vitro phosphorylation. Biochem. Biophys. Res. Commun. 211, 410-416.

Park, J.W., & Babior, B.M. (1997). Activation of the leukocyte NADPH oxidase subunit p47phox by protein kinase C. A phosphorylation-dependent change in the conformation of the C-terminal end of p47phox. Biochemistry 36, 7474-7480.

Park, J.W., Royal, C.R., El Benna, J., & Babior, B.M. (1997). Kinase-dependent activation of the leukocyte NADPH oxidase in a cell-free system. J. Biol. Chem. 272, 11035-11043.

Parker, P.J. (1995). Intracellular signalling. PI 3-kinase puts GTP on the Rac. Curr. Biol. 5, 577-579.

Parkos, C.A., Allen, R.A., Cochrane, C.G., & Jesaitis, A.J. (1987). Purified cytochrome b from human granulocyte plasma membrane is comprised of two polypeptides with relative molecular weights of 91,000 and 22,000. J Clin. Invest. 80, 732-742.

Parkos, C.A., Dinauer, M.C., Walker, L.E., Allen, R.A., Jesaitis, A.J., & Orkin, S.H. (1988). Primary structure and unique expression of the 22-kilodalton light chain of human neutrophil cytochrome b. Proc. Natl. Acad. Sci. USA 85, 3319-3323.

Philips, M.R., Feoktistov, A., Pillinger, M.H., & Abramson, S.B. (1995). Translocation of p21rac2 from cytosol to plasma membrane is neither necessary nor sufficient for neutrophil NADPH oxidase activity. J. Biol. Chem. 270, 11514-11521.

Pick, E., Kroizman, T., & Abo, A. (1989). Activation of the superoxide-forming NADPH oxidase of macrophages requires two cytosolic components—One of them is also present in certain nonphagocytic cells. J. Immunol. 143, 4180-4187.

Pilloud-Dagher, M., & Vignais, P.V. (1991). Purification and characterization of an oxidase activating factor of 63 kilodaltons from bovine neutrophils. Biochemistry 30, 2753-2760.

Ponting, C.P. (1996). Novel domains in NADPH oxidase subunits, sorting nexins, and PtdIns 3-kinases: binding partners of SH3 domains? Prot. Sci. 5, 2353-2357.

Prigmore, E., Ahmed, S., Best, A., Kozma, R., Manser, E., Segal, A.W., & Lim, L. (1995). A 68-kDa kinase and NADPH oxidase component p67phox are targets for Cdc42Hs and Rac1 in neutrophils. J. Biol. Chem. 270, 10717-10722.

Ptasznik, A., Prossnitz, E.R., Yoshikawa, D., Smrcka, A., Traynor Kaplan, A.E., & Bokoch, G.M. (1996). A tyrosine kinase signaling pathway accounts for the majority of phosphatidylinositol 3,4,5-trisphosphate formation in chemoattractant-stimulated human neutrophils. J. Biol. Chem. 271, 25204-25207.

Ptasznik, A., Traynor Kaplan, A., & Bokoch, G.M. (1995). G protein-coupled chemoattractant receptors regulate Lyn tyrosine kinase. Shc adapter protein signaling complexes [published erratum appears in J. Biol. Chem. (1995) 270, 24622]. J. Biol. Chem. 270, 19969-19973.

Qualliotine Mann, D., Agwu, D.E., Ellenburg, M.D., McCall, C.E., & McPhail, L.C. (1993). Phosphatidic acid and diacylglycerol synergize in a cell-free system for activation of NADPH oxidase from human neutrophils. J. Biol. Chem. 268, 23843-23849.

Quinn, M.T., Evans, T., Loetterle, L.R., Jesaitis, A.J., & Bokoch, G.M. (1993). Translocation of Rac correlates with NADPH oxidase activation. Evidence for equimolar translocation of oxidase components. J. Biol. Chem. 268, 20983-20987.

Rane, M.J., Carrithers, S.L., Arthur, J.M., Klein, J.B., & McLeish, K.R. (1997). Formyl peptide receptors are coupled to multiple mitogen-activated protein kinase cascades by distinct signal transduction pathways: role in activation of reduced nicotinamide adenine dinucleotide oxidase. J. Immunol. 159, 5070-5078.

Ravel, P., & Lederer, F. (1993). Affinity-labeling of an NADPH-binding site on the heavy subunit of flavocytochrome b558 in particulate NADPH oxidase from activated human neutrophils. Biochem. Biophys. Res Commun. 196, 543-552.

Ravetch, J.V. (1997). Fc receptors. Curr. Opin. Immunol. 9, 121-125.

Reinhold, S.L., Prescott, S.M., Zimmerman, G.A., & McIntyre, T.M. (1990). Activation of human neutrophil phospholipase D by three separable mechanisms. FASEB J. 4, 208-214.

Rhee, S.G., & Bae, Y.S. (1997). Regulation of phosphoinositide-specific phospholipase C isozymes. J. Biol. Chem. 272, 15045-15048.

Rickard, J.E., & Sheterline, P. (1988). Effect of ATP removal and inorganic phosphate on length redistribution of sheared actin filament populations. Evidence for a mechanism of end-to-end annealing. J. Mol. Biol. 201, 675-681.

Ridley, A.J., Paterson, H.F., Johnston, C.L., Diekmann, D., & Hall, A. (1992). The small GTP binding protein rac regulates growth factor induced membrane ruffling. Cell 70, 401-410.

Rodaway, A.R., Teahan, C.G., Casimir, C.M., Segal, A.W., & Bentley, D.L. (1990). Characterization of the 47-kilodalton autosomal chronic granulomatous disease protein: tissue-specific expression and transcriptional control by retinoic acid. Mol. Cell Biol. 10, 5388-5396.

Rodriguez-Viciana, P., Warne, P.H., Dhand, R., Vanhaesebroeck, B., Gout, I., Fry, M.J., Waterfield, M.D., & Downward, J. (1994). Phosphatidylinositol-3-OH kinase as a direct target of Ras. Nature 370, 527-532.

Roman, D.G., Dancis, A., Anderson, G.J., & Klausner, R.D. (1993). The fission yeast ferric reductase gene frp1+ is required for ferric uptake and encodes a protein that is homologous to the gp91-phox subunit of the human NADPH phagocyte oxidoreductase. Mol. Cell. Biol. 13, 4342-4350.

Roos, D., de Boer, M., Kuribayashi, F., Weening, R.S., Segal, A.W., Ahlin, A., Nemet, K., Hossle, J.P., Bernatowska-Matuszkiewicz, E., & Middleton-Price, H. (1996). Mutations in the X-linked and autosomal recessive forms of chronic granulomatous disease. Blood 87, 1663-1681.

Rotrosen, D., & Leto, T.L. (1990). Phosphorylation of neutrophil 47-kDa cytosolic factor. Translocation to membrane is associated with distinct phosphorylation events. J. Biol. Chem. 265, 19910-19915.

Royer-Pokora, B., Kunkel, L.M., Monaco, A.P., Goff, S.C., Newburger, P.E., Baehner, R.L., Cole, F.S., Curnutte, J.T., & Orkin, S.H. (1986). Cloning the gene for an inherited human disorder—chronic granulomatous disease—on the basis of its chromosomal location. Nature 322, 32-38.

Sathyamoorthy, M., de Mendez, I., Adams, A.G., & Leto, T.L. (1997). p40^{phox} down-regulates NADPH oxidase activity through interactions with its SH3 domain. J. Biol. Chem. 272, 9141-9146.

Sbarra, A.J., & Karnovsky, M.L. (1959). The biochemical basis of phagocytosis.1. Metabolic changes during the ingestion of particles by polymorphonuclear leukocytes. J. Biol. Chem. 234, 1355-1362.

Schonwasser, D.C., Marais, R.M., Marshall, C.J., & Parker, P.J. (1998). Activation of the mitogen-activated protein kinase/extracellular signal- regulated kinase pathway by conventional, novel, and atypical protein kinase C isotypes. Mol. Cell Biol. 18, 790-798.

Segal, A.W. (1987). Absence of both cytochrome b-245 subunits from neutrophils in X-linked chronic granulomatous disease. Nature 326, 88-91.

Segal, A.W., & Allison, A.C. (1978). Oxygen consumption by stimulated human neutrophils. Ciba. Found. Symp. 205-223.

Segal, A.W., & Jones, O.T. (1979). The subcellular distribution and some properties of the cytochrome b component of the microbicidal oxidase system of human neutrophils. Biochem. J. 182, 181-188.

Segal, A.W., & Meshulam, T. (1979). Production of superoxide by neutrophils: a reappraisal. FEBS. Lett. 100, 27-32.

Segal, A.W., Garcia, R.C., Harper, A.M., & Banga, J.P. (1983). Iodination by stimulated human neutrophils. Studies on its stoichiometry, subcellular localization and relevance to microbial killing. Biochem. J. 210, 215-225.

Segal, A.W., Geisow, M., Garcia, R., Harper, A., & Miller, R. (1981). The respiratory burst of phagocytic cells is associated with a rise in vacuolar pH. Nature 290, 406-409.

Segal, A.W., Heyworth, P.G., Cockcroft, S., & Barrowman, M.M. (1985). Stimulated neutrophils from patients with autosomal recessive chronic granulomatous disease fail to phosphorylate a Mr-44,000 protein. Nature 316, 547-549.

Segal, A.W., West, I., Wientjes, F.B., Nugent, J.H.A., Chavan, A.J., Haley, B., Garcia, R.C., Rosen, H., & Scrace, G. (1992). Cytochrome b-245 is a flavocytochrome containing FAD and the NADPH-binding site of the microbicidal oxidase of phagocytes. Biochem. J. 284, 781-788.

Selvaraj, R.J., & Sbarra, A.J. (1966). Relationship of glycolytic and oxidative metabolism to particle entry and destruction in phagocytosing cells. Nature 211, 1272-1276.

Sergeant, S., & McPhail, L.C. (1997). Opsonized zymosan stimulates the redistribution of protein kinase C isoforms in human neutrophils. J. Immunol. 159, 2877-2885.

Sha'ag, D., & Pick, E. (1990). Nucleotide binding properties of cytosolic components required for expression of activity of the superoxide generating NADPH oxidase. Biochim. Biophys. Acta 1037, 405-412.

Shatwell, K.P., Dancis, A., Cross, A.R., Klausner, R.D., & Segal, A.W. (1996). The FRE1 ferric reductase of Saccharomyces cerevisiae is a cytochrome b similar to that of NADPH oxidase. J. Biol. Chem. 271, 14240-14244.

Simon, M.I., Strathmann, M.P., & Gautam, N. (1991). Diversity of G proteins in signal transduction. Science 252, 802-808.

Singer, W.D., Brown, H.A., Jiang, X., & Sternweis, P.C. (1996). Regulation of phospholipase D by protein kinase C is synergistic with ADP-ribosylation factor and independent of protein kinase activity. J. Biol. Chem. 271, 4504-4510.

Someya, A., Nagaoka, I., & Yamashita, T. (1993). Purification of the 260 kDa complex involved in the superoxide production of guinea pig neutrophils. FEBS Lett. 330, 215-218.

Stephens, L., Smrcka, A., Cooke, F.T., Jackson, T.R., Sternweis, P.C., & Hawkins, P.T. (1994). A novel phosphoinositide 3 kinase activity in myeloid-derived cells is activated by G protein beta gamma subunits. Cell 77, 83-93.

Sue, A.K., Fialkow, L., Vlahos, C.J., Schelm, J.A., Grinstein, S., Butler, J., & Downey, G.P. (1997). Inhibition of neutrophil oxidative burst and granule secretion by wortmannin: potential role of MAP kinase and renaturable kinases. J. Cell Physiol. 172, 94-108.

Sumimoto, H., Hata, K., Mizuki, K., Ito, T., Kage, Y., Sakaki, Y., Fukumaki, Y., Nakamura, M., & Takeshige, K. (1996). Assembly and activation of the phagocyte NADPH oxidase. Specific interaction of the N-terminal Src homology 3 domain of p47phox with p22phox is required for activation of the NADPH oxidase. J. Biol. Chem. 271, 22152-22158.

Sumimoto, H., Kage, Y., Nunoi, H., Sasaki, H., Nose, T., Fukumaki, Y., Ohno, M., Minakami, S., & Takeshige, K. (1994). Role of Src homology domains in assembly and activation of the phagocyte NADPH oxidase. Proc. Natl. Acad. Sci. USA 91, 5345-5349.

Swain, S.D., Helgerson, S.L., Davis, A.R., Nelson, L.K., & Quinn, M.T. (1997). Analysis of activation-induced conformational changes in p47phox using tryptophan fluorescence spectroscopy. J. Biol. Chem. 272, 29502-29510.

Tanaka, T., Imajoh-Ohmi, S., Kanegasaki, S., Takagi, Y., Makino, R., & Ishimura, Y. (1990). A 63-kilodalton cytosolic polypeptide involved in superoxide generation in porcine neutrophils. J. Biol. Chem. 265, 18717-18720.

Tang, W.J., & Gilman, A.G. (1991). Type-specific regulation of adenylyl cyclase by G protein beta gamma subunits. Science 254, 1500-1503.

Tang, W.J., & Gilman, A.G. (1992). Adenylyl cyclases. Cell 70, 869-872.

Tang, W.J., Iniguez Lluhi, J.A., Mumby, S., & Gilman, A.G. (1992). Regulation of mammalian adenylyl cyclases by G-protein alpha and beta gamma subunits. Cold Spring Harb. Symp. Quant. Biol. 57, 135-144.

Taylor, W.R., Jones, D.T., & Segal, A.W. (1993). A structural model for the nucleotide binding domain of the cytochrome b_{-245} ß chain. Prot. Sci. 2, 1675-1685.

Teahan, C., Rowe, P., Parker, P., Totty, N., & Segal, A.W. (1987). The X-linked chronic granulomatous disease gene codes for the beta-chain of cytochrome b_{-245}. Nature 327, 720-721.

Teahan, C.G., Totty, N., Casimir, C.M., & Segal, A.W. (1990). Purification of the 47kDa phosphoprotein associated with the NADPH oxidase of human neutrophils. Biochem. J. 267, 485-489.

Thrasher, A.J., Keep, N.H., Wientjes, F., & Segal, A.W. (1994). Chronic granulomatous disease. Biochim. Biophys. Acta 1227, 1-24.

Tilton, B., Andjelkovic, M., Didichenko, S.A., Hemmings, B.A., & Thelen, M. (1997). G-protein-coupled receptors and Fcgamma-receptors mediate activation of Akt/protein kinase B in human phagocytes. J. Biol. Chem. 272, 28096-28101.

Toker, A., Meyer, M., Reddy, K.K., Falck, J.R., Aneja, R., Aneja, S., Parra, A., Burns, D.J., Ballas, L.M., & Cantley, L.C. (1994). Activation of protein kinase C family members by the novel polyphosphoinositides PtdIns-3,4-P2 and PtdIns-3,4,5-P3. J. Biol. Chem. 269, 32358-32367.

Torres, M., Hall, F.L., & O'Neill, K. (1993). Stimulation of human neutrophils with formyl-methionyl-leucyl-phenylalanine induces tyrosine phosphorylation and activation of two distinct mitogen-activated protein-kinases. J. Immunol. 150, 1563-1577.

Tsunawaki, S., Kagara, S., Yoshikawa, K., Yoshida, L.S., Kuratsuji, T., & Namiki, H. (1996). Involvement of p40phox in activation of phagocyte NADPH oxidase through association of its carboxyl-terminal, but not its amino-terminal, with p67phox. J. Exp. Med. 184, 893-902.

Tsunawaki, S., Mizunari, H., Nagata, M., Tatsuzawa, O., & Kuratsuji, T. (1994). A novel cytosolic component, p40phox, of respiratory burst oxidase associates with p67phox and is absent in patients with chronic granulomatous disease who Lack p67phox. Biochem. Biophys. Res. Commun. 199, 1378-1387.

Twomey, B., Muid, R.E., & Dale, M.M. (1990). The effect of putative protein kinase C inhibitors, K252a and staurosporine, on the human neutrophil respiratory burst activated by both receptor stimulation and post-receptor mechanisms. Br. J. Pharmacol. 100, 819-825.

Vanhaesebroeck, B., Leevers, S.J., Panayotou, G., & Waterfield, M.D. (1997a). Phosphoinositide 3-kinases: a conserved family of signal transducers. Trends Biochem. Sci. 22, 267-272.

Vanhaesebroeck, B., Welham, M.J., Kotani, K., Stein, R., Warne, P.H., Zvelebil, M.J., Higashi, K., Volinia, S., Downward, J., & Waterfield, M.D. (1997b). P110delta, a novel phosphoinositide 3-kinase in leukocytes. Proc. Natl. Acad. Sci. USA 94, 4330-4335.

Volpp, B.D., Nauseef, W.M., & Clark, R.A. (1988). Two cytosolic neutrophil oxidase components absent in autosomal chronic granulomatous disease. Science 242, 1295-1297.

Volpp, B.D., Nauseef, W.M., Donelson, J.E., Moser, D.R., & Clark, R.A. (1989a). Cloning of the cDNA and functional expression of the 47-kilodalton cytosolic component of human neutrophil respiratory burst oxidase. Proc. Natl. Acad. Sci. USA 86, 7195-7199.

Volpp, B.D., Nauseef, W.M., Donelson, J.E., Moser, D.R., & Clark, R.A. (1989b). Cloning of the cDNA and functional expression of the 47 kilodalton cytosolic component of human neutrophil respiratory burst oxidase. [Author's correction.] Proc. Natl. Acad. Sci. USA 86, 9563.

Voncken, J.W., Van Schaick, H., Kaartinen, V., Deemer, K., Coates, T., Landing, B., Pattengale, P., Dorseuil, O., Bokoch, G.M., Groffen, J., & Heisterkamp, N. (1995). Increased neutrophil respiratory burst in *bcr*-null mutants. Cell 80, 719-728.

Waite, K.A., Wallin, R., Qualliotine Mann, D., & McPhail, L.C. (1997). Phosphatidic acid-mediated phosphorylation of the NADPH oxidase component p47-phox. Evidence that phosphatidic acid may activate a novel protein kinase. J. Biol. Chem. 272, 15569-15578.

Wallach, T.M., & Segal, A.W. (1996). Stoichiometry of the subunits of flavocytochrome b558 of the NADPH oxidase of phagocytes. Biochem. J. 320, 33-38. Abstract.

Wallach, T.M., & Segal, A.W. (1997). Analysis of glycosylation sites on gp91phox, the flavocytochrome of the NADPH oxidase, by site directed mutagenesis and in vitro translation. Biochem. J. 321, 583-585.

Weiss, S.J. (1989). Tissue destruction by neutrophils. N. Engl. J. Med. 320, 365-376.

Wenzel Seifert, K., Schachtele, C., & Seifert, R. (1994). N-protein kinase C isoenzymes may be involved in the regulation of various neutrophil functions. Biochem. Biophys. Res Commun. 200, 1536-1543.

Wientjes, F.B., & Segal, A.W. (1995). NADPH oxidase and the respiratory burst. Sem. Cell Biol. 6, 357-365.

Wientjes, F.B., Hsuan, J.J., Totty, N.F., & Segal, A.W. (1993). p40phox, a third cytosolic component of the activation complex of the NADPH oxidase to contain SH3 domains. Biochem. J. 296, 557-562.

Wientjes, F.B., Panayotou, G., Reeves, E., & Segal, A.W. (1996). Interactions between cytosolic components of the NADPH oxidase—p40phox interacts with both p67phox and p47phox. Biochem. J. 317, 919-924.

Winterbourn, C.C., Garcia, R.C., & Segal, A.W. (1985). Production of the superoxide adduct of myeloperoxidase (compound III) by stimulated human neutrophils and its reactivity with hydrogen peroxide and chloride. Biochem. J. 228, 583-592.

Woodman, R.C., Ruedi, J.M., Jesaitis, A.J., Okamura, N., Quinn, M.T., Smith, R.M., Curnutte, J.T., & Babior, B.M. (1991). Respiratory burst oxidase and three of four oxidase-related polypeptides are associated with the cytoskeleton of human neutrophils. J. Clin. Invest. 87, 1345-1351.

Worthen, G.S., Avdi, N., Buhl, A.M., Suzuki, N., & Johnson, G.L. (1994). FMLP activates Ras and Raf in human neutrophils. Potential role in activation of MAP kinase [see comments]. J. Clin. Invest. 94, 815-823.

Yamashita, T., Someya, A., & Hara, E. (1985). Response of superoxide anion production by guinea pig eosinophils to various soluble stimuli: comparison to neutrophils. Arch. Biochem. Biophys. 241, 447-452.

Yu, L., Quinn, M.T., Cross, A.R., & Dinauer, M.C. (1998). $gp91^{phox}$ is the heme binding subunit of the superoxide-generating NADPH oxidase. Proc. Natl. Acad. Sci. U.S.A. 95, 7993-7998.

Zatti, M., & Rossi, F. (1965). Early changes of hexose monophosphate pathway activity and of NADPH oxidation in phagocytosing leucocytes. Biochim. Biophys. Acta 99, 557.

OXYGEN-INDEPENDENT ANTIMICROBIAL MECHANISMS OF PMN

Peter Elsbach

Advances in Cell and Molecular Biology of Membranes and Organelles
Volume 5, pages 485-512.

ISBN: 1-55938-999-0

I. INTRODUCTION

Throughout evolution ecologic interdependence has been maintained between prokaryotic and eukaryotic organisms. Of necessity mutual defenses had to be developed against disruption of this balance by overly aggressive partners. Prokaryotes employ extraordinarily effective adaptive mechanisms, both genetic and biochemical, to deal with hostile stresses in their environment, including the eukaryotic host (Qi et al., 1995; Groisman, 1996; Fallman et al., 1997; Guiney, 1997; Guo et al., 1997; Yamamoto et al., 1997). In turn, the host possesses a complex cytotoxic arsenal to fight off the enormous variety of potential microbial predators (for reviews see Elsbach and Weiss, 1992; Weiss, 1994; Levy, 1996; Elsbach et al., 1998). Among these antimicrobial elements are a range of polypeptides and proteins found all along the evolutionary scale in plants (Broekaert et al., 1995), insects (Boman, 1995; Hoffmann et al., 1996; Hoffmann and Reichhart, 1997), primitive animals, and mammals including humans (Zasloff, 1992; Boman, 1995; Levy, 1996; Elsbach et al., 1999), that are thought to serve as endogenous antibiotics. The origin of these agents can actually be traced to prokaryotes such as *Escherichia coli* that also employ potent cytotoxic proteins, bacteriocins, with target specificity for related bacterial strains to establish their supremacy in a competitive environment (Baba and Schneewind, 1998).

In recent years much progress has been made in the isolation from mammals, including humans, of polypeptides with antimicrobial activities *in vitro* (Elsbach and Weiss, 1992; Levy, 1996). Many of these have been cloned and several are now the subject of exploration of their potential use as therapeutic agents. A sense of urgency has been prompted by the progressive emergence of microbial infections resistant to conventional antibiotics and hence the need to find new ways of dealing with life-threatening drug-resistant microbes.

The effectiveness of endogenous antimicrobial defenses is evident from the fact that the normal host generally goes through most of life without major infections. Thus, the dissection of the host's complex antimicrobial equipment into its components and the study of their structural and functional characteristics as antimicrobial agents may yield clues that lead to ultimate clinical application.

This review provides an outline of the categories of polypeptides/proteins with antimicrobial properties that are recognized as constituents of phagocytes. While the focus is on the polymorphonuclear leukocytes (PMN) that among the professional phagocytes are the most prominent source of these antimicrobial polypeptides, it is now clear that many epithelia also produce similar agents that may contribute to their role as barriers against microbial invaders. For details not provided in this synopsis the reader will be referred to very recent and comprehensive reviews of the same topic that have recently appeared or are in press (Levy, 1996; Ganz and Lehrer, 1997; Ganz and Weiss, 1997; Elsbach, 1998a, b; Elsbach and Weiss, 1998; Elsbach et al., 1999).

II. ANTIMICROBIAL POLYPEPTIDES AND PROTEINS OF PHAGOCYTES

After the discovery of the antibacterial effects of the enzyme lysozyme, a constituent of phagocytes, epithelia and body fluids, nearly a century ago (Fleming, 1922), there was a long period with little progress in the identification of individual components that contribute to the antimicrobial capabilities of the host. The isolation and characterization during the past two decades of many other cytotoxic proteins and peptides from phagocytes, mainly the PMN, has given rise to new concepts of host-defense against infectious agents. The description and classification of these polypeptides rests on their molecular properties, their location within the cell and the spectrum of their cytotoxic activities.

The order of presentation of the various categories is inevitably arbitrary. In this review the order is based on what is known about the cytoplasmic granules of the PMN, the intracellular sites where most of the antimicrobial polypeptides are stored. Much has been learned about the assembly of these intracellular organelles during myeloid differentiation. Progress is also being made in the analysis of their contents and the roles these constituents play in cellular and extracellular events during various forms of inflammation (Borregaard et al., 1995; Gullberg et al., 1997).

Table 1 lists the principal antimicrobial polypeptides by molecular mass and indicates their association with either primary (azurophil) and/or secondary (specific) granules (Borregaard et al., 1995; Gullberg et al., 1997). The contents of granules are subject to release triggered by a wide range of (patho)physiologic and artificial stimuli (Henson et al., 1992). The contents of the primary granules are mainly released into newly formed phagocytic vacuoles and to a lesser extent extracellularly, i.e., serving a prominent intracellular function. In contrast, secondary granules have the characteristics of secretory granules and readily release their contents upon stimulation of the PMN into the extracellular (inflammatory) environment. How such segregation of the contents of different granule populations defines different host-defence roles is a subject of much investigation (Campbell et al., 1989; Owen et al., 1995; Sengelov et al., 1995; Borregaard, 1996).

III. CYTOTOXIC POLYPEPTIDES STORED IN PRIMARY GRANULES

The proteins and peptides stored in the primary granules are of two kinds: 1) those with cytotoxic but no known catalytic properties, including the bactericidal/permeability-increasing protein (BPI), azurocidin/CAP37, and defensins, and 2) a range of enzymes, encompassing a broad spectrum of degradative enzymes capable of contributing to the destruction of killed bacteria by digesting their macromolecules (Borregaard et al., 1995; Borregaard, 1996; Gullberg et al.,

Table 1. Antibiotic Proteins of PMN

Localization	*Protein*	*MW (kDa)*	*μg/10^6 PMN*	*Animal(s)*	*Activity*
1° Granule	Defensins	4	~4 H/12 R	H, R, G, Hm, Rt, C	B+, F, EV, m, M
	Serprocidins	30	1–2.5 (each)	H, C	B±, F, m, M
	BPI	50–60	1.5	H, R, C	B–
2° Granule	Cathelicidins	11–20 (2–7)[b]	2–3	H, R, M, C[c], S, P	B±, F
	Lactoferrin	78	3	H, R	B±, F
1° and 2° Granule	Lysozyme	14.5	2	H, E	B+, F
Granule (unspecified)					
	C3 (C6 and C7)	185 (110)	? (0.15, 0.06)	H, M	(B–)[a]
	PLA_2	14	0.05–0.5	H, R	B+
	AOAH	52–60	<01	H, R	B–
Cytosol	Calprotectin	14 and 8	5	H, M, P	B±, F

Note: Abbreviations: H, human; R, rabbit, M, mouse; G, guinea pig; Hm, hamster, Rt, rat; E, horse; C, cow; S, sheep; P, pig; B, bacteria (Gram staining ±); F, fungi; m, metazoan parasites; M, mammalian cells; EV, enveloped viruses.

[a] Bactericidal activity of complement requires assembly of the membrane attack complex.

[b] Most cathelicidins require proteolytic cleavage to generate active microbicidal peptides of 2–7 kDa.

[c] Bactenecins are located in the large granules of bovine PMN.

1997). Among these are lysozyme (a "peptidoglycanase"), proteinases, some with independent antimicrobial activity (elastase, cathepsin G), nucleases, and saccharidases. Enzymes degrading bacterial phospholipids (Wright et al., 1990b; Weiss et al., 1994; Madsen et al., 1996; Elsbach, 1997) and lipopolysaccharides (LPS) (Munford and Hunter, 1992; Luchi and Munford, 1993) are known to be granule-associated but have not yet been assigned to a defined subset of granules. The enzyme, myeloperoxidase, a component of the oxygen-dependent antimicrobial system of phagocytes, is also a constituent of the primary granules. The components of tertiary granules have been recently reviewed (Borregaard et al., 1995; Borregaard, 1996; Gullberg et al., 1997).

A. Bactericidal/Permeability-Increasing Protein

BPI is a 50–55 kDa cationic protein of 456 amino acid residues isolated and cloned from human, rabbit and cow PMN (Elsbach, 1994; Elsbach et al., 1994; Elsbach and Weiss, 1998). Synthesis of the protein has only been demonstrated in myeloid precursors. In addition to its presence in the primary granules, BPI has also been identified on the surface membrane of PMN (Weersink et al., 1993) and mononuclear cells (Dentener et al., 1996), and in a granule population, termed "specific," of eosinophils (Calafat et al., 1998). BPI is structurally related to a family of lipid and/or LPS-binding proteins that include the LPS-binding protein (LBP), cholesterolester exchange protein (CETP) and phospholipid-transfer protein (PLTP) (Beamer et al., 1997). The crystal structure of human BPI has recently been solved, revealing a highly elongated molecule composed of two distinct domains of similar size, separated by a proline-rich linker (Beamer et al., 1997). Each domain contains a hydrophobic phospholipid containing pocket that is also predicted to be present in the other members of this family. All four proteins also share the presence of a disulfide (cys 135–175) that is necessary for the stability and the functional integrity of these molecules. For further details of the structure-function relationships in BPI and their apparent relevance in LBP, CETP, and PLTP see (Beamer et al., 1997). Other evidence of the relatedness of these four proteins is indicated by a nearly identical genomic organization (Hubacek et al., 1997). The two-domain structure of BPI and LBP reflects bifunctional roles of these molecules (see below).

BPI is potently antibacterial at low nM concentrations both in artificial media and in biological fluids (Weiss et al., 1992; Weinrauch et al., 1995). Its cytotoxicity is highly selective for gram-negative bacteria (GNB), a target-specificity that reflects high binding affinity for the LPS in the outer membrane of the bacterial envelope (Weiss et al., 1980, 1984, 1992). The LPS-binding affinity is also evident toward cell-free LPS (apparent Kd: ≤ 5 nM). Complex formation between BPI and LPS blocks all endotoxin-mediated host cell responses examined (Marra et al., 1990, 1992; Gazzano-Santoro et al., 1992, 1994; Ammons et al., 1993, 1994). Because BPI targets the anionic sites clustered in the highly conserved lipid A portion of the LPS molecule (Gazzano-Santoro et al., 1995) the anti-LPS activity of BPI is manifest toward LPS from a wide range of GNB species, regardless of the highly variable composition of the polysaccharide chain.

The antibacterial actions of BPI against a broad range of GNB species (Elsbach and Weiss, 1992) cause initially reversible and subsequently irreversible growth inhibition. The former is linked to alterations in the outer membrane and the latter to damage to the cytoplasmic membrane (Mannion et al., 1990a, b). The antibacterial and anti-endotoxin activities of BPI do not require the holo-protein (Ooi et al., 1987). The N-terminal domain of BPI contains the determinants of interaction with LPS (Gazzano-Santoro et al., 1992). In LBP LPS binding, although with a ca. 70-fold lower apparent affinity, is also mediated by the N-terminal domain. This difference in affinity most likely reflects the high concentration of basic residues, unopposed by acidic residues, at the end of the N-terminal domain of BPI, promoting strong attraction to the negatively charged LPS (Beamer et al., 1997). The two structurally similar proteins are also functionally very different in other respects. The primary function of LBP, a plasma protein constitutively secreted by the liver and in increased amounts during the acute phase response, is to mobilize and amplify host cell responses to invading GNB by delivering LPS to CD14 acceptor molecules, present either on cells or in body fluids (Ulevitch and Tobias, 1994, 1995; Ulevitch et al., 1996). In contrast, binding of BPI or its N-terminal derivative to LPS in the bacterial envelope or after its release strongly inhibits host-responses (Weiss et al., 1992; Elsbach, 1994; Elsbach et al., 1994a, b). Moreover, whereas BPI, or its proteolytically prepared or recombinant N-terminal half, potently inhibit growth of GNB (Weiss et al., 1992; Elsbach, 1994; Elsbach et al., 1994a, b), LBP possesses no antibacterial capabilities. The functions associated with the C-terminal domains of BPI and LBP are also profoundly different. Delivery of LBP-bound LPS to CD14 depends on the C-terminal half of the molecule (Ulevitch et al., 1995). The C-terminal half of BPI lacks this function (Iovine et al., 1997). However, this region of BPI can also play a delivery role by promoting phagocytosis of BPI-coated encapsulated, phagocytosis-resistant, *E. coli* by human PMN (Iovine et al., 1997). Thus, the structurally similar LBP and BPI fulfill dissimilar and even opposite biologic functions in the complex pathways that lead to activation or inhibition of host responses to invading GNB and their endotoxic products.

The different interactions of BPI and LBP with LPS are further illustrated by studies of the physical properties of complexes formed between the two proteins with LPS. Whereas LBP pulls single molecules out of LPS aggregates in line with its role in delivery of LPS signals, BPI inserts into LPS aggregates, blocking LBP-induced disaggregation and may promote formation of larger aggregates (Tobias et al., 1995, 1997; Wiese et al., 1997a, b).

Role of BPI in Antibacterial Host-Defense

That BPI actually plays a role in antibacterial host defense is strongly suggested by the fact that the fates of GNB ingested by PMN or treated with isolated BPI are nearly identical (Mannion et al., 1990a, b). Moreover, anti-BPI sera abolish the potent antibacterial activities of whole PMN lysates directed at GNB (Weiss et al., 1982) and of inflammatory fluids *ex vivo* (Weinrauch et al., 1995), consistent with a prominent role of BPI as an endogenous antibiotic. However, while the antiserum effect implies that BPI is necessary for the anti-GNB activity of a complex host environment, BPI does not act alone; other cellular and extracellular constituents act synergistically with BPI (Levy et al., 1994; Elsbach et al., 1994a, 1998; Weinrauch et al., 1995; Levy, 1996). It should also be stressed that not all mammals need to rely on BPI for effective antibacterial host defense. The PMN of laboratory mice lack not only defensins (Lehrer et al., 1993), but also do not contain BPI (unpublished observations).

The potent antibacterial and anti-LPS activities of BPI and its N-terminal fragment have prompted the large-scale production of recombinant BPI derivatives in order to permit testing of their potential as therapeutic agents (Elsbach and Weiss, 1995). Protection of several animal species against lethal doses of LPS and bacterial inocula has been documented (Gazzano-Santoro et al., 1994; Kung et al., 1994;; Lin et al., 1994a, b). These results have led to multiple clinical trials, including two Phase III trials, currently in progress, involving fulminant pediatric meningococcal sepsis (Giroir et al., 1997) and hemorrhagic trauma. Other trials in progress include partial hepatectomy, severe peritoneal infections and cystic fibrosis. In none of more than 1000 normal or severely ill individuals receiving $rBPI_{21}$ (BPI-derived ca. 21 kDa N-terminal fragment) have issues of safety or immunogenicity been encountered.

B. Defensins

For more detailed reviews of the defensins the reader is referred to very recent reviews (Ganz and Lehrer, 1997; Ganz and Weiss, 1997). Briefly, defensins, ca. 4,000 kDa polypeptides, are the most abundant protein components of primary granules of the PMN of human and other mammals (but are absent in the PMN of laboratory mice, horse and pig), representing as much as 30% of total granule protein. Defensin-like peptides are also constituents of rabbit alveolar macrophages and of Paneth cells in the human as well as the murine intestinal tract (Ganz and Lehrer, 1997; Ganz and Weiss, 1997). The widespread occurrence of both α- and β-defensins in higher animals and more distant defensin relatives in plants (Broekaert et al., 1995) and insects (Hoffmann and Reichhart, 1997) is consistent with an early evolutionary origin of these agents (Ganz and Lehrer, 1997). Human defensin genes are clustered on chromosomal band 8p23 whether expressed in neutrophils or in epithelial cells (Sparkes et al., 1989; Ganz and Weiss, 1997). A characteristic three-dimensional structure that includes the positions of 3 disulfides is shared by the members of this extensive family of polypeptides. Defensins are produced as biologically inactive preproproteins that are converted into broadly cytotoxic peptides by proteolytic cleavage before storage in the azurophilic granules of the PMN (Lehrer et al., 1993). Human defensins (both classical α- and somewhat structurally divergent β-defensins) are also constituents of secretory granules of the intestinal Paneth cells (Jones and Bevins, 1992; Ouellette et al., 1992; Ganz and Weiss, 1997). In other mammalian epithelia of many organs, including trachea, tongue, kidney and skin, defensin-like peptides have been demonstrated that are induced in response to inflammatory stimuli (Schonwetter et al., 1995; Diamond et al., 1996; Ganz and Weiss, 1997).

The principal biologic role of the defensins, as their name implies, is generally considered to be a host-defense one, based on their antimicrobial activities in laboratory (low ionic strength) media. In physiologic and biologic fluids their activity toward most bacterial species is markedly reduced or even eliminated (Levy et al., 1995; Ganz and Lehrer, 1997). It must be stressed, however, that estimates of the molar concentrations of the extraordinarily abundant defensins upon their release into the phagosome support the view that they participate in the creation of a toxic intracellular antimicrobial environment (Levy et al., 1995; Ganz and Lehrer, 1997). Nonetheless, the possibility that the defensins and other polypeptides that are antimicrobial *in vitro*, may in fact serve other and even more important roles *in vivo* is supported by reports indicating that these molecules exhibit a range of biologic (hormone-like, tissue repair) activities, unrelated to antimicrobial action (reviewed in Levy et al., 1995; Ganz and Lehrer, 1997; Ganz and Weiss, 1997). Another reason to explore additional functions in inflammation is that *in vitro* defensins express these other activities at much lower concentrations than their activity against microbes.

The cytotoxic action of defensins, whether toward bacteria, *Candida*, or human tumor cells, appears to involve several steps that lead to membrane-permeabilization and lysis of the target cells (Ganz and Lehrer, 1997; Ganz and Weiss, 1997). Attempts have been made to relate evidence of pore formation by defensins in artificial and biologic membranes to defensin structures (White et al., 1995).

IV. CYTOTOXIC POLYPEPTIDES STORED IN SECONDARY GRANULES

A. Cathelicidins

For a more comprehensive treatment of these polypeptides, especially their molecular and gene structures, the reader is directed to several reviews of the most current literature (Zanetti et al., 1995; Levy, 1996; Ganz and Lehrer, 1997; Ganz and Weiss, 1997).

The cathelicidins are the most recently recognized family of PMN polypeptides with antimicrobial activity and common structural features. The name derives from the initial description of a 12 kDa protein isolated from pig neutrophils, termed cathelin (Turk and Bode, 1991). Subsequently many polypeptides present in neutrophils of a broad range of mammalian species have been reported that share amino acid sequence homology with cathelin in their N-terminal region. This negatively charged "cathelin-like" region is highly conserved among members of the family and represents the proregion of these molecules that is linked to a highly variable cationic C-terminal peptide (10–40 amino acids). This peptide, upon proteolytic release from the inactive precursor exhibits antimicrobial properties (reviewed in Zanetti et al., 1995; Levy, 1996; Ganz and Weiss, 1997). The antimicrobial activity of these polypeptides has given rise to the designation "cathelicidins" (Zanetti et al., 1995).

The number of cathelicidins varies greatly among different animal species. In a bovine genomic library a cluster of 11 distinct genes has been identified (Scocchi et al., 1997). More distantly related members have been isolated and cloned from rabbit PMN (Levy et al., 1993; Levy, 1996) that do not require proteolytic cleavage for biologic activity. These 15 kDa polypeptides (p15s) express antibacterial activity against GNB at relatively high concentrations (low μmolar) (Levy et al., 1994; Weinrauch et al., 1995). However, the p15s and also another antibacterial cathelicidin polypeptide, CAP18, found in rabbit and human neutrophils (Larrick et al., 1993, 1994a, b, 1995) when combined with BPI show potent synergy both in laboratory media and as natural constituents of inflammatory fluids (Levy et al., 1994; Weinrauch et al., 1995) and unpublished observations). While the p15s only share ca. 35% sequence identity with cathelin and other cathelicidins (Levy et al., 1993), the four cysteines common to the family are positioned similarly in the p15s.

So far the only cathelicidin found in human PMN is CAP18, also referred to as FALL39 and LL37, depending on the investigators determining its expression and translation products (Agerberth et al., 1995; Cowland et al., 1995; Larrick et al., 1995; Sorensen et al., 1997a, b). Human CAP18 is stored in unprocessed form in the secondary granules of PMN (Sorensen et al., 1997a). Expression of this cathelicidin is constitutive in PMN and testis but is inducible in keratinocytes (Frohm et al., 1997). The rabbit p15s have also been located to the secondary granules (Zorember et al., 1997), suggesting that this granule population is also the storage site for other cathelicidins. This raises questions about the proteolytic event that converts inactive precursors to the active peptide, if the cathelicidin proforms and the proteinases do not reside in the same granule population, unlike the defensins (Ganz and Lehrer, 1997). This issue has been well-studied in the case of the bovine cathelicidins, the bactenicins (Zanetti et al., 1995) The inactive probactenicins are stored in the large granules that are unique to the PMN of ruminant animals. Proteolysis by elastase, a constituent of the primary granules, and activation does not occur, therefore, until the PMN are activated and degranulate (Zanetti et al., 1995).

Other bovine cathelicidins are the proforms of disulfide containing dodecapeptide (Storici et al., 1992) and the tryptophan-rich indolicidin (Del Sal et al., 1992; Selsted et al., 1992). The pig PMN protegrins—cathelicidins that are similarly stored and processed—also release an antimicrobial C-terminal peptide that contains two disulfides (Harwig et al., 1996; Lehrer and Ganz, 1996; Panyutich et al., 1997). The mature protegrins are antibacterial in physiologic salt solutions and serum against both gram-positive and gram-negative bacterial species and also inhibit *in vitro* the growth of *Mycobacterium tuberculosis* at 50 μg/ml (Miyakawa et al., 1996).

It seems likely that the separation of proteinases and cathelicidins in different granules of the PMN of non-ruminants would also ensure protection from inopportune peptide activation until degranulation and extracellular release of granule contents are triggered. Since only minute concentrations of proteinase (elastase) suffice for processing of proforms, the much greater extracellular release of constituents of the secondary granules than those of the primary granules does not argue against such a control mechanism.

The appearance of bioactive concentrations of cathelicidins, such as p15s in the fluid phase of inflammatory exudates (Weinrauch et al., 1995; Zarember et al., 1997), CAP18 in human serum (Sorensen et al., 1997a, b) and PR39 in the circulation of pigs infected with *Salmonella* (Zhang et al., 1997) is consistent with extracellular functions of these agents in inflammation. As has been shown for defensins (Levy, 1996; Ganz and Lehrer, 1997), much recent evidence indicates that these functions are not limited to antimicrobial actions, especially in the highly complex natural host environment (Gallo et al., 1994; Shi et al., 1996; Li et al., 1997). It has not yet been determined what the functions are of defensins and human CAP18 peptide in serum/plasma where they exist as high molecular weight complexes (Panyutich et al., 1993; Sorensen et al., 1997b).

Despite the conserved primary structures of the polypeptides that are grouped as cathelicidins, these proteins/peptides assume very different molecular conformations, ranging from linear to more folded (e.g., disulfide mediated) members of this family (see Levy, 1996; Ganz and Lehrer, 1997; Ganz and Weiss, 1997). This implies that the mode of antimicrobial action of the different members is likely to be different, some acting by generating envelope alterations, others through effects on macromolecular synthesis and degradation (Boman et al., 1993; Ganz and Weiss, 1997).

Expression of cathelicidin genes is not restricted to myeloid cells, but includes keratinocytes, the spleen and the testis (Frohm et al., 1996, 1997; Scocchi et al., 1997).

It is becoming increasingly clear that the various families of structurally distinct polypeptides described here, notably the defensins and the cathelicidins, actually share some biologically important features. Thus, the growing evidence that antimicrobial proteins and polypeptides that were first recognized as constituents of leukocytes only are in fact widely distributed in epithelia also extends to the cathelicidins. Further, it can no longer be claimed that the reliance of many students of these polypeptides on antimicrobial assays *in vitro* adequately assesses the range of biologic roles these agents may fulfill *in vivo*. Both defensins and several cathelicidins are capable of other functions related to the inflammatory responses of the host.

B. Lactoferrin (Lf)

For a more extensive literature review of Lf see Ellison, 1994; Brock, 1995; Levy, 1996; Elsbach et al., 1999. Lf, an 80 kDa cationic protein that belongs to a family of iron binding proteins, is a constituent of secondary granules of PMN and mucosal secretions in all animal species studied. Iron-binding by Lf, by depriving bacteria dependent on this growth factor, accounts in part for its antibacterial activity. However, Lf also interacts directly with bacterial targets via a range of surface molecules, including Lf receptors (Gray-Owen and Schryvers, 1996; Miehlke et al., 1996; Dhaenens et al., 1997) and outer membrane LPS (Elass-Rochard et al., 1995). An N-terminal region has been identified that includes a stretch from position 28 to 36, containing 4 basic residues, that mediates high affinity binding to lipid A of isolated LPS (Appelmelk et al., 1994; Elass-Rochard et al., 1995; 1998). The crystal structure of Lf has revealed a bilobed molecule (Andersen et al., 1987), in which iron-binding involves cooperative interactions between the N-terminal and C-terminal domains (Ward et al., 1996). The combination of the iron-complexing and bacterial surface interactive properties of Lf may actually provide a source of iron for some (gram-negative) bacterial species rather than contributing to antibacterial host defense (Gray-Owen and Schryvers, 1996).

Attempts have been made to further identify molecular determinants of the antibacterial action of Lf by generating proteolytic fragments. One of these, lactoferricin, encompasses what is thought to be the primary N-terminal LPS-binding domain of Lf and includes a loop that has been claimed to be similar in other LPS-binding proteins, such as BPI, LBP and *Limulus* anti-LPS factor (Elass-Rochard et al., 1998). However, the recently solved 3-dimensional stucture of BPI does not support a closely similar loop region in that molcule (Beamer et al., 1997). The antimicrobial spectrum of lactoferricin differs from that of the parent protein (Yamauchi et al., 1993), a common finding when the activities of holoproteins and derived peptides are compared (Little et al., 1994).

At high concentrations ($\geq$ 20 μg/ml) human recombinant Lf inhibits LPS-binding to a differentiated monocytic (THP-1) cell line and, when coincubated with LBP in 10% serum, prevents LBP-mediated interaction with CD14 (Elass-Rochard et al., 1998). Since in the presence of serum/plasma the high affinity LPS-binding by Lf (apparent kDa ca. 4 nM) is inhibited (Wang et al., 1995), it remains to be determined whether Lf affects LPS signaling in complete biologic fluids and at physiologic concentrations.

Under various laboratory conditions high concentrations of Lf exhibit a range of activities in addition to acting as an antibacterial agent, including effects on virus-host cell interaction (Harmsen et al., 1995) and pleiotropic effects on the functions of many different mammalian host cells (reviewed in Brock, 1995; Levy, 1996; Elsbach et al., 1999). Thus, as pointed out above for the defensins and cathelicidins, further exploration in biologic settings of the ability of Lf to modulate the host environment appears warranted.

V. CYTOTOXIC DEGRADATIVE ENZYMES

The view that degradative enzymes can act as independent antimicrobial agents was initially based on the discovery of lysozyme by Fleming (1922). Lysozyme, whether isolated from mammalian body fluids and secretions or as a constituent of both primary and secondary granules of PMN (for more extensive reviews of lysozyme, see Jolles and Jolles, 1984; Irwin and Wilson, 1990; Takeuchi et al., 1993; Irwin et al., 1996; Levy, 1996), promptly lyses a number of saprophytic gram-positive bacterial species in which glycosidic bonds in the peptidoglycan network are readily accessible. However, it is still uncertain how much lysozyme as a constituent of the antibacterial arsenal of the mammalian host contributes to the killing/destruction of pathogenic microorganisms.

The notion that host enzymes capable of degrading microbial macromolecules are integral to host-defense is supported by the original description of the primary granules of (rabbit) PMN as lysosome-like cytoplasmic particles that contain a range of acid hydrolases (Hirsch, 1974). Subsequently, Ginsburg and collaborators have been strong advocates of the importance of hydrolases in combination with nonenzymatic host factors in antimicrobial host defense (Lahav et al., 1974, 1975; Ginsburg and Sela, 1976; Ginsburg et al., 1982; Ginsburg and Lahav, 1983; Ginsburg, 1988).

However, evaluation of the participation of host hydrolases in determining the fate/digestion of bacterial targets must take into consideration that activation of bacterial degradative enzymes is a common bacterial response to adverse conditions (Weiss and Elsbach, 1977; Elsbach, 1980). Further, bacteria have a remarkable ability to repair damage. Thus, biochemical evidence of degradation of bacterial macromolecules often inaccurately reflects actual damage, because compensatory biosynthesis may have replaced broken down components. Hence a meaningful assessment of the role of host hydrolases in overall antimicrobial activities of the host must include more than mere measurement of degradation of labeled bacterial constituents.

Such dissection of the antibacterial effects of one host enzyme, a 14 kDa phospholipase A_2 (PLA), has now yielded for the first time convincing evidence that this hydrolase qualifies as a biologically important independent endogenous antibiotic agent.

A. Inflammatory 14-kDa Phospholipase(s) A_2

Low molecular weight deacylating phospholipases represent a heterogeneous family of enzymes, widely distributed in nature, whose biologic functions are often not well-established (Tischfield, 1997). Members of a subcategory of this family, the group IIa 14-kDa PLA_2s, with shared structural characteristics (Weiss et al., 1991, 1994; Elsbach, 1997) are constituents of phagocytes (Wright et al., 1990a, b), bodyfluids and secretions (Kramer and Pepinsky, 1991; Nevalainen et al., 1994; Elsbach, 1997; Ganz and Weiss, 1997), and have been shown to participate in the destruction of gram-negative bacteria (Wright et al., 1990b; Weiss et al., 1994; Elsbach, 1997). By use of bacterial targets lacking activatable phospholipase, the catalytic involvement of these host phospholipases has been established (Wright et al., 1990b; Weiss et al., 1994; Madsen et al., 1996; Elsbach, 1997). Manifestation of the catalytic activity against the phospholipids of *E. coli* is entirely dependent on other host-defense agents (such as BPI and the membrane attack complex of complement Weiss et al., 1991, 1994; Madsen et al., 1996) that, by altering the outer envelope, promote binding of active but not inactive PLA2 family members (Weiss et al., 1991, 1994; Madsen et al., 1996). Both the cellular (PMN) and extracellular (inflammatory fluid) group IIa 14-kDa PLA_2s have been shown to participate in the attack on the phospholipids of *E. coli* (Wright et al., 1990a, b; Weiss et al., 1991; Weiss, 1994; Madsen et al., 1996; Elsbach, 1997).

In contrast to the inability of these inflammatory group IIa 14-kDa PLA_2s to target GNB in the absence of other surface-perturbing antimicrobial agents of the host, both the PLA_2 that is a natural constituent of inflammatory fluid elicited in the rabbit, and recombinant human synovial fluid PLA_2 act as potent independent bactericidal agents against a range of Gram-positive bacterial species (Weinrauch

Table 2. Antibacterial Activities of Mammalian 14 kDa Phospholipases A_2 in Inflammatory Settings

		Dependence on Other Antibacterial Agents
Targets	Gram-negative bacteria (GNB)	+ (BPI; MAC*)
	Gram-positive bacteria (GPS)	–
Location	Body fluids: Serum/plasma; tears; seminal and inflammatory fluids Cells: PMN; platelets; liver; others?	
Structural requirements for targeting bacteria	High net + charge	

Note: * MAC, membrane attack complex of compliments.

et al., 1996) (Table 2). This activity is evident in physiologic fluids and at the concentrations found there (Kramer and Pepinsky, 1991; Nevalainen et al., 1994; Weinrauch et al., 1998). The bactericidal action is linked to Ca^{2+}-dependent catalytic activity and is abolished by all procedures that block catalysis (Weinrauch et al., 1996b), verifying that the action of the highly cationic enzyme is not attributable to a noncatalytic interaction of the protein with the bacteria (Weinrauch et al., 1996b). The bactericidal action of inflammatory fluid PLA_2 extends to multiple-antibiotic resistant and encapsulated *Staphylococcus aureus* strains (Weinrauch et al., 1996b). Moreover, against highly antibiotic-resistant and moderately PLA_2-sensitive clinical isolates of *Enterococcus faecium,* combinations of low active or inactive concentrations of antibiotics and PLA_2 show dramatic synergy (Foreman-Wykert, 1999). In baboons that had received a lethal systemic inoculum of *E. coli*, circulating PLA_2 levels rise exponentially in the 24 hours after challenge, coincident with rising anti-staphylococcal activity that is blocked by anti-human group II 14-kDa PLA_2 antibodies (Weinrauch et al., 1999).

Of the many members of the large family of 14-kDa PLA_2s tested for antibacterial activity (Weiss et al., 1991, 1994; Elsbach, 1997), only a few possess this capability even though all show comparable catalytic activity toward isolated bacterial phospholipids or when these are made readily accessible by removal of the cell wall (protoplasts; manuscript in preparation) or autoclaving (Elsbach and Weiss, 1991). This signifies that noncatalytic properties of the PLA_2 determine the ability of a PLA_2 to reach bacterial targets. The primary determinants are the surface charge of the enzyme. Both the concentration of basic residues in discrete regions of the molecule (Weiss et al., 1991, 1994; Elsbach, 1997) as well as its overall cationicity (manuscript in preparation) have been shown to define interaction with the bacterial surface (binding) and subsequent antibacterial potency.

The cellular origin of the antibacterial 14-kDa group IIa PLA_2s found in the circulation and inflammatory fluids is still uncertain. Many cell types that participate in inflammatory events have been shown to secrete the 14-kDa PLA_2s, including PMN, platelets, articular cartilage and hepatocytes (Rosenthal et al., 1995; Kallajoki and Nevalainen, 1997). The growing evidence that this type of PLA_2 can act as a potent antibacterial agent *in vitro* and *in vivo,* combined with reports of very high constitutive PLA_2 levels not only in inflammatory fluids but also in tears (Nevalainen et al., 1994) and other secretions (Harwig et al., 1995) support the notion that this enzyme is involved in antimicrobial protection. This concept is important because of the common view that many of the subspecies of PLA_2s fulfill potentially harmful pro-inflammatory roles in various disease states that therefore need to be checked by use of inhibitors of catalytic activity (Pruzanski and Vadas, 1997). Whether one or both of these disparate functions may dominate in a given inflammatory setting remains to be explored.

B. Acyloxyacyl Hydrolase (AOAH)

The pathophysiology of GNB infections is closely linked to detection by the host of LPS (endotoxin) whether as a bacterial surface component or after shedding from the envelope. While recognition of the presence of GNB and their LPS ususally triggers effective protective host responses, these may also become excessive and self-destructive. Many of the molecular participants in these interactions have been identified. However, less is known about the means by which the host detoxifies and disposes of LPS, specifically of its potent bioactive lipid A moiety, once the initial LPS signal has been delivered. In rats, intravenously administered LPS forms complexes with various lipoproteins, including chylomicra (Harris et al., 1990, 1993; Read et al., 1993, 1995) before being traced to the liver. The radiolabeled LPS is detected first in Kupffer cells and granulocytes and then in hepatocytes (Freudenburg et al., 1984), to appear ultimately via the bile in the intestinal tract in a process that takes several weeks (Freudenburg et al., 1984). Evidence of slow release of fatty acids from the radiolabeled LPS at several tissue sites implied the presence of enzyme(s) capable of deacylating lipid A (Freudenburg et al., 1984). AOAH is the first such enzyme that has been isolated, cloned and characterized (Erwin and Munford, 1991; Munford and Hunter, 1992).

This lipolytic enzyme, found in PMN and mononuclear phagocytes and in circulating and inflammatory fluids, is the only well-defined mammalian hydrolase capable of deacylating and thereby detoxifying LPS (Erwin and Munford, 1991; Munford and Hunter, 1992). AOAH not only selectively cleaves the secondary (acyloxyacyl-linked) fatty acids from lipid A but also exhibits phospholipase, lysophospholipase, lipase and acyltransferase activities *in vitro* (Munford and Hunter, 1992). AOAH (Mr 52–60 kDa) is also structurally a highly unusual lipolytic enzyme, consisting of two disulfide linked subunits (Munford and Hall, 1989; Hagen et al., 1991). Both subunits are necessary for catalysis. The active site is thought to be located in the larger subunit; the smaller saposin-like subunit apparently stabilizes the enzyme and plays a role in its intracellular localization (Hangen et al., 1991; Staab et al., 1994). Human neutrophils have been shown to deacylate LPS (Luchi and Munford, 1993). The biologic participation *in vivo* of AOAH in the disposition of LPS is strongly suggested by the demonstration of increased expression of AOAH after LPS-challenge in rabbits (Erwin and Munford, 1991) and increased LPS deacylation after transient overexpression of AOAH in mice infected with a recombinant adenovirus carrying the AOAH gene (Coulthard et al., 1996). Until recently, it had not been established what is the potential contribution of AOAH to the disposal of LPS under pathophysiologic conditions. It has now been shown that isolated LPS undergoes rapid and extensive AOAH-like deacylation in a sterile inflammatory pertoneal exudate elicited in rabbits (Weinrauch et al., 1999). In this *ex vivo* inflammatory setting, the LPS of intact bacteria is also subject to subtantial AOAH-like deacylation (Katz et al., 1999). The mononuclear cells present in the exudate are the principal site of deacylating activity. The use of this model of inflammation has served to demonstrate for the first time that the mammalian host is capable of effectively detoxifying both cell-free and bacteria-associated LPS by AOAH-like deacylation. Because of the prominence of LPS as a trigger in disease caused by GNB, much remains to be learned about the relative importance of the removal of intact LPS complexed to lipoproteins en route to the intestinal tract and detoxification by degradative enzyme(s).

C. Other PMN-Associated Hydrolases with Antimicrobial Actvities

Among the many hydrolases stored in the primary granules of PMN are several serine proteases that are antimicrobial as isolated proteins *in vitro* and accordingly have been named serprocidins (Odenberg and Olsson, 1975; Campanelli et al., 1990a, b; Levy, 1996; Elsbach et al., 1999). The cytotoxicity toward bacteria of the active enzymes elastase, cathepsin G, and proteinase 3 can be dissociated, at least in part, from the hydrolytic activity (Odenberg and Olsson, 1975; Campanelli et al., 1990a, b; Levy, 1996; Elsbach et al., 1999). The antibacterial activity of the proteolytically inactive serprocidin analogue azurocidin/CAP37 argues most strongly for a noncatalytic mode of antimicrobial action of the serprocidins as has also been reported for lysozyme (Laible and Germaine, 1985). Attempts have been made to identify molecular determinants of antimicrobial activity of the serprocidin cathepsin G, a protein that is broadly cytotoxic *in vitro* against both Gram-positive and Gram-negative bacterial species, by preparing synthetic sequential peptides extending along the full length of the protein (Shafer et al., 1993, 1996). No obvious correlations were found between regions linked to catalysis and bioactivity of individual peptides. Overall, the insights gained from this dissection have been limited, mainly because the potency on a molar basis of the most active peptides is far less than that of the native protein.

Again the question may be asked whether the serprocidins should be viewed mainly as antimicrobial agents rather than as participants in a range of inflammatory events in which enzymes such as elastase and cathepsin G play regulatory roles by processing other proteins and polypeptides, for example in chemotaxis and in priming and activation of PMN and other cells (reviewed in Levy, 1996; Elsbach et al., 1999).

For a consideration of other hydrolases of phagocytes capable of degrading microbial macromolecular constituents, the reader is referred to earlier reviews (Elsbach, 1980; Elsbach and Weiss, 1992).

VI. CYTOSOLIC ANTIMICROBIAL POLYPEPTIDES

Two abundant anionic calcium and zinc-binding polypeptides that form a complex, termed calprotectin, in the cytosol of PMN and monocytes and some mucosal cells are broadly cytotoxic under various laboratory conditions (Sohnle et al., 1991a, b, 1996; Hessian et al., 1993; Yui et al., 1997). The ability of these polypeptides to complex zinc is thought to be sufficient to account for their cytotoxic effects (Sohnle et al., 1996). In addition to a possible contribution to the antimicrobial actions of phagocytes after these cells lyse at inflammatory sites (abscesses), calprotectins may have broader regulatory functions in inflammation (Hessian et al., 1993).

VII. PERSPECTIVES

Owing to recent technological advances the isolation and cloning of new proteins and polypeptides with antimicrobial actvities from many sources has become almost commonplace. However, as stressed repeatedly in this review the actual biologic role of many of these individual agents in overall antimicrobial host-defense remains to be established. The assumption that these newly isolated polypeptides, especially when highly cationic, are part of an antimicrobial arsenal is usually based upon antibacterial assays employing laboratory strains of bacteria in simple media. More critical appraisal of biologic function requires testing under conditions resembling the host environment, where none of these agents acts in isolation, but may be more or less active because of interactions with other host elements. Examples have been given above of both synergy among antimicrobial proteins and peptides and of profound inhibition by physiologic electrolyte concentrations or formation of inactive complexes.

Further, in the whole animal the mononuclear and polymorphonuclear phagocytes are highly interactive participants in multisystem responses to diverse stimuli and fulfill many functions beyond ingesting and killing microbial invaders. These cells secrete a multitude of bioactive molecules, including cytokines, complement components, coagulation-stimulatory tissue factor, adhesive molecules, lipid mediators and reactive oxygen derivatives. Most of these secretory products have no direct antimicrobial capabilities. Thus, the possibility should still be considered that the primary function in the inflammatory process of members of the families of proteins and polypeptides reviewed here has actually not yet been identified. On the other hand, the impressive success in implicating specific gene products in antimicrobial host defense that has been achieved in *Drosophila* plus the evidence that the regulatory pathways governing host responses in this insect bear remarkable resemblance to those in mammals, should point the way toward clearer identification of principal mammalian endogenous antibiotics as well (Hoffmann et al., 1996; Hoffmann and Reichhart, 1997). It should also be pointed out that among the proteins and polypeptides discussed here some clearly do fulfill the criteria of an endogenous antibiotic. Thus, protection against lethal bacterial inoculates and toxic bacterial products has been shown in various animal models. Mice and pigs can be protected against systemically administered LPS by CAP18 peptide (Larrick et al., 1995; VanderMeer et al., 1995; Fletcher et al., 1997). Mice are also protected against LPS by protegrins (Ganz and Weiss, 1997; Steinberg et al., 1997). Mice, rats, rabbits and pigs are protected against both live GNB and LPS by recombinant N-terminal fragments of BPI ($rBPI_{23}$ or $rBPI_{21}$) (Ammons et al., 1994; Kung et al., 1994; Lin et al., 1994a, 1996). In a rabbit model in which neither $rBPI_{21}$ nor a conventional antibiotic protected against gram-negative bacterial sepsis, the animals were protected when the two agents were combined (Lin et al., 1996).

So far only one of all the known mammalian proteins and polypeptides with antimicrobial activity has reached clinical testing. The BPI-derived $rBPI_{21}$ has now passed through phase I and phase II testing, showing no issues of safety either in normal humans or in patients with serious disease (de Winter et al., 1995; Van der Mohlen et al., 1995a, b; Giroir et al., 1997). Currently additional phase I/II placebo-controlled trials are in progress, including testing of rBPI21 in partial hepatectomy, serious abdominal infections and cystic fibrosis. Two phase III trials are in progress involving: 1) fulminant pediatric meningococcemia, extending a completed open phase II trial (Giroir et al., 1997) being carried out in the U.S., Canada and Great Britain, and 2) a 40-center trial in the US of $rBPI_{21}$ in hemorrhagic trauma, also extending two completed placebo-controlled phase II trials with 570 patients.

Obviously only a beginning has been made with the exploration of endogenous antibiotics as agents that can substitute or supplement the conventional antibiotics that progressively lose their efficacy in the treatment of serious bacterial infections (Levy, 1992; Zasloff, 1992; Jacob and Zasloff, 1994; Elsbach and Weiss, 1995). The further study of components of host-defense systems should include incorporation of new concepts concerning microbial adaptation to the host environment as a major determinant in pathogenesis. Recent evidence indicates that initially nonvirulent bacteria, after invading the host, can acquire resistance to host defenses through regulatory adjustment of gene expression rather than mutations (Qi et al., 1995; Groisman, 1996; Fallman et al., 1997; Guiney, 1997; Guo et al., 1997; Yamamoto et al., 1997). These findings further illustrate the need to assess the potential effectiveness of endogenous antibiotics as therapeutic agents under conditions that closely mimic those in the host.

Future investigations should also include the largely unexplored antimicrobial equipment of the mononuclear phagocyte. To what extent the essential role of this cell has in protection against a spectrum of pathogens against which the PMN is generally less or not effective, relies on O_2-independent antimicrobial proteins and peptides is entirely unknown.

ACKNOWLEDGMENTS

The work in this laboratory is supported by USPHS grants AM 05472 and AI 18571 and a grant from the XOMA Corporation, Berkeley, California.

REFERENCES

Agerberth, B., Gunne, H., Odeberg, J., Kogner, P., Boman, H.G., & Gudmundsson, G.H. (1995). FALL-39, a putative human peptide antibiotic, is cysteine-free and expressed in bone marrow and testis. Proc. Natl. Acad. Sci. USA 92, 195-199.

Ammons, W.S., & Kung, A.H. (1993). Recombinant amino terminal fragment of bactericidal/permeability-increasing protein prevents hemodynamic responses to endotoxin. Circ. Shock 41, 176-184.

Ammons, W.S., Kohn, F.R., & Kung, A.H.C. (1994). Protective effects of an N-terminal fragment of bactericidal/permeability-increasing protein in rodent models of Gram-negative sepsis: role of bactericidal properties. J. Infect. Dis. 170, 1473-1482.

Anderson, B.F., Baker, H.M., Dodson, E.J., Norris, G.E., Rumball, S.V., Waters, J.M., & Baker, E.N. (1987). Structure of human lactoferrin at 3.2-A resolution. Proc. Natl. Acad. Sci. USA 84, 1769-1773.

Appelmelk, B.J., An, Y.Q., Geerts, M., Thijs, B.G., de Boer, H.A., MacLaren, D.M., de Graaff, J., & Nuijens, J.H. (1994). Lactoferrin is a lipid A-binding protein. Infect. Immun. 62, 2628-2632.

Baba, T., & Schneewind, O. (1998). Instruments of microbial warfare: bacteriocin synthesis, toxicity and immunity. Trends Microbiol. 6, 66-71.

Beamer, L.J., Carroll, S.F., & Eisenberg, D. (1997). Crystal structure of human BPI and two bound phospholipids at 2.4 Angstrom resolution. Science. 276, 1881-1884.

Boman, H.G. (1995). Peptide antibiotics and their role in innate immunity. Annu. Rev. Immunol. 13, 61-92.

Boman, H.G., Agerberth, B., & Boman, A. (1993). Mechanisms of action on Escherichia coli of cecropin P1 and PR-39, two antibacterial peptides from pig intestine. Infect. Immun. 61, 2978-2984.

Borregaard, N. (1996). Current concepts about neutrophil granule physiology. Curr. Opin. Hematol. 3, 11-18.

Borregaard, N., Sehested, M., Nielsen, B.S., Sengelov, H., & Kjeldsen, L. (1995). Biosynthesis of granule proteins in normal human bone marrow cells. Gelatinase is a marker of terminal neutrophil differentiation. Blood 85, 812-817.

Brock, J. (1995). Lactoferrin: a multifunctional immunoregulatory protein? Immunol. Today 16, 417-419.

Broekaert, W.F., Terras, F.R., Cammue, B.P., & Osborn, R.W. (1995). Plant defensins: novel antimicrobial peptides as components of the host defense system. Plant Physiol. 108, 1353-1358.

Calafat, J., Janssen, H., Tool, A., Dentener, M., Knol, E.F., Rosenberg, H.F., & Egesten, A. (1998). The bactericidal/permeability-increasing protein is present in specific granules of human eosinophils. Blood. 91, 4770-4775.

Campanelli, D., Detmers, P.A., Nathan, C.F., & Gabay, J.E. (1990a). Azurocidin and a homologous serine protein from neutrophils. Differential antimicrobial and proteolytic properties. J. Clin. Invest. 85, 904-915.

Campanelli, D., Melchior, M., Fu, Y., Nakata, M., Shuman, H., & Gabay, J.E. (1990b). Cloning of cDNA for proteinase 3: a serien protease, antibiotic and autoantigen from human neutrophils. J. Exp. Med. 172, 1709-1715.

Campbell, E.J., Silverman, E.K., & Campbell, M.A. (1989). Elastase and cathepsin G of human monocytes. Quantification of cellular content, release in response to stimuli and heterogeneity in elastase-mediated proteolytic activity. J. Immunol. 143, 2961-2968.

Coulthard, M.G., Swindle, J., Munford, R.S., Gerard, R.D., & Meidell, R.S. (1996). Adenovirus-mediated transfer of a gene encoding acyloxyacyl hydrolase (AOAH) into mice increases tissue and plasma AOAH activity. Infect. Immun. 64, 1510-1515.

Cowland, J.B., Johnsen, A.H., & Borregaard, N. (1995). hCAP-18, a cathelin/probactenecin-like protein of human neutrophil specific granules. FEBS Lett. 368, 173-176.

de Winter, R.J., von der Mohlen, M.A., van Lieshout, H., Wedel, N., Nelson, B., Friedmann, N., Delemarre, B.J., & van Deventer, S.J. (1995). Recombinant endotoxin-binding protein (rBPI23) attenuates endotoxin-induced circulatory changes in humans. J. Inflamm. 45, 193-206.

Del Sal, G., Storici, P., Schneider, C., Romeo, D., & Zanetti, M. (1992). cDNA cloning of the neutrophil bactericidal peptide indolicidin. Biochem. Biophys. Res. Commun. 187, 467-472.

Dentener, M.A., Francot, G.J., & Buurman, W.A. (1996). Bactericidal/permeability-increasing protein, a lipopolysaccharide-specific protein on the surface of human peripheral blood monocytes. J. Infect. Dis. 173, 252-255.

Dhaenens, L., Szczebara, F., & Husson, M.O. (1997). Identification, characterization, and immunogenicity of the lactoferrin-binding protein from Helicobacter pylori. Infect. Immun. 65, 514-518.

Diamond, G., Russell, J.P., & Bevins, C.L. (1996). Inducible expression of an antibiotic peptide gene in lipopolysaccharide-challenged tracheal epithelial cells. Proc. Natl. Acad. Sci. USA 93, 5156-5160.

Elass-Rochard, E., Legrand, D., Salmon, V., Roseanu, A., Trif, M., Tobias, P.S., Mazurier, J., & Spik, G. (1998). Lactoferrin inhibits the endotoxin interaction with CD14 by competition with the lipopolysaccharide-binding protein. Infect. Immun. 66, 486-491.

Elass-Rochard, E., Roseanu, A., Legrand, D., Trif, M., Salmon, V., Motas, C., Montreuil, J., & Spik, G. (1995). Lactoferrin-lipopolysaccharide interaction: involvement of the 28-34 loop region of human lactoferrin in the high-affinity binding to Escherichia coli 055B5 lipopolysaccharide. Biochem. J. 312, 839-845.

Ellison, R.T. (1994). The effects of lactoferrin on gram-negative bacteria. In: Lactoferrin: Structure and Function, Vol. 357 (Hutchens, T.W., Rumball, S.V., & Lonnerdal, B., Eds.), pp. 71-90. Plenum Press, New York.

Elsbach, P. (1980). Degradation of microorganisms by phagocytic cells. Rev. Infect. Dis. 2, 106-128.

Elsbach, P. (1994). Bactericidal permeability-increasing protein in host defence against Gram-negative bacteria and endotoxin. In: Antimicrobial Peptides. Ciba Foundation Symposium 186 (Marsh, J., & Goode, J.A. Eds.), pp. 176-189. John Wiley & Sons Ltd, Chichester.

Elsbach, P. (1997). Determinants of the antimicrobial action of 14-kD phospholipases A2. In: Phospholipase A2. Basic and Clinical Aspects in Inflammatory Diseases. Vol. 24. (Uhl, W., Nevalainen T.J., & Buchler, M.W., Eds.), pp. 17-22. Karger, Basel.

Elsbach, P. (1998a). Recent advances in therapy of sepsis: focus on recombinant bactericidal/permeability-increasing protein (BPI). BioDrugs. 1, 435-442.

Elsbach, P. (1999). Bactericidal/permeability-increasing protein, p15s and phospholipases A2, endogenous antibiotics in host-defense against bacterial infections. In: *Endotoxin in Health and Disease.* (Brade H., Opal, S., Morrison, D.C., & Vogel, S., Eds.), pp. 369-377. Marcel Dekker, Inc., New York.

Elsbach, P., & Weiss, J. (1991). Utilization of labeled *Escherichia coli* as phospholipase substrate. Meth. Enzymol. 197, 24-31.

Elsbach, P., & Weiss, J. (1992). Oxygen-independent antimicrobial systems of phagocytes. In: *Inflammation: Basic Principles and Clinical Correlates, 2nd ed.* (Gallin, J.I., Goldstein, I.M., & Snyderman, R., Eds.), pp. 603-636. Raven Press, New York.

Elsbach, P., & Weiss, J. (1995). Prospects for the use of recombinant BPI in the treatment of gram-negative bacterial infections. Infect. Agents Dis. 4, 102-109.

Elsbach, P., & Weiss, J. (1998). Role of the bactericidal/permeability-increasing protein in host-defence. Curr. Opin. Immunol. 10, 45-49.

Elsbach, P., Weiss, J., & Levy, O. (1994a). Integration of antimicrobial host defenses: role of the bactericidal/permeability-increasing protein. Trends Microbiol. 2, 324-328.

Elsbach, P., Weiss, J., & Levy, O. (1999). Oxygen-independent antimicrobial systems of phagocytes. In: *Inflammation: Basic Principles and Clinical Correlates, 3rd ed.* (Gallin, J., Snyderman, R., Fearon, D., Haynes B., & Nathan, C., Eds.), pp. 801-818. Lippincott-Raven, Philadelphia.

Elsbach, P., Weiss, J., Doerfler, M., Shu, C., Kohn, F., Ammons, W.S., Kung, A.H., Meszaros, K.K., & Parent, J.B. (1994b). The bactericidal/permeability increasing protein of neutrophils is a potent antibacterial and anti-endotoxin agent in vitro and in vivo. [42 refs]. Prog. Clin. Biol. Res. 388, 41-51.

Erwin, A.L., & Munford, R.S. (1991). Plasma lipopolysaccharide-deacylating activity (acyloxyacyl hydrolase) increases after lipopolysaccharide administration to rabbits [see comments]. Lab. Invest. 65, 138-144.

Fallman, M., Persson, C., & Wolf-Watz, H. (1997). Yersinia proteins that target host cell signaling pathways. J. Clin. Invest. 99, 1153-1157.

Fleming, A. (1922). On a remarkable bacteriolytic element found in tissues and excretions. Proc. R. Soc. (London) 93, 306-310.

Fletcher, M.A., Kloczewiak, M.A., Loiselle, P.M., Ogata, M., Vermeulen, M.W., Zanzot, E.M., & Warren, H.S. (1997). A novel peptide-IgG conjugate, CAP18(106-138)-IgG, that binds and neutralizes endotoxin and kills gram-negative bacteria. J. Infect. Dis. 175, 621-632.

Foreman-Wykert, A.K., Weinrauch, Y., Elsbach, P., & Weiss, J. (1999). Cell-wall determinants of the bactericidal action of group II: A phospholipase A_2 against Gram-positive bacteria. J. Clin. Invest. 103, 715-721.

Freudenberg, M.A., Kleine, B., & Galanos, C. (1984). The fate of lipopolysaccharide in rats: evidence for chemical alteration in the molecule. Rev. Infect. Dis. 6, 483-487.

Frohm, M., Agerberth, B., Ahangari, G., Stahle-Backdahl, M., Liden, S., Wigzell, H., & Gudmundsson, G.H. (1997). The expression of the gene coding for the antibacterial peptide LL-37 is induced in human keratinocytes during inflammatory disorders. J. Biol. Chem. 272, 15258-15263.

Frohm, M., Gunne, H., Bergman, A.C., Agerberth, B., Bergman, T., Boman, A., Liden, S., Jornvall, H., & Boman, H.G. (1996). Biochemical and antibacterial analysis of human wound and blister fluid. Eur. J. Biochem. 237, 86-92.

Gallo, R.L., Ono, M., Povsic, T., Page, C., Eriksson, E., Klagsbrun, M., & Bernfield, M. (1994). Syndecans, cell surface heparan sulfate proteoglycans, are induced by a proline-rich antimicrobial peptide from wounds. Proc. Natl. Acad. Sci. USA 91, 11035-11039.

Ganz, T., & Lehrer, R.I. (1997). Antimicrobial peptides of leukocytes. Curr. Opin. Hematol. 4, 53-58.

Ganz, T., & Weiss, J. (1997). Antimicrobial peptides of phagocytes and epithelia. Sem. Hematol. 34, 342-354.

Gazzano-Santoro, H., Meszaros, K., Birr, C., Carrol, S.F., Theofan, G., Horwitz, A.H., Lim, E., Aberle, S., Kasler, H., & Parent, J.B. (1994). Competition between rBPI23, a recombinant fragment of bactericidal/permeability-increasing protein, and lipopolysaccharide (LPS)-binding protein for binding to LPS and Gram-negative bacteria. Infect. Immun. 62, 1185-1191.

Gazzano-Santoro, H., Parent, J.B., Conlon, P.J., Kasler, H.G., Tsai, C.M., Lill-Elghanian, D.A., & Hollingsworth, R.I. (1995). Characterization of the structural elements in lipid A required for binding of a recombinant fragment of bactericidal/permeability-increasing protein rBPI23. Infect. Immun. 63, 2201-2205.

Gazzano-Santoro, H., Parent, J.B., Grinna, L., Horwitz, A., Parsons, T., Theofan, G., Elsbach, P., Weiss, J., & Conlon, P.J. (1992). High-affinity binding of the bactericidal/permeability-increasing protein and a recombinant amino-terminal fragment to the lipid A region of lipopolysaccharide. Infect. Immun. 60, 4754-4761.

Ginsburg, I. (1988). The biochemistry of bacteriolysis: paradoxes, facts and myths. Microbiol. Sci. 5, 137-142.

Ginsburg, I., & Lahav, M. (1983). Lysis and biodegradation of microorganisms in infectious sites may involve cooperation between leukocyte, serum factors and bacterial wall autolysins: a working hypothesis. Eur. J. Clin. Microbiol. 2, 186-191.

Ginsburg, I., & Sela, M.N. (1976). The role of leukocytes and their hydrolases in the persistence of degradation and transport of bacterial constituents in tissues: relation to chronic inflammatory processes in staphylococcal, streptococcal and mycobacterial infections and in chronic peridontal disease. Crit. Rev. Microbiol. 4, 249-322.

Ginsburg, I., Lahav, M., & Giesbrecht, P. (1982). Effect of leukocyte hydrolases on bacteria XVI. Activation by leukocyte factors and cationic substances of autolytic enzymes in Staphylococcus aureus: modulation by anionic polyelectrolytes in relation to survival of bacteria in inflammatory exudates. Inflammation 6, 269-284.

Giroir, B.P., Quint, P.A., Barton, P., Kirsch, E.A., Kitchen, L., Goldstein, B., Margraf, D., Sastry, S., Orlowski, J.P., Bradley, J.S., Nelson, B.J., Wedel, N.I., White, M.L., Bauer, R.J., Carroll, S.F., & Scannon, P.J. (1997). Evaluation of rBPI21 (recombinant amino terminal fragment of human bactericidal/permeability-increasing protein) in children with severe meningococcal disease. Lancet 350, 1439-1443.

Gray-Owen, S.D., & Schryvers, A.B. (1996). Bacterial transferrin and lactoferrin receptors. Trends Microbiol. 4, 185-191.

Groisman, E.A. (1996). Bacterial responses to host-defense peptides. Trends Microb. 4, 127-128.

Guiney, D.G. (1997). Regulation of bacterial virulence gene expression by the host environment. J. Clin. Invest. 99, 565-569.

Gullberg, U., Andersson, E., Garwicz, D., Lindmark, A., & Olsson, I. (1997). Biosynthesis, processing and sorting of neutrophil proteins: insight into neutrophil granule development. Eur. J. Haematol. 58, 137-153.

Guo, L., Lim, K.B., Gunn, J.S., Bainbridge, B., Darveau, R.P., Hackett, M., & Miller, S.L. (1997). Regulation of lipid A modifications by Salmonella typhimurium virulence genes phoP-phoQ. Science 276, 250-253.

Hagen, F.S., Grant, F.J., Kuijper, J.L., Slaughter, C.A., Moomaw, C.R., Orth, K., O'Hara, P.J., & Munford, R.S. (1991). Expression and characterization of recombinant human acyloxyacyl hydrolase, a leukocyte enzyme that deacylates bacterial lipopolysaccharides. Biochemistry 30, 8415-8423.

Harmsen, M.C., Swart, P.J., Bethune, M.d., Pauwels, R., Clerq, E.D., The, T.H., & Meijer, D.K.F. (1995). Antiviral effects of plasma and milk proteins: lactoferrin shows potent activity against both human immunodeficiency virus and human cytomegalovirus replication in vitro. J. Infect. Dis. 172, 380-388.

Harris, H.W., Grunfeld, C., Feingold, K.R., & Rapp, J.H. (1990). Human very low density lipoproteins and chylomicrons can protect against endotoxin-induced death in mice. J. Clin. Invest. 86, 696-702.

Harris, H.W., Grunfeld, C., Feingold, K.R., Read, T.E., Kane, J.P., Jones, A.L., Eichbaum, E.B., Bland, G.F., & Rapp, J.H. (1993). Chylomicrons alter the fate of endotoxin, decreasing tumor necrosis factor release and preventing death. J. Clin. Invest. 91, 1028-1034.

Harwig, S.S., Tan, L., Qu, X.D., Cho, Y., Eisenhauer, P.B., & Lehrer, R.I. (1995). Bactericidal properties of murine intestinal phospholipase A2. J. Clin. Invest. 95, 603-610.

Harwig, S.S., Waring, A., Yang, H.J., Cho, Y., Tan, L., & Lehrer, R.I. (1996). Intramolecular disulfide bonds enhance the antimicrobial and lytic activities of protegrins at physiological sodium chloride concentrations. Eur. J. Biochem. 240, 352-357.

Henson, P.M., Henson, J.E., Fittschen, C., Bratton, D.L., & Riches, D.W.H. (1992). Degranulation and secretion by phagocytic cells. In: Inflammation: Basic Principles and Clinical Correlates (Gallin, J.I., Goldstein, I.M., & Snyderman, R., Eds.), pp. 511-540. Raven Press, New York.

Hessian, P.A., Edgeworth, J., & Hogg, N. (1993). MRP-8 and MRP-14, two abundant Ca2+-binding proteins of neutrophils and monocytes. J. Leukoc. Biol. 53, 197-204.

Hirsch, J.G. (1974). Neutrophil leukocytes. In: The Inflammatory Process, Vol. 1 (Zweifach, B.W., Grant, L., & Mc-Cluskey, R.T., Eds.), pp. 411-447. Academic Press, New York.

Hoffmann, J.A., & Reichhart, J. (1997). Drosophila immunity. Trends Cell Biol. 7, 309-316.

Hoffmann, J.A., Reichhart, J.M., & Hetru, C. (1996). Innate immunity in higher insects. Curr. Opin. Immunol. 8, 8-13.

Hubacek, J.A., Buchler, C., Aslanidis, C., & Schnitz, G. (1997). The genomic organization of the genes for human lipopolysaccharide binding protein (LBP) and bactericidal/permeability-increasing protein (BPI) is highly conserved. Biochem. Biophys. Res. Commun. 236, 427-430.

Iovine, N., Elsbach, P., & Weiss, J. (1997). An opsonic function of the neutrophil bactericidal/permeability-increasing protein depends on both its N- and C-terminal domains. Proc. Natl. Acad. Sci. USA 94, 10973-10978.

Irwin, D.M., & Wilson, A.C. (1990). Concerted evolution of ruminant stomach lysozymes. Characterization of lysozyme cDNA clones from sheep and deer. J. Biol. Chem. 265, 4944-4952.

Irwin, D.M., Yu, M., & Wen, Y. (1996). Isolation and characterization of vertebrate lysozyme genes. EXS 75, 225-241.

Jacob, L., & Zasloff, M. (1994). Potential therapeutic applications of magainins and other antimicrobial agents of animal origin. [91 refs]. In: Ciba Foundation Symposium, Vol. 186 (Boman, H., Marsh J., & Goode, J.A., Eds.), pp. 197-216. John Wiley & Sons, Chichester.

Jolles, P., & Jolles, J. (1984). What's new in lysozyme research? Mol. Cell. Biochem. 63, 165-189.

Jones, D.E., & Bevins, C.L. (1992). Paneth cells of the human small intestine express an antimicrobial peptide gene. J. Biol. Chem. 267, 23216-23225.

Kallajoki, M., & Nevalainen, T.J. (1997). Expression of group II phospholipases A2 in human tissues. In: *Phospholipases A2. Basic and Clinical Aspects in Inflammatory Diseases.* (Uhl, W. Nevalainen, T.J., & Buchler, M.W., Eds.) Vol. 24, pp. 8-16. Karger, Basel.

Katz, S.S., Weinrauch, Y., Munford, R.S., Elsbach, P., & Weiss, J. Deacylation of LPS in *E. coli* during destruction by cellular and extracellular components of a rabbit peritoneal inflammatory exudate. (Submitted for publication.)

Kramer, R.M., & Pepinsky, R.B. (1991). Assay and purification of phospholipase A2 from human synovial fluid in rheumatoid arthritis. Meth. Enzymol. 197, 373-381.

Kung, A.H., Ammons, W.S., Lin, Y., & Kohn, F.R. (1994). Efficacy of a recombinant amino terminal fragment of bactericidal/permeability increasing protein in rodents challenged with LPS or E. coli bacteria. Prog. Clin. Biol. Res. 388, 255-263.

Lahav, M., Ne'eman, N., Adler, E., & Ginsburg, I. (1974). Effect of leukocyte hydrolases on bacteria. I. Degradation of 14C- labeled Streptococcus and Staphylococcus by leukocyte lysates in vitro. J. Infect. Dis. 129, 528-537.

Lahav, M., Ne'eman, N., James, J., & Ginsburg, I. (1975). The effect of leukocyte hydrolases on bacteria. III. Bacteriolysis induced by extracts of different leukocyte populations and the inhibition of lysis by macromolecular substances. J. Infect. Dis. 131, 149-157.

Laible, N.J., & Germaine, G.R. (1985). Bactericidal activity of human lysozyme, muramidase-inactive lysozyme, and cationic polypeptides against Streptococcus sanguis and Streptococcus faecalis: inhibition by chitin oligosaccharides. Infect. Immun. 48, 720-728.

Larrick, J.W., Hirata, M., Balint, R.F., Lee, J., Zhong, J., & Wright, S.C. (1995). Human CAP18: a novel antimicrobial lipopolysaccharide-binding protein. Infect. Immun. 63, 1291-1297.

Larrick, J.W., Hirata, M., Shimomoura, Y., Yoshida, M., Zheng, H., Zhong, J., & Wright, S.C. (1993). Antimicrobial activity of rabbit CAP18-derived peptides. Antimicrob. Agents Chemother. 37, 2534-2539.

Larrick, J.W., Hirata, M., Shimomoura, Y., Yoshida, M., Zheng, H., Zhong, J., & Wright, S.C. (1994a). Rabbit CAP18 derived peptides inhibit gram negative and gram positive bacteria. Prog. Clin. Biol. Res. 388, 125-135.

Larrick, J.W., Hirata, M., Zheng, H., Zhong, J., Bolin, D., Cavaillon, J.M., Warren, H.S. & Wright, S.C. (1994b). A novel granulocyte-derived peptide with lipopolysaccharide-neutralizing activity. J. Immunol. 152, 231-240.

Lehrer, R.I., & Ganz, T. (1996). Endogenous vertebrate antibiotics. Defensins, protegrins, and other cysteine-rich antimicrobial peptides. Ann. NY Acad. Sci. 797, 228-239.

Lehrer, R.I., Lichtenstein, A.K., & Ganz, T. (1993). Defensins: antimicrobial and cytotoxic peptides of mammalian cells. Annu. Rev. Immunol. 11, 105-128.

Levy, O. (1996). Antibiotic proteins of polymorphonuclear leukocytes. Eur. J. Haem. 56, 263-277.

Levy, O., Ooi, C.E., Elsbach, P., Doerfler, M.E., Lehrer, R.I. & Weiss, J. (1995). Antibacterial proteins of granulocytes differ in interaction with endotoxin: comparison of bactericidal/permeability-increasing protein, p15s, and defensins. J. Immunol. 154, 5403-5410.

Levy, O., Ooi, C.E., Weiss, J., Lehrer, R.I., & Elsbach, P. (1994). Individual and synergistic effects of rabbit granulocyte proteins on Escherichia coli. J. Clin. Invest. 94, 672-682.

Levy, O., Weiss, J., Zarember, K., Ooi, C.E., & Elsbach, P. (1993). Antibacterial 15-kDa protein isoforms (p15s) are members of a novel family of leukocyte proteins. J. Biol. Chem. 268, 6058-6063.

Levy, S.B. (1992). In: The Antibiotic Paradox: How Miracle Drugs are Destroying a Miracle, p. 148. Plenum Press, New York.

Li, J., Brown, L.F., Laham, R.J., Volk, R., & Simons, M. (1997). Macrophage-dependent regulation of syndecan gene expression. Circ. Res. 81, 785-796.

Lin, Y., Ammons, W.S., Leach, W.J., & Kung, A.H. (1994a). Protective effects of a recombinant N-terminal fragment of bactericidal/permeability increasing protein on endotoxic shock in conscious rabbits. Shock 2, 324-331.

Lin, Y., Kohn, F.R., Kung, A.H., & Ammons, W.S. (1994b). Protective effect of a recombinant fragment of bactericidal/permeability increasing protein against carbohydrate dyshomeostasis and tumor necrosis factor-alpha elevation in rat endotoxemia. Biochem. Pharmacol. 47, 1553-1559.

Lin, Y., Leach, W.J., & Ammons, W.S. (1996). Synergistic effect of a recombinant N-terminal fragment of bactericidal/permeability-increasing protein and cefamandole in treatment of rabbit gram-negative sepsis. Antimicrob. Agents Chemother. 40, 65-69.

Little, R.G., Kelner, D.N., Lim, E., Burke, D.J., & Conlon, P.J. (1994). Functional domains of recombinant bactericidal/permeability-increasing protein (rBPI23). J. Biol. Chem. 268, 1865-1872.

Luchi, M., & Munford, R.S. (1993). Binding, internalization, and deacylation of bacterial lipopolysaccharide by human neutrophils. J. Immunol. 151, 959-969.

Madsen, L.M., Inada, M., & Weiss, J. (1996). Determinants of activation by complement of group II phospholipase A2 acting against Escherichia coli. Infect. Immun. 64, 2425-2430.

Mannion, B.A., Weiss, J., & Elsbach, P. (1990a). Separation of sublethal and lethal effects of polymorphonuclear leukocytes on Escherichia coli. J. Clin. Invest. 86, 631-641.

Mannion, B.A., Weiss, J., & Elsbach, P. (1990b). Separation of sublethal and lethal effects of the bactericidal/permeability increasing protein on Escherichia coli. J. Clin. Invest. 85, 853-860.

Marra, M.N., Wilde, C.G., Collins, M.S., Snable, J.L., Thornton, M.B., & Scott, R.W. (1992). The role of bactericidal/permeability-increasing protein as a natural inhibitor of bacterial endotoxin. J. Immunol. 148, 532-537.

Marra, M.N., Wilde, C.G., Griffith, J.E., Snable, J.L., & Scott, R.W. (1990). Bactericidal/permeability-increasing protein has endotoxin-neutralizing activity. J. Immunol. 144, 662-666.

Miehlke, S., Reddy, R., Osato, M.S., Ward, P.P., Conneely, O.M., & Graham, D.Y. (1996). Direct activity of recombinant human lactoferrin against Helicobacter pylori. J. Clin. Microbiol. 34, 2593-2594.

Miyakawa, Y., Ratnakar, P., Rao, A.G., Costello, M.L., Mathieu-Costello, O., Lehrer, R.I., & Catanzaro, A. (1996). In vitro activity of the antimicrobial peptides human and rabbit defensins and porcine leukocyte protegrin against Mycobacterium tuberculosis. Infect. Immun. 64, 926-932.

Munford, R.S., & Hall, C.L. (1989). Purification of acyloxyacyl hydrolase, a leukocyte enzyme that removes secondary acyl chains from bacterial lipopolysaccharides. J. Biol. Chem. 264, 15613-15619.

Munford, R.S., & Hunter, J.P. (1992). Acyloxyacyl hydrolase, a leukocyte enzyme that deacylates bacterial lipopolysaccharides, has phospholipase, lysophospholipase, diacylglycerollipase, and acyltransferase activities in vitro. J. Biol. Chem. 267, 10116-10121.

Nevalainen, T.J., Aho, H.J., & Peuravuori, H. (1994). Secretion of group 2 phospholipase A2 by lacrimal glands. Invest. Ophthalmol. Vis. Sci. 35, 417-421.

Odeberg, H., & Olsson, I. (1975). Antibacterial activity of cationic proteins from human granulocytes. J. Clin. Invest. 56, 1118-1124.

Ooi, C.E., Weiss, J., Elsbach, P., Frangione, B., & Mannion, B. (1987). A 25-kDa NH2-terminal fragment carries all the antibacterial activities of the human neutrophil 60-kDa bactericidal/permeability-increasing protein. J. Biol. Chem. 262, 14891-14894.

Ouellette, A.J., Miller, S.I., Henschen, A.H., & Selsted, M.E. (1992). Purification and primary structure of murine cryptdin-1, a Paneth cell defensin. FEBS Lett. 304, 146-148.

Owen, C.A., Campbell, M.A., Sannes, P.L., Boukedes, S.S., & Campbell, E.J. (1995). Cell surface-bound elastase and cathepsin G on human neutrophils: a novel, non-oxidative mechanism by which neutrophils focus and preserve catalytic activity of serine proteases. J. Cell Biol. 131, 775-789.

Panyutich, A., Shi, J., Boutz, P.L., Zhao, C., & Ganz, T. (1997). Porcine polymorphonuclear leukocytes generate extracellular microbicidal activity by elastase-mediated activation of secreted protegrins. Infect. Immun. 65, 978-985.

Panyutich, A.V., Panyutich, E.A., Krapivin, V.A., Baturevich, E.A., & Ganz, T. (1993). Plasma defensin concentrations are elevated in patients with septicemia or bacterial meningitis. J. Lab. Clin. Med. 122, 202-207.

Pruzanski, W., & Vadas, P. (1997). Secretory nonpancreatic phospholipase A2 (sPLA2) as a mediator of inflammation. In: *Phospholipase A2. Basic and Clinical Aspects in Inflammatory Diseases.* (Uhl, W., Nevalainen, T.J., & Buchler, M.W.), Vol. 24, pp. 38-42. Karger, Basel.

Qi, S.Y., Li, Y., Szyroki, A., Giles, I.G., Moir, A., & O'Connor, C.D. (1995). Salmonella typhimurium responses to a bactericidal protein from human neutrophils. Molec. Microbiol. 17, 523-531.

Read, T.E., Grunfeld, C., Kumwenda, Z.L., Calhoun, M.C., Kane, J.P., Feingold, K.R., & Rapp, J.H. (1995). Triglyceride-rich lipoproteins prevent septic death in rats. J. Exp. Med. 182, 267-272.

Read, T.E., Harris, H.W., Grunfeld, C., Feingold, K.R., Calhoun, M.C., Kane, J.P., & Rapp, J.H. (1993). Chylomicrons enhance endotoxin excretion in bile. Infect. Immun. 61, 3496-3502.

Rosenthal, M.D., Gordon, M.N., Buescher, E.S., Slusser, J.H., Harris, L.K., & Franson, R.C. (1995). Human neutrophils store type II 14-kDa phospholipase A2 in granules and secrete active enzyme in response to soluble stimuli. Biochem. Biophys. Res. Commun. 208, 650-656.

Schonwetter, B.S., Stolzenberg, E.D., & Zasloff, M.A. (1995). Epithelial antibiotics induced at sites of inflammation. Science 267, 1645-1648.

Scocchi, M., Wang, S., & Zanetti, M. (1997). Structural organization of the bovine cathelicidin gene family and identification of a novel member [In Process Citation]. FEBS Lett. 417, 311-315.

Selsted, M.E., Novotny, M.J., Morris, W.L., Tang, Y.Q., Smith, W., & Cullor, J.S. (1992). Indolicidin, a novel bactericidal tridecapeptide amide from neutrophils. J. Biol. Chem. 267, 4292-4295.

Sengelov, H., Follin, P., Kjeldsen, L., Lollike, K., Dahlgren, C., & Borregaard, N. (1995). Mobilization of granules and secretory vesicles during in vivo exudation of human neutrophils. J. Immunol. 154, 4157-4165.

Shafer, W.M., Hubalek, F., Huang, M., & Pohl, J. (1996). Bactericidal activity of a synthetic peptide (CG 117-136) of human lysosomal cathepsin G is dependent on arginine content. Infect Immun. 64, 4842-4845.

Shafer, W.M., Shepherd, M.E., Boltin, B., Wells, L., & Pohl, J. (1993). Synthetic peptides of human lysosomal cathepsin G with potent antipseudomonal activity. Infect. Immun. 61, 1900-1908.

Shi, J., Ross, C.R., Leto, T.L., & Blecha, F. (1996). PR-39, a proline-rich antibacterial peptide that inhibits phagocyte NADPH oxidase activity by binding to Src homology 3 domains of p47 phox. Proc. Natl. Acad. Sci. USA 93, 6014-6018.

Sohnle, P.G., Collins, L.C., & Wiessner, J.H. (1991a). The zinc-reversible antimicrobial activity of neutrophil lysates and abscess fluid supernatants. J. Infect. Dis. 164, 137-142.

Sohnle, P.G., Collins-Lech, C.. & Wiessner, J.H. (1991b). Antimicrobial activity of an abundant calcium-binding protein in the cytoplasm of human neutrophils. J. Infect. Dis. 163, 187-192.

Sohnle, P.G., Hahn, B.L., & Santhanagopalan, V. (1996). Inhibition of Candida albicans growth by calprotectin in the absence of direct contact with the organisms. J. Infect. Dis. 174, 1369-1372.

Sorensen, O., Arnljots, K., Cowland, J.B., Bainton, D.F., & Borregaard, N. (1997a). The human antibacterial cathelicidin, hCAP-18, is synthesized in myelocytes and metamyelocytes and localized to specific granules in neutrophils. Blood 90, 2796-2803.

Sorensen, O., Cowland, J.B., Askaa, J., & Borregaard, N. (1997b). An ELISA for hCAP-18, the cathelicidin present in human neutrophils and plasma. J. Immunol. Meth. 206, 53-59.

Sparkes, R.S., Kronenberg, M., Heinzmann, C., Daher, K.A., Klisak, I., Ganz, T., & Mohandas, T. (1989). Assignment of defensin gene(s) to human chromosome 8p23. Genomics 5, 240-244.

Staab, J.F., Ginkel, D.L., Rosenberg, G.B., & Munford, R.S. (1994). A saposin-like domain influences the intracellular localization, stability, and catalytic activity of human acyloxyacyl hydrolase. J, Biol, Chem. 269, 23736-23742.

Steinberg, D.A., Hurst, M.A., Fujii, C.A., Kung, A.H., Ho, J.F., Cheng, F.C., Loury, D.J., & Fiddes, J.C. (1997). Protegrin-1: a broad-spectrum, rapidly microbicidal peptide with in vivo activity. Antimicrob. Agents Chemother. 41, 1738-1742.

Storici, P., Del, S.G., Schneider, C., & Zanetti, M. (1992). cDNA sequence analysis of an antibiotic dodecapeptide from neutrophils. FEBS Lett. 314, 187-190.

Takeuchi, K., Irwin, D.M., Gallup, M., Shinbrot, E., Kai, H., Stewart, C., & Basbaum, C. (1993). Multiple cDNA sequences of bovine tracheal lysozyme. J. Biol. Chem. 268, 27440-27446.

Tischfield, J.A. (1997). A reassessment of the low molecular weight phospholipase A2 gene family in mammals. J. Biol. Chem. 272, 17247-17250.

Tobias, P.S., Soldau, K., Gegner, J.A., Mintz, D., & Ulevitch, R.J. (1995). Lipopolysaccharide binding protein-mediated complexation of lipopolysaccharide with soluble CD14. J. Biol. Chem. 270, 10482-10488.

Tobias, P.S., Soldau, K., Iovine, N., Elsbach, P., & Weiss, J. (1997). Lipopolysaccharide (LPS) binding proteins BPI and LBP form different complexes with LPS. J. Biol. Chem. 272, 18682-18685.

Turk, V., & Bode, W. (1991). The cystatins: protein inhibitors of cysteine proteinases. FEBS Lett. 285, 213-219.

Ulevitch, R.J., & Tobias, P.S. (1994). Recognition of endotoxin by cells leading to transmembrane signaling. [40 refs]. Curr. Opin. Immunol. 6, 125-130.

Ulevitch, R.J., & Tobias, P.S. (1995). Receptor-dependent mechanisms of cell stimulation by bacterial endotoxin. Annu. Rev. Immunol. 13, 437-457.

Ulevitch, R.J., Dunn, D.L., Fink, M.P., & Taylor, C.E. (1996). Endotoxin-related intracellular pathways: implications for therapeutic intervention. Shock 6, 1-2.

VanderMeer, T.J., Menconi, M.J., Zhuang, J., Wang, H., Murtaugh, R., Bouza, C., Stevens, P., & Fink, M.P. (1995). Protective effects of a novel 32-amino acid C-terminal fragment of CAP18 in endotoxemic pigs. Surgery 117, 656-662.

von der Mohlen, M.A., Kimmings, A.N., Wedel, N.I., Mevissen, M.L., Jansen, J., Friedmann, N., Lorenz, T.J., Nelson, B.J., White, M.L., Bauer, R., & et al. (1995a). Inhibition of endotoxin-induced cytokine release and neutrophil activation in humans by use of recombinant bactericidal/permeability-increasing protein. J. Infect. Dis. 172, 144-151.

von der Mohlen, M.A., van Deventer, S.J., Levi, M., van den Ende, B., Wedel, N.I., Nelson, B.J., Friedmann, N., & ten Cate, J.W. (1995b). Inhibition of endotoxin-induced activation of the coagulation and fibrinolytic pathways using a recombinant endotoxin-binding protein (rBPI23). Blood 85, 3437-3443.

Wang, D., Pabst, K.M., Aida, Y., & Pabst, M.J. (1995). Lipopolysaccharide-inactivating activity of neutrophils is due to lactoferrin. J. Leukoc. Biol. 57, 865-874.

Ward, P.P., Zhou, X., & Conneely, O.M. (1996). Cooperative interactions between the amino- and carboxyl-terminal lobes contribute to the unique iron-binding stability of lactoferrin. J. Biol. Chem. 271, 12790-12794.

Weersink, A.J., van Kessel, K.P., van den Tol, M.E., van Strijp, J.A., Torensma, R., Verhoef, J., Elsbach, P., & Weiss, J. (1993). Human granulocytes express a 55-kDa lipopolysaccharide-binding protein on the cell surface that is identical to the bactericidal/permeability-increasing protein. J. Immunol. 150, 253-263.

Weinrauch, Y., Abad, C., Ning-Sheng, L., P., Lowry, S., & Weiss, J. (1998). Mobilization of potent plasma bactericidal activity during systemic bacterial challenge: Role of group II phospholipase A2. J. Clin. Invest. 102, 633-638.

Weinrauch, Y., Elsbach, P., Madsen, L.M., Foreman, A., & Weiss, J. (1996b). The potent anti-Staphylococcus aureus activity of a sterile rabbit inflammatory fluid is due to a 14-kD phospholipase A2. J. Clin. Invest. 97, 250-257.

Weinrauch, Y., Foreman, A., Shu, C., Zarember, K., Levy, O., Elsbach, P., & Weiss, J. (1995). Extracellular accumulation of potently microbicidal bactericidal/permeability-increasing protein and p15s in an evolving sterile rabbit peritoneal inflammatory exudate. J. Clin. Invest. 95, 1916-1924.

Weinrauch, Y., Katz, S.S., Munford, R.S., Elsbach, P., & Weiss, J. (1999). Deacylation of purified lipopolysaccharides by cellular and extracellular components of a sterile rabbit peritoneal inflammatory exudate. Infect. Immun. 67, 3376-3382.

Weiss, J. (1994). Leukocyte-derived antimicrobial proteins. Curr. Opin. Hemat. 1, 78-84.

Weiss, J., & Elsbach, P. (1977). The use of a phospholipase A-less Escherichia coli mutant to establish the action of granulocyte phospholipase A on bacterial phospholipids during killing by a highly purified granulocyte fraction. Biochim. Biophys. Acta 466, 23-33.

Weiss, J., Beckerdite-Quagliata, S., & Elsbach, P. (1980). Resistance of gram-negative bacteria to purified bactericidal leukocyte proteins: relation to binding and bacterial lipopolysaccharide structure. J. Clin. Invest. 65, 619-628.

Weiss, J., Wright, S., Bekkers, A.C.A.P.A., van den Bergh, C.J., & Verheij, H.M. (1991). Conversion of pig pancreas phospholipase A2 by protein engineering into enzyme active against *Escherichia coli* treated with the bactericidal/permeability-increasing protein. J. Biol. Chem. 266, 4162-4167.

Weiss, J., Elsbach, P., Shu, C., Castillo, J., Grinna, L., Horwitz, A., & Theofan, G. (1992). Human bactericidal/permeability-increasing protein and a recombinant NH2-terminal fragment cause killing of serum-resistant gram-negative bacteria in whole blood and inhibit tumor necrosis factor release induced by the bacteria. J. Clin. Invest. 90, 1122-1130.

Weiss, J., Inada, M., Elsbach, P., & Crowl, R.M. (1994). Structural determinants of the action against Escherichia coli of a human inflammatory fluid phospholipase A2 in concert with polymorphonuclear leukocytes. J. Biol. Chem. 269, 26331-26337.

Weiss, J., Muello, K., Victor, M., & Elsbach, P. (1984). The role of lipopolysaccharides in the action of the bactericidal/permeability-increasing neutrophil protein on the bacterial envelope. J. Immunol. 132, 3109-3115.

Weiss, J., Victor, M., Stendhal, O., & Elsbach, P. (1982). Killing of gram-negative bacteria by polymorphonuclear leukocytes: role of an O2-independent bactericidal system. J. Clin. Invest. 69, 959-970.

White, S.H., Wimley, W.C., & Selsted, M.E. (1995). Structure, function, and membrane integration of defensins. Curr. Opin. Struct. Biol. 5, 521-527.

Wiese, A., Brandenburg, K., Carroll, S.F., Rietschel, E.T., & Seydel, U. (1997a). Mechanisms of action of bactericidal/permeability-increasing protein BPI on reconstituted outer membranes of gram-negative bacteria. Biochemistry 36, 10311-10319.

Wiese, A., Brandenburg, K., Lindner, B., Schromm, A.B., Carroll, S.F., Rietschel, E.T., & Seydel, U. (1997b). Mechanisms of action of the bactericidal/permeability-increasing protein BPI on endotoxin and phospholipid monolayers and aggregates. Biochemistry 36, 10301-10310.

Wright, G.C., Ooi, C.E., Weiss, J., & Elsbach, P. (1990a). Purification of a cellular (granulocyte) and an extracellular (serum) phospholipase A2 that participate in the destruction of Escherichia coli in a rabbit inflammatory exudate. J. Biol. Chem. 265, 6675-6681.

Wright, G.C., Weiss, J., Kim, K.S., Verheij, H., & Elsbach, P. (1990b). Bacterial phospholipid hydrolysis enhances the destruction of Escherichia coli ingested by rabbit neutrophils. Role of cellular and extracellular phospholipases. J. Clin. Invest. 85, 1925-1935.

Yamamoto, T., Hanawa, T., Ogata, S., & Kamiya, S. (1997). The Yersinia enterocolitica GsrA stress protein, involved in intracellular survival, is induced by macrophage phagocytosis. Infect. Immun. 65, 2190-2196.

Yamauchi, K., Tomita, M., Giehl, T.J., & Ellison, R.T. (1993). Antibacterial activity of lactoferrin and a pepsin-derived lactoferrin peptide fragment. Infect. Immun. 61, 719-728.

Yui, S., Mikami, M., Tsurumaki, K., & Yamazaki, M. (1997). Growth-inhibitory and apoptosis-inducing activities of calprotectin derived from inflammatory exudate cells on normal fibroblasts: regulation by metal ions. J. Leukoc. Biol. 61, 50-57.

Zanetti, M., Gennaro, R., & Romeo, D. (1995). Cathelicidins: a novel protein family with a common proregion and a variable C-terminal antimicrobial domain. FEBS Lett. 374, 1-5.

Zarember, K., Elsbach, P., Shin-Kim, K., & Weiss, J. (1997). p15s (15-kD antimicrobial proteins) are stored in the secondary granules of rabbit granulocytes: implications for antibacterial synergy with the bactericidal/permeability-increasing protein in inflammatory fluids. Blood 89, 672-679.

Zasloff, M. (1992). Antibiotic peptides as mediators of innate immunity. Curr. Opin. Immunol. 4, 3-7.

Zhang, G., Ross, C.R., Dritz, S.S., Nietfeld, J.C., & Blecha, F. (1997). Salmonella infection increases porcine antibacterial peptide concentrations in serum. Clin. Diagn. Lab. Immunol. 4, 774-777.

INDEX

www.ingramcontent.com/pod-product-compliance
Ingram Content Group UK Ltd.
Pitfield, Milton Keynes, MK11 3LW, UK
UKHW020225290926
15470UKWH00001B/380